AF721561

High-Throughput Plant Metabolomics

CABI BIOTECHNOLOGY SERIES

Biotechnology, in particular the use of transgenic organisms, has a wide range of applications including agriculture, forestry, food and health. There is evidence that it could make a major impact in producing plants and animals that are able to resist stresses and diseases, thereby increasing food security. There is also potential to produce pharmaceuticals in plants through biotechnology, and provide foods that are nutritionally enhanced. Genetically modified organisms can also be used in cleaning up pollution and contamination. However, the application of biotechnology has raised concerns about biosafety, and it is vital to ensure that genetically modified organisms do not pose new risks to the environment or health. To understand the full potential of biotechnology and the issues that relate to it, scientists need access to information that not only provides an overview of and background to the field, but also keeps them up to date with the latest research findings.

This series, which extends the scope of CABI's successful "Biotechnology in Agriculture" series, addresses all topics relating to biotechnology including transgenic organisms, molecular analysis techniques, molecular pharming, *in vitro* culture, public opinion, economics, development and biosafety. Aimed at researchers, upper-level students and policy makers, titles in the series provide international coverage of topics related to biotechnology, including both a synthesis of facts and discussions of future research perspectives and possible solutions.

Titles Available

1. **Animal Nutrition with Transgenic Plants**
   Edited by G. Flachowsky
2. **Plant-derived Pharmaceuticals: Principles and Applications for Developing Countries**
   Edited by K.L. Hefferon
3. **Transgenic Insects: Techniques and Applications**
   Edited by M.Q. Benedict
4. **Bt Resistance: Characterization and Strategies for GM Crops Producing *Bacillus thuringiensis* Toxins**
   Edited by Mario Soberón, Yulin Gao and Alejandra Bravo
5. **Plant Gene Silencing: Mechanisms and Applications**
   Edited by Tamas Dalmay
6. **Ethical Tensions from New Technology: The Case of Agricultural Biotechnology**
   Edited by Harvey James
7. **GM Food Systems and their Economic Impact**
   Tatjana Brankov and Koviljko Lovre
8. **Endophyte Biotechnology: Potential for Agriculture and Pharmacology**
   Edited by Alexander Schouten
9. **Forest Genomics and Biotechnology**
   Richard Meilan and Matias Kirst
10. **Transgenic Insects: Techniques and Applications, 2nd Edition**
    Edited by M.Q. Benedict and Maxwell J. Scott
11. **Plant Omics: Advances in Big Data Biology**
    Edited by Hajime Ohyanagi, Eiji Yamamoto, Ai Kitazumi and Kentaro Yano

12. **Next-generation Sequencing and Agriculture**
Edited by Phillip Bayer and Dave Edwards

13. **Aquaculture and Fisheries Biotechnology: Genetic Approaches**
Rex A. Dunham

14. **Algal Biotechnology**
Edited by Qiang Wang

15. **Genetic Engineering of Vegetable Crops**
Edited by Pritam Kalia

16. **Plant Breeding Technology: Future Trends and Challenges**
Edited by Zeba Khan

17. **Stress-resilient Crops: Coordinated Omics-CRISPR-Nanotechnology Strategies**
Edited by Jen-Tsung Chen

18. **High-Throughput Plant Metabolomics**
Edited by Jen-Tsung Chen

High-Throughput Plant Metabolomics

Edited by

Jen-Tsung Chen

CABI

CABI is a trading name of CAB International

CABI
Nosworthy Way
Wallingford
Oxfordshire OX10 8DE
UK

Tel: +44 (0)1491 832111
E-mail: info@cabi.org
Website: www.cabi.org

CABI
200 Portland Street
Boston
MA 02114
USA

Tel: +1 (617)682-9015
E-mail: cabi-nao@cabi.org

© CAB International 2025. All rights, including for text and data mining, AI training, and similar technologies, are reserved. No part of this publication may be reproduced in any form or by any means, electronically, mechanically, by photocopying, recording or otherwise, without the prior permission of the copyright owners.

The views expressed in this publication are those of the author(s) and do not necessarily represent those of, and should not be attributed to, CAB International (CABI). Any images, figures and tables not otherwise attributed are the author(s)' own. References to internet websites (URLs) were accurate at the time of writing.
CAB International and, where different, the copyright owner shall not be liable for technical or other errors or omissions contained herein. The information is supplied without obligation and on the understanding that any person who acts upon it, or otherwise changes their position in reliance thereon, does so entirely at their own risk. Information supplied is neither intended nor implied to be a substitute for professional advice. The reader/user accepts all risks and responsibility for losses, damages, costs and other consequences resulting directly or indirectly from using this information.

CABI's Terms and Conditions, including its full disclaimer, may be found at https://www.cabidigitallibrary.org/terms-and-conditions.

A catalogue record for this book is available from the British Library, London, UK.

ISBN-13: 9781800626898 (hardback)
9781800626904 (ePDF)
9781800626911 (ePub)

DOI: 10.1079/9781800626911.0000

Commissioning Editor: Ward Cooper
Editorial Assistant: Emma McCann
Production Editor: Shankari Wilford

Typeset by Straive, Pondicherry, India
Printed and bound by CPI Group (UK) Ltd, Croydon, CR0 4YY

Contents

Contributors xi

Preface xix

1. **Applied Metabolomics for Studying Plant-specialized Metabolites in Plant Health, Human Health, and Food Quality** 1
Harmonie M. Bettenhausen, Dhirendra Niroula, and Janak R. Joshi

2. **Applied Metabolomics for Plant Protection** 18
Carmen Grech, Julia Pérez-Pérez, Francisco Vázquez Prol, Ismael Rodrigo, Francisco Vera-Sirera, María Pilar López-Gresa, and Purificación Lisón

3. **High-throughput Plant Metabolomics and Predictive Modelling** 38
Millena C. Barros Santos, Cathleen Mirande-Ney, Thomas Dussarrat, Guillaume Marti, Alain Bouchereau, Younès Dellero, Sylvain Prigent, and Pierre Pétriacq

4. **Technological Advancements in Plant High-throughput Metabolomics** 67
Tharani Sureshkumar, Chithra Devi Balasundarm Saraswathy, Ramkumar Govindarajan, Ajay Samuel Siva Daniel, Sunitha Kumari Krishnankutty, Sarah Jaison, Muthukrishnan Arun, and Arumugam Vijaya Anand

5. **High-throughput Metabolomics for Uncovering Plant–Herbivore Interactions** 96
Debajit Das, Sangeeta Sarma, Dhrubo Jyoti Gogoi, Abhik Gogoi, Channakeshavaiah Chikkaputtaiah, and Dhanawantari L. Singha

6. **High-throughput Metabolomics for Studying Plant Stress Physiology Under a Changing Climate** 114
Mahmoud F. Seleiman, Awais Ahmad, Majed Alotaibi, Adel M. Al-Saif, Shivani Ranjan, Sumit Sow, and Harun H. Gitari

7. **Plant Metabolomics Based on Single-cell Technology** 132
Ekta Pandey, Rinkee Kumari, Shahla Faizan, Sayyada Bushra, Alisha Hussain, Parul Verma, and Mohammed Sayeed Akhtar

8. **Constructing Metabolomics Using Integrated Bioinformatics Tools for Advancing Plant Science** 153
Parkirti, Sandeep Kour, Mohd Ali, Deepak Kumar, Roohi Sharma, Vikram, Alisha Sharma, Anchita Singh, and Puja Ohri

9. **Applied Metabolomics for Plant Biotechnology** 176
Ilham Dehbi, Khadija Benamar, Rachid Ezzouggari, Mouna Janati, Slimane Khayi, Rachid Mentag, Zineb Belabess, Yunfei Jiang, and Rachid Lahlali

10. **Metabolomics and High-throughput Approaches to Assessing Drought-stress Tolerance in Crops** 201
Suman Gusain, Akshita Patial, Rakhi Chandoliya, Vikas Sharma, and Rohit Joshi

11. **Applied Metabolomics for Crop Improvement and Plant Breeding** 216
Saika Nabi, M. Altaf Wani, Asif. B. Shikari, Zahoor A. Dar, P.A. Sofi, Khazin Hussain, Asma Shakeel, Mehnaz Shakeel, Liset George, and Seelam Giri Uma Sankar Rao

12. **High-throughput Metabolomics for Plant Phenotyping** 236
Sarfraz Ahmad, Maksud Hasan Shah, Saniya, Mohammad Ghani Raghib, Vikas Belwal, Shaukat Ali, and Rubina Khan

13. **High-throughput Metabolomics for the Dissection of Plant Biotic and Abiotic Stress Responses** 252
Berivan Ozlem Gumus, Alberto Ribes-Navarro, Eva Cañizares, Juan Manuel Acién, Miguel González-Guzmán, and Vicent Arbona

14. **High-throughput Metabolomics for Assessing CRISPR-edited Plants** 276
Ajay Kumar Keot, Yogita N. Sarki, Riwandahun Marwein, Channakeshavaiah Chikkaputtaiah, and Dhanawantari L. Singha

15. **Metabolomics-assisted Breeding: A New Omics-integrated Avenue for Improved Stress Response in Plants** 292
Rajib Roychowdhury, Archana Prasad, Soumya Prakash Das, Priya Shah, Ajay Kumar, Umakanta Sarker, Sunil S. Gangurde, and Manish K Pandey

16. **Plant High-throughput Metabolomics: Recent Achievements, Applications, and Future Perspectives** 315
Lumbini Devi Darapu, Shivani Tiwari, and Prateek Gupta

17. **Integrating Advanced Metabolomics with Plant Functional Genomics** 335
Akash Shaikh, Varsha Toppo, Jameel M. Al-Khayri, and Praveen Nagella

18. **Insights into Medicinal Herbs Based on Metabolomics and High-throughput Technologies** 350
Akshita Patial, Rakhi Chandoliya, Suman Gusain, Vikas Sharma, and Rohit Joshi

19. **High-throughput Metabolomics for Medicinal Plant Research** 371
Gunna Sureshbabu Suruthi, Sivamani Ganesan, Ramakrishnan Aruna Raji, Shanmugam Velayuthaprabhu, Saradhadevi Muthukrishnan, Palanisamy Sampathkumar, Asirvatham Alwin Robert, and Arumugam Vijaya Anand

20. **High-throughput Metabolomics for the Exploration of Plant Secondary Metabolism** 388
Gunna Sureshbabu Suruthi, Galeel Rahman Rihana Begum, Murugesan Vinothini, Balasubramanian Balamuralikrishnan, Naif Abdullah Al-Dhabi, Mariadhas Valan Arasu, Vino Palani, and Arumugam Vijaya Anand

21. High-throughput Metabolomics for Plant Ecology and Biodiversity Research 405
Sarah Jaison, Chithra Devi Balasundarm Saraswathy, Tharani Sureshkumar, Ajay Samuel Siva Daniel, Sunitha Kumari Krishnankutty, Ramkumar Govindarajan, Muthukrishnan Arun, and Arumugam Vijaya Anand

22. High-throughput Metabolomics for Plant Natural Product Research 427
Gunna Sureshbabu Suruthi, Pandiarajan Anisha, Sivagaminathan Thangavalli, Balasubramanian Balamuralikrishnan, Naif Abdullah Al-Dhabi, Mariadhas Valan Arasu, Vino Palani, and Arumugam Vijaya Anand

Index 445

Contributors

Juan Manuel Acién, Departament de Biologia, Bioquímica i Ciències Naturals, Universitat Jaume I, Castelló de la Plana, Spain. Email: acien@uji.es

Awais Ahmad, Department of Plant Production, College of Food and Agriculture Sciences, King Saud University, Riyadh, Saudi Arabia. ORCID: 0000-0002-6197-2116. Email: awaisahmad1543@gmail.com

Sarfraz Ahmad, Department of Genetics and Plant Breeding, College of Agriculture Mandawa, Swami Keshwanand Rajasthan Agriculture University, Bikaner, Rajasthan, India. Email: sarfraz_firoz@rediffmail.com

Mohd. Sayeed Akhtar, Department of Botany, Gandhi Faiz-e-Aam College, Shahjahanpur 242001, Uttar Pradesh, India. ORCID: 0000-0003-0251-8279. Email: sayeedbot@gmail.com

Naif Abdullah Al-Dhabi, Department of Botany and Microbiology, College of Science, King Saud University, PO Box 2455, Riyadh 11451, Saudi Arabia. Email: naldhabi@ksu.edu.sa

Mohd Ali, Department of Zoology, Guru Nanak Dev University, Amritsar, Punjab 143005, India. ORCID: 0000-0001-7815-2344. Email: xyaansari774@gmail.com

Shaukat Ali, Department of Agronomy, Krishi Vigyan Kendra Chandgothi, Swami Keshwanand Rajasthan Agriculture University, Bikaner, Rajasthan, India. Email: shaukatagro@gmail.com

Jameel M. Al-Khayri, Department of Agricultural Biotechnology, College of Agriculture and Food Sciences, King Faisal University, Al-Ahsa 31982, Saudi Arabia. ORCID: 0000-0001-9507-0201. Email: jkhayri@kfu.edu.sa

Majed Alotaibi, Department of Plant Production, College of Food and Agriculture Sciences, King Saud University, Riyadh, Saudi Arabia. Email: malotaibi@KSU.EDU.sa

Adel M. Al-Saif, Department of Plant Production, College of Food and Agriculture Sciences, King Saud University, Riyadh, Saudi Arabia. Email: adelsaef@ksu.edu.sa

Arumugam Vijaya Anand, Department of Human Genetics and Molecular Biology, Bharathiar University, Coimbatore 641046, Tamil Nadu, India. Email: avahgmb@buc.edu.in

Pandiarajan Anisha, Department of Human Genetics and Molecular Biology, Bharathiar University, Coimbatore 641046, Tamil Nadu, India. Email: anisha.prajan@gmail.com

Mariadhas Valan Arasu, Department of Botany and Microbiology, College of Science, King Saud University, PO Box 2455, Riyadh 11451, Saudi Arabia. Email: mvalanarasu@gmail.com

Vicent Arbona, Departament de Biologia, Bioquímica i Ciències Naturals, Universitat Jaume I, Castelló de la Plana, Spain. Email: arbona@uji.es

Muthukrishnan Arun, Department of Biotechnology, Bharathiar University, Coimbatore 641046, Tamil Nadu, India. Email: arun@buc.edu.in

Balasubramanian Balamuralikrishnan, Department of Food Science and Biotechnology, College of Life Science, Sejong University, Seoul 05006, Republic of Korea. Email: bala.m.k@sejong.ac.kr

Galeel Rahman Rihana Begum, Department of Human Genetics and Molecular Biology, Bharathiar University, Coimbatore 641046, Tamil Nadu, India. Email: rihanabegum20062004@gmail.com

Zineb Belabess, Biotechnology Research Unit, Regional Center of Agricultural Research of Rabat, National Institute of Agricultural Research, Rabat, Morocco. Email: zineb.belabess@inra.ma

Vikas Belwal, Department of Tree Improvement, College of Forestry Ranichauri, VCSG Uttarakhand University of Horticulture and Forestry, Bharsar, Uttarakhand, India. Email: vikasbelwal@hotmail.com

Khadija Benamar, Department of Biology, Laboratory of Microbial Biotechnology and Bioactive Molecules, Sciences and Technologies Faculty, Sidi Mohamed Ben Abdellah University, PO Box 2202, Fez, Morocco. Email: khadija.benamar2@usmba.ac.ma

Harmonie M. Bettenhausen, James B. Beam Institute for Kentucky Spirits, University of Kentucky, Kentucky, USA. Email: hbettenhausen@uky.edu

Alain Bouchereau, Institute for Genetics, Environment and Plant Protection (IGEPP), National Research Institute for Agriculture, Food and Environment (INRAE), Institut Agro Rennes-Angers, Université Rennes, 35650 Le Rheu, France; Metabolic Profiling and Metabolomic Platform (P2M2), MetaboHUB-Grand-Ouest, Rennes, France. Email: alain.bouchereau@univ-rennes.fr

Sayyada Bushra, Department of Botany, Aligarh Muslim University, Aligarh 202002, Uttar Pradesh, India. Email: sbushra012@gmail.com

Eva Cañizares, Departament de Biologia, Bioquímica i Ciències Naturals, Universitat Jaume I, Castelló de la Plana, Spain. Email: ecanizar@uji.es

Rakhi Chandoliya, Biotechnology Division, CSIR-Institute of Himalayan Bioresource Technology (CSIR-IHBT), Palampur, Himachal Pradesh 176061, India; Department of Biotechnology, School of Bioengineering and Biosciences, Lovely Professional University, Phagwara 144411, India. Email: chandoliyarakhi76@gmail.com

Channakeshavaiah Chikkaputtaiah, Biological Sciences and Technology Division, CSIR-North East Institute of Science and Technology (NEIST), Jorhat 785 006, Assam, India; Academy of Scientific and Innovative Research (AcSIR), Ghaziabad 201 002, India. Email: channakeshav@neist.res.in

Ajay Samuel Siva Daniel, GRG BIRAC EYUVA Centre, PSGR Krishnammal College for Women, Coimbatore 641004, Tamil Nadu, India. Email: ajaysamuel@psgrkcw.ac.in

Zahoor A. Dar, Dryland Agriculture Research Station Rangreth, Sher-e-Kashmir University of Agricultural Sciences and Technology (SKUAST), Kashmir, India. Email: zahoorpbg@gmail.com

Lumbini Devi Darapu, Department of Biological Sciences, SRM University AP, Amaravati 522240, Andhra Pradesh, India. Email: lumbini_devidarapu@srmap.edu.in

Debajit Das, Department of Agricultural Biotechnology, Assam Agricultural University, Jorhat 785 013, Assam, India. Email: debajitbnc@gmail.com

Soumya Prakash Das, School of Life Sciences, Seacom Skills University, Bolpur 731236, West Bengal, India. ORCID: 0000-0002-5115-7679. Email: jptahbu@gmail.com

Ilham Dehbi, Phytopathology Unit, Department of Plant Protection, Ecole Nationale d'Agriculture de Meknès, Km 10, Rte Haj Kaddour, BP S/40, Meknès 50001, Morocco; Laboratory of Plant Biotechnologies and Valorization of Bio-Resources, Moulay Ismail University of Meknès, BP 11201, Zitoune, Meknès 50000, Morocco. Email: ilhamdehbi60@gmail.com

Younès Dellero, Institute for Genetics, Environment and Plant Protection (IGEPP), National Research Institute for Agriculture, Food and Environment (INRAE), Institut Agro Rennes-Angers, Université Rennes, 35650 Le Rheu, France; Metabolic Profiling and Metabolomic Platform (P2M2), MetaboHUB-Grand-Ouest, Rennes, France. Email: Younes.Dellero@inrae.fr

Thomas Dussarrat, Department of Chemical Ecology, Bielefeld University, Universitätsstrasse 25, 33615 Bielefeld, Germany. Email: thomas.dussarrat@uni-bielefeld.de

Rachid Ezzouggari, Phytopathology Unit, Department of Plant Protection, Ecole Nationale d'Agriculture de Meknès, Km 10, Rte Haj Kaddour, BP S/40, Meknès 50001, Morocco; Laboratory of

Biotechnology, Conservation, and Valorization of Natural Resources (LBCVNR), Faculty of Sciences Dhar El Mehraz, Sidi Mohamed Ben Abdallah University, Fez 30000, Morocco. Email: ezzouggarirachid@gmail.com

Shahla Faizan, Department of Botany, Aligarh Muslim University, Aligarh 202002, Uttar Pradesh, India. Email: sfaizan.bt.@amu.ac.in

Sivamani Ganesan, PG and Research Department of Zoology and Biotechnology, A.V.V.M. Sri Pushpam College (Autonomous), Poondi, Thanjavur 613503, Tamil Nadu, India. Email: ganesanmolbio@gmail.com

Sunil S. Gangurde, Center of Excellence in Genomics & Systems Biology (CEGSB), and Center for Pre-breeding Research (CPBR), International Crops Research Institute for the Semi-Arid Tropics (ICRISAT), Hyderabad 502324, Telangana, India. ORCID: 0000-0001-7119-8649. Email: Sunil.Gangurde@icrisat.org

Liset George, Division of Genetics and Plant Breeding, Faculty of Agriculture, Sher-e-Kashmir University of Agricultural Sciences and Technology (SKUAST), Kashmir 192301, India. Email: georgeliset@gmail.com

Harun H. Gitari, Department of Agricultural Science and Technology, School of Agriculture and Environmental Sciences, Kenyatta University, Nairobi, Kenya. Email: harun.gitari@ku.ac.ke

Abhik Gogoi, Department of Agricultural Biotechnology, Assam Agricultural University, Jorhat 785 013, Assam, India. Email: abhik.gogoi.adj21@aau.ac.in

Dhrubo Jyoti Gogoi, Biological Sciences and Technology Division, CSIR-North East Institute of Science and Technology (NEIST), Jorhat 785 006, Assam, India; Academy of Scientific and Innovative Research (AcSIR), Ghaziabad 201 002, India. Email: dhrubojyotigogoi111@gmail.com

Miguel González-Guzmán, Departament de Biologia, Bioquímica i Ciències Naturals, Universitat Jaume I, Castelló de la Plana, Spain. Email: mguzman@uji.es

Ramkumar Govindarajan, GRG BIRAC EYUVA Centre, PSGR Krishnammal College for Women, Coimbatore 641004, Tamil Nadu, India. Email: ramkumarg@psgrkcw.ac.in

Carmen Grech, Instituto de Biología Molecular y Celular de Plantas (IBMCP), Consejo Superior de Investigaciones Científicas (CSIC), Universitat Politècnica de València (UPV), Ciudad Politécnica de la Innovación (CPI) 8 E, Ingeniero Fausto Elio s/n, 46011 Valencia, Spain. ORCID: 0009-0003-1864-6969. E-mail: cgreher@ibmcp.upv.es

Berivan Ozlem Gumus, Departament de Biologia, Bioquímica i Ciències Naturals, Universitat Jaume I, Castelló de la Plana, Spain. Email: gumu@uji.es

Prateek Gupta, Department of Biological Sciences, SRM University AP, Amaravati 522240, Andhra Pradesh, India. Email: prateekg04@gmail.com

Suman Gusain, Biotechnology Division, CSIR-Institute of Himalayan Bioresource Technology (CSIR-IHBT), Palampur, Himachal Pradesh 176061, India; Academy of Scientific and Innovative Research (AcSIR), CSIR-HRDC Campus, Ghaziabad, Uttar Pradesh 201 002, India. Email: suman6gusain@gmail.com

Alisha Hussain, Department of Botany, Aligarh Muslim University, Aligarh 202002, Uttar Pradesh, India. Email: alishahussain302@gmail.com

Khazin Hussain, Division of Genetics and Plant Breeding, Faculty of Agriculture, Sher-e-Kashmir University of Agricultural Sciences and Technology (SKUAST), Kashmir 192301, India. Email: kainruk21@gmail.com

Sarah Jaison, Department of Botany, PSGR Krishnammal College for Women, Coimbatore 641004, Tamil Nadu, India. Email: sarah@psgrkcw.ac.in

Mouna Janati, Laboratory of Plant Biotechnologies and Valorization of Bio-Resources, Moulay Ismail University of Meknes, BP 11201, Zitoune, Meknès 50000, Morocco. Email: m.janati@edu.umi.ac.ma

Yunfei Jiang, Department of Plant, Food and Environmental Sciences, Faculty of Agriculture, Dalhousie University, Truro, Nova Scotia, Canada B2N 5E3. Email: yunfei.jiang@dal.ca

Janak R. Joshi, Department of Plant Sciences and Plant Pathology, Montana State University, Montana, USA. Email: janak.joshi@montana.edu

Rohit Joshi, Biotechnology Division, CSIR-Institute of Himalayan Bioresource Technology (CSIR-IHBT), Palampur, Himachal Pradesh 176061, India; Academy of Scientific and Innovative Research (AcSIR), CSIR-HRDC Campus, Ghaziabad, Uttar Pradesh, 201 002, India. Email: rohit-joshi@ihbt.res.in

Ajay Kumar Keot, Biological Sciences and Technology Division, CSIR-North East Institute of Science and Technology (CSIR-NEIST), Jorhat 785 006, Assam, India; Academy of Scientific and Innovative Research (AcSIR), Ghaziabad 201 002, India. Email: ajay.keot92@gmail.com

Rubina Khan, Department of Genetics and Plant Breeding, College of Agriculture Jobner, Sri Karan Narendra Agriculture University, Jobner, Rajasthan, India. Email: rkrubikhan7@gmail.com

Slimane Khayi, Plant Protection Laboratory, Regional Center of Agricultural Research of Meknès , National Institute of Agricultural Research, Km 13, Route Haj Kaddour, BP 578, Meknès 50001, Morocco. Email: slimane.khayi@inra.ma

Sandeep Kour, Department of Zoology, Guru Nanak Dev University, Amritsar, Punjab 143005, India. ORCID: 0000-0002-8241-8835. Email: koursandeep0309@gmail.com

Sunitha Kumari Krishnankutty, Department of Botany, PSGR Krishnammal College for Women, Coimbatore 641004, Tamil Nadu, India. Email: ksunitha@psgrkcw.ac.in

Ajay Kumar, Amity Institute of Biotechnology, Amity University, Noida, 201313, India. ORCID: 0000-0002-3260-0807. Email: ajaykumar_bhu@yahoo.com

Deepak Kumar, Department of Zoology, Guru Nanak Dev University, Amritsar, Punjab 143005, India. ORCID: 0000-0003-4313-4247. Email: deepdhaka415@gmail.com

Rinkee Kumari, Department of Botany, Aligarh Muslim University, Aligarh 202002, Uttar Pradesh, India. Email: krinki252@gmail.com

Rachid Lahlali, Phytopathology Unit, Department of Plant Protection, Ecole Nationale d'Agriculture de Meknès, Km 10, Rte Haj Kaddour, BP S/40, Meknès 50001, Morocco. Email: rlahlali@enameknes.ac.ma

Purificación Lisón, Instituto de Biología Molecular y Celular de Plantas (IBMCP), Consejo Superior de Investigaciones Científicas (CSIC), Universitat Politècnica de València (UPV), Ciudad Politécnica de la Innovación (CPI) 8 E, Ingeniero Fausto Elio s/n, 46011 Valencia, Spain. ORCID: 0000-0002-1662-8084. Email: plison@ibmcp.upv.es

María Pilar López-Gresa, Instituto de Biología Molecular y Celular de Plantas (IBMCP), Consejo Superior de Investigaciones Científicas (CSIC), Universitat Politècnica de València (UPV), Ciudad Politécnica de la Innovación (CPI) 8 E, Ingeniero Fausto Elio s/n, 46011 Valencia, Spain. ORCID: 0000-0001-9251-0160. E-mail: mplopez@ceqa.upv.es

Guillaume Marti, Laboratoire de Recherche en Sciences Végétales, Metatoul-AgromiX Platform, Université de Toulouse, CNRS, INP, 24 Chemin de Borde Rouge, Auzeville, Auzeville-Tolosane 31320, France. Email: guillaume.marti@univ-tlse3.fr

Riwandahun Marwein, Biological Sciences and Technology Division, CSIR-North East Institute of Science and Technology (CSIR-NEIST), Jorhat 785 006, Assam, India; Academy of Scientific and Innovative Research (AcSIR), Ghaziabad 201 002, India. Email: riwanmarwein@gmail.com

Rachid Mentag, Plant Protection Laboratory, Regional Center of Agricultural Research of Meknès, National Institute of Agricultural Research, Km 13, Route Haj Kaddour, BP 578, Meknès 50001, Morocco. Email: rachidmentag@yahoo.ca

Cathleen Mirande-Ney, University of Bordeaux, INRAE, UMR1332 BFP, 33140 Villenave d'Ornon, France. Email: cathleen.mirande-ney@inrae.fr

Saradhadevi Muthukrishnan, Department of Biochemistry, Bharathiar University, Coimbatore 641046, Tamil Nadu, India. Email: saradhadevi@buc.edu.in

Saika Nabi, Division of Genetics and Plant Breeding, Faculty of Agriculture, Sher-e-Kashmir University of Agricultural Sciences and Technology (SKUAST), Kashmir 192301, India. Email: saikanabi@skuastkashmir.ac.in

Praveen Nagella, Department of Life Sciences, School of Sciences, CHRIST (Deemed to be University), Hosur Road, Bengaluru 560029, India. ORCID: 0000-0001-6971-0527. Email: praveen.n@christuniversity.in

Dhirendra Niroula, Department of Plant Sciences and Plant Pathology, Montana State University, Montana, USA. Email: dhirendra.niroula@student.montana.edu

Puja Ohri, Department of Zoology, Guru Nanak Dev University, Amritsar, Punjab 143005, India. ORCID: 0000-0002-4528-2385. Email: ohri_puja_11@rediffmail.com

Vino Palani, Department of Botany, Sri Vidya Mandir Arts and Science College (Autonomous), Krishnagiri 636902, Tamil Nadu, India. Email: vinopalanibota@gmail.com

Ekta Pandey, Department of Botany, Aligarh Muslim University, Aligarh 202002, Uttar Pradesh, India. ORCID: 0000-0003-0251-8279. Email: epandey251@gmail.com

Manish K. Pandey, Center of Excellence in Genomics & Systems Biology (CEGSB), and Center for Pre-breeding Research (CPBR), International Crops Research Institute for the Semi-Arid Tropics (ICRISAT), Hyderabad 502324, Telangana, India. ORCID: 0000-0002-4101-6530. Email: manish.pandey@icrisat.org

Akshita Patial, Biotechnology Division, CSIR-Institute of Himalayan Bioresource Technology (CSIR-IHBT), Palampur, Himachal Pradesh 176061, India; Department of Molecular Biology and Genetic Engineering, School of Bioengineering and Biosciences, Lovely Professional University, Phagwara 144411, India. Email: patyal.akshita17@gmail.com

Parkirti, Department of Zoology, Guru Nanak Dev University, Amritsar, Punjab 143005, India. ORCID: 0000-0002-3255-7706. Email: Parkirti209@gmail.com

Julia Pérez-Pérez, Instituto de Biología Molecular y Celular de Plantas (IBMCP), Consejo Superior de Investigaciones Científicas (CSIC), Universitat Politècnica de València (UPV), Ciudad Politécnica de la Innovación (CPI) 8 E, Ingeniero Fausto Elio s/n, 46011 Valencia, Spain. ORCID: 0000-0001-9448-0700. E-mail: jperper@ibmcp.upv.es

Pierre Pétriacq, University of Bordeaux, INRAE, UMR1332 BFP, 33140 Villenave d'Ornon, France; Bordeaux Metabolome, MetaboHUB, PHENOME-EMPHASIS, Villenave d'Ornon, 33140, France. Email: pierre.petriacq@inrae.fr

Archana Prasad, Grape Physiology and Breeding Lab, Department of Fruit Tree Sciences, Institute of Plant Sciences, Agricultural Research Organization (ARO) – Volcani Institute, Rishon Lezion 7505101, Israel. ORCID: 0000-0003-3491-5893. Email: archanaambt@gmail.com

Sylvain Prigent, University of Bordeaux, INRAE, UMR1332 BFP, 33140 Villenave d'Ornon, France; Bordeaux Metabolome, MetaboHUB, PHENOME-EMPHASIS, Villenave d'Ornon, 33140, France. Email: sylvain.prigent@inrae.fr

Francisco Vázquez Prol, Instituto de Biología Molecular y Celular de Plantas (IBMCP), Consejo Superior de Investigaciones Científicas (CSIC), Universitat Politècnica de València (UPV), Ciudad Politécnica de la Innovación (CPI) 8 E, Ingeniero Fausto Elio s/n, 46011 Valencia, Spain. ORCID: 0000-0001-6044-6571. E-mail: fravazpr@posgrado.upv.es

Mohammad Ghani Raghib, Department of Horticulture, Faculty of Agriculture, Nangarhar University, Jalalabad, Afghanistan. Email: mohammadghaniraghib@gmail.com

Ramakrishnan Aruna Raji, Department of Medicine, UV Gullas College of Medicine, Banilad, Cebu 6014, Philippines. Email: rarunaraji2002@gmail.com

Shivani Ranjan, Department of Agronomy, Dr. Rajendra Prasad Central Agricultural University, Pusa, Samastipur, Bihar 848125, India. Email: ranjanshivani54@gmail.com

Seelam Giri Uma Sankar Rao, Division of Genetics and Plant Breeding, Faculty of Agriculture, Sher-e-Kashmir University of Agricultural Sciences and Technology (SKUAST), Kashmir 192301, India. Email: umasankarseelam18@gmail.com

Alberto Ribes-Navarro, Departament de Biologia, Bioquímica i Ciències Naturals, Universitat Jaume I, Castelló de la Plana, Spain. Email: alribes@uji.es

Asirvatham Alwin Robert, Department of Endocrinology and Diabetes, Prince Sultan Military Medical City, Ministry of Defense, Riyadh 11159, Saudi Arabia. Email: aalwinrobert@gmail.com

Ismael Rodrigo, Instituto de Biología Molecular y Celular de Plantas (IBMCP), Consejo Superior de Investigaciones Científicas (CSIC), Universitat Politècnica de València (UPV), Ciudad Politécnica de la Innovación (CPI) 8 E, Ingeniero Fausto Elio s/n, 46011 Valencia, Spain. ORCID: 0000-0002-8994-6112. E-mail: irodrig@ibmcp.upv.es

Rajib Roychowdhury, Agricultural Research Organization (ARO) – The Volcani Institute, Rishon Lezion 7505101, Israel; Center of Excellence in Genomics & Systems Biology (CEGSB), and Center for Pre-breeding Research (CPBR), International Crops Research Institute for the Semi-Arid Tropics (ICRISAT), Hyderabad 502324, Telangana, India. ORCID: 0000-0001-8051-7285. Email: rajibroychowdhury86@yahoo.com

Palanisamy Sampathkumar, Department of Chemistry and Biosciences, SASTRA Deemed University, Srinivasa Ramanujan Centre, Kumbakonam 612 001, Tamil Nadu, India. Email: pskumar16@yahoo.com

Saniya, Department of Horticulture, ICAR-National Research Centre for Grapes, Pune, Maharashtra, India. Email: saniyacg.17@gmail.com

Millena C. Barros Santos, University of Bordeaux, INRAE, UMR1332 BFP, 33140 Villenave d'Ornon, France; Bordeaux Metabolome, MetaboHUB, PHENOME-EMPHASIS, Villenave d'Ornon, 33140, France. Email: millena.barros-santos@inrae.fr

Chithra Devi Balasundarm Saraswathy, Department of Botany, PSGR Krishnammal College for Women, Coimbatore 641004, Tamil Nadu, India. Email: bchitradevi@psgrkcw.ac.in

Umakanta Sarker, Department of Genetics and Plant Breeding, Faculty of Agriculture, Gazipur Agricultural University, Gazipur, 1706, Bangladesh. ORCID: 0000-0002-6814-8816. Email: umakanta@gau.edu.bd

Yogita N. Sarki, Biological Sciences and Technology Division, CSIR-North East Institute of Science and Technology (CSIR-NEIST), Jorhat 785 006, Assam, India; Academy of Scientific and Innovative Research (AcSIR), Ghaziabad 201 002, India. Email: yogita.sarki3106@gmail.com

Sangeeta Sarma, Department of Botany, Jagiroad College, Jagiroad, Morigaon 782410, Assam, India. Email: sangeeta.sarma8@gmail.com

Mahmoud F. Seleiman, Department of Plant Production, College of Food and Agriculture Sciences, King Saud University, Riyadh, Saudi Arabia; Department of Crop Sciences, Faculty of Agriculture, Menoufia University, Shibin El-kom, Egypt. Email: mseleiman@ksu.edu.sa

Maksud Hasan Shah, Department of Agronomy, College of Agriculture Chandgothi, Swami Keshwanand Rajasthan Agriculture University, Bikaner, Rajasthan, India. Email: maksudhasanshah@gmail.com

Priya Shah, Center of Excellence in Genomics & Systems Biology (CEGSB), and Center for Pre-breeding Research (CPBR), International Crops Research Institute for the Semi-Arid Tropics (ICRISAT), Hyderabad 502324, Telangana, India. ORCID: 0000-0002-7566-1175. Email: Priya.Shah@icrisat.org

Akash Shaikh, Department of Life Sciences, School of Sciences, CHRIST (Deemed to be University), Hosur Road, Bengaluru – 560029, India. Email: akash.shaikh@res.christuniversity.in

Asma Shakeel, Division of Genetics and Plant Breeding, Faculty of Agriculture, Sher-e-Kashmir University of Agricultural Sciences and Technology (SKUAST), Kashmir 192301, India. ORCID: 0000-0001-5232-7089. Email: asmasm1319@skuastkashmir.ac.in

Mehnaz Shakeel, Division of Genetics and Plant Breeding, Faculty of Agriculture, Sher-e-Kashmir University of Agricultural Sciences and Technology (SKUAST), Kashmir 192301, India. Email: mehnazsm13@skuastkashmir.ac.in

Alisha Sharma, Department of Zoology, Guru Nanak Dev University, Amritsar, Punjab 143005, India. ORCID: 0009-0003-7858-5213. Email: sharmaalisha405@gmail.com

Roohi Sharma, Department of Zoology, Guru Nanak Dev University, Amritsar, Punjab 143005, India. ORCID: 0000-0001-8833-3296. Email: sharmaroohi77.rs@gmail.com

Vikas Sharma, Department of Molecular Biology and Genetic Engineering, School of Bioengineering and Biosciences, Lovely Professional University, Phagwara 144411, India. Email: vikas.25269@lpu.co.in

Asif. B. Shikari, Division of Genetics and Plant Breeding, Faculty of Agriculture, Sher-e-Kashmir University of Agricultural Sciences and Technology (SKUAST), Kashmir 192301, India. Email: asifshikari@skuastkashmir.ac.in

Anchita Singh, Department of Zoology, Guru Nanak Dev University, Amritsar, Punjab 143005, India. ORCID: 0009-0009-0041-9454. Email: anchitasinghthakur@gmail.com

Dhanawantari L. Singha, Department of Botany, Rabindranath Tagore University, Hojai 782435, Assam, India. Email: dhanawantarisingha@gmail.com

P.A. Sofi, Division of Genetics and Plant Breeding, Faculty of Agriculture, Sher-e-Kashmir University of Agricultural Sciences and Technology (SKUAST), Kashmir 192301, India. Email: parvazesofi@gmail.com

Sumit Sow, Department of Agronomy, Dr. Rajendra Prasad Central Agricultural University, Pusa, Samastipur, Bihar 848125, India. Email: sumitsow19@gmail.com

Tharani Sureshkumar, GRG BIRAC EYUVA Centre, PSGR Krishnammal College for Women, Coimbatore 641004, Tamil Nadu, India. Email: tharani@psgrkcw.ac.in

Gunna Sureshbabu Suruthi, Department of Human Genetics and Molecular Biology, Bharathiar University, Coimbatore 641046, Tamil Nadu, India. Email: sruthigunna11@gmail.com

Sivagaminathan Thangavalli, Department of Human Genetics and Molecular Biology, Bharathiar University, Coimbatore 641046, Tamil Nadu, India. Email: thangavallisuzo@gmail.com

Shivani Tiwari, Department of Biological Sciences, SRM University AP, Amaravati 522240, Andhra Pradesh, India. Email: shivani_tiwari@srmap.edu.in

Varsha Toppo, Department of Life Sciences, School of Sciences, CHRIST (Deemed to be University), Hosur Road, Bengaluru 560029, India. ORCID: 0000-0003-1967-3929. Email: varsha.toppo@res.christuniversity.in

Shanmugam Velayuthaprabhu, Department of Biotechnology, Bharathiar University, Bharathiar University, Coimbatore 641046, Tamil Nadu, India. Email: velayuthaprabhu@buc.edu.in

Francisco Vera-Sirera, Instituto de Biología Molecular y Celular de Plantas (IBMCP), Consejo Superior de Investigaciones Científicas (CSIC), Universitat Politècnica de València (UPV), Ciudad Politécnica de la Innovación (CPI) 8 E, Ingeniero Fausto Elio s/n, 46011 Valencia, Spain. ORCID: 0000-0002-1143-7692. Email: fravesi@upv.es

Parul Verma, Department of Botany, Aligarh Muslim University, Aligarh 202002, Uttar Pradesh, India. Email: vermaparul9955@gmail.com

Vikram, Department of Zoology, Guru Nanak Dev University, Amritsar, Punjab 143005, India. ORCID: 0009-0008-0724-2551. Email: vikramhomosapiens29@gmail.com

Murugesan Vinothini, Department of Human Genetics and Molecular Biology, Bharathiar University, Coimbatore 641046, Tamil Nadu, India. Email: vinothini49113@gmail.com

M. Altaf Wani, Division of Genetics and Plant Breeding, Faculty of Agriculture, Sher-e-Kashmir University of Agricultural Sciences and Technology (SKUAST), Kashmir 192301, India. Email: wanialtaf100@gmail.com

Preface

Functional genomics has been constructed intensively for decades to unravel the secrets of plant life and the complexity of plant biological networks. It has undoubtedly become the main investigating system for plant biology. Metabolomics, one of the essential components of plant functional genomics, continues to be upgraded to advance nearly all aspects of plant research. It can deal with primary and secondary metabolites to understand their role in plant growth, development, and adaptative mechanisms under stressful environments, which are promising to advance breeding programs for high-yield, high-quality, and stress-resilient crops.

Under stress, plant primary metabolisms have an array of responses to mitigate negative impacts through specific production patterns of amino acids, polyamines, carbohydrates, lipids, and so on. Using functional genomics, particularly metabolomics, to identify the genetic background underlying the diversity of plant primary metabolites will pave the way for molecular breeding to improve and develop plants' stress tolerance.

Plants sustain critical activities through the continuous production of secondary metabolites, which are essential players in the machinery of reproduction, defense, and competition in their environment. Interestingly and fortunately, a huge array of these constituents are well-recognized bioactive compounds that possess health benefits and are used as therapeutics for a wide range of diseases, many of them having been applied for thousands of years in traditional remedies or herbal medicines. Hence, the function and production of secondary metabolites derived from foods and medicinal plants have greatly interested scientists for decades. Additionally, research on secondary metabolites attracts much attention in plant protection due to their role in the mechanisms of defense against pests and pathogens.

This book summarizes the current knowledge on this topic by refining scientific literature in the research field of plant primary- and secondary-metabolite-based metabolomics, which is an essential component of plant functional genomics and uses powerful technology to profile metabolites in a high-throughput manner.

This book includes the current achievements of metabolomics in revealing the roles of both primary and secondary metabolisms of major crops and medicinal plants. It presents methods and applications of metabolomics for the exploration of stress responses and may pave the way for obtaining climate-smart and stress-tolerant crops to face biotic and abiotic stressors in the scenario of climate change. The proposed strategies based on cutting-edge approaches can advance the exploration of diverse subtopics of plant sciences as well as crop breeding for the development of future anti-stress and high-yield plants and achieve precision and sustainable agriculture. The knowledge and technologies of this book support the goals of the United Nations Sustainable Development

Goals, the Paris Agreement, and the vision of sustainable agriculture. As a reference for students, researchers, teachers, professors, and experts in the field of plant science, this book provides an overview of this crucial topic of plant science and may inspire researchers' ideas for future experiments.

The editor, Dr Jen-Tsung Chen, appreciates all chapter authors and staff of the publisher, CABI, for their valuable contributions to this book.

1 Applied Metabolomics for Studying Plant-specialized Metabolites in Plant Health, Human Health, and Food Quality

Harmonie M. Bettenhausen[1]*, Dhirendra Niroula[2], and Janak R. Joshi[2]
[1]*James B. Beam Institute for Kentucky Spirits, University of Kentucky, Kentucky, USA;* [2]*Department of Plant Sciences and Plant Pathology, Montana State University, Montana, USA*

Abstract
Metabolomics, a cross-disciplinary study emerging in the post-genomics era, is regarded as the bridge between genotypes and phenotypes. It involves studying and quantitatively analyzing metabolites and small molecules within biological systems. Plants, being rooted in place, require diverse metabolites to navigate biotic and abiotic challenges. The complexity of this metabolite matrix significantly challenges identifying and measuring specific metabolites. Enhanced access to analytical tools and databases has facilitated the integration of metabolomics into scientific domains, notably in agriculture affecting plant health, human health, and food quality. Metabolomics now aids in understanding biosynthesis pathways and provides insights into plant responses to stresses, early pathogen detection, and the quality of plant foods for human health and food processing.

Keywords: agriculture, bioactive compounds, food quality, mass spectrometry, metabolomics, plant health, plant–microbe interactions

1.1 Introduction

Metabolomics, emerging prominently in the post-genomics era (early 2000s), represents a transformative approach in modern biology, bridging the gap between genotypes and phenotypes through the comprehensive analysis of metabolites within a biological system. As the study of small molecules or metabolites, metabolomics offers a dynamic snapshot of the biochemical activities occurring in organisms at any given time. This scientific discipline has proven particularly crucial in agriculture, where understanding the metabolic responses of plants to various stresses directly impacts plant health, food quality, and human nutrition.

In agriculture, plants are immobile and face myriad biotic and abiotic challenges. The ability to navigate these challenges hinges significantly on their metabolic resilience. Metabolomics facilitates a deeper understanding of these complex biochemical networks, providing insights that are vital for improving crop resistance, enhancing food quality, and ensuring food safety.

The integration of advanced analytical tools, technologies, and databases has accelerated the adoption of metabolomics across various domains of agricultural science. This chapter delves into

* E-mail: hbettenhausen@uky.edu

© CAB International 2025. *High-Throughput Plant Metabolomics* (Jen-Tsung Chen ed.)
DOI: 10.1079/9781800626911.0001

how metabolomics is being utilized to revolutionize our approach to three pivotal areas: enhancing plant health through early detection of pathogens and stress responses, improving food quality by analyzing nutritional content and sensory attributes, and promoting human health through the discovery of bioactive compounds and personalized nutrition solutions. Each section not only explores the current applications and benefits but also addresses the challenges faced and anticipates future directions in this exciting field of study (Table 1.1).

1.2 Importance of Metabolomics in Plant Health

Plants generate a diverse array of metabolites that not only facilitate their fundamental cellular functions but also safeguard them from various biotic and abiotic threats. These metabolites can generally be classified into primary metabolites, secondary metabolites, and plant hormones. The science of metabolomics allows the identification and quantification of these metabolites within plants, providing valuable information

Table 1.1. Comprehensive overview of metabolomics.

Section	Key themes	Technologies	Applications
Metabolomics overview	Study of metabolites, bridge between genotypes and phenotypes	LC-MS, GC-MS, NMR, HPLC	Identifying and quantifying metabolites, understanding metabolic responses, biosynthesis pathways, early pathogen detection
Importance in plant health	Metabolic resilience, defense against biotic and abiotic stresses	LC-MS, 2D gel electrophoresis, GC-TOF-MS, DESI	Identifying defense metabolites, understanding stress responses, plant–microbe interactions
Agriculture	Enhancing crop resistance, food quality, and food safety	Metabolic profiling, bioinformatics	Improving crop resilience, enhancing nutritional content, ensuring food safety, biofortification
Human health	Bioactive compounds, personalized nutrition	NMR, LC-MS	Identifying health-promoting compounds, functional foods, dietary planning
Food processing	Quality control, flavor and aroma profiling, shelf-life enhancement	GC-MS, LC-MS, NMR, modified atmosphere packaging	Improving flavor and aroma profiles, extending shelf life, optimizing processing techniques, detecting contaminants
Plant–microbe interactions	Influence of plant metabolites on microbial communities	LC-MS/MS, DESI metagenomics	Shaping microbial communities, enhancing plant health through beneficial microbes
Analytical tools and techniques	Advanced analytical technologies for metabolomics	LC-MS, GC-MS, NMR, HPLC, bioinformatics	Comprehensive analysis of metabolites, identifying and quantifying small molecules, integrating with other omics technologies
Challenges and future directions	Data complexity, integration with other omics technologies, reproducibility	AI, machine learning, high-throughput metabolomics	Overcoming data analysis challenges, integrating with genomics and proteomics, improving reproducibility, developing personalized medicine and nutrition, advancing real-time metabolomics

2D, two dimensional; AI, artificial intelligence; GC, gas chromatography; HPLC, high-performance liquid chromatography; LC-MS, liquid chromatography–mass spectrometry; GC-TOF-MS, gas chromatography–time-of-flight–mass spectrometry; GC-MS, gas chromatography–mass spectrometry; NMR, nuclear magnetic resonance.

that can help clarify their roles in a variety of plant physiological processes, defense mechanisms, and biochemical pathways, as well as plant–microbe interaction.

1.2.1 Defense response against biotic challenges

Metabolomic research plays a crucial role in understanding how plants react to various phytopathogens. It has been demonstrated in numerous studies that analyzing metabolic profiles is key to pinpointing the metabolites that contribute to plant defense against biotic stressors. In one such study, liquid chromatography–mass spectrometry (LC-MS) was used to identify 496 metabolites that were more prevalent in barley cultivars resistant to *Fusarium graminearum* compared with susceptible ones (Bollina *et al.*, 2010). Metabolites like phenylpropanoids, flavonoids, fatty acids, and terpenoids were found to be more abundant in the resistant cultivars, suggesting their role in defending barley against the harmful *F. graminearum* (Bollina *et al.*, 2010). In the case of wheat's *Fusarium* head blight disease, resistance is due primarily to the activation of phenylpropanoid, terpenoid, and fatty acid metabolic pathways, as well as the detoxification of deoxynivalenol into deoxynivalenol glucoside (Lemmens *et al.*, 2007). Techniques such as two-dimensional gel electrophoresis and liquid chromatography–tandem mass spectrometry (LC-MS/MS) have been employed to study the metabolomics related to resistance to *Fusarium* head blight in wheat. Furthermore, the response of maize (*Zea mays*) to belowground invasion of the western corn rootworm, *Diabrotica virgifera virgifera*, and the defense response of the aboveground parts to another herbivore insect pest, *Spodoptera littoralis*, was studied. This led to monitoring of the defense-related metabolites that were more prevalent in *Spodoptera* spp. Metabolomic analysis of the leaf's metabolites using high-performance liquid chromatography with diode-array detection (HPLC-DAD) revealed the induction of 2,4-dihydroxy-7-methoxy-1,4-benzoxazin-3-one (DIMBOA) during root attack and an additional increase upon challenge with *S. littoralis* (Erb *et al.*, 2009).

In a similar way, three rice varieties, TN1, Kavya, and RP2068, were subjected to metabolic profiling using gas chromatography–mass spectrometry (GC-MS) after exposure to gall midge biotype 1 (GMB1). It was found that 16 fatty acids (including unsaturated linoleic acid) and two amino acids (glutamine and phenylalanine) were the major components that were more prevalent in resistant rice varieties (Goff *et al.*, 2002). Another metabolomics study of sensitive and resistant rice cultivars following infestation with the bacterial leaf blight-causing *Xanthomonas oryzae* pv. *oryzae* (Xoo) led to the identification of metabolites such as acetophenone, xanthophylls, alkaloids, carbohydrates, and lipids (Sana *et al.*, 2010). Lastly, a metabolomics study done using MS identified that flavone C-glycosides were induced in oat (*Avena sativa*) upon invasion with the nematode, and one of the flavonoids, *O*-methyl-apigenin-*C*-deoxyhexoside-*O*-hexoside, was found to be an effective protectant against two harmful nematodes of cereals, *Pratylenchus* and *Heterodera* spp. (Soriano *et al.*, 2007).

1.2.2 Elucidating protection against abiotic stresses

Metabolomics is a powerful tool for examining how plants react to a variety of abiotic stressors, including temperature fluctuations, salinity, osmotic stress, and exposure to sulfur, phosphorus, and heavy metals. In a study involving *Arabidopsis thaliana*, gas chromatography–time of flight–MS (GC-TOF-MS) was used following exposure to cold stress. The results highlighted the significant role of the C-repeat/dehydration-responsive element-binding factor 3 (CBF3) in the plant's response to cold stress (Shulaev *et al.*, 2008). In another study, a chilling-tolerant rice variety was found to exhibit increased levels of galactose and raffinose compared with a chilling-sensitive genotype when subjected to cold stress, as determined by quantitative HPLC (Morsy *et al.*, 2007). Furthermore, a comparative analysis of metabolites under 150 mM salt stress in *Thellungiella halophila* and *A. thaliana* was conducted using GC-MS profiling and microarray. The study found that certain metabolites, such as sucrose and proline, were equally increased in both species. However, *T. halophila* showed an additional increase in sugars, sugar alcohols, organic acids, proline, and phosphate (Gong *et al.*, 2005).

Metabolomics has also been employed to investigate the effect of heavy metals on plant metabolomes. A study using metabolic fingerprinting and nuclear magnetic resonance (NMR) spectroscopy, along with multivariate analysis, revealed metabolic changes in *Silene cucubalus* cell cultures treated with cadmium compared with a control group. Specifically, compounds such as malic acid and acetate were found to increase, while glutamate and branched-chain amino acids decreased in the cadmium-treated cell culture (Bailey *et al.*, 2003).

1.2.3 Exploring biochemical pathways

Numerous metabolomic studies have shed light on the regulation of various pathways under biotic and abiotic stresses. One such study examined the response of *Arabidopsis* cell culture to salt stress over a time-course experiment. Metabolomic profiling was performed using GC-MS and LC-MS at multiple time points (0.5, 1, 2, 4, 12, 24, 48, and 72 h) following a salt treatment of 100 mM NaCl. Bioinformatics analysis of the data using principal component analysis and batch-learning self-organizing mapping analysis revealed that the short-term response to salt stress involved the activation of the methylation cycle to provide methyl groups, the phenylpropanoid pathway to promote lignin synthesis, and the production of glycine. In contrast, the long-term effects included the simultaneous upregulation of glycolysis and sucrose metabolism, along with concurrent downregulation of the methylation cycle (Kim *et al.*, 2007).

In another study, researchers investigated the metabolic dynamics induced by oxidative stress caused by menadione in heterotrophic *Arabidopsis* cells. Using GC-MS profiling, the authors measured the levels of 50 polar metabolites after stress treatment and correlated these changes with alterations in mRNA levels within the same sample. The findings indicated that oxidative stress initially resulted in a significant inhibition of the tricarboxylic acid cycle (TCA) cycle and various aspects of amino acid metabolism (Baxter *et al.*, 2007). Moreover, another study revealed that under salinity, TCA cycle activity was reduced while secretion of osmoprotectants was increased in the halophyte *Salvadora persica*. The study also discovered that amino acids, sugars, and sugar alcohols function as osmoprotectants under salt stress (Kumari and Parida, 2018).

1.2.4 Metabolomics in plant–microbe interactions

A microbiome is a community of microorganisms that inhabits a specific environment. The microbiome of plants is crucial for their health, vitality, and productivity. The exudates of a plant and the composition of metabolites in its tissues can shape the microbial communities in the rhizosphere, endosphere, and phyllosphere. Numerous metabolomic studies have shown that plant metabolites can affect the microbial community, and conversely, the microbial community can influence the relative abundance of plant metabolites.

One study employed the LC-MS/MS detection method to track the structure of the microbial community and the regulation of metabolites in the rhizosphere soil of *Fritillaria unibracteata* over a 3-year cultivation period. The study discovered that the relative abundance of sucrose, palmitoleic acid, myristic acid, and oleic acid decreased with time, while the relative abundances of serotonin, L-hydroxyproline, and 2-deoxyuridine increased. The microbial community initially dominated by *Pseudomonas* and *Botrytis* spp. shifted to *Mycolicibacterium* and *Pseudogymnoascus* spp. by the end of the third year (Liu *et al.*, 2023).

In another study, researchers investigated how plant metabolites influence the microbial community and vice versa in bitter sugarcane (BS) and sweet sugarcane (SS). The study found that the soil bacterial community of the BS group had higher abundances of Bacteroidetes, Firmicutes, and Chloroflexi compared with lower abundances of Actinobacteria, Latescibacteria, and Verrucomicrobia relative to the SS group. The study also analyzed the composition of metabolites that influenced the microbial community. It was found that lamioside, primisulfuron, sitaxentan, crolibulin, 2-formylpyridine, and asparagine were overproduced in BS, while leucoside, isovitexin 2″-*O*-arabinoside, vaccarin, kuromanin, and robinin were less produced in BS compared with SS (Huang *et al.*, 2021).

Communication of plants and microbes often occurs at the rhizosphere, where plant roots secrete exudates. The chemical composition of these root exudates shapes the surrounding microbial community (Sasse *et al.*, 2018). A metabolomic analysis found that citric acid released from cucumber attracted *Bacillus amyloliquefaciens* SQR9, aiding in biofilm formation (Zhang *et al.*, 2014). Another metabolomic study used LC/MS and NMR to study two hybrid *Zea mays*, PR37Y15 and DK315, after their seeds were inoculated with *Azospirillum* spp. The study found qualitative and quantitative modifications in the contents of benzoxazinoids, a secondary metabolite, in the maize plants (Walker *et al.*, 2011). Yet another metabolomic study using the nitrogen-fixing bacteria *Herbaspirillum seropedicae* and *Azospirillum brasilense* (Nif^+) and their counterparts deficient in nitrogenase activity (Nif^-) to evaluate the impact of bacterial inoculation and nitrogen fixation on the root and leaf metabolic profiles in *Z. mays*. This study identified metabolites such as mannitol and, to a lesser extent, trehalose and isocitrate in nitrogen-fixing bacterial–plant interactions (Brusamarello-Santos *et al.*, 2017).

1.3 Importance of Metabolomics in Agriculture: Human Health and Food Processing

Metabolomics provides a unique lens through which the complex interactions within agricultural ecosystems can be understood. Particularly in the context of human health and food processing, this approach is pivotal for advancing our capacity to enhance food quality and optimize nutritional outcomes.

1.3.1 Human health

Metabolomics offers a direct pathway to enhance human health by enabling a detailed analysis of bioactive compounds in food. Foods contain myriad compounds that interact with human metabolic pathways, influencing health outcomes (Vernocchi *et al.*, 2020). By identifying and quantifying these compounds, metabolomics aids in the discovery of functional foods and the assessment of their health benefits. For instance, metabolomics has been used to identify specific phenolic compounds in berries associated with anti-inflammatory properties (Ellis, 2006). Additionally, the role of metabolomics in personalizing nutrition is emerging as a transformative approach to dietary planning, where individual metabolic responses to different foods can be systematically evaluated to tailor dietary recommendations that optimize health (Johnson *et al.*, 2016).

1.3.2 Food processing

In the realm of food processing, metabolomics has the potential to revolutionize quality control and enhancement. By profiling the metabolites that contribute to flavor and aroma, manufacturers can develop products that cater to specific consumer preferences while maintaining nutritional integrity (Pavagadhi and Swarup, 2020). For example, metabolomics has enabled the detailed characterization of volatile and non-volatile compounds in coffee, providing insights that are critical for flavor-enhancement strategies (Green *et al.*, 2024). Moreover, metabolomics aids in understanding how processing techniques affect the stability and availability of nutrients in food products. For example, Rehman and Shah (2005) demonstrated how thermal processing influences the antioxidant levels in legumes, guiding processing parameters to preserve these important compounds.

1.3.3 Quality and safety

Metabolomics also plays a crucial role in enhancing food safety and quality by detecting contaminants and adulterants that could pose health risks. The rapid screening abilities of metabolomic tools allow the detection of pesticide residues or harmful microbial metabolites in food products, ensuring compliance with safety standards and regulations (Zhao *et al.*, 2018). Furthermore, metabolomic studies can assess the impact of agricultural practices on the metabolic profile of crops, thereby providing data to support sustainable and health-promoting farming practices.

1.3.4 Bioactive compounds in plant identification and characterization

Plants produce an array of bioactive compounds that play crucial roles in health promotion and disease prevention. Metabolomics provides a powerful approach to identify and characterize these compounds by analyzing the complete set of metabolites within plant or plant-based foods. Technologies such as MS and NMR spectroscopy are commonly used in metabolomic studies for their precision in identifying molecular structures and concentrations of plant metabolites (Salem *et al.*, 2020). For instance, flavonoids, a diverse group of phytonutrients, have been identified in fruits and vegetables using metabolomic tools. These studies involve the separation of compounds using LC followed by their mass spectral analysis, which allows the detailed profiling of these substances (Pandey and Rizvi, 2009). Detailed metabolomic analyses aid in mapping out the metabolic pathways involved in flavonoid biosynthesis, offering insights into how these compounds are produced in plants. Metabolomics also enables the characterization of less common, novel bioactive compounds. For example, glucosinolates in cruciferous vegetables such as broccoli and kale have been studied for their cancer-preventive properties (Melim *et al.*, 2022). Through targeted and non-targeted metabolomic approaches, researchers have been able to identify specific glucosinolates responsible for anti-carcinogenic activity, providing valuable information for breeding strategies aimed at enhancing their content in crops (Le *et al.*, 2019).

1.3.5 Role in disease prevention and health promotion

The characterization of bioactive compounds in plants has a significant impact on human health, particularly in the realms of disease prevention and health promotion. The detailed metabolic profiles provided by metabolomic studies help in understanding the mechanisms through which these compounds use their beneficial effects. For example, polyphenols, which are abundant in berries and other fruits, are known for their antioxidant properties (Bouyahya *et al.*, 2022). Metabolomic analyses have shown that these compounds help reduce oxidative stress, a key factor in the development of chronic diseases such as cardiovascular disease and diabetes (Manach *et al.*, 2004). Such findings underscore the potential of diet-based interventions in health maintenance and disease risk reduction. Research has revealed the role of certain plant-derived metabolites in modulating the immune system. Saponins, found in high quantities in beans and legumes, have been shown to possess immunomodulatory properties. Recent studies have linked these compounds with an improved immune response, highlighting their role in disease prevention (Shi *et al.*, 2004; Gupta *et al.*, 2022). Metabolomics provides a detailed understanding of the bioactive compounds in plants, contributing significantly to the field of nutraceuticals and functional foods (Hu *et al.*, 2022). This knowledge not only promotes a better understanding of diet–health relationships but also supports the development of food products tailored for health promotion and disease prevention. The ability of metabolomics to unravel the biochemical narratives of plant foods enriches our dietary strategies and paves the way for future advancements in public health nutrition.

1.4 Diet-Related Metabolomics

1.4.1 Understanding dietary impacts on human metabolism

Diet-related metabolomics provides an insightful look into how the human body reacts to dietary inputs at the metabolic level. By analyzing the spectrum of metabolites present after dietary interventions, this field allows the examination of how nutrients and other dietary components influence both health and disease outcomes. The core of diet-related metabolomics is detailing the metabolic adjustments triggered by various foods or diets (Tebani and Bekri, 2019). This involves capturing the flux in metabolite profiles that occurs due to specific dietary inputs, thereby shedding light on the metabolic pathways influenced by nutritional components (Emwas *et al.*, 2022). A pivotal study at Colorado State University utilized metabolomics to examine the impact of fermented foods on gut health. This research demonstrated that regular consumption of fermented products significantly alters the gut microbiome and associated metabolites, enhancing gastrointestinal health and systemic immunity (Weir *et al.*, 2013, 2021).

1.4.2 Case studies linking diet and metabolic profiles

Metabolomics has been instrumental in connecting specific dietary practices with distinct metabolic signatures, providing insights into how these dietary elements can potentiate health benefits or predispose to certain diseases. Recent studies analyzed the effects of the Mediterranean diet on metabolic health, offering a wide range of health benefits (Papadaki *et al.*, 2020). Metabolomic analysis revealed significant elevations in metabolites associated with reduced inflammation and improved lipid metabolism, markers crucial for reducing the risk of cardiovascular disease and metabolic disorders (Bailey and Holscher, 2018; Papadaki *et al.*, 2020). Research into the consumption of green tea illustrates how specific components can modulate health-related metabolic pathways. Metabolomics studies have shown that catechins in green tea enhance metabolic profiles related to antioxidative capacity and weight management, providing a basis for its use in obesity and diabetes management (Janssens *et al.*, 2016; Taghizadeh *et al.*, 2017). The impact of ω-3 fatty acids on lipid metabolism has been studied extensively using metabolomics. Investigations have linked these fatty acids with beneficial effects on lipid profiles, including reductions in triglycerides and improvements in high-density lipoprotein cholesterol levels, pointing toward their protective role in cardiovascular diseases (Nigam *et al.*, 2014; Thota *et al.*, 2018). Through these case studies, metabolomics not only highlights the potential of dietary components in modulating disease risk but also underpins the development of personalized nutritional advice and therapeutic strategies. As such, metabolomics stands at the frontier of nutritional science, paving the way for diet-based interventions tailored to individual metabolic needs. This approach also offers profound contributions to our understanding of the complex interplay between diet and human metabolism. Through the analysis of metabolic changes induced by various diets and food components, this field enhances our ability to formulate precise nutritional strategies aimed at disease prevention and health optimization. The insights garnered from metabolomics research are invaluable, providing a scientific basis for tailored dietary recommendations that could lead to improved health outcomes across diverse populations. As the field evolves, it promises to deliver more nuanced understandings of nutrition and metabolism, further revolutionizing our approach to health and disease management.

1.5 Personalized Nutrition and Functional Foods

1.5.1 Tailoring diets based on metabolomic data

Personalized nutrition, tailored to individual metabolic responses, represents a cutting-edge application of metabolomics. By analyzing metabolomic profiles that reflect a person's unique biochemical environment, scientists can develop customized dietary recommendations that optimize health outcomes based on individual needs. Metabolomics allows for the identification of specific metabolic signatures that predict an individual's response to different foods. This ability is crucial in addressing metabolic disorders such as obesity, diabetes, and cardiovascular diseases, where dietary intervention can significantly impact the course of the disease (Dalle Grave *et al.*, 2010). Research has demonstrated how metabolomic profiling can distinguish between metabolically healthy and unhealthy individuals within the same body mass index range, suggesting different dietary needs for health optimization (German *et al.*, 2003). One practical application of metabolomics in personalized nutrition is seen in the development of nutrigenomic diets (Zeisel, 2007). These diets consider genetic variations and their interaction with diet, providing tailored nutritional advice that prevents or mitigates chronic diseases based on individual genetic profiles. Individuals with specific genetic markers linked to poor metabolism of fatty acids may benefit from customized diets richer in ω-3 fatty acids, tailored to their metabolic needs (Picó *et al.*, 2019; Hippe and Schiller, 2022).

1.5.2 Future of functional foods enhanced by metabolomic insights

Functional foods are those that contain bioactive compounds offering health benefits beyond basic nutrition. Metabolomics plays a key role in the development and optimization of functional foods by elucidating the mechanisms through which bioactive ingredients exert their effects (Wu *et al.*, 2022). The future of functional foods lies in enhancing their efficacy and targeting abilities through metabolomic insights. For example, the identification of bioactive peptides within dairy products that can help in hypertension management showcases how metabolomics can aid in designing foods with specific health benefits (Korhonen, 2009). Such targeted functional foods can revolutionize dietary approaches to health maintenance and disease prevention. Moreover, the future advancements in functional foods will likely see the integration of metabolomics with other omics technologies such as genomics and proteomics. This integration will enable a more comprehensive understanding of the interactions between dietary components and the human body, leading to the development of hyperfunctional foods customized not only to the metabolic needs of individuals but also to their genetic predispositions (Pinu *et al.*, 2019; Son *et al.*, 2020; Wu *et al.*, 2022). The field of personalized nutrition and functional foods is poised for transformation through the application of metabolomics. As the precision of metabolomic measurements improves and as data analytics becomes more sophisticated, the potential to develop highly individualized dietary recommendations and functionally enhanced foods becomes more tangible (Guasch-Ferré *et al.*, 2018). These advancements will undoubtedly contribute to more proactive approaches in health management and disease prevention, tailored to the unique biological signatures of individuals, heralding a new era in nutrition science.

1.6 Metabolomics in Food Quality and Processing

1.6.1 Techniques for measuring nutrient content

Metabolomics is revolutionizing the assessment of nutritional quality in food sciences, offering profound insights into nutrient content and composition that directly influence breeding and crop selection strategies. By leveraging advanced metabolomic techniques, researchers and food scientists can quantify and optimize the nutritional value of foods at a molecular level.

The application of metabolomics in measuring nutrient content involves sophisticated analytical techniques that provide detailed and comprehensive profiles of food components. Two of the most prevalent techniques in nutritional metabolomics are NMR spectroscopy and MS, often coupled with LC. NMR is invaluable for its precision in identifying and quantifying molecular structures without the need for prior separation (Selamat *et al.*, 2021). In the context of food quality, NMR can rapidly assess the content of key nutrients such as fats, sugars, and amino acids in a non-destructive manner, allowing whole-food analysis (Chowdhury *et al.*, 2023). MS, particularly when coupled with chromatographic techniques such as LC, provides sensitive and specific detection of small-molecule metabolites. This method is especially useful for profiling vitamins, antioxidants, and plant secondary metabolites, which are crucial for assessing nutritional quality. MS-based metabolomics has facilitated the detailed mapping of vitamin C in citrus fruits, highlighting its potential to guide biofortification and breeding programs aimed at enhancing nutritional profiles (Pujos-Guillot *et al.*, 2013; Perez de Souza *et al.*, 2019).

1.6.2 Impact on breeding and crop selection

Metabolomics not only aids in measuring nutrient content but also profoundly impacts the selection and breeding of crops with enhanced nutritional qualities (Sun *et al.*, 2020; Marchev *et al.*, 2021). By identifying key metabolites that correlate with desired traits, metabolomics guides the breeding of crop varieties that are nutritionally superior and more appealing to consumers. Metabolomics supports biofortification efforts by identifying genetic pathways associated with nutrient synthesis in plants. For instance, metabolomic studies have been instrumental in enhancing the micronutrient content of staple crops such as rice and wheat, aiding global efforts to combat malnutrition (Jamaloddin *et al.*, 2021). With an increased consumer demand for health-promoting foods, metabolomics helps breed crops that meet specific dietary needs. By analyzing the metabolomic profiles of fruits and vegetables, researchers can identify varieties that offer enhanced health benefits, such as higher antioxidant capacity or improved anti-inflammatory properties, thus guiding consumer choices and breeding strategies (Sun *et al.*, 2020; Hotea *et al.*, 2023). The role of metabolomics in the assessment of nutritional quality extends beyond simple measurement to influencing the very methods by which foods are bred and developed. This approach not only helps in optimizing the nutrient content of foods but also ensures that the crops developed are aligned with health directives and consumer preferences, ultimately enhancing the overall quality and appeal of agricultural products. As metabolomics continues to evolve, its integration into food quality assessment and crop development is expected to yield even more significant insights and tangible benefits in the field of food science.

1.7 Expanded Flavor and Aroma Profiling in Food and Beverage Industries

1.7.1 Advanced techniques for profiling flavor and aroma components

The application of metabolomics in flavor and aroma profiling involves sophisticated analytical tools that decode complex sensory attributes of food and beverages, influencing product development and consumer satisfaction. GC-MS is necessary for analyzing volatile organic compounds that define a product's aroma (Perez-Hurtado *et al.*, 2017). The ability of GC-MS to separate and identify these compounds provides a detailed aroma profile crucial for products such as coffee, where aroma plays a significant role in consumer preference. For instance, recent studies using GC-MS have identified over 800 volatile compounds in coffee, which are critical for tailoring roasting processes to enhance desired aromatics (Caporaso *et al.*, 2018; Yu *et al.*, 2021). LC-MS is adept at profiling non-volatile taste compounds such as phenolics, peptides, and other flavor precursors. The technique is indispensable in industries such as winemaking, where it is used to profile tannins and other polyphenols that affect the taste and mouthfeel of wine (Flamini *et al.*, 2015; Pinu, 2018; Pereira-Coelho *et al.*, 2023).

1.7.2 Integration with sensory science in fermented beverages

The integration of metabolomic data with sensory evaluation and consumer feedback is transforming product-development strategies. This approach allows a more nuanced understanding of how molecular composition affects sensory perceptions and consumer preferences. Combining sensory data with metabolomics provides a holistic view of how specific metabolites contribute to perceived flavors and aromas (Bettenhausen *et al.*, 2018, 2020, 2021). For example, research in craft beer has utilized sensory metabolomics to link flavor profiles with consumer preferences, guiding brewers in recipe adjustment to suit market trends (Bettenhausen *et al.*, 2020; Cavallini *et al.*, 2021; Schreurs *et al.*, 2024). Metabolomics enables manufacturers to innovate based on precise consumer preference data. In the distilling industry, metabolomics assists in crafting spirits with desired flavor profiles by analyzing the impact of different fermentation conditions, ageing processes, and raw materials (Lachenmeier *et al.*, 2005; Lahne, 2010; Kew *et al.*, 2017; Roullier-Gall *et al.*, 2018; Arnold, 2020).

The convergence of metabolomics with the brewing and distilling industries represents a significant advance in flavor science, enabling more precise control over the creation of complex flavor profiles in beverages such as beer and spirits. In brewing, metabolomics is used to dissect the biochemical pathways that lead to the formation of key flavor compounds during fermentation. For example, yeast metabolism not only influences ethanol production but also plays a crucial role in the formation of esters, fusel alcohols, and phenolic compounds, which contribute significantly to the aroma and taste of beer (Caudy *et al.*, 2017; Roullier-Gall *et al.*, 2018, 2020). Researchers are now able to identify specific yeast strains that can produce desired flavor profiles or modify existing ones to enhance certain sensory attributes. This tailored approach helps brewers meet consumer demands for both traditional and innovative flavors, ensuring consistency and quality in production. Similarly, in the distilling industry, metabolomics offers insights into the complex interactions between raw materials and fermentation conditions that define the character of distilled spirits. The composition of the mash, the type of yeast used, fermentation temperature, and duration are all variables that influence the metabolomic profile of the end product (Lachenmeier *et al.*, 2005; Power *et al.*, 2021). Distillers can use these data to optimize the production process to achieve a specific balance of flavors. For instance, the development of signature flavors in whiskey, such as those derived from lignin breakdown products during barrel ageing, can be finely tuned by understanding the metabolic pathways involved in their formation (Kelly *et al.*, 2023). Moreover, metabolomic profiling is increasingly employed to investigate the ageing process itself, identifying key compounds that contribute to the maturity of a spirit, which can lead to innovations in accelerating the ageing process without compromising quality (Awad *et al.*, 2017).

The application of metabolomics extends to quality control and authenticity verification in both brewing and distilling. By creating metabolic fingerprints of different beverages, manufacturers can not only ensure product consistency across different batches but also protect their brands against counterfeiting. This aspect is critical in premium product markets where authenticity plays a major role in consumer preference and pricing (Selamat *et al.*, 2021; Mialon *et al.*, 2023). The potential of metabolomics in these industries is also expanding into environmental sustainability and resource management. By analyzing the metabolic profiles of waste byproducts in brewing and distilling, companies can explore ways to repurpose these materials into valuable products such as biofuels, animal feed, or even water purification agents, thus reducing waste and enhancing the sustainability of their operations (Chattaraj *et al.*, 2024). Metabolomics is not simply enhancing the flavor and quality of alcoholic beverages but is also pioneering new frontiers in sustainability and production efficiency (Sarkar *et al.*, 2023). As this field evolves, it will continue to empower brewers and distillers to craft superior products that meet modern consumer tastes and environmental standards. This holistic approach will inevitably drive innovation, leading to new product development and enhanced market competitiveness.

1.7.3 Impact on food design and consumer preferences

Metabolomics is not just a tool for analysis; it is a driver of innovation in food design, directly influenced by evolving consumer demands for healthier and more flavorful options (Kim *et al.*, 2016). With health becoming a central consumer focus, metabolomics is instrumental in developing products that are both healthful and flavorful. Identifying specific metabolites that can enhance the sweetness of fruits without adding sugar can help address health concerns such as diabetes while satisfying sweetness cravings (Colantonio *et al.*, 2022). As global dietary trends shift toward plant-based diets, metabolomics helps enhance the flavor profiles of plant-based meats to make them more appealing to a broader audience. By identifying and tweaking metabolites responsible for meat-like flavors and aromas, manufacturers can improve the sensory attributes of plant-based products (Kaczmarska *et al.*, 2021; van Vliet *et al.*, 2021). Application of metabolomics in flavor and aroma profiling not only deepens our understanding of food composition but also drives innovation across the food and beverage industries. By aligning product development with precise consumer preferences and health trends, metabolomics fosters the creation of products that are delicious, healthy, and highly marketable. This science-based approach ensures that new food products not only meet but exceed consumer expectations in a competitive global market.

1.7.4 Influence on postharvest processes and how metabolomics can guide postharvest handling

The postharvest handling of agricultural products plays a critical role in maintaining food quality and safety, and extending shelf life. Metabolomics offers a detailed snapshot of the metabolic state of produce postharvest, which can be critical for determining the optimal conditions for storage and handling. Metabolomic analysis can identify stress responses in fruits and vegetables triggered by harvesting, which may lead to spoilage or quality degradation (Fatima *et al.*, 2021). By monitoring these stress-related metabolites, adjustments can be made to postharvest processes to minimize stress and maintain quality (Moing *et al.*, 2020; Yun *et al.*, 2022; Martínez-Rivas and Fernie, 2024). Metabolomics can also guide decisions regarding the appropriate temperature and atmosphere for storage (Yokota *et al.*, 2019). Different crops have unique metabolic profiles that dictate their storage needs to prevent over-ripening or decay. Metabolomics has been used to determine the optimal storage conditions for apples, identifying key metabolites that are affected by different temperatures and controlled atmospheres, thus informing storage protocols that extend freshness and reduce waste (Amyotte *et al.*, 2017; Xu *et al.*, 2020; Zenoni *et al.*, 2023).

1.7.5 Enhancing food preservation and shelf life

Beyond guiding handling and storage, metabolomics also enhances food preservation techniques and extends the shelf life of products. By analyzing how metabolic processes change during storage and under different preservation methods, metabolomics helps optimize food-processing techniques to retain nutritional quality and sensory attributes. One area where metabolomics has proven especially valuable is in the modification of packaging technologies. Metabolomics can assess the effectiveness of different packaging materials and technologies, such as modified atmosphere packaging, in preserving the desired metabolic profile of food products. This approach has been applied successfully to extend the shelf life of perishable products such as meat and dairy by optimizing the gas composition in packaging to slow down microbial growth and enzymatic activities that lead to spoilage (Ranjitha *et al.*, 2015). Furthermore, metabolomics facilitates the development of natural preservatives by identifying plant-derived compounds with antimicrobial and antioxidant properties. For example, research has demonstrated the potential of certain plant extracts to extend the shelf life of fresh produce by targeting and inhibiting specific spoilage-related metabolites (Salem *et al.*, 2020; Beya *et al.*, 2021; Sieniawska and Georgiev, 2022).

The integration of metabolomics into postharvest processes revolutionizes the way food quality and safety are managed from farm to table. By providing a deep understanding of the metabolic changes during postharvest handling and storage, metabolomics guides the optimization of conditions to enhance preservation and extend shelf life. As this field advances, its application in postharvest technology promises to reduce food loss, improve economic returns, and ensure that consumers have access to high-quality, fresh, and nutritious products. This science-driven approach not only aligns with sustainable food production goals but also meets the growing consumer demand for high-standard and minimally processed foods.

1.8 Challenges and Future Directions in Metabolomics: Current Limitations in the Field

Despite significant advancements, metabolomics faces several challenges that can impact efficacy and broader application. One major limitation is the complexity of metabolomic data, which often requires sophisticated tools and algorithms for analysis and interpretation. The sheer volume and diversity of metabolites present in biological samples can make data processing and accurate metabolite identification challenging, particularly when distinguishing isomers and dealing with highly complex mixtures (Scalbert *et al.*, 2009; Johnson and Gonzalez, 2012; Di Minno *et al.*, 2022;). Another limitation is the integration of metabolomics with other omics technologies such as genomics, proteomics, and transcriptomics.

While such integration promises a comprehensive understanding of biological systems, it requires seamless coordination and data integration across different scales of biological organization, which is not yet fully streamlined (Buescher and Driggers, 2016; Chu *et al.*, 2019; Yu *et al.*, 2022). Additionally, the reproducibility of metabolomic studies can be a concern due to variability in sample preparation, analytical conditions, and instrument sensitivity. Standardizing methodologies across laboratories remains a significant hurdle, which is essential for translating metabolomic findings from the bench to clinical and industrial applications (Fernie *et al.*, 2011).

1.9 Potential Breakthroughs and Future Research Areas

Despite these challenges, the future of metabolomics is bright, with potential breakthroughs on the horizon. One promising area is the development of artificial intelligence and machine learning models to handle complex metabolomic data more effectively. These technologies have the potential to revolutionize how data are analyzed, enabling faster and more accurate predictions of metabolic pathways and biomarker discovery (Galal *et al.*, 2022; Zhang *et al.*, 2023; Schreurs *et al.*, 2024).

Another exciting development is the expansion of metabolomics applications in personalized medicine and nutrition. As metabolic profiles can provide unique insights into individual health statuses, metabolomics is expected to play a crucial role in tailoring dietary and medical interventions to enhance personal health outcomes (Hippe and Schiller, 2022). Furthermore, advancements in high-throughput and real-time metabolomics are set to enhance the speed and efficiency of data collection, significantly impacting agricultural practices, food safety, and environmental monitoring (Fuhrer and Zamboni, 2015; Selamat *et al.*, 2021; Khanijou *et al.*, 2022; Wu *et al.*, 2022).

In conclusion, while metabolomics faces certain limitations that challenge its current applications, ongoing technological advancements and interdisciplinary research are opening new frontiers. These developments promise not only to overcome existing hurdles but also to expand the capabilities of metabolomics in revolutionizing health, nutrition, and environmental sciences.

1.10 References

Amyotte, B., Bowen, A.J., Banks, T., Rajcan, I., and Somers, D.J. (2017) Mapping the sensory perception of apple using descriptive sensory evaluation in a genome wide association study. *PLOS ONE* 12: e0171710.

Arnold, R. (2020) *The Terroir of Whiskey: A Distiller's Journey into the Flavor of Place.* Columbia University Press, New York.

Awad, P., Athès, V., Decloux, M.E., Ferrari, G., Snakkers, G., *et al.* (2017) Evolution of volatile compounds during the distillation of cognac spirit. *Journal of Agricultural and Food Chemistry* 65, 7736–7748.

Bailey, M.A. and Holscher, H.D. (2018) Microbiome-mediated effects of the mediterranean diet on inflammation. *Advances in Nutrition* 9, 193–206.

Bailey, N.J.C., Oven, M., Holmes, E., Nicholson, J.K., and Zenk, M.H. (2003) Metabolomic analysis of the consequences of cadmium exposure in *Silene cucubalus* cell cultures via ^{1}H NMR spectroscopy and chemometrics. *Phytochemistry* 62, 851–858.

Baxter, C.J., Redestig, H., Schauer, N., Repsilber, D., Patil, K.R., *et al.* (2007) The metabolic response of heterotrophic *Arabidopsis* cells to oxidative stress. *Plant Physiology* 143, 312–325.

Bettenhausen, H.M., Barr, L., Broeckling, C.D., Chaparro, J.M., Holbrook, C., *et al.* (2018) Influence of malt source on beer chemistry, flavor, and flavor stability. *Food Research International* 113, 487–504.

Bettenhausen, H.M., Benson, A., Fisk, S., Herb, D., Hernandez, J., *et al.* (2020) Variation in sensory attributes and volatile compounds in beers brewed from genetically distinct malts: an integrated sensory and non-targeted metabolomics approach. *Journal of the American Society of Brewing Chemists* 78, 136–152.

Bettenhausen, H.M., Barr, L., Omerigic, H., Yao, L., and Heuberger, A.L. (2021) Mass spectrometry metabolomics of hot steep malt extracts and association to sensory traits. *Journal of the American Society of Brewing Chemists* 79, 394–406.

Beya, M.M., Netzel, M.E., Sultanbawa, Y., Smyth, H., and Hoffman, L.C. (2021) Plant-based phenolic molecules as natural preservatives in comminuted meats: a review. *Antioxidants* 10: 263.

Bollina, V., Kumaraswamy, G.K., Kushalappa, A.C., Choo, T.M., Dion, Y., *et al.* (2010) Mass spectrometry-based metabolomics application to identify quantitative resistance-related metabolites in barley against *Fusarium* head blight. *Molecular Plant Pathology* 11, 769–782.

Bouyahya, A., Omari, N.E., El Hachlafi, N., Jemly, M.E., Hakkour, M., *et al.* (2022) Chemical compounds of berry-derived polyphenols and their effects on gut microbiota, inflammation, and cancer. *Molecules* 27: 3286.

Brusamarello-Santos, L.C., Gilard, F., Brulé, L., Quilleré, I., Gourion, B., *et al.* (2017) Metabolic profiling of two maize (*Zea mays* L.) inbred lines inoculated with the nitrogen fixing plant-interacting bacteria *Herbaspirillum seropedicae* and *Azospirillum brasilense*. *PLOS ONE* 12: e0174576.

Buescher, J.M. and Driggers, E.M. (2016) Integration of omics: more than the sum of its parts. *Cancer and Metabolism* 4: 4.

Caporaso, N., Whitworth, M.B., Cui, C., and Fisk, I.D. (2018) Variability of single bean coffee volatile compounds of *Arabica* and robusta roasted coffees analysed by SPME-GC-MS. *Food Research International* 108, 628–640.

Caudy, A.A., Mülleder, M., and Ralser, M. (2017) Metabolomics in yeast. *Cold Spring Harbor Protocols* 2017: pdb.top083576.

Cavallini, N., Savorani, F., Bro, R., and Cocchi, M. (2021) A metabolomic approach to beer characterization. *Molecules* 26: 1472.

Chattaraj, S., Mitra, D., Ganguly, A., Thatoi, H., and Das Mohapatra, P.K. (2024) A critical review on the biotechnological potential of brewers' waste: challenges and future alternatives. *Current Research in Microbial Sciences* 6: 100228.

Chowdhury, C.R., Kavitake, D., Jaiswal, K.K., Jaiswal, K.S., Reddy, G.B., *et al.* (2023) NMR-based metabolomics as a significant tool for human nutritional research and health applications. *Food Bioscience* 53: 102538.

Chu, S., Huang, M., Kelly, R., Benedetti, E., Siddiqui, J., *et al.* (2019) Integration of metabolomic and other omics data in population-based study designs: an epidemiological perspective. *Metabolites* 9: 117.

Colantonio, V., Ferrão, L.F.V., Tieman, D.M., Bliznyuk, N., Sims, C., *et al.* (2022) Metabolomic selection for enhanced fruit flavor. *Proceedings of the National Academy of Sciences* 119: e2115865119.

Dalle Grave, R., Calugi, S., Centis, E., Marzocchi, R., El Ghoch, M., and Marchesini, G. (2010) Lifestyle modification in the management of the metabolic syndrome: achievements and challenges. *Diabetes, Metabolic Syndrome and Obesity* 3, 373–385.

Di Minno, A., Gelzo, M., Caterino, M., Costanzo, M., Ruoppolo, M., and Castaldo, G. (2022) Challenges in metabolomics-based tests, biomarkers revealed by metabolomic analysis, and the promise of the application of metabolomics in precision medicine. *International Journal of Molecular Sciences* 23: 5213.

Ellis, D. (2006) Metabolic fingerprinting in disease diagnosis: biomedical applications of infrared and Raman spectroscopy. *Analyst* 131, 875–885.

Emwas, A.H., Szczepski, K., Al-Younis, I., Lachowicz, J.I., and Jaremko, M. (2022) Fluxomics—new metabolomics approaches to monitor metabolic pathways. *Frontiers in Pharmacology* 13: 805782.

Erb, M., Gordon-Weeks, R., Flors, V., Camañes, G., Turlings, T.C.J., and Ton, J. (2009) Belowground ABA boosts aboveground production of DIMBOA and primes induction of chlorogenic acid in maize. *Plant Signaling and Behavior* 4, 636–638.

Fatima, E.A., Moha, T., Said, W., Abdelilah, M., and Mohammed, R. (2021) Use of metabolomics data analysis to identify fruit quality markers enhanced by the application of an aminopolysaccharide. *RSC Advances* 11, 35514–35524.

Fernie, A.R., Aharoni, A., Willmitzer, L., Stitt, M., Tohge, T., *et al.* (2011) Recommendations for reporting metabolite data. *Plant Cell* 23, 2477–2482.

Flamini, R., de Rosso, M., and Bavaresco, L. (2015) Study of grape polyphenols by liquid chromatography-high-resolution mass spectrometry (UHPLC/QTOF) and suspect screening analysis. *Journal of Analytical Methods in Chemistry* 2015: 350259.

Fuhrer, T. and Zamboni, N. (2015) High-throughput discovery metabolomics. *Current Opinion in Biotechnology* 31, 73–78.

Galal, A., Talal, M., and Moustafa, A. (2022) Applications of machine learning in metabolomics: disease modeling and classification. *Frontiers in Genetics* 13: e2400305.

German, J.B., Roberts, M.A., and Watkins, S.M. (2003) Personal metabolomics as a next generation nutritional assessment. *Journal of Nutrition* 133, 4260–4266.

Goff, S.A., Ricke, D., Lan, T.H., Presting, G., Wang, R., *et al.* (2002) A draft sequence of the rice genome (*Oryza sativa* L. ssp. *japonica*). *Science* 296, 92–100.

Gong, Q., Li, P., Ma, S., Indu Rupassara, S., and Bohnert, H.J. (2005) Salinity stress adaptation competence in the extremophile *Thellungiella halophila* in comparison with its relative *Arabidopsis thaliana*. *Plant Journal* 44, 826–839.

Green, S., Fanning, E., Sim, J., Eyres, G. T., Frew, R., and Kebede, B. (2024) The potential of NIR spectroscopy and chemometrics to discriminate roast degrees and predict volatiles in coffee. *Molecules* 29: 318.

Guasch-Ferré, M., Bhupathiraju, S.N., and Hu, F.B. (2018) Use of metabolomics in improving assessment of dietary intake. *Clinical Chemistry* 64, 82–98.

Gupta, S., Schillaci, M., and Roessner, U. (2022) Metabolomics as an emerging tool to study plant–microbe interactions. *Emerging Topics in Life Sciences* 6, 175–183.

Hippe, B. and Schiller, M. (2022) Translational aspects in precision nutrition, personalization, biomarkers and healthy aging. In: Haslberger, A.G. (ed.) *Advances in Precision Nutrition, Personalization and Healthy Aging*. Springer International Publishing, Cham, Switzerland, pp. 279–299.

Hotea, I., Sirbu, C., Plotuna, A.M., Tîrziu, E., Badea, C., *et al.* (2023) Integrating (nutri-) metabolomics into the One Health tendency—the key for personalized medicine advancement. *Metabolites* 13: 800.

Hu, S., Liu, C., and Liu, X. (2022) Innovative application of metabolomics on bioactive ingredients of foods. *Foods* 11: 2974.

Huang, W., Sun, D., Chen, L., and An, Y. (2021) Integrative analysis of the microbiome and metabolome in understanding the causes of sugarcane bitterness. *Scientific Reports* 11: 6024.

Jamaloddin, M., Maliha, A., Gokulan, C.G., Gaur, N., and Patel, H.K. (2021) Metabolomics-assisted breeding for crop improvement: an emerging approach. In: *Omics Technologies for Sustainable Agriculture and Global Food Security, Vol. 1*. Springer, Singapore, pp. 241–279.

Janssens, P.L.H.R., Hursel, R., and Westerterp-Plantenga, M.S. (2016) Nutraceuticals for body-weight management: the role of green tea catechins. *Physiology and Behavior* 162, 83–87.

Johnson, C.H. and Gonzalez, F.J. (2012) Challenges and opportunities of metabolomics. *Journal of Cellular Physiology* 227, 2975–2981.

Johnson, C. H., Ivanisevic, J., and Siuzdak, G. (2016) Metabolomics: beyond biomarkers and towards mechanisms. *Nature Reviews Molecular Cell Biology* 17, 451–459.

Kaczmarska, K., Taylor, M., Piyasiri, U., and Frank, D. (2021) Flavor and metabolite profiles of meat, meat substitutes, and traditional plant-based high-protein food products available in Australia. *Foods* 10: 801.

Kelly, T.J., O'Connor, C., and Kilcawley, K.N. (2023) Sources of volatile aromatic congeners in whiskey. *Beverages* 9: 64.

Kew, W., Goodall, I., Clarke, D., and Uhrín, D. (2017) Chemical diversity and complexity of scotch whisky as revealed by high-resolution mass spectrometry. *Journal of the American Society for Mass Spectrometry* 28, 200–213.

Khanijou, J.K., Kulyk, H., Bergès, C., Khoo, L.W., Ng, P., *et al.* (2022) Metabolomics and modelling approaches for systems metabolic engineering. *Metabolic Engineering Communications* 15: e00209.

Kim, J.K., Bamba, T., Harada, K., Fukusaki, E., and Kobayashi, A. (2007) Time-course metabolic profiling in *Arabidopsis thaliana* cell cultures after salt stress treatment. *Journal of Experimental Botany* 58, 415–424.

Kim, S., Kim, J., Yun, E.J., and Kim, K.H. (2016) Food metabolomics: from farm to human. *Current Opinion in Biotechnology* 37, 16–23.

Korhonen, H. (2009) Milk-derived bioactive peptides: from science to applications. *Journal of Functional Foods* 1, 177–187.

Kumari, A. and Parida, A.K. (2018) Metabolomics and network analysis reveal the potential metabolites and biological pathways involved in salinity tolerance of the halophyte *Salvadora persica*. *Environmental and Experimental Botany* 148, 85–99.

Lachenmeier, D.W., Richling, E., López, M.G., Frank, W., and Schreier, P. (2005) Multivariate analysis of FTIR and ion chromatographic data for the quality control of tequila. *Journal of Agricultural and Food Chemistry* 53, 2151–2157.

Lahne, J. (2010) Aroma characterization of American rye whiskey by chemical and sensory assays. M.Sc. thesis, University of Illinois at Urbana-Champaign, Urbana, Illinois.

Le, T.N., Luong, H.Q., Li, H.P., Chiu, C.H., and Hsieh, P.C. (2019) Broccoli (*Brassica oleracea* L. var. *italica*) sprouts as the potential food source for bioactive properties: a comprehensive study on *in vitro* disease models. *Foods* 8: 532.

Lemmens, M., Scholz, U., Berthiller, F., Dall'Asta, C., Koutnik, A., *et al.* (2007) The ability to detoxify the mycotoxin deoxynivalenol colocalizes with a major quantitative trait locus for *Fusarium* head blight resistance in wheat. *Molecular Plant–Microbe Interactions* 18, 1318–1324.

Liu, C., Yu, J., Ying, J., Zhang, K., Hu, Z., *et al.* (2023) Integrated metagenomics and metabolomics analysis reveals changes in the microbiome and metabolites in the rhizosphere soil of *Fritillaria unibracteata*. *Frontiers in Plant Science* 14: 1223720.

Manach, C., Scalbert, A., Morand, C., Rémésy, C., and Jiménez, L. (2004) Polyphenols: food sources and bioavailability. *American Journal of Clinical Nutrition* 79, 727–747.

Marchev, A.S., Vasileva, L.V., Amirova, K.M., Savova, M.S., Balcheva-Sivenova, Z.P., and Georgiev, M.I. (2021) Metabolomics and health: from nutritional crops and plant-based pharmaceuticals to profiling of human biofluids. *Cellular and Molecular Life Sciences* 78, 6487–6503.

Martínez-Rivas, F.J. and Fernie, A.R. (2024) Metabolomics to understand metabolic regulation underpinning fruit ripening, development, and quality. *Journal of Experimental Botany* 75, 1726–1740.

Melim, C., Lauro, M.R., Pires, I.M., Oliveira, P.J., and Cabral, C. (2022) The role of glucosinolates from cruciferous vegetables (Brassicaceae) in gastrointestinal cancers: from prevention to therapeutics. *Pharmaceutics* 14: 190.

Mialon, N., Roig, B., Capodanno, E., and Cadiere, A. (2023) Untargeted metabolomic approaches in food authenticity: a review that showcases biomarkers. *Food Chemistry* 398: 133856.

Moing, A., Pétriacq, P., and Osorio, S. (2020) Special issue on fruit metabolism and metabolomics. *Metabolites* 10: 230.

Morsy, M.R., Jouve, L., Hausman, J.F., Hoffmann, L., and Stewart, J.M. (2007) Alteration of oxidative and carbohydrate metabolism under abiotic stress in two rice (*Oryza sativa* L.) genotypes contrasting in chilling tolerance. *Journal of Plant Physiology* 164, 157–167.

Nigam, A., Talajic, M., Roy, D., Nattel, S., Lambert, J., *et al.* (2014) Fish oil for the reduction of atrial fibrillation recurrence, inflammation, and oxidative stress. *Journal of the American College of Cardiology* 64, 1441–1448.

Pandey, K.B. and Rizvi, S.I. (2009) Plant polyphenols as dietary antioxidants in human health and disease. *Oxidative Medicine and Cellular Longevity* 2, 270–278.

Papadaki, A., Nolen-Doerr, E., and Mantzoros, C.S. (2020) The effect of the mediterranean diet on metabolic health: a systematic review and meta-analysis of controlled trials in adult. *Nutrients* 12: 3342.

Pavagadhi, S. and Swarup, S. (2020) Metabolomics for evaluating flavor-associated metabolites in plant-based products. *Metabolites* 10: 197.

Pereira-Coelho, M., Haas, I.C.D.S., Katrin Reinke, C., Dognini, J., Amboni, R.D.D.M.C., *et al.* (2023) A green analytical method for the determination of polyphenols in wine by dispersive pipette extraction and LC-MS/MS. *Food Chemistry* 405: 134860.

Perez de Souza, L., Alseekh, S., Naake, T., and Fernie, A. (2019) Mass spectrometry-based untargeted plant metabolomic. *Current Protocols in Plant Biology* 4: e20100.

Perez-Hurtado, P., Palmer, E., Owen, T., Aldcroft, C., Allen, M.H., *et al.* (2017) Direct analysis of volatile organic compounds in foods by headspace extraction atmospheric pressure chemical ionisation mass spectrometry. *Rapid Communications in Mass Spectrometry* 31, 1947–1956.

Picó, C., Serra, F., Rodríguez, A.M., Keijer, J., and Palou, A. (2019) Biomarkers of nutrition and health: new tools for new approaches. *Nutrients* 11: 1092.

Pinu, F.R. (2018) Grape and wine metabolomics to develop new insights using untargeted and targeted approaches. *Fermentation* 4: 92.

Pinu, F.R., Beale, D.J., Paten, A.M., Kouremenos, K., Swarup, S., *et al.* (2019) Systems biology and multi-omics integration: viewpoints from the metabolomics research community. *Metabolites* 9: 76.

Power, A.C., Jones, J., NiNeil, C., Geoghegan, S., Warren, S., *et al.* (2021) What's in this drink? Classification and adulterant detection in Irish Whiskey samples using near infrared spectroscopy combined with chemometrics. *Journal of the Science of Food and Agriculture* 101, 5256–5263.

Pujos-Guillot, E., Hubert, J., Martin, J.F., Lyan, B., Quintana, M., *et al.* (2013) Mass spectrometry-based metabolomics for the discovery of biomarkers of fruit and vegetable intake: citrus fruit as a case study. *Journal of Proteome Research* 12, 1645–1659.

Ranjitha, K., Sudhakar Rao, D.V., Shivashankara, K.S., and Roy, T.K. (2015) Effect of pretreatments and modified atmosphere packaging on the shelf life and quality of fresh-cut green bell pepper. *Journal of Food Science and Technology* 52, 7872–7882.

Rehman, Z.U. and Shah, W.H. (2005) Thermal heat processing effects on antinutrients, protein and starch digestibility of food legumes. *Food Chemistry* 91, 327–331.

Roullier-Gall, C., Signoret, J., Hemmler, D., Witting, M.A., Kanawati, B., *et al.* (2018) Usage of FT-ICR-MS metabolomics for characterizing the chemical signatures of barrel-aged whisky. *Frontiers in Chemistry* 6: 29.

Roullier-Gall, C., David, V., Hemmler, D., Schmitt-Kopplin, P., and Alexandre, H. (2020) Exploring yeast interactions through metabolic profiling. *Scientific Reports* 10: 6073.

Salem, M.A., Perez de Souza, L., Serag, A., Fernie, A.R., Farag, M.A., *et al.* (2020) Metabolomics in the context of plant natural products research: from sample preparation to metabolite analysis. *Metabolites* 10: 37.

Sana, T.R., Fischer, S., Wohlgemuth, G., Katrekar, A., Jung, K.H., *et al.* (2010) Metabolomic and transcriptomic analysis of the rice response to the bacterial blight pathogen *Xanthomonas oryzae* pv. *oryzae*. *Metabolomics* 6, 451–465.

Sarkar, S., Sha, S.P., and Ghatani, K. (2023) Metabolomics of ethnic fermented foods and beverages: understanding new aspects through omic techniques. *Frontiers in Sustainable Food Systems* 7: 1040567.

Sasse, J., Martinoia, E., and Northen, T. (2018) Feed your friends: do plant exudates shape the root microbiome? *Trends in Plant Science* 23, 25–41.

Scalbert, A., Brennan, L., Fiehn, O., Hankemeier, T., Kristal, B.S., *et al.* (2009) Mass-spectrometry-based metabolomics: limitations and recommendations for future progress with particular focus on nutrition research. *Metabolomics* 5, 435–458.

Schreurs, M., Piampongsant, S., Roncoroni, M., Cool, L., Herrera-Malaver, B., *et al.* (2024) Predicting and improving complex beer flavor through machine learning. *Nature Communications* 15: 2368.

Selamat, J., Rozani, N.A.A., and Murugesu, S. (2021) Application of the metabolomics approach in food authentication. *Molecules* 26: 7565.

Shi, J., Arunasalam, K., Yeung, D., Kakuda, Y., Mittal, G., and Jiang, Y. (2004) Saponins from edible legumes: chemistry, processing, and health benefits. *Journal of Medicinal Food* 7, 67–78.

Shulaev, V., Cortes, D., Miller, G., and Mittler, R. (2008) Metabolomics for plant stress response. *Physiologia Plantarum* 132, 199–208.

Sieniawska, E. and Georgiev, M.I. (2022) Metabolomics: towards acceleration of antibacterial plant-based leads discovery. *Phytochemistry Reviews* 21, 765–781.

Son, J.W., Shoaie, S., and Lee, S. (2020) Systems biology: a multi-omics integration approach to metabolism and the microbiome. *Endocrinology and Metabolism* 35, 507–514.

Soriano, I.R., Asenstorfer, R.E., Schmidt, O., and Riley, I.T. (2007) Inducible flavone in oats (*Avena sativa*) is a novel defense against plant-parasitic nematodes. *Phytopathology* 94, 1207–1214.

Sun, W., Chen, Z., Hong, J., and Shi, J. (2020) Promoting human nutrition and health through plant metabolomics: current status and challenges. *Biology* 10: 20.

Taghizadeh, M., Farzin, N., Taheri, S., Mahlouji, M., Akbari, H., *et al.* (2017) The effect of dietary supplements containing green tea, capsaicin and ginger extracts on weight loss and metabolic profiles in overweight women: a randomized double-blind placebo-controlled clinical trial. *Annals of Nutrition and Metabolism* 70, 277–285.

Tebani, A. and Bekri, S. (2019) Paving the way to precision nutrition through metabolomics. *Frontiers in Nutrition* 6: 41.

Thota, R.N., Ferguson, J.J.A., Abbott, K.A., Dias, C.B., and Garg, M.L. (2018) Science behind the cardio-metabolic benefits of omega-3 polyunsaturated fatty acids: biochemical effects *vs.* clinical outcomes. *Food and Function* 9, 3576–3596.

van Vliet, S., Bain, J.R., Muehlbauer, M.J., Provenza, F.D., Kronberg, S.L. *et al.* (2021) A metabolomics comparison of plant-based meat and grass-fed meat indicates large nutritional differences despite comparable nutrition facts panels. *Scientific Reports* 11: 13828.

Vernocchi, P., Del Chierico, F., and Putignani, L. (2020) Gut microbiota metabolism and interaction with food components. *International Journal of Molecular Sciences* 21: 3688.

Walker, V., Bertrand, C., Bellvert, F., Moënne-Loccoz, Y., Bally, R., and Comte, G. (2011) Host plant secondary metabolite profiling shows a complex, strain-dependent response of maize to plant growth-promoting rhizobacteria of the genus *Azospirillum*. *New Phytologist* 189, 494–506.

Weir, T.L., Manter, D.K., Sheflin, A.M., Barnett, B.A., Heuberger, A.L., and Ryan, E.P. (2013) Stool microbiome and metabolome differences between colorectal cancer patients and healthy adults. *PLOS ONE* 8: e70803.

Weir, T.L., Trikha, S.R.J., and Thompson, H.J. (2021) Diet and cancer risk reduction: the role of diet–microbiota interactions and microbial metabolites. *Seminars in Cancer Biology* 70, 53–60.

Wu, W., Zhang, L., Zheng, X., Huang, Q., Farag, M.A., *et al.* (2022) Emerging applications of metabolomics in food science and future trends. *Food Chemistry: X* 16, 100500.

Xu, J., Yan, J., Li, W., Wang, Q., Wang, C., *et al.* (2020) Integrative analyses of widely targeted metabolic profiling and transcriptome data reveals molecular insight into metabolomic variations during apple (*Malus domestica*) fruit development and ripening. *International Journal of Molecular Sciences* 21: 4797.

Yokota, Y., Akihiro, T., Boerzhijin, S., Yamada, T., and Makino, Y. (2019) Effect of the storage atmosphere on metabolomics of harvested tomatoes (*Solanum lycopersicum* L.). *Food Science and Nutrition* 7, 773–778.

Yu, J.M., Chu, M., Park, H., Park, J., and Lee, K.G. (2021) Analysis of volatile compounds in coffee prepared by various brewing and roasting methods. *Foods* 10: 1347.

Yu, C.T., Chao, B.N., Barajas, R., Haznadar, M., Maruvada, P., *et al.* (2022) An evaluation of the National Institutes of Health grants portfolio: identifying opportunities and challenges for multi-omics research that leverage metabolomics data. *Metabolomics* 18: 29.

Yun, Z., Gao, H., and Jiang, Y. (2022) Insights into metabolomics in quality attributes of postharvest fruit. *Current Opinion in Food Science* 45: 100836.

Zeisel, S.H. (2007) Nutrigenomics and metabolomics will change clinical nutrition and public health practice: insights from studies on dietary requirements for choline. *American Journal of Clinical Nutrition* 86, 542–548.

Zenoni, S., Savoi, S., Busatto, N., Tornielli, G.B., and Costa, F. (2023) Molecular regulation of apple and grape ripening: exploring common and distinct transcriptional aspects of representative climacteric and non-climacteric fruits. *Journal of Experimental Botany* 74, 6207–6223.

Zhang, J.D., Xue, C., Kolachalama, V.B., and Donald, W.A. (2023) Interpretable machine learning on metabolomics data reveals biomarkers for Parkinson's disease. *ACS Central Science* 9, 1035–1045.

Zhang, N., Wang, D., Liu, Y., Li, S., Shen, Q., and Zhang, R. (2014) Effects of different plant root exudates and their organic acid components on chemotaxis, biofilm formation and colonization by beneficial rhizosphere-associated bacterial strains. *Plant and Soil* 374, 689–700.

Zhao, X., Li, M., and Xu, Z. (2018) Detection of foodborne pathogens by surface enhanced Raman spectroscopy. *Frontiers in Microbiology* 9: 01236.

2 Applied Metabolomics for Plant Protection

Carmen Grech*, Julia Pérez-Pérez*, Francisco Vázquez Prol*, Ismael Rodrigo†, Francisco Vera-Sirera†, María Pilar López-Gresa†, and Purificación Lisón†‡
Instituto de Biología Molecular y Celular de Plantas (IBMCP), Consejo Superior de Investigaciones Científicas (CSIC), Universitat Politècnica de València (UPV), Ciudad Politécnica de la Innovación (CPI), Valencia, Spain

Abstract

This chapter discusses the principles, methods, and applications of high-throughput metabolomics in plant protection. By combining advanced analytical techniques with bioinformatics, metabolomics enables comprehensive analysis of plant metabolites, enhancing our understanding of metabolic pathways involved in plant defense. Technologies such as nuclear magnetic resonance (NMR) and mass spectrometry are employed, using ^{1}H-NMR for analyzing a broad range of abundant metabolites, liquid chromatography–mass spectrometry for polar and semi-polar compounds, and gas chromatography–mass spectrometry for volatiles. These technologies allow the identification of differential metabolites that may be characteristic of disease or resistance. Specific metabolite profiles linked to disease symptoms aid early detection and intervention to minimize crop losses. Metabolites associated with resistance can be used in breeding resistant crop varieties or in the development of genetic and pharmaceutical strategies to improve crop protection. This chapter reviews differential metabolites identified by metabolomics in response to viroids, viruses, bacteria, fungi, or insects, and their potential applications in crop protection.

Keywords: bacteria, fungi, GC-MS, ^{1}H-NMR, insects, LC-MS, metabolites, viroids, viruses

2.1 Introduction

In recent years, metabolomics has emerged as a powerful tool for unraveling the metabolic complexity of plants. This chapter provides a comprehensive overview of the principles, methods, and applications of high-throughput metabolomics in plant protection. By combining the power of current analytical techniques together with extensive statistical analysis performed by bioinformatics tools, metabolomics enables rapid and large-scale analysis of plant metabolites. The transformative approach of applying these novel technologies has revolutionized our understanding of plant metabolic pathways, and has numerous applications in agriculture, environmental sciences, and biotechnology (Fig. 2.1).

Several key technologies and methodologies are employed in metabolomics to enhance analytical performance and coverage. In this sense, the two most used techniques in data acquisition are nuclear magnetic resonance (NMR) and mass

* These authors contributed equally to this work.
† These authors contributed equally to this work.
‡ E-mail: plison@ibmcp.upv.es

© CAB International 2025. *High-Throughput Plant Metabolomics* (Jen-Tsung Chen ed.)
DOI: 10.1079/9781800626911.0002

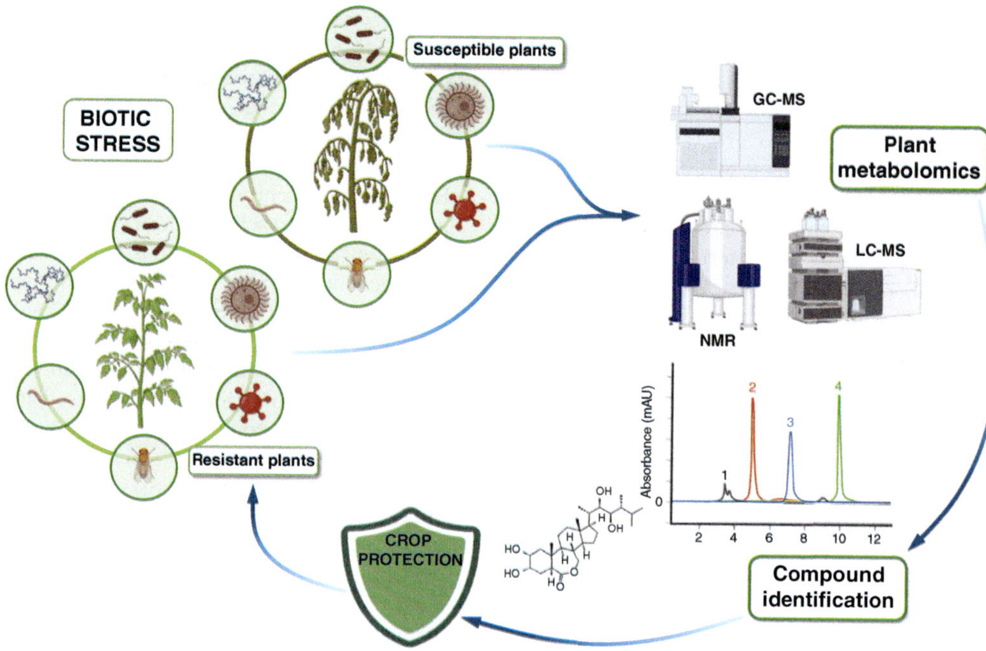

Fig. 2.1. Overview of high-throughput metabolomics in plant protection. This approach enables compound identification and the comparison of metabolic profiles in resistant and susceptible plants under biotic stress, advancing our understanding of plant defense mechanisms and supporting crop protection strategies. GC-MS, gas chromatography–mass spectrometry; LC-MS, liquid chromatography–mass spectrometry; NMR, nuclear magnetic resonance.

spectrometry (MS), with reproducibility and sensitivity being the advantages of each technology, respectively. Liquid chromatography–mass spectrometry (LC-MS) is widely used for the analysis of polar and semi-polar metabolites, while gas chromatography–mass spectrometry (GC-MS) is preferred for volatile and non-polar compounds. Furthermore, advances in high-resolution MS and data-processing algorithms have improved metabolite identification and quantification accuracy. Several other techniques, such as thin layer chromatography and Fourier-transform infrared spectroscopy, often offer complementary information and can be integrated into multi-platform metabolomic workflows for extensive metabolite profiling. These techniques allow the identification and quantification of a wide range of metabolites, providing a comprehensive view of plant metabolic profiles under different stresses (Fig. 2.1). Metabolomics emerges as a powerful approach for discovering differential compounds and is an invaluable tool for pinpointing end point metabolic effects resulting from external stimuli. It is widely applied to explore metabolite contents, identify key metabolites, and decipher metabolic pathways in plants, contributing to our understanding of plant responses to various stress factors.

2.1.1 Current analytical techniques in plant metabolomics

Due to the diverse range of plant metabolite structures, stabilities, and abundances, successful analysis requires selective extraction of plant tissues followed by parallel data acquisition on each extract. An appropriate selective extraction method provides a class of similar chemical compounds facilitating the use of analytical techniques specific for a given metabolite class. Different extraction methods are currently being employed or developed for primary and secondary plant metabolites including soluble sugars, amino acids, organic acids, alcohols, lipids, sterols, phenylpropanoids, flavonoids, lignins, and saponins (Dixon *et al.*, 2006). In addition, for volatile components, organic solvents or solid-phase extraction methods are described in the literature for their recovery. The extraction efficiency is critical as it represents the true state of the original sample.

Once the metabolites are extracted, they must be identified or characterized by conclusive structural elucidation techniques. The most widely used analytical techniques, as well as the experimental workflow usually followed in plant metabolomic studies, are as follows:

- *GC-MS.* Gas Chromatography is a powerful technique suitable for the separation of small molecules such as volatile organic compounds (VOC) like monoterpenes, sesquiterpenes, aldehydes, ketones, alcohols, esters, or phenolics, and non-volatile primary metabolites like monosaccharides, disaccharides, amino acids, or organic acids. Through a chemical derivatization process, these large, non-volatile molecules are converted into volatile and thermostable compounds that can be properly analyzed. GC-MS attached to electron impact (EI) single quadrupole (Q) is the most mature and cost-effective technology for rapid metabolite profiling (Shulaev, 2006). Nowadays, GC incorporating time-of-flight (TOF) mass analyzers offers an alternative to quadrupole and provides greater mass:charge ratio (m/z) accuracies despite cost increases. In addition, newer multi-dimensional chromatographic methods such as GC × GC-TOF-MS will improve compound separation and mass precision. The main advantages of GC-MS are the separation efficiencies by the GC and the excellent metabolite identification performed by the deconvolution software calculated from the mass spectra obtained from ionized molecules by EI. Many searchable mass spectral libraries are available to compare with your analytical EI mass spectra. The only requirement for resolving complex biological mixtures by GC is that samples should be volatile and thermostable (Aharoni *et al.*, 2023).
- *LC-MS.* This technique is highly suitable for the analysis of semi-polar and polar compounds including many secondary metabolites relevant to plant researchers. Chemical derivatization is not required for metabolite detection, although it may be beneficial to improve chromatographic resolution or to provide ionizable groups on metabolites that would otherwise be undetectable by a mass spectrometer (Dunn and Ellis, 2005). However, special chromatographic conditions must be applied to achieve correct separation of different classes of compounds. Compound identification is generally a time-consuming process and can sometimes be the real bottleneck in metabolomics, as mass spectra libraries are not always available due to the different ionization methods used in LC-MS, making co-elution of standards indispensable. Soft ionization methods such as electrospray ionization (ESI) and atmospheric pressure chemical ionization (APCI) are employed in this technology instead of fast atom bombardment or EI. ESI is the most common ionization method in plant metabolomics because it is well suited to ionize a wide range of polar metabolites including sugars, amino acids, organic acids, glucosides, and phenylpropanoids, while APCI is more appropriate to ionize non-polar compounds such as carotenoids. Diverse mass analyzers can be attached to the liquid chromatograph, with single quadrupole (Q), triple quadrupole (QQQ), TOF, and Orbitrap being the most frequently used in plant metabolomics (Chen *et al.*, 2019).
- *NMR spectroscopy.* The reproducibility of this technique makes it an ideal rapid and non-destructive tool for broad-range profiling of abundant metabolites and for metabolite fingerprinting in large sample sets (Dixon *et al.*, 2006). Despite its lower sensitivity, detecting 30–60 compounds per extract compared with the more than 200 detected by MS techniques, NMR still reveals significant metabolic trends and provides valuable structural data. In addition, the increased sensitivity and spectral resolution due to the ongoing accessibility of higher field strength magnets and the introduction of cryoprobes has led to an increase in the use of NMR spectroscopy for metabolic profiling studies. A ^{1}H-NMR spectrum of crude extract shows the protons of all the metabolites present in the sample extract within a few minutes, but it is sometimes difficult to

identify them because of overlapping chemical shifts. Therefore, additional two-dimensional (2D) NMR or MS analyses are often required to identify a metabolite of interest. The most useful 2D methods in NMR metabolomics are *J*-resolved spectroscopy and heteronuclear single quantum coherence (HSQC) spectroscopy for determining the presence of certain compounds. Other functional 2D-NMR experiments that are very common in metabolomics studies for metabolite identification are ^{1}H–^{1}H COSY (correlation spectroscopy), HMBC (heteronuclear multiple quantum coherence), and NOESY (nuclear Overhauser effect spectroscopy). ^{1}H-NMR is particularly effective for studying primary metabolites like amino acids, organic acids, and carbohydrates, as well as secondary metabolites present in higher concentrations.

Both MS and NMR provide detailed information on the molecular structure of metabolites, which is crucial for identifying and characterizing the specific compounds involved in the plant's defensive response to pathogens and allows real-time monitoring of metabolite changes over the course of infection. This helps to understand the temporal dynamics of the plant response and the stages of pathogen development. The integration of large-scale data obtained through these techniques requires the use of bioinformatics tools and multivariate data analysis. Data handling is particularly useful in the context of untargeted metabolomic profiling, where the goal is to explore and discover global metabolic changes without a specific prior hypothesis. Through plant metabolomics, it is possible to identify patterns, correlations, and significant differences in metabolic profiles, facilitating the identification of key metabolites involved in the plant defense response and opening up new ways for the development of more effective plant protection strategies (Castro-Moretti *et al.*, 2020).

2.1.2 The power of high-throughput metabolomics in crop protection

The outstanding power of metabolomics allows identification of even the most subtle variations in metabolic profiles under different physiological and developmental conditions. Consequently, metabolic analysis has become an essential tool to study the adaptations of plants to different environmental stresses, including biotic agents such as microorganisms and herbivores.

The unique characteristics of high-throughput metabolomics (Heuberger *et al.*, 2014; Castro-Moretti *et al.*, 2019; Chen *et al.*, 2019; Aharoni *et al.*, 2023; Dixon and Dickinson, 2024), that make this suite of technologies a powerful tool to apply in crop protection include:

- *Comprehensive analysis.* Metabolomics allows the simultaneous measurements and analysis of a wide range of small molecules within a biological system. This extensive approach provides a holistic view of the metabolic changes occurring in the plant and the pathogen or pest during their interaction. Furthermore, it can shed light on the coevolutionary relationships between plants and pathogens or pests by revealing how metabolic networks in both organisms interact and adapt to each other over time. This understanding is crucial for developing sustainable strategies for disease management and crop improvement.
- *Dynamic nature.* The metabolome is highly dynamic and responsive to environmental stimuli, including pathogen attack. By capturing these metabolic changes in real-time, metabolomics enables researchers to understand the dynamic responses of both the plant, the pathogen, and the pests during the interaction.
- *Identification of differential metabolites.* Metabolomics facilitates the identification of specific metabolites involved in plant–pathogen interactions. By identifying signature metabolite profiles associated with specific pathogens or disease symptoms, these biomarkers can be indicative of disease progression, resistance mechanisms, or biochemical pathways involved in host defense or pathogen virulence. Metabolomics can enable early detection and diagnosis of plant diseases, facilitating timely interventions to mitigate crop losses.
- *Unbiased approach.* Unlike targeted approaches that focus on specific metabolites or pathways, metabolomics offers an unbiased assessment of the entire metabolome. This approach

allows the discovery of unexpected metabolic changes or novel metabolites that may play critical roles in plant–pathogen interactions. In this way, researchers can unravel the biochemical mechanisms underlying host defense responses, pathogen adaptation strategies, and disease development.

In essence, metabolomics offers a powerful approach to untangling the intricate biochemical dialogues between plants, pathogens, and pests, providing crucial insights for understanding plant defense mechanisms, pathogen virulence strategies, and, ultimately, proposing strategies for disease management and crop protection.

2.1.3 Harnessing metabolomics for enhanced crop protection strategies

Crops are exposed to multiple environmental aggressions that cause a decrease in agricultural productivity, resulting in severe economic losses and an increase in production costs. The problems they provoke and the difficulty of controlling them, together with a trend toward greater social and environmental awareness, demonstrate the need to fight them through different sustainable strategies. A directly applied strategy to increase disease resistance lies in the identification of either metabolites capable of inducing the defensive response, or biomarkers that can be used for preventive actions. In this sense, advances in the knowledge of plant secondary metabolism are essential to uncover these molecules (Dixon *et al.*, 2024).

To implement the aforementioned high-throughput metabolomics technologies for identification of these metabolites, the biological starting materials (i.e. plants experiencing biotic stress compared with the corresponding non-stressed control plants) are chemically investigated to discover the specific compounds accumulated to survive. According to the chemical nature of the metabolites to be analyzed, different extractions will be performed, and the metabolomic studies will use the appropriate technique(s): ^{1}H-NMR, MS-GC, and/or LC-MS. Non-targeted metabolomic profilings are performed to identify differential metabolites and to get massive data. When the identified differential metabolites are characteristic of the disease, they can be used as biomarkers in diagnosis or monitoring to activate prevention plans (Fig. 2.2: A). In contrast, when the differential metabolites are characteristic of resistance, they can be used by breeders as biomarkers for resistant crop varieties (Fig. 2.2: B1). Alternatively, pharmaceutical (Fig. 2.2: B2) and genetic (Fig. 2.2: B3) approaches can also be developed, contributing to effective crop protection. These two approaches lead to the development of plant resistance inducers that could be applied to crops (Abdul Malik *et al.*, 2020), and to the generation of genetically modified resistant plants with redirected biosynthesis of these signals, respectively.

2.2 Applied Metabolomics for Crop Protection Against Virus and Viroid Infections

The complex interactions between viruses, viroids, and their plant hosts are central to studying various crop protection strategies. With the advent of advanced methodologies such as omics, recent studies have delved deeper into the complexities of infection mechanisms. Metabolomics has also emerged as a powerful tool in this field, providing a detailed snapshot of the biochemical activities occurring in response to infection. For example, metabolic changes in infected plants can reveal how viral and viroidal pathogens recruit the cellular machinery, evade immune responses, and spread within the host. Moreover, metabolomics holds potential for identifying novel biomarkers for early detection and diagnosis, as well as therapeutic targets for developing antiviral treatments. By integrating metabolomics with other omics approaches such as genomics, transcriptomics, and proteomics, a more holistic understanding of viral and viroid infections can be achieved. This chapter aims to provide a comprehensive overview of current methodologies, key findings, and future directions in the application of metabolomics to virology and plant pathology, underscoring its significance in advancing both basic and applied research in infectious diseases.

GC-MS has produced interesting results in plant–viroid interaction research. In this sense, the identification of differential metabolites could be used as biomarkers in the selection of resistant crop varieties (Fig. 2.2: B1). Bagherian

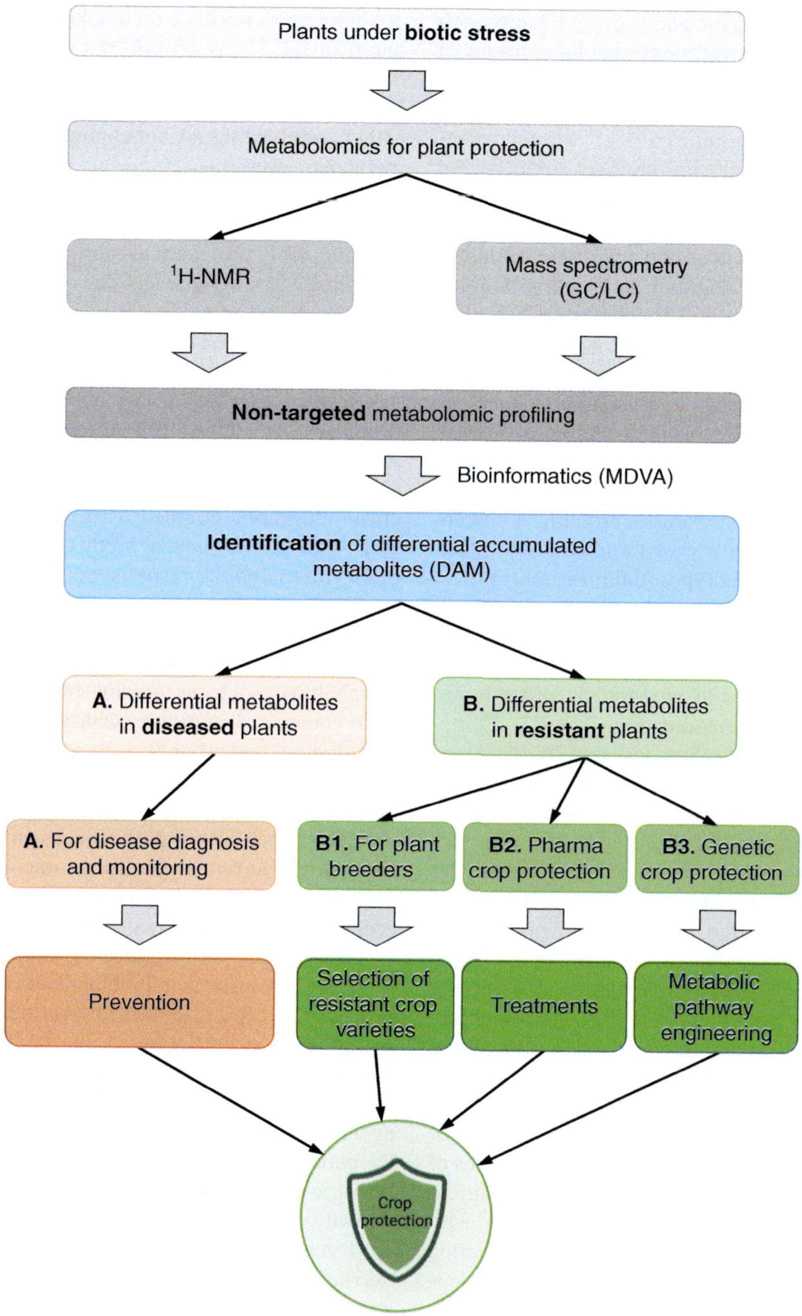

Fig. 2.2. Workflow of plant metabolomics for crop protection under biotic stress. The process involves metabolomic profiling using ^{1}H-nuclear magnetic resonance (^{1}H-NMR) or mass spectrometry (gas chromatography (GC) or liquid chromatography (LC)), followed by bioinformatics-based analysis to identify differentially accumulated metabolites. Differential metabolites in diseased plants (A) are used for disease diagnosis, monitoring, and prevention, while those in resistant plants (B) guide plant breeding, pharmaceutical crop protection, and genetic engineering. Together, these approaches enhance crop protection strategies. MDVA, multivariate data analysis.

et al. (2016) used a metabolic profiling approach using GC-MS to analyze metabolic changes in tomato plants infected with potato spindle tuber viroid (PSTVd). The study identified 79 metabolites in the tolerant tomato cultivar 'Moneymaker' that displayed significant changes upon PSTVd infection. The identified metabolites spanned various chemical groups, including fatty acids, sugars, alcohols, and amino acids. These findings provide valuable insights into how the 'Moneymaker' cultivar manages to tolerate PSTVd infection through specific metabolic adjustments and pathway regulations. A recent GC-MS-based metabolomic analysis identified monoterpenoids and methyl salicylate as key volatile involved in the salicylic acid-dependent defense response of tomato against Citrus Exocortis Viroid (CEVd), as demonstrated by comparative profiling of susceptible (NahG, low SA levels) and resistant (S5H, high SA levels) genotypes (Balanzá, 2025).

GC-MS was also used by Spanò *et al.* (2023) to identify VOCs in tomato plants infected with a recombinant strain of potato virus Y (PVYC-to), and the responses in susceptible and tolerant varieties were compared (Fig. 2.2: B1). Symptomatic infected plants produced VOCs enriched in alcohols, fatty acid derivates, benzenoids, and salicylic acid derivatives, while mock-inoculated plants had methyl ester compounds. Principal component analysis revealed closer VOC profiles between PVYC-to-infected and mock-inoculated grafted plants than symptomatic non-grafted plants. This suggests that VOC profiles are linked to viral RNA accumulation, disease severity, and graft-derived tolerance to PVYC-to infection. The metabolomic analysis helped to understand how grafting influences plant metabolism and enhances its defense capabilities.

Regarding genetic approaches, the work of Chang *et al.* (2021) analyzed the VOC profiles of both Cry1C transgenic rice expressing the *Bacillus thuringiensis* (*Bt*) gene and its parental rice with and without rice dwarf virus (RDV) infection (Fig. 2.2: B3). The analysis showed that the transgenic line did not cause significant differences in the volatilome compared with the parent under RDV-free conditions. However, a more significant response in secondary metabolites such as alkaloids, phenolics, terpenoids, and glycosides in the transgenic line upon infection with RDV was observed, indicating the role of the metabolome in this type of pathogenic interaction in the genetically modified and unmodified plants in response to stress.

The application of metabolomics platforms based on LC-MS is an optimum alternative to identify metabolites for crop protection. The LC-MS methodologies used by Bazzini *et al.* (2011) revealed significant changes in secondary metabolite levels (Fig. 2.2: A) at various stages of tomato mosaic virus (ToMV) infection in tobacco plants. Compounds such as chlorogenic acid, nicotine, and rutin were among the metabolites identified with altered levels. The data pointed to a biphasic response to ToMV infection: an early phase characterized by increased levels of several metabolites without detectable presence of the virus, and a later phase with both increases and decreases in metabolite levels correlating with virus accumulation. Rossouw *et al.* (2019) employed ultra-high-performance liquid chromatography coupled with mass spectrometry (UHPLC-MS)-based metabolomics to explore the metabolic responses of tomato plants to whitefly-mediated tomato curly stunt virus infection. The resistant tomato line (Fig. 2.2: B) produced a variety of defense-related secondary metabolites, such as several hydroxycinnamic acids conjugated to quinic acid, galactaric acid, and glucose, and other important secondary metabolites were identified including benzenoids, flavonoids, and steroidal glycoalkaloids. These secondary metabolites play a significant role in the plant defense mechanism, indicating their potential use in breeding programs to enhance crop resistance to viral infections.

Using ultra performance LC (UPLC) coupled with MS, a non-targeted metabolomics analysis by Z. Li *et al.* (2023) identified several secondary metabolites in bottle gourd plants affected by cucumber green mottle mosaic virus (CGMMV). The secondary metabolites that showed differential accumulation included those involved in pathways such as flavonoid biosynthesis, tyrosine metabolism, and jasmonic acid and its derivatives. The study highlighted an increase in the levels of 69 metabolites, while 7 metabolites decreased in their accumulation under CGMMV stress, indicating significant alterations in the metabolic profile. These differential metabolites are expected to be tested (Fig. 2.2: B2) for crop protection and could eventually be part of crop protection strategies. The most upregulated differentially accumulated metabolites were 4-caffeoylquinic acid, acteoside, and malvidin-3,5-*O*-diglucoside. The most downregulated metabolites were glucarate *O*-phosphoric acid, cyclic AMP, and scopoletin (7-hydroxy-5-methoxycoumarin).

Regarding plant–viroid interactions, a comprehensive metabolomic analysis was performed to assess the effects of apple scar skin viroid (ASSVd) infection on two varieties (Fig. 2.2: B1) of apple saplings (Li *et al.*, 2023a). Using an Agilent 1290 UHPLC system and a TripleTOF 5600 mass spectrometer, significant changes related to lipid metabolism, carbohydrate metabolism, and amino acid metabolism were identified including the biosynthesis of secondary metabolites such as L-2-aminobutyric acid, 6″-*O*-malonyldaidzin, D-xylose, and coumarin. The increase in certain metabolites such as L-2-aminobutyric acid and 6″-*O*-malonyldaidzin is representative of a stress response to viroid infection. Understanding the changes in secondary metabolites due to ASSVd infection on apple saplings provides valuable information for developing crop protection strategies aimed at mitigating the impact of viroid infections on apple production.

The metabolomics study described by Huang *et al.* (2024) involved metabolite profiling and analysis in rice lines genetically engineered using RNA interference (Fig. 2.2: B3), resistant to black-streaked dwarf virus (RBSDV). Using HPLC-MS, a total of 224 metabolites, including diterpenoids, phenolic glycosides, and α-amino acids and derivatives, were detected, providing new insights into understanding the metabolic changes in genetically engineered rice lines that confer resistance to RBSDV. Enhanced crop protection mechanisms were indicated by the upregulation of specific plant signaling hormone-related metabolites (e.g. salicylic acid, jasmonic acid), which suggested enhanced defense responses in genetically engineered rice. The identified metabolites play crucial roles in plant immune responses, pathogen resistance, and stress tolerance. These results are essential for understanding how genetic modifications can enhance crop protection and ensure the safety and sustainability of genetically engineered crops.

Several reports indicate the effectiveness of NMR-based metabolomics in identifying and quantifying metabolic changes in plants in response to virus infection. A metabolomic study on tobacco mosaic virus (TMV) infection in *Nicotiana tabacum* leaves described by Choi *et al.* (2006) used NMR to uncover the metabolic changes associated with the infection (Fig. 2.2: A). The authors observed a clear increase in certain metabolites such as 5-caffeoylquinic acid, α-linolenic acid analogs, and various sesquiterpenoids and diterpenoids, which might play a significant role in the defense mechanisms. The study of López-Gresa *et al.* (2009) revealed that glycosylated gentisic acid was the most important induced metabolite upon viroid infection in tomato plants, demonstrating the potential of NMR-based metabolomics to provide insights into plant–pathogen interactions (Fig. 2.2: A). Exogenous treatments with this phenolic compound have been described to induce resistance to RNA pathogens such as ToMV and citrus exocortis viroid (CEVd) in tomato and *Gynura auriantiaca*, respectively (Campos *et al.*, 2014a), indicating that the identification of differential metabolites can contribute to the development of crop protection against viruses and viroids (Fig. 2.2: B2).

Analyzing the response of tomato plants against ToMV, López-Gresa *et al.* (2012) identified various secondary metabolites using ^{1}H-NMR. This study provided a comprehensive view of the plant metabolome, allowing the detection and identification of metabolites involved in plant defense mechanisms, including phenylpropanoids, which are known to be involved in the defense response against various pathogens and flavonoids such as rutin that contribute to plant defense due to its antioxidant properties (Fig. 2.2: B). These metabolites can be used in biotechnological approaches to engineer crops with enhanced resistance to pathogens, reducing the need for chemical pesticides and contributing to sustainable agriculture.

The combination of two powerful techniques also can provide complementary information on the metabolome, increasing the possibility of discovering new compounds as crop protectors. In this way, the use of ^{1}H-NMR spectroscopy and GC-TOF-MS to investigate the metabolic changes induced in tomato plants by CEVd infection and chemical treatments with benzothiadiazole allowed the identification of several compounds that play a role in plant defense mechanisms (López-Gresa *et al.*, 2019). For instance, glycosylated phenolic compounds such as gentisic acid and ferulic acid, as well as amino acids such as phenylalanine, tyrosine, aspartate, glutamate, and asparagine, were found to accumulate in viroid-infected plants. Benzothiadiazole treatment led to the accumulation of γ-amino butyric acid (GABA), which was identified as a key metabolite contributing to the plant defense response (Fig. 2.2: B2). The study illustrated how metabolomics can be applied to understand and enhance crop protection by identifying critical metabolites and pathways involved in the plant defense response to viroid infection and chemical treatments.

2.3 Applied Metabolomics for Crop Protection Against Bacteria and Fungi

Beneficial microorganisms are pivotal in promoting plant growth. However, pathogenic bacteria and fungi cause significant plant diseases, resulting in compromised development, diminished productivity, and even plant collapse and death. Given the complexity of plant–microbe interactions, metabolomics emerges as a critical tool for identifying metabolites that facilitate these interactions or confer resistance to diseases. The use of this analytical platform is essential in identifying defensive secondary metabolites possessing antiviral, antibacterial, and antifungal properties, or signal molecules that activate plant defenses against pathogenic infections, being invaluable for recognizing resistance-related metabolites. The profile of metabolites synthesized and accumulated in plant tissues, organs, and cells varies in response to infections by different pathogens. Several metabolomics techniques facilitate the analysis of changes in metabolic profiles associated with pathogen attack, followed by identification through metabolomics databases and literature comparisons. Significant differences in metabolic profiles are observed in plants with varying resistance levels when challenged by pathogenic microorganisms. In plants with different resistance cultivars, it is possible to identify resistance-related metabolites by comparing the differential metabolite content after infection with the pathogen in resistant and susceptible varieties.

GC-MS metabolomics is invaluable for early detection, improvement of disease resistance, and evaluation of biocontrol strategies. The applications of this analytical platform in plant protection are very broad and encompass a wide range of approaches. GC-MS is employed to analyze VOCs emitted by plants infected with pathogens, by comparing the volatile profiles of healthy and diseased plants to identify specific compounds associated with pathogen infection. Specifically, the "aroma of resistance" in Rio Grande tomato plants was described, identifying VOCs associated with the immune response following infection with virulent or avirulent strains of *Pseudomonas syringae* pv. *tomato*. Infection with the virulent strain led to the production of monoterpenes and salicylic acid derivatives, which can be used for early pathogen detection (Fig. 2.2: A). Conversely, the tomato interaction with the avirulent strain resulted in the production of VOCs such as (Z)-3-hexenol esters and hydroxylated monoterpenes, indicative of a plant immune response (López-Gresa *et al.*, 2017). The use of these volatiles for crop protection has been described (Fig. 2.2: B2). In this sense, treatments with green leaf volatile esters activate plant defense responses against *Pseudomonas syringae* (López-Gresa *et al.*, 2018). In particular, (Z)-3-hexenyl butyrate (HB) has been described as a stoma closer, conferring enhanced resistance against *Phytophthora* spp. and *P. syringae* infection in potato and tomato plants, respectively, under open-field conditions. Additionally, exogenous treatments with this volatile alleviate water stress and improve fruit productivity in tomato plants (Payá *et al.*, 2024). The use of this volatile is extended to other species and uses. In *Vitis vinifera* L. grapevines, field experiments have investigated the impact of HB treatments (Fig. 2.2: B2). The application of HB around veraison induced a higher color intensity in berries, and vines treated at higher doses reached this stage earlier than the untreated controls (Payá *et al.*, 2020). All these results suggest the potential of these volatiles as a novel phytoprotector for sustainable agriculture. As far as hydroxylated monoterpenes are concerned, both pharmacological and genetic approaches have confirmed their use for crop protection, as treatments with these volatiles resulted in stomatal closure, Pathogenesis-related protein 1 (PR1) induction, and enhanced resistance against *P. syringae* infection in tomato plants (Pérez-Pérez *et al.*, 2024). In addition, transgenic tomato plants overexpressing the monoterpene synthase *MTS1* gene (Fig. 2.2: B3) induced *PR1* gene expression and stomatal closure in neighboring plants, indicating that these volatiles may play an important role not only in intra- but also in interplant immune signaling (Pérez-Pérez *et al.*, 2024).

GC-MS analysis facilitates the early detection of pathogens before visible symptoms appear.

By detecting changes in VOC emissions, the presence of pathogens can be identified, allowing timely intervention measures to be taken to prevent the spread of disease (Fig. 2.2: A). For example, in vineyards, GC has been used to detect volatile compounds associated with *Botrytis cinerea* infections before visible gray mold symptoms appear on grapes. VOCs such as 1,5-dimethyltetralin, 1,5-dimethylnaphthalene, phenylethyl alcohol, and 3-octanol were correlated with *B. cinerea* infection levels, with 2-octen-1-ol identified as a potential early differential metabolite of infection (Jiang *et al.*, 2023). Analysis of a compatible and incompatible interaction between tomato plants and *Fusarium oxysporum* f. sp. *lycopersici*, revealed the specific volatile chemical composition and plant signals associated with them, with susceptible plants characterized by the over-emission of methyl and ethyl salicylate, as well as some fatty acid derivatives (Hernández-Aparicio *et al.*, 2021). These volatiles could also be used as biomarkers for this fungal disease (Fig. 2.2: A).

In plant-breeding programs, GC-MS is used to select disease-resistant crop varieties. By analyzing the VOC profiles of different plant genotypes under pathogen attack, breeders can identify plants that emit fewer disease-promoting compounds or more disease-inhibiting compounds (Fig. 2.2: B1). For instance, GC analysis has been applied to identify wheat genotypes that emit lower levels of VOCs associated with *Fusarium* head blight (Buśko *et al.*, 2023). Similarly, in common beans, specific VOCs such as limonene, linalool, nonanal, methyl salicylate, and methyl jasmonate were identified in resistant varieties. These VOCs can trigger resistance responses in susceptible neighboring plants, making them overcome the disease in the same way as the resistant cultivar itself (Quintana-Rodriguez *et al.*, 2015). Volatiles identified by Hernández-Aparicio *et al.* (2021) could also be used to select disease-resistant crop varieties, as the resistant Movione and the susceptible Momor tomato varieties, carrying or lacking the *I* resistance gene, were used to identify differential VOCs.

LC-MS is suitable for studying plant–pathogen interactions due to its ability to detect a wide range of compounds with high sensitivity and specificity. This analytical technique can identify various metabolites produced by both plants and pathogens. This includes primary metabolites (e.g. sugars, amino acids, lipids) and secondary metabolites (e.g. phytoalexins, toxins). Understanding changes in metabolite profiles after an infection can provide insights into plant metabolic responses to pathogen attack and the biochemical pathways involved in defense mechanisms, and aids in the identification of new compounds. The identification of potent antifungal or antibacterial compounds via LC-MS can lead to the development of novel biocontrol agents. These compounds can be isolated and applied to crops as natural pesticides, reducing reliance on synthetic chemicals and promoting sustainable agriculture. For instance, Zacarés *et al.* (2007) and López-Gresa *et al.* (2011) characterized the phenylpropanoids produced by tomato plants when exposed to the bacterial pathogen *P. syringae* pv. *tomato* (Fig. 2.2: A). Using LC-MS and NMR spectroscopy, some of the most prominent differential metabolites were found to accumulate in response to the pathogen, including phytoalexins such as hydroxycinnamic acid amides (HCAAs) derived from dopamine, noradrenaline, and octopamine, chlorogenic acid, and rutin. Furthermore, chlorogenic acid induced the expression of the defense-related genes *PR1* and *P23*, while certain HCAA compounds exhibited potent antioxidant properties, suggesting their role in tomato plant defense against bacterial infection (Fig. 2.2: B2). Once the differential metabolite is identified, a genetic approach for plant protection can be developed. Consequently, tomato plants overexpressing tyramine *N*-hydroxycinnamoyltransferase, an enzyme involved in the last step of HCAA production, were generated (Fig. 2.2: B3), exhibiting both elevated HCAA levels and enhanced resistance to *P. syringae* (Campos *et al.*, 2014b).

In another study, Bollina *et al.* (2010) employed mass spectrometry-based metabolomics to identify 496 metabolites in barley associated with resistance against *Fusarium* head blight. Among these, 194 were deemed constitutive resistance-related metabolites, showing a higher abundance in resistant genotypes compared with the susceptible ones. Resistance-related metabolites were predominantly derived from phenylpropanoid, flavonoid, fatty acid, and terpenoid metabolic pathways. Selected resistance-related metabolites were tested *in vitro* for antifungal activity, demonstrating their ability to inhibit fungal biomass production (Fig. 2.2: B2). This study highlighted the utility of MS-based metabolomics in identifying *Fusarium* head blight resistance in plants and the potential of these resistance-related metabolites for screening resistant genotypes.

Pushpa *et al.* (2014) focused on understanding the mechanisms of quantitative resistance against *Phytophthora infestans*, the causal agent of late blight in potato, employing a non-targeted high-resolution LC-MS metabolomics analysis to compare resistant and susceptible potato cultivars (Fig. 2.2: B1). This study revealed a significant accumulation of HCAAs from the phenylpropanoid shunt pathway in the resistant genotype upon pathogen inoculation. HCAAs deposited on host cell walls inhibited pathogen colonization and acted as phytoalexins, reducing pathogen biomass and lesion expansion. The study suggests that these HCAAs could be used as biomarkers for late blight resistance.

Metabolomic insights gained from LC-MS studies can provide valuable information for breeding programs aimed at developing pathogen-resistant plant varieties. By identifying key resistance-indicative metabolites, breeders can select traits that enhance the plant's natural defense mechanisms. Furthermore, LC-MS can be used to monitor dynamic changes in the plant metabolome during pathogen infection. This real-time monitoring allows early detection of infection and timely application of treatments, improving crop management and yield. For example, UPLC coupled with Q-TOF MS was used to evaluate the potential of non-targeted metabolomics for distinguishing between *Fusarium*-infected and control barley samples. Metabolomic fingerprints were acquired from nine barley varieties infected with *Fusarium culmorum* and a control group without infection (Fig. 2.2: B1). Statistical evaluations demonstrated that this metabolomics approach could effectively differentiate barley responses to *F. culmorum* infection, identifying the mycotoxins deoxynivalenol and deoxynivalenol 3-glucoside as key resistance indicator metabolites in barley (Cajka *et al.*, 2014). Overall, the application of LC-MS metabolomics in studying plant–pathogen interactions provides a powerful tool for advancing our understanding of plant defense mechanisms and developing innovative strategies for crop protection.

It is well known that primary metabolism profiles change during different pathogenic infections. Primary metabolism adjustments, as plants reallocate their resources following biotic stress to bolster defense mechanisms, produce necessary signaling molecules, detoxify reactive oxygen species, and reinforce structural barriers against pathogens. This metabolic reprogramming is a crucial aspect of the plant adaptive response to biotic stress.

NMR spectroscopy is particularly well suited for the identification and characterization of major compounds such as primary metabolites and abundant secondary metabolites, including phenylpropanoids, flavonoids, glucosinolates, acyl sugars, saponins, alkaloids, polyketides, and carotenoids. These compounds are known for their antifeedant or antimicrobial activities, and can be clearly identified and quantified in plant extracts using NMR. Like the rest of the techniques, NMR is useful for the identification of metabolites related to resistance phenotypes in crops. This can contribute to the development of natural products with a positive impact on crop protection against pathogens. Using resistant cultivars is the most effective way to study the metabolic changes associated with pathogen attack. For example, *Ralstonia solanacearum* is a Gram-negative bacterial pathogen causing wilting, yield losses, and death in tomato plants. López-Gresa *et al.* 2009 described the detailed identification of key phenylpropanoids and flavonoids associated with the tomato–bacteria interaction, as revealed through NMR-based metabolomics, highlighting the metabolic specificity and complexity involved in plant defense responses. Using ^{1}H-NMR spectroscopy, Murti *et al.* (2021) identified key biochemical compounds involved in tomato defense against bacterial wilt. Leucine was described as a notable differential metabolite that distinguishes resistant from susceptible cultivars (Fig. 2.2: B1). Consequently, the insights gained from the correlation between leucine levels and cultivar resistance to bacterial wilt present promising knowledge for the development of innovative strategies aimed at enhancing crop protection. Interventions targeting leucine metabolism, or the application of leucine-based compounds, may hold considerable potential for bolstering the resistance of cultivars susceptible to *R. solanacearum* infection, thereby contributing to the advancement of cultivar protection methodologies.

NMR metabolomics is used to identify differential metabolites of disease resistance in agricultural crops. These metabolites can be used as biomarkers in the selection of resistant crop varieties and the development of sustainable agricultural practices, but they can also help monitor diseases before potential agricultural losses. For example, a study by Pontes *et al.* (2016) aimed to address the economic impact of huanglongbing (HLB; also known as citrus greening disease) on the citrus industry. HLB poses significant economic losses to the citrus industry worldwide. Early detection of the disease is crucial for minimizing these losses. A combined approach of ^{1}H-NMR spectroscopy and chemometrics was applied to identify differential metabolites in citrus leaves at different disease stages. Analysis of healthy and diseased leaf samples has shown some key metabolites that were enhanced in resistant plants: purine, *p*-coumaroyl alcohol, and 5-acetylamido-2-chloroaniline, as well as GABA, L-homocitrulline, L-lysine, and acetate. These metabolites can serve as HLB early detection markers (Fig. 2.2: A), thus enabling the adequate treatment of affected citrus trees and disease monitoring. Overall, NMR spectroscopy is a valuable tool in plant–pathogen interaction studies, providing deep insights into the biochemical changes that occur during infection and offering a comprehensive understanding of plant defense mechanisms.

Considering all these methodologies, it is evident that metabolomics provides a comprehensive toolkit for exploring the biochemical interactions between plants and microorganisms. By identifying key metabolites involved in growth and disease resistance, metabolomics can significantly contribute to the development of effective strategies for enhancing plant health and crop productivity.

2.4 Applied Metabolomics for Crop Protection Against Insects and Herbivores

In response to herbivory attacks or mechanical injury caused by herbivores, plants have evolved sophisticated defense mechanisms to mitigate damage and maintain homeostasis. Metabolomics has become an essential tool to study these responses, as it allows a detailed understanding of the metabolic changes associated with these stresses. Responses of plants to these specific stresses are critical components of their defense mechanisms, triggering rapid changes in gene expression, protein synthesis, and metabolite accumulation. High-throughput metabolomics has significantly advanced plant science research, allowing rapid and comprehensive analysis of the dynamic metabolic changes that occur in response to mechanical injuries and herbivore attacks. While traditional molecular biology techniques have offered valuable insights into the genetic regulation of these responses, metabolomics provides a more comprehensive view by analyzing the metabolic changes associated with plant injuries. This approach not only reveals metabolic adjustments crucial for plant defense but also identifies specific metabolites that play key roles in these processes.

Several studies have utilized metabolomics, including GC-MS, LC-MS, and NMR analytical techniques, to investigate the metabolic responses of plants to mechanical wounding or insect attack, shedding light on the complex interactions between plants and their environment. Metabolomics is emerging as an essential tool for understanding the complex metabolic interactions between plants and their environment during herbivory and injury responses.

GC-MS metabolomics has been used to elucidate herbivore-induced plant volatiles that the plants emit upon interaction with different species of insects and pests. The work by Wei *et al.* (2007) examined the emission of VOCs by plants to attract parasitic wasps for pest defense (Fig. 2.2: B2). To date, nearly 2000 volatile compounds, released from various plant organs in response to biotic activities, have been identified across over 90 plant families. These compounds play a crucial role in influencing the behavior of herbivores and their predators. Variations in volatile profiles were observed among healthy plants, mechanically damaged plants, and those infested with leaf miner larvae. These compounds include six-carbon alcohols, aldehydes, esters, monoterpenes, and sesquiterpenes such as (*E*)-2-hexenal, (*Z*)-3-hexenol, ocimene, and linalool. In a related investigation, 18 VOCs emitted by *Fallopia baldschuanica* flowers were analyzed using headspace-GC-MS. Compounds such as β-ocimene and limonene were identified as aphid alarm pheromones, effectively repelling them (Fig. 2.2: B2). This study illustrated the significant influence of volatile compounds on plant–insect interactions and their role in plant defense mechanisms (Salerno *et al.*, 2023). Furthermore, Nenaah and Ibrahim (2011) reported the insecticidal properties of various plant products, identifying active compounds using GC-MS, particularly terpenoids from essential oils of plants such as *Cinnamomum camphora*, *Ocimum basilicum*, and *Chenopodium ambrosioides*. Turlings and Erb (2018) explored tritrophic interactions mediated by herbivore-induced plant volatiles, emphasizing how plants release specific volatiles following herbivory to attract natural enemies such as parasitoids and predators. These interactions are pivotal in ecological communities and offer promising avenues for integrated pest management in agriculture. Wang *et al.* (2008) delved into the roles of lipoxygenase 3 (LOX3) and lipoxygenase 4 (LOX4) in the jasmonate signaling pathway. By using a genetic approach (Fig. 2.2: B3), they used *LOX3*- and *LOX4*-silenced plants, revealing distinct functions of these enzymes in direct and indirect plant defenses against herbivores. The study highlights contrasting effects on indirect defenses mediated by VOCs such as terpenoids and green leaf volatiles, showing reduced emission in *LOX3*-silenced plants and heightened emission in *LOX4*-silenced plants. This differentiation underscores the specialized roles of LOX3 and LOX4 in regulating VOC production crucial for indirect plant defenses. Together, these studies underscore the complex interplay between plant signaling molecules, volatiles, and defense mechanisms against herbivores, providing essential insights into ecological dynamics and offering practical implications for sustainable agricultural practices.

High-resolution GC-MS analysis of melon and cotton plants colonized by the fungus *Beauveria bassiana* revealed unique volatile emissions when infested by *Spodoptera* spp. and aphids compared with uncolonized controls. Benzaldehyde was identified as a potential marker for wild cotton simultaneously colonized and infested (Fig. 2.2: A). For melon plants similarly affected, 2-ethylhexyl nonyl sulfite was detected as a potential marker (González-Mas *et al.*, 2021). Another study on the integrated transcriptome and volatilome of *Tanacetum vulgare* showed increased terpenoid VOC emissions after infestation by *Spodoptera littoralis* and *Metopeurum fuscoviride* (Clancy *et al.*, 2020). After infestation by the tea green leafhopper (*Empoasca onukii*), *Camellia sinensis* plants were subjected to combined untargeted metabolome and transcriptome analysis, revealing significant volatile emissions such as 4-hydroxybenzaldehyde and an abundance of transcripts associated with a specific tea fragrance (Liu *et al.*, 2021). Research on parasitoid attraction revealed that *Peristenus spretus* was consistently drawn to a synthetic form of methyl salicylate and mixtures of cotton volatiles stimulated by insect feeding (Xu *et al.*, 2021).

The above studies and many others collectively underscore the critical role of GC-MS in identifying and analyzing volatile compounds involved in plant defense against herbivory pests. Volatilomics, as a burgeoning field, holds immense promise in advancing sustainable agricultural practices by elucidating complex plant–environment interactions and enhancing crop resistance and productivity (Kasote *et al.*, 2023).

The complex chemical interactions produced by herbivorous insects and nematodes in plants dramatically impact plant physiology, and consequently plant yield. An excellent analytical platform to explore the metabolic homeostasis sought by plants under this stress is metabolomics based on LC-MS. Employing non-targeted metabolomics analysis, Zhang *et al.* (2024) explored the effects of mechanical wounding on tea leaf metabolites. Their study revealed differential regulation of flavonoids (e.g. epicatechin, catechin) and phenolics (e.g. chlorogenic acid) in response to wounding. As described previously for pathogen-resistant plant varieties, LC-MS metabolomic studies can provide valuable information for breeding programs aimed at developing insect-resistant varieties (Fig. 2.2: B1). In this sense, potential metabolite markers for brown planthopper resistance such as phenylalanine and tryptophan were identified in different rice varieties (Uawisetwathana *et al.*, 2022).

Zhang *et al.* (2023) performed an untargeted metabolite profiling of *C. tetracocca* to examine its response to *Empoasca onukii* infestations (Fig. 2.2: A), using HPLC-MS. They identified 567 differentially abundant metabolites, with 280 upregulated and 287 downregulated, in both positive and negative ion modes. Among the most notable metabolites found at elevated levels in leaves affected by leafhoppers were indoleacrylic acid, shikimic acid, 8-*O*-acetyl shanzhiside methyl ester, medecassic acid, cimimanol D, coroloside, quercetin

3-*O*-rhamnosyl-(1→2)-rhamnosyl-(1→6)-glucoside, skimmianine, aginoside progenin, *cis*-3-hexenyl β-primeveroside, and theogallin. These findings suggest that these metabolites may contribute to the plant defense mechanisms against pests and hold potential for biocontrol applications.

NMR metabolomics is a powerful analytical tool used widely in plant biology to identify and quantify metabolites involved in defensive responses against herbivores and insects. By comparing the NMR spectra of plants before and after herbivore or insect attack, researchers can detect changes in metabolite profiles and pinpoint specific compounds that play crucial roles in plant defense. The non-destructive nature and high reproducibility of NMR make it an invaluable method for understanding the biochemical pathways activated during plant–insect interactions and for identifying potential biomarkers for plant resistance, ultimately contributing to the development of more resistant crop varieties. Shami *et al.* (2023) utilized NMR spectroscopy to identify various chemical compounds in different cotton varieties (Fig. 2.2: B1), including *Bt* cotton, a non-transgenic pest-resistant variety, and wild type. By employing multivariate data analysis to detect differential metabolites, principal component analysis was used to discern key differences, followed by the isolation and identification of these metabolites. The most prominent difference observed among cotton varieties was the presence of stigmasterol in non-transgenic pest-resistant varieties. Moreover, the signal intensity of valine, limonene, succinic acid, terpinolene, γ-aminobutyric acid, scylloinositol, proline, shikimic acid, chlorogenic acid, ferulic acid, arabinose, xylulose, trigonelline, and formate was found to be higher in *Bt* cotton plants. The use of NMR spectroscopy allowed a detailed chemical profile of the plant response to biotic stress to be obtained, which provided insights into the metabolic adjustments occurring in different tissues. Furthermore, the subsequent bioassays with identified metabolites elucidated their specific roles and the underlying signaling pathways involved in plant defense, highlighting the complexity and specificity of plant–herbivore interactions. This study therefore contributes to a broader understanding of plant defense strategies and the metabolic pathways activated in response to herbivory.

A study by Leiss *et al.* (2009), which used NMR metabolomics to assess *Chrysanthemum* spp. resistance to thrips, found that chlorogenic acid and feruloyl quinic acid were key metabolites distinguishing resistant from susceptible chrysanthemum varieties (Fig. 2.2: B1), as shown by partial least-squares discriminant analysis. These findings were confirmed by direct quantification using NMR and validated with HPLC.

2.5 Future Perspectives

The extended use of MS imaging will be particularly useful for crop protection. This innovative technique can identify and spatially map the distribution of natural metabolites and defense-related compounds within plants. Understanding where and how these compounds are produced and stored can help in breeding or engineering crops with enhanced resistance to pests and diseases. This can also provide early detection of infections or infestations, allowing timely intervention and management in a non-destructive and easy way, and, importantly, by analyzing the spatial distribution of both plant and pest metabolites we can understand the biochemical interactions between crops and pests. This knowledge can be used to develop targeted strategies to disrupt these interactions and protect crops. MS imaging data can be integrated with other omics data (genomics, transcriptomics, and proteomics) to provide a comprehensive view of plant biology and responses to stress. This holistic approach facilitates the development of more effective and targeted crop protection strategies. From the point of view of agrochemicals, MS imaging can track the fate and distribution of treatments within plant tissues and surrounding environments, helping to assess the environmental impact of crop protection practices and to develop more sustainable approaches. In summary, MS imaging offers a powerful tool for enhancing crop protection by providing detailed spatial and molecular insights into plant metabolism, pathogen interactions, and the effects of chemical treatments. This information is crucial for developing more effective, sustainable, and targeted approaches to protect crops from various stresses.

Artificial intelligence (AI) is also revolutionizing metabolomics by significantly enhancing

Table 2.1. Summary of differential accumulated metabolites identified through metabolomics in plant–pathogen/insect interactions. The table categorizes differential accumulated metabolites based on the type of pathogen or insect (viroids/viruses, bacteria/fungi, and insects), the host plant species, the analytical technique used, and relevant references.

	Pathogen/insect	Plant	Technique	Differential accumulated metabolites	Reference
Virus and viroids	Potyvirus necrotic infection	*Solanum lycopersicum*	GC-MS	Terpenes, alcohols, aldehydes, esters, ketones, and furan derivatives	Spanò *et al.* (2023)
	Rice dwarf virus	*Oryza sativa*	GC-MS	Gentiobiose, threonic acid, 2-deoxyerythritol, cytidine-monophosphate, glutamine, d-arabitol, 5-oxoproline, lactic acid, *N*-acetyl-l-leucine, uridine, glutaric acid	Chang *et al.* (2021)
	Potato spindle tuber viroid	*S. lycopersicum*	GC-MS, LC-MS	Glycosylated gentisic acid, glucose, malic acid, caffeic acid 4-*O*-β-d-glucoside, rutin (quercetin-gluco-rhamnoside), gentisic acid, eicosane, hexadecane, trichloroacetate, 3-methylbenzaldehyde, 1-*O*-methyl-β-d-glucuronate, 3-octadecenoic acid, methyl ester, sebacate, methyl 2-hydroxy-heptadecanoate, palmitoleate, d-fructose, galactose, erythritol, lignocerate, heptadecane, hexadecanol, octanol, viridiflorene, 1-octadecene, eicosatrienoate, pre-vitamin D3, uridine, 5α-cholestane-3β,7β,12β-triol, d-glucose, threose, cyclohexane-1-carboxylate, benzoate, palmitate, heptadecanoate, linoleate, myristate, behenate, γ-linolenate, myo-inositol, succinate, 2,4,6-trihydroxybenzophenone	Bagherian *et al.* *(2016)*
	Different plant–pathogen interactions	*S. lycopersicum*	GC-MS, LC-MS	Glycosylated gentisic acid, rutin, phenylpropanoids (e.g. ferulic acid glucoside), glucose, malic acid, sucrose, alanine, glycine, isoleucine, threonine, valine, acetic acid, aspartic acid, citric acid, glutamic acid, glyceric acid, β-glucose, α-glucose, rhamnose, fructose, γ-aminobutyric acid (GABA)	López-Gresa *et al.* (2009)
	Tobacco mosaic virus	*Arabidopsis*, *Nicotiana benthamiana*	LC-MS	Dehydroascorbate, gulonate, 2-oxo-gulonate, palmitate, myristate, pelargonate, stearate, benzoate, GABA, ribose, rhamnose, sorbose, malate, isocitrate, pyruvate, proline, alanine, phenylalanine, valine	Bazzini *et al.* (2011)
	Black-streaked dwarf virus	*O. sativa*	LC-MS	Luteolin, chrysin, steroids, thiamine, nicotinate, 5-l-glutamyl-l-alanine	Huang *et al.* (2024)
	Cucumber green mottle mosaic virus	*Lagenaria siceraria*	LC-MS	4-Caffeoylquinic acid, acteoside, malvidin-3,5-*O*-diglucoside, glucarate *O*-phosphoric acid, cyclic AMP, scopoletin (7-hydroxy-5-methoxycoumarin)	G. Li *et al.* (2023)

	Apple scar skin viroid	*Malus domestica*	LC-MS	L-2-aminobutyric acid, 6″-*O*-malonyldaidzin, d-xylose, coumarin	Z. Li *et al.* (2023)
	Citrus exocortis viroid	*S. lycopersicum*	LC-MS	Gentisic acid, ferulic acid, phenylalanine, tyrosine, aspartate, glutamate, asparagine, GABA, glutamine, adenosine, trigonelline	López-Gresa *et al.* (2019)
	Tomato curly stunt virus	*S. lycopersicum*	UHPLC-MS	Coumaric acid, caffeoyl putrescine, 5-*O*-caffeoylgalactaric acid (dimer), 5-*O*-(*E*)-feruloylquinic acid (dimer), 5-*O*-(*E*)-caffeoylquinic acid, sinapoylglycoside, quinic acid, galactaric acid, quercetin 3-rutinoside, quercetin-3-rutinoside-7-glycoside, quercetin-3-*O*-deoxyhexose-*O*-hexose-*O*-pentose, benzyl alcohol-hexose-pentose	Rossouw *et al.* (2019)
	Tobacco mosaic virus	*N. tabacum*	NMR	*R*-linolenic acid analogs, phenylpropanoids (3-*O*-, 4-*O*-, and 5-*O*-caffeoylquinic acids), inositol, sesquiterpenoids, diterpenoids (capsidiol and cembratriene-4,6-diol), sucrose, glucose, malic acid, citric acid, alanine, glutamine, proline	Choi *et al.* (2006)
	Tomato mosaic virus	*S. lycopersicum*	NMR, LC-MS	Rutin, glucose, sucrose, choline, GABA, glutamine, asparagine, phenylalanine, glutamic, aspartic, malic, citric, formic, fumaric acids, phenylpropanoids, tryptophan	López-Gresa *et al.* (2012)
Bacteria and fungus	*Pseudomonas syringae*	*S. lycopersicum*	GC-MS	(*Z*)-3-hexenol esters and hydroxylated monoterpenes	López-Gresa *et al.* (2017, 2018), Payá *et al.* (2020, 2024), Pérez-Pérez *et al.* (2024)
	Botrytis cinerea	*Vitis vinifera*	GC-MS	1,5-dimethyltetralin, 1,5-dimethylnaphthalene, phenylethyl alcohol, 3-octanol, 2-octenol	Jiang *et al.* (2023)
	Fusarium oxysporum	*S. lycopersicum*	GC-MS	Methyl and ethyl salicylate	Hernández-Aparicio *et al.* (2021)
	F. culmorum	*Triticum aestivum*	GC-MS	Methylcycloheptanol, 2-heptanone, 2(3H)-furanone, 5-hexyldihydro-2(3H)-furanone	Buśko *et al.* (2023)
	Colletotrichum lindemuthianum	*Phaseolus vulgaris*	GC-MS	Limonene, linalool, nonanal, methyl salicylate, methyl jasmonate	Quintana-Rodriguez *et al.* (2015)
	P. syringae	*S. lycopersicum*	LC-MS	Chlorogenic acid, rutin, hydroxycinnamic acid amides	Zacarés *et al.* (2007), López-Gresa *et al.* (2011), Campos *et al.* (2014)
	F. graminearum	*Hordeum vulgare*	LC-MS	Phenylpropanoids, flavonoids, fatty acids, terpenoids	Bollina *et al.* (2010
	Phytophthora infestans	*S. tuberosum*	LC-MS	Hydroxycinnamic acid amides	Pushpa *et al.* (2014)

Continued

Table 2.1. Continued.

	Pathogen/insect	Plant	Technique	Differential accumulated metabolites	Reference
	F. culmorum	*H. vulgare*	LC-MS	Mycotoxins deoxynivalenol, deoxynivalenol-3-glucoside	Cajka *et al.* (2014)
	Ralstonia solanacearum	*S. lycopersicum*	NMR	Leucine	Murti *et al.* (2021)
	Candidatus liberibacter spp.	*Citrus* spp.	NMR	Purine, *p*-coumaryl alcohol, 5-acetylamido-2-chloroaniline, GABA, l-homocitrulline, l-lysine, acetate	Pontes *et al.* (2016)
Insects and herbivores	*Opius dissitus*	*S. lycopersicum*	GC-MS	(*E*)-2-hexenal, (*Z*)-3-hexenol, ocimene, linalool	Wei *et al.* (2007)
	Choristoneura fumiferana	*Fallopia baldschuanica*	GC-MS	β-Ocimene and limonene	Salerno *et al.* (2023)
	Beauveria bassiana	*Gossypium hirsutum*	GC-MS	Benzaldehyde	González-Mas *et al.* (2021)
	Aphis gossypii	*Cucumis melo*	GC-MS	2-Ethylhexyl nonyl sulfite	González-Mas *et al.* (2021)
	Spodoptera littoralis, Metopeurum fuscoviride	*Tanacetum vulgare*	GC-MS	Terpenoids	Clancy *et al.* (2020)
	Empoasca onukii	*Camellia sinensis*	GC-MS	4-Hydroxybenzaldehyde	Liu *et al.* (2020)
	Peristenus spretus	*G. hirsutum*	GC-MS	Methyl salicylate	Xu *et al.* (2021)
	Wounding	*Camellia C.*	LC-MS	Flavonoids (epicatechin, catechin), phenolics (chlorogenic acid)	Zhang *et al.* (2014)
	Brown planthopper	*O. sativa*	LC-MS	Phenylalanine, tryptophan	Uawisetwathana *et al.* (2022)
	E. onukii	*C. tetracocca*	LC-MS	Indoleacrylic acid, shikimic acid, 8-*O*-acetyl shanzhiside methyl ester, medecassic acid, cimimanol D, coroloside, quercetin 3-*O*-rhamnosyl-(1→2)-rhamnosyl-(1→6)-glucoside, skimmianine, aginoside progenin, *cis*-3-hexenyl β-primeveroside, theogallin	Zhang *et al.* (2023)
	Herbivore	*G. hirsutum*	NMR	Valine, limonene, succinic acid, terpinolene, γ-aminobutyric acid, scylloinositol, proline, shikimic acid, chlorogenic acid, ferulic acid, arabinose, xylulose, trigonelline, formate	Shami *et al.* (2023)
	Frankliniella occidentalis	*Chrysanthemum* spp.	NMR	Chlorogenic acid, feruloyl quinic acid	Leiss *et al.* (2009)

GC-MS, gas chromatography–mass spectrometry; LC-MS, liquid chromatography–mass spectrometry; NMR, nuclear magnetic resonance; UHPLC-MS, ultra-high-performance liquid chromatography–mass spectrometry.

the identification and characterization of metabolites. Machine learning algorithms streamline the analysis of complex metabolomic datasets, improving the accuracy of metabolite identification and facilitating the discovery of novel compounds. AI techniques, including supervised and unsupervised learning, are employed to recognize patterns in spectral data from MS and NMR, which speeds up the process of identifying known metabolites and discovering new ones.

Furthermore, AI plays a critical role in elucidating differential metabolites by detecting significant changes in metabolite levels across different conditions or treatments. This ability helps in understanding biochemical pathways and uncovering mechanisms underlying diseases. AI also integrates metabolomic data with genomic and proteomic information, providing a holistic view of biological systems. By enhancing data analysis, visualization, and interpretation, AI accelerates scientific discoveries and deepens insights into metabolic processes.

In conclusion, the expanded integration of AI into metabolomics will offer transformative potential for advancing the field. By improving the efficiency and accuracy of metabolite analysis and integrating multi-omics data, AI not only accelerates discoveries but also provides a more comprehensive understanding of metabolic functions, ultimately advancing research and applications in various scientific fields.

2.6 Concluding Remarks

Metabolomics provides a set of powerful tools that have become essential in understanding the defensive response of plants to biotic stress. Table 2.1 summarizes differential metabolites identified by metabolomics in response to viroids or viruses, bacteria or fungi, and insects. The insights derived from these advances not only expand our knowledge of plant defense mechanisms, but also offer new opportunities for developing agricultural and biotechnological strategies aimed at enhancing crop resistance to environmental challenges.

2.7 References

Abdul Malik, N.A., Kumar, I.S., and Nadarajah, K. (2020) Elicitor and receptor molecules: orchestrators of plant defence and immunity. *International Journal of Molecular Sciences* 21: 963.

Aharoni, A., Goodacre, R., and Fernie, A.R (2023) Plant and microbial sciences as key drivers in the development of metabolomics research. *Proceedings of the National Academy of Sciences USA* 120: e2217383120.

Bagherian, S.A.A., Hamzehzarghani, H., Izadpanah, K., and Djavaheri, M. (2016) Effects of potato spindle tuber viroid infection on tomato metabolic profile. *Journal of Plant Physiology* 201, 42–53.

Balanzá, M., Vázquez-Prol, F., Rodrigo, I., Bellés, J.M., Vera-Sirera, F. *et al.* (2025) Salicylic acid modulates volatile organic compound profiles during CEVd infection in tomato plants. *Metabolites* 15: 102. DOI: 10.3390/metabo15020102.

Bazzini, A.A., Manacorda, C.A., Tohge, T., Conti, G., and Rodriguez, M.C. (2011) Metabolic and miRNA profiling of TMV infected plants reveals biphasic temporal changes. *PLOS ONE* 6: 28466.

Bollina, V., Kumaraswamy, G.K., Kushalappa, A.C., Choo, T.M., Dion, Y., *et al.* (2010) Mass spectrometry-based metabolomics application to identify quantitative resistance-related metabolites in barley against *Fusarium* head blight. *Molecular Plant Pathology* 11, 769–782.

Buśko, M., Gracka, A., Jeleń, H., Szablewska, K.S., Przybylska-Balcerek, A., *et al.* (2023) The effect of organic and conventional cultivation systems on the profile of volatile organic compounds in winter wheat grain, including susceptibility to *Fusarium* head blight. *Metabolites* 13: 1045.

Cajka, T., Vaclavikova, M., Dzuman, Z., Vaclavik, L., Ovesna, J., and Hajslova, J. (2014) Rapid LC-MS-based metabolomics method to study the *Fusarium* infection of barley. *Journal of Separation Science* 37, 912–919.

Campos, L., Granell, P., Tárraga, S., López-Gresa, P., Conejero, V., *et al.* (2014a) Salicylic acid and gentisic acid induce RNA silencing-related genes and plant resistance to RNA pathogens. *Plant Physiology and Biochemistry* 77, 35–43.

Campos, L., Lisón, P., López-Gresa, M.P., Rodrigo, I., Zacarés, L., *et al.* (2014b) Transgenic tomato plants overexpressing tyramine *N*-hydroxycinnamoyltransferase exhibit elevated hydroxycinnamic acid amide levels and enhanced resistance to *Pseudomonas syringae*. *Molecular Plant–Microbe Interactions* 27, 1159–1169.

Castro-Moretti, F.R., Gentzel, I.N., Mackey, D., and Alonso, A.P. (2020) Metabolomics as an emerging tool for the study of plant–pathogen interactions. *Metabolites* 10: 52.

Chang, X., Ning, D., Mao, L., Wang, B., Fang, Q., *et al.* (2021) Metabolic analysis reveals *Cry1C* gene transformation does not affect the sensitivity of rice to rice dwarf virus. *Metabolites* 11: 209.

Chen, F., Ma, R., and Chen, X. L. (2019) Advances of metabolomics in fungal pathogen–plant interactions. *Metabolites* 9, 169.

Choi, Y.H., Kim, H.K., Linthorst, H.J.M., Hollander, J.G., Lefeber, A.W.M., *et al.* (2006) NMR metabolomics to revisit the tobacco mosaic virus infection *in Nicotiana tabacum* leaves. *Journal of Natural Products* 69, 742–748.

Clancy, M.V., Haberer, G., Jud, W., Niederbacher, B., Niederbacher, S., *et al.* (2020) Under fire—simultaneous volatilome and transcriptome analysis unravels fine-scale responses of tansy chemotypes to dual herbivore attack. *BMC Plant Biology* 20: 551.

Dixon, R.A. and Dickinson, A.J. (2024) A century of studying plant secondary metabolism—from “what?” to “where, how, and why?” *Plant Physiology* 195, 48–66.

Dixon R.A., Gang D.R., Charlton A.J., Fiehn O., Kuiper H.A., *et al.* (2006) Applications of metabolomics in agriculture. *Journal of Agricultural and Food Chemistry* 54, 8984–8994.

Dunn, W.B. and Ellis, D.I. (2005) Metabolomics: current analytical platforms and methodologies. *Trends in Analytical Chemistry* 24, 285–294.

González-Mas, N., Gutiérrez-Sánchez, F., Sánchez-Ortiz, A., Grandi, L., Turlings, T.C.J., *et al.* (2021) Endophytic colonization by the entomopathogenic fungus *Beauveria bassiana* affects plant volatile emissions in the presence or absence of chewing and sap-sucking insects. *Frontiers in Plant Science* 12: 660460.

Hernández-Aparicio, F., Lisón, P., Rodrigo, I., Bellés, J.M., and López-Gresa, M.P. (2021) Signaling in the tomato immunity against *Fusarium oxysporum*. *Molecules* 26: 1818.

Heuberger, A.L., Robison, F.M., Lyons, S.M., Broeckling, C.D., and Prenni, J.E. (2014) Evaluating plant immunity using mass spectrometry-based metabolomics workflows. *Frontiers in Plant Science* 24, 291.

Huang, C., Wang, Z., Zhu, P., Xu, W., Du, Z., *et al.* (2024) RNAi-based genetically engineered rice resistant to black-streaked dwarf virus does not show adverse genetic effects: a multi-omics analysis. *Plants, People, Planet* 6, 622–639.

Jiang, L., Qiu, Y., Dumlao, M.C., Donald, W.A., Steel, C.C., and Schmidtke, L.M. (2023) Detection and prediction of *Botrytis cinerea* infection levels in wine grapes using volatile analysis. *Food Chemistry* 421: 136120.

Kasote, D., Lee, J., and Sreenivasulu, N. (2023) Volatilomics in plant and agricultural research: recent trends. *Frontiers in Plant Science* 14: 1289998.

Leiss, K.A., Maltese, F., Choi, Y.H., Verpoorte, R., and Klinkhamer, P.G.L. (2009) Identification of chlorogenic acid as a resistance factor for thrips in chrysanthemum. *Plant Physiology* 150, 1567–1575.

Li, G., Li, J., Zhang, H., Li, J., Jia, L., *et al.* (2023a) ASSVd infection inhibits the vegetative growth of apple trees by affecting leaf metabolism. *Frontiers in Plant Science* 14: 1137630.

Li, Z., Tang, Y., Lan, G., Yu, L., Ding, S., *et al.* (2023b) Transcriptome and metabolome analyses reveal that jasmonic acids may facilitate the infection of cucumber green mottle mosaic virus in bottle gourd. *International Journal of Molecular Sciences* 24: 16566.

Liu, H., Li, S., Xiao, G., and Wang, Q. (2021) Formation of volatiles in response to tea green leafhopper (*Empoasca onukii* Matsuda) herbivory in tea plants: a multi-omics study. *Plant Cell Reports* 40, 753–766.

López-Gresa, M.P., Maltese, F., Bellés, J.M., Conejero, V., Kim, H.K., *et al.* (2009) Metabolic response of tomato leaves upon different plant–pathogen interactions. *Phytochemical Analysis* 21, 89–94.

López-Gresa, M.P., Torres, C., Campos, L., Lisón, P., Rodrigo, I., *et al.* (2011) Identification of defence metabolites in tomato plants infected by the bacterial pathogen *Pseudomonas syringae*. *Environmental and Experimental Botany* 74, 216–228.

López-Gresa, M.P., Lisón, P., Kim, H.K., Choi, Y.H., Verpoorte, R., *et al.* (2012) Metabolic fingerprinting of tomato mosaic virus infected *Solanum lycopersicum*. *Journal of Plant Physiology* 169, 1586–1596.

López-Gresa, M.P., Lisón, P., Campos, L., Rodrigo, I., Rambla, J.L., *et al.* (2017) A non-targeted metabolomics approach unravels the VOCs associated with the tomato immune response against *Pseudomonas syringae*. *Frontiers in Plant Science* 8: 1188.

López-Gresa, M.P., Payá, C., Ozáez, M., Rodrigo, I., Conejero, V., *et al.* (2018) A new role for green leaf volatile esters in tomato stomatal defence against *Pseudomonas syringe* pv. *tomato*. *Frontiers in Plant Science* 18: 1855.

López-Gresa, M.P., Payá, C., Rodrigo, I., Bellés, J.M., Barceló, S., *et al.* (2019) Effect of benzothiadiazole on the metabolome of tomato plants infected by citrus exocortis viroid. *Viruses* 11: 437.

Murti, R.H., Afifah, E.N., and Nuringtyas, T.R. (2021) Metabolomic response of tomatoes (*Solanum lycopersicum* L.) against bacterial wilt (*Ralstonia solanacearum*) using ^{1}H-NMR spectroscopy. *Plants* 10: 1143.

Nenaah, G.E. and Ibrahim, S.I.A. (2011) Chemical composition and the insecticidal activity of certain plants applied as powders and essential oils against two stored-products coleopteran beetles. *Journal of Pest Science* 84, 393–402.

Payá, C., López-Gresa, M.P. Intrigliolo, D.S. Rodrigo, I. Bellés, J.M. Lisón, P. (2020) (*Z*)-3-hexenyl butyrate induces stomata closure and ripening in *Vitis vinifera*. *Agronomy* 10: 1122.

Payá, C., Belda-Palazón, B., Vera-Sirera, F., Pérez-Pérez, J., Jordá, L., *et al.* (2024) Signalling mechanisms and agricultural applications of (*Z*)-3-hexenyl butyrate-mediated stomatal closure. *Horticulture Research* 11: uhad248.

Pérez-Pérez, J., Minguillón, S., Kabbas-Piñango, E., Payá, C., Campos, L., *et al.* (2024) Metabolic crosstalk between hydroxylated monoterpenes and salicylic acid in tomato defence response against bacteria. *Plant Physiology* 195, 2323–2338.

Pontes, J.G.M., Ohashi, W.Y., Brasil, A.J.M., Filgueiras, P.R., Espíndola, A.P.D.M., *et al.* (2016) Metabolomics by NMR spectroscopy in plant disease diagnostic: huanglongbing as a case study. *ChemistrySelect* 1, 1176–1178.

Pushpa, D., Yogendra, K.N., Gunnaiah, R., Kushalapp, A.C., and Murphy, A. (2014) Identification of late blight resistance-related metabolites and genes in potato through nontargeted metabolomics. *Plant Molecular Biology Reporter* 32, 584–595.

Quintana-Rodriguez, E., Morales-Vargas, A.T., Molina-Torres, J., Adame-Alvarez, R.M., Acosta-Gallegos, J.A., and Heil, M. (2015) Plant volatiles cause direct, induced and associational resistance in common bean to the fungal pathogen *Colletotrichum lindemuthianum*. *Journal of Ecology* 103, 250–260.

Rossouw, L.T., Madala, N.E., Tugizimana, F., Steenkamp, P.A., Esterhuizen, L.L., and Dubery, I.A. (2019) Deciphering the resistance mechanism of tomato plants against whitefly-mediated tomato curly stunt virus infection through ultra-high-performance liquid chromatography coupled to mass spectrometry (UHPLC-MS)-based metabolomics approaches. *Metabolites* 9: 60.

Salerno, G., Rebora, M., and Gorb, S. (2023) Mechanoecology and chemoecology: physical and chemical interactions between insects and plants. *Insects* 14, 10–12.

Shami, A.A., Akhtar, M.T., Mumtaz, M.W., Mukhtar, H., Tahir, A., *et al.* (2023) NMR-based metabolomics: a new paradigm to unravel defence-related metabolites in insect-resistant cotton variety through different multivariate data analysis approaches. *Molecules* 28: 1763.

Shulaev, V. (2006) Metabolomics technology and bioinformatics. *Briefings in Bioinformatics* 7, 128–139.

Spanò, R., Mastrochirico, M., Longobardi, F., Cervellieri, S., Lippolis, V., and Mascia, T. (2023) Characterization of volatile organic compounds in grafted tomato plants upon potyvirus necrotic infection. *Journal of Integrative Agriculture* 22, 2426–2440.

Turlings, T.C.J. and Erb, M. (2018) Tritrophic interactions mediated by herbivore-induced plant volatiles: mechanisms, ecological relevance, and application potential. *Annual Review of Entomology* 63, 433–452.

Uawisetwathana, U., Jamboonsri, W., Bamrungthai, J., Jitthiang, P., Nookaew, I., and Karoonuthaisiri, N. (2022) Metabolite profiles of brown planthopper-susceptible and resistant rice (*Oryza sativa*) varieties associated with infestation and mechanical stimuli. *Phytochemistry* 194; 113044.

Wang, L., Allmann, S., Wu, J., and Baldwin, I.T. (2008) Comparisons of LIPOXYGENASE3- and JASMONATE-RESISTANT 4/6-silenced plants reveal that jasmonic acid and jasmonic acid–amino acid conjugates play different roles in herbivore resistance of *Nicotiana attenuata*. *Plant Physiology* 146, 904–915.

Wei, J., Wang, L., Zhu, J., Zhang, S., Nandi, O.I., and Kang, L. (2007) Plants attract parasitic wasps to defend themselves against insect pests by releasing hexenol. *PLOS ONE* 2: e852.

Xu, M., Jiang, Y., Chen, S., Chen, F., and Chen, F. 2021 Herbivory-induced emission of volatile terpenes in *Chrysanthemum morifolium* functions as an indirect defence against *Spodoptera litura* larvae by attracting natural enemies. *Journal of Agricultural and Food Chemistry* 69, 9743–9753.

Zacarés, L., López-Gresa, M.P., Fayos, J., Primo, J., Bellés, J.M., and Conejero, V. (2007) Induction of *p*-coumaroyldopamine and feruloyldopamine, two novel metabolites, in tomato by the bacterial pathogen *Pseudomonas syringae*. *Molecular Plant–Microbe Interactions* 20, 1439–1448.

Zhang, N., Tan, W., Luo, G., Pu, T., Wang, J., *et al.* (2023) Untargeted metabolite profiling of *Camellia tetracocca*'s response to an *Empoasca onukii* attack using GC-MS and LC-MS. *Agronomy* 13: 791.

Zhang, M., Zhang, L., Zhou, C., Xu, K., Chen, G., *et al.* (2024) Metabolite profiling reveals the dynamic changes in non-volatiles and volatiles during the enzymatic-catalyzed processing of Aijiao Oolong tea. *Plants* 13: 1249.

3 High-throughput Plant Metabolomics and Predictive Modelling

Millena C. Barros Santos[1,6*†], Cathleen Mirande-Ney[1*], Thomas Dussarrat[2], Guillaume Marti[3], Alain Bouchereau[4,5], Younès Dellero[4,5], Sylvain Prigent[1,6], and Pierre Pétriacq[1,6†]

[1]University of Bordeaux, INRAE, Villenave d'Ornon, France; [2]Department of Chemical Ecology, Bielefeld University, Bielefeld, Germany; [3]Laboratoire de Recherche en Sciences Végétales, Metatoul-AgromiX Platform, Université de Toulouse, CNRS, INP, Auzeville, Auzeville-Tolosane, France; [4]Institute for Genetics, Environment and Plant Protection (IGEPP), National Research Institute for Agriculture, Food and Environment (INRAE), Institut Agro Rennes-Angers, Université Rennes, Le Rheu, France; [5]Metabolic Profiling and Metabolomic Platform (P2M2), MetaboHUB-Grand-Ouest, Rennes, France; [6]Bordeaux Metabolome, MetaboHUB, PHENOME-EMPHASIS, Villenave d'Ornon, France

Abstract

This chapter explores advances and methodologies in high-throughput metabolic phenotyping through metabolomics and predictive modeling to enhance the understanding of plant metabolism. Key techniques, data analysis tools, and applications in plant science research are discussed. The potential of predictive modeling to identify new metabolic pathways and markers associated with plant performance and improve crop traits is highlighted. Future directions and challenges in the field are also examined.

Keywords: bioinformatics, data science, high-throughput phenotyping, metabolomics, predictive modeling

3.1 Introduction

The metabolome is often seen as the ultimate cellular phenotype, resulting from all upstream biochemical processes taking place in an organism. Metabolomics—the detection and quantitation of the metabolome—is a highly effective analytical tool in crop research, providing comprehensive insights into plant metabolism and physiology (Tian *et al.*, 2016). Metabolomics has made significant strides, playing a pivotal role in bridging the phenotype–genotype gap, informing genotype and environment interactions, and aiding in genome sequence annotation (Hall *et al.*, 2022). It holds immense potential in agricultural applications, including genetic improvement of crops, evaluation of transgenic varieties, and characterization of cultivars (Ibarra-Estrada *et al.*, 2016). The importance and role of metabolomics in these different fields are made possible by advances in analytical facilities and bioinformatics, which have substantially and sequentially

* Joint first authors.

† E-mail: millena.barros-santos@inrae.fr or pierre.petriacq@inrae.fr

© CAB International 2025. *High-Throughput Plant Metabolomics* (Jen-Tsung Chen ed.)
DOI: 10.1079/9781800626911.0003

addressed the challenges faced by metabolomics (Mendez *et al.*, 2019). Although efforts in metabolic analyses could be observed by the end of the 20th century, the shift toward metabolomics and so-called "untargeted" approaches relied on the dramatically increased sensitivity and resolution of analytical tools such as liquid chromatography–mass spectrometry (LC-MS) (Jones *et al.*, 2013; Alseekh and Fernie, 2018). The recent ability to detect tens of thousands of features in a single analysis has created a demand for new processing tools and statistical methods to address noisy and highly collinear data sets (Saccenti *et al.*, 2014; Mendez *et al.*, 2019). The development of various data-processing tools such as XCMS, MS-DIAL, and mzmine, as well as the development of multivariate statistics and, more recently, modeling packages and artificial neural networks, have greatly improved data analyses and reproducibility, as well as analyte annotation (Pluskal *et al.*, 2010; Tsugawa *et al.*, 2015; Chong and Xia, 2018; Mendez *et al.*, 2019). In particular, artificial intelligence (AI), including machine learning (ML) and deep learning (DL), has remarkable capabilities for big-data analysis as it uses computational modeling to approximate human intellect. AI has demonstrated remarkable proficiency in recognizing hidden patterns, discerning informative features from noisy backgrounds, and delivering accurate predictions. These capabilities make it particularly well suited for metabolomic data analysis (Chi *et al.*, 2024). Indeed, innovative AI algorithms have been developed to process raw (mass spectrometry (MS)) data, detect chromatogram peaks, identify metabolites, discover biomarkers, build predictive models, and construct pathways, as well as integrate multi-omics data analysis, thereby advancing the field of plant metabolomics (Greener *et al.*, 2022; Chi *et al.*, 2024). Although certain analytical steps are still limiting, such as feature annotation, other aspects such as phenotyping and metadata acquisition represent current and future challenges for high-throughput metabolomics (Hall *et al.*, 2022). A literature search showed that high-throughput metabolomics studies are on the rise and currently account for 8% of the 3477 publications involving plant sciences (WebOfScience, September 2024: www.webofscience.com; accessed 16 January 2025). Altogether, these major advances have shaped high-throughput metabolomics and its versatility.

This chapter offers an updated description of the approaches and tools used to meet high-throughput demands. We cover the whole metabolomics workflow, including sampling, sample preparation, data acquisition, data processing, and interpretation. We also discuss the critical role of bioinformatics in the development and advancement of metabolomics, as well as recent studies benefiting from high-throughput plant metabolomics.

3.2 High-throughput Plant Metabolomics Techniques

Traditional field phenotyping methods, such as measuring plant height, biomass, and macroscopic trait capture, are being outpaced by modern, large-scale, high-precision techniques (Hall *et al.*, 2022). Nowadays, non-destructive, computational, and imaging tools are crucial in developing new cultivars. They are faster, more precise, and less resource intensive, making them game changers in the field.

High-throughput metabolomics techniques have become increasingly important in plant sciences (Hall *et al.*, 2022). Prospectively, integrating metabolomics into high-throughput plant phenotyping platforms holds significant potential as metabolites define and support important physiological and agronomic traits in plants. Primary metabolites are direct contributors to nutrition, growth, and development processes, while secondary (specialized) metabolites are distinguished in their ecological functions, and all these metabolites contribute to establishing the quality of plant products for food or non-food purposes. Moreover, high-throughput metabolomics offers a promising avenue for bridging the genotype–phenotype gap and understanding plant responses to environmental changes (Hall *et al.*, 2022). The most pressing challenge, however, is developing high-throughput metabolomics methods without compromising data and information quality (Hajjar *et al.*, 2023).

In a recent review, Vervoort *et al.* (2021) differentiated high-throughput experimentation (HTE) and high-throughput analytical techniques (HTA). Briefly, HTE involves simultaneously carrying out workflow/experiments, maximizing efficiency with automation and miniaturization, and considering different steps of the metabolomics workflow. HTA refers to the technical analysis used to generate data from these experiments and reduce processing time to keep up with the high volume of HTE.

In this section, we will discuss the metabolomic workflow from high-throughput perspectives that have yet to be explored in the context of plant research (Fig. 3.1).

3.2.1 Sample preparation and processing

The chemical complexity of the plant matrices requires careful sample preparation and processing from the field to data acquisition. Sample preparation steps are critical for accurate results and involve harvesting, drying, storage, weighing, extraction, and filtration/purification (Kim and Verpoorte, 2010). In the context of high-throughput plant metabolomics, developing simplified protocols and automating the steps to analyze a large number of samples is crucial but remains challenging.

Recent studies have shown the benefits of AI and robots in agroecology and agriculture (Pandu *et al.*, 2022; Pappalardo and Andrade,

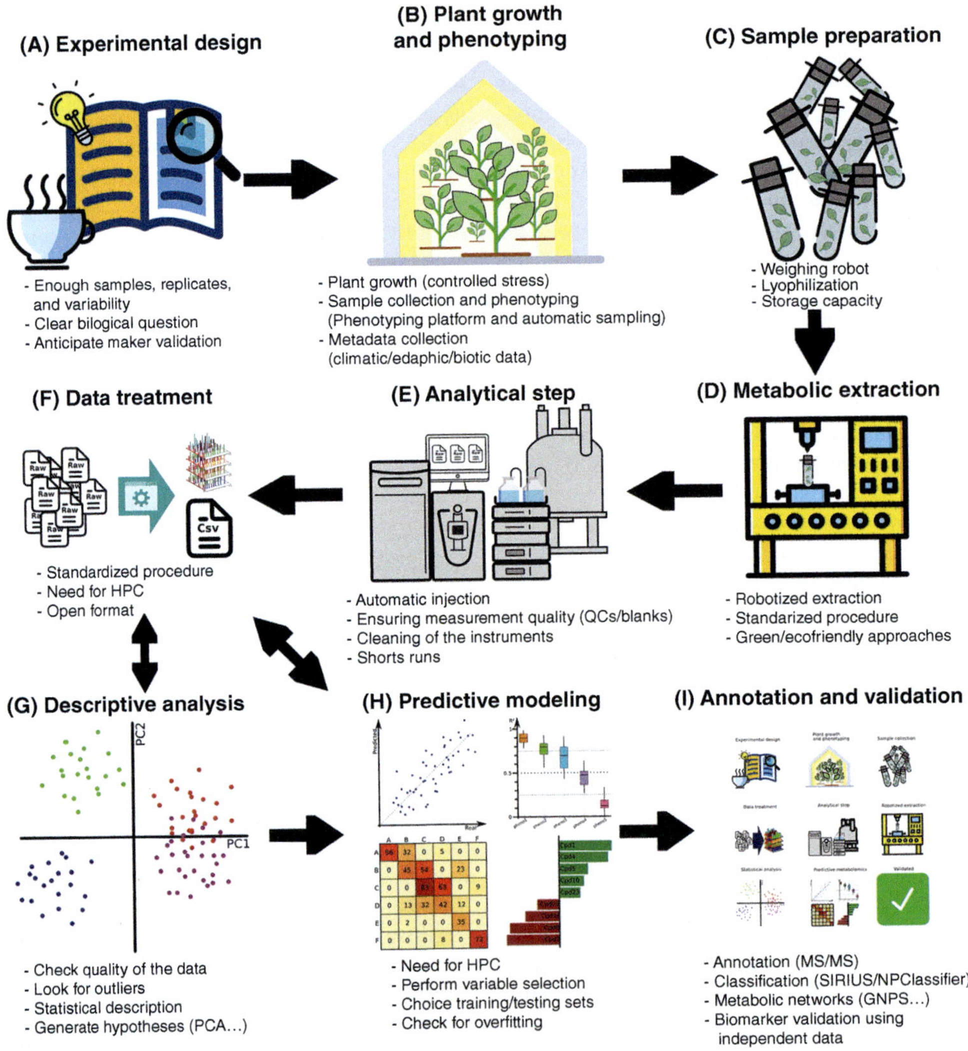

Fig. 3.1. Workflow for high-throughput plant metabolomics and predictive modeling. HPC, high-performance computing; QC, quality control; PCA, principal component analysis; MS, mass spectrometry; GNPS, Global Natural Product Social Molecular Networking.

2022). Robots/drones are used for mapping and spatialization in agricultural and forest environments (Aguiar *et al.*, 2020), for crop cultivation, monitoring, and harvesting (Wakchaure *et al.*, 2023), and for direct phenotyping of harvested products (Jahnke *et al.*, 2016). These facilities are also used to collect samples in the field or in greenhouses in an automated, reproducible, and efficient way while minimizing waste (Onishi *et al.*, 2019). Further advancements are necessary to automate the weighing process, such as using prototypes capable of weighing numerous samples in a few-hours cycle. Overall, these significant advances in phenotyping have accelerated the early steps in the metabolomics workflow and improved reproducibility. Other significant advances in plant metabolomics are found in sample preparation, now allowing standardized high-yield extraction of more than 300 samples in just hours from lower amounts of dry material (Luna *et al.*, 2020; Dussarrat *et al.*, 2022). This success has been achieved by developing specialized robots designed to perform specific tasks such as pipetting, homogenization, and centrifugation. In addition, regarding untargeted metabolomic analysis, adopting environmentally friendly extraction solvents, such as ethanol, has further increased the efficiency of automated plant extract preparation (Hajjar *et al.*, 2023). At this stage of the preparation of the extracts, the operations of internal standardization and design of quality controls are absolutely decisive for the robustness of the high-throughput metabolomics workflow (Hall *et al.*, 2022).

Looking ahead, the rise of automated plant phenotyping facilities has sparked significant interest in integrating rapid, sensitive, cost-effective, and automated metabolomics for comprehensive phenotyping (Hall *et al.*, 2022).

3.2.2 Analytical tools and acquisition methods

In the last decade, metabolomics studies have been presented as an indispensable technique for analyzing biomolecules and components of plant primary and secondary metabolisms. Due to their different chemical properties, different analytical platforms must be combined to cover the astonishing diversity of plant metabolites (Bijttebier *et al.*, 2016). Essential analytical methods include coupled techniques such as LC-MS, gas chromatography–mass spectrometry (GC-MS), and nuclear magnetic resonance (NMR) spectroscopy (Hall *et al.*, 2022). In this section, we will focus on MS-based metabolomics.

Unrivaled plant chemical diversity cannot be captured using a single analytical method (Allwood *et al.*, 2021). For instance, LC-MS is suitable for analyzing polar/semi-polar/apolar, non-volatile and thermally labile compounds, while GC-MS is widely used to analyze volatile and thermally stable compounds. LC-MS has emerged as a pivotal technique in high-throughput plant metabolomics, enabling the analysis of various metabolites, including specialized metabolites and polar compounds (Hill and Roessner, 2015; Allwood *et al.*, 2021). The development of ultra-high-performance liquid chromatography (UHPLC) and high-resolution mass spectrometry (HRMS)-based technologies (Orbitrap and TOF systems) has provided partial solutions to high-throughput analysis, while enabling complete metabolite profiling with better analyte resolution and greater mass accuracy (Allwood *et al.*, 2021). UHPLC systems can withstand higher pressure, allowing optimum mobile phase velocity to meet the requirements of smaller, more compact columns, resulting in faster injections to meet the concept of high throughput. However, studies are still employing a 10–25 min elution, which is considered a moderate-throughput method (Fuhrer and Zamboni, 2015). HTA counts as lower run times, allowing more than 1000 daily injections (Plumb *et al.*, 2023). However, recent high-throughput studies are applying these advancements to provide faster chromatography analysis, thus allowing gradient runs of less than 10 min with very short columns (Plumb *et al.*, 2023). Overall, some limitations must be considered because they can reflect on coelution, intensified matrix effects, decreased sensitivity due to poorer ionization at higher flow rates, and asymmetric peaks in the case of hydrophilic interaction liquid chromatography (HILIC) columns.

Also, recent techniques in MS, such as ion mobility systems, have enhanced metabolite separation and mass measurement accuracy, providing several benefits (Plumb *et al.*, 2023), such as allowing the separation of the several isomers present in plants. In the high-throughput context, incorporating ion mobility systems with LC-MS offers increased resolution and analysis speed, with the potential to separate and annotate structurally similar metabolites not achievable with standard LC-MS (Hall *et al.*, 2022), making it invaluable for plant functional genomics and biochemical composition studies (Zhou *et al.*, 2012).

Offering a sensitive and comprehensive analysis of primary metabolites, GC-MS is a historically powerful tool for high-throughput plant metabolomics (Wase *et al.*, 2022). It is also particularly tailored for profiling volatile metabolites and can detect over 400 compounds in a single analysis (Jonsson *et al.*, 2005; Kueffer, 2019). Several studies have compared different techniques (e.g. two-dimensional GC-MS, Fourier-transform–ion cyclotron resonance–mass spectrometry (FT-ICR-MS)), and despite recent advancements in GC-MS technology, such as GC-Orbitrap-MS and HRMS-time-of-flight, which have improved metabolite resolution and mass accuracy (Wase *et al.*, 2022), acquisition runs still require 20–40 min per sample. In addition, new advances incorporating low-thermal-mass devices can provide rapid heating temperature programs and cooling rates for capillary columns. Fialkov *et al.* (2020) compared standard GC-MS and fast GC-MS and showed a 4-fold gain in speed and a 3.5-fold loss in separation, going from a 27 min run to less than 1 min. Other benefits, such as narrower peaks, improved sensitivity, and elimination of column bleed, were demonstrated and showed promise for high-throughput analysis (Fialkov *et al.*, 2020). Key advancements in chromatographic and spectrometric techniques such as UHPLC, GC, thin-layer chromatography, superficially porous particles and open tubular columns, supercritical fluid chromatography, multiple injections in a single experimental run (MISER), and ON-CHIP chromatography and MS can support faster and more efficient workflows (Vervoort *et al.*, 2021).

Although chromatographic techniques provide a significant amount of information in an experiment, their throughput is often insufficient to meet the demands of high-throughput workflow, in which many samples need to be analyzed per day. Significant advances have been made to reduce analytical cycle times, such as chromatography-free approaches through direct infusion, flow-injection methods, and laser desorption/ionization (Han *et al.*, 2009). Although this could cause the loss of metabolome coverage, this strategy allows the acquisition of thousands of samples per day (Fuhrer and Zamboni, 2015), and may be very efficient for metabolic signature sorting and comparison. Some platforms can achieve rapid analysis of up to 0.5 s per sample using desorption electrospray ionization (DESI) and nanospray DESI and matrix-assisted laser desorption/ionization (MALDI) techniques (de Raad *et al.*, 2016).

Emerging technologies such as microfluidics, and miniaturized separation techniques (e.g. nanoscale LC-electrospray ionization), hold promise for achieving both rapid and comprehensive metabolite analysis (de Raad *et al.*, 2016), potentially addressing some of the challenges faced by untargeted metabolomics (Hill and Roessner, 2015). Despite remaining challenges, ongoing technological innovations are expected to facilitate the full integration of high-throughput metabolomics into phenotyping pipelines in the near future (Hall *et al.*, 2022).

3.2.3 Data processing, annotation, contextualization, and storage

Identifying small molecules, such as metabolites, is crucial in metabolomics and plant sciences. MS-based metabolomics still faces challenges, including compound annotation (in untargeted approaches) and difficulties contextualizing metabolites within biological pathways (Alvarez and Naldrett, 2021). Four levels of identification are described in the Metabolomics Standards Initiative (MSI): (i) identified compounds (with standards); (ii) putatively annotated compounds (based on physicochemical properties and/or spectral similarity with public/commercial spectral libraries); (iii) putatively characterized compound classes (based on characteristic physicochemical properties of a chemical class of compounds, or by spectral similarity to known compounds of a chemical class); and (iv) unknown compounds (Sumner *et al.*, 2007; Schymanski *et al.*, 2014).

Metabolomics data processing and analysis require specialized bioinformatics tools to handle the complexity and volume of data generated by high-throughput techniques (Lazar *et al.*,

2015; Misra *et al.*, 2017). The workflow typically involves data acquisition, data processing, annotation, data interpretation, and integration (Misra and van der Hooft, 2016; Cambiaghi *et al.*, 2017). Fundamental processing steps include noise filtering, peak picking, deconvolution, alignment, and creating a data matrix for statistical analysis. Various open-source tools and databases have been developed and are constantly updated to support these processes, such as MS-DIAL and mzmine, including those using R language (e.g. XCMS, MetaboAnalyst). For NMR-based metabolomics known for generating gold-standard quantitative metabolomic data, several tools have been developed and are constantly evolving, both in the area of preprocessing and statistical analysis and in data normalization, including Bayesian methods, statistical correlation analysis, statistical association networks, and data standards. Misra and Mohapatra (2019) discussed the available tools (R package-based and web-based) for MS- and NMR-based metabolomics research users. A summary of open-source software for data processing, annotation, and storage can be found in Table 3.1.

As the field rapidly evolves, there is a growing need for data standardization and improved tools to handle increasing data complexity and size. Some practices, such as quality control, are essential to guarantee quality reporting and community confidence (Broeckling *et al.*, 2023). Quality control allows the handling of systematic and random variations induced during experimental and analytical workflows. For high-throughput and large-scale studies, the acquisitions are susceptible to batch effects, and quality control is essential for correcting intra- and inter-batch drifts and data-normalization strategies, and is also useful for metabolite annotation strategies.

The annotation is a challenging step, and multiple improvements have been made to speed up the annotation process and enhance its reliability. Distinct annotation strategies can be used depending on the biological question (Fig. 3.1). The most widely used approach annotates only features that show some importance from previous statistical/descriptive analyses, but new bioinformatics advancements allow the annotation of all features (Dührkop *et al.*, 2019). Several databases have been established for plant metabolomics, enabling accurate metabolite annotation and pathway elucidation (e.g. METLIN, KEGG, PhenolExplorer, PLASMA, FragHUB) (Dablanc *et al.*, 2024). Also, *in silico* fragmentation methods combine compound database searching and fragmentation prediction for small-molecule identification, going through algorithmic and scoring criteria refinements (e.g. MetFrag). In this case, metabolite candidates can be filtered or scored based on factors such as substructures, specific elements, or presence on the 'suspect list' (Ruttkies *et al.*, 2016), in addition to scoring based on fragment structures learned in a training step (Ruttkies *et al.*, 2019). In addition, retention time can be calculated/predicted from user-supplied data or added separately for candidate ranking (Ruttkies *et al.*, 2016). Another computational approach for molecular structure identification is SIRIUS (ML tools), which allows the analysis of a full data set and integrates high-resolution isotope pattern analysis and fragmentation trees (Dührkop *et al.*, 2019). These tools hold great promise for high-throughput plant metabolomics, due to the possibility of analyzing a large number of samples. This facilitates data contextualization, with biomarker identification, allowing identification of the chemical family with predicting compound classes directly from MS/MS (e.g. CANOPUS-SIRIUS) and provides insights for several applications, such as crop improvement, interactions between plants and environment, and plant resilience to climate change. In a recent study, Peters *et al.* (2023) distinguished chemotaxonomy based on specialized metabolic markers at different complexity levels, from molecules to compound classes, superclasses, and pathways, as well as molecular descriptors using some of these bioinformatics tools (Peters *et al.*, 2023).

As mentioned earlier, one of the challenges in plant metabolomics is to accurately identify the features that hold valuable information about a biological system. Therefore, it is crucial to develop and have access to public databases containing information on compounds, metabolic pathways, spectra, and experimental procedures. Additionally, standardized protocols should be established to streamline information exchange between laboratories. In this context, FAIR principles were created in 2016 to make scientific data and research software more *F*indable, *A*ccessible, *I*nteroperable, and *R*eusable (Wilkinson *et al.*, 2016). Since then, new tools have been developed to make research metabolomics

Table 3.1. Open-source software for metabolomics data processing and annotation.

Software	Metabolomics data			Comment	Reference(s)
	Processing	Annotation	Repository		
Workflow4Metabolomics (W4M)	✓	✓		Integrates XCMS, CAMERA, and other tools within a Galaxy framework Provides standardized workflows and quality control tools	Giacomoni *et al.* (2015); Guitton *et al.* (2017)
XCMS	✓			Widely used for peak detection, alignment, and quantification Comprehensive support for LC/MS data Integrates with R for flexible data analysis	Smith *et al.* (2006); Tautenhahn *et al.* (2008); Benton *et al.* (2010)
MS-DIAL	✓	✓		Supports untargeted metabolomics, lipidomics, and proteomics Integrates with a wide range of mass spectrometry data Built-in spectral library for annotation	Tsugawa *et al.* (2015, 2019, 2020); Lai *et al.* (2018)
mzmine	✓	✓		Modular and flexible for LC-MS data analysis Real-time visualization of processing steps Supports integration with external databases for annotation	Katajamaa *et al.* (2006); Pluskal *et al.* (2010)
enviMass	✓			Based on the R environment including a graphical user interface in a web browser Versatile data-mining workflow for the automatized detection of patterns and compounds in hyphenated (LC or GC) or direct-injection and mass spectrometry (MS, MS/MS) measurement sequences	Blosloos/enviMass version 3.5 LC-HRMS trend detection workflow: R package, Zenodo
MetaboAnalyst	✓	✓		Web-based and user-friendly interface Offers a wide range of statistical and functional analysis tools Supports multi-omics data integration	Chong and Xia (2018); Pang *et al.* (2021)
OpenMS	✓			Extensive set of tools for MS data processing Supports targeted and untargeted workflows Interoperable with other bioinformatics tools (e.g. KNIME, Galaxy)	Röst *et al.* (2016); Pfeuffer *et al.* (2017)
MetFrag		✓		*In silico* fragmentation tool for metabolite identification Integrates with various public databases for compound search	Wolf *et al.* (2010); Ruttkies *et al.* (2016)
SIRIUS		✓		Advanced computational tools for molecular formula determination Supports combinatorial fragmentation trees Integrate with CANOPUS for chemical class identification Can integrate with MetFrag for enhanced annotation	Böcker *et al.* (2009); Dührkop *et al.* (2019)

PeakForest	✓	✓	Platform for annotating metabolomics data from NMR and MS Intuitive and user-friendly interface for managing spectral data Enables efficient spectral storage, ensuring data are organized and accessible for future studies	Paulhe *et al.* (2022)
GNPS (Global Natural Products Social Molecular Networking)	✓	✓	Specialized in natural products and metabolomics data sharing Provides community-contributed spectral libraries for compound identification	Wang *et al.* (2016); Nothias *et al.* (2020)
METASPACE	✓	✓	Annotation platform for spatial metabolomics based on MS imaging Repository for high-resolution MS imaging data sets	Palmer *et al.* (2017)
Metabolights		✓	Public repository for metabolomics data Supports a wide range of file formats and metadata standards Cross-species (allows a search for species-specific metabolites)	Haug *et al.* (2013); Kale *et al.* (2016)
Metabolomics Workbench		✓	Repository for metabolomics data and metadata, metabolite standards, protocols, tutorials and training, and analysis tools	Sud *et al.* (2016)
MetabolomeXchange		✓	A consortium that connects different metabolomics repositories Facilitates data sharing and retrieval across multiple platforms	metabolomexchange.org
Datagouv		✓	Provides free and open access to government data sets, promoting transparency and innovation Covers diverse domains including health, environment, and economics, facilitating multidisciplinary research	data.gouv.fr
Zenodo		✓	Open repository for European Union-funded research outputs from Horizon Europe, Euratom, and earlier framework programs Automatically generates DOIs for data sets, ensuring proper citation and long-term accessibility	zenodo.org

DIMs, direct injection mass spectrometry; GC, gas chromatography; LC-MS, liquid chromatography–mass spectrometry; MS, mass spectrometry.

processes more consistent and reusable (Zulfiqar *et al.*, 2024).

Hajjar *et al.* (2023) discussed several important FAIR aspects/examples: (i) making data *findable* by assigning them unique, persistent identifiers such as the InChIKey or SPLASH, which enable easy retrieval; (ii) ensuring data *accessibility* through standardized protocols, and possibly adding authentication mechanisms, such as ELIXIR, MetaboLights, and Metabolomics Workbench; (iii) making data *interoperable* by formatting and structuring it for use across different systems, often using ontologies and standard formats such as mzML, nmrML, the ISA data model, and the ontologies 4Chem and BioPortal; and (iv) ensuring data *reproducibility* by providing well-documented metadata and appropriate licensing, such as Creative Commons licenses, and adherence to guidelines such as the MSI.

Global Natural Product Social Molecular Networking (GNPS) is nowadays the most used platform for annotation by similarity and data storage. The GNPS and MassIVE integration allows access to a huge amount of metabolomics data. In 2021, there were over 1800 public data sets, which included more than 490,000 MS files and over 1.2 billion tandem mass spectra with users from over 160 countries (Leao *et al.*, 2021). By bringing together data, knowledge, and tools, GNPS maximizes the potential for the scientific community to reuse the resources. GNPS also provides unique tools for matching data to public libraries and discovering related metabolites without needing a special label.

The adoption of FAIR principles is crucial for plant metabolomics to facilitate data sharing, integration with other omics fields, and HTA, which are essential for generating and reusing knowledge.

3.3 Predictive Modeling in Metabolomics

As plant metabolomics often involves the analysis of a large amount of data, high-throughput metabolomics data pose a variety of analytical challenges (see section 3.2.3). As a result, it is essential to design well-structured analytical workflows to manage both data preprocessing and processing. After metabolomics data normalization and transformation, various univariate and multivariate statistics can be conducted (Bartel *et al.*, 2013). In this chapter, we focus on the particular task of predictive modeling (PM).

In plant metabolomics, PM has emerged as a powerful tool for uncovering the intricate link between metabolites and plant traits such as growth, yield, stress tolerance, or disease resistance, as well as for identifying key biomarkers (Worley and Powers, 2013; Khakimov *et al.*, 2014; Kisiel *et al.*, 2023). PM is a key benefit of data science, enabling the development of models that can simulate and predict the behavior of plant metabolism under different conditions (Ghosh *et al.*, 2020; García-Pérez *et al.*, 2021). According to Geisser (1993), PM is defined as “the process by which a model is created or chosen to try to best predict the probability of an outcome.” In plant sciences, models could be defined as a simplification of reality used to predict a specific outcome. This predictive outcome can either represent classes (e.g. species) or quantitative values (e.g. biomass, elevation). These models offer valuable applications in agriculture, plant breeding, and environmental monitoring, guiding crop improvement and sustainable farming practices (Ghosh *et al.*, 2020; García-Pérez *et al.*, 2021).

Although prediction research has traditionally used a Bayesian framework approach and statistical techniques such as regression models, AI techniques such as ML, artificial neural networks (ANNs), and DL algorithms are being used with increasing frequency (Waljee *et al.*, 2014). ML has become a widely adopted technology for identifying patterns and rules within large data sets (Chlingaryan *et al.*, 2018), including metabolomics data. Furthermore, the rapid growth of computing capabilities and the availability of cloud computing resources have led to the widespread adoption of automated ML (Sheikh *et al.*, 2024). Automated ML is now seen as an attractive alternative to traditional manual ML methods. It has the potential to deliver high-performance, all-encompassing ML processes that cover various stages, including data preparation, feature engineering, model creation, and model assessment (Sheikh *et al.*, 2024). Furthermore, the increasing capacity of machines to process and analyze plant metabolomics has also sparked interest in high-throughput plant metabolomics.

3.3.1 Methodologies and algorithms

ML techniques can be categorized according to the type of learning (supervised, unsupervised, semi-supervised, and reinforcement learning) and the learning models they employ (classification, regression, clustering, and dimensionality reduction) (Greener *et al.*, 2022). Figure 3.2 provides a summary of these categories. The most common form of ML used in PM is supervised learning. Supervised learning involves both input data (x) and known output labels (y), and the algorithm is trained to identify the function that links x to y (Zhang, 2010; Janiesch *et al.*, 2021). A variety of supervised ML approaches and algorithms are involved in PM, including regression analysis and classification (Ghosh *et al.*, 2020).

Regression finds the relationship between dependent and real or continuous values (e.g. temperature, yield) and one or more independent variables (features). Regression algorithms include linear regression, partial least squares, and support vector machines (SVMs), as well as other subtypes (Maulud and Abdulazeez, 2020). Classification algorithms predict categorical output variables (e.g. "disease" or "no disease") by labeling pieces of input data. Classification algorithms include random forests, decision trees and SVMs, among others (Sen *et al.*, 2020). In this section, the most commonly used supervised regression and classification learning methods (linear regression, partial least squares-discriminant analysis, SVMs, decision trees, and random forests) (Galal *et al.*, 2022) will be described briefly:

- Linear regression is a fundamental ML algorithm used for predicting outcomes based on linear relationships between variables (Maulud and Abdulazeez, 2020). It assumes that the output can be calculated from a linear combination of inputs (features); i.e. each input feature is multiplied by a

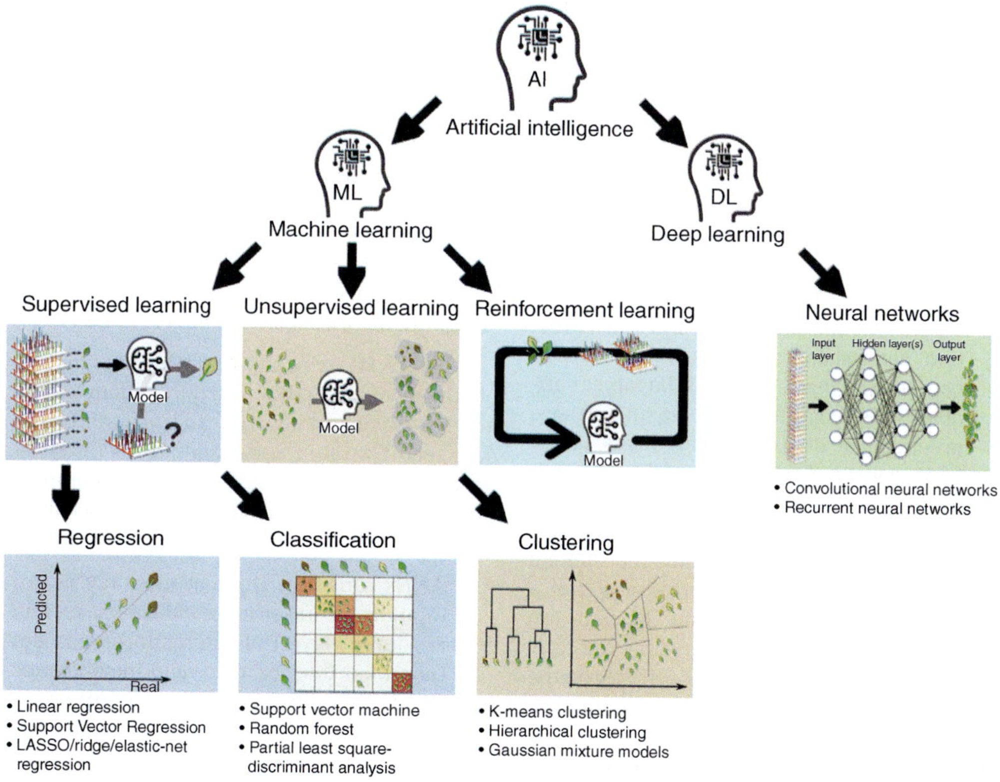

Fig. 3.2. Machine learning and deep learning approaches.

single parameter and these values are added. Linear regression predictions are easy to interpret. The objective is to find the best-fit line that minimizes the difference between the predicted output values and the real output values, usually by minimizing a loss function. To prevent overfitting, especially when the number of features is large, regularization methods are commonly used including ridge, LASSO, and elastic-net regression (Obi and Jecinta, 2023).

- Partial least squares-discriminant analysis (PLS-DA) is a supervised technique widely used in metabolomics studies. PLS-DA identifies latent variables (components) that optimize the distinction between predefined classes (e.g. healthy and diseased) in the data. By projecting the data onto these latent variables, PLS-DA highlights the variance and correlation structure that is most important for distinguishing between the groups (Barker and Rayens, 2003). Moreover, PLS-DA can simultaneously perform dimension reduction and classification, helping researchers identify patterns associated with different experimental conditions or sample classes (Lottering *et al.*, 2020).
- Developed in the 1990s by Cortes and Vapnik (1995), SVMs are referred to as the set of supervised learning methods applied in classification (support vector classification, SVC) and regression (support vector regression, SVR) problems (Gunn, 1997; Brereton and Lloyd, 2010). SVRs use continuous values and non-linear data and find a function that best approximates the relationship between the input data and the target variable. SVCs separate a data set into classes, producing a discrete output (class label). SVMs consist of finding the optimal boundary (hyperplane) between different classes. Although the SVM classification process may appear simple, selecting the optimal hyperplane is not. In fact, this notion of a boundary presupposes that the data are linearly separable, which is rarely the case. To address this, SVMs frequently rely on the usage of "kernels". These mathematical functions allow data to be separated by projecting them into a feature space (a higher-dimensional vector space). Therefore, the optimal hyperplane can be selected with an orientation that maximizes the margin constructed from two additional hyperplanes (parallel to the main one and at a tangent to the first two vectors) (Centorame *et al.*, 2024), ensuring a more accurate classification. The margin-maximization technique, in contrast, guarantees greater robustness to noise, and therefore a more generalizable model.
- Decision trees are a set of algorithms that fall under non-parametric supervised ML techniques. Briefly, the decision tree algorithm splits data into subsets based on certain criteria, such as temperature or soil type, and builds a tree-like model. Decision trees are simple and interpretable, and are often used in plant disease prediction (Centorame *et al.*, 2024). They can be classified into two groups: (i) a classification tree where the predicted outcome is a categorical variable, representing two or more classes to which the observation belongs; and (ii) a regression tree where the predicted value is a continuous variable.
- Random forests are widely used in metabolomics for both classification and feature selection. They can handle high-dimensional data and provide interpretable results by ranking the importance of metabolites. A random forest is an extension of a decision tree: it builds multiple trees and aggregates their predictions. It is a powerful and robust algorithm for tasks such as predicting crop yields, classifying plant species, and detecting plant diseases (Centorame *et al.*, 2024).

Unlike ML, which typically involves the use of statistical methods to learn from data, ANNs and DL use neural networks to learn from large data sets (Sharifani and Amini, 2023). Neural networks are increasingly used for complex metabolomics data sets where non-linear relationships exist between metabolites and plant traits:

- ANNs consist of three or more layers (an input layer, one or more hidden layers, and an output layer) of interconnected nodes (neurons) that learn to map metabolomic inputs to predicted outcomes (Centorame *et al.*, 2024). More precisely, the input layer receives data and passes it to the neurons in the hidden layer. After processing the data, the neurons produce an output that can either be sent as the final results of the computations or used as input by the neurons in the next layer. Thus, the fundamental elements of ANNs are neuron mathematical functions, which take data as input, process the data, and produce an output. Each connection has a connection weight, and each neuron has a threshold value and an activation function (Dastres and Soori, 2021). ANNs are used in metabolomics to model non-linear relationships between metabolite profiles and phenotypic traits. Different basic neural networks exist, such as multi-layer perceptrons, convolutional neural networks, recurrent neural networks, graph convolutional networks, and autoencoders (Greener *et al.*, 2022).

- While ANN usually consists of a few layers, DL models are characterized by having way more hidden layers, enabling them to capture more intricate patterns. This means that DL is suited for high-throughput analysis and extraction of useful knowledge from both large amounts of data and data collected from different sources. These models are particularly useful for complex tasks such as predicting plant metabolic responses to environmental changes (Grapov *et al.*, 2018; Pomyen *et al.*, 2020). DL has been used to group several algorithms including convolutional neural networks and recurrent neural networks (Grapov *et al.*, 2018).

3.3.2 Model training and validation

The concept of "learning" in ML broadly refers to the process of acquiring knowledge, skills, and understanding through instruction or experience (McQueen *et al.*, 1995). In ML, the learning process is based on a "training data set," where the model learns from examples to perform a specific task (Benos *et al.*, 2021). Thus, ML typically involves dividing the data into training and test sets. The training data is used to build a candidate model, while the test data is used to evaluate the model's performance. The simplest and perhaps most common strategy to split a data set and obtain the corresponding training set (and test set) is to sample a fraction (usually 70–80%) of the data set randomly. This strategy is referred to as random splitting. Other sampling techniques exist such as cluster-based, stratified, time series, and adversarial or biased sampling (Berrar, 2019; Søgaard *et al.*, 2021; Fontanari *et al.*, 2022). When dealing with imbalanced data sets, stratified sampling ensures that the class distribution is consistent across training, validation, and test sets. For time series data, time series splitting maintains the chronological sequence of the data while it is being partitioned.

A general restriction of supervised methods is the possibility of overfitting, which refers to the integration of noise into the statistical model, such as through excessive learning on a training data set (Broadhurst and Kell, 2006). To overcome this issue, validation techniques such as cross-validation (Westerhuis *et al.*, 2008) or bootstrapping (Wehrens *et al.*, 2000) exist. K-fold cross-validation separates the data set into "k" equally sized folds, permitting several rounds of training and validation. K-fold cross-validation is an effective method for evaluating the performance and generalization ability of ML models. Leave-one-out cross-validation is a comprehensive cross-validation strategy, particularly suitable for small data sets. In this method, each of the n samples is taken as a test sample, while the remaining $n - 1$ samples serve as the training set. This process is repeated for each sample, resulting in n classifiers and n test results. The performance of the model is then assessed by averaging these n results.

The ability to interpret a model depends on the ML method used and the input data. There are a few potential metrics to assess a prediction model. In fact, for classification models, several classification metrics are used to assess the performance of a model, particularly when the classes are imbalanced or when we need specific insights about different aspects of the prediction (Ghosh *et al.*, 2020):

- *Accuracy* indicates the proportion of correctly predicted instances compared with the total number of instances.

$$\text{Accuracy} = \frac{\text{Number of correct predictions}}{\text{Total number of predictions}} = \frac{\text{TP} + \text{TN}}{\text{TP} + \text{TN} + \text{FP} + \text{FN}}$$

where TP (true positive) corresponds to the correctly predicted positive instances, TN (true negative) corresponds to the correctly predicted negative instances, FP (false positive) corresponds to the incorrectly predicted positive instances (type I error), and FN (false negative) corresponds to the incorrectly predicted negative instances (type II error).

- *Precision* provides insights into the reliability of positive predictions. It measures how many of the predicted positive instances are actually positive. It is useful when the cost of false positives is high. A high precision means that the model has a low rate of false positives.

$$\text{Precision} = \frac{\text{TP}}{\text{TP}+\text{FP}}$$

- *Sensitivity* (or recall) provides insights into the ability to capture all positive instances. It measures how many actual positive instances are correctly predicted by the model. It is useful when the cost of false negatives is high. A high recall means the model correctly identifies most of the positive instances.

$$\text{Sensitivity} = \frac{\text{TP}}{\text{TP}+\text{FN}}$$

- *Specificity* provides insights into the ability to capture all negative instances. It measures how many of the actual negative instances are correctly predicted by the model. It is useful when the focus is on reducing false positives. Specificity is complementary to recall. In a balanced situation, a good classifier will have high values for both specificity and recall.

$$\text{Specificity} = \frac{\text{TN}}{\text{TN}+\text{FP}}$$

For regression models, metrics provide insights into the accuracy, efficiency, and robustness of the model's predictions in predicting continuous outcomes. The choice of metrics depends on the problem domain, whether we need to penalize large errors, how we handle outliers, and whether we need relative comparisons (Tatachar, 2021). Some relevant metrics that are commonly used are listed below. In the following definitions, y_i is the actual value, $\hat{y}_i$ is the predicted value, $\overline{y}$ is the mean of the actual values, and n is the number of data points.

- *Mean absolute error (MAE)* is the average of the absolute differences between actual values and predicted values. It measures the magnitude of errors in a set of predictions, without considering their direction.

$$MAE = \frac{1}{n}\sum_{i=1}^{n}\left|y_i - \hat{y}_i\right|$$

- *Mean squared error (MSE)* is the average of the squared differences between the predicted values and the actual values. It penalizes larger errors more heavily due to the squaring of the residuals. The closer the MSE is to 0, the more accurate is the prediction.

$$MSE = \frac{1}{n}\sum_{i=1}^{n}\left(y_i - \hat{y}_i\right)^2$$

- *Root mean squared error (RMSE)* is the square root of the MSE. It is in the same units as the target variable, making it easier to interpret than MSE. In other words, the RMSE is simply the standard deviation of the errors.

$$RMSE = \sqrt{\frac{1}{n}\sum_{i=1}^{n}\left(y_i - \hat{y}_i\right)^2}$$

- R^2 measures the proportion of variance in the dependent variable that is predictable from the independent variables. It ranges from 0 to 1, with 1 indicating perfect predictions and 0 indicating that the model explains none of the variance in the data.

$$R^2 = 1 - \frac{\sum_{i=1}^{n}\left(y_i - \hat{y}_i\right)^2}{\sum_{i=1}^{n}\left(y_i - \overline{y}_i\right)^2}$$

- *Mean absolute percentage error (MAPE)* measures the prediction accuracy by calculating the percentage difference between actual and predicted values. It is widely used in time series forecasting.

$$MAPE = \frac{1}{n}\sum_{i=1}^{n}\left|\frac{y_i - \hat{y}_i}{y_i}\right| \times 100$$

3.4 Applications of High-throughput Metabolomics and Predictive Modeling

As stated earlier, implementing high-throughput strategies in metabolomics workflows enables the generation of metabolomic data for hundreds of samples in just a few days. These experiments result in thousands of metabolic features, providing a detailed depiction of the cell/tissue/organ state at a specific time. This state serves as a reliable proxy for the past, present, and future of a given tissue, encompassing the outcomes of previous developmental stages, the current status, and the building blocks that facilitate future cell development. Additionally, subjecting an organism to various stresses can expand the diversity of the metabolome, thereby revealing broader characteristics.

Through the combination of metabolomic and phenotypic data and the application of ML methodologies, several complex phenotypes, such as relative growth rate, grafting success, plant elevation, or resistance to pathogens across diverse plant matrices, can be predicted. Furthermore, these techniques enable the prediction of these complex phenotypes and allow the identification of biomarkers that provide insights into the mechanisms regulating these phenotypes (Fig. 3.3).

In this section, we will review recent case studies conducted in high-throughput metabolomics and PM in plant research (Table 3.2).

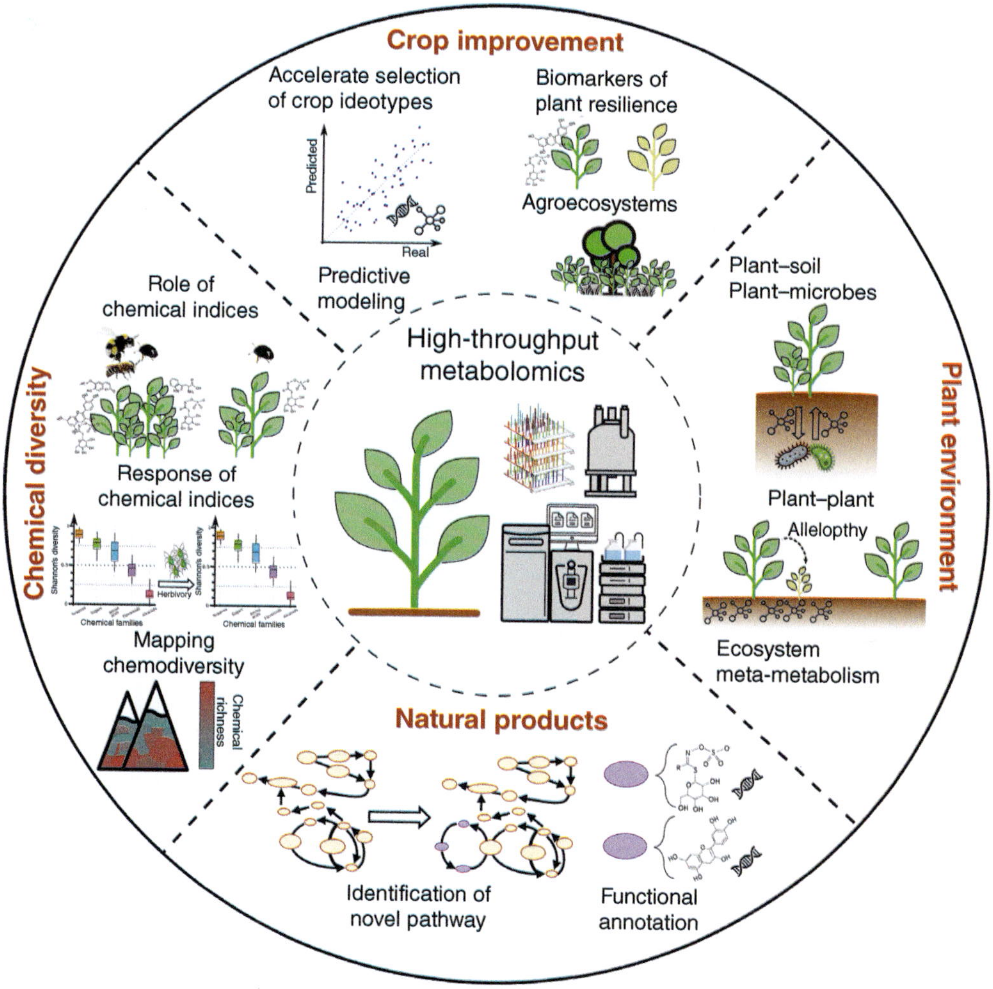

Fig. 3.3. Key applications of high-throughput metabolomics.

Table 3.2. List of applications of high-throughput metabolomics and predictive modeling.

Model used	Species	Variables predicted	Reference
BLUP	Wheat	Hybrid performance	Zhao *et al.* (2015)
BLUP	Oat	Seventeen phenotypic traits, including plant height, seed morphology, seed yield, and seed fatty acid composition	Hu *et al.* (2021)
BLUP	Canola	Seedling emergence and seed quality	Knoch *et al.* (2021)
PLS-DA	Rice	Yield heterosis	Dan *et al.* (2020)
PLS-DA	Rice	Rice 1000-grain weight	Dan *et al.* (2019)
PLS	Rice	Plant height, heading date, and grain yield	Dan *et al.* (2016)
BLUP, GlmNet-LASSO, SVM-R(EPS), SVM-R(NU), SVM-P(EPS), SVM-P(NU), and PLS	Rice	Yield, 1000-grain weight, number of grains per panicle, and number of tillers per plant	Wang *et al.* (2019a)
SVR, random forest, XGBoost, OLS, and BLUP	Rice	Grain-size phenotypes (grain length (major axis), grain width (minor axis), and grain perimeter) and genotypes under high night temperature conditions	Bi *et al.* (2023)
SR, rrBLUP, PLS, LASSO, elastic net, SVR, MARS, BART, GBM, LightGBM, random forest, ANN, DNNGP, and GaussprRadial	Maize	Yield	Wu *et al.* (2024)
SVM	Maize	Field hybrid performance	de Abreu E. Lima *et al.* (2018)
Random forest	Potato	Drought tolerance	Sprenger *et al.* (2018)
Random forest	Soil	Bacterial diversity	Liu *et al.* (2020)
LASSO, elastic net, and ridge regression	Several plant species	Elevation levels	Dussarrat *et al.* (2022)
Random forest, recursive partitioning, and XGBoost	Several plant species	Phytochemical diversity	Defossez *et al.* (2021)
LASSO, elastic net, and ridge regression	Tansy	Parental genotype	Dussarrat *et al.* (2023)
Simple linear regressions	Oak	Latitude	Moreira *et al.* (2018)
LASSO, elastic net, and ridge regression	*Atriplex imbricata* and *Maihueniopsis camachoi*	Plant–plant interaction pattern	Díaz *et al.* (2024)
LASSO, elastic net, and ridge regression	Several plant species	Relative growth rate	Roch *et al.* (2020)
Linear regressions	Several plant species	Bacterial diversity	Zhang *et al.* (2024)

ANN, artificial neural network; BART, Bayesian additive regression trees; BLUP, best linear unbiased prediction; DNNGP, deep neural network for genomic prediction; GaussprRadial, Gaussian process with Radial basis function kernel; GBM, generalized boosted regression models; glmnet, LASSO and elastic-net regularized generalized linear models; LASSO, least absolute shrinkage and selection operator regression; LightGBM, light gradient boosting machine; MARS, multivariate adaptive regression splines; OLS, ordinary least square; PLS, partial least squares; PLS-DA, partial least squares with discriminant analysis; rrBLUP, ridge regression and other kernels for genomic selection; SR, stepwise regression; SVM-P(EPS), support vector machine using the polynomial kernel function and epsilon regressions; SVM-P(NU), support vector machine using the polynomial kernel function and nu regressions; SVM-R(EPS), support vector machine using the radial basis function and epsilon regressions; SVM-R(NU), support vector machine using the radial basis function and nu regressions; SVR, support vector regression; XGBoost, extreme gradient boosting.

3.4.1 Plant metabolic pathway analysis

Pathway analysis in plants is a critical tool for unraveling the complexity of plant metabolism, revealing the biochemical routes that underlie plant physiology and stress responses. The diversity of plant metabolites, estimated to range from hundreds of thousands to over a million compounds, poses a significant challenge for accurate pathway elucidation (Medema *et al.*, 2021). Plant metabolic pathways are involved in key processes such as nutrition, growth, reproduction, defense, and adaptation to environmental stresses, and their analysis provides insights into both primary and specialized metabolite biosynthesis. Despite its importance, pathway analysis faces considerable challenges, from metabolite identification to integrating large-scale multi-omics data.

One of the primary challenges is the complexity and structural diversity of plant metabolites. Many natural products remain uncharacterized, particularly specialized metabolites that vary significantly between tissues, developmental stages, species, and environmental conditions (Wang *et al.*, 2019a; Tsugawa *et al.*, 2021). In addition, plant metabolic pathways are often encoded by biosynthetic gene clusters that can be scattered across genomes, making it difficult to link genes to their metabolic products (Blin *et al.*, 2019; van der Hooft *et al.*, 2020). Another major issue is metabolite misidentification, which can lead to erroneous pathway mappings. Studies have shown that even minor misannotations can lead to false-positive or false-negative results in pathway-enrichment analyses (Wieder *et al.*, 2021).

In addition, integrating multi-omics data poses both an opportunity and a challenge. As plants exhibit complex regulatory mechanisms involving genes, proteins, and metabolites, integrating data from genomics, epigenomics, transcriptomics, proteomics, and metabolomics is essential to obtain a holistic view of metabolic pathways. However, differences in data types and scales, such as varying temporal and spatial resolutions, make integration difficult (Noor *et al.*, 2019; van der Hooft *et al.*, 2020; Lee *et al.*, 2025).

To address these challenges, current bioinformatics developments are being made in the context of the overall field of metabolomics and are especially applicable to high throughput, which will require throughput and computational power in order to rapidly process a large number of samples. In the case of pathway analysis, different computational and experimental approaches have been developed. One prominent method is over-representation analysis, which assesses whether specific pathways are over-represented in a list of differentially abundant metabolites. However, over-representation analysis is sensitive to input parameters such as background sets and metabolite identification, requiring careful curation of data to avoid false results (Wieder *et al.*, 2021). Another powerful technique is genome mining, particularly using tools such as antiSMASH, which identifies biosynthetic gene clusters responsible for the production of specialized metabolites (Blin *et al.*, 2019). Recently, the latest version of antiSMASH has been enhanced to detect a broader range of biosynthetic gene cluster types, providing valuable insights into plant biosynthetic pathways.

Metabologenomics has emerged as an integrative approach, combining genomic and metabolomic data to map specialized metabolic pathways in plants (Fig. 3.3) (van der Hooft *et al.*, 2020). By linking biosynthetic gene clusters with corresponding metabolites, this approach bridges the gap between genotype and phenotype, facilitating the discovery of novel metabolic pathways (Medema *et al.*, 2021). The Paired Omics Data Platform (PoDP) is another interesting tool based on public repositories providing a community-driven resource for integrating MS and genomic data, enhancing the ability to connect metabolite profiles with their biosynthetic origins (Schorn *et al.*, 2021). Additionally, ML methods, such as DIABLO (*d*ata *i*ntegration *a*nalysis for *b*iomarker discovery using *l*atent c*o*mponents) are increasingly being used to integrate multi-omics data sets, providing more robust predictions of pathway involvement and broader contextualization (Singh *et al.*, 2019).

Several case studies illustrate the success of these approaches in plant pathway analysis. For example, Jeon *et al.* (2020) identified a pathogen-responsive gene cluster in tomato responsible for the biosynthesis of falcarindiol, a highly modified fatty acid involved in plant defense. Using CRISPR (clustered regularly interspaced short palindromic repeats) mutants, the study confirmed the role of this cluster in mediating pathogen resistance. This highlights how integrating genomic tools with pathway analysis can identify novel defense-related metabolites.

Stable isotope labeling is also of particular interest in improving pathway analysis using fully ^{13}C-labeled plant tissues. By integrating stable isotope labeling with a cheminformatics workflow in MS-DIAL, Tsugawa *et al.* (2019) identified 1092 metabolites across 31 plant tissues from 12 species. This technique allows precise determination of carbon numbers in metabolites, enhancing the annotation of unknown compounds and providing insights into novel metabolic pathways.

Plant pathway analysis remains a complex yet essential field for understanding plant metabolism and its environmental interactions. With multi-omics data integration, advanced computational tools, and experimental validation, the field is moving closer to fully characterizing the vast diversity of plant metabolic pathways. However, challenges such as metabolite identification, pathway mapping, and data integration remain ongoing hurdles that must be overcome through improved methodologies and collaborative platforms.

3.4.2 Crop improvement

Despite the utilization of marker-assisted selection in crop improvement programs since the 1990s, the majority of agronomically relevant traits are frequently complex and regulated by multiple minor-effect loci (Desta and Ortiz, 2014). To address this challenge, genomic selection has been introduced to estimate the genetic merits of individuals with respect to specific agronomic traits (highest estimated breeding values). Among the prediction models developed, linear mixed models with the best linear unbiased prediction (BLUP) have been the preferred option for single-trait genomic selection (Tong and Nikoloski, 2021). Notably, genomic-BLUP (G-BLUP) enables the compression of genome-wide information into a genomic relationship matrix (kinship coefficients between individuals) without a concomitant reduction in prediction accuracy in comparison with other models (Morota and Gianola, 2014; Wang *et al.*, 2019b). However, genomic prediction may not capture all complex gene interactions and downstream regulatory processes. The use of endophenotypes such as transcripts, metabolites, or multi-spectral reflectance has been proposed to improve the prediction of complex traits, as they capture the variability across genotypes more closely than genomic data (Mackay *et al.*, 2009; Civelek and Lusis, 2014; Tong and Nikoloski, 2021).

In recent years, the accuracy of metabolomic-based predictions for agronomic traits in different plant species has been evaluated using the metabolomic BLUP (M-BLUP) approach. This modeling approach has also allowed assessment of the interest in multi-omic predictions by combining kinship matrices derived from different omic data. In many cases, M-BLUP models have provided relatively equivalent efficiencies compared with predictions based solely on genomic data (G-BLUP). In some cases, models integrating both genomic and metabolomic data outperformed individual models, underscoring the potential of metabolomics for crop improvement. In wheat, metabolomics enabled the prediction of hybrid performance, encompassing grain yield, six abiotic and biotic stress traits, and seven quality characteristics. The associated M-BLUP model yielded a prediction accuracy of approximately 75% in the T_2 progeny derived from two parental lines (Zhao *et al.*, 2015). However, models based on genomic data (G- and GM-BLUP) exhibited prediction accuracy at 90%. In oat, M-BLUP models constructed exploiting seed metabolome data from 630 lines provided the highest prediction accuracy for seed fatty acid composition compared with transcriptomic-based (T-BLUP) and G-BLUP models when considering a single environment. This result is likely to reflect the close connection between the phenotypic traits selected and certain biochemical pathways. However, the GM-BLUP and GTM-BLUP models showed the highest prediction accuracy for plant height, seed morphology, and seed yield (Hu *et al.*, 2021). In canola, some BLUP models were constructed using genomic, transcriptomic, and metabolomic data from a panel of 477 hybrids to predict seedling emergence and seed quality (Knoch *et al.*, 2021). The prediction accuracies of the single and combined models were found to be relatively similar, except for the M-BLUP model (154 metabolites), which exhibited lower accuracy. Nevertheless, the prediction accuracies for seedling emergence and seed yield were only 30%, while those for other traits related to seed quality were between 50% and 60%. The combination of other endophenotypic data, such as phenomics, with genomic data did improve the ability to predict seed germination-related traits in oilseed rape (Laurençon *et al.*, 2024). In rice, three metabolomics studies conducted on seedlings from diverse hybrid populations were integrated to construct PLS regression models to predict hybrid performance, 1000-grain weight, plant height, and heading date (Dan *et al.*, 2016, 2019, 2020). The prediction accuracies were relatively high, ranging between 70% and 85%, but were found to be highly dependent on the structure of the training set (population and environmental representation) and the number of predictive analytes selected. Another study compared the efficacy of multi-omic models for predicting yield, 1000-grain weight, number of grains per panicle, and number of tillers per plant in a large panel of hybrid rice progeny (Wang *et al.*, 2019b). Despite the superior predictive accuracy demonstrated by M-BLUP models in comparison with G-BLUP models, the overall prediction performance of all models remained relatively low. Using a set of 73 metabolites detected in rice grains of 152 genotypes, random forest and extreme gradient boosting methods classified control and high night temperature conditions with 98% accuracy (Bi *et al.*, 2023). However, genomic-based predictions outperformed metabolomic-based predictions in at least 87–98% of the cross-validation resampling runs for grain length, width, and perimeter. In maize, the combination of metabolomic, genomic, and phenomic data from seedlings and leaves of 163 genotypes provided superior yield predictions in comparison with single models (Wu *et al.*, 2024). However, the prediction accuracies remained low (40–50%) (Wu *et al.*, 2024). Nevertheless, the use of the maize root metabolome from 381 genotypes cultivated *in vitro* (49 parental lines and 332 hybrids) permitted the identification of parental analytes that most accurately predicted the metabolic inheritance patterns in hybrids. The predictive accuracy of different models based on these analytes ranged between 60% and 79% for key agronomic traits such as hybrid performance and biomass in field conditions (Feher *et al.*, 2014; de Abreu E. Lima *et al.*, 2018). In potato, drought tolerance has been modeled using metabolomic and transcriptomic data from 32 cultivars grown across 16 independent agronomic field trials (Sprenger *et al.*, 2018). The random forest model demonstrated superior performance when metabolomic data were employed, attaining up to 90% prediction accuracies.

Overall, metabolomic data appear to be a highly effective alternative to genomic data for predicting a range of crucial agronomic traits in different crops. Combining endophenotypes with genomic data could also be an interesting strategy to increase the stability and the wide-ranging applicability of predictive models. Notably, this will allow integrating knowledge on regulatory networks and genotype × environment interactions. However, the efficacy of these methodologies will be contingent upon the structure of the screened populations, the quality of metabolomic analyses, and the selection of models.

3.4.3 Environmental interactions and ecosystem dynamics

Chemically mediated relationships between plants and microbes influence plant fitness, probably since plants first colonized the Earth (Sasse *et al.*, 2018; Redecker *et al.*, 2000). High-throughput metabolomics has been used to explore positive (e.g. mycorrhizal fungi) and negative (e.g. pathogenic bacteria) interactions (Fig. 3.3). Although crucial to enhance our understanding of below-ground processes, these studies have also highlighted the complexity of their analysis. For instance, metabolomics applied to plant–microbe interactions in natural environments is limited by its inability to differentiate between plant and microbial metabolites, unless these compounds are exclusive to one of the kingdoms (Allwood *et al.*, 2021). In addition, high chemical and biological diversity and species specificity of plant root exudates have been noted (Trivedi *et al.*, 2020). At a species level, studies have reported an influence of phenolic acids, benzoxazinoids, and flavonoids in *Rehmannia glutinosa* monoculture, *Zea mays*, and *Triadica sebifera*, respectively (Wu *et al.*, 2015; Cotton *et al.*, 2019; Tian *et al.*, 2021). Studies performed at the ecosystem level have allowed exploration of the interplay between root exudates from multiple species and microbial communities, as well as the impact of climate and edaphic parameters on these relationships (Jing *et al.*, 2023). For example, high-throughput metabolomics have shown that the soil metabolome, and, more precisely, several sugars, organic acid, flavonoids, and strigolactones, were correlated to changes in bacterial and fungi communities (Liu *et al.*, 2020; Yu *et al.*, 2022). Inoculation of arbuscular mycorrhizal or ectomycorrhizal fungi affected root and leaf metabolism (Xia *et al.*, 2023), and mycorrhizas influenced plant adaptation to drought (Cosme, 2023). Even more thrilling, multi-species analyses have shed light on potential converging rhizochemical strategies for influencing the soil microbiome (Vieira *et al.*, 2020). These convergent patterns are congruent and are supported by large-scale studies showing a significant but limited effect of plant phylogeny (around 11–14%) on variation in bacterial and fungi communities (Yang *et al.*, 2019; Gong *et al.*, 2023), and similar convergent patterns in aerial plant parts have been reported (Dussarrat *et al.*, 2022). Overall, metabolomics has considerably improved our understanding of plant–microbe interactions, highlighting the role of various chemical families and identifying both dissimilarities and similarities between plant species and environments, offering promising perspectives for agriculture, particularly in the context of intercropping and push–pull systems (Vishwakarma *et al.*, 2020; Wang *et al.*, 2021; Schuman, 2023).

Plant performances are also determined by plant–plant interactions (Weir *et al.*, 2004). Metabolomics has emerged as an ideal tool for exploring phytotoxic compounds and their role in plant growth and ecosystem functioning (Fig. 3.3) (Su *et al.*, 2022; Tuomi *et al.*, 2024). Allelopathy, the process by which released compounds interact with putative competitors, has been widely characterized in invasive plants (Yuan *et al.*, 2013; Zhu *et al.*, 2021), as well as in certain crops such as canola (Asaduzzaman *et al.*, 2015). For instance, phenolic acids, *p*-coumaric acid, sesquiterpenes, and alkaloids could be described as allelochemicals (Vivanco *et al.*, 2004; Rice *et al.*, 2021; Su *et al.*, 2022; Xie *et al.*, 2023). Recently, metabolomics has also highlighted the effect of climatic parameters (e.g. CO_2 levels) on toxin production and the need for a better understanding of the intraspecific variability of chemically mediated plant–plant interactions (Jandová *et al.*, 2015; Rice *et al.*, 2021). Although essential for explaining community structure, metabolomics studies on plant–plant positive interactions are even rarer (Becker *et al.*, 2023; Díaz *et al.*, 2024). Analyses of the positive and negative chemical dialogues are therefore essential to improving our understanding of plant–plant interactions and their impact at the ecosystem level. Ultimately, high-throughput metabolomics, combined with genetic analyses, should: (i) provide sufficient data to feed ecological models to determine the best-performing plant assemblages; and (ii) predict their response to climate change (Becker *et al.*, 2023).

Plant metabolism can be studied from the metabolite level to chemical indices (Fig. 3.3). Recently, phytochemical traits have been accepted as major components and actors of ecosystem functioning (Müller and Junker, 2022). Chemical diversity (chemodiversity) is estimated through compound richness, evenness, and disparity, which are used to assess indices such as phytochemical richness, Shannon diversity, and functional Hill diversity (Wetzel and Whitehead, 2020; Petrén *et al.*, 2023). Although most studies are performed at the compound level (Gargallo-Garriga *et al.*, 2020; Dussarrat *et al.*, 2022), chemodiversity indices have shown great potential in governing other components of biodiversity and are essential for a better understanding of ecosystem dynamics. First, phytochemicals define major plant functional traits such as leaf mass per area, leaf nitrogen content, and fruit biomass (Roch *et al.*, 2020; Walker *et al.*, 2022; Díaz *et al.*, 2024). In addition, generic responses of chemodiversity are observed in response to climatic and edaphic variations worldwide. Previous studies have shown the influence of climatic conditions on phytochemical diversity (Bakhtiari *et al.*, 2021; Sedio *et al.*, 2021), while several patterns arose when exploring the response of chemical diversity across elevation gradients (Walker *et al.*, 2022). This intricate link between environmental parameters and phytochemistry has made it possible to define a chemical map in the Alpine mountains (Defossez *et al.*, 2021). The existence of intraspecific chemodiversity is also puzzling, as its maintenance despite its inherent costs in highly chemodiverse species suggests evolutionary functions (Dussarrat *et al.*, 2023; Thon *et al.*, 2024; Wittmann and Bräutigam, 2024). Hence, these results strongly support the role of phytochemical traits in specialization and adaptation to environmental conditions, their status of niche dimension, and their role in defining plant community composition (Sedio, 2017; Müller and Junker, 2022).

Chemical indices can also be used to explore and understand multi-trophic interactions, for instance predicting plant–insect interactions (Schuman *et al.*, 2016). Higher phytochemical diversity coincided with greater herbivory damage in multiple trees across elevation or latitude gradients (Moreira *et al.*, 2018; Sun *et al.*, 2024). A positive relationship between phytochemical diversity and insect community diversity was also highlighted when comparing the metabolism of 112 *Piper* spp. worldwide (Richards *et al.*, 2015). In addition, floral scent has been shown to influence pollinator behaviors (Junker and Blüthgen, 2010), while root chemodiversity shaped bacterial communities across a climatic gradient (Zhang *et al.*, 2024). Overall, the development of high-throughput metabolomics has unlocked the large-scale analysis of a fundamental aspect of biodiversity: plant chemical diversity. This opportunity is the source of a better mechanistic understanding of plant functioning and a better assessment of the chemical signals that underpin ecological processes and ecosystem dynamics.

3.5 Future Directions and Concluding Remarks

High-throughput plant metabolomics is rapidly advancing, presenting new opportunities across various applications, from plant development and crop improvement to stress response studies. Technological advancements already enable true high-throughput metabolomics, facilitating large-scale studies of genetically diverse individuals and time-dependent responses to environmental stimuli (Zampieri *et al.*, 2017). None the less, further efforts are required to speed the throughput. The high-throughput plant metabolomics approach is increasingly combined with other omics techniques (see examples in section 3.4). Data integration to establish links between genes, transcripts, proteins, metabolites, and phenotypes remains challenging but is greatly improved by multivariate statistics and modeling. For instance, statistical associations between metabolites and transcripts through correlation-based or multivariate-based integrations provide a better biological understanding and better predictions of the response variable (Cavill *et al.*, 2016). Chemical analyses can be linked to quantitative genetics and genome-wide association studies to identify functional genes and develop trait-specific markers for plant breeding and whose performance is clearly linked to the large number of individuals to be phenotyped (Kumar *et al.*, 2017). Although more challenging as it requires complete and high-quality genomics and metabolomics data, pathway-based integrations can be used to combine metabolomics with phylogenetic data, yielding a unique opportunity to unravel adaptive metabolites to biotic or abiotic threats as well as unknown groups of natural products of interest (Chervin *et al.*, 2019). In ecology, the quest for general trends also requires an effective large-scale phenotyping strategy. In this context, advances in remote sensing have allowed the measurement of multiple functional traits through spectral data, which can then be correlated with metabolite abundances (Asner and Martin, 2011; Fine *et al.*, 2021). Although combining spectranomics, phylogenetics, and metabolomics represents a significant challenge, its success could represent a decisive breakthrough for agriculture and chemical ecology.

The deployment of high-throughput predictive metabolomic approaches will offer a means of understanding even more complex systems, such as agroecosystems. This integration can be crucial for expanding investigations into, for example, plant–pathogen interactions to elucidate defense mechanisms (Kumari *et al.*, 2024), or identifying predictive metabolic markers for crop responses to global changes, helping create more resistant plants, and addressing food production and environmental sustainability mitigating the impacts of climate change (Leisner, 2020; Janni *et al.*, 2024). The multi-faceted integration of heterogeneous and multi-compartmental data should provide essential information on the functioning of agroecosystems to improve the sustainability of future agriculture, which is directly impacted by climate change. Rising temperatures, changing rainfall patterns, and extreme weather events are expected to affect productivity and, consequently, the nutritional value of crops (Anderson *et al.*, 2020; Leisner, 2020). In addition, to meet the growing demand for food, agricultural production must potentially double by 2050 (Smith and Gregory, 2013; Janni *et al.*, 2024). Adaptation strategies include changing cultivation and land practices, developing improved crop varieties, and modifying food consumption patterns (Anderson *et al.*, 2020). In addition, integrating more complex modeling to embed a higher realism of phenotypes is also key to future advances in high-throughput plant metabolomics (Fig. 3.2). To enhance accuracy and create a robust knowledge base, models should incorporate various system levels and appropriately manage their respective formalisms (Bardini *et al.*, 2017). For instance, multi-directional models that are both multi-level and hybrid meet these criteria, making them highly valuable tools in computational systems biology. Overcoming these limitations will deepen our understanding of plant metabolism and accelerate crop improvement efforts.

In conclusion, it is an understatement to say that high-throughput metabolomics has a promising future ahead, both due to its exciting technical and methodological developments and the scientific breakthroughs it will enable, particularly in relation to plant research and the agriculture of tomorrow. More particularly, the advent of predictive approaches, thanks to the new tools developed and deployed in AI, and to collective data management and processing/mining initiatives (e.g. the GLOMICAVE consortium), will accelerate the discovery of new metabolic mechanisms underlying the life of plants.

3.6 Acknowledgments

This work was supported by the MetaboHUB infrastructure funded by the Agence Nationale de la Recherche under the France 2030 program (MetaboHUB ANR-11-INBS-0010 ; MetEx+ ANR-21-ESRE-0035; MetaboHUB (JVCE) ANR-24-INBS-0012).

3.7 References

Aguiar, A.S., dos Santos, F.N., Cunha, J.B., Sobreira, H., and Sousa, A.J. (2020) Localization and mapping for robots in agriculture and forestry: a survey. *Robotics* 9: 97.

Allwood, J.W., Williams, A., Uthe, H., van Dam, N.M., Mur, L.A.J., *et al.* (2021) Unravelling plant responses to stress—the importance of targeted and untargeted metabolomics. *Metabolites* 11: 558.

Alseekh, S. and Fernie, A.R. (2018) Metabolomics 20 years on: what have we learned and what hurdles remain? *Plant Journal* 94, 933–942.

Alvarez, S. and Naldrett, M.J. (2021) Mass spectrometry based untargeted metabolomics for plant systems biology. *Emerging Topics in Life Sciences* 5, 189–201.

Anderson, R., Bayer, P.E., and Edwards, D. (2020) Climate change and the need for agricultural adaptation. *Current Opinion in Plant Biology* 56, 197–202.

Asaduzzaman, M., Pratley, J.E., An, M., Luckett, D.J., and Lemerle, D. (2015) Metabolomics differentiation of canola genotypes: toward an understanding of canola allelochemicals. *Frontiers in Plant Science* 5: 765.

Asner, G.P. and Martin, R.E. (2011) Canopy phylogenetic, chemical and spectral assembly in a lowland Amazonian forest. *New Phytologist* 189, 999–1012.

Bakhtiari, M., Glauser, G., Defossez, E., and Rasmann, S. (2021) Ecological convergence of secondary phytochemicals along elevational gradients. *New Phytologist* 229, 1755–1767.

Bardini, R., Politano, G., Benso, A., and di Carlo, S. (2017) Multi-level and hybrid modelling approaches for systems biology. *Computational and Structural Biotechnology Journal* 15, 396–402.

Barker, M. and Rayens, W. (2003) Partial least squares for discrimination. *Journal of Chemometrics* 17, 166–173.

Bartel, J., Krumsiek, J., and Theis, F.J. (2013) Statistical methods for the analysis of high-throughput metabolomics data. *Computational and Structural Biotechnology Journal* 4: e201301009.

Becker, C., Berthomé, R., Delavault, P., Flutre, T., Fréville, H., *et al.* (2023) The ecologically relevant genetics of plant–plant interactions. *Trends in Plant Science* 28, 31–42.

Benos, L., Tagarakis, A.C., Dolias, G., Berruto, R., Kateris, D., and Bochtis, D. (2021) Machine learning in agriculture: a comprehensive updated review. *Sensors* 21: 3758.

Benton, H.P., Want, E.J., and Ebbels, T.M.D. (2010) Correction of mass calibration gaps in liquid chromatography–mass spectrometry metabolomics data. *Bioinformatics* 26, 2488–2489.

Berrar, D. (2019) Cross-validation. In: Ranganathan, S., Gribskov, M., Nakai, K., and Schönbach, C. (eds) *Encyclopedia of Bioinformatics and Computational Biology*, Elsevier, pp. 542–545.

Bi, Y., Yassue, R.M., Paul, P., Dhatt, B.K., Sandhu, J., *et al.* (2023) Evaluating metabolic and genomic data for predicting grain traits under high night temperature stress in rice. *G3 (Bethesda)* 13: jkad052.

Bijttebier, S., van der Auwera, A., Foubert, K., Voorspoels, S., Pieters, L., and Apers, S. (2016) Bridging the gap between comprehensive extraction protocols in plant metabolomics studies and method validation. *Analytica Chimica Acta* 935, 136–150.

Blin, K., Shaw, S., Steinke, K., Villebro, R., Ziemert, N., *et al.* (2019) antiSMASH 5.0: updates to the secondary metabolite genome mining pipeline. *Nucleic Acids Research* 47, W81–W87.

Böcker, S., Letzel, M.C., Lipták, Z., and Pervukhin, A. (2009) SIRIUS: decomposing isotope patterns for metabolite identification. *Bioinformatics* 25, 218–224.

Brereton, G. and Lloyd, R.G. (2010) Support vector machines for classification and regression. *Analyst* 135, 230–267.

Broadhurst, D.I. and Kell, D.B. (2006) Statistical strategies for avoiding false discoveries in metabolomics and related experiments. *Metabolomics* 2, 171–196.

Broeckling, C.D., Beger, R.D., Cheng, L.L., Cumeras, R., Cuthbertson, D.J., *et al.* (2023) Current practices in LC-MS untargeted metabolomics: a scoping review on the use of pooled quality control samples. *Analytical Chemistry* 95, 18645–18654.

Cambiaghi, A., Ferrario, M., and Masseroli, M. (2017) Analysis of metabolomic data: tools, current strategies and future challenges for omics data integration. *Briefings in Bioinformatics* 18, 498–510.

Cavill, R., Jennen, D., Kleinjans, J., and Briedé, J.J. (2016) Transcriptomic and metabolomic data integration. *Briefings in Bioinformatics* 17, 891–901.

Centorame, L., Gasperini, T., Ilari, A., Del Gatto, A., and Foppa Pedretti, E. (2024) An overview of machine learning applications on plant phenotyping, with a focus on sunflower. *Agronomy* 14: 719.

Chervin, J., Talou, T., Audonnet, M., Dumas, B., Camborde, L., *et al.* (2019) Deciphering the phylogeny of violets based on multiplexed genetic and metabolomic approaches. *Phytochemistry* 163, 99–110.

Chi, J., Shu, J., Li, M., Mudappathi, R., Jin, Y., *et al.* (2024) Artificial intelligence in metabolomics: a current review. *Trends in Analytical Chemistry* 178: 117852.

Chlingaryan, A., Sukkarieh, S., and Whelan, B. (2018) Machine learning approaches for crop yield prediction and nitrogen status estimation in precision agriculture: a review. *Computers and Electronics in Agriculture* 151, 61–69.

Chong, J. and Xia, J. (2018) MetaboAnalystR: an R package for flexible and reproducible analysis of metabolomics data. *Bioinformatics* 34, 4313–4314.

Civelek, M. and Lusis, A.J. (2014) Systems genetics approaches to understand complex traits. *Nature Reviews Genetics* 15, 34–48.

Cortes, C. and Vapnik, V. (1995) Support-vector networks. *Machine Learning* 20, 273–297.

Cosme, M. (2023) Mycorrhizas drive the evolution of plant adaptation to drought. *Communications Biology* 6: 346.

Cotton, T.E.A., Pétriacq, P., Cameron, D.D., Meselmani, M.A., Schwarzenbacher, R., *et al.* (2019) Metabolic regulation of the maize rhizobiome by benzoxazinoids. *ISME Journal* 13, 1647–1658.

Dablanc, A., Hennechart, S., Perez, A., Cabanac, G., Guitton, Y., *et al.* (2024) FragHub: a mass spectral library data integration workflow. *Analytical Chemistry* 96, 12489–12496.

Dan, Z., Hu, J., Zhou, W., Yao, G., Zhu, R., *et al.* (2016) Metabolic prediction of important agronomic traits in hybrid rice (*Oryza sativa* L.). *Scientific Reports* 6: 21732.

Dan, Z., Chen, Y., Xu, Y., Huang, J., Huang, J., *et al.* (2019) A metabolome-based core hybridisation strategy for the prediction of rice grain weight across environments. *Plant Biotechnology Journal* 17, 906–913.

Dan, Z., Chen, Y., Zhao, W., Wang, Q., and Huang, W. (2020) Metabolome-based prediction of yield heterosis contributes to the breeding of elite rice. *Life Science Alliance* 3: e201900551.

Dastres, R. and Soori, M. (2021) Artificial neural network systems. *International Journal of Imaging and Robotics* 1, 13–25.

de Abreu E. Lima, F., Willmitzer, L., and Nikoloski, Z. (2018) Classification-driven framework to predict maize hybrid field performance from metabolic profiles of young parental roots. *PLOS ONE* 13: e0196038.

de Raad, M., Fischer, C.R., and Northen, T.R. (2016) High-throughput platforms for metabolomics. *Current Opinion in Chemical Biology* 30, 7–13.

Defossez, E., Pitteloud, C., Descombes, P., Glauser, G., Allard, P.-M., *et al.* (2021) Spatial and evolutionary predictability of phytochemical diversity. *Proceedings of the National Academy of Sciences USA* 118: e2013344118.

Desta, Z.A. and Ortiz, R. (2014) Genomic selection: genome-wide prediction in plant improvement. *Trends in Plant Science* 19, 592–601.

Díaz, F.P., Dussarrat, T., Carrasco-Puga, G., Colombié, S., Prigent, S., *et al.* (2024) Ecological and metabolic implications of the nurse effect of *Maihueniopsis camachoi* in the Atacama Desert. *New Phytologist* 241, 1074–1087.

Dührkop, K., Fleischauer, M., Ludwig, M., Aksenov, A.A., Melnik, A.V., *et al.* (2019) SIRIUS 4: a rapid tool for turning tandem mass spectra into metabolite structure information. *Nature Methods* 16, 299–302.

Dussarrat, T., Prigent, S., Latorre, C., Bernillon, S., Flandin, A., *et al.* (2022) Predictive metabolomics of multiple Atacama plant species unveils a core set of generic metabolites for extreme climate resilience. *New Phytologist* 234, 1614–1628.

Dussarrat, T., Schweiger, R., Ziaja, D., Nguyen, T.T.N., Krause, L., *et al.* (2023) Influences of chemotype and parental genotype on metabolic fingerprints of tansy plants uncovered by predictive metabolomics. *Scientific Reports* 13: 11645.

Feher, K., Lisec, J., Römisch-Margl, L., Selbig, J., Gierl, A., *et al.* (2014) Deducing hybrid performance from parental metabolic profiles of young primary roots of maize by using a multivariate diallel approach. *PLOS ONE* 9: e85435.

Fialkov, A.B., Lehotay, S.J., and Amirav, A. (2020) Less than one minute low-pressure gas chromatography–mass spectrometry. *Journal of Chromatography A* 1612: 460691.

Fine, P.V.A., Salazar, D., Martin, R.E., Metz, M.R., Misiewicz, T.M., and Asner, G.P. (2021) Exploring the links between secondary metabolites and leaf spectral reflectance in a diverse genus of Amazonian trees. *Ecosphere* 12: e03362.

Fontanari, T., Fróes, T.C., and Recamonde-Mendoza, M. (2022) Cross-validation strategies for balanced and imbalanced datasets. In: Xavier-Junior, J.C. and Rios, R.A. (eds) *Intelligent Systems*. Springer International Publishing, Cham, Switzerland, pp. 626–640.

Fuhrer, T. and Zamboni, N. (2015) High-throughput discovery metabolomics. *Current Opinion in Biotechnology* 31, 73–78.

Galal, A., Talal, M., and Moustafa, A. (2022) Applications of machine learning in metabolomics: disease modeling and classification. *Frontiers in Genetics* 13: 1017340.

García-Pérez, P., Zhang, L., Miras-Moreno, B., Lozano-Milo, E., Landin, M., *et al.* (2021) The combination of untargeted metabolomics and machine learning predicts the biosynthesis of phenolic compounds in *Bryophyllum* medicinal plants (genus *Kalanchoe*). *Plants* 10: 2430.

Gargallo-Garriga, A., Sardans, J., Granda, V., Llusià, J., Peguero, G., *et al.* (2020) Different "metabolomic niches" of the highly diverse tree species of the French Guiana rainforests. *Scientific Reports* 10: 6937.

Geisser, S. (1993) *Predictive Inference: An Introduction.* CRC Press, Boca Raton, Florida.

Ghosh, T., Zhang, W., Ghosh, D., and Kechris, K. (2020) Predictive modeling for metabolomics data. *Methods in Molecular Biology* 2104, 313–336.

Giacomoni, F., Le Corguillé, G., Monsoor, M., Landi, M., Pericard, P., *et al.* (2015) Workflow4Metabolomics: a collaborative research infrastructure for computational metabolomics. *Bioinformatics* 31, 1493–1495.

Gong, X., Feng, Y., Dang, K., Jiang, Y., Qi, H., and Feng, B. (2023) Linkages of microbial community structure and root exudates: evidence from microbial nitrogen limitation in soils of crop families. *Science of the Total Environment* 881: 163536.

Grapov, D., Fahrmann, J., Wanichthanarak, K., and Khoomrung, S. (2018) Rise of deep learning for genomic, proteomic, and metabolomic data integration in precision medicine. *OMICS* 22, 630–636.

Greener, J.G., Kandathil, S.M., Moffat, L., and Jones, D.T. (2022) A guide to machine learning for biologists. *Nature Reviews Molecular Cell Biology* 23, 40–55.

Guitton, Y., Tremblay-Franco, M., Le Corguillé, G., Martin, J.-F., Pétéra, M., *et al.* (2017) Create, run, share, publish, and reference your LC–MS, FIA–MS, GC–MS, and NMR data analysis workflows with the Workflow4Metabolomics 3.0 Galaxy online infrastructure for metabolomics. *International Journal of Biochemistry and Cell Biology* 93, 89–101.

Gunn, S. (1997) *Support Vector Machines for Classification and Regression.* Technical Report, University of Southampton, Southampton, UK.

Hajjar, G., Barros Santos, M.C., Bertrand-Michel, J., Canlet, C., Castelli, F., *et al.* (2023) Scaling-up metabolomics: current state and perspectives. *Trends in Analytical Chemistry* 167: 117225.

Hall, R.D., d'Auria, J.C., Silva Ferreira, A.C., Gibon, Y., Kruszka, D., *et al.* (2022) High-throughput plant phenotyping: a role for metabolomics? *Trends in Plant Science* 27, 549–563.

Han, J., Datla, R., Chan, S., and Borchers, C.H. (2009) Mass spectrometry-based technologies for high-throughput metabolomics. *Bioanalysis* 1, 1665–1684.

Haug, K., Salek, R.M., Conesa, P., Hastings, J., de Matos, P., *et al.* (2013) MetaboLights—an open-access general-purpose repository for metabolomics studies and associated meta-data. *Nucleic Acids Research* 41, D781–786.

Hill, C.B. and Roessner, U. (2015) Advances in high-throughput untargeted LC–MS analysis for plant metabolomics. In: Parker, C. and Borchers, C. (eds) *Advanced LC-MS Applications in Metabolomics*. Future Science Ltd, University of Victoria Proteomics Centre, Victoria, Canada, pp. 58–71.

Hu, H., Campbell, M.T., Yeats, T.H., Zheng, X., Runcie, D.E., *et al.* (2021) Multi-omics prediction of oat agronomic and seed nutritional traits across environments and in distantly related populations. *Theoretical and Applied Genetics* 134, 4043–4054.

Ibarra-Estrada, E., Palma-Tenango, R.M.S.-H., Ibarra-Estrada, E., and Palma-Tenango, R.M.S.-H (2016) Metabolomics as a tool in agriculture. In: *Metabolomics – Fundamentals and Applications*. IntechOpen, London. Available at: www.intechopen.com/chapters/53228 (accessed 28 January 2025).

Jahnke, S., Roussel, J., Hombach, T., Kochs, J., Fischbach, A., *et al.* (2016) *pheno*Seeder—a robot system for automated handling and phenotyping of individual seeds. *Plant Physiology* 172, 1358–1370.

Jandová, K., Dostál, P., Cajthaml, T., and Kameník, Z. (2015) Intraspecific variability in allelopathy of *Heracleum mantegazzianum* is linked to the metabolic profile of root exudates. *Annals of Botany* 115, 821–831.

Janiesch, C., Zschech, P., and Heinrich, K. (2021) Machine learning and deep learning. *Electron Markets* 31, 685–695.

Janni, M., Maestri, E., Gullì, M., Marmiroli, M., and Marmiroli, N. (2024) Plant responses to climate change, how global warming may impact on food security: a critical review. *Frontiers in Plant Science* 14: 1297569.

Jeon, J.E., Kim, J.-G., Fischer, C.R., Mehta, N., Dufour-Schroif, C., *et al.* (2020) A pathogen-responsive gene cluster for highly modified fatty acids in tomato. *Cell* 180, 176–187.e19.

Jing, H., Wang, H., Wang, G., Liu, G., and Cheng, Y. (2023) The mechanism effects of root exudate on microbial community of rhizosphere soil of tree, shrub, and grass in forest ecosystem under N deposition. *ISME Communications* 3: 120.

Jones, O.A.H., Maguire, M.L., Griffin, J.L., Dias, D.A., Spurgeon, D.J., and Svendsen, C. (2013) Metabolomics and its use in ecology. *Austral Ecology* 38, 713–720.

Jonsson, P., Johansson, A.I., Gullberg, J., Trygg, J., Grung, B., *et al.* (2005) High-throughput data analysis for detecting and identifying differences between samples in GC/MS-based metabolomic analyses. *Analytical Chemistry* 77, 5635–5642.

Junker, R.R. and Blüthgen, N. (2010) Floral scents repel facultative flower visitors, but attract obligate ones. *Annals of Botany* 105, 777–782.

Kale, N.S., Haug, K., Conesa, P., Jayseelan, K., Moreno, P., *et al.* (2016) MetaboLights: an open-access database repository for metabolomics data. *Current Protocols in Bioinformatics* 53, 14.13.1–14.13.18.

Katajamaa, M., Miettinen, J., and Orešič, M. (2006) MZmine: toolbox for processing and visualization of mass spectrometry based molecular profile data. *Bioinformatics* 22, 634–636.

Khakimov, B., Bak, S., and Engelsen, S.B. (2014) High-throughput cereal metabolomics: current analytical technologies, challenges and perspectives. *Journal of Cereal Science* 59, 393–418.

Kim, H.K. and Verpoorte, R. (2010) Sample preparation for plant metabolomics. *Phytochemical Analysis* 21, 4–13.

Kisiel, A., Krzemińska, A., Cembrowska-Lech, D., and Miller, T. (2023) Data science and plant metabolomics. *Metabolites* 13: 454.

Knoch, D., Werner, C.R., Meyer, R.C., Riewe, D., and Abbadi, A. (2021) Multi-omics-based prediction of hybrid performance in canola. *Theoretical and Applied Genetics* 134, 1147–1165.

Kueffer, C. (2019) Inheritors of the Earth: how nature is thriving in an age of extinction. *Basic and Applied Ecology* 35, 13–17.

Kumar, R., Bohra, A., Pandey, A.K., Pandey, M.K., and Kumar, A. (2017) Metabolomics for plant improvement: status and prospects. *Frontiers in Plant Science* 8: 1302.

Kumari, M., Yagnik, K.N., Gupta, V., Singh, I.K., Gupta, R., *et al.* (2024) Metabolomics-driven investigation of plant defense response against pest and pathogen attack. *Physiologia Plantarum* 176: e14270.

Lai, Z., Tsugawa, H., Wohlgemuth, G., Mehta, S., Mueller, M., *et al.* (2018) Identifying metabolites by integrating metabolome databases with mass spectrometry cheminformatics. *Nature Methods* 15, 53–56.

Laurençon, M., Legrix, J., Wagner, M.-H., Demilly, D., Baron, C., *et al.* (2024) Genomic and phenomic predictions help capture low-effect alleles promoting seed germination in oilseed rape in addition to QTL analyses. *Theoretical and Applied Genetics* 137: 156.

Lazar, A., Romanciuc, F., Socaciu, M.A., and Socaciu, C. (2015) Bioinformatics tools for metabolomic data processing and analysis using untargeted liquid chromatography coupled with mass spectrometry. *Bulletin of University of Agricultural Sciences and Veterinary Medicine Cluj-Napoca. Animal Science and Biotechnologies* 72, 103–115.

Leao, T.F., Clark, C.M., Bauermeister, A., Elijah, E.O., Gentry, E.C., *et al.* (2021) Quick-start infrastructure for untargeted metabolomics analysis in GNPS. *Nature Metabolism* 3, 880–882.

Lee, K.S., Su, X. and Huan, T. (2025) Metabolites are not genes — avoiding the misuse of pathway analysis in metabolomics. *nature metabolism* 7, 858–861.

Leisner, C.P. (2020) Climate change impacts on food security—focus on perennial cropping systems and nutritional value. *Plant Science* 293: 110412.

Liu, K., Ding, X., and Wang, J. (2020) Soil metabolome correlates with bacterial diversity and co-occurrence patterns in root-associated soils on the Tibetan Plateau. *Science of the Total Environment* 735: 139572.

Lottering, R.T., Govender, M., Peerbhay, K., and Lottering, S. (2020) Comparing partial least squares (PLS) discriminant analysis and sparse PLS discriminant analysis in detecting and mapping *Solanum mauritianum* in commercial forest plantations using image texture. *ISPRS Journal of Photogrammetry and Remote Sensing* 159, 271–280.

Luna, E., Flandin, A., Cassan, C., Prigent, S., Chevanne, C., *et al.* (2020) Metabolomics to exploit the primed immune system of tomato fruit. *Metabolites* 10: 96.

Mackay, T.F.C., Stone, E.A., and Ayroles, J.F. (2009) The genetics of quantitative traits: challenges and prospects. *Nature Reviews Genetics* 10, 565–577.

Maulud, D. and Abdulazeez, A. (2020) A review on linear regression comprehensive in machine learning. *Journal of Applied Science and Technology Trends* 1, 140–147.

McQueen, R.J., Garner, S.R., Nevill-Manning, C.G., and Witten, I.H. (1995) Applying machine learning to agricultural data. *Computers and Electronics in Agriculture* 12, 275–293.

Medema, M.H., de Rond, T., and Moore, B.S. (2021) Mining genomes to illuminate the specialized chemistry of life. *Nature Reviews Genetics* 22, 553–571.

Mendez, K.M., Broadhurst, D.I., and Reinke, S.N. (2019) The application of artificial neural networks in metabolomics: a historical perspective. *Metabolomics* 15: 142.

Misra, B.B. and Mohapatra, S. (2019) Tools and resources for metabolomics research community: a 2017–2018 update. *Electrophoresis* 40, 227–246.

Misra, B.B. and van der Hooft, J.J.J. (2016) Updates in metabolomics tools and resources: 2014–2015. *Electrophoresis* 37, 86–110.

Misra, B.B., Fahrmann, J.F., and Grapov, D. (2017) Review of emerging metabolomic tools and resources: 2015–2016. *Electrophoresis* 38, 2257–2274.

Moreira, X., Castagneyrol, B., Abdala-Roberts, L., Berny-Miery Teran, J.C., Timmermans, B.G.H., *et al.* (2018) Latitudinal variation in plant chemical defences drives latitudinal patterns of leaf herbivory. *Ecography* 41, 1124–1134.

Morota, G. and Gianola, D. (2014) Kernel-based whole-genome prediction of complex traits: a review. *Frontiers in Genetics* 5: 00363.

Müller, C. and Junker, R.R. (2022) Chemical phenotype as important and dynamic niche dimension of plants. *New Phytologist* 234, 1168–1174.

Noor, E., Cherkaoui, S., and Sauer, U. (2019) Biological insights through omics data integration. *Current Opinion in Systems Biology* 15, 39–47.

Nothias, L.-F., Petras, D., Schmid, R., Dührkop, K., Rainer, J., *et al.* (2020) Feature-based molecular networking in the GNPS analysis environment. *Nature Methods* 17, 905–908.

Obi, J.C. and Jecinta, I.C. (2023) A review of techniques for regularization. *International Journal of Research in Engineering and Science* 11, 360–367.

Onishi, Y., Yoshida, T., Kurita, H., Fukao, T., Arihara, H., and Iwai, A. (2019) An automated fruit harvesting robot by using deep learning. *Robomech Journal* 6: 13.

Palmer, A., Phapale, P., Chernyavsky, I., Lavigne, R., Fay, D., *et al.* (2017) FDR-controlled metabolite annotation for high-resolution imaging mass spectrometry. *Nature Methods* 14, 57–60.

Pandu, U., Desai, K., Krupashree, R., Theerthana, T., and Raddy, G. (2022) A review on artificial intelligence and robots in agriculture. *Mysore Journal of Agricultural Sciences* 56, 1–13.

Pang, Z., Chong, J., Zhou, G., de Lima Morais, D.A., Chang, L., *et al.* (2021) MetaboAnalyst 5.0: narrowing the gap between raw spectra and functional insights. *Nucleic Acids Research* 49, W388–W396.

Pappalardo, S.E. and Andrade, D. (2022) Drones for good: UAS applications in agroecology and organic farming. In: *Drones and Geographical Information Technologies in Agroecology and Organic Farming*. CRC Press, Boca Raton, Florida.

Paulhe, N., Canlet, C., Damont, A., Peyriga, L., Durand, S., *et al.* (2022) PeakForest: A multi-platform digital infrastructure for interoperable metabolite spectral data and metadata management. *Metabolomics* 18: 40.

Peters, K., Blatt-Janmaat, K.L., Tkach, N., van Dam, N.M., and Neumann, S. (2023) Untargeted metabolomics for integrative taxonomy: metabolomics, DNA marker-based sequencing, and phenotype bioimaging. *Plants* 12: 881.

Petrén, H., Köllner, T.G., and Junker, R.R. (2023) Quantifying chemodiversity considering biochemical and structural properties of compounds with the R package chemodiv. *New Phytologist* 237, 2478–2492.

Pfeuffer, J., Sachsenberg, T., Alka, O., Walzer, M., Fillbrunn, A., *et al.* (2017) OpenMS—a platform for reproducible analysis of mass spectrometry data. *Journal of Biotechnology* 261, 142–148.

Plumb, R.S., Gethings, L.A., Rainville, P.D., Isaac, G., Trengove, R., *et al.* (2023) Advances in high throughput LC/MS based metabolomics: a review. *Trends in Analytical Chemistry* 160: 116954.

Pluskal, T., Castillo, S., Villar-Briones, A., and Orešič, M. (2010) MZmine 2: modular framework for processing, visualizing, and analyzing mass spectrometry-based molecular profile data. *BMC Bioinformatics* 11: 395.

Pomyen, Y., Wanichthanarak, K., Poungsombat, P., Fahrmann, J., Grapov, D., and Khoomrung, S. (2020) Deep metabolome: applications of deep learning in metabolomics. *Computational and Structural Biotechnology Journal* 18, 2818–2825.

Redecker, D., Kodner, R., and Graham, L.E. (2000) Glomalean fungi from the Ordovician. *Science* 289, 1920–1921.

Rice, C., Wolf, J., Fleisher, D.H., Acosta, S.M., Adkins, S.W., *et al.* (2021) Recent CO_2 levels promote increased production of the toxin parthenin in an invasive *Parthenium hysterophorus* biotype. *Nature Plants* 7, 725–729.

Richards, L.A., Dyer, L.A., Forister, M.L., Smilanich, A.M., Dodson, C.D., *et al.* (2015) Phytochemical diversity drives plant–insect community diversity. *Proceedings of the National Academy of Sciences USA* 112, 10973–10978.

Roch, L., Prigent, S., Klose, H., Cakpo, C.-B., Beauvoit, B., *et al.* (2020) Biomass composition explains fruit relative growth rate and discriminates climacteric from non-climacteric species. *Journal of Experimental Botany* 71, 5823–5836.

Röst, H.L., Sachsenberg, T., Aiche, S., Bielow, C., Weisser, H., *et al.* (2016) OpenMS: a flexible open-source software platform for mass spectrometry data analysis. *Nature Methods* 13, 741–748.

Ruttkies, C., Schymanski, E.L., Wolf, S., Hollender, J., and Neumann, S. (2016) MetFrag relaunched: incorporating strategies beyond *in silico* fragmentation. *Journal of Cheminformatics* 8: 3.

Ruttkies, C., Neumann, S., and Posch, S. (2019) Improving MetFrag with statistical learning of fragment annotations. *BMC Bioinformatics* 20: 376.

Saccenti, E., Hoefsloot, H.C.J., Smilde, A.K., Westerhuis, J.A., and Hendriks, M.M.W.B. (2014) Reflections on univariate and multivariate analysis of metabolomics data. *Metabolomics* 10, 361–374.

Sasse, J., Martinoia, E., and Northen, T. (2018) Feed your friends: do plant exudates shape the root microbiome? *Trends Plant Science* 23, 25–41.

Schorn, M.A., Verhoeven, S., Ridder, L., Huber, F., Acharya, D.D., *et al.* (2021) A community resource for paired genomic and metabolomic data mining. *Nature Chemical Biology* 17, 363–368.

Schuman, M.C. (2023) Where, when, and why do plant volatiles mediate ecological signaling? The answer is blowing in the wind. *Annual Review of Plant Biology* 74, 609–633.

Schuman, M.C., van Dam, N.M., Beran, F., and Harpole, W.S. (2016) How does plant chemical diversity contribute to biodiversity at higher trophic levels? *Current Opinion in Insect Science* 14, 46–55.

Schymanski, E.L., Jeon, J., Gulde, R., Fenner, K., Ruff, M., *et al.* (2014) Identifying small molecules via high resolution mass spectrometry: communicating confidence. *Environmental Science and Technology* 48, 2097–2098.

Sedio, B.E. (2017) Recent breakthroughs in metabolomics promise to reveal the cryptic chemical traits that mediate plant community composition, character evolution and lineage diversification. *New Phytologist* 214, 952–958.

Sedio, B.E., Spasojevic, M.J., Myers, J.A., Wright, S.J., Person, M.D., *et al.* (2021) Chemical similarity of co-occurring trees decreases with precipitation and temperature in North American forests. *Frontiers in Ecology and Evolution* 9: 679638.

Sen, P.C., Hajra, M., and Ghosh, M. (2020) Supervised classification algorithms in machine learning: a survey and review. In: Mandal, J.K. and Bhattacharya, D. (eds) *Emerging Technology in Modelling and Graphics*. Springer, Singapore, pp. 99–111.

Sharifani, K. and Amini, M. (2023) Machine learning and deep learning: a review of methods and applications. *World Information Technology and Engineering Journal* 10, 3897–3904.

Sheikh, M., Iqra, F., Ambreen, H., Pravin, K.A., Ikra, M., and Chung, Y.S. (2024) Integrating artificial intelligence and high-throughput phenotyping for crop improvement. *Journal of Integrative Agriculture* 23, 1787–1802.

Singh, A., Shannon, C.P., Gautier, B., Rohart, F., Vacher, M., *et al.* (2019) DIABLO: an integrative approach for identifying key molecular drivers from multi-omics assays. *Bioinformatics* 35, 3055–3062.

Smith, C.A., Want, E.J., O'Maille, G., Abagyan, R., and Siuzdak, G. (2006) XCMS: Processing mass spectrometry data for metabolite profiling using nonlinear peak alignment, matching, and identification. *Analytical Chemistry* 78, 779–787.

Smith, P. and Gregory, P.J. (2013) Climate change and sustainable food production. *Proceedings of the Nutrition Society* 72, 21–28.

Søgaard, A., Ebert, S., Bastings, J., and Filippova, K. (2021) We need to talk about random splits. In: Merlo, P., Tiedemann, J. and Tsarfaty, R. (eds) *Proceedings of the 16th Conference of the European Chapter of the Association for Computational Linguistics: Main Volume*. Association for Computational Linguistics, pp. 1823–1832. Available at: https://aclanthology.org/2021.eacl-main.156.pdf (accessed 28 January 20250.

Sprenger, H., Erban, A., Seddig, S., Rudack, K., Thalhammer, A., *et al.* (2018) Metabolite and transcript markers for the prediction of potato drought tolerance. *Plant Biotechnology Journal* 16, 939–950.

Su, P., Liu, X., Wang, R., Liu, T., Zhao, W., *et al.* (2022) Autotoxicity of *Ambrosia artemisiifolia* and *Ambrosia trifida* and its significance for the regulation of intraspecific populations density. *Scientific Reports* 12: 17424.

Sud, M., Fahy, E., Cotter, D., Azam, K., Vadivelu, I., *et al.* (2016) Metabolomics Workbench: an international repository for metabolomics data and metadata, metabolite standards, protocols, tutorials and training, and analysis tools. *Nucleic Acids Research* 44, D463–D470.

Sumner, L.W., Amberg, A., Barrett, D., Beale, M.H., Beger, R., *et al.* (2007) Proposed minimum reporting standards for chemical analysis. *Metabolomics* 3, 211–221.

Sun, L., He, Y., Cao, M., Wang, X., Zhou, X., *et al.* (2024) Tree phytochemical diversity and herbivory are higher in the tropics. *Nature Ecology and Evolution* 8, 1426–1436.

Tatachar, A.V. (2021) Comparative assessment of regression models based on model evaluation metrics. *International Research Journal of Engineering and Technology* 8, 853–860.

Tautenhahn, R., Böttcher, C., and Neumann, S. (2008) Highly sensitive feature detection for high resolution LC/MS. *BMC Bioinformatics* 9: 504.

Thon, F.M., Müller, C., and Wittmann, M.J. (2024) The evolution of chemodiversity in plants—from verbal to quantitative models. *Ecology Letters* 27: e14365.

Tian, B., Pei, Y., Huang, W., Ding, J., and Siemann, E. (2021) Increasing flavonoid concentrations in root exudates enhance associations between arbuscular mycorrhizal fungi and an invasive plant. *ISME Journal* 15, 1919–1930.

Tian, H., Lam, S.M., and Shui, G. (2016) Metabolomics, a powerful tool for agricultural research. *International Journal of Molecular Sciences* 17: 1871.

Tong, H. and Nikoloski, Z. (2021) Machine learning approaches for crop improvement: leveraging phenotypic and genotypic big data. *Journal of Plant Physiology* 257: 153354.

Trivedi, P., Leach, J.E., Tringe, S.G., Sa, T., and Singh, B.K. (2020) Plant–microbiome interactions: from community assembly to plant health. *Nature Reviews Microbiology* 18, 607–621.

Tsugawa, H., Cajka, T., Kind, T., Ma, Y., Higgins, B., *et al.* (2015) MS-DIAL: data-independent MS/MS deconvolution for comprehensive metabolome analysis. *Nature Methods* 12, 523–526.

Tsugawa, H., Nakabayashi, R., Mori, T., Yamada, Y., Takahashi, M., *et al.* (2019) A cheminformatics approach to characterize metabolomes in stable-isotope-labeled organisms. *Nature Methods* 16, 295–298.

Tsugawa, H., Ikeda, K., Takahashi, M., Satoh, A., Mori, Y., *et al.* (2020) A lipidome atlas in MS-DIAL 4. *Nature Biotechnology* 38, 1159–1163.

Tsugawa, H., Rai, A., Saito, K., and Nakabayashi, R. (2021) Metabolomics and complementary techniques to investigate the plant phytochemical cosmos. *Natural Product Reports* 38, 1729–1759.

Tuomi, M.W., Utsi, T.A., Yoccoz, N.G., Armstrong, C.W., Gonzalez, V., *et al.* (2024) The increase of an allelopathic and unpalatable plant undermines reindeer pasture quality and current management in the Norwegian tundra. *Communications Earth and Environment* 5: 414.

van der Hooft, J.J.J., Mohimani, H., Bauermeister, A., Dorrestein, P.C., Duncan, K.R., and Medema, M.H. (2020) Linking genomics and metabolomics to chart specialized metabolic diversity. *Chemical Society Reviews* 49, 3297–3314.

Vervoort, N., Goossens, K., Baeten, M., and Chen, Q. (2021) Recent advances in analytical techniques for high throughput experimentation. *Analytical Science Advances* 2, 109–127.

Vieira, S., Sikorski, J., Dietz, S., Herz, K., Schrumpf, M., *et al.* (2020) Drivers of the composition of active rhizosphere bacterial communities in temperate grasslands. *ISME Journal* 14, 463–475.

Vishwakarma, K., Kumar, N., Shandilya, C., Mohapatra, S., Bhayana, S., and Varma, A. (2020) Revisiting plant–microbe interactions and microbial consortia application for enhancing sustainable agriculture: a review. *Frontiers in Microbiology* 11: 560406.

Vivanco, J.M., Bais, H.P., Stermitz, F.R., Thelen, G.C., and Callaway, R.M. (2004) Biogeographical variation in community response to root allelochemistry: novel weapons and exotic invasion. *Ecology Letters* 7, 285–292.

Wakchaure, M., Patle, B.K., Mahindrakar, A.K., 2023. Application of AI techniques and robotics in agriculture: a review. *Artificial Intelligence in the Life Sciences* 3: 100057.

Waljee, A.K., Higgins, P.D.R., and Singal, A.G. (2014) A primer on predictive models. *Clinical and Translational Gastroenterology* 5: e44.

Walker, T.W.N., Alexander, J.M., Allard, P.-M., Baines, O., Baldy, V., *et al.* (2022) Functional Traits 2.0: the power of the metabolome for ecology. *Journal of Ecology* 110, 4–20.

Wang, J., Long, X., Chern, M., and Chen, X. (2021) Understanding the molecular mechanisms of trade-offs between plant growth and immunity. *Science China Life Sciences* 64, 234–241.

Wang, M., Carver, J.J., Phelan, V.V., Sanchez, L.M., Garg, N., *et al.* (2016) Sharing and community curation of mass spectrometry data with global natural products social molecular networking. *Nature Biotechnology* 34, 828–837.

Wang, S., Alseekh, S., Fernie, A.R., and Luo, J. (2019a) The structure and function of major plant metabolite modifications. *Molecular Plant* 12, 899–919.

Wang, S., Wei, J., Li, R., Qu, H., Chater, J.M., *et al.* (2019b) Identification of optimal prediction models using multi-omic data for selecting hybrid rice. *Heredity* 123, 395–406.

Wase, N., Abshire, N., and Obata, T. (2022) High-throughput profiling of metabolic phenotypes using high-resolution GC-MS. In: Lorence, A. and Medina Jimenez, K. (eds.) *High-throughput Plant Phenotyping: Methods and Protocols*. Springer, New York, pp. 235–260.

Wehrens, R., Putter, H., and Buydens, L.M.C. (2000) The bootstrap: a tutorial. *Chemometrics and Intelligent Laboratory Systems* 54, 35–52.

Weir, T.L., Park, S.-W., and Vivanco, J.M. (2004) Biochemical and physiological mechanisms mediated by allelochemicals. *Current Opinion in Plant Biology* 7, 472–479.

Westerhuis, J.A., Hoefsloot, H.C.J., Smit, S., Vis, D.J., Smilde, A.K., *et al.* (2008) Assessment of PLSDA cross validation. *Metabolomics* 4, 81–89.

Wetzel, W.C. and Whitehead, S.R. (2020) The many dimensions of phytochemical diversity: linking theory to practice. *Ecology Letters* 23, 16–32.

Wieder, C., Frainay, C., Poupin, N., Rodríguez-Mier, P., Vinson, F., *et al.* (2021) Pathway analysis in metabolomics: recommendations for the use of over-representation analysis. *PLOS Computational Biology* 17: e1009105.

Wilkinson, M.D., Dumontier, M., Aalbersberg, I.J., Appleton, G., Axton, M., *et al.* (2016) The FAIR guiding principles for scientific data management and stewardship. *Scientific Data* 3: 160018.

Wittmann, M.J. and Bräutigam, A. (2024) How does plant chemodiversity evolve? Testing five hypotheses in one population genetic model. *New Phytologist* 245, 1302–1314.

Wolf, S., Schmidt, S., Müller-Hannemann, M., and Neumann, S. (2010) *In silico* fragmentation for computer assisted identification of metabolite mass spectra. *BMC Bioinformatics* 11: 148.

Worley, B. and Powers, R. (2013) Multivariate analysis in metabolomics. *Current Metabolomics* 1, 92–107.

Wu, C., Luo, J., and Xiao, Y. (2024) Multi-omics assists genomic prediction of maize yield with machine learning approaches. *Molecular Breeding* 44: 14.

Wu, L., Wang, J., Huang, W., Wu, H., Chen, J., *et al.* (2015) Plant–microbe rhizosphere interactions mediated by *Rehmannia glutinosa* root exudates under consecutive monoculture. *Scientific Reports* 5: 15871.

Xia, M., Suseela, V., McCormack, M.L., Kennedy, P.G., and Tharayil, N. (2023) Common and lifestyle-specific traits of mycorrhizal root metabolome reflect ecological strategies of plant–mycorrhizal interactions. *Journal of Ecology* 111, 601–616.

Xie, Z., Zhao, S., Li, Y., Deng, Y., Shi, Y., *et al.* (2023) Phenolic acid-induced phase separation and translation inhibition mediate plant interspecific competition. *Nature Plants* 9, 1481–1499.

Yang, T., Tedersoo, L., Soltis, P.S., Soltis, D.E., Gilbert, J.A., *et al.* (2019) Phylogenetic imprint of woody plants on the soil mycobiome in natural mountain forests of eastern China. *ISME Journal* 13, 686–697.

Yu, H., He, Y., Zhang, W., Chen, L., Zhang, J., *et al.* (2022) Greater chemical signaling in root exudates enhances soil mutualistic associations in invasive plants compared to natives. *New Phytologist* 236, 1140–1153.

Yuan, Y., Wang, B., Zhang, S., Tang, J., Tu, C., *et al.* (2013) Enhanced allelopathy and competitive ability of invasive plant *Solidago canadensis* in its introduced range. *Journal of Plant Ecology* 6, 253–263.

Zampieri, M., Sekar, K., Zamboni, N., and Sauer, U. (2017) Frontiers of high-throughput metabolomics. *Current Opinion in Chemical Biology* 36, 15–23.

Zhang, Y. (2010) *New Advances in Machine Learning*. IntechOpen, London. Available at: www.perlego.com/book/2024136/new-advances-in-machine-learning-pdf (accessed 28 January 2025).

Zhang, Y., Worthy, S.J., Xu, S., He, Y., Wang, X., *et al.* (2024) Phytochemical diversity and their adaptations to abiotic and biotic pressures in fine roots across a climatic gradient. *Science of the Total Environment* 927: 172051.

Zhao, Y., Li, Z., Liu, G., Jiang, Y., Maurer, H.P., *et al.* (2015) Genome-based establishment of a high-yielding heterotic pattern for hybrid wheat breeding. *Proceedings of the National Academy of Sciences USA* 112, 15624–15629.

Zhou, B., Xiao, J.F., Tuli, L., and Ressom, H.W. (2012) LC-MS-based metabolomics. *Molecular BioSystems* 8, 470–481.

Zhu, X., Yi, Y., Huang, L., Zhang, C., and Shao, H. (2021) Metabolomics reveals the allelopathic potential of the invasive plant *Eupatorium adenophorum*. *Plants* 10: 1473.

Zulfiqar, M., Crusoe, M.R., König-Ries, B., Steinbeck, C., Peters, K., and Gadelha, L. (2024) Implementation of FAIR practices in computational metabolomics workflows—a case study. *Metabolites* 14: 118.

4 Technological Advancements in Plant High-throughput Metabolomics

Tharani Sureshkumar[1], Chithra Devi Balasundarm Saraswathy[2*], Ramkumar Govindarajan[1], Ajay Samuel Siva Daniel[1], Sunitha Kumari Krishnankutty[2], Sarah Jaison[2], Muthukrishnan Arun[3], and Arumugam Vijaya Anand[4]

[1]GRG BIRAC EYUVA Centre, PSGR Krishnammal College for Women, Coimbatore, Tamil Nadu, India; [2]Department of Botany, PSGR Krishnammal College for Women, Coimbatore, Tamil Nadu, India; [3]Department of Biotechnology, Bharathiar University, Coimbatore 641046, Tamil Nadu, India; [4]Department of Human Genetics and Molecular Biology, Bharathiar University, Coimbatore, Tamil Nadu, India

Abstract

Advancements in high-throughput metabolomics have dramatically enhanced our ability to study plant metabolism, offering new understandings of the complex biochemical processes that underlie plant growth, development, and environmental responses. Central to these advancements are sophisticated mass spectrometry (MS) technologies, including quadrupole time-of-flight MS, Orbitrap MS, and tandem MS. These technologies provide high-resolution, accurate mass measurements and detailed structural information, enabling the comprehensive recognition and quantification of a wide range of plant metabolites. The ability to analyze complex mixtures with high sensitivity and specificity has made MS indispensable in plant metabolomics, allowing researchers to uncover metabolic networks, identify biomarkers, and explore the roles of primary and secondary metabolites in plant physiology and adaptation. High-throughput metabolomics has facilitated large-scale studies, enabling the concurrent investigation of thousands of metabolites across multiple samples, thus offering a holistic view of plant metabolism. These advancements have been particularly valuable in comprehending how plants respond to various biotic and abiotic stresses, revealing key metabolic paths concerned with stress tolerance and resilience. Additionally, incorporating metabolomics with additional omics technologies, such as genomics and transcriptomics, has provided a more comprehensive understanding of plant systems biology. As the field progresses, emerging technologies such as single-cell and spatial metabolomics hold the prospect of further refining our experience of plant metabolic processes at the cellular and tissue levels. These technological innovations are expected to drive future research, leading to discoveries in plant biology, and enhancing crop improvement strategies and contributing to sustainable agricultural practices. The ongoing advancements in high-throughput metabolomics are poised to make a lasting impact on plant science and its applications in agriculture and biotechnology.

Keywords: chromatography, high-throughput analysis, mass spectrometry, metabolite identification, nuclear magnetic resonance spectroscopy, plant metabolomics

* E-mail: bchitradevi@psgrkcw.ac.in

© CAB International 2025. *High-Throughput Plant Metabolomics* (Jen-Tsung Chen ed.)
DOI: 10.1079/9781800626911.0004

4.1 Introduction

Plant metabolomics is a rapidly evolving field dedicated to thoroughly analyzing metabolites within plant systems. Metabolites are small molecules that play essential roles in diverse biochemical processes, serving as substrates, intermediates, and products in metabolic pathways (Sangwan *et al.*, 2015). These molecules are integral to plant development and environmental interaction. The study of these metabolites therefore offers invaluable insights into plant physiology, adaptation mechanisms, and responses to both abiotic and biotic stresses (Anzano *et al.*, 2022). Understanding the metabolic profile of plants can reveal the intricacies of their biochemical networks and physiological states. This is especially important because metabolites can be biomarkers for specific physiological conditions or responses (Arbona *et al.*, 2013). For instance, specific or metabolite profiles can indicate how a plant adapts to drought, salinity, pathogen attack, and nutrient availability. Furthermore, metabolites are often the end products of gene expression and protein activity, providing an explicit link between the phenotype and genotype (Jorge and António, 2018).

Metabolomics has its roots in classical biochemistry, where individual metabolites were studied in isolation (Hollywood *et al.*, 2006). Early metabolomic studies were constrained by the technological limitations of the time, resulting in low-throughput and limited metabolome coverage (Oldiges *et al.*, 2007). However, the past decades have seen notable advancements in analytical methods, markedly enhancing the scope and scale of metabolomic studies (Wolfender *et al.*, 2013). Initially, gas chromatography–mass spectrometry (GC-MS) and liquid chromatography–mass spectrometry (LC-MS) were the primary tools for metabolite analysis. These methods were powerful but had limitations regarding throughput and the range of detectable metabolites (Gika *et al.*, 2014). The advent of high-resolution mass spectrometry (MS), such as quadrupole time-of-flight–mass spectrometry (Q-TOF-MS) and Orbitrap MS, has revolutionized the field, offering unprecedented sensitivity, resolution, and accuracy (Law and Zhang, 2019). Additionally, advancements in nuclear magnetic resonance (NMR) spectroscopy have complemented MS techniques by providing detailed structural information and enabling non-destructive analysis (Marshall and Powers, 2017). The integration of high-throughput technologies into plant metabolomics has marked a paradigm shift, enabling the analysis of hundreds to thousands of metabolites simultaneously (Li and Gaquerel, 2021).

Automated systems for sample extraction, purification, and preparation have significantly increased throughput and reproducibility. These systems minimize human error and variability, critical for large-scale studies (Allwood *et al.*, 2011). Improvements in MS and NMR, along with innovations in chromatography, have expanded the range of detectable metabolites and enhanced data quality (Plumb *et al.*, 2023). Modern metabolomics generates vast amounts of data, necessitating sophisticated software and bioinformatics tools for data acquisition, processing, and interpretation (Monteiro *et al.*, 2013). High-throughput data acquisition systems coupled with real-time processing capabilities allow the immediate analysis of complex data sets (Liu *et al.*, 2014). Combining metabolomics with other omics technologies provides a more comprehensive understanding of plant biology (Yang *et al.*, 2021). These integrative approaches facilitate the correlation of metabolic changes with genetic and proteomic alterations, offering a holistic view of plant function and regulation (Tohge *et al.*, 2015).

High-throughput plant metabolomics has broad applications across various fields. Metabolomic profiling helps elucidate the biochemical changes underlying plant growth and development, providing insights into processes including germination, flowering, and senescence (Kumar *et al.*, 2017). By analyzing the metabolic responses of plants to various stresses (e.g. drought, salinity, pathogens), researchers can identify crucial metabolites involved in stress toleration and resistance mechanisms (Khan *et al.*, 2017). Metabolomics aids in developing crops with enhanced traits, such as improved nutritional content, increased yield, and resistance to pests and diseases. Metabolite profiling can guide metabolic engineering and breeding programs (Jamaloddin *et al.*, 2021). Understanding how plants interact with their environment, including soil microbiomes and pollutants, can inform strategies for sustainable agriculture and environmental remediation.

Despite significant progress, the field of plant metabolomics faces several challenges. Ensuring reproducibility and standardization in metabolomic studies is crucial, as is the development of robust bioinformatics tools for data integration and interpretation (Cambiaghi *et al.*, 2017). Future advancements are expected to focus on single-cell metabolomics, spatial metabolomics, and the integration of metabolomics with other omics technologies to provide deeper insights into plant biology (de Souza *et al.*, 2020). Advances in single-cell metabolomics will enable the analysis of metabolites at the cellular level, providing insights into cellular heterogeneity and the specific metabolic activities of different cell types within a plant (de Souza *et al.*, 2020). Spatial metabolomics involves mapping the distribution of metabolites within tissues, offering a spatial perspective on metabolic processes. Spatial metabolomics can reveal how metabolites are distributed within different parts of a plant, shedding light on tissue-specific functions and interactions (Sumner *et al.*, 2011; Feenstra *et al.*, 2017). Integrative omics approaches can uncover the complex regulatory networks that govern plant metabolism and physiology, facilitating the discovery of new metabolic pathways and regulatory mechanisms (Tohge *et al.*, 2015).

High-throughput metabolomics has transformed our ability to study plant metabolism on a large scale (Fuhrer and Zamboni, 2015). The continuous technological advancements in this field promise to further enhance our understanding of plant biology, paving the way for innovations in agriculture, biotechnology, and environmental science (National Academy of Sciences, 2019). As we continue to refine and expand our metabolomic capabilities, the integration of metabolomics with other omics technologies and the development of new analytical and computational tools will be essential in addressing the complex challenges of plant science (Sumner *et al.*, 2015). The future of plant metabolomics is bright, with the prospect of revolutionizing our approach to plant research and contributing to the sustainable development of agriculture and the preservation of our environment (Razzaq *et al.*, 2019).

4.2 Technologies in High-throughput Plant Metabolomics

4.2.1 Mass spectrometry

MS is a cornerstone technology in high-throughput plant metabolomics, playing an essential role in the recognition and quantification of metabolites based on their mass-to-charge ratios (m/z) (Marchev *et al.*, 2021). It offers high perceptiveness, explicitness, and the ability to analyze complex mixtures of metabolites simultaneously. Several advanced MS technologies are employed in plant metabolomics, each with unique capabilities suited for different analytical needs (Table 4.1) (Zhang *et al.*, 2012):

- Q-TOF-MS combines the benefits of a quadrupole mass analyzer with a TOF mass analyzer (Allen and McWhinney, 2019). This combination allows both high-resolution and accurate mass measurements, making Q-TOF-MS particularly effective for the analysis of complex plant metabolite mixtures (Ernst *et al.*, 2014; Alvarez-Rivera *et al.*, 2019). Q-TOF-MS provides high-resolution mass spectra that allow the precise determination of the m/z of ions (Lai and Wang, 2023). This high resolution is crucial for distinguishing between metabolites with similar masses, which is often necessary in plant metabolomics where many compounds may have similar or overlapping masses. The ability of Q-TOF-MS to analyze complex mixtures with high accuracy makes it ideal for profiling diverse plant metabolomes (Glauser *et al.*, 2013; Zhu *et al.*, 2013). It allows the identification of a wide range of metabolites, from primary metabolites such as amino acids and sugars to secondary metabolites including flavonoids and alkaloids (Farag *et al.*, 2014; Alvarez-Rivera *et al.*, 2019). Q-TOF-MS is sensitive enough to detect low-abundance metabolites, which is essential for comprehensive metabolic profiling in plants. This sensitivity helps in identifying metabolites that are critical for plant responses to environmental changes or stress circumstances (Jorge *et al.*, 2016b).
- Orbitrap MS is known for its exceptional mass resolution and accuracy, which are

Table 4.1. Key high-throughput mass spectrometry (MS) technologies used in plant metabolomics.

Technology	Features	Advantages	Applications in plant metabolomics	References
Q-TOF-MS	Combines quadrupole mass filter with time-of-flight analyzer High resolution and mass accuracy	High resolution for complex mixtures Accurate mass measurements Suitable for untargeted analysis	Metabolic profiling of complex plant samples Identification of diverse metabolites Biomarker discovery	Ibáñez *et al*. (2014); Cajka and Fiehn (2016)
Orbitrap MS	Electrostatic ion trapping Measures ion oscillation frequency	Exceptional mass resolution High mass accuracy Wide mass range	Detailed metabolic profiling Quantification of metabolites Analysis of isomeric compounds	Alves, *et al*., (2013); Glauser *et al*. (2013)
MS/MS	Multiple stages of mass analysis Fragmentation of selected ions	Provides structural information Enhanced sensitivity Helps in the identification of unknown metabolites	Structural elucidation of metabolites Discovery of new metabolites Analysis of metabolic pathways	Ma *et al*. (2006); Krettler and Thallinger (2021); Neagu *et al*. (2022)
GC-MS	Separation of volatile metabolites Identification of metabolites based on mass:charge ratio (*m*/*z*)	Enhanced sensitivity for low-abundance metabolites Development of advanced chromatographic columns	Analysis of primary metabolites (e.g. sugars, organic acids) Profiling of volatile organic compounds	Stashenko and Martínez (2014); Kaur and Sharma (2018); Makoś *et al*. (2019)
LC-MS	Separation of non-volatile metabolites Coupling of LC with MS	Improved separation techniques High-resolution MS integration	Profiling of metabolites including flavonoids and alkaloids Quantitative and qualitative analysis	de Vos *et al*. (2007); Allwood and Goodacre (2010); Hansen and Reubsaet (2015); McCalley (2023)
NMR spectroscopy	Non-destructive analysis Delivers structural information based on nuclear spin properties	Higher sensitivity with advanced NMR techniques Improved resolution and data processing	Structural elucidation of complex metabolites Quantification of metabolites in plant extracts	Krishnan *et al*. (2005); Deborde *et al*. (2017); Selegato *et al*. (2019)
Direct infusion MS	Rapid analysis without chromatographic separation Useful for profiling complex mixtures	Development of high-resolution ionization techniques Enhanced data-acquisition speed	Quick metabolic profiling Identification of metabolites in crude extracts	Bedair and Sumner (2008); Han *et al*. (2009); Southam *et al*. (2017)
Metabolite imaging MS	Spatial distribution of metabolites within plant tissues Mapping of metabolic patterns	Advanced imaging techniques Integration with high-resolution MS for detailed spatial analysis	Visualization of metabolite distribution in plant tissues Study of spatial variations in metabolite levels	Matros and Mock (2013); Sturtevant *et al*. (2016); Ho *et al*. (2017)
HPLC	High-resolution separation of metabolites Versatile separation methods for various compounds	Improved separation efficiency Development of new stationary phases	Separation of complex plant extracts Preparation for subsequent MS analysis	Türkan and Kalay (2006); Perez de Souza *et al*. (2021)

GC-MS, gas chromatography–mass spectrometry; HPLC, high-performance liquid chromatography; LC-MS, liquid chromatography–mass spectrometry; MS, mass spectrometry; MS/MS, tandem mass spectrometry; NMR, nuclear magnetic resonance; Q-TOF-MS, quadrupole time-of-flight–mass spectrometry.

achieved through its unique ion-trapping mechanism (Senyuva *et al.*, 2015). In Orbitrap MS, ions are trapped in an electrostatic field and oscillate at frequencies proportional to their *m/z*, which are then analyzed to produce high-resolution mass spectra (Perry *et al.*, 2008). Orbitrap MS provides extremely high resolution and mass accuracy, enabling the precise determination of molecular weights and the identification of metabolites with high confidence (Yan *et al.*, 2022). This is particularly useful in plant metabolomics for differentiating metabolites with very similar masses and for detailed structural elucidation (Stettin *et al.*, 2020). Orbitrap MS can detect a broad range of metabolites, from small molecules to larger, more complex compounds. This versatility is valuable in plant studies, where the metabolome includes a diverse array of chemical entities (Vinaixa *et al.*, 2016). The high resolution of Orbitrap MS allows detailed analysis of complex plant metabolite spectra, providing insights into metabolite structures and facilitating the identification of unknown compounds. This ability is crucial for understanding plant biochemical pathways and interactions (Rojas-Cherto *et al.*, 2012).

- Tandem MS (MS/MS) concerns multiple stages of mass analysis, where ions generated in the first stage are fragmented in a collision cell and then examined in a second mass analyzer (Neagu *et al.*, 2022). This technique provides valuable structural information about metabolites (Ma *et al.*, 2006). The fragmentation of ions in the collision cell produces fragment ions that can be used to extrapolate the structure of the parent compound. This is especially useful for identifying complex plant metabolites and understanding their chemical structures (Krettler and Thallinger, 2021). By analyzing the fragmentation patterns of metabolites, MS/MS helps in distinguishing between isomers and closely related compounds, which is crucial for accurate metabolite identification in plant studies (Watson, 2013; Oppermann *et al.*, 2012). This detailed structural information aids in confirming the identity of metabolites and elucidating their roles in plant biology. MS/MS can also improve quantitative analysis by providing more specific detection of metabolites (Xiao *et al.*, 2012). The ability to focus on specific precursor ions and their fragment ions enhances the sensitivity and accuracy of quantification, which is important for assessing changes in metabolite levels under different experimental conditions (Ma *et al.*, 2006).

MS technologies such as Q-TOF-MS, Orbitrap MS, and MS/MS each offer unique capabilities that enhance high-throughput plant metabolomics. Q-TOF-MS excels in high-resolution analysis of complex mixtures, Orbitrap MS provides exceptional mass accuracy and resolution for a broad range of metabolites, and MS/MS delivers detailed structural information essential for metabolite identification and quantification (Yu *et al.*, 2021). Together, these advanced MS technologies enable comprehensive metabolic profiling, facilitating in-depth studies of plant metabolism and contributing to advancements in agricultural research and crop improvement (Fig. 4.1).

4.2.2 NMR spectroscopy

NMR spectroscopy is a pivotal technology in plant metabolomics, offering unique advantages such as non-destructive analysis and the provision of detailed structural information (Deborde *et al.*, 2017). Unlike other analytical methods, NMR does not require extensive sample preparation or chemical derivatization, preserving the natural state of the metabolites (Selegato *et al.*, 2019). This characteristic is particularly valuable in plant metabolomics, where complex samples are often analyzed.

Cryogenic probes

Cryogenic probes are a significant advancement in NMR spectroscopy that greatly enhance its performance (Kovacs *et al.*, 2005). These probes operate at extremely low temperatures, typically around 100 K, which significantly increases their sensitivity and resolution (Moskau *et al.*, 2003).

- *Enhanced sensitivity*. Cryogenic probes utilize superconducting magnets and low-temperature techniques to reduce thermal

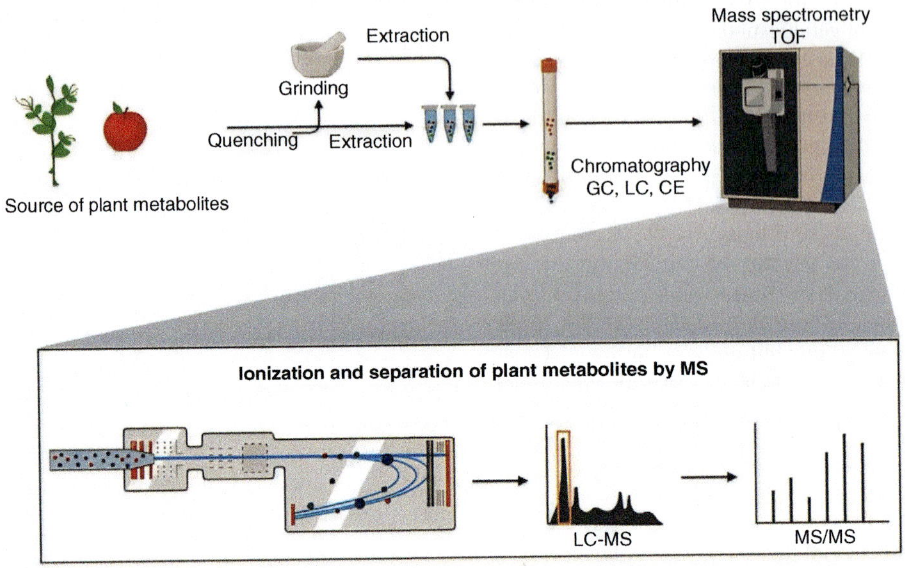

Fig. 4.1. An overview of plant high-throughput metabolomics. CE, capillary electrophoresis; GC, gas chromatography; LC, liquid chromatography; LC-MS, liquid chromatography–mass spectrometry; MS, mass spectrometry; MS/MS, tandem mass spectrometry; TOF, time-of-flight.

noise and enhance the signal:noise ratio (Ramaswamy *et al.*, 2007). This increased sensitivity permits the identification of low-abundance metabolites that might be missed with conventional NMR probes (Spraul *et al.*, 2003). In plant metabolomics, this ability is crucial for analyzing metabolites present in trace amounts, such as rare secondary metabolites or those produced under specific stress conditions (Tian *et al.*, 2016).

- *Improved resolution.* By reducing noise and enhancing signal detection, cryogenic probes provide better spectral resolution (Cloarec *et al.*, 2007). This improvement enables the precise separation of closely related metabolites and the acquisition of clearer spectra, which is essential for detailed metabolic profiling and the identification of complex plant metabolites (Seger and Sturm, 2007).

High-field NMR spectroscopy

High-field NMR spectroscopy refers to the use of powerful magnetic fields, typically greater than 600 MHz, which greatly enhances the resolution and sensitivity of NMR measurements (Moser *et al.*, 2017). High-field NMR instruments offer superior resolution by increasing the power of the magnetic field, which enhances the separation of resonance signals from different metabolites (Salem *et al.*, 2020). This high resolution is crucial for resolving overlapping peaks and distinguishing between metabolites with similar chemical shifts, providing clearer and more detailed spectra (Rolin *et al.*, 2013). The increased magnetic field strength also improves the sensitivity of NMR measurements, allowing the detection of metabolites at lower concentrations (Moser *et al.*, 2017). This sensitivity is especially beneficial for comprehensive metabolomic analyses in plants, where the concentration of some metabolites can be quite low, especially those involved in secondary metabolic pathways or specific developmental stages (Leiss *et al.*, 2011).

High-field NMR facilitates more detailed and accurate metabolomic analyses by providing better spectral resolution and sensitivity (Selegato *et al.*, 2019). This ability enables researchers to obtain more precise metabolic profiles and to recognize and quantify a broader range of metabolites in plant samples (Shachar-Hill, 2002). High-field NMR is instrumental in studying complex metabolic networks and understanding how plants adapt to various environmental conditions or genetic modifications (Shachar-Hill, 2002). NMR spectroscopy, with advancements such as cryogenic probes and high-field NMR, provides powerful tools for high-throughput plant metabolomics. Cryogenic probes enhance perceptiveness and resolution, allowing the identification of low-abundance metabolites, while high-field NMR improves resolution and sensitivity for more detailed and comprehensive metabolomic analyses. These advancements enable researchers to gain deeper insights into plant metabolism, contributing to advances in crop improvement, stress response studies, and overall agricultural research.

4.2.3 Chromatography techniques

Chromatography is a fundamental technique in plant metabolomics, often employed in conjunction with MS or NMR spectroscopy to separate and analyze complex mixtures of metabolites (Marshall and Powers, 2017). Each chromatography technique offers unique advantages suited to different types of metabolites and analytical goals.

Gas chromatography

Gas chromatography (GC) is a powerful method for separating volatile and semi-volatile compounds (Mondello *et al.*, 2008). In GC, the sample is vaporized and taken through a chromatographic column by an inert gas (usually helium or nitrogen), with separation achieved based on the interaction of the analytes with the stationary phase of the column (Kaur and Sharma, 2018). GC is particularly effective for analyzing small, volatile metabolites such as fatty acids, essential oils, and certain secondary metabolites such as terpenes (Kalidasu *et al.*, 2017; Wolfender *et al.*, 2015). In plant metabolomics, GC is used to profile essential oils and other volatiles that play key roles in plant aroma, flavor, and defense mechanisms (Stashenko and Martínez, 2012). GC offers high resolution and sensitivity for the separation of volatile compounds (Qualley and Dudareva, 2009). This high resolution allows for the detection and quantification of intricate mixtures with many overlapping peaks, which is critical for accurate metabolite profiling (Shellie *et al.*, 2005). For compounds that are not naturally volatile, derivatization can be employed to convert them into more volatile forms. This technique is often used in GC to enhance the analysis of metabolites such as fatty acids or sugars that require chemical modification to be analyzed effectively (Fiehn, 2016; Villas-Bôas *et al.*, 2011).

Liquid chromatography

Liquid chromatography (LC) is a versatile technique employed to separate a wide range of metabolites, including those that are non-volatile, polar, or thermally labile (Yergey *et al.*, 2013). In LC, the sample is dissolved in a liquid mobile phase and pumped through a column packed with a solid stationary phase. Separation is achieved based on differential interactions with the stationary phase (Galea *et al.*, 2015). LC can separate a wide variety of metabolites, including large biomolecules such as peptides, nucleotides, and complex carbohydrates, as well as smaller compounds (Wolfender *et al.*, 2015). This versatility makes LC a cornerstone technique in plant metabolomics for analyzing a comprehensive range of plant metabolites. LC is often coupled with MS (LC-MS) to enhance detection and quantification. LC-MS provides detailed information about the mass and structure of metabolites, facilitating the detection and analysis of complex mixtures (Xiao *et al.*, 2012). This combination is invaluable for profiling complex plant metabolomes and understanding metabolic changes under various conditions. High-performance liquid chromatography (HPLC) is a common LC method that uses high pressure to push the mobile phase through the column, improving separation efficiency and resolution (Hansen and Reubsaet, 2015). HPLC is widely used in plant metabolomics for its ability to handle complex samples and provide high-resolution separation of metabolites (Allwood and Goodacre, 2010).

Capillary electrophoresis

Capillary electrophoresis (CE) is a method that separates compounds based on their m/z as they migrate through a capillary tube under the influence of an electric field (Poole, 2018). CE is known for its high separation efficiency and resolution (Ramautar *et al.*, 2006). CE provides excellent separation efficiency due to the narrow capillary tubes and the high voltage applied (Aupiais and Chartier, 2020). This high efficiency allows for the precise separation of closely related metabolites, making CE suitable for analyzing charged metabolites such as amino acids and polar compounds (Kok *et al.*, 2014; Monton and Soga, 2007). CE requires very small amounts of sample, which is advantageous when working with limited quantities of plant material. This feature is particularly useful in plant metabolomics, where sample availability can be a constraint (Khakimov *et al.*, 2014). CE can be coupled with MS (CE-MS) to combine the high separation efficiency of CE with the detailed detection and quantification capabilities of MS (Týčová *et al.*, 2017). This integration enhances the analysis of complex plant samples and improves the resolution of metabolite profiles (Waris *et al.*, 2022). GC, LC, and CE each offer distinct advantages for separating and analyzing plant metabolites. GC is ideal for volatile compounds, LC provides versatility for a broad range of metabolites, and CE excels in separating charged metabolites with high efficiency. When combined with MS or NMR, these chromatography techniques facilitate comprehensive metabolic profiling and contribute to a deeper understanding of plant biology, improving crop quality, yield, and resilience (Kueffer, 2019).

4.3 Sample Preparation and Data Acquisition

4.3.1 Sample preparation

Efficient and reproducible sample preparation is essential for high-throughput metabolomics, as it directly influences the accurateness and reliability of the resulting data (Salem *et al.*, 2020). Proper sample preparation ensures that the metabolite profiles obtained are reflective of the biological state of the samples and that comparisons between samples are meaningful (Monteiro *et al.*, 2013). This section discusses the key considerations for sample preparation in high-throughput metabolomics.

Extraction methods

The selection of extraction process is crucial and should be tailored to the nature of the target metabolites (Ernst *et al.*, 2014). Different metabolites have varying solubilities and polarities, so the extraction method must effectively isolate the metabolites of interest while minimizing the extraction of interfering substances (Foito and Stewart, 2018).

- *Polar metabolites.* For polar metabolites such as amino acids, organic acids, and sugars, aqueous or aqueous-organic solvent systems (e.g. water, methanol, acetonitrile) are commonly used (Rodrigues *et al.*, 2021). Techniques such as liquid–liquid extraction or solid-phase extraction may also be employed to concentrate and purify these metabolites (Perez de Souza *et al.*, 2021).
- *Non-polar metabolites.* Non-polar metabolites such as lipids and hydrophobic compounds often require non-polar solvents such as hexane or chloroform for extraction (Aldana *et al.*, 2020). Methods such as Soxhlet extraction or accelerated solvent extraction can be used to ensure comprehensive extraction of these metabolites (Mushtaq *et al.*, 2014; Roopashree and Naik, 2019).

Quenching metabolism

To accurately capture the metabolite profile of a sample, it is vital to rapidly halt ongoing metabolic processes at the moment of sampling (Lämmerhofer and Weckwerth, 2013). This is achieved through quenching methods that quickly stop enzyme activity and metabolic reactions, thereby preserving the metabolic state of the sample (Álvarez-Sánchez *et al.*, 2010). Common quenching techniques include the following:

- *Rapid freezing.* Flash-freezing the sample in liquid nitrogen effectively stops metabolic activity and preserves the metabolite profile (Rodrigues *et al.*, 2019).
- *Acidification.* Adding acidic solvents (e.g. cold perchloric acid or hydrochloric acid) can halt enzymatic reactions by denaturing enzymes and stabilizing metabolites (Lu *et al.*, 2017).

Sample standardization

Consistency in sample preparation is critical for reliable and comparable results across different samples and experimental conditions (Rodrigues *et al.*, 2019). Standardization involves several key practices:

- *Homogenization.* Homogenizing plant tissues or other biological samples ensures uniform sample consistency. This helps in achieving uniform extraction and reduces variability between samples (Salem *et al.*, 2020).
- *Internal standards.* Adding internal standards or isotopically labeled compounds to samples helps to account for variability in sample preparation and extraction efficiency. These standards provide a reference point for quantification and help in normalizing data (Salem *et al.*, 2020).
- *Protocol consistency.* It is important to use standardized protocols for sample handling, extraction, and processing to minimize variation and ensure reproducibility. This includes maintaining consistent conditions such as temperature, pH, and extraction times across all samples (Kuligowski and Quintás, 2021).

Efficient sample preparation is fundamental to high-throughput metabolomics. Tailoring extraction methods to the target metabolites, effectively quenching metabolism to capture the true metabolome, and ensuring sample standardization are critical considerations that enhance the dependability and reproducibility of metabolomic investigations (Allwood *et al.*, 2011). By carefully addressing these aspects, researchers can obtain accurate and meaningful metabolite profiles, facilitating robust and insightful analyses in plant metabolomics and beyond.

4.3.2 Data acquisition

Modern high-throughput metabolomics is characterized by its ability to generate and handle vast quantities of complex data, driven by substantial advancements in data-acquisition technologies (Guo *et al.*, 2022). These advancements enhance the efficiency, accuracy, and scope of metabolomic analyses. Key advancements in data acquisition include automated sample handling, high-resolution data acquisition, and real-time analysis.

Automated sample handling

Automated sample handling systems have revolutionized high-throughput metabolomics by significantly reducing human error and increasing the overall throughput of analyses. Automation minimizes manual intervention in sample preparation, injection, and processing, leading to higher consistency and reproducibility across experiments. Robotic systems handle tasks such as liquid dispensing, sample mixing, and injection into analytical instruments with precision and speed, reducing variability and potential errors (Bertić *et al.*, 2023).

Automated systems enable the processing of a considerable number of samples in a shorter time frame. This increased throughput is crucial for large-scale metabolomic studies, such as those involving extensive plant sample collections or high-resolution profiling of metabolic changes across different conditions (Zampieri *et al.*, 2017; Malinowska *et al.*, 2022). Automated sample handling is often integrated with data-acquisition systems, allowing seamless operation from sample preparation to data collection (Fuhrer and Zamboni, 2015). This integration ensures efficient workflow management and real-time data acquisition, optimizing the overall analytical process (Malinowska *et al.*, 2022).

High-resolution data acquisition

High-resolution data acquisition is a cornerstone of modern metabolomics, providing detailed and accurate information about metabolite structures and concentrations (Junot *et al.*, 2014; Jones, 2018). High-resolution MS and high-field NMR are examples of technologies that deliver high-resolution data (Vinay *et al.*, 2021). These technologies allow the precise measurement of metabolite masses and the detection of minor peaks in complex mixtures. This level of detail is essential for distinguishing among metabolites with similar masses and for identifying novel or low-abundance metabolites (Junot *et al.*, 2014; Jones, 2018).

High-resolution data acquisition enhances the accuracy of metabolite quantification by providing better peak resolution and reducing interference from overlapping signals (Evans *et al.*, 2014). This accuracy is critical for reliable metabolic profiling and for drawing meaningful conclusions from the data (Ma and Chowdhury, 2013). It also enhances the ability to obtain structural information about metabolites (Wagner *et al.*, 2013). For instance, MS/MS can provide fragmentation patterns that aid in identifying and characterizing complex metabolites, while high-resolution NMR can reveal detailed molecular structures (Wang *et al.*, 2017).

Real-time analysis

Real-time analysis refers to the ability to process and interpret data as they are being acquired, allowing immediate feedback and adjustments during experiments (Majchrzak *et al.*, 2020). Real-time analysis systems can process data while in progress, providing instant results and enabling researchers to monitor their ongoing experiments. This ability is beneficial for dynamic studies where rapid changes in metabolite levels need to be tracked, such as in response to environmental stresses or experimental treatments (Heinemann *et al.*, 2014).

Real-time data feedback allows researchers to make on-the-spot adjustments to experimental conditions or analytical parameters (Wang *et al.*, 2024). For example, if preliminary results indicate unexpected metabolite profiles, researchers can modify sample preparation or analysis protocols to address these findings and refine their experimental approach. Immediate access to data helps in making informed decisions during the experimental phase, enhancing the efficiency of the research strategy and potentially leading to more rapid discoveries and insights (Uthe *et al.*, 2021). Advancements in automated sample handling, high-resolution data acquisition, and real-time analysis have significantly enhanced the abilities of high-throughput metabolomics (Zampieri *et al.*, 2017; Hall *et al.*, 2022). These advancements contribute to increased efficiency, accuracy, and insight in metabolic analyses, enabling researchers to handle large-scale studies, achieve detailed metabolite profiling, and make timely adjustments to experimental protocols. Together, these technologies support the comprehensive exploration of metabolic networks and contribute to advancements in plant metabolomics and other fields.

4.4 Data Analysis and Interpretation

4.4.1 Metabolite identification

Accurate detection of metabolites is a fundamental characteristic of high-throughput metabolomics, as it enables researchers to link observed data to specific biochemical compounds and gain insights into metabolic processes (Peters *et al.*, 2018). The identification process typically involves comparing experimental data with established references and utilizing computational tools to interpret complex data (Weckwerth, 2008; Wishart, 2009a,b). Key tools and databases for metabolite identification are discussed below.

Metabolite databases

Metabolite databases are essential resources that provide comprehensive reference information for determining metabolites established on experimental data (Bais *et al.*, 2010). These databases include the following:

- The Human Metabolome Database (HMDB) is a valuable resource having detailed information on human metabolites, including their chemical structures, biological functions, and spectral data (Forsythe and Wishart, 2009). While focused primarily on human metabolism, it also contains data relevant to plant and microbial metabolites (Wishart *et al.*, 2009). The HMDB provides reference spectra and detailed profiles that assist the detection of metabolites observed in experimental samples (Neumann and Böcker, 2010).
- METLIN is a comprehensive metabolite database that includes a vast array of metabolite spectra from different organisms, including plants (Sangwan *et al.*, 2015). It offers reference mass spectra, fragmentation patterns, and chemical properties of metabolites (Smith *et al.*, 2005). METLIN is especially useful for matching experimental data with known metabolites and identifying compounds based on their mass and fragmentation patterns (Guijas *et al.*, 2018).
- The Kyoto Encyclopedia of Genes and Genomes (KEGG) is a widely employed database that delivers knowledge on metabolic paths,

enzyme functions, and compound structures (Kanehisa *et al.*, 2016). It includes data on various metabolites and their roles in biological processes (Kanehisa *et al.*, 2014). KEGG's pathway maps and compound annotations facilitate the identification of metabolites and help in understanding their functional context within metabolic networks (Altman *et al.*, 2013).

In silico *prediction tools*

In silico prediction tools complement experimental data by providing computational approaches to metabolite identification, particularly for novel or unknown compounds (Blaženović *et al.*, 2018). These tools use algorithms and databases to predict metabolite structures and properties based on known chemical information (Chagas-Paula *et al.*, 2015; Krettler and Thallinger, 2021). Key features of *in silico* prediction tools include the following:

- *Chemical property prediction.* Tools such as ChemDraw or ChemSketch use chemical structure prediction algorithms to generate possible structures of unknown metabolites established on their observed mass spectra and fragmentation patterns (Jones *et al.*, 2021; Sherlin *et al.*, 2022). These tools can suggest potential candidates and guide further experimental validation (Nahar and Sarker, 2024).
- *Spectral matching. In silico* tools can compare experimental spectra with theoretical spectra generated from chemical structure databases (Krettler and Thallinger, 2021). This comparison helps in identifying metabolites by matching observed fragmentation patterns, *m/z*, and retention times with predicted values (Neumann and Böcker, 2010).
- *Machine learning algorithms.* Advanced machine learning and artificial intelligence algorithms are increasingly employed to predict metabolite identities from complex data (Chi *et al.*, 2024). These algorithms can analyze large data sets to identify patterns and correlations, offering predictions of metabolite identities based on known chemical properties and spectral data (Tsugawa *et al.*, 2021).
- *Database-integrated search engines.* Tools such as METLIN's search engine or Global Natural Products Social Molecular Networking (GNPS) integrate multiple databases and prediction algorithms to offer comprehensive search capabilities (Misra, 2021). These platforms allow researchers to query experimental data against extensive metabolite libraries and prediction models, enhancing the accuracy of metabolite identification (Misra and Mohapatra, 2019).

The accurate identification of metabolites in high-throughput metabolomics relies on a combination of comprehensive metabolite databases and advanced *in silico* prediction tools. Metabolite databases such as HMDB, METLIN, and KEGG provide valuable reference information and spectral data for matching and identifying known metabolites, while *in silico* tools offer predictive capabilities for novel or unknown compounds. Together, these resources enable researchers to effectively interpret complex metabolomic data, identify key metabolites, and gain insights into metabolic processes.

4.4.2 Quantification and normalization

Quantification and normalization are critical steps in metabolomics for accurately measuring metabolite concentrations and ensuring that the data are comparable across different samples and experimental conditions (de Livera *et al.*, 2012; Wu and Li, 2016). These steps are essential for deriving meaningful biological insights and ensuring the reliability and reproducibility of metabolomic studies. Key approaches for quantification and normalization include the use of internal standards and normalization techniques.

Internal standards

Internal standards are chemicals added to samples in known quantities to account for variations in sample preparation, extraction, and analytical procedures (Salem *et al.*, 2020). They serve as reference points to correct for technical variability and enhance the accuracy of metabolite quantification (Mattoli *et al.*, 2023). Key aspects of using internal standards include the following:

- *Selection of appropriate standards.* Internal standards should be chemically identical to the metabolites of interest but not noted in the biological sample. This ensures that they undergo similar extraction, ionization, and detection processes as the target metabolites, providing a reliable basis for correction (de Vos *et al.*, 2007; Roberts *et al.*, 2012).
- *Spike-in method.* Internal standards are added to each sample at a consistent concentration before sample preparation begins (de Vos *et al.*, 2007). By comparing the measured concentration of the internal standard in each sample with its known concentration, researchers can correct for losses and variations during sample handling and analysis (Barnes *et al.*, 2016; Wu and Li, 2016).
- *Correction of analytical variability.* Internal standards help correct for fluctuations in instrument performance, such as variations in ionization efficiency, detector sensitivity, and chromatographic retention times (Ernst *et al.*, 2014). This correction ensures that the measured metabolite concentrations accurately reflect their true abundance in the samples (Alseekh *et al.*, 2021).

Normalization techniques

Normalization techniques are employed to correct for systematic variations between samples that may arise from differences in sample concentration, extraction efficiency, or instrument performance (Dudzik *et al.*, 2018). Effective normalization ensures that the data are comparable across different samples and conditions (de Livera *et al.*, 2012). Key normalization techniques include the following:

- *Total ion chromatogram (TIC) normalization.* TIC normalization involves adjusting the measured metabolite concentrations based on the total ion current detected in each sample (Wu and Li, 2016; Misra, 2020). The TIC represents the sum of all ion intensities in a given chromatogram or mass spectrum (Zhang and Zhao, 2014). By normalizing metabolite intensities to the TIC, variations in sample injection volumes, instrument sensitivity, and overall sample concentration are corrected, making the data more comparable across samples (Rambla *et al.*, 2015).
- *Normalization to biomass or tissue weight.* In plant metabolomics, normalizing metabolite concentrations to the biomass or tissue weight from which the metabolites were extracted is a common practice (Jorge *et al.*, 2016a). This approach ensures that variations in the amount of biological material used for extraction do not affect the measured metabolite concentrations, allowing for more accurate comparisons between samples (Wu and Li, 2016).
- *Normalization to housekeeping metabolites.* Similar to the use of housekeeping genes in transcriptomics, certain metabolites that are consistently present at stable levels across samples can be used as internal references for normalization (Creydt and Fischer, 2020). These housekeeping metabolites serve as benchmarks to adjust for variations in sample processing and analysis, ensuring that changes in metabolite levels reflect true biological differences (Wu and Li, 2016).
- *Quantile normalization.* This statistical technique involves adjusting the distribution of metabolite intensities across samples to achieve a common distribution (Ren *et al.*, 2015). Quantile normalization is particularly useful when comparing metabolomic data from different batches or experimental conditions, as it helps to reduce batch effects and technical variability (Reinhold *et al.*, 2019).

4.4.3 Statistical analysis

High-throughput metabolomics generates complex and large-scale data sets that require robust statistical methods to interpret the data accurately and derive meaningful biological insights (Hall *et al.*, 2022). The key statistical methods used in metabolomics include multivariate analysis and machine learning methods.

Multivariate analysis

Multivariate analysis encompasses a range of statistical techniques designed to handle data sets with multiple variables simultaneously (Worley and Powers, 2013). In metabolomics, these methods are essential for identifying patterns, correlations, and differences in metabolite profiles across samples.

- *Principal component analysis (PCA)*. PCA is an unsupervised multivariate approach that decreases the dimensionality of the data while preserving the variance (Weckwerth, 2008). It transforms the actual variables into a new set of uncorrelated variables named principal components. PCA assists in visualizing the overall structure of the data, identifying outliers, and revealing underlying patterns, such as the clustering of samples based on their metabolic characteristics (Ren *et al.*, 2015).
- *Partial least squares discriminant analysis (PLS-DA)*. PLS-DA is a supervised technique that extends PLS regression to classify samples based on their metabolomic profiles (Ren *et al.*, 2015; Ghosh *et al.*, 2020). It identifies the variables (metabolites) that give most to the separation between predefined groups, such as treated versus control samples (Okada *et al.*, 2010). PLS-DA is particularly useful for biomarker discovery and understanding the metabolic differences between conditions (Kalivodová *et al.*, 2015).

Machine learning

Machine learning techniques are increasingly used in metabolomics to build predictive models, classify samples, and uncover complex patterns in the data (Liebal *et al.*, 2020). These algorithms can handle high-dimensional data and capture non-linear relationships between variables (Labory *et al.*, 2024).

- *Random forests*. Random forests are ensemble learning techniques that create considerable decision trees during training and output the mode of the classes for classification (Liebal *et al.*, 2020). They are robust to overfitting and can handle large data sets with many variables, making them suitable for identifying important metabolites and classifying samples based on metabolomic profiles (Liebal *et al.*, 2020).
- *Support vector machines* (SVMs). SVMs comprise a governed learning algorithm that locates the optimal hyperplane that separates classes in a high-dimensional space (Galal *et al.*, 2022). It is adequate for handling non-linear relationships through the use of kernel functions (Mendez *et al.*, 2019). SVMs are used in metabolomics for sample classification, biomarker identification, and predicting outcomes based on metabolomic data (Liebal *et al.*, 2020).

4.4.4 Pathway analysis

Pathway analysis is crucial for interpreting metabolomic data within a biological context. It helps understand how changes in metabolite levels affect metabolic pathways and overall cellular functions (Bhalla *et al.*, 2005). Key tools and approaches for pathway investigation include the following:

- *Pathway enrichment analysis*. This identifies metabolic paths that are greatly affected in the data set. This approach involves comparing the observed metabolite changes with comprehended metabolic pathways to determine which pathways are over-represented or under-represented (Janz *et al.*, 2010).
- *Metabolite set enrichment analysis (MSEA)*. MSEA assesses whether predefined sets of metabolites (metabolite sets) are particularly enriched in the data (Persicke *et al.*, 2012). By comparing the observed data with reference metabolite sets associated with specific pathways, MSEA highlights pathways that are impacted by the experimental conditions (Lamichhane *et al.*, 2018).
- *Over-representation analysis*. This tests whether specific pathways contain more differentially abundant metabolites than expected by chance (Johnson *et al.*, 2015). This method helps pinpoint pathways that are significantly affected, providing insights into the biological processes underlying the observed metabolomic changes (Eicher *et al.*, 2020).

- *Network analysis*. Network analysis visualizes and interprets the relationships between metabolites and other molecular entities, such as genes, proteins, and enzymes (Toubiana *et al*., 2013). It helps elucidate the complex relationships within the metabolome and integrates metabolomic data with other omics data sets (Toubiana *et al*., 2013).
- *Metabolic network reconstruction*. This approach involves building comprehensive maps of metabolic pathways based on experimental data and known biochemical reactions (Töpfer *et al*., 2015). Metabolic network reconstruction provides a detailed view of the metabolic processes and how metabolites interact within the cell (Stitt *et al*., 2010).
- *Correlation networks*. These represent the relationships between metabolites based on their co-abundance patterns (McClure *et al*., 2020). Nodes denote metabolites and edges symbolize noteworthy correlations between them. Correlation networks help identify co-regulated metabolites and potential regulatory mechanisms (Allen *et al*., 2010).

Summary

Robust statistical analysis and pathway analysis are essential components of high-throughput metabolomics (Bartel *et al*., 2013). Multivariate analysis techniques such as PCA and PLS-DA help identify patterns and correlations in complex data sets, while machine learning algorithms such as random forests and SVMs classify samples and predict outcomes based on metabolomic profiles (Liebal *et al*., 2020). Pathway enrichment analysis and network analysis provide biological context to the data, highlighting affected pathways and visualizing interactions between metabolites and other molecular entities (Barupal *et al*., 2012). Together, these techniques enable a comprehensive understanding of metabolic changes and their biological implications.

4.5 Applications of High-throughput Plant Metabolomics

High-throughput metabolomics is transforming plant research by providing deep insights into plant physiology, stress responses, crop improvement, and environmental interactions (Razzaq *et al*., 2022). Some of the key applications of this powerful technology in plant research are discussed in this section.

4.5.1 Plant physiology and development

High-throughput metabolomics is pivotal in understanding the intricate metabolic changes that underpin plant growth and development (Allwood *et al*., 2021). Specific applications include analysis of hormone profiling and nutrient metabolism.

- *Hormone profiling*. Phytohormones, including auxins, abscisic acid, gibberellins, ethylene, and cytokinins, play an essential function in regulating plant development and responses to environmental stimuli (Alferez *et al*., 2021). Metabolomics enables the comprehensive profiling of these hormones, helping researchers decipher their biosynthesis, metabolism, and signaling pathways (Šimura *et al*., 2018). By analyzing phytohormone levels and their dynamic changes during different developmental stages or in response to environmental cues, investigators can gain insights into their roles in processes such as germination, flowering, fruiting, and senescence (Šimura *et al*., 2018; Chang and Lv, 2024).
- *Nutrient metabolism*. Nutrient availability and allocation are critical for plant growth and development (Ericsson, 1995). Metabolomics allows a detailed investigation of how plants absorb, transport, and utilize essential nutrients such as nitrogen, potassium, phosphorus, and trace elements (Lai *et al*., 2020). By profiling primary and secondary metabolites, researchers can elucidate the metabolic pathways involved in nutrient assimilation and the interplay between nutrient status and metabolic processes (Sharma *et al*., 2021). This information is vital for optimizing fertilization practices and improving crop yields.

4.5.2 Stress responses

Plants encounter a wide range of abiotic and biotic stresses, and metabolomics can reveal the metabolic adjustments they make to cope with these challenges. Key applications include the following:

- *Drought and salinity.* Water scarcity and high salinity are major abiotic stresses that limit plant productivity (Majeed and Muhammad, 2019). Metabolomics helps identify the metabolites involved in osmoprotection, such as osmolytes (e.g. proline, glycine betaine) and antioxidants (e.g. ascorbate, glutathione) (Arbona *et al.*, 2013; Anzano *et al.*, 2022). It also sheds light on stress signaling molecules, including abscisic acid and reactive oxygen species, and their roles in activating stress response pathways (Khan *et al.*, 2017). Understanding these metabolic adaptations can inform breeding techniques for producing drought- and salt-tolerant crop varieties.
- *Pathogen resistance.* Plants have evolved sophisticated defense mechanisms against pathogens involving a complex network of metabolic changes (Castro-Moretti *et al.*, 2020). Metabolomics enables the characterization of these changes, including the assembly of defense-related metabolites such as phytoalexins, pathogenesis-related proteins, and signaling molecules such as salicylic acid and jasmonic acid (Allwood *et al.*, 2021). By determining the metabolic profiles of resistant and susceptible plant varieties, investigators can identify key metabolites and pathways associated with disease resistance, guiding the development of resistant crops (Arbona and Gomez-Cadenas, 2016).

4.5.3 Crop improvement

Metabolomics is a valuable tool for crop improvement, offering insights into metabolic traits that can be harnessed to enhance crop quality and productivity (Razzaq *et al.*, 2022). Key applications include metabolic engineering and phenotyping.

- *Metabolic engineering.* Metabolic engineering involves manipulating metabolic pathways to improve the production of desirable compounds, including vitamins, antioxidants, and secondary metabolites with health benefits (Dixit *et al.*, 2021). Metabolomics provides a detailed map of the metabolic networks, allowing researchers to identify target pathways and assess the influence of genetic modifications on the metabolome (Fiehn, 2001). This approach can direct the development of crops with improved nutritional profiles, enhanced flavors, and increased resistance to pests and diseases (Kumar *et al.*, 2017).
- *Phenotyping.* High-throughput metabolomics facilitates the phenotyping of large populations of plants by providing comprehensive metabolic profiles (Hall *et al.*, 2022). These profiles can be used to identify biomarkers related to desirable traits, such as yield, stress tolerance, and other quality attributes (Dawid and Hille, 2018). By integrating metabolomic data with genomic and phenotypic information, breeders can select plants with optimal metabolic traits, accelerating the breeding method and improving the efficiency of crop progress programs (Litvinov *et al.*, 2021).

4.5.4 Environmental interactions

Plants interact with their environment in complex ways, and metabolomics can help unravel these interactions, contributing to our understanding of plant ecology and environmental sustainability (van Dam and van der Meijden, 2011). Examples include interactions with the soil microbiome and pollutants.

- *Soil microbiome.* Plants release a combination of metabolites into the soil, influencing the composition and activity of soil microbial communities (Chen *et al.*, 2021). Metabolomics allows the characterization of these root exudates and their impact on microbial communities, shedding light on plant–microbe interactions (Manriquez *et al.*, 2021). This knowledge can inform strategies to promote beneficial microbes, enhance nutrient uptake, and improve soil health, leading to more sustainable agricultural practices (Bertola *et al.*, 2021).
- *Pollution and phytoremediation.* Plants can uptake and metabolize pollutants from the environment, a process known as

phytoremediation (Yadav *et al.*, 2023). Metabolomics helps identify the metabolic paths concerned in the detoxification and sequestration of pollutants such as heavy metals, organic contaminants, and excess nutrients (Singh *et al.*, 2016). By understanding these processes, researchers can develop plants with enhanced phytoremediation capabilities, contributing to environmental cleanup and pollution mitigation (Agarwal and Rani, 2022).

4.6. Future Directions in Plant Metabolomics

The field of plant metabolomics is growing rapidly, driven by technological advancements that are opening new frontiers for research and application. This section discusses some of the key areas where these advancements are poised to make significant impacts.

4.6.1 Technological advancements

Technological innovations are at the forefront of the next wave of breakthroughs in plant metabolomics (Aharoni *et al.*, 2023). These improvements are enhancing our ability to analyze, visualize, and integrate metabolic data with other biological information, providing deeper insights into plant biology and paving the way for new applications (Sharma *et al.*, 2021).

Single-cell metabolomics

Single-cell metabolomics represents a cutting-edge approach to understanding cellular heterogeneity by analyzing the metabolic profiles of individual cells (Prasad *et al.*, 2022). This technology is crucial for uncovering cell-specific metabolic responses that are often masked in bulk tissue analyses (Dolatmoradi *et al.*, 2022). In plant research, single-cell metabolomics can reveal the metabolic diversity among different cell types within a tissue (de Souza *et al.*, 2020). It uncovers unique metabolic pathways active in specialized cells, such as guard cells, trichomes, or root hair cells, and enhances our understanding of cell differentiation and development processes by linking metabolic changes to gene expression and protein synthesis (Dong *et al.*, 2023). Technologies such as MS imaging and advanced single-cell isolation techniques (e.g. laser-capture microdissection, microfluidics) are pivotal in pushing the boundaries of single-cell metabolomics (Taylor *et al.*, 2021).

Spatial metabolomics

Spatial metabolomics involves mapping the dispersal of metabolites within plant tissues, providing a precise view of metabolic dynamics in their anatomical context (Feenstra *et al.*, 2017). This approach can identify tissue-specific metabolic pathways and their regulation. It reveals the spatial organization of metabolic networks, which is critical for understanding processes such as nutrient transport, signaling, and defense mechanisms, and aids in visualizing how environmental factors, such as light, temperature, and stress, influence metabolic activities at the tissue level (Katam *et al.*, 2022). Advanced imaging techniques, such as MS imaging, desorption electrospray ionization (DESI), and matrix-assisted laser desorption/ionization (MALDI), are instrumental in spatial metabolomics (Siyuan Ma *et al.*, 2023). These techniques enable high-resolution visualization of metabolite distribution, enhancing our understanding of plant metabolism.

Integration with other omics

Integrating metabolomics with other omics technologies such as genomics, transcriptomics, and proteomics offers a holistic view of plant biology (Yang *et al.*, 2021). This multi-omics approach can provide an extensive understanding of the regulatory networks governing plant metabolism and facilitate the identification of key genes, transcripts, and proteins related to specific metabolic traits (Martin *et al.*, 2024). This approach enhances the discovery of biomarkers for plant breeding and stress resilience by linking metabolic changes to genetic and proteomic data (Raza, 2022).

Bioinformatics tools and systems biology approaches are crucial for managing and integrating multi-omics data (Subramanian *et al.*, 2020). This integration can lead to more accurate models of metabolic pathways and networks, driving innovations in plant science and agriculture (Kumar *et al.*, 2016).

4.6.2 Data integration and management

The high-throughput nature of modern metabolomics creates vast amounts of complex data. Efficient handling, integration, and interpretation of these data sets are crucial for deriving meaningful biological insights (Zhang *et al.*, 2012). Future efforts in this area will focus on the use of bioinformatics and data sharing.

Bioinformatics tools

Developing more sophisticated bioinformatics tools is essential for managing the complexity of metabolomic data (Cambiaghi *et al.*, 2017). These tools will need to address various challenges, including the following:

- *Data preprocessing.* This comprises methods for noise reduction and peak alignment to improve data quality (Wei *et al.*, 2012; Vettukattil, 2015). Advanced algorithms and machine learning techniques will play a crucial role in automating these processes, making them more efficient and accurate (Sarker, 2021).
- *Feature extraction and annotation.* Improving algorithms for identifying and annotating metabolites from raw data is critical (Chaleckis *et al.*, 2019). Leveraging advances in machine learning and artificial intelligence can help automate the identification of metabolic features, reduce manual intervention, and increase the accuracy and speed of annotation (Ahmed *et al.*, 2020).
- *Integration with other omics.* Creating platforms that seamlessly integrate metabolomic information with genomic, transcriptomic, and proteomic data is vital for comprehensive multi-omics analyses (O'Connor *et al.*, 2023). Such integration will facilitate a holistic experience of biological systems by linking metabolic changes to underlying genetic and proteomic alterations (Sweetlove and Fernie, 2005).

Data sharing

Standardized formats and centralized repositories are critical for facilitating data sharing and collaboration within the scientific community. Future efforts will include the following:

- *Standardized data formats.* It will be important to develop and adopt universal data formats that ensure consistency and interoperability across different studies and platforms (González Morales and Orrell, 2018). This standardization will make it more comfortable to compare and integrate data from various sources, enhancing the robustness of metabolomic studies (Rocca-Serra *et al.*, 2016).
- *Public repositories.* Expanding and enhancing repositories such as MetaboLights, HMDB, and other public databases will be important to store and share metabolomic data (Haug *et al.*, 2017). These repositories promote transparency, reproducibility, and allow researchers to validate findings and build on each other's work (Stodden *et al.*, 2016).
- *Collaborative platforms.* Platforms need to be established that enable researchers to collaboratively analyze and interpret metabolomic data (Sud *et al.*, 2016). These platforms will foster open science and innovation by providing tools for data visualization, statistical investigation, and pathway mapping, enabling researchers to share insights and develop new hypotheses collaboratively (Levine *et al.*, 2020).

Efficient data integration and management are foundational for advancing the field of metabolomics. By developing sophisticated bioinformatics tools, standardizing data formats, and enhancing data-sharing practices, the scientific community can maximize the influence of metabolomic investigation, leading to deeper biological insights and more robust applications in plant science and beyond.

4.6.3 Standardization and reproducibility

Ensuring reproducibility and standardization is crucial for the credibility and reliability of metabolomic studies (Rodrigues *et al.*, 2019). By focusing on standardization and reproducibility, the field of plant metabolomics can produce more reliable and comparable results, advancing our understanding of plant biology and improving applications in agriculture (Saito and Matsuda, 2010). Standardized protocols and rigorous quality control (QC) measures are essential for building a solid foundation for future research and ensuring that findings are robust and reproducible across different studies and laboratories (Fiehn *et al.*, 2008).

Standard operating procedures

Developing and widely adopting standard operating procedures for every stage of metabolomic research is essential.

- *Sample preparation.* This involves standardizing methods for sample collection, storage, and extraction to minimize variability and ensure consistency. This involves clearly defined procedures for handling different types of plant tissues, the use of specific solvents for extraction, and the timing and conditions for sample storage (Salem *et al.*, 2020).
- *Data acquisition.* It is important to harmonize procedures for instrument calibration, data-acquisition settings, and QC to enhance data comparability (Mosley *et al.*, 2024). Regular calibration of instruments such as mass spectrometers and NMR spectrometers ensures that data are consistent over time and across different laboratories (Deborde *et al.*, 2017).
- *Data analysis.* Establishing standardized workflows for data processing, normalization, and statistical analysis will ensure reproducible results (Uthe *et al.*, 2021). This includes setting clear guidelines for the use of software tools, algorithms for peak detection and alignment, and statistical methods for data interpretation (Sumner *et al.*, 2007).

Quality control

Implementing rigorous QC measures is vital for maintaining the integrity of metabolomic data. These measures include the following:

- *Internal standards.* It is important to use internal standards to monitor and update for variations in sample preparation and tool performance (van der Kloet *et al.*, 2009). Internal standards are substances that are added in known quantities to samples, blanks, and calibration standards to track and correct for any variations during the analytical process (Gonzalez-Riano *et al.*, 2019).
- *QC samples.* QC samples should be regularly analyzed alongside experimental samples to assess data reliability and instrument stability (Fiehn *et al.*, 2008). QC samples are typically prepared by pooling a small portion of each sample in the study to create a representative mixture that is run at regular intervals to monitor instrument performance (Dudzik *et al.*, 2018).
- *Data validation.* Applying robust validation techniques, such as cross-validation and external validation with independent data sets, will ensure the accuracy of metabolite identification and quantification (Hageman *et al.*, 2017; Rha *et al.*, 2021). Cross-validation concerns dividing the data into subsets, training the model on some subsets, and testing it on others to ensure robustness (Hoekenga, 2008). External validation requires using completely independent data sets to confirm findings (Rey *et al.*, 2020).

4.7 Conclusion

The field of plant metabolomics has experienced significant growth and transformation over recent decades, largely due to the integration and advancement of high-throughput technologies. As plants produce a vast array of metabolites that are central to their growth, development, and adaptation to environmental stresses, the ability to comprehensively analyze these small molecules has become increasingly important. The technological advancements in high-throughput metabolomics, particularly in MS and its associated techniques, have opened new frontiers in plant science, allowing researchers to explore plant metabolism with unprecedented detail and scale.

MS has emerged as a cornerstone technology in plant metabolomics, offering powerful abilities for the identification and quantification of metabolites. The development of Q-TOF-MS,

Orbitrap MS, and MS/MS has particularly improved our ability to analyze complex plant metabolomes. Q-TOF-MS, with its combination of high resolution and mass accuracy, is ideal for the analysis of complex mixtures, allowing the precise identification of a wide range of metabolites in plant tissues. Orbitrap MS, known for its exceptional mass resolution and accuracy, enables the detection of even the most challenging metabolites, including isomers and other closely related compounds. MS/MS provides critical structural information that aids in the detection of unknown metabolites, offering insights into the intricate biochemical pathways within plants.

These advancements have not only improved the accuracy and resolution of metabolomic analyses but have also increased throughput, enabling large-scale studies that were previously unimaginable. High-throughput metabolomics allows the simultaneous analysis of hundreds to thousands of metabolites across multiple samples, providing a comprehensive view of plant metabolism. This is particularly important for understanding how plants respond to various stresses, such as drought, salinity, and pathogen attack, as well as for investigating the metabolic changes that occur during different stages of plant growth and development. The integration of these advanced technologies has also facilitated the finding of new metabolites and metabolic paths, contributing to our knowledge of plant biology at a molecular level. By enabling the detailed analysis of plant secondary metabolites, which are often involved in defense mechanisms and ecological interactions, high-throughput metabolomics has provided new insights into the evolutionary adaptations of plants. Moreover, the ability to link metabolomic data with other omics data, such as genomics, transcriptomics, and proteomics, has paved the way for a more holistic knowledge of plant systems biology.

However, while considerable progress has been created, challenges remain in the field of plant metabolomics. Ensuring the reproducibility and standardization of metabolomic studies is critical, as variations in sample preparation, acquisition of data, and analysis can lead to discrepancies in results. The development of robust bioinformatics tools and databases is essential for the accurate understanding of metabolomic data and for the integration of metabolomics with omics techniques. The emergence of single-cell metabolomics and spatial metabolomics shows great promise for understanding the heterogeneity of plant tissues and the spatial distribution of metabolites within plants. These approaches will provide new insights into the cellular and tissue-specific metabolic processes that drive plant growth, development, and adaptation.

In conclusion, the technological advancements in high-throughput plant metabolomics have revolutionized our ability to study plant metabolism on a large scale. By providing detailed, accurate, and comprehensive analyses of plant metabolomes, these technologies have deepened our understanding of plant biology and opened new avenues for research in plant science, agriculture, and biotechnology. As the field continues to evolve, the integration of metabolomics with other omics technologies and the development of new analytical tools will be crucial in addressing the complex challenges of plant research and in contributing to the sustainable development of agriculture and environmental management.

4.8 Acknowledgment

All the authors are thankful to their respective universities and institutions for their valuable support.

4.9 References

Agarwal, P. and Rani, R. (2022) Strategic management of contaminated water bodies: omics, genome-editing and other recent advances in phytoremediation. *Environmental Technology and Innovation* 27: 102463.

Aharoni, A., Goodacre, R., and Fernie, A.R. (2023) Plant and microbial sciences as key drivers in the development of metabolomics research. *Proceedings of the National Academy of Sciences USA* 120: e2217383120.

Ahmed, Z., Mohamed, K., Zeeshan, S., and Dong, X. (2020) Artificial intelligence with multi-functional machine learning platform development for better healthcare and precision medicine. *Database* 2020: baaa010.

Aldana, J., Romero-Otero, A., and Cala, M.P. (2020) Exploring the lipidome: current lipid extraction techniques for mass spectrometry analysis. *Metabolites* 10: 231.

Alferez, F., de Carvalho, D.U., and Boakye, D. (2021) Interplay between abscisic acid and gibberellins, as related to ethylene and sugars, in regulating maturation of non-climacteric fruit. *International Journal of Molecular Sciences* 22: 669.

Allen, D.R. and McWhinney, B.C. (2019) Quadrupole time-of-flight mass spectrometry: a paradigm shift in toxicology screening applications. *Clinical Biochemist Reviews* 40: 135.

Allen, E., Moing, A., Ebbels, T.M., Maucourt, M., Tomos, A.D., *et al.* (2010) Correlation network analysis reveals a sequential reorganization of metabolic and transcriptional states during germination and gene-metabolite relationships in developing seedlings of *Arabidopsis*. *BMC Systems Biology* 4: 62.

Allwood, J.W. and Goodacre, R. (2010) An introduction to liquid chromatography–mass spectrometry instrumentation applied in plant metabolomic analyses. *Phytochemical Analysis* 21, 33–47.

Allwood, J.W., de Vos, R.C., Moing, A., Deborde, C., Erban, A., *et al.* (2011) Plant metabolomics and its potential for systems biology research: background concepts, technology, and methodology. *Methods in Enzymology* 500, 299–336.

Allwood, J.W., Williams, A., Uthe, H., van Dam, N.M., Mur, L.A., *et al.* (2021) Unravelling plant responses to stress—the importance of targeted and untargeted metabolomics. *Metabolites* 11: 558.

Alseekh, S., Aharoni, A., Brotman, Y., Contrepois, K., d'Auria, J., *et al.* (2021) Mass spectrometry-based metabolomics: a guide for annotation, quantification and best reporting practices. *Nature Methods* 18, 747–756.

Altman, T., Travers, M., Kothari, A., Caspi, R., and Karp, P.D. (2013) A systematic comparison of the MetaCyc and KEGG pathway databases. *BMC Bioinformatics* 14: 112.

Alvarez-Rivera, G., Ballesteros-Vivas, D., Parada-Alfonso, F., Ibañez, E., and Cifuentes, A. (2019) Recent applications of high resolution mass spectrometry for the characterization of plant natural products. *Trends in Analytical Chemistry* 112, 87–101.

Álvarez-Sánchez, B., Priego-Capote, F., and de Castro, M.L. (2010) Metabolomics analysis II. Preparation of biological samples prior to detection. *Trends in Analytical Chemistry* 29, 120–127.

Alves, S., Rathahao-Paris, E., and Tabet, J.C. (2013) Potential of Fourier transform mass spectrometry for high-throughput metabolomics analysis. *Advances in Botanical Research* 67, 219–302.

Anzano, A., Bonanomi, G., Mazzoleni, S., and Lanzotti, V. (2022) Plant metabolomics in biotic and abiotic stress: a critical overview. *Phytochemistry Reviews* 21, 503–524.

Arbona, V. and Gomez-Cadenas, A. (2016) Metabolomics of disease resistance in crops. *Current Issues in Molecular Biology* 19, 13–30.

Arbona, V., Manzi, M., de Ollas, C., and Gómez-Cadenas, A. (2013) Metabolomics as a tool to investigate abiotic stress tolerance in plants. *International Journal of Molecular Sciences* 14, 4885–4911.

Aupiais, J. and Chartier, F. (2020) Capillary electrophoresis. In: Beauchemin, D. (ed.) *Sample Introduction Systems in ICPMS and ICPOES*. Elsevier, Amsterdam, the Netherlands pp. 299–356.

Bais, P., Moon, S.M., He, K., Leitao, R., Dreher, K., *et al.* (2010) PlantMetabolomics.org: a web portal for plant metabolomics experiments. *Plant Physiology* 152, 1807–1816.

Barnes, S., Benton, H.P., Casazza, K., Cooper, S. J., Cui, X., *et al.* (2016) Training in metabolomics research. I. Designing the experiment, collecting and extracting samples and generating metabolomics data. *Journal of Mass Spectrometry* 51, 461–475.

Bartel, J., Krumsiek, J., and Theis, F.J. (2013) Statistical methods for the analysis of high-throughput metabolomics data. *Computational and Structural Biotechnology Journal* 4: e201301009.

Barupal, D.K., Haldiya, P.K., Wohlgemuth, G., Kind, T., Kothari, S.L., *et al.* (2012) MetaMapp: mapping and visualizing metabolomic data by integrating information from biochemical pathways and chemical and mass spectral similarity. *BMC Bioinformatics* 13: 99.

Bedair, M. and Sumner, L.W. (2008) Current and emerging mass-spectrometry technologies for metabolomics. *Trends in Analytical Chemistry* 27, 238–250.

Bertić, M., Zimmer, I. Andrés-Montaner, D., Rosenkranz, M., Kangasjärvi, J., *et al.* (2023) Automatization of metabolite extraction for high-throughput metabolomics: case study on transgenic isoprene-emitting birch. *Tree Physiology* 43, 1855–1869.

Bertola, M., Ferrarini, A., and Visioli, G. (2021) Improvement of soil microbial diversity through sustainable agricultural practices and its evaluation by-omics approaches: a perspective for the environment, food quality and human safety. *Microorganisms* 9: 1400.

Bhalla, R., Narasimhan, K., and Swarup, S. (2005) Metabolomics and its role in understanding cellular responses in plants. *Plant Cell Reports* 24, 562–571.

Blaženović, I., Kind, T., Ji, J., and Fiehn, O. (2018) Software tools and approaches for compound identification of LC-MS/MS data in metabolomics. *Metabolites* 8: 31.

Cajka, T. and Fiehn, O. (2016) Toward merging untargeted and targeted methods in mass spectrometry-based metabolomics and lipidomics. *Analytical Chemistry* 88, 524–545.

Cambiaghi, A., Ferrario, M., and Masseroli, M. (2017) Analysis of metabolomic data: tools, current strategies and future challenges for omics data integration. *Briefings in Bioinformatics* 18, 498–510.

Castro-Moretti, F.R., Gentzel, I.N., Mackey, D., and Alonso, A.P. (2020) Metabolomics as an emerging tool for the study of plant–pathogen interactions. *Metabolites* 10: 52.

Chagas-Paula, D.A., Oliveira, T.B., Zhang, T., Edrada-Ebel, R., and Da Costa, F.B. (2015) Prediction of anti-inflammatory plants and discovery of their biomarkers by machine learning algorithms and metabolomic studies. *Planta Medica* 81, 450–458.

Chaleckis, R., Meister, I., Zhang, P., and Wheelock, C.E. (2019) Challenges, progress and promises of metabolite annotation for LC–MS-based metabolomics. *Current Opinion in Biotechnology* 55, 44–50.

Chang, Y. and Lv, G. (2024) Key role of hormone signal transduction and lipid metabolism in the development of *Nitraria sibirica* leaves: an integrated metabolomic and transcriptomic analysis. *Industrial Crops and Products* 212: 118322.

Chen, L., Schwier, M., Krumbach, J., Kopriva, S., and Jacoby, R.P. (2021) Metabolomics in plant–microbe interactions in the roots. *Advances in Botanical Research* 98, 133–161.

Chi, J., Shu, J., Li, M., Mudappathi, R., and Jin, Y. (2024) Artificial intelligence in metabolomics: a current review. *Trends in Analytical Chemistry* 178: 117852.

Cloarec, O., Campbell, A., Tseng, L.-H., Braumann, U., Spraul, M., *et al.* (2007) Virtual chromatographic resolution enhancement in cryoflow LC–NMR experiments via statistical total correlation spectroscopy. *Analytical Chemistry* 79, 3304–3311.

Creydt, M. and Fischer, M. (2020) Mass-spectrometry-based food metabolomics in routine applications: a basic standardization approach using housekeeping metabolites for the authentication of asparagus. *Journal of Agricultural and Food Chemistry* 68, 14343–14352.

Dawid, C. and Hille, K. (2018) Functional metabolomics—a useful tool to characterize stress-induced metabolome alterations opening new avenues towards tailoring food crop quality. *Agronomy* 8: 138.

de Livera, A.M., Dias, D.A., de Souza, D., Rupasinghe, T., Pyke, J., *et al.* (2012) Normalizing and integrating metabolomics data. *Analytical Chemistry* 84, 10768–10776.

de Souza, L. P., Borghi, M., and Fernie, A. (2020) Plant single-cell metabolomics—challenges and perspectives. *International Journal of Molecular Sciences* 21: 8987.

de Vos, R.C., Moco, S., Lommen, A., Keurentjes, J.J., Bino, R.J., and Hall, R.D. (2007) Untargeted large-scale plant metabolomics using liquid chromatography coupled to mass spectrometry. *Nature Protocols* 2, 778–791.

Deborde, C., Moing, A., Roch, L., Jacob, D., Rolin, D., and Giraudeau, P. (2017) Plant metabolism as studied by NMR spectroscopy. *Progress in Nuclear Magnetic Resonance Spectroscopy* 102, 61–97.

Dixit, S., Shukla, A., Singh, V., and Upadhyay, S.K. (2021) Engineering of plant metabolic pathway for nutritional improvement: recent advances and challenges. In: Upadhyay, S.K. (ed.) *Genome Engineering for Crop Improvement*. Wiley, Hoboken, New Jersey, pp. 351–379.

Dolatmoradi, M., Samarah, L.Z., and Vertes, A. (2022) Single-cell metabolomics by mass spectrometry: opportunities and challenges. *Analysis and Sensing* 2: e202100032.

Dong, Y., Li, S., Wu, H., Gao, Y., Feng, Z., *et al.* (2023) Advances in understanding epigenetic regulation of plant trichome development: a comprehensive review. *Horticulture Research* 10: uhad145.

Dudzik, D., Barbas-Bernardos, C., García, A., and Barbas, C. (2018) Quality assurance procedures for mass spectrometry untargeted metabolomics. a review. *Journal of Pharmaceutical and Biomedical Analysis* 147, 149–173.

Eicher, T., Kinnebrew, G., Patt, A., Spencer, K., Ying, K., *et al.* (2020) Metabolomics and multi-omics integration: a survey of computational methods and resources. *Metabolites* 10: 202.

Ericsson, T. (1995) Growth and shoot:root ratio of seedlings in relation to nutrient availability. *Plant and Soil* 168, 205–214.

Ernst, M., Silva, D.B., Silva, R.R., Vêncio, R.Z., and Lopes, N.P. (2014) Mass spectrometry in plant metabolomics strategies: from analytical platforms to data acquisition and processing. *Natural Product Reports* 31, 784–806.

Evans, A.M., Bridgewater, B., Liu, Q., Mitchell, M., Robinson, R., *et al.* (2014) High resolution mass spectrometry improves data quantity and quality as compared to unit mass resolution mass spectrometry in high-throughput profiling metabolomics. *Metabolomics* 4: 2.

Farag, M.A., Mohsen, M., Heinke, R., and Wessjohann, L.A. (2014) Metabolomic fingerprints of 21 date palm fruit varieties from Egypt using UPLC/PDA/ESI–qTOF-MS and GC–MS analyzed by chemometrics. *Food Research International* 64, 218–226.

Feenstra, A.D., Alexander, L.E., Song, Z., Korte, A.R., Yandeau-Nelson, M.D., *et al.* (2017) Spatial mapping and profiling of metabolite distributions during germination. *Plant Physiology* 174, 2532–2548.

Fiehn, O. (2001) Combining genomics, metabolome analysis, and biochemical modelling to understand metabolic networks. *Comparative and Functional Genomics* 2, 155–168.

Fiehn, O. (2016) Metabolomics by gas chromatography–mass spectrometry: combined targeted and untargeted profiling. *Current Protocols in Molecular Biology* 114, 30.34. 31–30.34. 32.

Fiehn, O., Wohlgemuth, G., Scholz, M., Kind, T., Lee, D.Y., *et al.* (2008) Quality control for plant metabolomics: reporting MSI-compliant studies. *Plant Journal* 53, 691–704.

Foito, A. and Stewart, D. (2018) Metabolomics: a high-throughput screen for biochemical and bioactivity diversity in plants and crops. *Current Pharmaceutical Design* 24, 2043–2054.

Forsythe, I.J. and Wishart, D.S. (2009) Exploring human metabolites using the human metabolome database. *Current Protocols in Bioinformatics* 25, 14.18.11–14.18.45.

Fuhrer, T. and Zamboni, N. (2015) High-throughput discovery metabolomics. *Current Opinion in Biotechnology* 31, 73–78.

Galal, A., Talal, M., and Moustafa, A. (2022) Applications of machine learning in metabolomics: disease modeling and classification. *Frontiers in Genetics* 13: 1017340.

Galea, C., Mangelings, D., and Vander Heyden, Y. (2015) Characterization and classification of stationary phases in HPLC and SFC—a review. *Analytica Chimica Acta* 886, 1–15.

Ghosh, T., Zhang, W., Ghosh, D., and Kechris, K. (2020) Predictive modeling for metabolomics data. *Computational Methods and Data Analysis for Metabolomics* 2104, 313–336.

Gika, H.G., Theodoridis, G.A., Plumb, R.S., and Wilson, I.D. (2014) Current practice of liquid chromatography–mass spectrometry in metabolomics and metabonomics. *Journal of Pharmaceutical and Biomedical Analysis* 87, 12–25.

Glauser, G., Veyrat, N., Rochat, B., Wolfender, J.-L., and Turlings, T.C. (2013) Ultra-high pressure liquid chromatography–mass spectrometry for plant metabolomics: a systematic comparison of high-resolution quadrupole-time-of-flight and single stage Orbitrap mass spectrometers. *Journal of Chromatography A* 1292, 151–159.

González Morales, L.G., and Orrell, T. (2018) *Data Interoperability: A Practitioner's Guide to Joining Up Data in the Development Sector*. United Nations Statistics Division, New York. Available at: https://repository.oceanbestpractices.org/handle/11329/1971 (accessed 28 January 2025).

Gonzalez-Riano, C., Dudzik, D., Garcia, A., Gil-de-la-Fuente, A., Gradillas, A., *et al.* (2019) Recent developments along the analytical process for metabolomics workflows. *Analytical Chemistry* 92, 203–226.

Guijas, C., Montenegro-Burke, J.R., Domingo-Almenara, X., Palermo, A., Warth, B., *et al.* (2018) METLIN: a technology platform for identifying knowns and unknowns. *Analytical Chemistry* 90, 3156–3164.

Guo, J., Yu, H., Xing, S., and Huan, T. (2022) Addressing big data challenges in mass spectrometry-based metabolomics. *Chemical Communications* 58, 9979–9990.

Hageman, J., Engel, B., de Vos, R.C., Mumm, R., Hall, R.D., *et al.* (2017) Robust and confident predictor selection in metabolomics. In: Datta, S. and Mertens, B. (eds) *Statistical Analysis of Proteomics, Metabolomics, and Lipidomics Data Using Mass Spectrometry*. Springer, Cham, Switzerland, pp. 239–257.

Hall, R.D., d'Auria, J.C., Ferreira, A.C.S., Gibon, Y., Kruszka, D., *et al.* (2022) High-throughput plant phenotyping: a role for metabolomics? *Trends in Plant Science* 27, 549–563.

Han, J., Datla, R., Chan, S., and Borchers, C.H. (2009) Mass spectrometry-based technologies for high-throughput metabolomics. *Bioanalysis* 1, 1665–1684.

Hansen, S.H., and Reubsaet, L. (2015) High-performance liquid chromatography (HPLC) and high-performance liquid chromatography–mass spectrometry (LC-MS). In: *Bioanalysis of Pharmaceuticals: Sample Preparation, Separation Techniques, and Mass Spectrometry*. Wiley, Chichester, UK, pp. 123–172.

Haug, K., Salek, R.M., and Steinbeck, C. (2017) Global open data management in metabolomics. *Current Opinion in Chemical Biology* 36, 58–63.

Heinemann, J., Noon, B., Mohigmi, M.J., Mazurie, A., Dickensheets, D.L., and Bothner, B. (2014) Real-time digitization of metabolomics patterns from a living system using mass spectrometry. *Journal of the American Society for Mass Spectrometry* 25, 1755–1762.

Ho, Y.N., Shu, L.J., and Yang, Y.L. (2017) Imaging mass spectrometry for metabolites: technical progress, multimodal imaging, and biological interactions. *Wiley Interdisciplinary Reviews: Systems Biology and Medicine* 9: e1387.

Hoekenga, O.A. (2008) Using metabolomics to estimate unintended effects in transgenic crop plants: problems, promises, and opportunities. *Journal of Biomolecular Techniques* 19: 159.

Hollywood, K., Brison, D.R., and Goodacre, R. (2006) Metabolomics: current technologies and future trends. *Proteomics* 6, 4716–4723.

Ibáñez, M., Sancho, J.V., Bijlsma, L., van Nuijs, A.L., Covaci, A., and Hernández, F. (2014) Comprehensive analytical strategies based on high-resolution time-of-flight mass spectrometry to identify new psychoactive substances. *Trends in Analytical Chemistry* 57, 107–117.

Jamaloddin, M., Maliha, A., Gokulan, C., Gaur, N., and Patel, H.K. (2021) Metabolomics-assisted breeding for crop improvement: an emerging approach. *Omics Technologies for Sustainable Agriculture and Global Food Security* 1, 241–279.

Janz, D., Behnke, K., Schnitzler, J.-P., Kanawati, B., Schmitt-Kopplin, P., and Polle, A., (2010) Pathway analysis of the transcriptome and metabolome of salt sensitive and tolerant poplar species reveals evolutionary adaption of stress tolerance mechanisms. *BMC Plant Biology* 10: 150.

Johnson, C.H., Ivanisevic, J., Benton, H.P., and Siuzdak, G. (2015) Bioinformatics: the next frontier of metabolomics. *Analytical Chemistry* 87, 147–156.

Jones, D. (2018) Recent technical advancement in high-resolution metabolomics. In: *ISEE Conference Abstracts*, Vol. 2018. Available at: https://ehp.niehs.nih.gov/doi/10.1289/isesisee.2018.S01.02.07 (accessed 28 January 2025).

Jones, M.R., Pinto, E., Torres, M.A., Dörr, F., Mazur-Marzec, H., *et al.* (2021) CyanoMetDB, a comprehensive public database of secondary metabolites from cyanobacteria. *Water Research* 196: 117017.

Jorge, T.F. and António, C. (2018) Plant metabolomics in a changing world: metabolite responses to abiotic stress combinations. In: Andjelkovic, V. (ed.) *Plant, Abiotic Stress and Responses to Climate Change*. IntechOpen, London. Available at: www.intechopen.com/chapters/57908 (accessed 28 January 2025).

Jorge, T.F., Mata, A.T., and António, C. (2016a) Mass spectrometry as a quantitative tool in plant metabolomics. *Philosophical Transactions of the Royal Society A: Mathematical, Physical and Engineering Sciences* 374: 20150370.

Jorge, T.F., Rodrigues, J.A., Caldana, C., Schmidt, R., van Dongen, J.T., *et al.* (2016b) Mass spectrometry-based plant metabolomics: metabolite responses to abiotic stress. *Mass Spectrometry Reviews* 35, 620–649.

Junot, C., Fenaille, F., Colsch, B., and Bécher, F. (2014) High resolution mass spectrometry based techniques at the crossroads of metabolic pathways. *Mass Spectrometry Reviews* 33, 471–500.

Kalidasu, G., Reddy, G.S., Kumari, S.S., Kumari, A.L., and Sivasankar, A. (2017) Secondary volatiles and metabolites from *Nigella sativa* L. seed. *Indian Journal of Natural Products and Resources* 8, 151–158.

Kalivodová, A., Hron, K., Filzmoser, P., Najdekr, L., Janečková, H., and Adam, T. (2015) PLS-DA for compositional data with application to metabolomics. *Journal of Chemometrics* 29, 21–28.

Kanehisa, M., Goto, S., Sato, Y., Kawashima, M., Furumichi, M., and Tanabe, M. (2014) Data, information, knowledge and principle: back to metabolism in KEGG. *Nucleic Acids Research* 42, D199–D205.

Kanehisa, M., Sato, Y., Kawashima, M., Furumichi, M., and Tanabe, M. (2016) KEGG as a reference resource for gene and protein annotation. *Nucleic Acids Research* 44, D457–D462.

Katam, R., Lin, C., Grant, K., Katam, C.S., and Chen, S. (2022) Advances in plant metabolomics and its applications in stress and single-cell biology. *International Journal of Molecular Sciences* 23: 6985.

Kaur, G. and Sharma, S., (2018) Gas chromatography—a brief review. *International Journal of Information and Computing Science* 5, 125–131.

Khakimov, B., Bak, S., and Engelsen, S.B. (2014) High-throughput cereal metabolomics: current analytical technologies, challenges and perspectives. *Journal of Cereal Science* 59, 393–418.

Khan, N., Ali, S., Shahid, M.A., and Kharabian-Masouleh, A. (2017) Advances in detection of stress tolerance in plants through metabolomics approaches. *Plant Omics* 10, 153–163.

Kok, M.G., Somsen, G.W., and de Jong, G.J. (2014) The role of capillary electrophoresis in metabolic profiling studies employing multiple analytical techniques. *Trends in Analytical Chemistry* 61, 223–235.

Kovacs, H., Moskau, D., and Spraul, M. (2005) Cryogenically cooled probes—a leap in NMR technology. *Progress in Nuclear Magnetic Resonance Spectroscopy* 46, 131–155.

Krettler, C.A. and Thallinger, G.G. (2021) A map of mass spectrometry-based *in silico* fragmentation prediction and compound identification in metabolomics. *Briefings in Bioinformatics* 22: bbab073.

Krishnan, P., Kruger, N., and Ratcliffe, R. (2005) Metabolite fingerprinting and profiling in plants using NMR. *Journal of Experimental Botany* 56, 255–265.

Kueffer, C. (2019) Book review: *Inheritors of the Earth: How Nature is Thriving in an Age of Extinction*, by Chris D. Thomas. *Basic and Applied Ecology* 35, 13–17.

Kuligowski, J. and Quintás, G. (eds) (2021) *Sample Preparation in Metabolomics*. MDPI, Basel, Switzerland.

Kumar, R., Bohra, A., Pandey, A.K., Pandey, M.K., and Kumar, A. (2017) Metabolomics for plant improvement: status and prospects. *Frontiers in Plant Science* 8: 1302.

Kumar, V., Baweja, M., Singh, P.K., and Shukla, P. (2016) Recent developments in systems biology and metabolic engineering of plant–microbe interactions. *Frontiers in Plant Science* 7: 1421.

Labory, J., Njomgue-Fotso, E., and Bottini, S. (2024) Benchmarking feature selection and feature extraction methods to improve the performances of machine-learning algorithms for patient classification using metabolomics biomedical data. *Computational and Structural Biotechnology Journal* 23, 1274–1287.

Lai, J.-L., Liu, Z.-W., and Luo, X.-G. (2020) A metabolomic, transcriptomic profiling, and mineral nutrient metabolism study of the phytotoxicity mechanism of uranium. *Journal of Hazardous Materials* 386: 121437.

Lai, Y.H. and Wang, Y.S. (2023) Advances in high-resolution mass spectrometry techniques for analysis of high mass-to-charge ions. *Mass Spectrometry Reviews* 42, 2426–2445.

Lamichhane, S., Sen, P., Dickens, A.M., Hyötyläinen, T., and Orešič, M. (2018) An overview of metabolomics data analysis: current tools and future perspectives. *Comprehensive Analytical Chemistry* 82, 387–413.

Lämmerhofer, M. and Weckwerth, W. (eds) (2013) *Metabolomics in Practice: Successful Strategies to Generate and Analyze Metabolic Data*. Wiley, Hoboken, New Jersey.

Law, K.P. and Zhang, C.L. (2019) Current progress and future trends in mass spectrometry-based archaeal lipidomics. *Organic Geochemistry* 134, 45–61.

Leiss, K.A., Choi, Y.H., Verpoorte, R., and Klinkhamer, P.G. (2011) An overview of NMR-based metabolomics to identify secondary plant compounds involved in host plant resistance. *Phytochemistry Reviews* 10, 205–216.

Levine, R.M., Fogaren, K.E., Rudzin, J.E., Russoniello, C.J., Soule, D.C., and Whitaker, J.M. (2020) Open data, collaborative working platforms, and interdisciplinary collaboration: building an early career scientist community of practice to leverage ocean observatories initiative data to address critical questions in marine science. *Frontiers in Marine Science* 7: 593512.

Li, D. and Gaquerel, E. (2021) Next-generation mass spectrometry metabolomics revives the functional analysis of plant metabolic diversity. *Annual Review of Plant Biology* 72, 867–891.

Liebal, U.W., Phan, A.N., Sudhakar, M., Raman, K., and Blank, L.M. (2020) Machine learning applications for mass spectrometry-based metabolomics. *Metabolites* 10: 243.

Litvinov, D.Y., Karlov, G.I., and Divashuk, M.G. (2021) Metabolomics for crop breeding: general considerations. *Genes* 12: 1602.

Liu, X., Iftikhar, N., and Xie, X. (2014) *Survey of Real-time Processing Systems for Big Data*. In: *Proceedings of the 18th International Database Engineering and Applications Symposium*. Association for Computing Machinery, New York, pp. 356–361.

Lu, W., Su, X., Klein, M.S., Lewis, I.A., Fiehn, O., and Rabinowitz, J.D. (2017) Metabolite measurement: pitfalls to avoid and practices to follow. *Annual Review of Biochemistry* 86, 277–304.

Ma, S. and Chowdhury, S.K. (2013) Data acquisition and data mining techniques for metabolite identification using LC coupled to high-resolution MS. *Bioanalysis* 5, 1285–1297.

Ma, S., Chowdhury, S.K., and Alton, K. (2006) Application of mass spectrometry for metabolite identification. *Current Drug Metabolism* 7, 503–523.

Ma, S., Leng, Y., Li, X., Meng, Y., Yin, Z., and Hang, W. (2023) High spatial resolution mass spectrometry imaging for spatial metabolomics: advances, challenges, and future perspectives. *Trends in Analytical Chemistry* 159: 116902.

Majchrzak, T., Wojnowski, W., Rutkowska, M., and Wasik, A. (2020) Real-time volatilomics: a novel approach for analyzing biological samples. *Trends in Plant Science* 25, 302–312.

Majeed, A. and Muhammad, Z. (2019) Salinity: a major agricultural problem—causes, impacts on crop productivity and management strategies. In: Hasanuzzaman, M., Hakeem, K., Nahar, K., and Alharby, H. (eds) *Plant Abiotic Stress Tolerance*. Springer, Cham, Switzerland, pp. 83–99.

Makoś, P., Przyjazny, A., and Boczkaj, G. (2019) Methods of assaying volatile oxygenated organic compounds in effluent samples by gas chromatography—a review. *Journal of Chromatography A* 1592, 143–160.

Malinowska, J.M., Palosaari, T., Sund, J., Carpi, D., Lloyd, G.R., *et al.* (2022) Automated sample preparation and data collection workflow for high-throughput *in vitro* metabolomics. *Metabolites* 12: 52.

Manriquez, B., Muller, D., and Prigent-Combaret, C. (2021) Experimental evolution in plant-microbe systems: a tool for deciphering the functioning and evolution of plant-associated microbial communities. *Frontiers in Microbiology* 12: 619122.

Marchev, A.S., Vasileva, L.V., Amirova, K.M., Savova, M.S., Balcheva-Sivenova, Z.P., and Georgiev, M.I. (2021) Metabolomics and health: from nutritional crops and plant-based pharmaceuticals to profiling of human biofluids. *Cellular and Molecular Life Sciences* 78, 6487–6503.

Marshall, D.D. and Powers, R. (2017) Beyond the paradigm: combining mass spectrometry and nuclear magnetic resonance for metabolomics. *Progress in Nuclear Magnetic Resonance Spectroscopy* 100, 1–16.

Martin, J.J., Song, Y., Hou, M., Zhou, L., Liu, X., *et al.* (2024) Multi-omics approaches in oil palm research: a comprehensive review of metabolomics, proteomics, and transcriptomics based on low-temperature stress. *International Journal of Molecular Sciences* 25: 7695.

Matros, A. and Mock, H.-P. (2013) Mass spectrometry based imaging techniques for spatially resolved analysis of molecules. *Frontiers in Plant Science* 4: 89.

Mattoli, L., Gianni, M., and Burico, M. (2023) Mass spectrometry-based metabolomic analysis as a tool for quality control of natural complex products. *Mass Spectrometry Reviews* 42, 1358–1396.

McCalley, D.V. (2023) Understanding and managing peak shape for basic solutes in reversed-phase high performance liquid chromatography. *Chemical Communications* 59, 7887–7899.

McClure, R.S., Lee, J.-Y., Chowdhury, T.R., Bottos, E.M., White, R.A., III, *et al.* (2020) Integrated network modeling approach defines key metabolic responses of soil microbiomes to perturbations. *Scientific Reports* 10: 10882.

Mendez, K.M., Broadhurst, D.I., and Reinke, S.N. (2019) The application of artificial neural networks in metabolomics: a historical perspective. *Metabolomics* 15: 142.

Misra, B.B. (2020) Data normalization strategies in metabolomics: current challenges, approaches, and tools. *European Journal of Mass Spectrometry* 26, 165–174.

Misra, B.B. (2021) New software tools, databases, and resources in metabolomics: updates from 2020. *Metabolomics* 17: 49.

Misra, B.B. and Mohapatra, S. (2019) Tools and resources for metabolomics research community: a 2017–2018 update. *Electrophoresis* 40, 227–246.

Mondello, L., Tranchida, P.Q., Dugo, P., and Dugo, G. (2008) Comprehensive two-dimensional gas chromatography-mass spectrometry: a review. *Mass Spectrometry Reviews* 27, 101–124.

Monteiro, M., Carvalho, M., Bastos, M., and Guedes de Pinho, P. (2013) Metabolomics analysis for biomarker discovery: advances and challenges. *Current Medicinal Chemistry* 20, 257–271.

Monton, M.R.N. and Soga, T. (2007) Metabolome analysis by capillary electrophoresis–mass spectrometry. *Journal of Chromatography A* 1168, 237–246.

Moser, E., Laistler, E., Schmitt, F., and Kontaxis, G. (2017) Ultra-high field NMR and MRI—the role of magnet technology to increase sensitivity and specificity. *Frontiers in Physics* 5: 33.

Moskau, D., Ritcher, C., Kovaks, H., Salzmann, M., Baselgia, L., and Haeberli, M. (2003) Highest sensitivity for cutting-edge NMR applications: 600 MHz CryoProbes. *Spectra 2000 Analyse* 32, 39–41.

Mosley, J.D., Schock, T.B., Beecher, C.W., Dunn, W.B., Kuligowski, J., *et al.* (2024) Establishing a framework for best practices for quality assurance and quality control in untargeted metabolomics. *Metabolomics* 20: 20.

Mushtaq, M.Y., Choi, Y.H., Verpoorte, R., and Wilson, E.G. (2014) Extraction for metabolomics: access to the metabolome. *Phytochemical Analysis* 25, 291–306.

Nahar, L. and Sarker, S.D. (2024) Computational phytochemistry: an overview. In: *Computational Phytochemistry*, 2nd edn. Elsevier, Amsterdam, the Netherlands, pp. 1–58.

Neagu, A.-N., Jayathirtha, M., Baxter, E., Donnelly, M., Petre, B.A., and Darie, C.C. (2022) Applications of tandem mass spectrometry (MS/MS) in protein analysis for biomedical research. *Molecules* 27: 2411.

Neumann, S. and Böcker, S. (2010) Computational mass spectrometry for metabolomics: identification of metabolites and small molecules. *Analytical and Bioanalytical Chemistry* 398, 2779–2788.

O'Connor, L.M., O'Connor, B.A., Lim, S.B., Zeng, J., and Lo, C.H. (2023) Integrative multi-omics and systems bioinformatics in translational neuroscience: a data mining perspective. *Journal of Pharmaceutical Analysis* 13, 836–850.

Okada, T., Mochamad Afendi, F., Altaf-Ul-Amin, M., Takahashi, H., Nakamura, K., and Kanaya, S. (2010) Metabolomics of medicinal plants: the importance of multivariate analysis of analytical chemistry data. *Current Computer-Aided Drug Design* 6, 179–196.

Oldiges, M., Lütz, S., Pflug, S., Schroer, K., Stein, N., and Wiendahl, C. (2007) Metabolomics: current state and evolving methodologies and tools. *Applied Microbiology and Biotechnology* 76, 495–511.

Oppermann, M., Damoc, N.E., Crone, C., Moehring, T., Muenster, H., and Hornshaw, M. (2012) High precision measurement and fragmentation analysis for metabolite identification. *Plant Metabolomics: Methods and Protocols* 860, 145–156.

Perez de Souza, L., Alseekh, S., Scossa, F., and Fernie, A.R. (2021) Ultra-high-performance liquid chromatography high-resolution mass spectrometry variants for metabolomics research. *Nature Methods* 18, 733–746.

Perry, R.H., Cooks, R.G., and Noll, R.J. (2008) Orbitrap mass spectrometry: instrumentation, ion motion and applications. *Mass Spectrometry Reviews* 27, 661–699.

Persicke, M., Rückert, C., Plassmeier, J., Stutz, L.J., Kessler, N., *et al.* (2012) MSEA: metabolite set enrichment analysis in the MeltDB metabolomics software platform: metabolic profiling of *Corynebacterium glutamicum* as an example. *Metabolomics* 8, 310–322.

Peters, K., Worrich, A., Weinhold, A., Alka, O., Balcke, G., *et al.* (2018) Current challenges in plant eco-metabolomics. *International Journal of Molecular Sciences* 19: 1385.

Plumb, R.S., Gethings, L.A., Rainville, P.D., Isaac, G., Trengove, R., *et al.* (2023) Advances in high throughput LC/MS based metabolomics: a review. *Trends in Analytical Chemistry* 160: 116954.

Poole, C.F. (ed.) (2018) *Capillary Electro Migration Separation Methods*. Elsevier, Amsterdam, the Netherlands.

Prasad, M., Ghosh, M., and Kumar, R. (2022) Single cell metabolomics. In: Troisi, J. (ed.) *Metabolomics Perspectives*. Elsevier, London, pp. 457–513.

Qualley, A.V. and Dudareva, N. (2009) Metabolomics of plant volatiles. *Methods in Molecular Biology* 553, 329–343.

Ramaswamy, V., Hooker, J.W., Withers, R.S., Nast, R.E., Edison, A.S., and Brey, W.W. (2007) Microsample cryogenic probes: technology and applications. *EMagRes* 2. Available at: https://onlinelibrary.wiley.com/doi/full/10.1002/9780470034590.emrstm1315 (accessed 28 January 2025).

Ramautar, R., Demirci, A., and de Jong, G.J. (2006) Capillary electrophoresis in metabolomics. *Trends in Analytical Chemistry* 25, 455–466.

Rambla, J.L., López-Gresa, M., Bellés, J., and Granell, A. (2015) Metabolomic profiling of plant tissues. *Methods in Molecular Biology* 1284, 221–235.

Raza, A. (2022) Metabolomics: a systems biology approach for enhancing heat stress tolerance in plants. *Plant Cell Reports* 41, 741–763.

Razzaq, A., Sadia, B., Raza, A., Khalid Hameed, M., and Saleem, F. (2019) Metabolomics: a way forward for crop improvement. *Metabolites* 9: 303.

Razzaq, A., Wishart, D.S., Wani, S.H., Hameed, M.K., Mubin, M., and Saleem, F. (2022) Advances in metabolomics-driven diagnostic breeding and crop improvement. *Metabolites* 12: 511.

Reinhold, D., Pielke-Lombardo, H., Jacobson, S., Ghosh, D., and Kechris, K. (2019) Pre-analytic considerations for mass spectrometry-based untargeted metabolomics data. *Methods in Molecular Biology* 1978, 323–340.

Ren, S., Hinzman, A.A., Kang, E.L., Szczesniak, R.D., and Lu, L.J. (2015) Computational and statistical analysis of metabolomics data. *Metabolomics* 11, 1492–1513.

Rey, M.-D., Valledor, L., Castillejo, M.Á., Sánchez-Lucas, R., López-Hidalgo, C., *et al.* (2020) Recent advances in MS-based plant proteomics: proteomics data validation through integration with other classic and -omics approaches. *Progress in Botany* 81, 77–101.

Rha, C.-S., Jang, E.K., Hong, Y.D., and Park, W.S. (2021) Supervised statistical learning prediction of soybean varieties and cultivation sites using rapid UPLC-MS separation, method validation, and targeted metabolomic analysis of 31 phenolic compounds in the leaves. *Metabolites* 11: 884.

Roberts, L.D., Souza, A.L., Gerszten, R.E., and Clish, C.B. (2012) Targeted metabolomics. *Current Protocols in Molecular Biology* 98, 30.32. 31–30.32. 24.

Rocca-Serra, P., Salek, R.M., Arita, M., Correa, E., Dayalan, S., *et al.* (2016) Data standards can boost metabolomics research, and if there is a will, there is a way. *Metabolomics* 12: 14.

Rodrigues, A.M., Ribeiro-Barros, A.I., and António, C. (2019) Experimental design and sample preparation in forest tree metabolomics. *Metabolites* 9: 285.

Rodrigues, A.M., Miguel, C., Chaves, I., and António, C. (2021) Mass spectrometry-based forest tree metabolomics. *Mass Spectrometry Reviews* 40, 126–157.

Rojas-Cherto, M., Peironcely, J.E., Kasper, P.T., van der Hooft, J.J., de Vos, R.C., *et al.* (2012) Metabolite identification using automated comparison of high-resolution multistage mass spectral trees. *Analytical Chemistry* 84, 5524–5534.

Rolin, D., Deborde, C., Maucourt, M., Cabasson, C., Fauvelle, F., *et al.* (2013) High-resolution 1H-NMR spectroscopy and beyond to explore plant metabolome. *Advances in Botanical Research* 67, 1–66.
Roopashree, K. and Naik, D. (2019) Advanced method of secondary metabolite extraction and quality analysis. *Journal of Pharmacognosy and Phytochemistry* 8, 1829–1842.
Saito, K. and Matsuda, F. (2010) Metabolomics for functional genomics, systems biology, and biotechnology. *Annual Review of Plant Biology* 61, 463–489.
Salem, M.A., Perez de Souza, L., Serag, A., Fernie, A.R., Farag, M.A., *et al.* (2020) Metabolomics in the context of plant natural products research: from sample preparation to metabolite analysis. *Metabolites* 10: 37.
Sangwan, N.S., Tiwari, P., Mishra, S.K., Yadav, R.K., Tripathi, S., *et al.* (2015) Plant metabolomics: an overview of technology platforms for applications in metabolism. In: Barh, D., Khan, M., Davies, E. (eds) *PlantOmics: The Omics of Plant Science*. Springer, New Delhi, India, pp. 257–298.
Sarker, I.H. (2021) Machine learning: algorithms, real-world applications and research directions. *SN Computer Science* 2: 160.
National Academy of Sciences (2019) *Science Breakthroughs to Advance Food and Agricultural Research by 2030*. National Academies Press, Washington, DC.
Seger, C. and Sturm, S. (2007) Analytical aspects of plant metabolite profiling platforms: current standings and future aims. *Journal of Proteome Research* 6, 480–497.
Selegato, D.M., Pilon, A.C., and Carnevale Neto, F. (2019) Plant metabolomics using NMR spectroscopy. *Methods in Molecular Biology* 2037, 345–362.
Senyuva, H.Z., Gökmen, V., and Sarikaya, E.A. (2015) Future perspectives in Orbitrap™-high-resolution mass spectrometry in food analysis: a review. *Food Additives and Contaminants: Part A* 32, 1568–1606.
Shachar-Hill, Y. (2002) Nuclear magnetic resonance and plant metabolic engineering. *Metabolic Engineering* 4, 90–97.
Sharma, V., Gupta, P., Kagolla, P., Hangargi, B., Veershetty, A., *et al.* (2021) Metabolomics intervention towards better understanding of plant traits. *Cells* 10: 346.
Shellie, R.A., Welthagen, W., Zrostliková, J., Spranger, J., Ristow, M., *et al.* (2005) Statistical methods for comparing comprehensive two-dimensional gas chromatography-time-of-flight mass spectrometry results: metabolomic analysis of mouse tissue extracts. *Journal of Chromatography A* 1086, 83–90.
Sherlin, R.A., Annadurai, P., Jebastin, T., Srinivasan, S., Arockiasamy, P., *et al.* (2022) Workflow and methods of *in silico* analysis of compounds isolated from medicinal plants. In: Singh, S., Datta, R., Johri, P., and Trvedi, M. (eds) *Phytopharmaceuticals and Biotechnology of Herbal Plants*. CRC Press, Boca Raton, Florida.
Šimura, J., Antoniadi, I., Široká, J., Tarkowská, D.E., Strnad, M., *et al.* (2018) Plant hormonomics: multiple phytohormone profiling by targeted metabolomics. *Plant Physiology* 177, 476–489.
Singh, S., Parihar, P., Singh, R., Singh, V.P., and Prasad, S.M. (2016) Heavy metal tolerance in plants: role of transcriptomics, proteomics, metabolomics, and ionomics. *Frontiers in Plant Science* 6: 1143.
Smith, C.A., O'Maille, G., Want, E.J., Qin, C., Trauger, S.A., *et al.* (2005) METLIN: a metabolite mass spectral database. *Therapeutic Drug Monitoring* 27, 747–751.
Southam, A.D., Weber, R.J., Engel, J., Jones, M.R., and Viant, M.R. (2017) A complete workflow for high-resolution spectral-stitching nanoelectrospray direct-infusion mass-spectrometry-based metabolomics and lipidomics. *Nature Protocols* 12, 310–328.
Spraul, M., Freund, A., Nast, R., Withers, R., Maas, W., and Corcoran, O. (2003) Advancing NMR sensitivity for LC-NMR-MS using a cryoflow probe: application to the analysis of acetaminophen metabolites in urine. *Analytical Chemistry* 75, 1536–1541.
Stashenko, E. and Martínez, J. (2012) GC-MS analysis of volatile plant secondary metabolites. In: Salih, B. and Çelikbıça, O. (eds) *Gas Chromatography in Plant Science, Wine Technology, Toxicology and Some Specific Applications*. IntechOpen, London. Available at: www.intechopen.com/chapters/30592 (accessed 28 January 2025).
Stashenko, E. and Martínez, J.R. (2014) Gas chromatography-mass spectrometry. In: Guo, X. (ed.) *Advances in Gas Chromatography*. IntechOpen, London. Available at: www.intechopen.com/chapters/46209 (accessed 28 January 2025).
Stettin, D., Poulin, R.X., and Pohnert, G. (2020) Metabolomics benefits from orbitrap GC-MS—comparison of low-and high-resolution GC–MS. *Metabolites* 10: 143.
Stitt, M., Sulpice, R., and Keurentjes, J. (2010) Metabolic networks: how to identify key components in the regulation of metabolism and growth. *Plant Physiology* 152, 428–444.
Stodden, V., McNutt, M., Bailey, D.H., Deelman, E., Gil, Y., *et al.* (2016) Enhancing reproducibility for computational methods. *Science* 354, 1240–1241.

Sturtevant, D., Lee, Y.-J., and Chapman, K.D. (2016) Matrix assisted laser desorption/ionization-mass spectrometry imaging (MALDI-MSI) for direct visualization of plant metabolites in situ. *Current Opinion in Biotechnology* 37, 53–60.

Subramanian, I., Verma, S., Kumar, S., Jere, A., and Anamika, K. (2020) Multi-omics data integration, interpretation, and its application. *Bioinformatics and Biology Insights* 14: 1177932219899051.

Sud, M., Fahy, E., Cotter, D., Azam, K., Vadivelu, I., *et al.* (2016) Metabolomics Workbench: an international repository for metabolomics data and metadata, metabolite standards, protocols, tutorials and training, and analysis tools. *Nucleic Acids Research* 44, D463–D470.

Sumner, L.W., Urbanczyk-Wochniak, E., and Broeckling, C.D. (2007) Metabolomics data analysis, visualization, and integration. In: Edwards, D. (ed.) *Plant Bioinformatics. Methods in Molecular Biology*. Humana Press, Totowa, New Jersey, pp. 409–436.

Sumner, L.W., Yang, D.S., Bench, B.J., Watson, B.S., Li, C., and Jones, A.D. (2011) Spatially resolved plant metabolomics. *Annual Plant Reviews* 43, 343–366.

Sumner, L.W., Lei, Z., Nikolau, B.J., and Saito, K. (2015) Modern plant metabolomics: advanced natural product gene discoveries, improved technologies, and future prospects. *Natural Product Reports* 32, 212–229.

Sweetlove, L.J. and Fernie, A.R. (2005) Regulation of metabolic networks: understanding metabolic complexity in the systems biology era. *New Phytologist* 168, 9–24.

Taylor, M.J., Lukowski, J.K., and Anderton, C.R. (2021) Spatially resolved mass spectrometry at the single cell: recent innovations in proteomics and metabolomics. *Journal of the American Society for Mass Spectrometry* 32, 872–894.

Tian, H., Lam, S.M., and Shui, G. (2016) Metabolomics, a powerful tool for agricultural research. *International Journal of Molecular Sciences* 17: 1871.

Tohge, T., Scossa, F., and Fernie, A.R. (2015) Integrative approaches to enhance understanding of plant metabolic pathway structure and regulation. *Plant Physiology* 169, 1499–1511.

Töpfer, N., Kleessen, S., and Nikoloski, Z. (2015) Integration of metabolomics data into metabolic networks. *Frontiers in Plant Science* 6: 49.

Toubiana, D., Fernie, A.R., Nikoloski, Z., and Fait, A. (2013) Network analysis: tackling complex data to study plant metabolism. *Trends in Biotechnology* 31, 29–36.

Tsugawa, H., Rai, A., Saito, K., and Nakabayashi, R. (2021) Metabolomics and complementary techniques to investigate the plant phytochemical cosmos. *Natural Product Reports* 38, 1729–1759.

Türkan, A. and Kalay, Ş. (2006) Monitoring lipase-catalyzed methanolysis of sunflower oil by reversed-phase high-performance liquid chromatography: elucidation of the mechanisms of lipases. *Journal of Chromatography A* 1127, 34–44.

Týčová, A., Ledvina, V., and Klepárník, K. (2017) Recent advances in CE-MS coupling: instrumentation, methodology, and applications. *Electrophoresis* 38, 115–134.

Uthe, H., van Dam, N.M., Hervé, M.R., Sorokina, M., Peters, K., and Weinhold, A. (2021) A practical guide to implementing metabolomics in plant ecology and biodiversity research. *Advances in Botanical Research* 98, 163–203.

van Dam, N.M., and van der Meijden, E. (2011) A role for metabolomics in plant ecology. *Annual Plant Reviews* 43, 87–107.

van der Kloet, F.M., Bobeldijk, I., Verheij, E.R., and Jellema, R.H. (2009) Analytical error reduction using single point calibration for accurate and precise metabolomic phenotyping. *Journal of Proteome Research* 8, 5132–5141.

Vettukattil, R. (2015) Preprocessing of raw metabonomic data. *Methods in Molecular Biology* 1277, 123–136.

Villas-Bôas, S.G., Smart, K.F., Sivakumaran, S., and Lane, G.A. (2011) Alkylation or silylation for analysis of amino and non-amino organic acids by GC-MS? *Metabolites* 1, 3–20.

Vinaixa, M., Schymanski, E.L., Neumann, S., Navarro, M., Salek, R.M., and Yanes, O. (2016) Mass spectral databases for LC/MS-and GC/MS-based metabolomics: state of the field and future prospects. *Trends in Analytical Chemistry* 78, 23–35.

Vinay, C.M., Udayamanoharan, S.K., Prabhu Basrur, N., Paul, B., and Rai, P.S. (2021) Current analytical technologies and bioinformatic resources for plant metabolomics data. *Plant Biotechnology Reports* 15, 561–572.

Wagner, D.S., Pirhalla, J.L., and Bowers, G.D. (2013) Metabolite structure analysis by high-resolution MS: supporting drug-development studies. *Bioanalysis* 5, 463–479.

Wang, C., He, L., Li, D.-W., Bruschweiler-Li, L., Marshall, A.G., and Brüschweiler, R. (2017) Accurate identification of unknown and known metabolic mixture components by combining 3D NMR with Fourier transform ion cyclotron resonance tandem mass spectrometry. *Journal of Proteome Research* 16, 3774–3786.

Wang, L., Tolok, G., Fu, Y., Xu, L., Li, L., *et al.* (2024) Application and research progress of laser-induced breakdown spectroscopy in agricultural product inspection. *ACS Omega* 9, 2403–2418.

Waris, M., Kocak, E., Gonulalan, E.M., Demirezer, L.O., Kır, S., and Nemutlu, E. (2022) Metabolomics analysis insight into medicinal plant science. *Trends in Analytical Chemistry* 157: 116795.

Watson, D.G. (2013) A rough guide to metabolite identification using high resolution liquid chromatography mass spectrometry in metabolomic profiling in metazoans. *Computational and Structural Biotechnology Journal* 4: e201301005.

Weckwerth, W. (2008) Integration of metabolomics and proteomics in molecular plant physiology–coping with the complexity by data-dimensionality reduction. *Physiologia Plantarum* 132, 176–189.

Wei, X., Shi, X., Kim, S., Zhang, L., Patrick, J.S., *et al.* (2012) Data preprocessing method for liquid chromatography–mass spectrometry based metabolomics. *Analytical Chemistry* 84, 7963–7971.

Wishart, D.S. (2009a) Computational approaches to metabolomics. *Methods in Molecular Biology* 593, 283–313.

Wishart, D.S. (2009b) Computational strategies for metabolite identification in metabolomics. *Bioanalysis* 1, 1579–1596.

Wishart, D.S., Knox, C., Guo, A.C., Eisner, R., Young, N., *et al.* (2009) HMDB: a knowledgebase for the human metabolome. *Nucleic Acids Research* 37 (Suppl. 1), D603–D610.

Wolfender, J.-L., Rudaz, S., Hae Choi, Y., and Kyong Kim, H. (2013) Plant metabolomics: from holistic data to relevant biomarkers. *Current Medicinal Chemistry* 20, 1056–1090.

Wolfender, J.-L., Marti, G., Thomas, A., and Bertrand, S. (2015) Current approaches and challenges for the metabolite profiling of complex natural extracts. *Journal of Chromatography A* 1382, 136–164.

Worley, B. and Powers, R. (2013) Multivariate analysis in metabolomics. *Current Metabolomics* 1, 92–107.

Wu, Y. and Li, L. (2016) Sample normalization methods in quantitative metabolomics. *Journal of Chromatography A* 1430, 80–95.

Xiao, J.F., Zhou, B., and Ressom, H.W. (2012) Metabolite identification and quantitation in LC-MS/MS-based metabolomics. *Trends in Analytical Chemistry* 32, 1–14.

Yadav, R., Singh, G., Santal, A.R., and Singh, N.P. (2023) Omics approaches in effective selection and generation of potential plants for phytoremediation of heavy metal from contaminated resources. *Journal of Environmental Management* 336: 117730.

Yan, X.-T., Zhang, Y., Zhou, Y., Li, G.-H., and Feng, X.-S. (2022) Technical overview of orbitrap high resolution mass spectrometry and its application to the detection of small molecules in food (update since 2012). *Critical Reviews in Analytical Chemistry* 52, 593–626.

Yang, Y., Saand, M.A., Huang, L., Abdelaal, W.B., Zhang, J., *et al.* (2021) Applications of multi-omics technologies for crop improvement. *Frontiers in Plant Science* 12, 563953.

Yergey, A.L., Edmonds, C.G., Lewis, I.A., and Vestal, M.L. (2013) *Liquid Chromatography/Mass Spectrometry: Techniques and Applications*. Springer Science and Business Media, New York.

Yu, Y., Yao, C., and Guo, D.-A. (2021) Insight into chemical basis of traditional Chinese medicine based on the state-of-the-art techniques of liquid chromatography–mass spectrometry. *Acta Pharmaceutica Sinica B* 11, 1469–1492.

Zampieri, M., Sekar, K., Zamboni, N., and Sauer, U. (2017) Frontiers of high-throughput metabolomics. *Current Opinion in Chemical Biology* 36, 15–23.

Zhang, A., Sun, H., Wang, P., Han, Y., and Wang, X. (2012) Modern analytical techniques in metabolomics analysis. *Analyst* 137, 293–300.

Zhang, W. and Zhao, P.X. (2014) Quality evaluation of extracted ion chromatograms and chromatographic peaks in liquid chromatography/mass spectrometry-based metabolomics data. *BMC Bioinformatics* 15 (Suppl. 11): S5.

Zhu, Z.-J., Schultz, A.W., Wang, J., Johnson, C.H., Yannone, S.M., *et al.* (2013) Liquid chromatography quadrupole time-of-flight mass spectrometry characterization of metabolites guided by the METLIN database. *Nature Protocols* 8, 451–460.

5 High-throughput Metabolomics for Uncovering Plant–Herbivore Interactions

Debajit Das[1], Sangeeta Sarma[2], Dhrubo Jyoti Gogoi[3,4], Abhik Gogoi[1], Channakeshavaiah Chikkaputtaiah[3,4*], and Dhanawantari L. Singha[5*]

[1]Department of Agricultural Biotechnology, Assam Agricultural University, Jorhat, Assam, India; [2]Department of Botany, Jagiroad College, Jagiroad, Morigaon, Assam, India; [3]Biological Sciences and Technology Division, CSIR-North East Institute of Science and Technology (NEIST), Jorhat, Assam, India; [4]Academy of Scientific and Innovative Research (AcSIR), Ghaziabad, India; [5]Department of Botany, Rabindranath Tagore University, Hojai, Assam, India

Abstract

Plant–herbivore interactions create a link between the organisms in various environments, including freshwater, marine, and terrestrial ecosystems. These connections are the essential link that generates animal biomass from simple sunlight in everything from simple organisms to complex organisms. Thus, a comprehensive study of the ecology and evolution of almost every ecosystem depends on these relationships. Metabolomics can be adopted to gain more insights and to study this plant–herbivore interaction in depth. Metabolomics, through the production of a profile of small molecules obtained from cellular metabolism, offers knowledge regarding numerous processes of plant physiology by showing the complex networks formed through various biochemical events in the cell. This chapter provides a guidance to studies of the current changes in the food chain, thus helping us to understand the environment better.

Keywords: metabolomics, plant–herbivore interactions, NMR spectroscopy, mass spectroscopy, gas chromatography–mass spectrometry

5.1 Introduction

The interactions between plants and herbivores, either directly or indirectly, assist them in their survival and are of great ecological importance. These interactions are very important for the energy flow in the food chain and ultimately in nutrient cycles (Gong and Zhang, 2014). Over millions of years, the interactions between them have evolved into a complex relationship wherein plants have developed strategies to defend themselves from herbivores, while organisms that feed on plants have made counteradaptations to plant defenses (Rioja *et al.*, 2017). These strategies involve biochemical, morphological, and molecular mechanisms that counter the scenario of plants being consumed by their attackers. Responding to the damage inflicted on them, plants activate dynamic biochemical reactions and produce certain chemicals that may retard the overall growth of herbivores. Regarding morphological strategies, they incorporate

* E-mail: channakeshav@neist.res.in or dhanawantarisingha@gmail.com

© CAB International 2025. *High-Throughput Plant Metabolomics* (Jen-Tsung Chen ed.)
DOI: 10.1079/9781800626911.0005

direct defenses through physical barriers by developing thorns, spines, hairs, and trichomes (War *et al.*, 2012). Molecular mechanisms involve pathways where several signaling compounds such as salicylic acid, jasmonic acid, abscisic acid, and ethylene control the chemicals related to defense (Vinod and Sabah, 2018). In response to the evolving defense strategies of plants, herbivores have equally evolved themselves to decrease the negative effects of these defenses. They have adapted to bypass these defenses by avoiding the store regions of the defensive compounds or metabolizing them if ingested. They may even invade the plant's defense system and suppress the plant's resistance. Even antinutritional defenses can be overcome by salivary effectors produced by herbivores (Kant *et al.*, 2015; Perkovich and Ward, 2022). Hence, these interactions become an intricate dance between plants and herbivores as a continuous arms race, evolving and maneuvering each other and ultimately leading to a dynamic ecological balance.

5.1.1 Role of metabolites in mediating plant–herbivore interactions

The primary metabolites of plants are responsible for their basic functions, whereas secondary metabolites are responsible for providing defensive protection by regulating the signaling pathways of herbivore attack (Divekar *et al.*, 2022). The latter include nitrogen-containing alkaloids, steroids, saponins, polyphenolics such as tannins, cyanogenic glucosides, γ-amino butyric acid, lectins, and sporamin. The presence of these compounds makes it repulsive for herbivores to attack them and may even cause death following ingestion. It is not necessary for a single compound to be harmful on its own; rather, a combination of these compounds can produce a toxic effect. These plant metabolites, together with plant protection mechanisms, play a multi-faceted role in orchestrating complex ecological interactions, and therefore a deeper understanding of plant–herbivore interactions is required (Mithöfer and Maffei, 2016). This understanding has been accomplished using traditional approaches including field-based approaches and biochemical analysis. However, these traditional approaches have significant shortcomings. Field-based observations or ecological observations require a longer period of time (Estarague *et al.*, 2023) and may provide only surface-level information without revealing the molecular details of the interactions. Biochemical analyses have focused on known metabolites and have generally overlooked the broad groups of chemicals involved in the interaction. As a consequence, a holistic understanding of the complex defense mechanisms of plants has been lacking. In contrast, metabolomic studies have included both targeted and untargeted metabolites, which has improved our understanding of these complex mechanisms (Allwood *et al.*, 2021). This has allowed researchers to discover and better understand the metabolomic pathways and networks associated with these interactions and to provide a larger picture of the biochemical processes involved. The ability of metabolomic approaches to unravel these untargeted metabolites will allow a deeper understanding of these chemical dialogues and provides a significant advantage over traditional approaches (Manickam *et al.*, 2023). Metabolomic studies have generated a vast array of data sets comprising information on chemical changes in plants in response to herbivore attacks and how plants communicate these signals to other organisms (Maag *et al.*, 2015). This data set, when integrated with other omics studies, will provide a deeper understanding of the molecular mechanisms involved in plant defense and adaptation (Kamali and Singh, 2023).

5.2 Unveiling the Metabolome

The various steps and types of metabolomics are presented in Fig. 5.1 and discussed in detail here. Harvesting, storing, extracting, and quantifying metabolites from plant samples is the first step in plant metabolomics research. Data interpretation comes next. Plant metabolomics relies heavily on sample preparation as this may alter the number of metabolites in the sample. Thus, plant sample collection and preservation should occur quickly to minimize modifications to the biochemical response occurring inside the plant cells (Tanna and Mishra, 2018). The most frequent cause of bias in plant metabolomic research is improper handling during sample collection (Salem *et al.*, 2020). Plant metabolomics

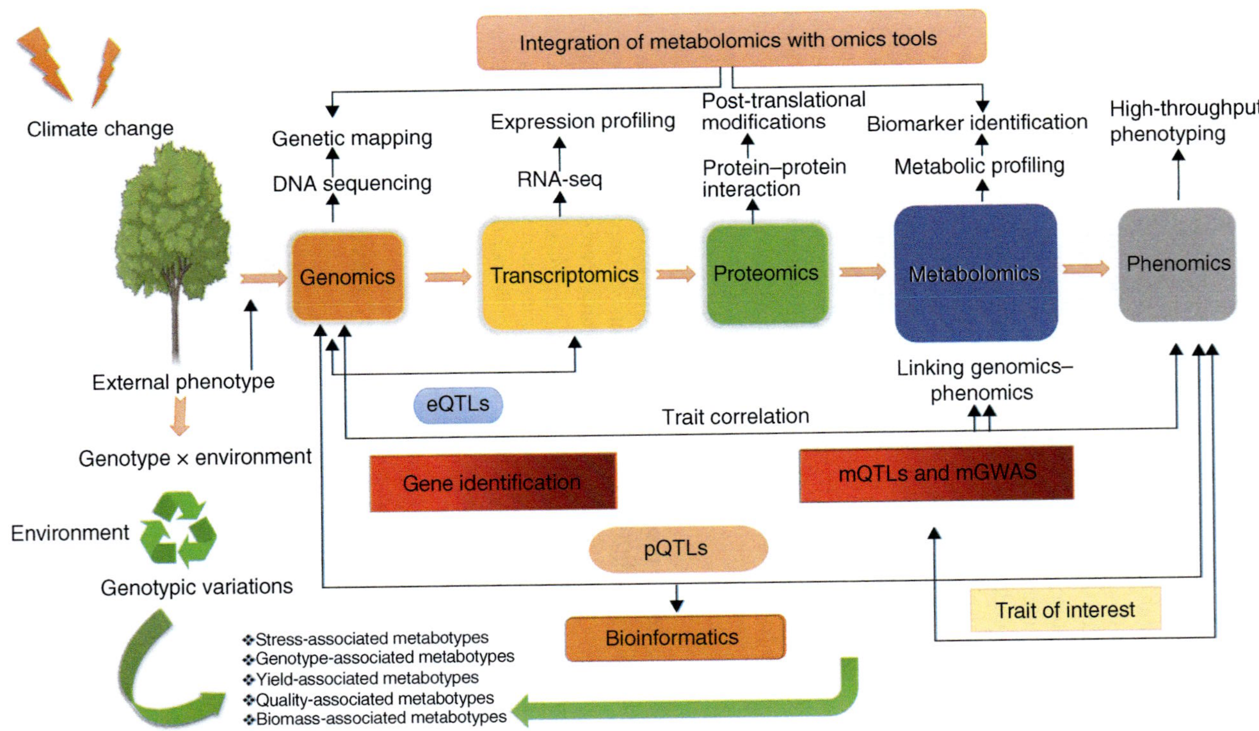

Fig. 5.1. Steps and types of plant metabolomics studies. eQTLs, epigenomic quantitative trait loci; mGWAS, metabolic genome-wide association studies; mQTLs, metabolic quantitative trait loci; pQTLs, proteomic quantitative trait loci; RNA-seq, RNA sequencing.

investigations should be justified by adhering to the Metabolomics Standards Initiative (MSI) for sample collection, storage, and extract preparation (Sumner *et al.*, 2007).

5.2.1 Sample collection and storage

Plant metabolomics typically involves four main steps: sample analysis, extraction, storage, and harvesting. Harvesting plant samples has to be done carefully as the plant's metabolome is susceptible to enzymatic responses that may break down various metabolites. Furthermore, metabolites change according to the stage of growth, age of the plant, and time of sample gathering (Patel *et al.*, 2016). Similarly, processing metabolomics requires a variety of storage methods, including air-, oven-, and freeze-drying (Harbourne *et al.*, 2009; Salem *et al.*, 2020).

5.2.2 Sample preparation: critical concerns and best practices

In the context of metabolomic analysis, quenching, mechanical, and ultrasonic extraction techniques show promise (Parida *et al.*, 2018). Furthermore, combining the ultrasonic extraction process with mechanical grinding may give good quality, yield, and chemical flexibility (Gong *et al.*, 2017). In addition to the extraction techniques, selecting the right solvents is essential as different types of metabolites (e.g. polar or non-polar) cannot be extracted using the same solvent. Chloroform, methanol, and water make up a solvent solution that can be used to isolate diverse metabolites (Silva-Navas *et al.*, 2016; Corrales *et al.*, 2017), and this solvent system is often used for various metabolites encompassing hydrophilic, non-polar, and polar substances. Nearly all metabolite types can be recovered using a single approach thanks to the development of a specialized solvent gradient extraction technique (Yuliana *et al.*, 2011). Sample preparation involves a number of techniques, including the Swiss rolling method (Velickovic *et al.*, 2020), enzyme-assisted extraction (Zuorro *et al.*, 2013), microwave-assisted extraction (Teo *et al.*, 2013), and ultrasound-assisted extraction (Altemimi *et al.*, 2016).

In metabolomics research, targeted detection and measurement of metabolites are the primary methods employed (Zhou *et al.*, 2016). Metabolites of interest are isolated from various plant parts, such as leaves, stems, and roots. Targeted metabolomics quantification aims to maximize metabolite coverage by analyzing specific metabolites (Cajka and Fiehn, 2016). Various extraction techniques, such as liquid–liquid extraction (Cajka and Fiehn, 2016), solid-phase extraction (Bojko *et al.*, 2014), and the use of varying ratios of organic solvents (Giavalisco *et al.*, 2011), are utilized to extract these targeted metabolites. To enhance analytical reliability, one or more internal standards may be added during sample preparation (Ciccimaro and Blair, 2010). The final step in preparing a sample for liquid chromatography–mass spectrometry (LC-MS) analysis involves evaporating the solvents and dissolving the sample in an appropriate solvent (Cajka and Fiehn, 2016). Targeted metabolite quantification is regarded as a crucial technique due to its consistent stability, sensitivity, and accuracy (Zhou and Lin, 2016). However, this approach is typically limited to measuring a predetermined set of recognized analytes and does not identify new or unknown metabolites. To broaden the scope of targeted metabolomics research, new methods employing high-resolution tools such as parallel reaction monitoring have been developed (Zhou and Lin, 2016). In the realm of plant metabolomics, novel extraction techniques are continually evolving, influenced by the properties of the components and the chosen analytical frameworks.

5.3 Analytical Techniques for Metabolomics Study

5.3.1 Nuclear magnetic resonance spectroscopy

An increasingly used analytical method in metabolomics research, especially in plant biology, is nuclear magnetic resonance (NMR) spectroscopy. By using the magnetic characteristics of atomic nuclei, this technique can provide comprehensive details on the molecular structure, dynamics, and surroundings of metabolites in a biological specimen. Within the field of plant biology, NMR spectroscopy has emerged as a crucial instrument for clarifying metabolic pathways, detecting biomarkers, and understanding the complex biochemical networks that regulate plant operations. The behavior of nuclei in a magnetic field is the fundamental idea behind NMR spectroscopy. Nuclear elements with non-zero spins, such as ^{13}C or ^{1}H, align with a strong magnetic field. These nuclei are brought out of equilibrium by the application of a radio frequency pulse. Upon returning to a state of equilibrium, they release energy that is picked up and converted into a spectrum. The "chemical shifts," or positions of the peaks, in the NMR spectrum provide information about the molecule structure by reflecting the electrical environment around the nuclei. Because it is non-destructive and can analyze complex biological mixtures without requiring substantial sample preparation, NMR spectroscopy is a highly regarded technique in the field of plant metabolomics. These features are essential for researching plant tissues as obtaining physiologically meaningful data often requires maintaining the tissues in their original condition. Determining the metabolic pathways of plants is one of the main uses of NMR spectroscopy. By following ^{13}C-labeled substrates through metabolic networks, for example, NMR has been used to map carbon flux in photosynthetic processes. These investigations have clarified how carbon is distributed across multiple metabolic pathways, illuminating the subtleties of plant metabolism in a range of physiological settings (Ratcliffe and Shachar-Hill, 2001). Key metabolites implicated in stress reactions have also been identified thanks in part to NMR-based metabolomics. The metabolome of plants is significantly altered when they are subjected to environmental stresses such as salt, drought, or pathogen attack. Researchers may find possible indicators of stress tolerance by detecting and quantifying these metabolomic alterations using NMR spectroscopy (Obata and Fernie, 2012). NMR spectroscopy has been used to investigate the function of secondary metabolites in plant defense systems in addition to elucidating pathways. Plant defense against diseases, herbivores, and environmental stressors is greatly aided by secondary metabolites such as phenolic compounds, terpenoids, and alkaloids. The structural variety and functional relevance of these chemicals in plant defense have been revealed via the comprehensive characterization made possible by NMR. For instance, knowledge of the function of certain phenolic chemicals in preventing pathogen development and discouraging herbivores has been gained by identifying them in various plant species (Dixon and Paiva, 1995). Additionally, the study of plant–microbe interactions has benefitted from the use of NMR spectroscopy. Plant health and nutrient uptake depend heavily on the symbiotic interactions that exist between plants and microbes such as nitrogen-fixing bacteria and mycorrhizal fungi, and the metabolic exchanges between plants and their microbial companions have been studied using NMR-based metabolomics. Researchers have improved their knowledge of the chemical communication underlying these relationships by examining the metabolites released by both the bacteria and plants. This information is essential for creating plant–microbe partnerships that are favorable to agriculture and are sustainable (Pritchard and Rogers, 2000). The investigation of metabolic diversity across various plant species is further aided by NMR spectroscopy. Plants create a wide variety of metabolites known as the "plant metabolome," which varies greatly across species due to their adaptation to different ecological niches. It is possible to determine distinctive metabolic fingerprints and learn about the evolutionary tactics used by plants to adapt to a variety of situations by comparing the NMR spectra of various plant species. This methodology has been very beneficial in the investigation of medicinal plants as the identification of bioactive chemicals has the potential to provide novel therapeutic medicines (Kim *et al.*, 2010). The use of NMR spectroscopy in plant metabolomics has been further strengthened in recent years by its combination with other analytical methods, such as mass spectrometry (MS).

In summary, NMR spectroscopy has become a vital component of plant metabolomics studies. The technique is very beneficial for examining metabolic pathways, stress responses, plant–microbe interactions, and metabolic diversity due to its capacity to provide comprehensive, repeatable, and non-destructive examination of plant metabolites. NMR spectroscopy is expected to become increasingly important in the study of plant metabolism as technical developments continue to improve its sensitivity and resolution. This will provide fresh insights into the biochemical complexity of the plant kingdom.

5.3.2 Mass spectrometry

MS is a very effective analytical method used extensively in the study of metabolomics in plant biology. It provides valuable information on the physiological condition of an organism. The presence of MS in this field has greatly expanded because of its heightened sensitivity, specificity, and adaptability, enabling the identification of metabolites even at very low levels. The fundamental concept of MS is based on the process of ionizing chemical compounds to produce charged molecules or fragments of molecules, which are then quantified by their mass:charge (m/z) ratio. The method starts with ionization of the sample, often achieved using techniques such as electrospray ionization or matrix-assisted laser desorption/ionization (MALDI), which are very efficient for analyzing a diverse array of metabolites. After ionization, the metabolites are put in a mass analyzer, where they are sorted according to their m/z ratio. The main mass analyzers used in plant metabolomics are time-of-flight (TOF), quadrupole (Q), and Orbitrap analyzers. These analyzers provide exceptional resolution and mass accuracy, which are essential for accurately identifying metabolites. Plant biology utilizes MS-based metabolomics to study metabolic pathways, understand the biochemical mechanisms behind plant responses to environmental challenges, and analyze the changes in plant metabolism under different situations. For example, MS has played a crucial role in revealing the metabolic alterations in plants exposed to drought stress. Research has shown that plants accumulate osmoprotectants such as proline and soluble sugars as a response to drought. These substances can be identified and measured using MS. The incorporation of MS with other analytical methods has broadened its use in plant metabolomics. Integrating MS with chromatographic techniques such as liquid chromatography (LC-MS) or gas chromatography (GC-MS) enables the precise isolation and identification of intricate mixtures of metabolites. This combination is especially advantageous for investigating volatile organic compounds (VOCs) in plants, which often play a role in plant defense and communication. In addition, MS is increasingly used with isotopic labeling methods to track the metabolic destiny of certain molecules inside plants. The stable isotope-resolved metabolomics (SIRM) technique enables researchers to monitor the integration of labeled atoms from precursors such as ^{13}C-glucose into subsequent metabolites. The combination of SIRM and MS has been used to accurately track the flow of carbon across key metabolic pathways in plants, providing a comprehensive understanding of the dynamics of the metabolic network (Ratcliffe and Shachar-Hill, 2001). Metabolic engineering greatly benefits from this, as it relies on comprehending the carbon flow in biosynthetic pathways to optimize the creation of certain molecules. The introduction of high-resolution MS equipment in recent years has expanded the limits of plant metabolomics. State-of-the-art devices such as the Orbitrap and Fourier-transform ion cyclotron resonance mass spectrometers provide exceptional precision and resolution in measuring mass. The increasing accessibility of metabolite databases and bioinformatics tools has also simplified the analysis of MS data in plant metabolomics. Databases such as METLIN, the Human Metabolome Database (HMDB), and MassBank provide reference spectra for a number of metabolites, enabling rapid identification of substances discovered using MS. In addition, the use of bioinformatics tools allows the integration of MS data with genomic and transcriptome information, enabling a comprehensive knowledge of plant metabolism at a systems level (Fukushima and Kusano, 2014). In plant biology, MS has significantly transformed metabolomics research by offering a full and intricate understanding of the plant metabolome. The uses of this technology range from understanding metabolic pathways and studying how plants respond to environmental challenges to discovering novel bioactive chemicals and advancing plant metabolic engineering. The continuing evolution of MS technology means that it will maintain its leading position in the field of plant metabolomics, and will drive the emergence of new discoveries and contribute to a deeper knowledge of plant biology.

5.3.3 Liquid chromatography–mass spectrometry

With its remarkable sensitivity and specificity, LC-MS provides extensive profiling of a wide variety of metabolites, making it a vital tool in the study of plant metabolomics. This technique enables the identification and quantification of complex plant metabolite mixtures by combining the potent detection powers of MS with the separation abilities of LC (Dunn, 2008). Understanding plant responses to environmental stresses and clarifying metabolic pathways have both benefitted greatly from LC-MS. For the purpose of pharmaceutical and nutraceutical research, its use is essential in the discovery of novel bioactive compounds. The analytical abilities of LC-MS have been strengthened by technological developments such as tandem MS (MS/MS) and high-resolution MS, which have enabled the identification of trace-level metabolites and the deciphering of their structures (Tsugawa *et al.*, 2015). Moreover, the use of bioinformatics tools in conjunction with LC-MS has significantly improved the examination of metabolomics data, enabling comprehensive metabolic fingerprinting and cross-species comparisons of plants (Fernie and Schauer, 2009). Thus, LC-MS is a key method in the field of plant metabolomics, contributing to our comprehension of plant biochemistry and encouraging its use in many scientific fields.

5.3.4 Gas chromatography–mass spectrometry

GC-MS is a powerful analytical method that integrates the functionalities of GC and MS to detect and identify various substances within a sample (Smith, 2004). This technique is widely recognized for its high sensitivity, specificity, and adaptability in analyzing intricate mixtures. In this process, the GC component is responsible for separating volatile and semi-volatile compounds according to their boiling points and polarity, whereas the MS component provides structural insights by ionizing these separated compounds and measuring their m/z (Adams, 2007). The resulting data provide detailed information about the molecular composition of the sample (McLafferty and Turecek, 1993). In the field of plant metabolomics, GC-MS is used extensively due to its ability to analyze a broad spectrum of primary and secondary metabolites, including sugars, amino acids, organic acids, fatty acids, and VOCs (Fiehn, 2002). This technique plays a vital role in understanding plant physiology, biochemistry, and interactions with the environment. As highlighted by Fiehn (2002), accurate identification and quantification of plant metabolites can significantly enhance our understanding of metabolic pathways and their regulation.

The typical workflow for GC-MS in plant metabolomics involves several essential steps, beginning with sample preparation. Plant tissues are usually homogenized, and the metabolites are extracted using solvents (Roessner *et al.*, 2001). After extraction, the metabolites are derivatized to increase their volatility and stability, which is essential for their separation and detection within the GC-MS system (Lisec *et al.*, 2006). The derivatized samples are then injected into the GC machine, where they are vaporized and carried by an inert gas, such as helium, through a capillary column containing a stationary phase (Schauer and Fernie, 2006). The metabolites are separated based on their interactions with the stationary phase as they move through the column. After separation, the metabolites enter the mass spectrometer, where they are ionized—usually by electron impact ionization—resulting in fragmentation of the compounds into characteristic ions (Kopka *et al.*, 2005). These ions are detected and recorded as mass spectra, which are then either matched against spectral libraries or identified based on their fragmentation patterns (Kind *et al.*, 2009). The data obtained provide comprehensive metabolic profiles, which are subsequently analyzed using bioinformatics tools to compare the metabolic states under different conditions (Weckwerth, 2003).

GC-MS has contributed to numerous breakthroughs in plant metabolomics. For example, it has been employed to profile metabolic changes in plants subjected to abiotic stresses such as drought, salinity, and extreme temperatures (Kaplan *et al.*, 2004). Metabolomic studies utilizing GC-MS have shown that plants accumulate specific metabolites in response to these stresses, which play key roles in stress tolerance mechanisms (Rizhsky *et al.*, 2004). Furthermore, GC-MS has proven instrumental in exploring plant–pathogen interactions by identifying metabolites involved in plant defense responses (Chong *et al.*, 2005). Research has demonstrated that infected plants produce secondary metabolites such as phytoalexins, which are crucial for pathogen resistance (Halket *et al.*, 2005). Additionally, GC-MS has been valuable in studying plant–microbe interactions in the rhizosphere by analyzing root exudates and their effects on soil microbial communities (Badri and Vivanco, 2009). The technique has also been applied to unravel the metabolic basis of plant development and growth. For instance, GC-MS-based metabolomic analyses of *Arabidopsis thaliana* have shed light on the metabolic networks that regulate flowering and seed development (Fait *et al.*, 2006).

In summary, GC-MS is an indispensable tool in plant metabolomics, offering unmatched abilities for analyzing a diverse range of metabolites. Its application has significantly advanced our understanding of plant metabolism, stress responses, development, and interactions with other organisms. The detailed metabolic data obtained through GC-MS analyses continue to drive innovations and discoveries in plant science, underscoring the technique's critical role in modern metabolomic research.

5.4 Data Processing, Analysis and Pathway Identification

5.4.1 Data preprocessing

To extract metabolite signals from raw data (e.g. chromatograms and NMR), peak detection and chromatogram alignment algorithms are essential. Numerous software tools and packages have been developed to streamline compound detection. Some of the widely used tools include XCMS (Smith *et al.*, 2006), mzMine2 (Katajamaa *et al.*, 2006), MAVEN (Metabolomic Analysis and Visualization Engine; Clasquin *et al.*, 2012), OpenMS (Sturm *et al.*, 2008), MS-DIAL (Mass Spectrometry-Data Independent Analysis; Tsugawa *et al.*, 2015), ADAP, and apLCMS (adaptive processing of LC-MS; Yu *et al.*, 2009). According to Kessner *et al.* (2008) and Martens *et al.* (2011), all major software applications adhere to open formats such as mzML, facilitating feature extraction and quantification. Once the data is preprocessed, it is crucial to interpret the results with accurate metabolite annotation.

5.4.2 Annotation tools and databases for metabolites

The integration of unprocessed analytical data with biological understanding via targeted or untargeted methods of profiling is known as metabolite annotation. The method of identifying metabolites or annotation, however, is often time-consuming. If reference chemicals are not accessible, the identification of metabolites is often accomplished by using fragmentation data collected from MS/MS targeted against online MS/MS databases. To identify and deconvolute mass spectra automatically from GC-MS data, the NIST spectral library is used (Ausloos *et al.* 1999). Another open-source program for spectrum searching is the Golm Metabolite Database (GMD; Kopka *et al.*, 2005). Many spectral databases, such as METLIN (Smith *et al.*, 2005), MoNA (http://mona.fiehnlab.ucdavis.edu/, accessed 17 January 2025), mzCloud (www.mzcloud.org/, accessed 17 January 2025), MassBank (Horai *et al.* 2010), and Global Natural Products Social Molecular Networking database (GNPS; Wang *et al.* 2016), are used for LC-MS metabolite annotation and spectral searches. Annotating metabolites in people and plants using mass spectrum and NMR data may be facilitated by the HMDB (Wishart *et al.*, 2009) and the Madison Metabolomics Consortium Database (MMCDB) (Markley *et al.*, 2007; Cui *et al.*, 2008). For metabolite annotation, a number of software tools are available, such as CAMERA (Kuhl *et al.*, 2012), RAMclust, MetAssign (Daly *et al.*, 2014), and xMSannotator (Uppal *et al.*, 2017). Many compound repositories that are available for metabolite annotation are also required.

5.4.3 Statistical analyses

The metabolite characteristics and the dependent variable are compared using statistical analysis. Appropriate statistical techniques are often used for metabolomics data, based on the research goal. Both multivariate analysis and chemometric approaches are employed for metabolomics data because metabolomics studies from the analytical methods introduce multiple biases into the data, which render comprehension difficult and demanding. There are both supervised and unsupervised approaches used in multivariate analysis. Principal component analysis (PCA) is one of the unsupervised approaches used in metabolomics investigation. By projecting and grouping the samples in a data table, PCA explains the systematic differences in the metabolites of each sample. PCA has been reported to be used in research for statistical analysis of metabolomics data (Catchpole *et al.*, 2005). Metabolomics data processing uses additional statistical methods such as orthogonal partial least squares (OPLS; Trygg and Wold, 2002), partial least squares discriminant analysis (PLS-DA; Barker and Rayens, 2003), hierarchical clustering (Murtagh, 1983), and *k*-means clustering (Murtagh, 1983). Statistical analysis is performed using commercial programs such as PLS-Toolbox (www.eigenvector.com/software/pls_toolbox.htm; accessed 17 January 2025) and SIMCA-P (http://umetrics.com/products/simca; accessed 17 January 2025). Statistical Metabolomics Analysis—An R Tool (SMART; Liang *et al.*, 2016) and specmine (Costa *et al.*, 2016) are two software tools created in R that are also being utilized.

5.4.4 Data interpretation

Once the metabolomics data have been statistically analyzed, the selected compounds need to be associated with certain metabolic pathways. Ad hoc software tools for metabolic pathway analysis include metPA (Xia *et al.*, 2011), MetExplore (Cottret *et al.*, 2018), Metabolite Set Enrichment Analysis (MSEA; Xia and Wishart, 2010), and metaP-server (Kastenmüller *et al.*, 2011), mapping to openly accessible biochemical pathway databases, such as KEGG (Ogata *et al.*, 1998). Furthermore, there are databases accessible for particular plant metabolomics pathways, such as Arabidopsis Reactome (Tsesmetzis *et al.*, 2008), MapMan (Thimm *et al.*, 2004), KaPPA-View (Tokimatsu *et al.*, 2005), and Plant Reactome (Naithani *et al.*, 2020). An example of one of these databases is PlantCyc. Although the information provided by these biochemical pathways is significant, the majority of metabolic pathways in plants are redundant and extensively intertwined, making it difficult to interpret the data from the metabolome biologically. A previous identification of metabolites is necessary for the analytic methods used to find perturbations in pathways. According to Heuberger *et al.* (2014), analysis using an untargeted metabolomics approach has gained more traction recently.

5.5 Case Studies

More than 200,000 metabolites are thought to exist in plants, and each species or cultivated variety of plants may have a unique pattern of chemotypic expression (Dixon and Strack, 2003). Hence, the use of traditional chemical assays to ascertain the modifications in plant metabolome status caused by insect herbivory is exceedingly difficult. MS- and NMR-based metabolomics techniques offer broader and non-targeted perception of metabolite profiles in host plants under biotic stress because of the swift advancement of omic technology and chemometric methodology (Fiehn, 2002). Plant–insect interactions have been studied extensively using the metabolomics approach (Jansen *et al.*, 2009; Marti *et al.*, 2013; Lu *et al.*, 2018).

5.5.1 Model systems

Metabolomic studies are being used extensively to understand plant–herbivore interactions in many model systems such as rice–planthopper interactions. For instance, metabolomic analysis of the leaf sheath of YHY15 rice plants revealed a distinct total of 26 metabolites. However, analysis of the honeydew produced following feeding by the brown planthopper, *Nilaparvata lugens* (Hemiptera: Delphacidae), revealed 33 metabolites. Metabolomics analysis suggested that amino acid absorption in YHY15 plants increased following brown planthopper infestation (Peng *et al.*, 2016). A comparative metabolomic investigation was also performed to understand the varied metabolomic patterns in susceptible and resistant rice varieties against brown planthopper (Kang *et al.*, 2019).

A metabolomic analysis examined the molecular mechanism behind rice's *Bph30*-mediated resistance to brown planthoppers and offered theoretical recommendations for developing brown planthopper-resistant rice cultivators (Shi *et al.*, 2023). High-performance liquid chromatography (HPLC) analysis revealed the production of ovicidal benzyl benzoate by *japonica* rice varieties, which was induced by elicitors in female white-backed planthoppers, *Sogatella furcifera* (Horvath), and killed the eggs of the plant hopper on the rice plants (Yang *et al.*, 2013). More recently, the secondary metabolite tryptamine and additionally two terpenoids and 11 unknown compounds were observed in rice plants infested by the brown planthopper (Gu *et al.*, 2024). The defense mechanism employed by monocots against insects has also been linked to the salicylic acid pathway (Qi *et al.*, 2018). Gibberellins, in addition to jasmonic acid, salicylic acid, and ethylene, have been proposed as regulators of herbivore resistance in rice (Qi *et al.*, 2018). A study was conducted using HPLC-MS/MS, which reported that by modifying the transcript levels of the rice genes *OsMPK3* and *OsWRKY24/45/53/70*, the rice DELLA gene *OsSLR1* positively regulates brown planthopper-induced levels of jasmonic acid, ethylene, and hydrogen peroxide (Zhang *et al.*, 2017).

5.5.2 Unravelling defense mechanisms

Significant alterations in plant primary and secondary metabolites as a result of pest infection have been identified by metabolomics analysis (Allwood *et al.*, 2008). Plants employ a variety of tactics to activate defense mechanisms against pest attacks (Manickam *et al.*, 2023). However, plant cells contain a wide variety of metabolites, making it difficult to decipher a species' whole metabolome (Tenenboim and Brotman, 2016). Herbivores with specialized metabolites that are tailored to the particular habitat of plant communities may have distinct effects on their host plants. For instance, when cabbage butterflies (*Pieris brassicae*) and flea beetles (*Phyllotreta nemorum*) infest white cabbage (*Brassica oleraceae* var. *capitata*), there is a rise in phenolic acids and flavonoids, particularly quercetin (Kovalikova *et al.*, 2019). An investigation demonstrated that, in comparison with null mutants lacking 3-deoxyanthocynadin, a flavonoid found in wild-type sorghum, there was significantly increased mortality and decreased population growth in maize leaf aphids (*Rhopalosiphum maidis*). Similarly, pericarp extracts rich in polyphenols from purple maize had a deleterious effect on the growth, development, and adult fitness features of tobacco hornworms (*Manduca sexta*) (Tayal *et al.*, 2020). These findings are supported by a substantial amount of research showing that several polyphenolic groups work together to defend a majority of plant species against a broad variety of insects. Plants have developed natural defense mechanisms to fend against diseases and insect pests by releasing glucosinolates, which are naturally occurring repellents. Brassicales produce three different forms of glucosinolates: aliphatic, indolic, and benzoic. When insects attack, the glucosinolates and the enzymes that hydrolyze them (myrosinases) emit harmful degradation products that repel generalist insects while drawing in specialists with some of the degradation products (Chhajed *et al.*, 2020). Certain herbivores' foraging signals have also been found to be guided by sugars and benzoxazinoids (Machado *et al.*, 2021). Plants produce secondary metabolites to protect themselves from herbivores by regulating defense signaling pathways and performing defensive functions. Damage caused by herbivores starts intricate processes that eventually result in plant secondary metabolite synthesis and buildup. The production of these metabolites is controlled by the interaction of signaling molecules called phytohormones. A metabolomic study revealed that application of oral sections from *Helicoverpa amigera* and *Spodoptera litura* strongly induced the jasmonic acid signaling pathway in cotton (Si *et al.*, 2020). Another metabolomic study reported hormone signaling following brown plant hopper infestation in rice (Zhang *et al.*, 2022). Hence, metabolomic analysis combined with other omics technologies have been used extensively to elucidate phytohormone signal transduction induced by plant–herbivore interactions.

5.6 Applications and Future Directions

5.6.1 Potential applications

The potential applications of high-throughput metabolomics in agriculture, pest management, and plant breeding are discussed below.

Agriculture

Different aspects of agriculture such as the development and differentiation processes of crops, fruit ripening, and resistance to adverse environmental conditions, stress-related challenges, pathogens, and pest attacks are closely associated with the content and composition of metabolites. Numerous analytical methods can be used to study a wide range of compounds, some of which are substance specific.

Rice (*Oryza sativa*) is one of the most important crops globally, especially in Asia. It has been studied extensively, with much of the research focusing on identifying cultivars through metabolic profiling of different plant parts, primarily leaves, using MS and capillary electrophoresis (Sato *et al.*, 2008). Moreover, rice has been a key subject in functional genomics, which integrates genetic and metabolic studies using techniques such as GC-flame ionization detection, GC-MS, and GC-TOF-MS (Gong *et al.*, 2013).

Numerous metabolomic investigations have been conducted on maize (*Zea mays*), many of which are related to genetic research. For instance, the metabolomic profile of its leaves was determined in order to use GC-MS for genetic mapping (Riedelsheimer *et al.*, 2012). Conversely, a study involving 702 genotypes gathered from various geographic zones and growing conditions examined the genetic basis of the metabolic diversity in its grains, quantifying 983 metabolites (Wen *et al.*, 2014). In another study, GC-MS was used to analyze heredity, various geographic locations, and environmental conditions (Röhlig *et al.*, 2009). The authors were able to distinguish the chemical composition of the genotypes under analysis, which helped them identify noteworthy chemical characteristics that varied depending on the region of origin.

Potato (*Solanum tuberosum*) is a crop that has been studied extensively in many fields, including metabolomics. These studies have had a variety of goals, from characterizing certain cultivars and elucidating the metabolic profile, to determining the chemical features of the tubers by GC-MS, to characterizing the relationship between composition and quality parameters (Carreno-Quintero *et al.*, 2012). Utilizing various techniques such as GC-TOF-MS and flow injection electrospray-MS, differences and similarities can be determined between related species or cultivars utilizing a chemotaxonomic categorization obtained in agriculture through the application of metabolomics.

The volatiles from fruit, peel, and pulp have been extensively researched and analyzed using ultra-performance liquid chromatography (UPLC)-Q-TOF-MS technology (Mintz-Oron *et al.*, 2008). Proton NMR has also been used to characterize the carotenoid component profile of the fruit. Through LC-MS, chemical profiles of tomato plants at various phases of growth have been obtained (Moco *et al.*, 2006). Moco *et al.* (2008) also conducted a similar study in which they used LC-MS and ^{1}H-NMR to obtain databases from the fruit metabolome. Using MALDI/TOF-MS, the basic characterization of fruits from several cultivars was accomplished.

Pest management

Some compounds directly harm herbivores and act as protective measures, such as toxic chemicals and substances that inhibit digestion. In contrast, indirect defense mechanisms involve releasing volatile signals that attract predators of the herbivores. Among the most common classes of chemicals that are important for indirect plant protection are VOCs. VOCs can be released immediately after a herbivore does harm, or they can be produced and released several hours later (Brilli *et al.*, 2019). After being infested by chewing herbivores (*Spodoptera* spp.) and sucking herbivores (*Aphis gossypii*), melon and cotton plants colonized by the entomopathogenic fungus *Beauveria bassiana* emitted distinct blends of volatile compounds, identified through high-resolution GC-MS analysis. In contrast, the control plants, which were not colonized by *B. bassiana*, did not exhibit these changes. This study suggested that the aromatic compound benzaldehyde could potentially serve as a marker for wild cotton that is simultaneously colonized by the fungus and attacked by pests (González-Mas *et al.*, 2021). Additionally, transcriptome and volatilome analysis of cotton plants showed an increase in transcripts associated with VOC synthesis and the jasmonate and salicylic acid pathways when *Apolygus lucorum* was fed on the plants. Similarly, a synthetic version of methyl salicylate and blends of cotton volatiles induced by insect feeding consistently attracted *Peristenus spretus* parasitoids (Huang *et al.*, 2022). Increased VOC discharge and phenolic chemical abundance were found in the volatilome and metabolome of *Populus × canescens* after *Laccaria bicolor* pretreatment and *Chrysomela populi* infestation, according to LC-MS analysis (Sivaprakasam Padmanaban *et al.*, 2022). By using headspace solid-phase microextraction coupled with GC-MS and UPLC-MS/MS to quantify VOCs and various metabolites, researchers found that *Tuta absoluta* feeding on tomato and eggplant showed differences in these regulatory patterns (Chen *et al.*, 2023).

Plant breeding

Metabolite profiling is a valuable tool that can steer the breeding process in the direction of identifying desirable characteristics during the early phases of selection. It can also be used to evaluate genotypic variation before developing molecular tools specific to a given species. Over the past 10 years, considerable progress in designing software tools and instrumentation innovation has been made possible by high-throughput metabolome analysis. Biotechnology, functional genomics, precision plant breeding, and disease diagnostics are just a few of the research areas that have profited from the applications of metabolomics and the movement toward translational metabolomics (Guijas *et al.*, 2018). Additionally, its use opens the door for translational metabolomics in plant breeding. Postgenomic approaches have sped up the screening process, and integrating metabolomics with other high-throughput tools will shorten the time required to develop elite crop varieties with enhanced resistance to biotic and abiotic stresses. Metabolomics has a great deal of potential to offer a comprehensive analysis of the diagnosis of various metabolites and phenotyping of plants (Fernie and Schauer, 2009). In 524 varieties of rice, 840 different metabolite units have been identified, which may be useful for future crop breeding (Chen *et al.*, 2014). Researchers are employing holistic approaches such as transcriptomics, proteomics, and metabolomics to map quantitative traits and examine genetic variations at the levels of mRNA, proteins, and metabolites. These methods are referred to as epigenomic quantitative trait loci (eQTL), proteomic QTL (pQTL), and metabolic QTL (mQTL) analyses. The metabolic traits of plants can be detected through mQTLs and genome-wide association studies (GWAS) combined with metabolomics techniques (mGWAS) (Gong *et al.*, 2013). In metabolic engineering, the important alleles found in crops including tomato, wheat, rice, sesame, broccoli, mustard, and *Arabidopsis* have been employed.

A paradigm shift in the breeding process will result from easy access to the many omics platforms, which will make it easier to pick plants based on genome-scale data produced at various stages of biological processes. Plant breeders will eventually adopt these advancements, which will enable them to make more educated choices. More recently, mGWAS have become a useful technique for elucidating the underlying genetic basis of various metabolic alterations in the plant metabolome. Phenotypic GWAS have effectively identified novel candidate genes in rice. Hundreds of distinct chemicals were discovered during the discovery of biomarker metabolites in *Arabidopsis* wild type and mutants using an integrated approach to metabolic profiling (Manickam *et al.*, 2023).

5.6.2 Promising future research avenues

Recent developments in plant metabolomics have made it possible to precisely choose desired features and have created opportunities for the study of plants with metabolite modifications. The technological shift from individual metabolite analysis to high-throughput assays that provide fingerprints of multiple compounds simultaneously has made it easier to identify strong biomarkers and develop improved models for metabolite networks. Over the past 10 years, the application of metabolomics in combination with other omics technologies has not only revealed a multitude of previously unidentified metabolites but has also made it possible to ascertain precisely how these metabolites contribute to the improvement of important plant characteristics such as yield, quality, and shelf life. In order to clarify the architecture of metabolic characteristics, high-throughput metabolomic techniques such as genotyping/sequencing technologies based on next-generation sequencing technology have proven to be a great help at a reasonable cost. A major boost to metabolomics-assisted breeding could come from recently established techniques such as GWAS, GS, and epigenome-wide association studies that enable the effective integration of metabolite profiling.

Metabolic marker identification to study different metabolic pathways in plants and to predict the type and extent of various stresses (biotic/abiotic) are potential future uses for metabolomics. Metabolism-assisted breeding is one method that will be widely used in agricultural improvement efforts to produce crop varieties that are climate smart and high yielding while tolerant to stress. Additionally, utilizing advanced genome-editing technologies such as clustered regularly interspaced short palindromic repeats (CRISPR)/Cas9, metabolic profiling of genome-edited plants can be done using metabolomics techniques to address regulatory concerns and risk assessment associated with genetically modified crops. Speed breeding is another fascinating area where metabolomics might provide a significant contribution toward crop improvement.

5.7 Conclusion

High-throughput metabolomics has significantly deepened our understanding of the intricate interactions between plants and herbivores. By analyzing the changes in the plant's metabolome before and after herbivore interaction, scientists have identified key metabolites that play critical roles in deterring herbivores or attracting natural enemies of the herbivores. This knowledge paves the way for developing pest-resistant crops, potentially reducing our reliance on chemical pesticides, and leading to the discovery of new, plant-derived natural products for pest control. By uncovering the hidden chemical language that governs these interactions, metabolomics has revolutionized our approach to plant defense mechanisms, with promising implications for ecological research, agriculture, and sustainable pest management. Although the complexity of metabolomes and the technical challenges involved present obstacles, the potential of high-throughput metabolomics to drive future discoveries is vast. As technology and computational methods continue to advance, this field holds the promise of uncovering novel metabolites, enhancing our understanding of biological processes, and contributing to personalized medicine, agriculture, and environmental science. With each advancement, we move closer to a future where ecological management is informed by a deep understanding of the natural processes that sustain life.

5.8 References

Adams, R.P. (2007) *Identification of Essential Oil Components by Gas Chromatography–Mass Spectrometry*. Allured Publishing Corporation, Carol Stream, Illinois.

Allwood, J.W., Ellis, D.I., and Goodacre, R. (2008) Metabolomic technologies and their application to the study of plants and plant–host interactions. *Physiologia Plantarum* 132, 117–135.

Allwood, J.W., Williams, A., Uthe, H., van Dam, N.M., Mur, L.A., *et al.* (2021) Unravelling plant responses to stress—the importance of targeted and untargeted metabolomics. *Metabolites* 11: 558.

Altemimi, A., Watson, D.G., Choudhary, R., Dasari, M.R., and Lightfoot, D.A. (2016) Ultrasound assisted extraction of phenolic compounds from peaches and pumpkins. *PLOS ONE* 11: e0148758.

Ausloos, P., Clifton, C.L., Lias, S.G., Mikaya, A.I., Stein, S.E., *et al.* (1999) The critical evaluation of a comprehensive mass spectral library. *Journal of the American Society for Mass Spectrometry* 10, 287–299.

Badri, D.V. and Vivanco, J.M. (2009) Regulation and function of root exudates. *Plant, Cell and Environment* 32, 666–681.

Barker, M. and Rayens, W. (2003) Partial least squares for discrimination. *Journal of Chemometrics* 17, 166–173.

Bojko, B., Reyes-Garcés, N., Bessonneau, V., Goryński, K., Mousavi, F., *et al.* (2014) Solid-phase microextraction in metabolomics. *Trends in Analytical Chemistry* 61, 168–180.

Brilli, F., Loreto, F., and Baccelli, I. (2019) Exploiting plant volatile organic compounds (VOCs) in agriculture to improve sustainable defense strategies and productivity of crops. *Frontiers in Plant Science* 10: 264.

Cajka, T. and Fiehn, O. (2016) Toward merging untargeted and targeted methods in mass spectrometry-based metabolomics and lipidomics. *Analytical Chemistry* 88, 524–545.

Carreno-Quintero, N., Acharjee, A., Maliepaard, C., Bachem, C.W., Mumm, R., *et al.* (2012) Untargeted metabolic quantitative trait loci analyses reveal a relationship between primary metabolism and potato tuber quality. *Plant Physiology* 158, 1306–1318.

Catchpole, G.S., Beckmann, M., Enot, D.P., Mondhe, M., Zywicki, B., *et al.* (2005) Hierarchical metabolomics demonstrates substantial compositional similarity between genetically modified and conventional potato crops. *Proceedings of the National Academy of Sciences USA* 102: 14458–14462.

Chen, T., Chen, L., Wang, J., Cheng, J., Yi, S., *et al.* (2023) Development of attractants and repellents for *Tuta absoluta* based on plant volatiles from tomato and eggplant. *Frontiers in Sustainable Food Systems* 7: 1155317.

Chen, W., Gao, Y., Xie, W., Gong, L., Lu, K., *et al.* (2014) Genome-wide association analyses provide genetic and biochemical insights into natural variation in rice metabolism. *Nature Genetics* 46, 714–721.

Chhajed, S., Mostafa, I., He, Y., Abou-Hashem, M., El-Domiaty, M., and Chen, S. (2020) Glucosinolate biosynthesis and the glucosinolate–myrosinase system in plant defense. *Agronomy* 10: 11.

Chong, J., Baltz, R., Schmitt, C., Beffa, R., Fritig, B., and Saindrenan, P. (2005) Downregulation of a pathogen-responsive tobacco UDP-Glc:phenylpropanoid glucosyltransferase reduces scopoletin glucoside accumulation, enhances oxidative stress, and weakens virus resistance. *Plant Cell* 14, 1093–1107.

Ciccimaro, E. and Blair, I.A. (2010) Stable-isotope dilution LC–MS for quantitative biomarker analysis. *Bioanalysis* 2, 311–341.

Clasquin, M.F., Melamud, E., and Rabinowitz, J.D. (2012) LC-MS data processing with MAVEN: a metabolomic analysis and visualization engine. *Current Protocols in Bioinformatics* 37, 14.11.1–14.11.23.

Corrales, A.R., Carrillo, L., and Lasierra, P. (2017) Multifaceted role of cycling DOF factor 3 (CDF3) in the regulation of flowering time and abiotic stress responses in *Arabidopsis*. *Plant, Cell and Environment* 40, 748–764.

Costa, C., Maraschin, M., and Rocha, M. (2016) An R package for the integrated analysis of metabolomics and spectral data. *Computer Methods and Programs in Biomedicine* 129, 117–124.

Cottret, L., Frainay, C., Chazalviel, M., Cabanettes, F., Gloaguen, Y., *et al.* (2018) MetExplore: collaborative edition and exploration of metabolic networks. *Nucleic Acids Research* 46, W495–W502.

Cui, Q., Lewis, I.A., Hegeman, A.D., Anderson, M.E., Li, J., *et al.* (2008) Metabolite identification via the Madison Metabolomics Consortium Database. *Nature Biotechnology* 26, 162–164.

Daly, R., Rogers, S., Wandy, J., Jankevics, A., Burgess, K.E., *et al.* (2014) MetAssign: probabilistic annotation of metabolites from LC–MS data using a Bayesian clustering approach. *Bioinformatics* 30, 2764–2771.

Divekar, P.A., Narayana, S., Divekar, B.A., Kumar, R., Gadratagi, B.G., *et al.* (2022) Plant secondary metabolites as defense tools against herbivores for sustainable crop protection. *International Journal of Molecular Sciences* 23: 2690.

Dixon, R.A. and Paiva, N.L. (1995) Stress-induced phenylpropanoid metabolism. *Plant Cell* 7: 1085.

Dixon, R.A. and Strack, D. (2003) Phytochemistry meets genome analysis, and beyond. *Phytochemistry* 62, 815–816.

Dunn, W.B. (2008) Current trends and future requirements for the mass spectrometric investigation of microbial, mammalian and plant metabolomes. *Physical Biology* 5: 011001.

Estarague, A., Violle, C., Vile, D., Hany, A., Martino, T., *et al.* (2023) Plant–herbivore interactions: experimental demonstration of genetic variability in plant–plant signalling. *Evolutionary Applications* 16, 772–780.

Fait, A., Fromm, H., Walter, D., Galili, G., and Fernie, A.R. (2006) Highway or byway: the metabolic role of the GABA shunt in plants. *Trends in Plant Science* 13, 14–19.

Fernie, A.R. and Schauer, N. (2009) Metabolomics-assisted breeding: a viable option for crop improvement? *Trends in Genetics* 25, 39–48.

Fiehn, O. (2002) Metabolomics—the link between genotypes and phenotypes. *Plant Molecular Biology* 48, 155–171.

Fukushima, A. and Kusano, M. (2014) Recent progress in the development of metabolome databases for plant systems biology. *Frontiers in Plant Science* 4: 73.

Giavalisco, P., Li, Y., and Matthes, A. (2011) Elemental formula annotation of polar and lipophilic metabolites using C, N and S isotope labelling, in combination with high-resolution mass spectrometry. *Plant Journal* 68, 364–376.

Gong, B. and Zhang, G. (2014) Interactions between plants and herbivores: a review of plant defense. *Acta Ecologica Sinica* 34, 325–336.

Gong, L., Chen, W., Gao, Y., Liu, X., Zhang, H., *et al.* (2013) Genetic analysis of the metabolome exemplified using a rice population. *Proceedings of the National Academy of Sciences USA* 110, 20320–20325.

Gong, Z.-G., Hu, J., Wu, X., and Xu, Y.-J. (2017) The recent developments in sample preparation for mass spectrometry-based metabolomics. *Critical Reviews in Analytical Chemistry* 347, 325–331.

González-Mas, N., Gutiérrez-Sánchez, F., Sánchez-Ortiz, A., Grandi, L., Turlings, T.C.J., *et al.* (2021) Endophytic colonization by the entomopathogenic fungus *Beauveria Bassiana* affects plant volatile emissions in the presence or absence of chewing and sap-sucking insects. *Frontiers in Plant Science* 12: 660460.

Gu, C., Zhang, Y., Wang, M., Lin, Y., Zeng, B., *et al.* (2024) Metabolomic profiling reveals the anti-herbivore mechanisms of rice (*Oryza sativa*). *International Journal of Molecular Sciences* 25: 11.

Guijas, C., Montenegro-Burke, J.R., Warth, B., Spilker, M.E., and Siuzdak, G. (2018) Metabolomics activity screening for identifying metabolites that modulate phenotype. *Nature Biotechnology* 36, 316–320.

Halket, J.M., Waterman, D., Przyborowska, A.M., Patel, R.K.P., Fraser, P.D., and Bramley, P.M. (2005) Chemical derivatization and mass spectral libraries in metabolic profiling by GC/MS and LC/MS/MS. *Journal of Experimental Botany* 56, 219–243.

Harbourne, N., Marete, E., Jacquier, J.C., and O'Riordan, D. (2009) Effect of drying methods on the phenolic constituents of meadowsweet (*Filipendula ulmaria*) and willow (*Salix alba*). *LWT—Food Science and Technology* 42, 1468–1473.

Heuberger, A.L., Robison, F.M., Lyons, S.M.A., Broeckling, C.D., and Prenni, J.E. (2014) Evaluating plant immunity using mass spectrometry-based metabolomics workflows. *Frontiers in Plant Science* 5: 291.

Horai, H., Arita, M., Kanaya, S., Nihei, Y., Ikeda, T., *et al.* (2010) MassBank: a public repository for sharing mass spectral data for life sciences. *Journal of Mass Spectrometry* 45, 703–714.

Huang, X., Kou, J., Jing, W., Han, X., Liu, D., *et al.* (2022) Transcriptomic and metabolomic reprogramming in cotton after *Apolygus lucorum* feeding implicated in enhancing recruitment of the parasitoid *Peristenus spretus*. *Journal of Pest Science* 95, 249–262.

Jansen, J.J., Allwood, J.W., Marsden-Edwards, E., van der Putten, W.H., Goodacre, R., and van Dam, N.M. (2009) Metabolomic analysis of the interaction between plants and herbivores. *Metabolomics* 5, 150–161.

Kamali, S. and Singh, A. (2023) Genomic and transcriptomic approaches to developing abiotic stress-resilient crops. *Agronomy* 13: 2903.

Kang, K., Yue, L., Xia, X., Liu, K., and Zhang, W. (2019) Comparative metabolomics analysis of different resistant rice varieties in response to the brown planthopper *Nilaparvata lugens* Hemiptera: Delphacidae. *Metabolomics* 15: 62.

Kant, M.R., Jonckheere, W., Knegt, B., Lemos, F., Liu, J., *et al.* (2015) Mechanisms and ecological consequences of plant defence induction and suppression in herbivore communities. *Annals of Botany* 115, 1015–1051.

Kaplan, F., Kopka, J., Sung, D.Y., Zhao, W., Popp, M., *et al.* (2004) Transcript and metabolite profiling during cold acclimation of *Arabidopsis* reveals an intricate relationship of cold-regulated gene expression with modifications in metabolite content. *Plant Journal* 37, 869–890.

Kastenmüller, G., Römisch-Margl, W., Wägele, B., Altmaier, E., and Suhre, K. (2011) Meta P-server: a web-based metabolomics data analysis tool. *Journal of Biomedicine and Biotechnology* 2011: 839862.

Katajamaa, M., Miettinen, J., and Orešič, M. (2006) MZmine: toolbox for processing and visualization of mass spectrometry based molecular profile data. *Bioinformatics* 22, 634–636.

Kessner, D., Chambers, M., Burke, R., Agus, D., and Mallick, P. (2008) ProteoWizard: open source software for rapid proteomics tools development. *Bioinformatics* 24, 2534–2536.

Kim, H.K., Choi, Y.H., and Verpoorte, R. (2010) NMR-based metabolomic analysis of plants. *Nature Protocols* 5, 536–549.

Kind, T., Wohlgemuth, G., Lee, D.Y., Lu, Y., Palazoglu, M., *et al.* (2009) FiehnLib: mass spectral and retention index libraries for metabolomics based on quadrupole and time-of-flight gas chromatography/mass spectrometry. *Analytical Chemistry* 81, 10038–10048.

Kopka, J., Schauer, N., Krueger, S., Birkemeyer, C., Usadel, B., *et al.* (2005) GMD@CSB.DB: the Golm metabolome database. *Bioinformatics* 21, 1635–1638.

Kovalikova, Z., Kubes, J., Skalicky, M., Kuchtickova, N., Maskova, L., *et al.* (2019) Changes in content of polyphenols and ascorbic acid in leaves of white cabbage after pest infestation. *Molecules* 24: 14.

Kuhl, C., Tautenhahn, R., Böttcher, C., Larson, T.R., and Neumann, S. (2012) CAMERA: an integrated strategy for compound spectra extraction and annotation of liquid chromatography/mass spectrometry data sets. *Analytical Chemistry* 84, 283–289.

Liang, Y.-J., Lin, Y.-T., Chen, C.-W., Lin, C.-W., Chao, K.-M., *et al.* (2016) SMART: statistical metabolomics analysis—an R tool. *Analytical Chemistry* 88, 6334–6341.

Lisec, J., Schauer, N., Kopka, J., Willmitzer, L., and Fernie, A.R. (2006) Gas chromatography mass spectrometry–based metabolite profiling in plants. *Nature Protocols* 1, 387–396.

Lu, H., Zhang, S., Luo, J., Wang, C., Lv, L., *et al.* (2018) Integrative proteomic and metabolomic analysis reveals the cotton plant defense mechanisms induced by insect (*Adelphocoris suturalis Jakovlev*) feeding. Available at: https://doi.org/10.20944/preprints201809.0058.v1 (accessed 29 January 2025).

Maag, D., Erb, M., and Glauser, G. (2015) Metabolomics in plant–herbivore interactions: challenges and applications. *Entomologia Experimentalis et Applicata* 157, 18–29.

Machado, R.A.R., Theepan, V., Robert, C.A.M., Züst, T., Hu, L., *et al.* (2021) The plant metabolome guides fitness-relevant foraging decisions of a specialist herbivore. *PLOS Biology* 19: e3001114.

Manickam, S., Rajagopalan, V.R., Kambale, R., Rajasekaran, R., Kanagarajan, S., and Muthurajan, R. (2023) Plant metabolomics: current initiatives and future prospects. *Current Issues in Molecular Biology* 45: 8894–8906.

Markley, J.L., Anderson, M.E., Cui, Q., Eghbalnia, H.R., Lewis, I.A., *et al.* (2007) New bioinformatics resources for metabolomics. *Pacific Symposium on Biocomputing* 12, 157–168.

Martens, L., Chambers, M., Sturm, M., Kessner, D., Levander, F., *et al.* (2011) mzML—a community standard for mass spectrometry data. *Molecular and Cellular Proteomics* 10: R110.000133.

Marti, G., Erb, M., Boccard, J., Glauser, G., Doyen, G.R., *et al.* (2013) Metabolomics reveals herbivore-induced metabolites of resistance and susceptibility in maize leaves and roots. *Plant, Cell and Environment* 36, 621–639.

McLafferty, F.W. and Turecek, F. (1993) *Interpretation of Mass Spectra*. University Science Books, Sausalito, California.

Mintz-Oron, S., Mandel, T., Rogachev, I., Feldberg, L., Lotan, O., *et al.* (2008) Gene expression and metabolism in tomato fruit surface tissues. *Plant Physiology* 147, 823–851.

Mithöfer, A. and Maffei, M.E. (2016) General mechanisms of plant defense and plant toxins. In: Carlini, C. and Ligabue-Braun, R. (eds) *Plant Toxins*. Springer, Dordrecht, the Netherlands, pp. 3–24.

Moco, S., Bino, R.J., Vorst, O., Verhoeven, H.A., de Groot, J., *et al.* (2006) A liquid chromatography-mass spectrometry-based metabolome database for tomato. *Plant Physiology* 141, 1205–1218.

Moco, S., Forshed, J., de Vos, R.C., Bino, R.J., and Vervoort, J. (2008) Intra-and inter-metabolite correlation spectroscopy of tomato metabolomics data obtained by liquid chromatography-mass spectrometry and nuclear magnetic resonance. *Metabolomics* 4, 202–215.

Murtagh, F. (1983) A survey of recent advances in hierarchical clustering algorithms. *Computer Journal* 26, 354–359.

Naithani, S., Gupta, P., Preece, J., d'Eustachio, P., Elser, J.L., *et al.* (2020) Plant Reactome: a knowledge-base and resource for comparative pathway analysis. *Nucleic Acids Research* 48, D1093–D1103.

Obata, T. and Fernie, A.R. (2012) The use of metabolomics to dissect plant responses to abiotic stresses. *Cellular and Molecular Life Sciences* 69, 3225–3243.

Ogata, H., Goto, S., Fujibuchi, W., and Kanehisa, M. (1998) Computation with the KEGG pathway database. *Biosystems* 47, 119–128.

Parida, A.K., Panda, A., and Rangani, J. (2018) Metabolomics-guided elucidation of abiotic stress tolerance mechanisms in plants. In: Ahmad, P., Ahanger, M.A., Singh, V.P., Tripathi, D.K., Alam, P., and Alyemeni, M.N. (eds) *Plant Metabolites and Regulation Under Environmental Stress*. Academic Press, Cambridge, Massachusetts, pp. 89–131.

Patel, M.K., Mishra, A., and Jha, B. (2016) Untargeted metabolomics of halophytes. In: Kim, S. (ed.) *Marine Omics: Principles and Applications*. CRC Press, Boca Raton, Florida, pp. 309–325.

Peng, L., Zhao, Y., Wang, H., Zhang, J., Song, C., *et al.* (2016) Comparative metabolomics of the interaction between rice and the brown planthopper. *Metabolomics* 12: 132.

Perkovich, C. and Ward, D. (2022) Differentiated plant defense strategies: herbivore community dynamics affect plant–herbivore interactions. *Ecosphere* 13: e3935.

Pritchard, S.G. and Rogers, H.H. (2000) Spatial and temporal deployment of crop roots in CO_2-enriched environments. *New Phytologist* 147, 55–71.

Qi, J., Malook, S., Shen, G., Gao, L., Zhang, C., *et al.* (2018) Current understanding of maize and rice defense against insect herbivores. *Plant Diversity* 40, 189–195.

Ratcliffe, R.G. and Shachar-Hill, Y. (2001) Probing plant metabolism with NMR. *Annual Review of Plant Biology* 52, 499–526.

Riedelsheimer, C., Lisec, J., Czedik-Eysenberg, A., Sulpice, R., Flis, A., *et al.* (2012) Genome-wide association mapping of leaf metabolic profiles for dissecting complex traits in maize. *Proceedings of the National Academy of Sciences USA* 109, 8872–8877.

Rioja, C., Zhurov, V., Bruinsma, K., Grbic, M., and Grbic, V. (2017) Plant–herbivore interactions: a case of an extreme generalist, the two-spotted spider mite *Tetranychus urticae*. *Molecular Plant–Microbe Interactions* 30, 935–945.

Rizhsky, L., Liang, H., and Mittler, R. (2004) The water–water cycle is essential for chloroplast protection in the absence of stress. *Journal of Biological Chemistry* 279, 38921–38925.

Roessner, U., Luedemann, A., Brust, D., Fiehn, O., Linke, T., *et al.* (2001) Metabolic profiling allows comprehensive phenotyping of genetically or environmentally modified plant systems. *Plant Cell* 13, 11–29.

Röhlig, R.M., Eder, J., and Engel, K.H. (2009) Metabolite profiling of maize grain: differentiation due to genetics and environment. *Metabolomics* 5, 459–477.

Salem, M.A., Perez de Souza, L., Serag, A., Fernie, A.R., Farag, M.A., *et al.* (2020) Metabolomics in the context of plant natural products research: from sample preparation to metabolite analysis. *Metabolites* 10: 37.

Sato, S., Arita, M., Soga, T., Nishioka, T., and Tomita, M. (2008) Time-resolved metabolomics reveals metabolic modulation in rice foliage. *BMC Systems Biology* 2: 51.

Schauer, N. and Fernie, A.R. (2006) Plant metabolomics: towards biological function and mechanism. *Trends in Plant Science* 11, 508–516.

Shi, S., Zha, W., Yu, X., Wu, Y., Li, S., *et al.* (2023) Integrated transcriptomics and metabolomics analysis provide insight into the resistance response of rice against brown planthopper. *Frontiers in Plant Science* 14: 1213257.

Si, H., Liu, H., Sun, Y., Xu, Z., Liang, S., *et al.* (2020) Transcriptome and metabolome analysis reveal that oral secretions from *Helicoverpa armigera* and *Spodoptera litura* influence wound-induced host response in cotton. *Crop Journal* 8, 929–942.

Silva-Navas, J., Moreno-Risuenom, M.A., and Manzanom, C. (2016) Flavonols mediate root phototropism and growth through regulation of proliferation-to differentiation transition. *Plant Cell* 28, 1372–1387.

Sivaprakasam Padmanaban, P.B., Rosenkranz, M., Zhu, P., Kaling, M., Schmidt, A., *et al.* (2022) Mycorrhiza-tree-herbivore interactions: alterations in poplar metabolome and volatilome. *Metabolites* 12: 93.

Smith, C.A., O'Maille, G., Want, E.J., Qin, C., Trauger, S.A., *et al.* (2005) METLIN: a metabolite mass spectral database. *Therapeutic Drug Monitoring* 27, 747–751.

Smith, C.A., Want, E.J., O'Maille, G., Abagyan, R., and Siuzdak, G. (2006) XCMS: processing mass spectrometry data for metabolite profiling using nonlinear peak alignment, matching, and identification. *Analytical Chemistry* 78, 779–787.

Smith, R.M. (2004) *Understanding Mass Spectra: A Basic Approach*. Wiley, Hoboken, New Jersey.

Sturm, M., Bertsch, A., Gröpl, C., Hildebrandt, A., Hussong, R., *et al.* (2008) OpenMS—an open-source software framework for mass spectrometry. *BMC Bioinformatics* 9: 163.

Sumner, L.W., Amberg, A., Barrett, D., Beale, M.H., Beger, R., *et al.* (2007) Proposed minimum reporting standards for chemical analysis Chemical Analysis Working Group (CAWG) Metabolomics Standards Initiative (MSI). *Metabolomics* 3, 211–221.

Tanna, B. and Mishra, A. (2018) Metabolomics of seaweeds: tools and techniques. In: Ahmad, P., Ahanger, M.A., Singh, V.P., Tripathi, D.K., Alam, P., and Alyemeni, M.N. (eds) *Plant Metabolites and Regulation Under Environmental Stress*. Academic Press, Cambridge, Massachusetts, pp. 37–52.

Tayal, M., Somavat, P., Rodriguez, I., Martinez, L., and Kariyat, R. (2020) Cascading effects of polyphenol-rich purple corn pericarp extract on pupal, adult, and offspring of tobacco hornworm (*Manduca sexta L.*). *Communicative and Integrative Biology* 13, 43–53.

Tenenboim, H. and Brotman, Y. (2016) Omic relief for the biotically stressed: metabolomics of plant biotic interactions. *Trends in Plant Science* 21, 781–791.

Teo, C.C., Chong, W.P.K., and Ho, Y.S. (2013) Development and application of microwave-assisted extraction technique in biological sample preparation for small molecule analysis. *Metabolomics* 9, 1109–1128.

Thimm, O., Bläsing, O., Gibon, Y., Nagel, A., Meyer, S., *et al.* (2004) MAPMAN: a user-driven tool to display genomics data sets onto diagrams of metabolic pathways and other biological processes. *Plant Journal* 37, 914–939.

Tokimatsu, T., Sakurai, N., Suzuki, H., Ohta, H., Nishitani, K., *et al.* (2005) KaPPA-view: a web-based analysis tool for integration of transcript and metabolite data on plant metabolic pathway maps. *Plant Physiology* 138, 1289–1300.

Trygg, J. and Wold, S. (2002) Orthogonal projections to latent structures (O-PLS). *Journal of Chemometrics* 16, 119–128.

Tsesmetzis, N., Couchman, M., Higgins, J., Smith, A., Doonan, J.H., *et al.* (2008) *Arabidopsis* reactome: a foundation knowledgebase for plant systems biology. *Plant Cell* 20, 1426–1436.

Tsugawa, H., Cajka, T., Kind, T., Ma, Y., Higgins, B., *et al.* (2015) MS-DIAL: data-independent MS/MS deconvolution for comprehensive metabolome analysis. *Nature Methods* 12, 523–526.

Uppal, K., Walker, D.I., and Jones, D.P. (2017) xMSannotator: an R package for network-based annotation of high-resolution metabolomics data. *Analytical Chemistry* 89, 1063–1067.

Velickovic, D., Chu, R.K., Myers, G.L., Ahkami, A.H., and Anderton, C.R. (2020) An approach for visualizing the spatial metabolome of an entire plant root system inspired by the Swiss-rolling technique. *Journal of Mass Spectrometry* 55: 4363.

Vinod, K. and Sabah, A. (2018) Plant defense against pathogens: the role of salicylic acid. *Research Journal of Biotechnology* 13, 97–103.

Wang, M., Carver, J.J., Phelan, V.V., *et al.* (2016) Sharing and community curation of mass spectrometry data with global natural products social molecular networking. *Nature Biotechnology* 34, 828–837.

War, A.R., Paulraj, M.G., Ahmad, T., Buhroo, A.A., Hussain, B., *et al.* (2012) Mechanisms of plant defense against insect herbivores. *Plant Signaling and Behavior* 7, 1306–1320.

Weckwerth, W. (2003) Metabolomics in systems biology. *Annual Review of Plant Biology* 54, 669–689.

Wen, W., Li, D., Li, X., Gao, Y., Li, W., *et al.* (2014) Metabolome-based genome-wide association study of maize kernel leads to novel biochemical insights. *Nature Communications* 5: 3438.

Wishart, D.S., Knox, C., Guo, A.C., Eisner, R., Young, N., *et al.* (2009) HMDB: a knowledgebase for the human metabolome. *Nucleic Acids Research* 37: D603–D610.

Xia, J. and Wishart, D.S. (2010) MSEA: a web-based tool to identify biologically meaningful patterns in quantitative metabolomic data. *Nucleic Acids Research* 38, W71–W77.

Xia, J., Wishart, D.S., and Valencia, A. (2011) MetPA: a web-based metabolomics tool for pathway analysis and visualization. *Bioinformatics* 26, 2342–2344.

Yang, J.-O., Nakayama, N., Toda, K., Tebayashi, S., and Kim, C.-S. (2013) Elicitor(s) in *Sogatella furcifera* (Horváth) causing the Japanese rice plant (*Oryza sativa* L.) to induce the ovicidal substance, benzyl benzoate. *Bioscience, Biotechnology, and Biochemistry* 77, 1258–1261.

Yu, T., Park, Y., Johnson, J.M., and Jones, D.P. (2009) apLCMS—adaptive processing of high-resolution LC/MS data. *Bioinformatics* 25, 1930–1936.

Yuliana, N.D., Khatib, A., Verpoorte, R., and Choi, Y.H. (2011) Comprehensive extraction method integrated with NMR metabolomics: a new bioactivity screening method for plants, adenosine A1 receptor binding compounds in *Orthosiphon stamineus* Benth. *Analytical Chemistry* 83, 6902–6906.

Zhang, J., Luo, T., Wang, W., Cao, T., Li, R., and Lou, Y. (2017) Silencing *OsSLR1* enhances the resistance of rice to the brown planthopper. *Plant, Cell and Environment* 40, 2147–2159.

Zhang, Q., Li, T., Gao, M., Ye, M., Lin, M., *et al.* (2022) Transcriptome and metabolome profiling reveal the resistance mechanisms of rice against brown planthopper *Nilaparvata lugens*. *International Journal of Molecular Sciences* 23, 4083.

Zhou, J. and Yin, Y. (2016) Strategies for large-scale targeted metabolomics quantification by liquid chromatography-mass spectrometry. *Analyst* 141, 6362–6373.

Zhou, J., Liu, H., Liu, Y., Liu, J., Zhao, X., and Yin, Y. (2016) Development and evaluation of a parallel reaction monitoring strategy for large-scale targeted metabolomics quantification. *Analytical Chemistry* 88: 4478.

Zuorro, A., Lavecchia, R., Medici, F., and Piga, L. (2013) Enzyme-assisted production of tomato seed oil enriched with lycopene from tomato pomace. *Food and Bioprocess Technology* 6, 3499–3509.

6 High-throughput Metabolomics for Studying Plant Stress Physiology Under a Changing Climate

Mahmoud F. Seleiman[1,2*], Awais Ahmad[1], Majed Alotaibi[1], Adel M. Al-Saif[1], Shivani Ranjan[3], Sumit Sow[3], and Harun H. Gitari[4]

[1]*Department of Plant Production, College of Food and Agriculture Sciences, King Saud University, Riyadh, Saudi Arabia;* [2]*Department of Crop Sciences, Faculty of Agriculture, Menoufia University, Shibin El-kom, Egypt;* [3]*Department of Agronomy, Dr. Rajendra Prasad Central Agricultural University, Pusa, Samastipur, Bihar, India;* [4]*Department of Agricultural Science and Technology, School of Agriculture and Environmental Sciences, Kenyatta University, Nairobi, Kenya*

Abstract

Climate change is an aggregative outcome of numerous interlinked meteorological anomalies and physical processes altering gradually over a long period of time. In the past few decades. Hyper-anthropogenic activities have disrupted the natural equilibrium of energy flow, accelerated pollution, and led to the accumulation of greenhouse gasses in the Earth's atmosphere. The consequent hyperthermal period, if unchecked, will lead to a hothouse Earth with profound implications at both planetary and societal levels. Changing climate is amplifying extreme weather events such as unpredicted spells of heat, drought, cold, and salinity around the world. Moreover, associated slow processes such as loss of soil fertility and biodiversity, pest infestation, genetic modifications, and alteration of seasonal periods are challenging sustainable agricultural production and threatening global food security. Thus, quantification of the impact of climate change on agricultural production has become inevitable, and in the past couple of decades efforts have been made in various scientific domains including omics studies. Climate change-associated stresses impact plants at cellular and metabolic levels, and thus an understanding of plant physiological responses and the underlying molecular mechanisms could help to mitigate and manage upcoming threats to crop production. This chapter covers the current status of climate change and future predictions in association with the impact on plant physiological processes. Advances in plant metabolomics and high-throughput analytical tools may facilitate our understanding of plant metabolite synthesis and cross-talk in response to external stimuli. The integration of plant metabolomics with other omics techniques such as transcriptomics and proteomics, with the assistance of specialized data-processing tools, may shape the future of stress computation. The amalgamation of existing knowledge with metabolomics may lead us to "metabolite biological markers" accessible through high-throughput analytical techniques with sufficient precision, reliability, and time effectiveness.

Keywords: climate change, high-throughput metabolomics, plant stress

* E-mail: mseleiman@ksu.edu.sa

© CAB International 2025. *High-Throughput Plant Metabolomics* (Jen-Tsung Chen ed.)
DOI: 10.1079/9781800626911.0006

6.1 Introduction

Climate change has occurred throughout the history of our planet, even before the existence of human civilization, in a periodic fashion of ice and warm cycles. However, over the last 10,000 years, climatic conditions have been quite stable to support life. Thus, about 5000 years ago, humans moved out of caves and created modern civilizations through an evolutionary process greatly influenced by sustainable climatic conditions. These stable climatic conditions gave birth to biodiversity, seasonal cycles, and thus the domestication of agriculture, which consequently promoted the domination of humankind in the ecosystem and led to unchecked population growth. In the last few decades, since the advent of the Industrial Revolution, the climate crisis has been greatly exacerbated as a consequence of anthropogenic activities (Massey, 2023). Ongoing climatic changes are occurring at a magnitude and speed unknown to us throughout human history, resulting in a sudden exposure to rapid climatic shift (Notz, 2020). For instance, the average global temperature has increased at a rate of 0.15–0.20°C per decade since 1975 and is expected to increase up to 5.8°C by the end of the current century, if unchecked (Arora *et al.*, 2005; Malhi *et al.*, 2021). Therefore, changes in climatic processes have attained significant attention from the scientific community due to the scale of their occurrence, magnitude of impact, and frequency of observable anomalies.

Climate change, generally defined as significant changes in meteorological elements over a long period, has become one of the biggest challenges to the survival of life on this planet in recent times (Malhi *et al.*, 2021). In the last few decades, enhanced human activity has resulted in an altered composition of the global atmosphere majorly contributed by anthropogenic hyper-emission of greenhouse gasses since the Industrial Revolution. The rising greenhouse gasses, mainly carbon dioxide, ozone, and water vapors, absorb energy emitted by the Earth and thus increase global warming and consequent extreme weather irregularities (Malhi *et al.*, 2021). It is expected that the average global temperature will rise by 2°C, even if anthropogenic emission of greenhouse gasses ceased immediately (Hare and Meinshausen, 2006). The climatic projections advocate the development of extreme heat, droughts, salinity, and cold stress spells in the near future, and such an undesirable complexity of stresses will significantly impose water shortages, soil fertility losses, pest infestations, and diseases in plants (Malhi *et al.*, 2021). Climate change is known to have adverse effects on agricultural production and is predicted to reduce global cereal production by 5.5% for wheat and 3.8% for maize by the end of 2050 (Lobell *et al.*, 2011). Therefore, to develop mitigation strategies for agriculture, an in-depth understanding of metabolic activities, physiological processes, and signal transduction mechanisms in plants under a changing climate is of vital importance. However, a highly efficient and functionally effective utilization of plant stress physiology coupled with modern high-tech sensing and analyzing techniques will be required for an advanced response and management in crop production.

Biology is an informational science (Hood *et al.*, 2004). As such, biologists are benefiting greatly from the recently developed and information-rich functional genomics technologies. These technologies, which are ultimately aimed at giving us full coverage of, for example, the total genome sequence or complete transcriptional analysis of an organism, are already expanding research horizons well beyond those envisaged just 5–10 years ago. The capacity to generate data on the molecular organization of living things has never been so great and, particularly with data coming from complementary, multivariate sources, this has resulted in a paradigm shift in our way of thinking regarding how we tackle complex biological questions. Multi-disciplinary approaches have become the order of the day. Divergent proficiencies are being combined to enable us to gain wholly new insights into, for example, cellular behavior, how plants are organized, and how they respond to genetic or environmental stimuli. Techniques such as DNA microarrays and proteomics have already greatly expanded our knowledge. Metabolomics will add to this, and indeed will benefit considerably from the additional advantage of not being reliant on having species-specific genomic information available for effective data analysis. However, while this switch from a reductionist to a more holistic approach has created many new opportunities, it has also raised a number of significant new challenges. This is particularly the case concerning the newest field of functional genomics—metabolomics—where attention is specifically focused on the biochemical complement of cells and tissues.

Since its establishment as an ambitious concept, metabolomics, in a short period of time, has rapidly grown to become a valuable technology to illustrate the molecular organization of multicellular organisms including plants. In the last few decades, scientific advances in various dimensions including analytical techniques, high-throughput capabilities, bioinformatics, genetic sequencing, molecular modeling, data mining, and dedicated statistical programs have achieved remarkable progress. Thus, understanding how plants react, respond, and manage an external stimulus has become considerably more precise and effective than ever before; however, integration with other techniques is needed for effective outcomes (Hall, 2006). Metabolomics offers a potential bridge between the phenotype and genotype of an organism and thus may significantly assist us in linking genes with their function in response to internal or external promoters or suppressants. Moreover, this advancement creates unique opportunities to understand functional biochemicals, metabolic pathways, and integration of existing knowledge of still-unexplored biodiversity in order to better mitigate the deleterious effects of changing climate with respect to sustainable food security.

6.2 Changing Climate: Current Scenarios and Future Projections

Since the creation of this planet, the climatic factors have been changing repeatedly, ultimately making the Earth habitable to support life. During this journey of about 5 billion years, the Earth went through a number of climatic transformations until it achieved periodic Ice Age and warm temperatures some million years ago (Notz, 2020). Interestingly, over the past 10,000 years, the climatic conditions became quite stable and Earth's ecosystem was able to absorb slow changes through natural phenomena, which led to agricultural developments and modern civilizations of humankind. However, in the past few decades, anthropogenic activities such as burning of fossil fuels, deforestation, and synthetic chemicals have resulted in a devastating imbalance of the natural equilibrium and consequent rapid climatic changes (Fig. 6.1). Hyper-rapid anthropogenic greenhouse gas release is the driving force of the current "hyperthermal" period, which, if unchecked, will lead to a "hothouse Earth" with profound implications both at planetary and societal levels (Zalasiewicz and Williams, 2021).

The Framework Convention on Climate Change (FCCC) in 1992 concluded that there was a significant association between the rapidly changing climate and anthropogenic activities (Dandotiya and Sharma, 2022). The subsequent impact on natural ecosystems, human civilization, and suitability of this planet could be disastrous. Hyperaccumulation of greenhouse gasses in the atmosphere alters the planet's energy equilibrium by increasing net influx and leading to progressive hyperthermal conditions also referred to as "global warming" (Karoly *et al.*, 2003; Mann and Jones, 2003; Massey, 2023). Moreover, such a natural disturbance results in unpredictable and frequent disasters including extreme heatwaves, droughts, floods, wildfires, and weather anomalies. In addition, average global temperature, melting of ice, rising sea levels, and shifts in climatic patterns have become significantly visible over the last century, considerably impacting the human population both directly and indirectly (Dandotiya and Sharma, 2022). According to the Intergovernmental Panel on Climate Change (IPCC), the average global temperature has increased during the 20th century ending with the 1990s as the warmest decade on record. Furthermore, the decline in ice cover, rising sea levels, and high heat contents were recurrently reported in the last few decades, which consequently resulted in changing high and low temperatures, cloud cover, precipitation, drought, and severe flooding. The IPCC has projected a 1.4–5.8°C increase in average global temperatures for the current century (2000–2100) (Fankhauser and Tol, 1997; Trenberth, 2001; Parry, 2007; O'neill *et al.*, 2017).

"Climate," by definition, is an aggregate of interconnected meteorological elements and physical processes of a particular region over a long period of time. Therefore, a number of natural phenomena and anthropogenic activities may contribute to climate change, and thus quantifying climate change as an individual factor or myriad climate processes can be used in

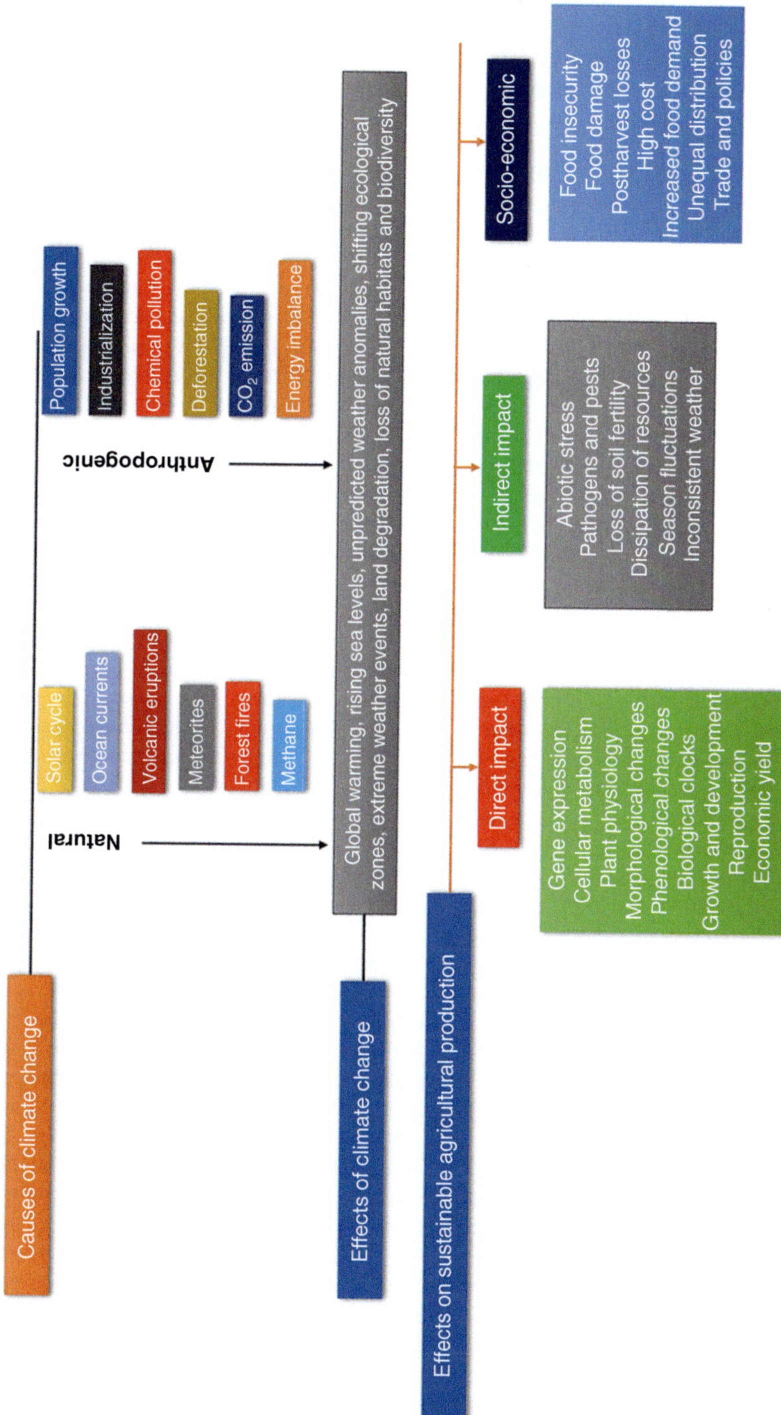

Fig. 6.1. Natural and anthropogenic activities causing climate change and their impact on sustainable agricultural production.

numerical modeling. Realization of climate change is a relatively new concept, but in the last few decades, enormous development has been made in weather-sensing technologies and data modeling, which is not limited to recording current climatic changes but also includes the reconstruction of past and the prediction of future climatic anomalies (Lyon *et al.*, 2022). The IPCC has developed a number of working groups, global projects, and policy directions focused on global climate change (Duchêne *et al.*, 2022). However, most scientific efforts are being utilized to understand and quantify climate change with high accuracy, and less attention is paid to its predicted impact. One of the fundamental reasons for this is the complexity of interlinked climatic processes, the lack of feedback detection, and the adaptability of the ecosystem and living components to climate change (Christensen *et al.*, 2022). In daily life, we directly or indirectly rely on ecosystem services, and therefore a natural equilibrium and the survival of such ecosystems is indispensable to drive sustainable societies. However, the changing climate poses momentous pressure on ecosystem services and thus needs to be addressed adequately to manage and mitigate the detrimental impact appropriately (Bakure *et al.*, 2022). The likelihood of intensified climatic changes in the near future has compelled humanity to chart dynamic pathways to survive through adaptations and mitigation strategies interconnected to the nexus of local priorities, vulnerable needs, and assurance of human well-being in a sustainable ecosystem (Anukwonke *et al.*, 2022).

6.3 Significance of Plant Stress Physiology Under a Changing Climate

Climate change threatens the survival of planet Earth, and more than 14,000 scientists have advocated it as an "emergency" (Gardner and Bullock, 2021). The consequent impact of climate change alters biodiversity, agriculture, and global food production, where plants—being immobile in nature—are significantly vulnerable. The interlinked physical processes of climate directly affect plant phenology. Different climatic components affecting plants include increasing average temperature, altered rainfall patterns, seasonal variations, frequent natural disasters, prevailing drought, and heat waves (Muluneh, 2021). Both experimental and hypothetical models prove that, in response to changing climate, most plants potentially migrate to new territories relatively more similar to their current habitats. Thus, some results have shown some plant species tentatively shifting to higher altitudes at the rate of 11.00 m to 16.90 km per decade; however, this assertion still needs more evidence (Muluneh, 2021). Moreover, most plant species may not have a chance to migrate due to finite phenotypic plasticity and are on the verge of extension (Bhadra *et al.*, 2022).

A constantly increasing human population and consequent deterioration of the Earth's ecosystem due to man-made contaminants are leading to drastic changes in weather elements, termed climate change (Pascual *et al.*, 2022). Such a climatic imbalance threatens life, potentially leading to hunger, unrest, and mass extinction (Sage, 2020). The increasing global temperatures and changing weather patterns are creating a wide range of abiotic stresses in plants, affecting agricultural production and global food security. For instance, the growing seasons of many crop plants have decreased by 10–25% in various geographical regions around the globe (Mgbemene *et al.*, 2016; Adamo *et al.*, 2021). Heat waves, forest fires, profound droughts, irregular precipitations, floods, and prevailing soil salinity are just a few examples of abiotic stresses representing the extremities of climate change (Chivangulula *et al.*, 2023). Overproduction of greenhouse gasses caused by human influence in the natural ecosystem has negative repercussions on crop production due to the susceptibility of plants to shifting weather patterns (Ratnayake *et al.*, 2023). Warm temperature spells reduced wheat crop output considerably from 1962 to 2002 (Lobell and Field, 2007), and similar findings were also confirmed by Gourdji *et al.* (2013) from 1980 to 2011 across Central Asia, South Asia, and South America. Several scientific observations have reported a significant decrease in grain and biomass production in wheat with respect to increasing temperatures in various ecological zones (Marcos-Barbero *et al.*, 2021; Karatayev *et al.*, 2022; Xiao *et al.*, 2022; Farhad *et al.*, 2023). Warming nights are predicted to significantly reduce crop yield including rice in the near future (Ma *et al.*,

2023). Changing climate patterns, increased day temperatures, and heat waves directly affect seed set by hastening blooming, shortening flowering duration, and lowering pollen sprouting in a number of agricultural plants (Ayyenar *et al.*, 2023; Liu *et al.*, 2023; Mehmood *et al.*, 2023). Furthermore, weather extremes impose corrosive effects on plant' phenology, biological signaling, metabolism, and biochemical processes including photosynthesis, genetic expression, signal transduction, and even reproduction (Farhad *et al.*, 2023). Discoloration of betel leaves, a reddish appearance of lemon leaves, and root decay in a pineapple under cold conditions are just a few examples of climatic extremes (Pautasso *et al.*, 2012; Rahman and Alam, 2016; Lim *et al.*, 2023). Moreover, under changing climate conditions, plants are becoming more vulnerable to different kinds of insects, pests, and pathogens. Likewise, climate change is also affecting evolution, survivability, and genetic modifications in other biological components of the ecosystem including microorganisms (Hamann *et al.*, 2021). Therefore, the combination of abiotic and biotic stresses is causing a deterioration of our agriculture systems, and their impact is projected to increase in the coming years (Zandalinas *et al.*, 2021). Plants in general, and especially those that have been domesticated as crops, have a certain threshold beyond which growth and yield are compromised when exposed to a changing climate. Agricultural production is predicted to decrease by 30% by 2050 in Africa (Muluneh, 2021). Agriculture is the primary source of food and is of immense importance to ensure food security, market prices, and the socio-economic health of a significant portion of the global population.

6.4 Basic Concept of Metabolomics

Near the end of the 20th century, technological advances gained enough pace to reshape scientific studies by adding various new dimensions to analytical approaches. The rapid developments in transcriptomes, proteomics, and genome sequencing technologies along with high-throughput analytical techniques such as mass spectrometry (MS) and nuclear magnetic resonance (NMR) in the late 1990s created a vacuum of complementary technology to comprehend metabolic actions in plants, especially in response to external stimuli (Hall, 2006). Thus, the term "metabolome" was coined in 1998 to describe the metabolite complement of living tissues (Oliver *et al.*, 1998). Since then, the desire to have more detailed knowledge of the biochemical composition of potential biomolecules and their interactive activities to stimulate and control functional genomics has gained immense scientific attention.

Metabolomics is a relatively new research field but is expanding exponentially in various biological fields including medicine, plants, pharmaceuticals, biotechnology, single-cell physiology, synthetic biology, and environmental sciences. Although several definitions of "metabolomics" have been proposed, nearly all of them share a common fundamental concept of "the analysis of small molecular mass compounds in a given biological condition" (de Raad *et al.*, 2016). Molecules larger than 1500 Da including sugar polymers, typical amino acids, and peptide chains are generally excluded due to larger molecular size and limitations in extraction and detection (Adams, 2003). The broad scope of metabolomics covers distinct analytical approaches including metabolic profiling, metabolic fingerprinting, targeted metabolomics, and exometabolomics.

6.5 Plant Metabolomics and Environmental Stresses

Being sessile obligate autotrophs, plants have evolved as multicellular structures that have developed divergent functions in order to withstand predominantly hostile environments and continuously changing meteorological elements (Gai *et al.*, 2020). Plant tissues such as parenchyma, epidermis, root hairs, and guard cells have their own dedicated set of metabolic mechanisms in order to perform their desired functions effectively (Hall, 2006). A plant's response to seasonal perturbations, temporal anomalies, and a changing local environment is predominantly facilitated by metabolic mechanisms largely depending on the genome. Moreover, plants have developed a complex of functionally active biomolecules, versatile in function and flexible in nature, to deal with both biotic (e.g. drought, heat, salinity, light, cold) and abiotic (e.g. pathogens, parasites, herbivores) external stresses. Some of these compounds exist commonly in the plant kingdom, but many exist in specialized genera and even species (Ghatak *et al.*, 2018). Speculatively, the estimated number of such biomolecules including secondary metabolites throughout the plant kingdom is a considerably big number but the real number most likely ranges between 100,000 and 200,000 (Oksman-Caldentey and Inzé, 2004). However, this complexity and wide range of compounds can be classified at every level of functionality, cell, tissue, organism, phenotype, and genotype. Consequently, a complete insight into metabolic molecules, their cross-talk, and their activity in various pathways may allow us to exploit them for interactive responses and provide predictive and diagnostic value (Table 6.1).

To date, there is a large gap in understanding the chemical nature/composition of biomolecules, their regulation, and their role in metabolic pathways (Jander *et al.*, 2004; Ghatak *et al.*, 2018). Much of the plant's biochemistry is still defined as "unidentified," both structurally and functionally, most likely due to the inevitable focus on tightly predefined questions in the past. The complete picture of a metabolic pathway has recently been described as a piecemeal elucidation of causal relationships within a circumscribed subset of cellular pathways (Fridman and Pichersky, 2005; Singh *et al.*, 2020). Clearly, increasing demand for other areas of molecular biology and changing climatic conditions are fueling the need for a broader understanding of plant metabolism that can be applied to a wide range of applications. Thus, an essentially non-targeted approach to metabolomics is potentially suitable to design future research.

Metabolomics appeared as a change in scientific philosophy and was embraced by the agricultural community due to the wide metabolic diversity within plants (Trethewey, 2004; Goodacre, 2005; Bagal *et al.*, 2023). Having more detailed information regarding the biochemical composition of compounds and their role in plant tissues has huge potential in both scientific and applied domains (Hall *et al.*, 2005). Moreover, the information obtained in metabolomics reflects much more of a biological endpoint compared with proteomics and transcriptomics, as it enhances our understanding of how plants respond, function, and survive in their environment. Metabolomics is of particular relevance considering the timeline of physiological events such as gene expression, protein synthesis, enzymatic function, and metabolite production (Nicholson *et al.*, 2004; Warren *et al.*, 2021).

The quality of crop plants, such as physical attributes, flavor, shelf life, color, and fragrance, are associated with the direct function of metabolic events taking place at that time along with the molecular profile of the plants (Memelink, 2005). Modern advances in metabolomics has shown it to be a complementary technology to proteomics and transcriptomics. Plant metabolomics, undoubtedly, has become an influential research tool to understand the biochemical mechanisms linked to plant growth and development under abiotic stresses particularly drought, extreme temperatures, and salinity. In fact, together with other omics technologies, metabolomics has accelerated our understanding of complex interactive relationships existing among biological molecules (Saito and Matsuda, 2010; Jorge *et al.*, 2016a). However, compared with other omics techniques, metabolomics does not require organism-specific genome data for analysis (Kopka, 2006).

The prime objective of plant metabolomics is to deliver non-biased characterization of the entire metabolite pool of a plant tissue or even a cell in response to its external environment. This metabolite pool, especially under abiotic stresses, consists of a wide range of chemical compounds diverse in nature both physically and chemically. Generally, the metabolite profile includes organic acids, amino acids, inorganic compounds, hydrophilic carbohydrates, and hydrophilic derivatives of lipids ranging from femtomolar to millimolar concentrations (Fernie, 2003). Largely, metabolites are divided into two major categories: (i) primary metabolites, which are fundamentally related to plant growth and development, being highly conserved in molecular structure and existing throughout the plant kingdom; and (ii) secondary metabolites, which aid in plant communications between the organism and the external environment, being dependent on environmental stimuli. This complexity and wide range of metabolites means that current plant metabolomics often combines with multiple analytical tools aiming for an effective outcome from a complex plant sample. Some of the most powerful analytical tools in metabolomics currently in practice include

Table 6.1. Some recent studies using various high-throughput analytical tools for metabolite profiling in plants under different environmental stresses.

Environmental stress	Plant species	Analytical tool	Metabolites	Reference
Drought	*Medicago sativa* *Medicago truncatula*	LC-TQMS	Abscisic acid, proline, flavonoids, carbohydrates	Echeverria *et al.* (2021)
	Arabidopsis thaliana	GC-MS LC/UPLC-MS	Phytohormone including; abscisic acid, jasmonic acid, salicylic acid, *trans*-zeatin, indole-3-acetic acid, brassinolide,	Salem *et al.* (2020)
	Oryza sativa	GC-MS LC	Benzoic acid, proline, citric acid, aconitic acids, carbohydrates, norvaline	Auler *et al.* (2021)
Salinity	*Hordeum vulgare*	LC-MS/MS	Fatty acyls, sphingolipids, polyketide, glycerolipids, pernol lipids	Sarabia *et al.* (2018)
	Zea mays	LC-MS	Flavonoids, amino acids, benzoxazinoids, lipids, terpenoids	Liang *et al.* (2021)
	Oryza sativa	GC-TQMS	Sugars, mannitol, organic acids	Chang *et al.* (2019)
Heat/ temperature	*Arabidopsis thaliana*	GC-MS	Trehalose, glycine, glutamine, maltose, glyceric acid	Casatejada-Anchel *et al.* (2021)
	Arabidopsis thaliana	LC-MS/MS	Phenylalanine-based metabolites	Simpson *et al.* (2021)
	Oryza sativa	GC-TOF-MS	Phenylpropanoids, organic acids, sugars, amino acids, phosphates, polyhydroxy acids	Glaubitz *et al.* (2017)
Drought and heat	*Arabidopsis thaliana*	FT-ICR-MS GC-TOF-MS	Fatty acids, coniferyl alcohol, sucrose	Georgii *et al.* (2017)
Salinity and cold	*Physcomitrella patens*	GC-MS	Amino acids, sugars, organic acids	Arif *et al.* (2018)
High light and cold	*Arabidopsis thaliana*	GC-TOF MS LC-MS/MS	Citrate, organic acids, gluconic acid, amino acids, hexose, sugars	Küstner *et al.* (2019)

FT-ICR-MS, Fourier-transform–ion cyclotron resonance–mass spectrometry; GC-MS, gas chromatography–mass spectrometry; GC-TOF-MS, gas chromatography–time-of-flight–mass spectrometry; GC-TQMS, gas chromatography–triple quadrupole mass spectrometry; LC, liquid chromatography; LC-MS, liquid chromatography–mass spectrometry; LC-MS/MS, liquid chromatography–tandem mass spectrometry; LC-TQMS, liquid chromatography–triple quadrupole mass spectrometry; LC/UPLC-MS, ultra-performance liquid chromatography–mass spectrometry.

NMR, MS, liquid chromatography (LC), and gas chromatography (GC) (Kim *et al.*, 2010). Recently, comprehensive information in plants under stress has been obtained by coupling more than one metabolomic tool, such as LC-MS and GC-MS (Obata and Fernie, 2012; Jorge *et al.*, 2016a). However, the effectiveness of choosing an analytical tool depends not only on the selection or sequence of analysis but also on the experimental design and sampling technique (Jorge *et al.*, 2016b). It becomes even more important when dealing with plant metabolomics under environmental stresses. Therefore, a well-studied experimental design should effectively address different environmental and experimental variables such as the metabolite extraction method, metabolic quenching, and plant tissue harvest. Furthermore, replication, randomization, and sample size are of immense importance in plant metabolomics under environmental stresses.

6.6 Technological Advances in High-throughput Plant Metabolomics

Despite the considerably fast progress made in metabolomics in the last decade, obtaining an adequate complement as achieved by other omics studies (i.e. genomics, transcriptomics, and proteomics), remains a significant challenge (Guijas *et al.*, 2018). The diverse nature of chemical structures, an unknown number of metabolites, and dynamic metabolic conversions are a few of the primary factors contributing to challenges in metabolomics. For instance, only 14,000 metabolites out of potentially 1 million compounds throughout the plant kingdom have been measured using advanced analytical tools (Alseekh and Fernie, 2018). Recent advances in hyphenated technologies have shown the utility of metabolomics in enhancing our understanding to a wide range and depth (Geng *et al.*, 2016, 2017; Zhu *et al.*, 2020). Untargeted metabolomics with the help of high-resolution MS has shown a significant compendium of different biological compounds, especially when compared with controls, in diseased and stressed plants. This method primarily targets all metabolites in a sample including known and unknown chemical compounds, whereas targeted metabolomics often focuses on relative or absolute quantification of a known metabolite or only a small number of metabolites at once using triple quadrupole MS (TQMS) as an analytical tool. Targeted metabolomics works as a comparative technique using authentic standards of specific metabolites (Katam *et al.*, 2022). For example, LC-MS/MS QTrap was used for *Triticum aestivum* and *Hordeum vulgare* genotypes to identify unique profiles of weed suppression for 6-methoxy-benzoxazolin-2-one, benzoxazolin-2-one, 2-hydroxy-1,4-benzoxazin-3-one, and benzoxazinoids (Mwendwa *et al.*, 2016).

Targeted metabolomics often has associated limitations due to the unavailability of authentic standards for most of the metabolites. In contrast, non-targeted metabolomics has no such limitations but does have a low rate of metabolite identification. Therefore, the latest development in using "semi-targeted metabolomics" offers a mid-way technique enabling the quantification of hundreds of metabolites of similar chemical structures using one calibration curve (Glover *et al.*, 2019; Dreier *et al.*, 2021). For example, in field metabotyping of downy mildew of *Vitis vinifera*, 32 metabolites were identified at level 1 confidence and another 15 were found as biomarkers of infection at level 2 confidence (Billet *et al.*, 2020). The emergence of a new metabolomic method known as "pseudo-targeted" has merged the advantages of both targeted and non-targeted metabolomics into one approach. This novel method, using Multiple Reaction Monitoring (MRM)-Ion Pair Finder software for TQMS, automatically defines ion pairs for metabolites of multi-reaction monitoring without using authentic chemical standards, unlike targeted metabolomics (Zheng *et al.*, 2020).

Advanced analytical tools in metabolomics can be divided into two groups comprising LC and GC in front-end separation and MS and NMR in spectroscopy. These techniques are complementary to each other and thus the pairing of such analytical tools has been used effectively in numerous metabolomic profiling studies including plants under abiotic stress caused by environmental changes (Katam *et al.*, 2022). Coupling MS with LC or GC has the advantage of separation through both chromatography and mass analyzers (Gowda and Djukovic, 2014). GC-LC separates a complex mixture of compounds based on the differential partitioning of molecules between mobile and stationary phases, whereas MS-NMR analyzes molecules based on mass:charge (m/z) ratio in an electric field followed by resonance in a magnetic field.

Although GC-MS has a low coverage of about 200 metabolites, due to the usual requirement of chemical derivations, it is still a popular technique for detecting volatile metabolites and hydrocarbons (Fiehn *et al.*, 2000; Stettin *et al.*, 2020; Zeki *et al.*, 2020). For example, 14 antimicrobial and anticancer compounds were detected from *Picrorhiza kurroa* roots using GC-MS, while 15 volatile metabolites were extracted from *Kigelia africana* fruit using GC-time-of-flight (TOF) coupled with MS (Fagbohun *et al.*, 2021; Tabassam *et al.*, 2021). The combination of GC-MS with isotopic labeling has enabled the study of glucogenesis in guard cells (Feizi *et al.*, 2021; Lima *et al.*, 2021), and a recently developed GC-Orbitrap-MS technique allowed identification of metabolic biomarkers for quality control and processing authentication in fruits and has been tested for *Piper nigrum* and *Solanum lycopersicum* (Romera-Torres *et al.*, 2019; Rivera-Perez *et al.*, 2021).

Likewise, LC-MS is the most widely used analytic approach in metabolomics due to its high versatility and wide range. In plant metabolomics, ultra-high pressure liquid chromatography (UPLC) prevailed for the last decade because of high separation efficiency (Geng *et al.*, 2016, 2017; Zhu *et al.*, 2020; Perez de Souza *et al.*, 2021). Furthermore, the development of microflow LC and nanoLC (nLC) has potentially empowered single-cell metabolomics with high sensitivity (Wickremsinhe *et al.*, 2006; Hilhorst *et al.*, 2014). nLC-MS is widely used in proteomics but has recently been reported with potential application in plant metabolomics. For instance, nLC-MS has recently been used in the quantification of different cannabinoids in hemp inflorescences (Žampachová *et al.*, 2021). The very small sample amount required for nLC-MS makes it an effective technique in plant metabolomics with limited quantity samples or low-concentration metabolites such as secondary metabolites in response to external abiotic stresses (Žampachová *et al.*, 2021; Katam *et al.*, 2022). Although the concept of metabolomics is relatively new compared with other omic studies, its promising potential has triggered the development of highly precise, sensitive, more reliable, and targeted analytical tools in recent years such as ion mobility spectrometry (IMS), which includes field asymmetric waveform IMS, drift tube IMS, trapped IMS, and traveling wave IMS. The first collision cross-section using trapped IMS has recently been developed for 146 plant metabolites of *Medicago truncatula* (Schroeder *et al.*, 2019). Considering the existence of thousands of plant metabolites, collision cross-section libraries will potentially become popular in the near future.

Progressively, the development of high-throughput instruments with high sensitivity, low sample requirements, and high speed is paving the way to single-cell metabolomics, and in the last 5 years, the techniques of microdissection, free-flow electrophoresis, and flow cytometry have been developed (Jehan, 2019). Although most of these advanced methods are utilized for animal cells, plant cell protoplasts can be isolated for single-cell metabolomics (Misra *et al.*, 2014). However, the use of enzymatic reactions to remove the cell wall in protoplast preparation may lead to alteration of the chemical properties of plant cells. Direct sampling of plant cells using micrometer tips is trending in single-cell metabolomics. These techniques have been applied to phloem-associated cells (Yamamoto *et al.*, 2019), trichome cells (Nakashima *et al.*, 2016), mesophyll cells (Fujii *et al.*, 2015), and epidermal cells (Shimizu *et al.*, 2015). MS imaging in another powerful metabolomic tool that provides *in situ* analysis without tagging, and can simultaneously analyze a wide range of metabolites (Katam *et al.*, 2022). Matrix-assisted laser desorption ionization (MALDI) is the most advanced method, as it covers a wide range of molecules. This technique was recently applied to *Zea mays* roots (Righetti *et al.*, 2021), green algae (Shanta *et al.*, 2021), and *S. lycopersicum* (Dong *et al.*, 2020). Conclusively, the pace of technological developments, and especially of analytical tools, will potentially lead to new dimensions of omics studies. Such scientific developments may contribute to an in-depth understanding of plant metabolic responses to environmental stresses and changing climate, and thus can potentially be utilized to improve crop production.

6.7 Integration of Metabolomics with Other High-throughput Techniques to Improve Plant Resilience to a Changing Climate

High-throughput technologies permit a wide range of omics data sets for phenomics, proteomics, transcriptomics, genomics, and metabolomics analysis using the latest computer facilities. The advances in integrating omics-related data and evolving bioinformatics accompanied by statistical programs and target-specific computer models may provide critical information and analysis of plant molecular systems and functions (Choi, 2019). These technological advances in gene function, regulation factors, and adaptive nature have significantly improved research in plant omics. Consequently, they contribute to the recovery of substantial plant biodiversity, environmental conservation, genetic improvement, crop production, and food security (Kumar *et al.*, 2021). Thus, the integration of multi-level biological data for various plant omics may facilitate a comprehensive investigation and insight into molecular aspects governing the plant response to changing climate and associated environmental stresses (Mora-Poblete *et al.*, 2023).

Advanced cutting-edge technologies in plant metabolomics can be used to enhance plant resilience against environmental challenges. Dwivedi *et al.* (2022) employed high-throughput phenomics to select reproductive-stage drought-stress tolerance in near-isogenic lines of *Oryza sativa* plants. The results revealed a significant relationship between quantitative trait loci and 11 of the studied phenomics traits. Thus, high-throughput phenomics-based analysis enables rapid and efficient selection of stress-tolerance genotypes in changing climates. Similarly, the integration of proteomics and transcriptomics is of supreme importance in understanding the complex response of biological systems to environmental stimuli. Li *et al.* (2022) performed proteomic and transcriptomic analyses of *Myricaria laxiflora* roots during flooding and post-flooding recovery periods. The transcriptomic data showed downregulation of genes related to calcium signaling, cell walls, and auxin during the early stage of flooding, whereas there was significant upregulation of genes associated with mitochondrial metabolism, root development, and reactive oxygen species scavenging in the post-flooding period. Protein groups related to redox reactions, G-protein signaling, carbohydrate metabolism, and glycolysis were differentially expressed in six groups under flooding stress, whereas transcription factor proteins, hormones, enzymes, and gene expression-associated proteins appeared in the post-flooding period. These outcomes have facilitated our understanding of flooding stress mechanisms and future directions in metabolomics for further insight and fast-track elucidation of how plants respond to flood.

Integrated metabolomics and transcriptomics analysis in *Populus tomentosa* was performed to examine metabolite profiles and their regulatory role in response pathways under low nitrogen stress (Chen *et al.*, 2021). Integrated metabolomics and transcriptomics showed co-expression of genes and metabolites. Nitrogen is an essential component for plant growth and development, and thus these findings provide an insight into metabolic and molecular transductions and an understanding of the underlying mechanisms in plants under nitrogen deficiency. Likewise, Ponce *et al.* (2022) and Sajad *et al.* (2022) conducted transcriptomic studies for *Solanum tuberosum* and *Brassica oleracea* under imposed drought and cold stresses to explore the genetic and molecular bases of plant responses. The findings provided directions for targeted metabolomics to bridge the knowledge gap between gene expression and metabolic pathways under various biotics stresses including drought and cold.

The integrative analysis of metabolomics and transcriptomics includes combining the processing of data from metabolomics (metabolite profile) and gene expression (transcriptomic profile) to achieve a comprehensive understanding of the biological process taking place in living organisms (Mora-Poblete *et al.*, 2023). Liu *et al.* (2021) studied the ultraviolet (UV)-B adaptive mechanism in *Leptobryum pyriforme* moss using interactive analysis of transcriptomics and metabolomics. The results showcased the co-expression of genes and metabolites synergistically and helped to develop an understanding of the underlying mechanisms for jasmine signaling, DNA repair, adaptation, UVR8-mediated signaling, flavonoid biosynthesis pathways, and hormone signal transduction. Likewise, Malviya *et al.* (2022) and Xu *et al.* (2023) used high-throughput metabolomic techniques in *Saccharum officinarum* aiming for breeding improvements. In short, utilization of advanced high-throughput metabolomic technologies integrated with other omics studies in plants may facilitate insight knowledge of plant metabolism at the molecular level, which potentially helps the development of resilience in plants against changing climate (Veley *et al.*, 2017; Gogolev *et al.*, 2021).

6.8 Conclusion

The recent anomalies in global climate and projected changes are challenging the very existence of life on Earth. Since the advent of the Industrial Revolution, the climate crisis has been greatly exacerbated as a consequence of anthropogenic activities. Meteorological irregularities and extreme spells of heat, drought, cold, and salinity along with the complexity of diseases, pest infestation, and loss in soil fertility are posing a serious threat to sustainable agricultural production. Therefore, to cope with the changing climate, an in-depth understanding of the physiological response in plants including metabolic activities, signal transduction, genetic expression, and homeostatic mechanisms is vital. The recently developed concept of metabolomics offers integration of modern technologies with advanced high-throughput analytical facilities, statistical approaches, and computer programming to provide an understanding of the effect of climate change on agriculture and the plant's response to such a stimulus.

The current achievements in plant metabolomics are nothing compared with its potential. The continuation of technological advances in high-throughput tools, integration of different omics profiles, and increasingly sophisticated bioinformatics have significantly empowered plant metabolomics, especially in data mining and analysis. However, special attention should be focused on the primary goal of advancing scientific knowledge rather than producing move data. The metabolomic objectives have already reached single-cell and even subcellular levels. Moreover, the application of plant metabolomics is expanding to a number of subsets such as ecology, biodiversity, plant diseases, symbiotic relationships, biochemistry, and climate change. This wide range of applications and diverse utilization of plant metabolomics provide a full-scale discovery platform that will support and feed complementary fields of knowledge to advance our understanding of multi-layer plant response systems.

Metabolomics has proven itself a potential diagnostic and analytic multi-dimensional approach to understanding plant mechanisms triggered by external stimuli such as diseases and abiotic stresses. Morphological to physiological mechanisms at the organism level are being studied as major analytical and response characteristics in plants under unfavorable environmental conditions. However, metabolomics offers quantification of the plant response to its changing climate at the molecular level, complementing existing knowledge to understand the biochemical pathways underlying these responses. Therefore, the utilization of metabolomics in developing physiological biomarkers may assist us with a time-effective, reliable, and relatively precise approach to understanding the impact of climate change on crop plants. Potentially, such a cascade of information from climate change (stimulus) to physiological response and hidden mechanisms at the molecular level (metabolomics) could contribute significantly to selecting more resilient genomes, developing mitigation strategies, and improving crop breeding and management for sustainable crop production in the future.

6.9 References

Adamo, N., Al-Ansari, N., and Sissakian, V. (2021) Review of climate change impacts on human environment: past, present and future projections. *Engineering* 13, 605–630.

Adams, A. (2003) Metabolomics: small-molecule 'omics: research advances in the bid to map biofluid metabolites on a system-wide scale. *Scientist* 17, 38–41.

Alseekh, S. and Fernie, A.R. (2018) Metabolomics 20 years on: what have we learned and what hurdles remain? *Plant Journal* 94, 933–942.

Anukwonke, C.C., Tambe, E.B., Nwafor, D.C., and Malik, K.T. (2022) Climate change and interconnected risks to sustainable development. In: Bandh, S.A. (ed.) *Climate Change: The Social and Scientific Construct*. Springer, Cham, Switzerland, pp. 71–86.

Arif, M.A., Alseekh, S., Harb, J., Fernie, A., and Frank, W. (2018) Abscisic acid, cold and salt stimulate conserved metabolic regulation in the moss *Physcomitrella patens*. *Plant Biology* 20, 1014–1022.

Arora, M., Goel, N.K., and Singh, P. (2005) Evaluation of temperature trends over India. *Hydrological Sciences Journal* 50, 81–93.

Auler, P.A., Souza, G.M., da Silva Engela, M.R.G., do Amaral, M.N., Rossatto, T., *et al.* (2021) Stress memory of physiological, biochemical and metabolomic responses in two different rice genotypes under drought stress: the scale matters. *Plant Science* 311: 110994.

Ayyenar, B., Kambale, R., Duraialagaraja, S., Manickam, S., Mohanavel, V., *et al.* (2023) Developing early morning flowering version of rice variety CO 51 to mitigate the heat-induced yield loss. *Agriculture* 13: 553.

Bagal, D., Chowdhary, A.A., Mehrotra, S., Mishra, S., Rathore, S., and Srivastava, V. (2023) Metabolic engineering in hairy roots: an outlook on production of plant secondary metabolites. *Plant Physiology and Biochemistry* 201: 107847.

Bakure, B.Z., Hundera, K., and Abara, M. (2022) Review on the effect of climate change on ecosystem services *IOP Conference Series: Earth and Environmental Science* 1016: 012055.

Bhadra, P., Maitra, S., Shankar, T., Hossain, A., Praharaj, S., and Aftab, T. (2022) Climate change impact on plants: plant responses and adaptations. In: Aftab, T. and Roychoudhury, A. (eds) *Plant Perspectives to Global Climate Changes*. Academic Press, London, pp. 1–24.

Billet, K., Malinowska, M.A., Munsch, T., Unlubayir, M., Adler, S., *et al.* (2020) Semi-targeted metabolomics to validate biomarkers of grape downy mildew infection under field conditions. *Plants* 9: 1008.

Casatejada-Anchel, R., Muñoz-Bertomeu, J., Rosa-Téllez, S., Anoman, A.D., Nebauer, S.G., *et al.* (2021) Phosphoglycerate dehydrogenase genes differentially affect *Arabidopsis* metabolism and development. *Plant Science* 306: 110863.

Chang, J., Cheong, B.E., Natera, S., and Roessner, U. (2019) Morphological and metabolic responses to salt stress of rice (*Oryza sativa* L.) cultivars which differ in salinity tolerance. *Plant Physiology and Biochemistry* 144, 427–435.

Chen, M., Yin, Y., Zhang, L., Yang, X., Fu, T., *et al.* (2021) Metabolomics and transcriptomics integration of early response of *Populus tomentosa* to reduced nitrogen availability. *Frontiers in Plant Science* 12: 769748.

Chivangulula, F.M., Amraoui, M., and Pereira, M.G. (2023) The drought regime in Southern Africa: a systematic review. *Climate* 11: 147.

Choi, H.K. (2019) Translational genomics and multi-omics integrated approaches as a useful strategy for crop breeding. *Genes and Genomics* 41, 133–146.

Christensen, O.B., Kjellström, E., Dieterich, C., Gröger, M., and Meier, H.E.M. (2022) Atmospheric regional climate projections for the Baltic Sea region until 2100. *Earth System Dynamics* 13, 133–157.

Dandotiya, B. and Sharma, H.K. (2022) Climate change and its impact on terrestrial ecosystems. In: *Research Anthology on Environmental and Societal Impacts of Climate Change*. IGI Global, Hershey, Pennsylvania, pp. 88–101.

de Raad, M., Fischer, C.R., and Northen, T.R. (2016) High-throughput platforms for metabolomics. *Current Opinion in Chemical Biology* 30, 7–13.

Dong, Y., Sonawane, P., Cohen, H., Polturak, G., Feldberg, L., *et al.* (2020) High mass resolution, spatial metabolite mapping enhances the current plant gene and pathway discovery toolbox. *New Phytologist* 228, 1986–2002.

Dreier, D.A., Nouri, M.Z., Denslow, N.D., and Martyniuk, C.J. (2021) Lipidomics reveals multiple stressor effects (temperature × mitochondrial toxicant) in the zebrafish embryo toxicity test. *Chemosphere* 264: 128472.

Duchêne, F., Hamdi, R., van Schaeybroeck, B., Caluwaerts, S., de Troch, R., *et al.* (2022) Downscaling ensemble climate projections to urban scale: Brussels's future climate at 1.5°C, 2°C, and 3°C global warming. *Urban Climate* 46: 101319.

Dwivedi, P., Ramawat, N., Raju, D., Dhawan, G., Gopala Krishnan, S., *et al.* (2022) Drought tolerant near isogenic lines of pusa 44 pyramided with qDTY2. 1 and qDTY3. 1, show accelerated recovery response in a high throughput phenomics based phenotyping. *Frontiers in Plant Science* 12: 752730.

Echeverria, A., Larrainzar, E., Li, W., Watanabe, Y., Sato, M., *et al.* (2021) *Medicago sativa* and *Medicago truncatula* show contrasting root metabolic responses to drought. *Frontiers in Plant Science* 12: 612.

Fagbohun, O.F., Olawoye, B., Ademakinwa, A.N., Oriyomi, O.V., Fagbohun, O.S., *et al.* (2021) UHPLC/GC-TOF-MS metabolomics, MTT assay, and molecular docking studies reveal physostigmine as a new anticancer agent from the ethyl acetate and butanol fractions of *Kigelia africana* (Lam.) Benth. fruit extracts. *Biomedical Chromatography* 35: e4979.

Fankhauser, S. and Tol, R.S. (1997) The social costs of climate change: the IPCC second assessment report and beyond. *Mitigation and Adaptation Strategies for Global Change* 1, 385–403.

Farhad, M., Kumar, U., Tomar, V., and Hossain, A. (2023) Heat stress in wheat: a global challenge to feed billions in the current era of the changing climate. *Frontiers in Sustainable Food Systems* 7: 1203721.

Feizi, N., Hashemi-Nasab, F.S., Golpelichi, F., Saburouh, N., and Parastar, H. (2021) Recent trends in application of chemometric methods for GC-MS and GCŨGC-MS-based metabolomic studies. *Trends in Analytical Chemistry* 138: 116239.

Fernie, A.R. (2003) Metabolome characterisation in plant system analysis. *Functional Plant Biology* 30, 111–120.

Fiehn, O., Kopka, J., Dörmann, P., Altmann, T., Trethewey, R.N., and Willmitzer, L. (2000) Metabolite profiling for plant functional genomics. *Nature Biotechnology* 18, 1157–1161.

Fridman, E. and Pichersky, E. (2005) Metabolomics, genomics, proteomics, and the identification of enzymes and their substrates and products. *Current Opinion in Plant Biology* 8, 242–248.

Fujii, T., Matsuda, S., Tejedor, M.L., Esaki, T., Sakane, I., *et al.* (2015) Direct metabolomics for plant cells by live single-cell mass spectrometry. *Nature Protocols* 10, 1445–1456.

Gai, Z., Wang, Y.U., Ding, Y., Qian, W., Qiu, C., *et al.* (2020) Exogenous abscisic acid induces the lipid and flavonoid metabolism of tea plants under drought stress. *Scientific Reports* 10: 12275.

Gardner, C.J. and Bullock, J.M. (2021) In the climate emergency, conservation must become survival ecology. *Frontiers in Conservation Science* 2: 659912.

Geng, S., Misra, B.B., de Armas, E., Huhman, D.V., Alborn, H.T., *et al.* (2016) Jasmonate-mediated stomatal closure under elevated CO_2 revealed by time-resolved metabolomics. *Plant Journal* 88, 947–962.

Geng, S., Yu, B., Zhu, N., Dufresne, C., and Chen, S. (2017) Metabolomics and proteomics of *Brassica napus* guard cells in response to low CO_2. *Frontiers in Molecular Biosciences* 4: 51.

Georgii, E., Jin, M., Zhao, J., Kanawati, B., Schmitt-Kopplin, P., *et al.* (2017) Relationships between drought, heat and air humidity responses revealed by transcriptome–metabolome co-analysis. *BMC Plant Biology* 17: 120.

Ghatak, A., Chaturvedi, P., and Weckwerth, W. (2018) Metabolomics in plant stress physiology. *Plant Genetics and Molecular Biology* 164, 187–236.

Glaubitz, U., Li, X., Schaedel, S., Erban, A., Sulpice, R., *et al.* (2017) Integrated analysis of rice transcriptomic and metabolomic responses to elevated night temperatures identifies sensitivity-and tolerance-related profiles. *Plant, Cell and Environment* 40, 121–137.

Glover, S.C., Nouri, M.Z., Tuna, K.M., Alvarez, L.B.M., Ryan, L.K., *et al.* (2019) Lipidomic analysis of urinary exosomes from hereditary α-tryptasemia patients and healthy volunteers. *FASEB BioAdvances* 1: 624.

Gogolev, Y.V., Ahmar, S., Akpinar, B.A., Budak, H., Kiryushkin, A.S., *et al.* (2021) Omics, epigenetics, and genome editing techniques for food and nutritional security. *Plants* 10, 1423.

Goodacre, R. (2005) Making sense of the metabolome using evolutionary computation: seeing the wood with the trees. *Journal of Experimental Botany* 56, 245–254.

Gourdji, S.M., Sibley, A.M., and Lobell, D.B. (2013) Global crop exposure to critical high temperatures in the reproductive period: historical trends and future projections. *Environmental Research Letters* 8: 024041.

Gowda, G.N. and Djukovic, D. (2014) Overview of mass spectrometry-based metabolomics: opportunities and challenges. *Methods in Molcular Biology* 1198, 3–12.

Guijas, C., Montenegro-Burke, J.R., Warth, B., Spilker, M.E., and Siuzdak, G. (2018) Metabolomics activity screening for identifying metabolites that modulate phenotype. *Nature Biotechnology* 36, 316–320.

Hall, R.D. (2006) Plant metabolomics: from holistic hope, to hype, to hot topic. *New Phytologist* 169, 453–468.

Hall, R.D., de Vos, C.R., Verhoeven, H.A., and Bino, R.J. (2005) Metabolomics for the assessment of functional diversity and quality traits in plants. In: Vaidyanathan, S., Harrigan, G.G., and Goodacre, R. (eds) *Metabolome Analyses: Strategies for Systems Biology*. Springer, Boston, Massachusetts, pp. 31–44.

Hamann, E., Blevins, C., Franks, S.J., Jameel, M.I., and Anderson, J.T. (2021) Climate change alters plant–herbivore interactions. *New Phytologist* 229, 1894–1910.

Hare, B. and Meinshausen, M. (2006) How much warming are we committed to and how much can be avoided? *Climatic Change* 75, 111–149.

Hilhorst, M., Briscoe, C., and van de Merbel, N. (2014) Sense and nonsense of miniaturized LC–MS/MS for bioanalysis. *Bioanalysis* 6, 3263–3265.

Hood, L., Heath, J.R., Phelps, M.E., and Lin, B. (2004) Systems biology and new technologies enable predictive and preventative medicine. *Science* 306, 640–643.

Jander, G., Norris, S.R., Joshi, V., Fraga, M., Rugg, A., *et al.* (2004) Application of a high-throughput HPLC-MS/MS assay to *Arabidopsis* mutant screening; evidence that threonine aldolase plays a role in seed nutritional quality. *Plant Journal* 39, 465–475.

Jehan, Z. (2019) Single-cell omics: an overview. In: Barh, D. and Azavedo, V. (eds) *Single-Cell Omics, Vol.* 1. Academic press, London, pp. 3–19.

Jorge, T.F., Rodrigues, J.A., Caldana, C., Schmidt, R., van Dongen, J.T., *et al.* (2016a) Mass spectrometry-based plant metabolomics: metabolite responses to abiotic stress. *Mass Spectrometry Reviews* 35, 620–649.

Jorge, T.F., Mata, A.T., and António, C. (2016b) Mass spectrometry as a quantitative tool in plant metabolomics. *Philosophical Transactions of the Royal Society A: Mathematical, Physical and Engineering Sciences* 374: 20150370.

Karatayev, M., Clarke, M., Salnikov, V., Bekseitova, R., and Nizamova, M. (2022) Monitoring climate change, drought conditions and wheat production in Eurasia: the case study of Kazakhstan. *Heliyon* 8: e08660.

Karoly, D.J., Braganza, K., Stott, P.A., Arblaster, J.M., Meehl, G.A., *et al.* (2003) Detection of a human influence on North American climate. *Science* 302, 1200–1203.

Katam, R., Lin, C., Grant, K., Katam, C.S., and Chen, S. (2022) Advances in plant metabolomics and its applications in stress and single-cell biology. *International Journal of Molecular Sciences* 23: 6985.

Kim, H.K., Choi, Y.H., and Verpoorte, R. (2010) NMR-based metabolomic analysis of plants. *Nature Protocols* 5, 536–549.

Kopka, J. (2006) Current challenges and developments in GC–MS based metabolite profiling technology. *Journal of Biotechnology* 124, 312–322.

Kumar, A., Anju, T., Kumar, S., Chhapekar, S.S., Sreedharan, S., *et al.* (2021) Integrating omics and gene editing tools for rapid improvement of traditional food plants for diversified and sustainable food security. *International Journal of Molecular Sciences* 22: 8093.

Küstner, L., Fürtauer, L., Weckwerth, W., Nägele, T., and Heyer, A.G. (2019) Subcellular dynamics of proteins and metabolites under abiotic stress reveal deferred response of the *Arabidopsis thaliana* hexokinase-1 mutant *gin2-1* to high light. *Plant Journal* 100, 456–472.

Li, L., Huang, G., Xiang, W., Zhu, H., Zhang, H., *et al.* (2022) Integrated transcriptomic and proteomic analyses uncover the regulatory mechanisms of *Myricaria laxiflora* under flooding stress. *Frontiers in Plant Science* 13: 924490.

Liang, X., Liu, S., Wang, T., Li, F., Cheng, J., *et al.* (2021) Metabolomics-driven gene mining and genetic improvement of tolerance to salt-induced osmotic stress in maize. *New Phytologist* 230, 2355–2370.

Lim, J.A., Yaacob, J.S., Mohd Rasli, S.R.A., Eyahmalay, J.E., El Enshasy, H.A., and Zakaria, M.R.S. (2023) Mitigating the repercussions of climate change on diseases affecting important crop commodities in Southeast Asia, for food security and environmental sustainability—a review. *Frontiers in Sustainable Food Systems* 6: 1030540.

Lima, V.F., Erban, A., Daubermann, A.G., Freire, F.B.S., Porto, N.P., *et al.* (2021) Establishment of a GC-MS-based 13C-positional isotopomer approach suitable for investigating metabolic fluxes in plant primary metabolism. *Plant Journal* 108, 1213–1233.

Liu, S., Fang, S., Liu, C., Zhao, L., Cong, B., and Zhang, Z. (2021) Transcriptomics integrated with metabolomics reveal the effects of ultraviolet-B radiation on flavonoid biosynthesis in Antarctic moss. *Frontiers in Plant Science* 12: 788377.

Liu, X., Gu, M., Lv, X., Sheng, D., Wang, X., *et al.* (2023) High temperature defense pathways mediate lodicule expansion and spikelet opening in maize tassels. *Journal of Experimental Botany* 74, 3684–3699.

Lobell, D.B. and Field, C.B. (2007) Global scale climate–crop yield relationships and the impacts of recent warming. *Environmental Research Letters* 2: 014002.

Lobell, D.B., Schlenker, W., and Costa-Roberts, J. (2011) Climate trends and global crop production since 1980. *Science* 333, 616–620.

Lyon, C., Saupe, E.E., Smith, C.J., Hill, D.J., Beckerman, A.P., *et al.* (2022) Climate change research and action must look beyond 2100. *Global Change Biology* 28, 349–361.

Ma, M., Huang, D., and Hossain, S.S. (2023) Opportunities or risks: economic impacts of climate change on crop structure adjustment in ecologically vulnerable regions in China. *Sustainability* 15: 6211.

Malhi, G.S., Kaur, M., and Kaushik, P. (2021) Impact of climate change on agriculture and its mitigation strategies: a review. *Sustainability* 13: 1318.

Malviya, M.K., Li, C.N., Lakshmanan, P., Solanki, M.K., Wang, Z., *et al.* (2022) High-throughput sequencing-based analysis of rhizosphere and diazotrophic bacterial diversity among wild progenitor and closely related species of sugarcane (*Saccharum* spp. inter-specific hybrids). *Frontiers in Plant Science* 13: 829337.

Mann, M.E. and Jones, P.D. (2003) Global surface temperatures over the past two millennia. *Geophysical Research Letters* 30, 5–4.

Marcos-Barbero, E.L., Pérez, P., Martínez-Carrasco, R., Arellano, J.B., and Morcuende, R. (2021) Screening for higher grain yield and biomass among sixty bread wheat genotypes grown under elevated CO_2 and high-temperature conditions. *Plants* 10: 1596.

Massey, D.S. (2023) The shape of things to come: international migration in the twenty-first century. In: Lerpold, L., Sjöberg, O., and Wennberg, K. (eds.) *Migration and Integration in a Post-Pandemic World: Socioeconomic Opportunities and Challenges*. Springer, Cham, Switzerland, pp. 29–81.

Mehmood, M., Qamar, R., and Joyia, F.A. (2023) Effect of high temperature stress on pollen grains in sunflower (*Helianthus annuus* L.) inbred lines. *Brazilian Archives of Biology and Technology* 66: e23220927.

Memelink, J. (2005) Tailoring the plant metabolome without a loose stitch. *Trends in Plant Science* 10, 305–307.

Mgbemene, C.A., Nnaji, C.C., and Nwozor, C. (2016) Industrialization and its backlash: focus on climate change and its consequences. *Journal of Environmental Science and Technology* 9, 301–316.

Misra, B.B., Assmann, S.M., and Chen, S. (2014) Plant single-cell and single-cell-type metabolomics. *Trends in Plant Science* 19, 637–646.

Mora-Poblete, F., Heidari, P., and Fuentes, S. (2023) Integrating advanced high-throughput technologies to improve plant resilience to environmental challenges. *Frontiers in Plant Science* 14: 1218691.

Muluneh, M.G. (2021) Impact of climate change on biodiversity and food security: a global perspective—a review article. *Agriculture and Food Security* 10: 36.

Mwendwa, J.M., Weston, P.A., Fomsgaard, I., Laursen, B.B., Brown, W.B., *et al.* (2016) Metabolic profiling for benzoxazinoids in weed-suppressive and early vigour wheat genotypes. In: Randall, R., Lloyd, S., and Borger, C. (eds.) *20th Australasian weeds Conference: Proceedings*. Weeds Society of Western Australia, Perth, Australia, pp. 11–15.

Nakashima, T., Wada, H., Morita, S., Erra-Balsells, R., Hiraoka, K., and Nonami, H. (2016) Single-cell metabolite profiling of stalk and glandular cells of intact trichomes with internal electrode capillary pressure probe electrospray ionization mass spectrometry. *Analytical Chemistry* 88, 3049–3057.

Nicholson, J.K., Holmes, E., Lindon, J.C., and Wilson, I.D. (2004) The challenges of modeling mammalian biocomplexity. *Nature Biotechnology* 22, 1268–1274.

Notz, D. (2020) A short history of climate change. *EPJ Web of Conferences* 246: 00002.

Obata, T. and Fernie, A.R. (2012) The use of metabolomics to dissect plant responses to abiotic stresses. *Cellular and Molecular Life Sciences* 69, 3225–3243.

Oksman-Caldentey, K.M. and Inzé, D. (2004) Plant cell factories in the post-genomic era: new ways to produce designer secondary metabolites. *Trends in Plant Science* 9, 433–440.

Oliver, S.G., Winson, M.K., Kell, D.B., and Baganz, F. (1998) Systematic functional analysis of the yeast genome. *Trends in Biotechnology* 16, 373–378.

O'Neill, B.C., Oppenheimer, M., Warren, R., Hallegatte, S., Kopp, R.E., *et al.* (2017) IPCC reasons for concern regarding climate change risks. *Nature Climate Change* 7, 28–37.

Parry, M.L. (ed.) (2007) *Climate Change 2007—Impacts, Adaptation and Vulnerability. Working Group II Contribution to the Fourth Assessment Report of the IPCC, Vol.* 4. Cambridge University Press, New York.

Pascual, L.S., Segarra-Medina, C., Gómez-Cadenas, A., López-Climent, M.F., Vives-Peris, V., and Zandalinas, S.I. (2022) Climate change-associated multifactorial stress combination: a present challenge for our ecosystems. *Journal of Plant Physiology* 276: 153764.

Pautasso, M., Döring, T.F., Garbelotto, M., Pellis, L., and Jeger, M.J. (2012) Impacts of climate change on plant diseases—opinions and trends. *European Journal of Plant Pathology* 133, 295–313.

Perez de Souza, L., Alseekh, S., Scossa, F., and Fernie, A.R. (2021) Ultra-high-performance liquid chromatography high-resolution mass spectrometry variants for metabolomics research. *Nature Methods* 18, 733–746.

Ponce, O.P., Torres, Y., Prashar, A., Buell, R., Lozano, R., *et al.* (2022) Transcriptome profiling shows a rapid variety-specific response in two Andigenum potato varieties under drought stress. *Frontiers in Plant Science* 13: 1003907.

Rahman, M.H. and Alam, K. (2016) Forest dependent indigenous communities' perception and adaptation to climate change through local knowledge in the protected area—a Bangladesh case study. *Climate* 4: 12.

Ratnayake, S.S., Reid, M., Larder, N., Kadupitiya, H.K., Hunter, D., *et al.* (2023) Impact of climate change on paddy farming in the village Tank Cascade Systems of Sri Lanka. *Sustainability* 15: 9271.

Righetti, L., Bhandari, D.R., Rolli, E., Tortorella, S., Bruni, R., *et al.* (2021) Unveiling the spatial distribution of aflatoxin B1 and plant defense metabolites in maize using AP-SMALDI mass spectrometry imaging. *Plant Journal* 106, 185–199.

Rivera-Perez, A., Romero-Gonzalez, R., and Garrido Frenich, A. (2021) Feasibility of applying untargeted metabolomics with GC-Orbitrap-HRMS and chemometrics for authentication of black pepper (*Piper nigrum* L.) and identification of geographical and processing markers. *Journal of Agricultural and Food Chemistry* 69, 5547–5558.

Romera-Torres, A., Arrebola-Liebanas, J., Vidal, J.L.M., and Frenich, A.G. (2019) Determination of calystegines in several tomato varieties based on GC-Q-Orbitrap analysis and their classification by ANOVA. *Journal of Agricultural and Food Chemistry* 67, 1284–1291.

Sage, R.F. (2020) Global change biology: a primer. *Global Change Biology* 26, 3–30.

Saito, K. and Matsuda, F. (2010) Metabolomics for functional genomics, systems biology, and biotechnology. *Annual Review of Plant Biology* 61, 463–489.

Sajad, S., Jiang, S., Anwar, M., Dai, Q., Luo, Y., *et al.* (2022) Genome-wide study of Hsp90 gene family in cabbage *(Brassica oleracea* var. *capitata* L.) and their imperative roles in response to cold stress. *Frontiers in Plant Science* 13: 908511.

Salem, M.A., Yoshida, T., Perez de Souza, L., Alseekh, S., Bajdzienko, K., *et al.* (2020) An improved extraction method enables the comprehensive analysis of lipids, proteins, metabolites and phytohormones from a single sample of leaf tissue under water-deficit stress. *Plant Journal* 103, 1614–1632.

Sarabia, L.D., Boughton, B.A., Rupasinghe, T., van de Meene, A.M., Callahan, D.L., *et al.* (2018) High-mass-resolution MALDI mass spectrometry imaging reveals detailed spatial distribution of metabolites and lipids in roots of barley seedlings in response to salinity stress. *Metabolomics* 14: 63.

Schroeder, M., Meyer, S.W., Heyman, H.M., Barsch, A., and Sumner, L.W. (2019) Generation of a collision cross section library for multi-dimensional plant metabolomics using UHPLC-trapped ion mobility-MS/MS. *Metabolites* 10: 13.

Shanta, P.V., Li, B., Stuart, D.D., and Cheng, Q. (2021) Lipidomic profiling of algae with microarray MALDI-MS toward ecotoxicological monitoring of herbicide exposure. *Environmental Science and Technology* 55, 10558–10568.

Shimizu, T., Miyakawa, S., Esaki, T., Mizuno, H., Masujima, T., *et al.* (2015) Live single-cell plant hormone analysis by video-mass spectrometry. *Plant and Cell Physiology* 56, 1287–1296.

Simpson, J.P., Wunderlich, C., Li, X., Svedin, E., Dilkes, B., and Chapple, C. (2021) Metabolic source isotopic pair labeling and genome-wide association are complementary tools for the identification of metabolite–gene associations in plants. *Plant Cell* 33, 492–510.

Singh, D., Geat, N., Mehriya, M., Rajawat, M.V.S., Prasanna, R., *et al.* (2020) Omics (genomics, proteomics, metabolomics, etc.) tools to study the environmental microbiome and bioremediation. In: Kashyap, B., Solanki, M.K., Kamboj, D.V., and Randey, A.K. (eds.) *Waste to Energy: Prospects and Applications*. Springer, Singapore, pp. 235–260.

Stettin, D., Poulin, R.X., and Pohnert, G. (2020) Metabolomics benefits from orbitrap GC–MS—comparison of low-and high-resolution GC–MS. *Metabolites* 10: 143.

Tabassam, Q., Mehmood, T., Ahmed, S., Saeed, S., Raza, A.R., and Anwar, F. (2021) GC-MS metabolomics profiling and HR-APCI-MS characterization of potential anticancer compounds and antimicrobial activities of extracts from *Picrorhiza kurroa* roots. *Journal of Applied Biomedical* 19, 26–39.

Trenberth, K.E. (2001) Stronger evidence of human influences on climate: the 2001 IPCC Assessment. *Environment: Science and Policy for Sustainable Development* 43, 8–19.

Trethewey, R.N. (2004) Metabolite profiling as an aid to metabolic engineering in plants. *Current Opinion in Plant Biology* 7, 196–201.

Veley, K.M., Berry, J.C., Fentress, S.J., Schachtman, D.P., Baxter, I., and Bart, R. (2017) High-throughput profiling and analysis of plant responses over time to abiotic stress. *Plant Direct* 1: e00023.

Warren, D.L., Dornburg, A., Zapfe, K., and Iglesias, T.L. (2021) The effects of climate change on Australia's only endemic Pokémon: measuring bias in species distribution models. *Methods in Ecology and Evolution* 12, 985–995.

Wickremsinhe, E.R., Singh, G., Ackermann, B.L., Gillespie, T.A., and Chaudhary, A.K. (2006) A review of nanoelectrospray ionization applications for drug metabolism and pharmacokinetics. *Current Drug Metabolism* 7, 913–928.

Xiao, L., Asseng, S., Wang, X., Xia, J., Zhang, P., *et al.* (2022) Simulating the effects of low-temperature stress on wheat biomass growth and yield. *Agricultural and Forest Meteorology* 326: 109191.

Xu, Z., Kong, R., An, D., Zhang, X., Li, Q., *et al.* (2023) Evaluation of a sugarcane (*Saccharum* spp.) hybrid F_1 population phenotypic diversity and construction of a rapid sucrose yield estimation model for breeding. *Plants* 12, 647.

Yamamoto, K., Takahashi, K., Caputi, L., Mizuno, H., Rodriguez-Lopez, C.E., *et al.* (2019) The complexity of intercellular localisation of alkaloids revealed by single-cell metabolomics. *New Phytologist* 224, 848–859.

Zalasiewicz, J. and Williams, M. (2021) Climate change through Earth history. In: Letcher, T.M. (ed.) *Climate Change*, 3rd edn. Elsevier, Amsterdam, the Netherlands, pp. 49–65.

Žampachová, L., Aturki, Z., Mariani, F., and Bednář, P. (2021) A rapid nano-liquid chromatographic method for the analysis of cannabinoids in *Cannabis sativa* L. extracts. *Molecules* 26: 1825.

Zandalinas, S.I., Sengupta, S., Fritschi, F.B., Azad, R.K., Nechushtai, R., and Mittler, R. (2021) The impact of multifactorial stress combination on plant growth and survival. *New Phytologist* 230, 1034–1048.

Zeki, Ö.C., Eylem, C.C., Reçber, T., Kır, S., and Nemutlu, E. (2020) Integration of GC–MS and LC–MS for untargeted metabolomics profiling. *Journal of Pharmaceutical and Biomedical Analysis* 190: 113509.

Zheng, F., Zhao, X., Zeng, Z., Wang, L., Lv, W., *et al.* (2020) Development of a plasma pseudotargeted metabolomics method based on ultra-high-performance liquid chromatography–mass spectrometry. *Nature Protocols* 15, 2519–2537.

Zhu, M., Geng, S., Chakravorty, D., Guan, Q., Chen, S., and Assmann, S.M. (2020) Metabolomics of red-light-induced stomatal opening in *Arabidopsis thaliana*: coupling with abscisic acid and jasmonic acid metabolism. *Plant Journal* 101, 1331–1348.

7 Plant Metabolomics Based on Single-cell Technology

Ekta Pandey[1], Rinkee Kumari[1], Shahla Faizan[1], Sayyada Bushra[1], Alisha Hussain[1], Parul Verma[1], and Mohammed Sayeed Akhtar[2*]
[1]*Department of Botany, Aligarh Muslim University, Aligarh 202002, Uttar Pradesh, India;* [2]*Department of Botany, Gadhi Faiz-e-Aam College, Shahjahanpur 242001, Uttar Pradesh, India*

Abstract

Single-cell technology has revolutionized plant metabolomics, allowing researchers to analyze metabolites in individual cells at the nanoliter scale. This high-throughput technique allows a more comprehensive understanding of plant metabolism, revealing cellular variability even within the same tissue. Single-cell metabolomics, a relatively new field, is based on advances in microfluidics, single-cell sequencing, and mass spectrometry. These technologies enable researchers to measure the metabolic profiles of individual cells with unprecedented resolution, accuracy, and throughput. This chapter discusses the current approaches to plant metabolomics using single-cell technology, and the challenges and future directions in this rapidly evolving field.

Keywords: high-throughput technology, metabolic profiling, microfluidics, plant metabolism, single-cell sequencing

7.1 Introduction

Metabolomics, as the latest functional genomics challenge, follows the establishment of methods for genomics (high-throughput DNA sequencing), transcriptomics (analysis of gene expression), and proteomics (analysis of proteins). Using high-throughput techniques, metabolomics is the careful and unbiased study of complex mixtures of metabolites that are often found in plant extracts (Hollywood *et al.*, 2006). Its origins can be traced back to the early metabolic studies of the 1950s and 1960s, which concentrated on isolated metabolites through methods such as gas chromatography (GC). The term "metabolomics" was coined in the late 1990s, aligning with advancements in high-throughput technologies such as mass spectrometry (MS) and nuclear magnetic resonance (NMR) spectroscopy (Griffiths and Wang, 2009). These advancements enabled the comprehensive profiling of metabolites, sparking exponential growth in the field and its applications in deciphering various physiological and pathological conditions in plants. Presently, approximately 200,000 metabolites have been identified in plants. Each distinct plant species typically harbors around 7000–15,000 of these metabolites, with 3000–5000 of them being present in the leaves (d'Auria and Gershenzon, 2005; Fernie, 2007). The intricate and diverse characteristics of these extensive chemical groups, which exhibit

* E-mail: sayeedbot@gmail.com

© CAB International 2025. *High-Throughput Plant Metabolomics* (Jen-Tsung Chen ed.)
DOI: 10.1079/9781800626911.0007

temporal and spatial variations, pose significant challenges for contemporary analytical technology when juxtaposed with the relatively straightforward polymeric structures of DNA molecules composed of four nucleotides.

Plant metabolomics has traditionally relied on bulk plant material, providing a general overview of the metabolic status of entire tissues or organs. Despite its informative nature, this approach often overlooks the inherent diversity within plant tissues, where individual cells exhibit unique metabolic patterns. In contrast, single-cell metabolomics dissects this intricate mosaic by analyzing metabolites at the cellular level, revealing unprecedented insights into cellular functions, interactions, and responses (Sumner *et al.*, 2011; Gautam *et al.*, 2024). Recent advancements in single-cell metabolomics have unveiled ground-breaking discoveries, such as the diverse metabolic states present in plant tissues. Being multicellular organisms, higher plant species, such as rice, with its 40 distinct cell types, have attracted considerable research attention (Jiao *et al.*, 2009). Studies focusing on single-cell metabolomics have thus utilized refined samples from various cell types across different plant genera. Unlike animals, plants exhibit remarkable adaptability in their metabolic activities, resulting in varied metabolic profiles among different cell types (Zenobi, 2013; Weng *et al.*, 2021). For instance, root cells engaging with the soil environment exhibit metabolic patterns distinct from those of photosynthetically active leaf cells. Through single-cell metabolomics, researchers can delineate these metabolic variations and elucidate cell-specific functions in processes such as nutrient absorption, stress responses, and growth regulation.

Furthermore, there is increasing clarity on the dynamic interplay between plants and their environment. Plants continuously adjust their metabolic pathways in response to environmental fluctuations, such as temperature, humidity, and nutrient availability (Dusenge *et al.*, 2019; Gilmore *et al.*, 2019). Single-cell metabolomics has revealed that these adaptations occur at the cellular level, rather than uniformly across tissues. These findings underscore the intricate nature of plant adaptation mechanisms and their implications for enhancing crop resilience against climate change and various stresses. This chapter discusses the current approaches to plant metabolomics using single-cell technology, and the challenges and future directions in this rapidly evolving field.

7.2 Plant Metabolomics

A major extra dimension is supplied by metabolites to the data flow that happens through DNA to RNA to protein. This flow of information brings about a considerable change. The development of sophisticated instruments and a more rapid comprehension of the activities that metabolites play in organisms have both contributed to this advancement (Schauer and Fernie, 2006). Plants produce hundreds of distinct metabolites. These metabolites serve as natural compounds that help plants attract pollinators, repel herbivores, fight microbial attacks, and protect themselves from environmental challenges. Two major categories may be used to classify plant metabolites: primary and specialized (secondary) metabolites (Castro-Moretti *et al.*, 2020). For plants to be able to withstand plant parasite infections as well as biotic and abiotic stresses, the plant must produce secondary or specialized metabolites. In contrast, primary metabolites comprise components vital to the plant's growth and reproduction (Fig. 7.1). Notably, the metabolites generated by fundamental metabolic biochemical pathways, such as the EMP (Embden–Meyerhof–Parnas) pathway, the pentose phosphate pathway, and the tricarboxylic acid cycle, also serve as the fundamental components of secondary metabolic pathways. This serves to emphasize the inherent connection that exists among these pathways. Amino acids, for instance, play a part in the process of sequestering nitrogen and also act as intermediates in the production of a wide range of specialized compounds, including pigments and phytohormones. Two kinds of biocidal secondary metabolites, also known as phytoalexins and phytoanticipins, are produced by plants as a means of protecting themselves against the damage caused by pathogens and pests (Khan *et al.*, 2023). One of the main goals of this new field of study, called single-cell plant metabolomics, is to thoroughly examine each plant cell's metabolic profile. At the most basic level of biological structure, this method offers a strong tool

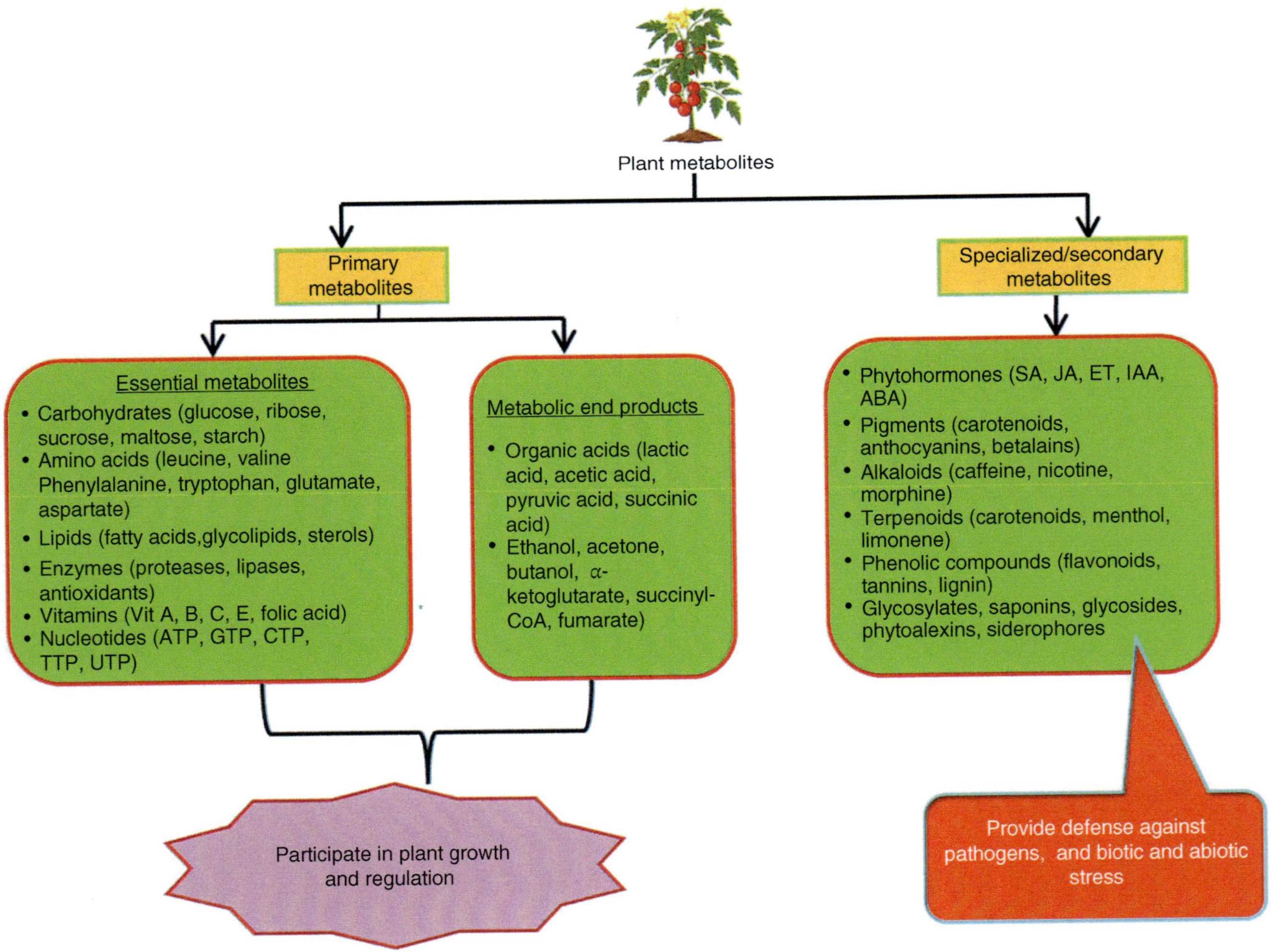

Fig. 7.1. Schematic representation of different categories of plant metabolites and their function in plant growth and development and defense mechanisms. ABA, abscisic acid; ET, ethylene; IAA, indole-3-acetic acid; JA, jasmonic acid; SA, salicylic acid.

for providing an understanding of the complex and ever-changing nature of plant metabolism (Katam *et al.*, 2022). In single-cell plant metabolomics, advanced MS techniques such as liquid chromatography–mass spectrometry (LC-MS) and gas chromatography–mass spectrometry (GC-MS) are used, along with other analytical platforms such as NMR spectroscopy. These methods can accurately identify and quantify several metabolites in small sample amounts (Kueger *et al.*, 2012). In addition to analytical obstacles, single-cell plant metabolomics demands specific sample preparation and data interpretation. It is difficult to isolate plant cells without affecting their metabolic integrity, and these investigations create huge data sets that need advanced bioinformatics and computational methods to draw conclusions.

7.3 Single-cell Sampling Technologies and Types

The accuracy of single-cell metabolomics data relies heavily on the precise isolation of each distinct target cell from tissues or the collection of single cells from a cell suspension. The primary methods used in the area of single-cell isolation include laser-capture microdissection, micromanipulation, microfluidics, single-cell aspiration, fluorescence-based cell sorting, and limiting dilution (Table 7.1). The techniques used depends on the source of cells being isolated. Several types of cell plants are used for the study of metabolomics.

7.4 Recent Methods in Single-cell Metabolomics Techniques

Apart from a few prominent metabolites such as sugars or pigments such as anthocyanins (Lindström and Andersson-Svahn, 2010), the small, limited volume of single cells (at the femtoliter to picoliter range) has made their sampling and analysis challenging (Zhang and Vertes, 2018). Single-cell metabolites are diverse and modest in content, and the concentration of each component varies significantly. Moreover, due to the absence of an amplification phase, metabolite analysis requires more sensitive detection techniques than DNA and RNA analysis (Eirew *et al.*, 2015; Comi *et al.*, 2017; Villani *et al.*, 2017). Single-cell metabolite types and amounts alter with constant changes in cellular metabolism and environmental conditions. Thus, for single-cell metabolite analysis, it is important to create analytical techniques that are highly sensitive, selective, and efficient in terms of resolution and throughput (Armbrecht and Dittrich, 2017). MS is the most common approach for detecting and analyzing single-cell metabolites because of its high accuracy, its capacity to detect many compounds simultaneously, and its qualitative capabilities. Table 7.2 details the primary fundamental techniques employed in single-cell metabolomics, emphasizing their primary characteristics, benefits, and restrictions. Various analytical techniques are utilized, with multiple modifications based on the specific aim of the analysis.

7.4.1 MS-based methods for single-cell metabolomics

MS is widely considered to be the most precise technique for identifying the diverse range of metabolites found at extremely low concentrations (femtomole to attomole) produced in a single cell (Minakshi *et al.*, 2019). In addition, the highly informative and label-free nature of MS analysis, along with its ability to work in numerous array methods, has established it as the predominant technique for single-cell metabolomics analysis. It also enables the collection of data from several metabolomics databases, along with the ability to correlate metabolomic information with other omics databases for valuable biological interpretations (Minakshi *et al.*, 2019; Bianchi and Careri, 2021; Prasad *et al.*, 2022). After the digestion of a single target cell, the cellular metabolites can be separated by a microfluidic system in a variety of ways, including electrophoretically via capillary electrophoresis (CE), or chromatographically through LC and GC. In order to perform identification, quantification, or downward analysis, isolated metabolites can be transferred directly to a connected MS platform (DeLaney *et al.*, 2018). A mass spectrometer comprises an ionization source, a mass analyzer that separates ions based on their mass:charge (m/z) ratio, and a detector that

Table 7.1. Summary of the different single-cell isolation sampling methods.

Sampling technology	Principle	Advantages	Limitations	Applications	Reference
Micromanipulation	Physical isolation of individual cells using fine-tipped glass micropipettes or microneedles	High precision, versatile, minimal tissue disruption	Low throughput, labor-intensive, and operator expertise necessary	Particular cell type isolation, cellular injection, and omics research	Ungai-Salánki *et al.* (2016)
Microfluidics	Utilizing microfluidic chambers and channels to separate and manipulate individual cells	Advanced single-cell analysis with high-throughput and integrated sample preparation and analysis	Requires fabrication, potential technical complexity	Metabolomics, single-cell sequencing, cell sorting	Sukovich *et al.* (2017)
Patch-clamp technique	Use of a glass pipette to create a high-resistance seal on the cell membrane	High resolution, suitable for live cells, can measure ionic currents	Complex setup requires skill, limited to electrophysiologically active cells	Study of cell signaling, neuroscience, and intracellular metabolomics	Gradov and Gradova (2015)
Laser-capture microdissection	Using a laser beam to selectively cut and capture individual cells or small cell groupings	Maintains spatial context and is compatible with both fresh and fixed tissues	Low throughput, risk of contamination or sample loss	Histology, proteomics, genomes, and metabolomics	Zeb *et al.* (2019)
Microcapillary extraction	Using glass capillaries to mechanically isolate individual cells	High isolation cell purity, little contamination risk, compatibility with further analyses	Labor intensive, low throughput, requiring care and expertise	Transcriptomics, proteomics, and metabolomics	Zhu *et al.* (2010)
Flow cytometry and fluorescence-activated cell sorting	Use of laser and fluorescence signals to sort and isolate single cells	High recovery and purity rates, appropriate for live cells, versatile marker use	Requires fluorescent labeling, potential cell damage due to sorting	Single-cell omics, population analysis, and cell sorting	Minakshi *et al.* (2019)

Table 7.2. Key analytical methods used for single-cell metabolomics in plants.

Method	Ionization technique	Instrument type	Advantages	Limitations	Typical metabolite size range	Reference
Single-cell LC-MS	ESI, APCI	Mass analyzer (quadrupole, TOF, Orbitrap)	Able to separate complex mixtures, high resolution, and compatible with a broad variety of metabolites	Time-consuming, requires specialist knowledge for data interpretation	100–2000 Da	Gika *et al.* (2014)
Single-cell GC-MS	EI, CI	Mass analyzer (quadrupole, TOF, Orbitrap)	High sensitivity for volatile compounds	Restricted to volatile and thermally stable compounds	50–800 Da	Fiehn (2016)
Single-cell CE-MS	ESI	Mass analyzer (Q-TOF, Orbitrap)	Excellent for charged metabolites, high separation efficiency	Limited throughput, lower reproducibility	50–2000 Da	Ramautar *et al.* (2019)
ESI-MS	ESI (API)	Mass analyzer (FT-ICR, TOF, Orbitrap)	Allows analysis of intact, labile, high-molecular-mass molecules and is suitable for polar and ionic compounds	Requires liquid sample, can have ion-suppression issues	100–2000 Da	Zhu *et al.* (2021)
LAESI-MS	Laser electrospray ionization	Mass analyzer (FT-ICR, TOF, Orbitrap)	*In situ* and spatially resolved study of tissues, cells, and polar and non-polar compounds	Lower sensitivity, matrix effects, and ion suppression	100–2000 Da	Kang *et al.* (2016)
DESI-MS	Desorption electrospray ionization	Mass analyzer (Orbitrap, FT-ICR)	Minimal sample preparation, ambient ionization conditions, suitable for *in situ* analysis of plant tissues	Limited to surface metabolites, poorer spatial resolution than MALDI	100–2000 Da	Takats (2012)

Continued

Table 7.2 Continued.

Method	Ionization technique	Instrument type	Advantages	Limitations	Typical metabolite size range	Reference
MALDI-MS	Matrix-assisted laser desorption/ ionization	Mass analyzer (Orbitrap, TOF)	Spatial resolution for imaging, suitable for large biomolecules (e.g. lipids, secondary metabolites)	Matrix interference, lower sensitivity for small metabolites	500–5000 Da	Ma *et al.* (2023)
SIMS	Ion bombardment	Mass analyzer (TOF)	High spatial resolution, surface sensitivity	Chemical fragmentation, limited to surface metabolites	50–1000 Da	Lockyer *et al.* (2024)
NMR spectroscopy	NA (non-ionization)	NMR spectrometer	Non-destructive, no extensive sample preparation	Lower sensitivity, large sample volume needed	Typically small molecules (50–1000 Da)	Emwas *et al.* (2022)
Raman spectroscopy	NA (vibrational spectroscopy)	Raman spectrometer	Non-destructive, minimal sample preparation	Lower sensitivity, water interference in samples	NA (depends on the chemical bonds)	Schie and Huser (2013)

APCI, atmospheric pressure chemical ionization; API, atmospheric pressure ionization; CE-MS, capillary electrophoresis–mass spectrometry; CI, chemical ionization; DESI-MS, desorption electrospray ionization–mass spectrometry; EI, electron ionization; ESI, electrospray ionization; ESI-MS, electrospray ionization–mass spectrometry; FT-ICR, Fourier-transform–ion cyclotron resonance; GC-MS, gas chromatography–mass spectrometry; LAESI-MS, laser ablation electrospray ionization–mass spectrometry; LC-MS, liquid chromatography–mass spectrometry; NA, not applicable; NMR, nuclear magnetic resonance; Q-TOF, quadrupole time-of-flight; SIMS, secondary ion mass spectrometry; TOF, time-of-flight.

records the ions or Fourier transform of the image current. The most widely used ionization source techniques for single-cell analysis are Fourier-transform–ion cyclotron resonance (FT-ICR), electrospray ionization (ESI), desorption ESI (DESI), matrix-assisted laser desorption/ionization (MALDI), laser desorption/ionization (LDI), and secondary ion mass spectrometry (SIMS) (Misra *et al.*, 2014; Hansen and Lee, 2018; Maia *et al.*, 2023). Utilizing several ionization sources combined with MS has shown the extensive array of methodologies accessible in MS. Figure 7.2 depicts the workflow for single-cell metabolomics derived from a number of cell types.

Single-cell LC-MS

The advancement of LC-MS techniques has dramatically transformed the field of plant metabolism at the single-cell level. The emergence of this potent analytical approach has become essential for deciphering the complex metabolic environments within individual plant cells, offering an unparalleled understanding of the spatial and temporal changes in plant metabolism. In the field of single-cell plant metabolomics, LC-MS provides an efficient and sensitive platform for detecting and identifying a diverse range of metabolites, from primary compounds such as amino acids and sugars to secondary metabolites with complex structures (Taylor *et al.*, 2021; Katam *et al.*, 2022). In LC-MS, the structural elucidation and sensitive detection of MS are combined with the high-resolution separation capabilities of LC to allow the thorough profiling of metabolites within individual plant cells. LC-MS allows researchers to separate, detect, and structurally elucidate these metabolites, providing an exhaustive overview of the metabolic profiles within single plant cells (Theodoridis *et al.*, 2012).

The procedure for single-cell plant metabolomics using LC-MS usually starts by extracting targeted single cells or small clusters of cells from plant tissues. This can be accomplished using various methods, including micromanipulation, laser-capture microdissection, and fluorescence-activated cell sorting (Misra, 2020). After the cells have been separated, various polar and non-polar metabolites are extracted from them, frequently using a mix of solvents such as acetonitrile, methanol, and water. The metabolites are then separated by LC. This technique uses a combination of stationary phases and mobile phases to resolve the complex mixture of compounds. Common methods employed in single-cell plant metabolomics include reverse-phase LC, which separates molecules according to their hydrophobicity (Diederen *et al.*, 2021), and hydrophilic interaction LC, which targets metabolites with a more polar geometry (Navarro-Reig *et al.*, 2017). After separation, the metabolites are loaded into the mass spectrometer, where they are ionized using techniques such as atmospheric pressure chemical ionization (APCI) or ESI (Theodoridis *et al.*, 2012). The mass spectrometer provides useful information about the molecular mass and structural features of the ionized metabolites by measuring their m/z ratios (Ren *et al.*, 2018). Yamamoto *et al.* (2019) identified the vindoline biosynthetic pathway in *Catharanthus roseus* leaf tissue using LC-MS and demonstrate that vindoline accumulates in the idioblast and laticifer cells of the elongating leaf. Yamamoto *et al.* (2016) also found different terpenoid indole alkaloids in the whole-stem tissue of *C. roseus*, which included catharanthine, ajmalicine, vindoline, serpentine, loganic acid, and loganin.

Single-cell GC-MS

Single-cell GC-MS capitalizes on the potential of MS and GC. GC separates volatile compounds by differentiating their interactions with a stationary phase within a column. In contrast, MS provides comprehensive structural information by ionizing these compounds and determining their m/z ratio. When coupled, GC-MS provides excellent specificity, sensitivity, and the capacity to evaluate small quantities of complex biological mixtures (Majchrzak *et al.*, 2018). Electron ionization (EI) and chemical ionization (CI) are the most common ionization sources used in GC-MS. However, compared with EI, CI has not been used as much in metabolomics. The reason for this may be that CI is a “soft” ionization technique, which results in fragments that are less sensitive and will not assist in the identification process as effectively as EI (Warren, 2013; Amirav *et al.*, 2020). The intricacy of the data and the requirement for advanced computational tools to handle, analyze, and display the enormous volumes of data produced provide a

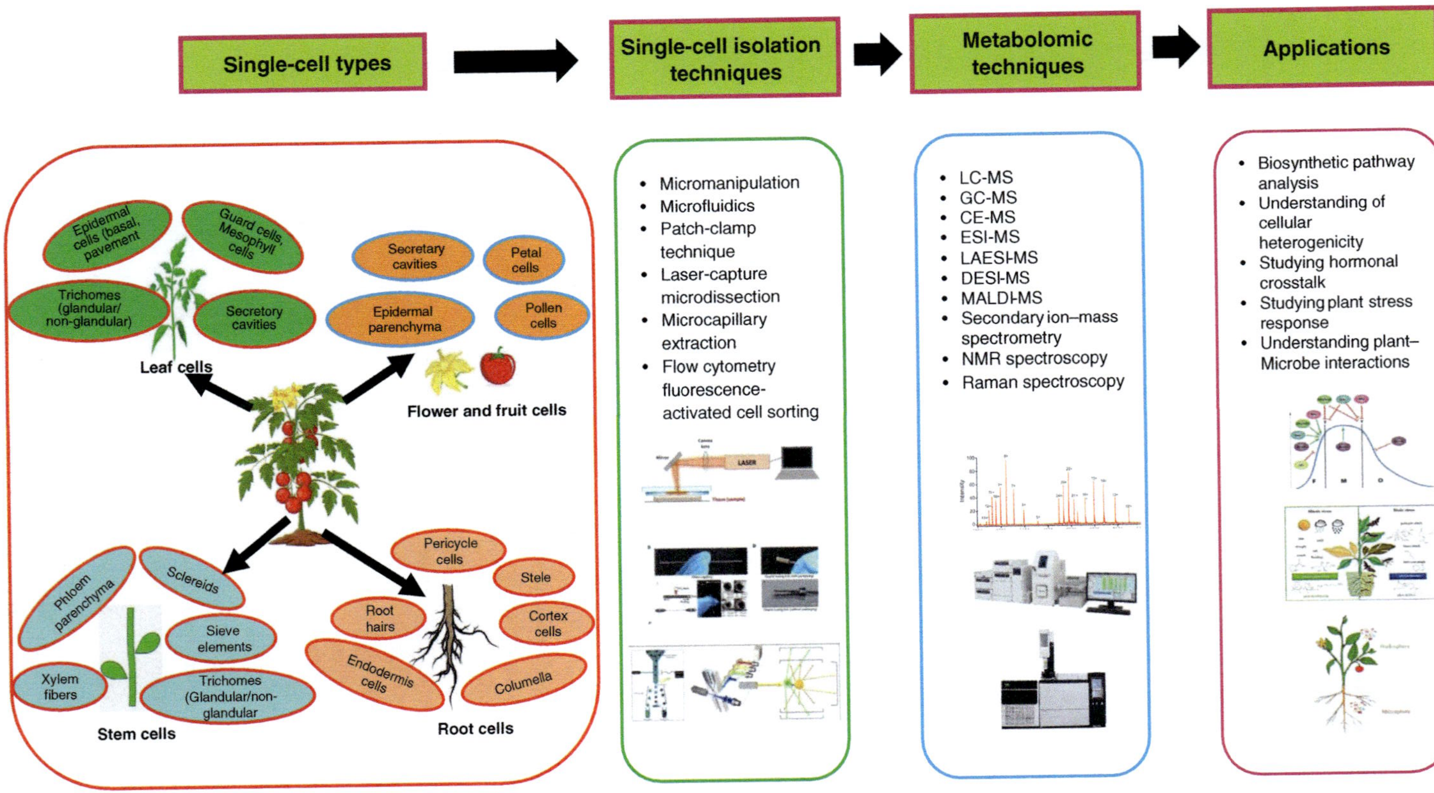

Fig. 7.2. Representation of a workflow chart for single-cell metabolomics using a variety of cell types. CE-MS, capillary electrophoresis–mass spectrometry; DESI-MS, desorption electrospray ionization–mass spectrometry; ESI-MS, electrospray ionization–mass spectrometry; GC-MS, gas chromatography–mass spectrometry; LAESI-MS, laser ablation electrospray ionization–mass spectrometry; LC-MS, liquid chromatography–mass spectrometry; MALDI-MS, matrix-assisted laser desorption/ionization–mass spectrometry; NMR, nuclear magnetic resonance.

significant challenge. Steinbrenner *et al.* (2011) used GC-MS to examine how the metabolism of individual cells in tomato leaves reacts to herbivore attacks and demonstrated how plant defense chemicals move around in space and time.

Single-cell CE-MS

CE-MS is an analytical method based on separation that involves lysing captured cells to release complex metabolites. Following a capillary channel separation process based on their physicochemical properties, the metabolites are placed into a MS analyzer for characterization. Electrophoresis in capillaries, typically nanoliter volumes, is ideal for single-cell metabolomics. Metabolite separations often use two-dimensional electrophoresis. In the first dimension, isoelectric focusing is used to separate metabolites based on charge, and in the second dimension, separation is by molecular mass. Based on diverse ionization techniques, different types of CE have been used for MS analysis, including CE-ESI, CE–inductively coupled plasma (CE-ICP), CE-APCI, and CE–atmospheric pressure photoionization (CE-APPI) (Minaksh *et al.*, 2019). CE-MS has been used to quantify carbohydrates, amino acids, and primary metabolites in reaction to high CO_2 levels (Watanabe *et al.*, 2014). Using CE and CE-MS, multiple studies have identified the main metabolites in rice. For example, Maruyama *et al.* (2014) discovered cold- and dehydration-responsive metabolites and phytohormones in rice. This research identified multiple genes that are upregulated in starch breakdown, the glyoxylate cycle, and sucrose metabolism. This analysis also helped to observe the correlation between glyoxylate cycle regulation and glucose buildup in rice, which is not present in *Arabidopsis*.

ESI-MS

Soft ionization techniques such as ESI have been utilized to analyze metabolites in single cells using ESI-MS-based techniques. Many researchers have created novel ESI methods and there have been significant advancements in this area of individual cell analysis to improve ESI-MS performance (Yin *et al.*, 2019). Masuda *et al.* (2018), for instance, created a live single-cell MS method. This procedure usually delivers a liquid sample (the cell's cytoplasm) into the mass spectrometer via a capillary or nozzle. The liquid sample is exposed to a strong electric field at the capillary or nozzle tip, usually between 2 and 5 kV (Zhang *et al.*, 2023). ESI is the mechanism by which charged droplets develop due to this electric field. Depending on the voltage generated by the ESI nozzle, the analytes can be either deprotonated (M^-, negative mode) or protonated (M^+, positive mode) (van Dam and Bouwmeester, 2016; Khan *et al.*, 2023). The charged droplets experience a sequence of Coulomb fission and solvent evaporation processes, which produce gas-phase ions containing the analytes found in the sample. Skimmers, lenses, and ion guides focus and convey ESI gas-phase ions into the mass analyzer. This stage guarantees effective ion transport from the atmospheric pressure ESI source to the high-vacuum mass analyzer. The gas-phase ions are separated by m/z ratio utilizing quadrupole, time-of-flight (TOF), ion trap, or orbitrap mass analyzers. Mass spectra from mass analyzers show the sample's ion abundance (Cubero Herrera, 2007). The advantage of this method is that ESI-MS can directly analyze samples in liquid form, such as solutions, extracts, or biological fluids, without extensive sample preparation. Because ESI-MS can identify analytes at femtomolar (10^{-15} M) concentrations, it is a good tool for analyzing tiny amounts of chemicals.

Laser ablation ESI-MS

In the field of ionization, laser ablation ESI (LAESI) is a relatively new technique that is particularly useful for assessing samples containing water. To stimulate the water molecules present in the materials under investigation, this technique uses a focused mid-infrared laser beam with a wavelength of 2.94 μm. During the procedure, gaseous metabolite particles are produced. These particles are then ionized by releasing charged droplets from the ESI laser. The ionized metabolite particles are then analyzed using MS. The laser beam passes through a germanium dioxide (GeO_2) glass fiber trim tip, which has a surface area of around 30 μm^2. With this configuration, spatial resolution is achievable down to 30 μm (Flanigan and Levis, 2014). Thus, this approach is best for analyzing large plant or animal cell clusters. LAESI is a technology with a reduced chemical background, easy sampling, and enhanced sensitivity, making it suitable for single-cell analysis. LAESI-MS has been used in several single-cell studies (e.g. Hieta *et al.*, 2020). Shrestha *et al.* (2011) used LAESI-MS to analyze single-cell metabolic profiles *in situ*. This study used small cell cultures of *Allium cepa*, *Narcissus pseudonarcissus* bulb epidermis, and *Lytechinus pictus* single eggs as research models.

DESI-MS

The DESI technique was developed by Takats *et al.* (2004). The idea behind DESI is to use ESI to create charged droplets and ions from a solvent, which are then directed toward the surface of the targeted analyte. The charged particles deposited on to the surface produce gaseous ions in conjunction with the analytes. After the sample mixture is applied to a polytetrafluoroethylene plate and allowed to dry, a high voltage (4 kV) is used to spray methanol–water (1:1) containing either 0.1% aqueous acetic acid or 1% acetic acid solution (Takats, 2012). DESI can ionize a larger spectrum of target compounds than other ionizing techniques such as MALDI and nanostructure-initiator-MS because, with the right derivatization, it can ionize lipids and free fatty acids, carbohydrates, lipid-soluble vitamins, and cholesterol. High spatial resolution, which is necessary for single-cell analysis, may be achieved using DESI-MS (Yin *et al.*, 2019). This makes it possible to determine differences in metabolism between single cells or between various areas of a single cell. Due to its versatility, DESI-MS may be used on a variety of materials, such as living cells, dried samples, and fresh tissues.

7.4.2 NMR spectroscopy

The quantified nature and wide range of metabolic information in NMR spectroscopy make it useful for single-cell metabolomics. The principle behind single-cell NMR spectroscopy is to use the magnetic characteristics of atomic nuclei, such as carbon (^{13}C) and hydrogen (^{1}H), to learn more about a material's chemical makeup and structure. In this instance, the sample is a single cell that is separated and examined to identify its chemical makeup and metabolic characteristics. ^{1}H-NMR helps identify the arrangement of hydrogen atoms in a molecule, whereas ^{13}C-NMR shows the arrangement of carbon atoms in an analytic sample (Emwas *et al.*, 2022). In metabolomics, NMR analyzes polar compounds. While less exact (micromolar range) than MS-based methods, it is more secure in terms of verification and consistency. NMR, an impartial and unbiased approach, reliably detects the molecular composition of substances. Metabolites have been analyzed extensively using quantitative and qualitative NMR techniques. The key challenges in single-cell NMR spectroscopy are the small sample volumes, which necessitate specialized instrumentation and high-sensitivity detection systems. Additionally, the sample preparation and handling of single cells without introducing artifacts or disrupting their natural state can be technically challenging. These volumes typically range from nanoliters to picoliters (Edison *et al.*, 2020).

7.4.3 MALDI-MS

MALDI-MS is commonly employed to investigate large molecules with minimal dissolution. MALDI-MS observations generally commence by combining the desired sample with a matrix, which is subsequently irradiated with an ultraviolet (UV) laser beam under vacuum conditions. The analytes are ionized and transported to a mass spectrometer analyzer, typically TOF-MS device (Weidner, 2012). The initial phase requires isolating individual cells from a larger plant tissue or sample. This can be accomplished through the use of microfluidic devices, micromanipulation, or laser-capture microdissection. The isolated single cells are subsequently coated with a specialized matrix compound, typically an organic acid (e.g. α-cyano-4-hydroxycinnamic acid or 2,5-dihydroxybenzoic acid). The matrix compound facilitates ionization of the metabolites present in the cell. The sample is exposed to an intense laser beam, typically a UV laser (e.g. nitrogen laser or neodymium-doped yttrium aluminum garnet (Nd:YAG) laser). This beam targets the matrix-coated single cells. Upon absorption of the laser energy, the matrix compound undergoes rapid ionization and vaporization. This procedure, known as laser desorption, discharges ionized, intact metabolite molecules from the individual cell (Earnshaw *et al.*, 2007). After ionization, the metabolites are propelled on to a TOF or FT-ICR mass spectrometer. Ionized metabolites are sorted by *m/z* ratio in the mass spectrometer, with lighter ions traveling faster than heavier ions. The mass spectrometer detects the separated ions and generates a mass spectrum that shows the molecular mass and relative abundance of the cell's metabolites (Chang *et al.*, 2007). This method has been utilized extensively to examine plant growth, stress responses, and secondary metabolite spatial organization, revealing the intricate and dynamic nature of plant metabolism in single cells.

7.4.4 SIMS

SIMS was developed in the 1960s and is often referred to as ion probe analysis. SIMS is a technology for analyzing very sensitive surface composition. A primary ion beam (e.g. O_2^+, Ar^+, N_2^+, O^-, N^-, F^-, Cs^+, or Ga^+) is used to bombard the surface of the material, causing the atoms on the surface of the material to be released as a spray of highly charged ions. Subsequently, a mass spectrometer is utilized to analyze the *m/z* ratio of the compound ions to ascertain the sample's composition (Wu *et al.*, 2017; Yin *et al.*, 2019). A two-dimensional or even three-dimensional picture of the sample's chemical composition may be formed by rastering the main ion beam over the sample surface and collecting the secondary ions at each pixel, allowing generation of the image. The spatial resolution that may be achieved using MALDI-MS imaging is less than 1 μm, but SIMS has a higher resolution (50–10 nm) (Lockyer *et al.*, 2024). Therefore, nano-SIMS is suited for smaller applications in biology, such as the study of calcium channeling, investigation of subcellular metabolite allocation, and analysis of microbial metabolites via the use of stable isotope labeling technology.

7.4.5 Raman spectroscopy

Raman spectroscopy is a non-destructive method for detecting and quantifying cell metabolites. The approach involves using visible, near-infrared, or near-UV lasers to illuminate cells and tag them with fluorescent or radioactive staining. Single-cell Raman spectroscopy is a potent, non-invasive, and label-free method for testing the chemical composition and molecular structure of individual cells. Cells are immobilized on a glass slide or microfluidic device to keep the isolated single cells still during Raman measurement (Georgakoudi *et al.*, 2008; Schie and Huser, 2013). A microscope objective focuses a visible or near-infrared monochromatic laser on the cell. The Raman effects occur when a tiny proportion of laser photons scatter inelastically with cell molecules. Raman scattering is shifted in frequency relative to laser light, unique to cell molecular vibrations. A Raman spectrometer differentiates its wavelengths after the microscope objective collects Raman-scattered light. A sensitive charge-coupled device (CCD) or electron-multiplying CCD camera detects the Raman signals (Golcuk *et al.*, 2006). Raman signals are processed to create a Raman spectrum, which shows the intensity of Raman-shifted light as a function of frequency/Raman shift (or wavenumber):

$$\text{Raman shift}\left(\text{cm}^{-1}\right)=\left(\frac{1}{\lambda^{\text{laser}}}-\frac{1}{\lambda^{\text{scattered}}}\right)\times 10^{7}$$

where λ^{laser} is the wavelength of the incident laser light (in nm) and $\lambda^{\text{scattered}}$ is the wavelength of the Raman-scattered light (in nm). The factor of 10^7 is used to convert the units from nanometers to wavenumbers (cm^{-1}).

As it reflects the vibrational modes of biomolecules, including proteins, lipids, and nucleic acids, the Raman spectrum gives the organism a unique molecular fingerprint. With single-cell Raman spectroscopy's high spatial resolution, certain subcellular components such as the nucleus, mitochondria, or lipid droplets can be examined, offering new perspectives on the spatial organization and heterogeneity of the cell.

7.4.6 Applications of single-cell metabolomics in plant biology

Single-cell metabolomics has resulted in an explosion of information on the roles of cells, metabolic pathways, and interactions between cells. It has been shown to be useful in the detection of persistent metabolite modifications that

Table 7.3. Single-cell analysis in plants using metabolomics approaches.

Plant species	Tissue or cell type	Analytic method	Metabolites or number of metabolites	Reference
Cannabis sativa	Trichome cells (sessile, stalked, and bulbous)	LC-MS	Cannabinoids (targeted)	Livingston *et al.* (2020)
Oryza sativa	Suspension culture cells	GC-MS	134 metabolites (main metabolites: methionine, serine, succinate, proline ornithine, glucose, fructose, galactose, trehalose, tyrosine, GABA, cysteine, leucine, asparagine, alanine, and phenylalanine)	Liu *et al.* (2013)
Solanum spp.	Trichome types 1, 4, 6, and 7	LC-MS	119 metabolites (main metabolites: fatty acids, flavonoids, terpenoids, acyl sugars, and alkaloids)	McDowell *et al.* (2011)
Artemisia annua	Glandular trichome cells	GC-MS	12 metabolites (main metabolites: mono- and sesquiterpenoids)	Bertea *et al.* (2005)
Oryza sativa	Cell cultures	CE-MS	84 targeted metabolites (main metabolites: glucose 6-phosphate, dihydroxyacetone phosphate, 3-phosphoglyceric acid, pyruvate, glucose 1-phosphate, glutamyl-l-cysteine, fructose 6-phosphate, phosphoenolpyruvate, and *O*-acetyl-l-serine)	Ishikawa *et al.* (2010)
Celtis australis	Single internodal cells	CE-MS	125 metabolites (main metabolites: pyruvate, GABA, malate, fumarate, succinate, AMP, CMP, and *N*-acetyl-l-cysteine)	Oikawa *et al.* (2011)
Glycine max	Mitochondrial cell fractions	CE-MS	Pyruvate, citrate, aconitate, NAD, succinate, NADH, UDP-glucose, glyceraldehyde 3-phosphate, GABA, G1P, and 6-phosphogluconate	Komatsu *et al.* (2011)
Allium cepa	Epidermal cells	LAESI-MS	Cyanidin (targeted)	Shrestha *et al.* (2011)
Egeria densa	Leaf blade cells	Fiber-based LAESI-MS	Aspartate, malate, ascorbate, glucosides, and triterpene saponins (e.g. azukisaponin I and medicoside I)	Stopka *et al.* (2018)
Viola cornuta	Petal cells	MALDI TOF-MS	Lavonol 3-*O*-glycosides and violanin (targeted)	Sugahara *et al.* (2018)
Zea mays	Root epidermal and cortex cells	MALDI imaging	Tricarboxylic acid cycle metabolites	Hansen and Lee (2018)
Closterium acerosum	Cell culture	MALDI-MS	Targeted (UDP-glucose, ADP, ATP, and GTP)	Amantonico *et al.* (2010)
Pelargonium zonale	Leaf, petal, and stem cells	Nano-ESI-MS	Terpenoids and isoprenoids	Lorenzo Tejedor *et al.* (2012)
Scrippsiella trochoidea	Cell culture	ESI-MS	588 metabolites (main metabolites: phosphoethanolamine, phosphatidylglycerol, phosphatidylserine, triglyceride, and phosphatic acid)	Sun *et al.* (2018)
Lotus japonicus	Flower bud and petal cells	NMR	Flavonoids (gossypetin and quercetagetin)	Hiraga *et al.* (2021)

CE-MS, capillary electrophoresis–mass spectrometry; ESI-MS, electrospray ionization–mass spectrometry; GABA, γ-aminobutyric acid; GC-MS, gas chromatography–mass spectrometry; LAESI-MS, laser ablation electrospray ionization–mass spectrometry; LC-MS, liquid chromatography–mass spectrometry; MALDI-TOF-MS, matrix-assisted laser desorption/ionization–time-of flight–-mass spectrometry; MALDI-MS, matrix-assisted laser desorption/ionization–mass spectrometry; NMR, nuclear magnetic resonance.

depict cellular signaling and metabolic responses, as well as the phenotypes and characteristics of organisms. The metabolome has a high degree of dynamic activity. Table 7.3 highlights some recent studies of single-cell metabolomics using different analytic approaches.

The bioactivity profile of *Fucus vesiculosus*, which was collected on a monthly basis over a period of 1 year, revealed that the concentration of several metabolites varied. The total ion count of phlorotannin was found to be at its maximum during the summer, but the concentrations of chlorophyll pigments, lipids, and carotenoids were at their highest during the winter and spring seasons (Heavisides *et al.*, 2018). The metabolic flexibility of xylem lignification was discovered by untargeted metabolomics in transgenic poplar trees with varied genetic origins. This finding raises the possibility of a genotype-specific metabolism that is connected to certain phenotypic characteristics. Additionally, the primary and secondary metabolic profiles of citrus rootstocks provided an explanation for the resistance of scions to bacteria and other diseases (Killiny *et al.*, 2018; Robinson *et al.*, 2018). A further metabolomic investigation conducted on pepper found that the shift of polyphenols happened throughout the ripening process of the fruit in various components of the fruit, and the abundance of these polyphenols varied depending on the cultivar (Calumpang *et al.*, 2020). Single-cell metabolomics thus has a broad variety of applications in the field of plant biology, such as the study of cellular heterogeneity, and the discovery of new insights into plant growth, stress responses, and disease resistance.

Understanding cellular heterogeneity

Plant tissues have various cell types, each with distinct roles and specific metabolic profiles. Single-cell metabolomics is a very effective method for understanding the differences between cells and within cells, allowing us to gain valuable knowledge about the many metabolic processes in plants. Laser microsampling and nanoflow LC-ESI-MS were used to conduct an analysis on a single-cell basis to determine the molecular components of the petal colors of *Torenia hybrida*, and the number of anthocyanins and the fraction of molecular components in three sections of the petal were examined at the individual cell position (Kajiyama *et al.*, 2006). According to Sugahara *et al.* (2018), the resolution of MS is improving and becoming increasingly more accurate. These authors analyzed the flavonol 3-*O*-glycosides expressed in the petals of *Viola cornuta* and found flavonol 3-O-glycosides that were not violanine. In addition, they investigated whether or not the flavonoids acted such as co-pigments for the bluish color of the flower petals. At the pixel level, they constructed a comprehensive map based on the chemical makeup and quantity of flavonoids obtained from MALDI TOF-MS imaging. This was done to elucidate the spatial distribution of the flavonoids based on the data. The flavonoids that have an effect on the coloring mechanism of viola were discovered through the use of small variations in coloration and the quantities of flavonoids comprising them. They also discovered that the violanin quinonoidal form is surrounded by an excessive amount of flavonol 3-*O*-glycoside molecules, and pigmentation is suppressed by this instead of the pH value in the vacuole (Sugahara *et al.*, 2018). Shrestha *et al.* (2011) demonstrated that it is feasible to do cell-by-cell imaging with LAESI *in situ* without causing any disruption to the epidermal cells of onion.

Exploring plant stress responses

Numerous aspects of plant growth and development are negatively impacted by abiotic and biotic stressors, which have become a persistent threat to agricultural production. These stressors reduce the germination process, the absorption of nutrients, photosynthesis, yield, quality, and primary and secondary metabolites. Plants produce a greater quantity of metabolites in comparison with both mammals and microbes (Brunetti *et al.*, 2013; Carrington *et al.*, 2018). Because plants are sessile creatures, they have developed systems that allow them to react and adapt to the changing circumstances of their environment. Flavonoids, glucosinolates, and alkaloids are examples of specialized metabolites that are produced as part of defense-related metabolites. This is one of the processes that contribute to defense-related activities. Signaling pathways, protein kinases, lipids, phytohormones, reactive oxygen species, and redox homeostasis are all altered as a result of plant responses (Viant and Sommer, 2013). Researchers can detect specific metabolic alterations in response to environmental stressors such as salt, drought, or pathogen invasion by analyzing a single cell. Certain cells may show distinct defense-related metabolites, such as those in roots or shoots. Plant immunity may be better understood by identifying these cellular metabolic defenses via single-cell analysis.

It is feasible to simultaneously see the relative position of hundreds or even thousands of chemical metabolites inside a single tissue segment with the use of MS imaging. Plants use their roots as belowground organs to react to environmental pressures, absorb water and nutrients, and attach themselves to the soil. Sarabia *et al.* (2018) observed and distinguished molecules in longitudinal root cell sections of *Hordeum vulgare* and compared ions, metabolites, and lipid profiles before and after treatment with salt stress. Salt stress altered phosphatidylcholine (PC) profiles, with PC 34:n decreasing in salt-treated roots of *H. vulgare*. MALDI-MS imaging showed root elemental content increases in Na^+ and decreases in K^+ (Sarabia *et al.*, 2018). In *Vicia faba* and *Arabidopsis thaliana* guard cells, several metabolomic studies have shown that flavonoids, reactive oxygen species, abscisic acid, nitric oxide, and auxin are major components of the signaling cascade regulating stomatal fluctuations in exposure to osmotic stresses and pathogenic organisms such as *Pseudomonas syringae* (Jin *et al.*, 2013; Ou *et al.*, 2014). Single-cell metabolomics may reveal unique metabolic responses to environmental stress. This information may reveal the defensive mechanisms and adaptation techniques of distinct plant cell types, which vary by location, function, and stress proximity. Understanding the metabolic fingerprints of stress-tolerant and -sensitive cell types might help improve plant stress resistance.

Studying hormonal signaling pathways

Plant hormones, also known as phytohormones, are tiny organic compounds essential for controlling how plants grow, develop, and react to their surroundings. They typically have extremely precise effects on individual cells and work at very low doses. Although conventional analytical methods often evaluate bulk tissue samples and average the responses of different cell populations, they have nevertheless yielded important insights into hormone signaling networks. By examining metabolite profiles at the level of individual cells, single-cell metabolomics overcomes this constraint and offers a more detailed knowledge of hormone signaling networks and their temporal and spatial dynamics within the plant (de Souza *et al.*, 2020). The spatial distribution patterns of many hormones, including auxins, cytokinins, gibberellins, abscisic acid, and ethylene, are each unique. Single-cell metabolomics has the potential to shed light on the manner in which these hormones accumulate and work in certain cell types, such as root hairs, guard cells, and vascular tissues (McReynolds, 2022). Individual cell analysis lets researchers see how hormone concentrations fluctuate inside tissue cells. This may illuminate local hormone biosynthesis, transport, and breakdown. For example, the glucosinolate, phenylpropanoid, and fatty acid metabolism in the *Arabidopsis* plant root metabolome was altered as a result of increases in both ethylene and auxin. These alterations had a negative correlation with the data extracted from the matching transcriptome (Hildreth *et al.*, 2020). Metabolite levels alter rapidly during hormonal signaling. These alterations may be resolved temporally using single-cell metabolomics to examine hormone signaling processes dynamically. As an example, the response of guard cells to abscisic acid in the presence of drought stress includes rapid changes in metabolite levels, which ultimately result in the closing of stomatal pores (Sirichandra *et al.*, 2009). The temporal metabolite profiles of these cells may reveal the timing and sequence of metabolic processes in response to hormone cues. Root development is strongly controlled by auxins and cytokinins. In root tip cells, single-cell metabolomics may show how hormone gradients affect cell proliferation, elongation, and differentiation (Skalický *et al.*, 2018). Researchers may find root hormonal regulatory nodes and downstream effectors by comparing the metabolomes of individual cells in wild-type and mutant plants with altered hormone signaling pathways.

Investigating symbiotic and pathogenic interactions

Plants interact with a variety of microbes, both helpful (symbionts) and harmful (pathogens). Metabolomics at the single-cell level with focused analysis of interaction sites, such as root nodules or infection sites, assists in elucidating metabolic exchanges between plant cells and microbes, as well as tracking metabolites produced or consumed during symbiosis or pathogenic attack, providing insights into the molecular dialog between plants and microorganisms (Khan *et al.*, 2023). Symbiotic relationships, such as those between plants and nitrogen-fixing bacteria or mycorrhizal fungi, are essential for nutrient acquisition, stress tolerance, and general plant health (Soumare *et al.*, 2015). Single-cell metabolomics provides a unique chance to investigate the metabolic exchanges and signaling pathways that underpin these mutually beneficial relationships. Lambert *et al.* (2020) utilized GC-MS to measure systemic nitrogen signaling-related soluble metabolites in control, nitrogen-limited, and nitrogen-satiated nodules of the model symbiotic association *Medicago truncatula*/*Sinorhizobium* after 1–4 days and identified 107 different chemicals in the nodules. The majority comprised soluble carbohydrates, proteins, and organic acids. In response to systemically nitrogen signaling, the plant nitrogen deficit increased global soluble metabolite content, whereas plant nitrogen satiety lowered it. In response to *Bradyrhizobium japonicum* inoculation of soybean, 166 amino acids, flavonoids, carboxylic acids, fatty acids, and trehalose were found to be strongly regulated. The crucial involvement of *B. japonicum* in nodulation and soybean symbiotic bacterial growth has been thoroughly characterized (Streeter and Gomez, 2006). Plant cells fight pathogens by producing antimicrobial chemicals, activating signaling cascades, and inducing programmed cell death. These defensive responses have cell-specific metabolic markers that single-cell metabolomics can identify. Pathogens use the host's metabolic mechanism to develop and multiply. Single-cell metabolomics may reveal pathogen-induced cell-specific metabolic alterations, revealing pathogenic processes.

7.5 Conclusion and Future Directions

Combining single-cell technology with plant metabolomics is a significant advancement in the field of plant sciences. While classical metabolomics provides comprehensive chemical profiles of bulk tissues, it frequently masks the variability observed within individual cells. This restriction is overcome by single-cell metabolomics, which makes it possible to conduct an in-depth cellular investigation and reveal the complexity, variety, and dynamism of plant metabolism. Single-cell isolation, high-throughput analytics, and bioinformatics developments have recently made it possible to investigate a single plant cell's metabolome. This discovery makes it easier to comprehend how cells respond to external cues, change throughout development, and interact with other living things. Research at this level of detail may clarify metabolic pathways unique to certain cell types, providing information on functions such as photosynthesis, the creation of secondary metabolites, and stress reactions. Therefore, by offering previously unheard-of clarity and accuracy, single-cell metabolomics is set to transform industries, including breeding, biotechnology, and ecological studies.

Future studies will need to concentrate on combining single-cell metabolomics with other omics technologies, such as transcriptomics, proteomics, and genomics. Plant biology will become more holistic due to this multi-omics approach, which will provide a more thorough knowledge of cellular connections and processes. Developing single-cell metabolomics methods that are more sensitive and economical and have high throughput is imperative. Improvements in data collection, processing, and sample preparation techniques will improve the precision and resolution of metabolic profiling.

7.6 References

Amantonico, A., Urban, P.L., Fagerer, S.R., Balabin, R.M., and Zenobi, R. (2010) Single-cell MALDI-MS as an analytical tool for studying intrapopulation metabolic heterogeneity of unicellular organisms. *Analytical Chemistry* 82, 7394–7400.

Amirav, A., Fialkov, A., Margolin-Eren, K., Neumark, B., Elkabets, O., *et al.* (2020) Gas chromatography–mass spectrometry (GC–MS) with cold electron ionization (EI): bridging the gap between GC–MS and LC–MS. *Spectroscopy Supplements* 18, 5–15.

Armbrecht, L. and Dittrich, P.S. (2017) Recent advances in the analysis of single cells. *Analytical Chemistry* 89, 2–21.

Bertea, C.M., Freije, J.R., van der Woude, H., Verstappen, F.W.A., Perk, L., *et al.* (2005) Identification of intermediates and enzymes involved in the early steps of artemisinin biosynthesis in *Artemisia annua*. *Planta Medica* 71, 40–47.

Bianchi, F. and Careri, M. (2021) Mass spectrometry in food authenticity and traceability. In: Galanakis, C.M. (ed.) *Food Authentication and Traceability*. Academic Press, Cambridge, Massachusetts, pp. 101–130.

Brunetti, C., George, R.M., Tattini, M., Field, K., and Davey, M.P. (2013) Metabolomics in plant environmental physiology. *Journal of Experimental Botany* 64, 4011–4020.

Calumpang, C.L.F., Saigo, T., Watanabe, M., and Tohge, T. (2020) Cross-species comparison of fruit-metabolomics to elucidate metabolic regulation of fruit polyphenolics among solanaceous crops. *Metabolites* 10: 209.

Carrington, Y., Guo, J., Le, C.H., Fillo, A., Kwon, J., *et al.* (2018) Evolution of a secondary metabolic pathway from primary metabolism: shikimate and quinate biosynthesis in plants. *Plant Journal* 95, 823–833.

Castro-Moretti, F.R., Gentzel, I.N., Mackey, D., and Alonso, A.P. (2020) Metabolomics as an emerging tool for the study of plant–pathogen interactions. *Metabolites* 10: 52.

Chang, W.C., Huang, L.C.L., Wang, Y.S., Peng, W.P., Chang, H.C., *et al.* (2007) Matrix-assisted laser desorption/ionization (MALDI) mechanism revisited. *Analytica Chimica Acta* 582, 1–9.

Comi, T.J., Do, T.D., Rubakhin, S.S., and Sweedler, J.V. (2017) Categorizing cells on the basis of their chemical profiles: progress in single-cell mass spectrometry. *Journal of the American Chemical Society* 139, 3920–3929.

Cubero Herrera, L. (2007) Studies on atmospheric pressure chemical ionization and tandem mass spectrometry of sulfur-containing compounds from petroleum. PhD thesis, Dalhousie University, Canada.

d'Auria, J.C. and Gershenzon, J. (2005) The secondary metabolism of *Arabidopsis thaliana*: growing like a weed. *Current Opinion in Plant Biology* 8, 308–316.

de Souza, L.P., Borghi, M., and Fernie, A. (2020) Plant single-cell metabolomics—challenges and perspectives. *International Journal of Molecular Sciences* 21: 8987.

DeLaney, K., Sauer, C.S., Vu, N.Q., and Li, L. (2018) Recent advances and new perspectives in capillary electrophoresis-mass spectrometry for single cell "omics". *Molecules* 24: 42.

Diederen, T., Delabrière, A., Othman, A., Reid, M.E., and Zamboni, N. (2021) Metabolomics. In: Nielsen, J., Stephanopoulos, G., and Lee, S.Y. (eds) *Metabolic Engineering: Concepts and Applications*. Wiley, Weinheim, Germany, pp. 259–299.

Dusenge, M.E., Duarte, A.G., and Way, D.A. (2019) Plant carbon metabolism and climate change: elevated CO_2 and temperature impacts on photosynthesis, photorespiration and respiration. *New Phytologist* 221, 32–49.

Earnshaw, C.J., Atkinson, S.J., Burrell, M., and Clench, M.R. (2007) Matrix assisted laser desorption ionisation mass spectrometric imaging—principles and applications. In: Griffiths, W.J. (ed.) *Metabolomics, Metabonomics and Metabolite Profiling*. Royal Society of Chemistry, London, pp. 234–253.

Edison, A.S., Colonna, M., Gouveia, G.J., Holderman, N.R., Judge, M.T., *et al.* (2020) NMR: unique strengths that enhance modern metabolomics research. *Analytical Chemistry* 93, 478–499.

Eirew, P., Steif, A., Khattra, J., Ha, G., Yap, D., *et al.* (2015) Dynamics of genomic clones in breast cancer patient xenografts at single-cell resolution. *Nature* 518, 422–426.

Emwas, A.H., Szczepski, K., Poulson, B.G., McKay, R., Tenori, L., *et al.* (2022) Nuclear magnetic resonance in metabolomics. In: Troisi, J. (ed.) *Metabolomics Perspectives*. Academic Press, Cambridge, Massachusetts, pp. 149–218.

Fernie, A.R. (2007) The future of metabolic phytochemistry: larger numbers of metabolites, higher resolution, greater understanding. *Phytochemistry* 68, 2861–2880.

Fiehn, O. (2016) Metabolomics by gas chromatography–mass spectrometry: combined targeted and untargeted profiling. *Current Protocols in Molecular Biology* 114, 30.4.1–30.4.32.

Flanigan, P. and Levis, R. (2014) Ambient femtosecond laser vaporization and nanosecond laser desorption electrospray ionization mass spectrometry. *Annual Review of Analytical Chemistry* 7, 229–256.

Gautam, D.K., Chimata, A.V., Gutti, R.K., and Paddibhatla, I. (2024) Investigative tools to study blood cells: a focus on single cell isolation and analysis. In: Gutti, R.K. (ed.) *Hematopoiesis*. Apple Academic Press, New York, pp. 19–49.

Georgakoudi, I., Rice, W.L., Hronik-Tupaj, M., and Kaplan, D.L. (2008) Optical spectroscopy and imaging for the noninvasive evaluation of engineered tissues. *Tissue Engineering Part B: Reviews* 14, 321–340.

Gika, H.G., Theodoridis, G.A., Plumb, R.S., and Wilson, I.D. (2014) Current practice of liquid chromatography–mass spectrometry in metabolomics and metabonomics. *Journal of Pharmaceutical and Biomedical Analysis* 87, 12–25.

Gilmore, I.S., Heiles, S., and Pieterse, C.L. (2019) Metabolic imaging at the single-cell scale: recent advances in mass spectrometry imaging. *Annual Review of Analytical Chemistry* 12, 201–224.

Golcuk, K., Mandair, G.S., Callender, A.F., Finney, W.F., Sahar, N., *et al.* (2006) Rapid Raman spectroscopy of musculoskeletal tissue using a visible laser and an electron-multiplying CCD (EMCCD) detector. In: Mahadevan-Jansen, A. and Wolfgang H. Petrich, W.H. (eds) *Biomedical Vibrational Spectroscopy III: Advances in Research and Industry, Vol.* 6093. Society of Photo Optical, Washington, DC, pp. 269–275.

Gradov, O.V. and Gradova, M.A. (2015) "MS-patch-clamp" or the possibility of mass spectrometry hybridization with patch-clamp setups for single cell metabolomics and channelomics. *Advances in Biochemistry* 3, 66–71.

Griffiths, W.J. and Wang, Y. (2009) Mass spectrometry: from proteomics to metabolomics and lipidomics. *Chemical Society Reviews* 38, 1882–1896.

Hansen, R.L. and Lee, Y.J. (2018) High-spatial resolution mass spectrometry imaging: toward single cell metabolomics in plant tissues. *Chemical Record* 18, 65–77.

Heavisides, E., Rouger, C., Reichel, A.F., Ulrich, C., Wenzel-Storjohann, A., *et al.* (2018) Seasonal variations in the metabolome and bioactivity profile of *Fucus vesiculosus* extracted by an optimised, pressurised liquid extraction protocol. *Marine Drugs* 16: 503.

Hieta, J.P., Kopra, J., Räikkönen, H., Kauppila, T.J., and Kostiainen, R. (2020) Sub-100 μm spatial resolution ambient mass spectrometry imaging of rodent brain with laser ablation atmospheric pressure photoionization (LAAPPI) and laser ablation electrospray ionization (LAESI). *Analytical Chemistry* 92, 13734–13741.

Hildreth, S.B., Foley, E.E., Muday, G.K., Helm, R.F., and Winkel, B.S. (2020) The dynamic response of the *Arabidopsis* root metabolome to auxin and ethylene is not predicted by changes in the transcriptome. *Scientific Reports* 10: 679.

Hiraga, Y., Shimada, N., Nagashima, Y., Suda, K., Kanamori, T., *et al.* (2021) Identification of a flavin monooxygenase-like flavonoid 8-hydroxylase with gossypetin synthase activity from *Lotus japonicus*. *Plant and Cell Physiology* 62, 411–423.

Hollywood, K., Brison, D.R., and Goodacre, R. (2006) Metabolomics: current technologies and future trends. *Proteomics* 6, 4716–4723.

Ishikawa, T., Takahara, K., Hirabayashi, T., Matsumura, H., Fujisawa, S., *et al.* (2010) Metabolome analysis of response to oxidative stress in rice suspension cells overexpressing cell death suppressor Bax inhibitor-1. *Plant and Cell Physiology* 51, 9–20.

Jiao, Y., Lori Tausta, S., Gandotra, N., Sun, N., Liu, T., *et al.* (2009) A transcriptome atlas of rice cell types uncovers cellular, functional and developmental hierarchies. *Nature Genetics* 41, 258–263.

Jin, X., Wang, R.S., Zhu, M., Jeon, B.W., Albert, R., *et al.* (2013) Abscisic acid–responsive guard cell metabolomes of *Arabidopsis* wild-type and gpa1 G-protein mutants. *Plant Cell* 25, 4789–4811.

Kajiyama, S.I., Harada, K., Fukusaki, E., and Kobayashi, A. (2006) Single cell-based analysis of *Torenia* petal pigments by a combination of ArF excimer laser micro sampling and nano-high performance liquid chromatography (HPLC)–mass spectrometry. *Journal of Bioscience and Bioengineering* 102, 575–578.

Kang, N., Lee, J.M., Jeon, A., Oh, H.B., and Moon, B. (2016) Design and synthesis of new mass tags for matrix-free laser desorption ionization mass spectrometry (LDI-MS) based on 6, 11-dihydrothiochromeno [4, 3-b] indole. *Tetrahedron* 72, 5612–5619.

Katam, R., Lin, C., Grant, K., Katam, C.S., and Chen, S. (2022) Advances in plant metabolomics and its applications in stress and single-cell biology. *International Journal of Molecular Sciences* 23: 6985.

Khan, F., Pandey, E., Fatima, S., Khan, A., Zeb, S.Z., and Ahmad, F. (2023) Prospects for the use of metabolomics engineering in exploring and harnessing chemical signaling in root galls. In: Ahmad, F. and Blázquez, G.N. (eds) *Root-Galling Disease of Vegetable Plants*. Springer, Singapore, pp. 309–338.

Killiny, N., Valim, M.F., Jones, S.E., and Hijaz, F. (2018) Effect of different rootstocks on the leaf metabolite profile of 'Sugar Belle' mandarin hybrid. *Plant Signaling and Behavior* 13: e1445934.

Komatsu, S., Yamamoto, A., Nakamura, T., Nouri, M.Z., Nanjo, Y., *et al.* (2011) Comprehensive analysis of mitochondria in roots and hypocotyls of soybean under flooding stress using proteomics and metabolomics techniques. *Journal of Proteome Research* 10, 3993–4004.

Kueger, S., Steinhauser, D., Willmitzer, L., and Giavalisco, P. (2012) High-resolution plant metabolomics: from mass spectral features to metabolites and from whole-cell analysis to subcellular metabolite distributions. *Plant Journal* 70, 39–50.

Lambert, I., Pervent, M., Le Queré, A., Clément, G., Tauzin, M., *et al.* (2020) Responses of mature symbiotic nodules to the whole-plant systemic nitrogen signaling. *Journal of Experimental Botany* 71, 5039–5052.

Lindström, S. and Andersson-Svahn, H. (2010) Overview of single-cell analyses: microdevices and applications. *Lab on a Chip* 10, 3363–3372.

Liu, D., Ford, K.L., Roessner, U., Natera, S., Cassin, A.M., *et al.* (2013) Rice suspension cultured cells are evaluated as a model system to study salt responsive networks in plants using a combined proteomic and metabolomic profiling approach. *Proteomics* 13, 2046–2062.

Livingston, S.J., Quilichini, T.D., Booth, J.K., Wong, D.C., Rensing, K.H., *et al.* (2020) *Cannabis* glandular trichomes alter morphology and metabolite content during flower maturation. *Plant Journal* 101, 37–56.

Lockyer, N.P., Aoyagi, S., Fletcher, J.S., Gilmore, I.S., van der Heide, P.A., *et al.* (2024) Secondary ion mass spectrometry. *Nature Reviews Methods Primers* 4: 32.

Lorenzo Tejedor, M., Mizuno, H., Tsuyama, N., Harada, T., and Masujima, T. (2012) *In situ* molecular analysis of plant tissues by live single-cell mass spectrometry. *Analytical Chemistry* 84, 5221–5228.

Ma, S., Leng, Y., Li, X., Meng, Y., Yin, Z., *et al.* (2023) High spatial resolution mass spectrometry imaging for spatial metabolomics: advances, challenges, and future perspectives. *Trends in Analytical Chemistry* 159: 116902.

Maia, M., Figueiredo, A., Cordeiro, C., and Sousa Silva, M. (2023) FT-ICR-MS-based metabolomics: a deep dive into plant metabolism. *Mass Spectrometry Reviews* 42, 1535–1556.

Majchrzak, T., Wojnowski, W., Lubinska-Szczygeł, M., Różańska, A., Namieśnik, J., *et al.* (2018) PTR-MS and GC-MS as complementary techniques for analysis of volatiles: a tutorial review. *Analytica Chimica Acta* 1035, 1–13.

Maruyama, K., Urano, K., Yoshiwara, K., Morishita, Y., Sakurai, N., *et al.* (2014) Integrated analysis of the effects of cold and dehydration on rice metabolites, phytohormones, and gene transcripts. *Plant Physiology* 164, 1759–1771.

Masuda, K., Abouleila, Y., Ali, A., Yanagida, T., and Masujima, T. (2018) Live single-cell mass spectrometry (LSC-MS) for plant metabolomics. *Methods in Molecular Biology* 1778, 269–282.

McDowell, E.T., Kapteyn, J., Schmidt, A., Li, C., Kang, J.H., *et al.* (2011) Comparative functional genomic analysis of *Solanum glandular* trichome types. *Plant Physiology* 155, 524–539.

McReynolds, M.R. (2022) Exploring the functions of hormone signaling using omics profiling. PhD thesis, Iowa State University, Ames, Iowa.

Minakshi, P., Ghosh, M., Kumar, R., Patki, H.S., Saini, H.M., *et al.* (2019) Single-cell metabolomics: technology and applications. In: Barh, D. and Azevedo, V. (eds) *Single-Cell Omics*. Academic Press, Cambridge, Massachusetts, pp. 319–353.

Misra, B.B. (2020) Open-source software tools, databases, and resources for single-cell and single-cell-type metabolomics. *Methods in Molecular Biology* 2064, 191–217.

Misra, B.B., Assmann, S.M., and Chen, S. (2014) Plant single-cell and single-cell-type metabolomics. *Trends in Plant Science* 19, 637–646.

Navarro-Reig, M., Ortiz-Villanueva, E., Tauler, R., and Jaumot, J. (2017) Modelling of hydrophilic interaction liquid chromatography stationary phases using chemometric approaches. *Metabolites* 7: 54.

Oikawa, A., Matsuda, F., Kikuyama, M., Mimura, T., and Saito, K. (2011) Metabolomics of a single vacuole reveals metabolic dynamism in an alga *Chara australis*. *Plant Physiology* 157, 544–551.

Ou, X., Gan, Y., Chen, P., Qiu, M., Jiang, K., *et al.* (2014) Stomata prioritize their responses to multiple biotic and abiotic signal inputs. *PLOS ONE* 9: e101587.

Prasad, M., Ghosh, M., and Kumar, R. (2022) Single cell metabolomics. In: Troisi, J. (ed.) *Metabolomics Perspectives*. Academic Press, Cambridge, Massachusetts, pp. 457–513.

Ramautar, R., Somsen, G.W., and de Jong, G.J. (2019) CE-MS for metabolomics: developments and applications in the period 2016–2018. *Electrophoresis* 40, 165–179.

Ren, J.L., Zhang, A.H., Kong, L., and Wang, X.J. (2018) Advances in mass spectrometry-based metabolomics for investigation of metabolites. *RSC Advances* 8, 22335–22350.

Robinson, A.R., Dauwe, R., and Mansfield, S.D., (2018) Assessing the between-background stability of metabolic effects arising from lignin-related transgenic modifications, in two *Populus* hybrids using non-targeted metabolomics. *Tree Physiology* 38, 378–396.

Sarabia, L.D., Boughton, B.A., Rupasinghe, T., van de Meene, A.M., Callahan, D.L., *et al.* (2018) High-mass-resolution MALDI mass spectrometry imaging reveals detailed spatial distribution of metabolites and lipids in roots of barley seedlings in response to salinity stress. *Metabolomics* 14: 63.

Schauer, N. and Fernie, A.R. (2006) Plant metabolomics: towards biological function and mechanism. *Trends in Plant Science* 11, 508–516.

Schie, I.W. and Huser, T. (2013) Methods and applications of Raman microspectroscopy to single-cell analysis. *Applied Spectroscopy* 67, 813–828.

Shrestha, B., Patt, J.M., and Vertes, A. (2011) *In situ* cell-by-cell imaging and analysis of small cell populations by mass spectrometry. *Analytical Chemistry* 83, 2947–2955.

Sirichandra, C., Wasilewska, A., Vlad, F., Valon, C., and Leung, J. (2009) The guard cell as a single-cell model towards understanding drought tolerance and abscisic acid action. *Journal of Experimental Botany* 60, 1439–1463.

Skalický, V., Kubeš, M., Napier, R., and Novák, O. (2018) Auxins and cytokinins-the role of subcellular organization on homeostasis. *International Journal of Molecular Sciences* 19: 3115.

Soumare, A., Diop, T., Manga, A., and Ndoye, I. (2015) Role of arbuscular mycorrhizal fungi and nitrogen fixing bacteria on legume growth under various environmental stresses. *International Journal of Biosciences* 7, 31–46.

Steinbrenner, A.D., Gómez, S., Osorio, S., Fernie, A.R., and Orians, C.M. (2011) Herbivore-induced changes in tomato (*Solanum lycopersicum*) primary metabolism: a whole plant perspective. *Journal of Chemical Ecology* 37, 1294–1303.

Stopka, S.A., Khattar, R., Agtuca, B.J., Anderton, C.R., Paša-Tolić, L., *et al.* (2018) Metabolic noise and distinct subpopulations observed by single cell LAESI mass spectrometry of plant cells *in situ*. *Frontiers in Plant Science* 9: 1646.

Streeter, J.G. and Gomez, M.L. (2006) Three enzymes for trehalose synthesis in *Bradyrhizobium* cultured bacteria and in bacteroids from soybean nodules. *Applied and Environmental Microbiology* 72, 4250–4255.

Sugahara, K., Kitao, K., Watanabe, T., and Yamagaki, T. (2018) Imaging mass spectrometry analysis of flavonoids in blue viola petals and their enclosure effects on violanin during color expression. *Analytical Chemistry* 91, 896–902.

Sukovich, D.J., Kim, S.C., Ahmed, N., and Abate, A.R. (2017) Bulk double emulsification for flow cytometric analysis of microfluidic droplets. *Analyst* 142, 4618–4622.

Sumner, L.W., Yang, D.S., Bench, B.J., Watson, B.S., Li, C., *et al.* (2011) Spatially resolved plant metabolomics. In: Hall, R.D. (ed.) *Annual Plant Reviews, Vol. 43: Biology of Plant Metabolomics*. Wiley, Ames, Iowa, pp. 343–366.

Sun, M., Yang, Z., and Wawrik, B. (2018) Metabolomic fingerprints of individual algal cells using the single-probe mass spectrometry technique. *Frontiers in Plant Science* 9: 571.

Takats, Z. (2012) Desorption electrospray ionization mass spectrometry. In: Lebedev, A.T. (ed.) *Comprehensive Environmental Mass Spectrometry*. ILM Publications, Hertforshire, UK, pp. 167–183.

Takats, Z., Wiseman, J.M., Gologan, B., and Cooks, R.G. (2004) Mass spectrometry sampling under ambient conditions with desorption electrospray ionization. *Science* 306, 471–473.

Taylor, M.J., Lukowski, J.K., and Anderton, C.R. (2021) Spatially resolved mass spectrometry at the single cell: recent innovations in proteomics and metabolomics. *Journal of the American Society for Mass Spectrometry* 32, 872–894.

Theodoridis, G.A., Gika, H.G., Want, E.J., and Wilson, I.D. (2012) Liquid chromatography–mass spectrometry based global metabolite profiling: a review. *Analytica Chimica Acta* 711, 7–16.

Ungai-Salánki, R., Gerecsei, T., Fürjes, P., Orgovan, N., Sándor, N., *et al.* (2016) Automated single cell isolation from suspension with computer vision. *Scientific Reports* 6: 20375.

van Dam, N.M. and Bouwmeester, H.J. (2016) Metabolomics in the rhizosphere: tapping into belowground chemical communication. *Trends in Plant Science* 21, 256–265.

Viant, M.R. and Sommer, U. (2013) Mass spectrometry based environmental metabolomics: a primer and review. *Metabolomics* 9 (Suppl. 1), 144–158.

Villani, A.C., Satija, R., Reynolds, G., Sarkizova, S., Shekhar, K., *et al.* (2017) Single-cell RNA-seq reveals new types of human blood dendritic cells, monocytes, and progenitors. *Science* 356: eaah4573.

Warren, C.R. (2013) Use of chemical ionization for GC–MS metabolite profiling. *Metabolomics* 9, 110–120.

Watanabe, C.K., Sato, S., Yanagisawa, S., Uesono, Y., Terashima, I., *et al.* (2014) Effects of elevated CO_2 on levels of primary metabolites and transcripts of genes encoding respiratory enzymes and their diurnal patterns in *Arabidopsis thaliana*: possible relationships with respiratory rates. *Plant and Cell Physiology* 55, 341–357.

Weidner, S.M. (2012) Mass spectrometry: MALDI (matrix-assisted laser desorption/ ionization) and ESI (electrospray ionization). In: Matyjaszewski, K. and Möller, M. (eds) *Polymer Science: A Comprehensive Reference*. Elsevier, Amsterdam, the Netherlands, pp. 93–109.

Weng, J.K., Lynch, J.H., Matos, J.O., and Dudareva, N. (2021) Adaptive mechanisms of plant specialized metabolism connecting chemistry to function. *Nature Chemical Biology* 17, 1037–1045.

Wu, K., Jia, F., Zheng, W., Luo, Q., Zhao, Y., *et al.* (2017) Visualization of metallodrugs in single cells by secondary ion mass spectrometry imaging. *Journal of Biological Inorganic Chemistry* 22, 653–661.

Yamamoto, K., Takahashi, K., Mizuno, H., Anegawa, A., Ishizaki, K., *et al.* (2016) Cell-specific localization of alkaloids in *Catharanthus roseus* stem tissue measured with imaging MS and single-cell MS. *Proceedings of the National Academy of Sciences USA* 113, 3891–3896.

Yamamoto, K., Takahashi, K., Caputi, L., Mizuno, H., Rodriguez-Lopez, C.E., *et al.* (2019) The complexity of intercellular localisation of alkaloids revealed by single-cell metabolomics. *New Phytologist* 224, 848–859.

Yin, L., Zhang, Z., Liu, Y., Gao, Y., and Gu, J. (2019) Recent advances in single-cell analysis by mass spectrometry. *Analyst* 144, 824–845.

Zeb, Q., Wang, C., Shafiq, S., and Liu, L. (2019) An overview of single-cell isolation techniques. In: Barh, D. and Azevedo, V. (eds) *Single-Cell Omics, Technological Advances and Applications, Vol.* 1. Academic Press, Cambridge, Massachusetts, pp. 101–135.

Zenobi, R. (2013) Single-cell metabolomics: analytical and biological perspectives. *Science* 342: 1243259.

Zhang, C., Le Dévédec, S. E., Ali, A., and Hankemeier, T. (2023) Single-cell metabolomics by mass spectrometry: READY for primetime? *Current Opinion in Biotechnology* 82: 102963.

Zhang, L. and Vertes, A. (2018) Single-cell mass spectrometry approaches to explore cellular heterogeneity. *Angewandte Chemie International Edition* 57, 4466–4477.

Zhu, G., Shao, Y., Liu, Y., Pei, T., Li, L., *et al.* (2021) Single-cell metabolite analysis by electrospray ionization mass spectrometry. *Trends in Analytical Chemistry* 143: 116351.

Zhu, Y., Chidekel, A., and Shaffer, T.H. (2010) Cultured human airway epithelial cells (Calu-3): a model of human respiratory function, structure, and inflammatory responses. *Critical Care Research and Practice* 2010: 394578.

8 Constructing Metabolomics Using Integrated Bioinformatics Tools for Advancing Plant Science

Parkirti, Sandeep Kour, Mohd Ali, Deepak Kumar, Roohi Sharma, Vikram, Alisha Sharma, Anchita Singh, and Puja Ohri*

Department of Zoology, Guru Nanak Dev University, Amritsar, Punjab, India

Abstract

Metabolomics is the study of profiles of cell metabolites and its understanding is essential to comprehend most cellular and molecular systems underlying various plant functions, omics, and bioinformatics. In order to achieve much deeper insights, integrated approaches that combine more advanced bioinformatic tools with transcriptomics, proteomics, metabolomics, and other data handling, processing, analysis, and integration techniques along with strong computational tools are indispensable. This approach is essential for conducting mechanistic research on plant metabolites. Bioinformatics databases and software applications needed for plant research are available for this. Plant science can be improved by combining powerful bioinformatics tools with omics techniques. An integrated approach typically completes a better research task, such as mechanistic studies on plant metabolism and plant authentication. This chapter will provide an objective description of some integrated metabolomic data analysis and bioinformatic tools for constructive metabolomics.

Keywords: metabolomics, omics, plants, bioinformatics, computational tools

8.1. Introduction

There are a number of recently developed resources that belong to a systems approach for plant science omics and bioinformatics analyses. To comprehend plant functions, systems analyses based on omics have been employed where many findings on bioinformatics and plant omics are available (Mochida and Shinozaki, 2011). In recent years, metabolomics data sets have expanded and become more complex. Thus, it is now more challenging to determine the clinical significance, biological significance, and interpretation of changes more efficiently at metabolite levels. Therefore, a combined performance of statistics, enrichment, visualization, and contextualization independent of the computational approach, along with integrated workflow for particular study designs of the metabolomic data set has been developed to obtain meaningful comprehensive information (Barupal *et al.*, 2018). For example, a number of Rp1-D-mediated defense response metabolites, such as chlorogenic acid, caffeic acid, ferulic acid, flavonoids, terpenoids, kauralexins, and zealexins, as well as genes involved in salicylic acid biosynthesis, including calmodulin-binding protein 60G and systemic acquired resistance deficient 1

* E-mail: ohri_puja_11@rediffmail.com

© CAB International 2025. *High-Throughput Plant Metabolomics* (Jen-Tsung Chen ed.)
DOI: 10.1079/9781800626911.0008

(SARD 1), and several transcription factors, including WRKY53, BZIP84, NKD1, BHLH124, and MYB100, were identified by integrated metabolomic, transcriptomic, and gene regulatory network analyses of common rust (*Puccinia sorghi*) resistance in maize (Kim *et al.*, 2021). Among omics approaches, metabolomics is one of the fields that is growing fast and has the most promising potential for crop improvement initiatives (Mochida and Shinozaki, 2011). Utilizing cutting-edge metabolomics platforms, researchers are able to decipher intricate biological processes and uncover latent regulatory networks governing crop health and growth. Numerous integrated technologies and approaches, such as those based on mass spectrometry (MS), are employed to analyze complex mixtures of plant extracts on a large scale (Patel *et al.*, 2021). Processing tools such as Met-Align, MET-XAlign, XCMS, PlantMAT, MET-COFEA, and ChromaTOF handle the data generated by these techniques. Following the identification of metabolites through statistical analysis, a combination of a variety of online resources and software options are available, such as principal component analysis, orthogonal partial least squares, partial least squares discriminant analysis, high-content screening, heatmaps, and correlation analysis as examples of multivariate analysis, while analysis of variance, the Mann–Whitney *U* test, Student's *t*-test, and others are used for univariate analyses (Raza, 2022). Due to recent technological advancements and omics-based approaches, the complex global biological systems that underpin different plant functions can now be addressed. By integrating knowledge across plant species, these technological advancements have also facilitated translational research and accelerated the development of genome-scale resources in emerging and applied model plant species (Mochida and Shinozaki, 2011).

8.2 Plant Metabolomics: An Overview

The final recipient of biological information needed to produce the intended phenotype is known as the metabolome. Generally speaking, a molecule smaller than 1.5 kDa is referred to as a metabolite (Wakayama *et al.*, 2015). Over the past 10 years, plant metabolomics has progressed from a completely theoretical concept to a highly recommended and commonly applied methodology. Studies on plant metabolomics have often been used to examine the metabolites of various crop plants. Owing to the fact that many metabolic approaches lack specificity, further research in this field can provide a comprehensive representation of both primary and specialized metabolites (van Dam and van der Meijden, 2011). In general, around 200,000 metabolites are found in plants; of these, 7000–15,000 are found in specific species (Fernie *et al.*, 2004) and 3000–5000 are found in all plant leaves (Kim *et al.*, 2010), and these numbers are rising with improved analytical methods (Last *et al.*, 2007). Selective metabolite profiling, aside from basic identification, is used to determine the biological properties of plants. These include: (i) ecotypes for taxonomic or biochemical information and species fingerprinting; (ii) examining the reaction of metabolites to physical stimuli and exogenous substances; and (iii) studies of developmental process, symbiotic relationship, and metabolite content comparison between transgenic and wild-type plants (Templer *et al.*, 2017; Alseekh and Fernie, 2018). In these studies, metabolome profiles are paired with additional omics tools, to examine intricate regulatory networks that are regulated during stress, changes in protein expression, global gene expression, and metabolite configurations (Weckwerth, 2011). Apart from individual cell metabolomes, there has been analysis of root hairs, pollen tissues, trichomes, and guard cells, with metabolomics thus playing a crucial role in plant research (Nägele *et al.*, 2017). These studies also indicate that the usefulness of metabolome analysis can differ among different cell types. A plant's phenotypic character is determined by the concentration and synthesis of metabolites in its various organs at different developmental stages; hence, the characteristics of the tissues and organs are related to their metabolite nature (Roldan *et al.*, 2014). Metabolite profiling is combined with markers to explore chemical differences among plant species (Schauer *et al.*, 2006). Because metabolic processes differ greatly at the cellular and subcellular levels in crops, the use of different metabolomics approaches with different protocols has increased significantly. In general, metabolomics can aid in the identification of new gene functions and provide clarification regarding the metabolisms that regulate a metabolome network.

8.3 Plant Metabolome Analysis Workflow

Metabolomics is a developing field within the realm of "omics" that focuses on identifying and quantifying metabolites and chemical traces of cellular regulatory processes across various biological species (Razzaq *et al.*, 2019). Effective experimental design along with proper sample preparation are important stages in metabolomics-related experiments, significantly impacting the biological understanding of outcomes interlinked to the biological system (de Souza *et al.*, 2024; Mardoc *et al.*, 2024). During a plant metabolomics experiment, the initial step involves crafting a well-defined hypothesis, i.e. a biological question (Fig. 8.1). Without a precise biological question, observed changes may be misconstrued, potentially obscuring important information regarding responses to environmental changes and determination of the active elements in different species with significance in pharmacology, agriculture, environment, or industry (Feng *et al.*, 2020). Hence, it is essential to comprehend the biological system under investigation, enabling the selection of suitable tissues for analysis and appropriate controls (Rodrigues *et al.*, 2019). Following conceptualization of the biological question, the initial critical measure in metabolomics workflow is planning the experimental design. This encompasses comprehensive planning, from the conditions of plant growth and treatment to sample preparation conditions, analytical platforms, and statistical treatments (Fig. 8.1) (Mead, 2017). The designs must clearly delineate the experimental design of the study, encompassing environmental conditions such as nurseries, growth chambers, greenhouses, or field conditions (Jacyna *et al.*, 2019). To address qualitative and quantitative variations in metabolomics analyses, biological replicates are indispensable for robust statistical analysis and dependable biological interpretation of the findings (Pei *et al.*, 2023). Randomization plays a critical role in minimizing biological variability and experimental error (Wang *et al.*, 2023). Another crucial aspect of plant metabolomics studies is the sample preparation workflow, which entails numerous steps preceding the extraction and analysis of metabolites. This is especially pertinent for trees, as their tissues pose challenges due to the presence of various interferents such as oleoresins and cellulose (Eshawu and Ghalsasi, 2024). To ensure a high-quality metabolomics analysis in plants, particular emphasis should be placed on the initial steps of the sample preparation workflow, which encompass experimental design, sampling, quenching, metabolite extraction, and conditions for harvesting tissues (Dominguez *et al.*, 2024). To enhance the statistical study significantly, determining replicate numbers required can be achieved through power analysis based on the analytical variance range within the populations (Shi *et al.*, 2024). The sample preparation protocol significantly influences the analytical data obtained. This workflow entails prompt quenching of metabolism and harvesting of biological material, followed by storage prior to sample homogenization and metabolite extraction. It needs to be meticulously planned to pinpoint potential sources of experimental variation and errors that could jeopardize data analysis, data reusability, or biological interpretation of the results (Jeppesen and Powers, 2023). In metabolomics, quenching is often achieved by sample homogenization of plants, typically utilizing a ball mill or pestle and mortar to break down plant cell walls, followed by sample weighing. These procedures are consistently carried out under liquid nitrogen to prevent the tissue thawing (Boness *et al.*, 2023). Metabolite extraction protocol selection is crucial in metabolomics studies as it can directly impact metabolite coverage and concentration (Salam *et al.*, 2023). Various factors should be taken into account, including the selection of a suitable solvent system, solvent solubility, solvent:sample ratio, extraction duration, and temperature (Duportet *et al.*, 2012). The choice of suitable solvent depends not only on the properties of the metabolites to be extracted but also on the specific requirements of the analytical platform to be used, such as gas chromatography–mass spectrometry (GC-MS) or liquid chromatography–mass spectrometry (LC-MS) (Oh *et al.*, 2023).

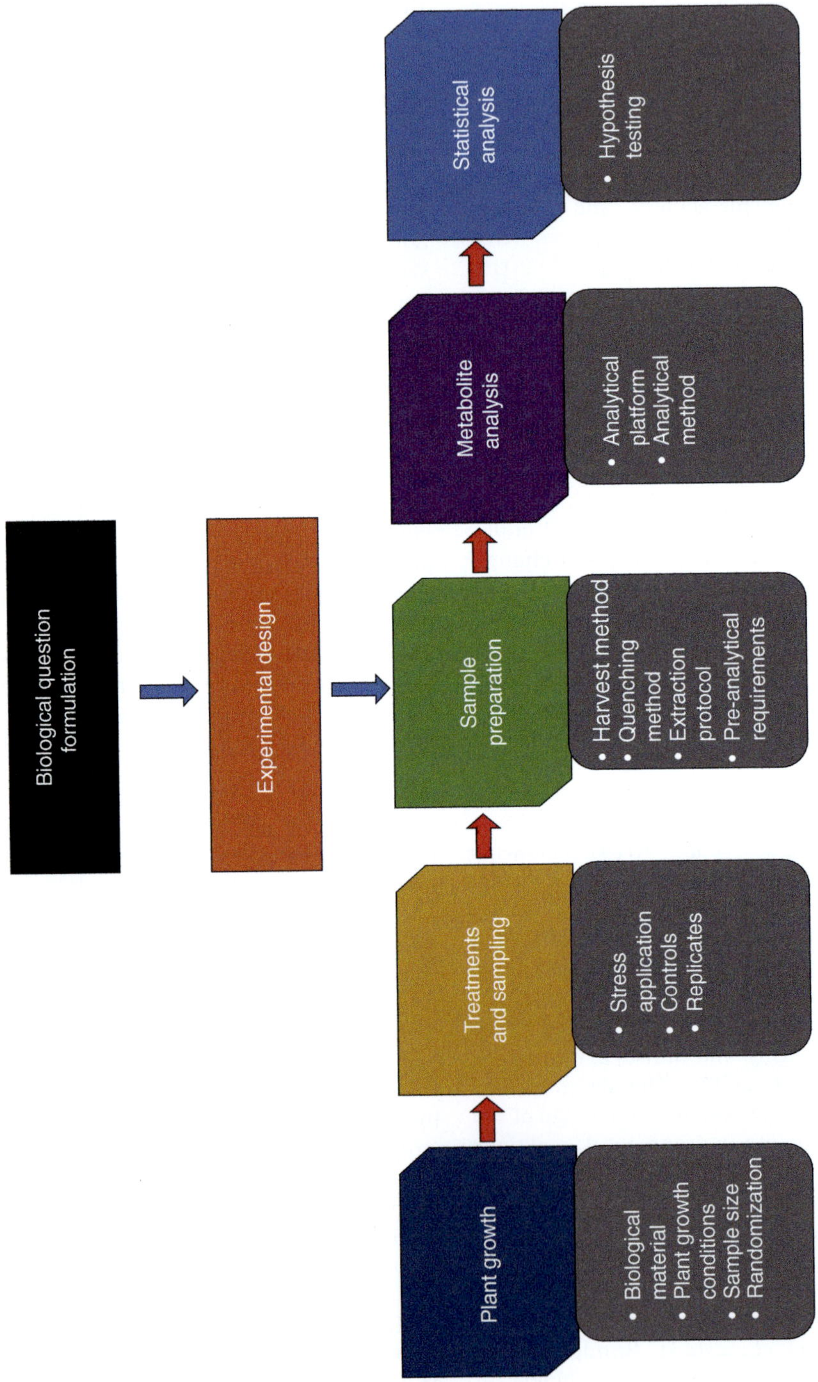

Fig. 8.1. General workflow for sample preparation for metabolomic analysis.

8.4 Metabolomics and Plant Development Under Stress Conditions

Plant metabolomics has emerged as a popular tool for studying plant stress (Piasecka *et al.*, 2019). Different types of abiotic and biotic stresses negatively affect the growth and development of plants and reduce crop yield. For sustainable productivity and successful management of these stresses, extensive research has been focused on comprehending the molecular basis of stress responses and the development of stress-tolerant plants (Ambrosino *et al.*, 2020). Plant metabolomics has played a huge role in this regard. Certain plant metabolites such as phytohormones, organic acids, amino acids, and sugars display stress-specific responses, while others have been observed to increase or decrease under multiple stress conditions. These findings imply that plants might have common strategies for coping with environmental stressors generally, as well as strategies for combating particular stresses. Usually, metabolomics comprises two methods for comprehensive metabolic profiling: targeted and untargeted approaches (Gorrochategui *et al.*, 2016; Ribbenstedt *et al.*, 2018). In the untargeted approach, processing of mass chromatograms from an experimental analysis is done using tools that allow comparison and retrieval of diversely amassed ions in the various samples being studied. In contrast, the targeted approach involves the utilization of public, customized, or specific metabolomics data sets that facilitate the recognition and assessment of known primary and secondary metabolites (Ambrosino *et al.*, 2020).

In order to withstand biotic and abiotic stressors and survive in harsh environmental conditions, plants have developed a number of adaptive mechanisms, including metabolic variations (Ashapkin *et al.*, 2020; Anzano *et al.*, 2022). Various plant metabolites act as osmolytes and osmoprotectants to ameliorate the severe impact of abiotic stresses such as salt, drought, and water stress (Teklić *et al.*, 2021). These plant metabolites include dimethylsulfoniopropionate and glycine betaine, amino acids such as ectoine and proline; sugar, such as sucrose, fructan and trehalose, and certain metabolites of mannitol, polyols and sorbitol (Dikilitas *et al.*, 2021; Kebert *et al.*, 2022). For example, a study conducted to determine the role of proline during salt stress in transgenic plants revealed that the overexpression of pyrroline-5-carboxylate synthetase, an enzyme involved in the production of proline, enhanced proline content and hence salt stress tolerance in transgenic plants (Meena *et al.*, 2019). One extensively researched molecule that is effective in maintaining the water equilibrium between the plant cell and its surroundings under water-deficit conditions is glycine betaine. Glycine betaine also helps in the stabilization of macromolecules, detoxification of reactive oxygen radicals, shielding of photosynthesis, and acts as an osmoprotectant (Salam *et al.*, 2023). As a source of energy and carbon in cells, soluble carbohydrates play a key role in plant metabolic processes (Salam *et al.*, 2023). As carbohydrate accumulation is linked with photosynthesis, various stress conditions might affect the levels of these soluble sugars. Rosa *et al.* (2009) revealed that some soluble sugars such as hexoses and sucrose downregulated the genes related to stress and upregulated the genes related to growth, thus increasing stress tolerance in plants. Plant metabolites are also key components in pathogen-infected plants. Treatment of *Solanum tuberosum* with potato virus Y resulted in elevated amino acid as well as organic acid levels, whereas the levels of sucrose were observed to decrease (Kogovsek *et al.*, 2016). However, amino acids such as arginine, alanine, tryptophan, serine, and glycine were reduced in soybean plants infected with soybean mosaic virus (Díaz-Cruz and Cassone, 2018).

Primary metabolites such as ascorbic acid, anthocyanins, glutamine, carotene and α-tocopherol protect plants from oxidative stress by scavenging oxygen free radicals (Hüdig *et al.*, 2018). Phytochemicals and phytohormones such as methyl jasmonate and jasmonic acid, methyl salicylate and salicylic acid, and several other molecules generated during stress conditions, aid in combating a number of environmental stresses (Hou *et al.*, 2016; Li *et al.*, 2019). Besides primary metabolites that have common physiological roles in plants, certain plant metabolites are produced under specific conditions for a specific purpose such as protection against pathogens, shielding from harmful ultraviolet (UV) radiations, and improved resistance to plant stressors. These metabolites include phenolic compounds such as flavonoids, isoflavonoids, lignin, anthocyanin, tannins and coumarins, terpenes, and sulfur- and nitrogen-containing compounds, and are known as secondary metabolites. The precursor of a number of phenolic compounds (i.e. phenylalanine) is involved in the resistance mechanism of plants in response to various biotic and abiotic environmental stresses. These phenolic compounds are produced by the phenylpropanoid pathway, which is regulated by abiotic as well as biotic stress factors such as salinity, drought stress, extreme temperatures, pathogen attack, wounding, treatment by herbicides, nutrient deficiencies, and UV radiation, resulting in the amassing of various phenolic compounds (Valifard *et al.*, 2015; Sharma *et al.*, 2019). For instance, flavonoids were observed to significantly improve resistance to water stress in *Arabidopsis thaliana* (Nakabayashi *et al.*, 2014). It has also been indicated that during drought stress, polyphenols were involved in seizing free radicals and maintaining osmotic potential in plants (Rodziewicz *et al.*, 2014).

The role of plant phenolics during biotic stress has also been studied through metabolomics analysis. Miotto-Vilanova *et al.* (2019) reported that flavonoids were accumulated in *Vitis vinifera* leaves treated with *Paraburkholderia phytofirmans* PsJN. Similar findings were documented by Gutiérrez-Albanchez *et al.* (2020) who tested the effect of *Bacillus amyloliquefaciens* QV15 on *Rubus* spp. Owing to their reactive oxygen species scavenging and antioxidant properties, anthocyanin content was found to be enhanced in plants during cold and drought stress, which provided protection to plants (Ancillotti *et al.*, 2015; Xiang *et al.*, 2021). Nitrogen-containing plant secondary metabolites such as alkaloids, glucosinolates, and cyanogenic glycosides, as well as non-protein amino acids, offer defense against invading pathogens, herbivores, and abiotic stressors. Alkaloids have been found to play significant roles against drought, salt stress, microbial pathogens, and insect pests (Salam *et al.*, 2023). Alkaloid content was found to be enhanced in *Lupinus termis* under activated yeast treatment and drought stress (Taha *et al.*, 2020).

8.5 Current Techniques for Analysis of Plant Metabolomics

In the metabolomics of plants, identification of all metabolites in plant samples cannot be carried out using a single analytical tool. The various tools of metabolomics include techniques such as liquid chromatography (LC), Fourier transform ion cyclotron resonance, ion mobility spectrometry (IMS), GC-MS, capillary electrophoresis–mass spectrometry (CE-MS), matrix-assisted laser desorption/ionization (MALDI), and nuclear magnetic resonance (NMR) for quantitative and qualitative analyses of specific metabolites (Patel *et al.*, 2021). Following sample preparation, a combination of these different analytical methods is used to identify, quantify, and separate the different metabolite groups in plant samples, as each technique has its own specificity. During the selection of appropriate analytical methods, the type of approach (targeted or untargeted) and the sample's physical and chemical properties need to be taken into consideration (Vinay *et al.*, 2021).

8.5.1 Gas chromatography–mass spectrometry

For the identification and measurement of small metabolites (less than 500 Da), GC-MS is the perfect method. Chemical derivatization is used primarily to identify these molecules, which include amines, fatty acids, alcohols, hydroxyl acids, sugars, sterols, and amino acids. The process makes the molecules volatile so that gas chromatography can be used. Electron ionization and chemical ionization are the two main types of ionization used in GC-MS. Currently, electron ionization is used in most GC-MS metabolomics techniques. The most standardized, effective, and productive metabolomics technique in plant metabolomics is GC-MS, which is also regarded as the most flexible structure for metabolite analysis (Patel *et al.*, 2021). According to a study by Benkeblia (2023), GC-MS, LC-MS and ultra-performance LC (UPLC)-MS are key technologies that help identification and quantification of oligosaccharides in plants.

8.5.2 Liquid chromatography–mass spectrometry

LC-MS is an additional platform for analyzing the plant metabolome. Compared with GC-MS, it is more appropriate for analyzing polar and non-volatile metabolites. It makes secondary metabolite and lipid analysis available. For metabolomics, lipidomics, and proteomics analyses, LC-MS systems with various instrumental settings, such as regular, nano-, and micro-LC systems, are helpful (Waris *et al.*, 2022). In a study by Yang *et al.* (2020), a LC-MS-based metabolomic method was used to study 11 anthraquinones and 60 flavonoids in buckwheat seed groats and hulls, with some differential metabolites in hulls found to be more concentrated than in groats.

8.5.3 Capillary electrophoresis–mass spectrometry

CE-MS is useful for metabolomics research. Charged metabolites first undergo size- and charge-based separation by CE, and then are ultimately selectively detected by MS. The two main benefits of CE-MS are its high resolution and its ability to analyze almost any charged species using both cationic and anionic methods (Wakayama *et al.*, 2015). Therefore, this is an effective method for analyzing charged metabolites with low molecular mass. Numerous CE-MS techniques have been developed over the years and have significantly advanced metabolomics research across a range of fields. Large-scale metabolomics techniques depend on GC, LC, and CE separation techniques usually coupled with MS to identify numerous metabolites, but their throughputs are relatively low (Soga, 2023). In a study by Hamid *et al.* (2019), water-soluble metabolites were evaluated in two green, five red, and four brown algae using LC-MS and CE-MS-like techniques. Sucrose, fructose and glucose were found in green algae, sorbitol was found in red algae, and mannitol was found in brown algae.

8.5.4 Ion mobility spectrometry

Because of the three-dimensional separation of metabolites and their isomers, IMS has become a sophisticated analytical method. The mass fragmentation pattern and collision cross-section values of the separated metabolites are used to identify them (Joshi *et al.*, 2024). Over the past two to three decades, IMS-MS has become an improved method for small-molecule analysis, providing both mass and structural information for detected ions in milliseconds. It is a gas-phase electrophoretic method that makes it possible to quickly separate ions according to their charge state, size, and shape. This is accomplished when ions pass through a drift cell in the presence of a specific electric field and collide with neutral ions in a buffer gas (Zhang *et al.*, 2018). In a study by Riboni *et al.* (2023), the metabolomic response of treated durum wheat was evaluated using techniques such as UPLC-IMS-high-resolution MS. The results detected 88 compounds and 48 discriminant ones with nutritional, biological, and technological implications and thus suggested safety levels for use in treated wheat cultivation.

8.5.5 NMR spectroscopy

NMR spectroscopy is highly influential in metabolomics and has numerous other interdisciplinary uses. The analytical ability of NMR spectroscopy provides data for a simple qualitative and quantitative evaluation of the endogenous and exogenous metabolites found in biological samples. NMR spectroscopy is a frequently utilized instrument in chemical profiling and plant metabolomics due to its numerous advantages over alternative methods, including its non-destructiveness, sample recovery, and relative quantification without the need for internal standards. NMR spectroscopy requires less work to prepare samples and offers a macroscopic view of the metabolome (Kumar, 2016). In a study by Fortier *et al.* (2023), metabolomic analysis was performed on the root exudate of pea plants using the NMR approach. The technique was helpful in possible discrimination between faba bean and pea depending on their exudate composition and provided characteristic metabolic fingerprinting of pea root exudate.

8.5.6 Matrix-assisted laser desorption/ionization

MALDI, secondary ion mass spectrometry (SIMS), and desorption electrospray ionization MS (DESI-MS) are among the most well-known MS imaging methods. For molecular imaging, MALDI is the most widely utilized method. One of the special contributions of MS imaging to bioanalytical analysis is its ability to analyze the distribution of hundreds of plant metabolites throughout a sample without requiring extensive sample preparation. The well-established MALDI-MS is used to analyze small metabolites, proteins, peptides, and inorganic ions (Heyman and Dubery, 2016). In a study by Sun *et al.* (2020), MALDI-MS was used to visualize the spectrum of various metabolites in *Salvia miltiorrhia* Burge. In this method, 1,1′-binaphthyl-2,2′-diamine and 1,5-diaminonaphthalene were amended as MALDI matrices in both positive and negative ion modes. This method ultimately imaged various molecules such as fatty acids, cholines, tanshinones, phospholipids, oligosaccharides, and amino acids. Similarly, the metabolomic profile of *Ginko biloba* was analyzed using MALDI-MS and other techniques. As a result, various rare flavonoids were detected, mainly from the leaf epidermis, with unequal distribution on the upper and lower sides (Li *et al.*, 2018).

8.6 Multi-omics Data Analysis Technologies

Biology is currently undergoing a rapid and extensive revolution of "multi-omics" technologies. Multi-omics approaches have been well documented in crop science and play a principal role in elucidating senescence, yield, growth, and responses to biotic and abiotic stress in various crops (Yang *et al.*, 2021). Different omics approaches (genomics, transcriptomics, metabolomics, proteomics, phenomics, and ionomics) generate a huge amount of data that can be used to study the novel chemical and genetic elements controlling physiological processes in plants (Cohen *et al.*, 2017). Different omics branches can be used individually or in an integrated manner to understand the collective characterization and quantification of pools of biological molecules (Fig. 8.2).

8.6.1 Genomics

Genomics is the branch of molecular biology that is concerned with the structure, function, evolution, and mapping of the genome. The potential of parallel sequencing of billions of DNA fragments using cheap and fast technologies has made genome sequencing feasible and accessible globally. Genome sequencing has evolved from sequencing of a finite number or only a reference model species to releasing a multitude of genome sequences from a large variety of plant species (Ambrosino *et al.*, 2018). For example, in roses, the ideal genes for recurrent blooming were identified using comparative genome analysis for high-quality rose genomes (Raymond *et al.*, 2018). By using a combination of short reads, long reads, and various scaffolding technologies, the sequence of wild allotetraploid wheat was assembled and annotated (Avni *et al.*, 2017), while Liao *et al.* (2018) revealed that selection-driven genes escape from centromeric regions by comparing genomes of *Oryza sativa* and *Oryza brachyantha*.

8.6.2 Transcriptomics

In order to understand gene functionality and depict gene-expression patterns in various tissues, a comparison of transcriptome data can be exploited (Mykles *et al.*, 2016). Transcriptome sequencing is a high-sensitivity, high-resolution, and high-throughput technique that can be utilized to study non-model organisms and the genomes of medicinal plants, and can provide critical insights into gene structure, expression, and regulation (Guo *et al.*, 2021). Transcriptomics permits the comparison of gene-expression patterns across samples and facilitates the discovery of new genes (Tyagi *et al.*, 2022). In *Saussurea lappa*, the leaf transcriptome was sequenced to identify transcripts responsible for flavonoid and sesquiterpene production (Bains *et al.*, 2019). Using transcriptome sequencing in *Lantana camara*, 229 and 943 genes were found in leaves and roots, respectively, that are responsible for the production of phenylpropanoic acid (Shah *et al.*, 2020).

8.6.3 Proteomics

Proteomics is an important tool that focuses on exploring post-translational modification of

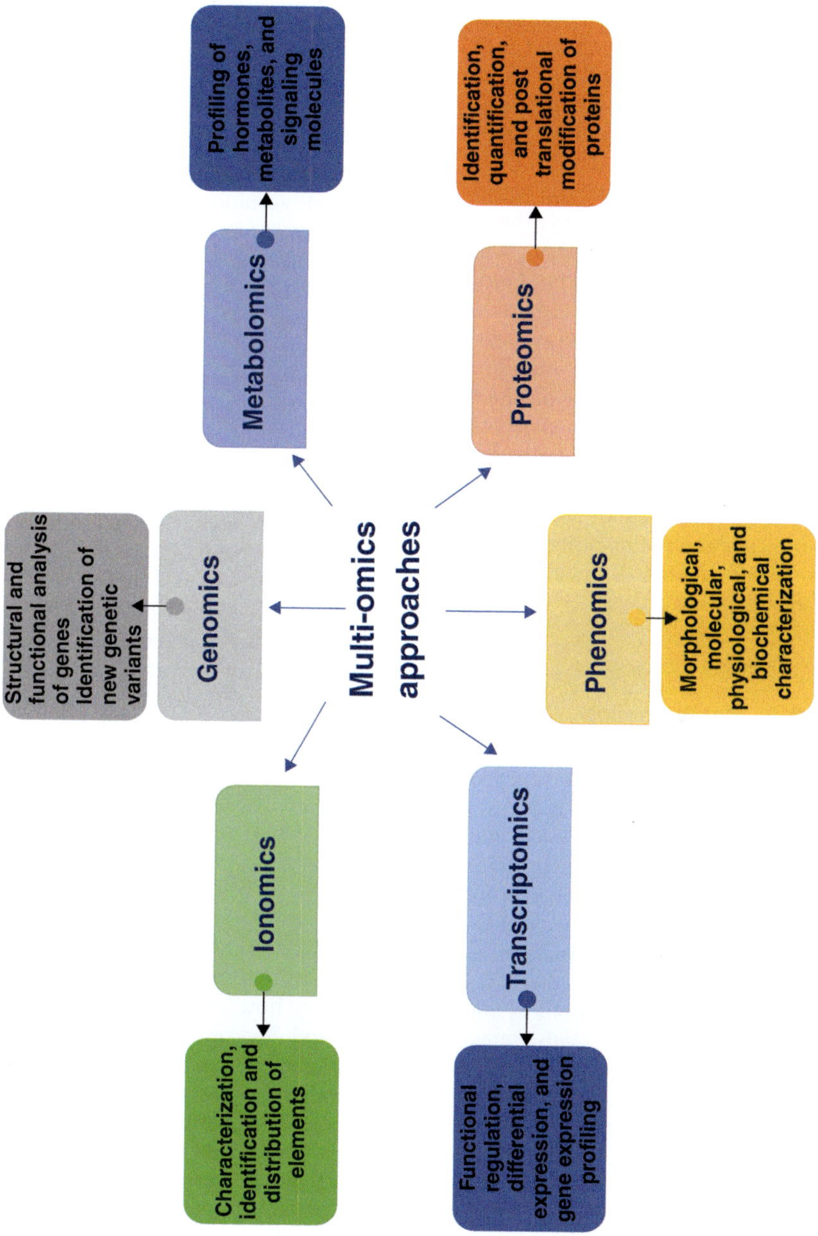

Fig. 8.2. Different omics branches used in plant sciences.

proteins and protein–protein interactions and unraveling the molecular effects of impaired plant metabolism (Meena *et al.*, 2017). Proteomic studyies of plants such as *Arabidopsis thaliana*, *Brassica napus*, *Glycine max*, *Solanum tuberosum*, *Solanum lycopersicum*, *Triticum aestivum*, and *Zea mays* have identified various proteins and their isoforms involved in signaling pathways, transcription factors, and regulatory pathways, and their importance in the production of stress-related chemicals (Kosová *et al.*, 2018). Chen *et al.* (2015) revealed that in *Medicago sativa*, most of the proteins involved in photosynthesis, protein metabolism, energy metabolism, and stress also aid in adaptation to cold stress. Using the shotgun proteomic approach, the response of two varieties of *Solanum tuberosum* to nitrogen deficiency was studied. The difference was found to be strongly associated with defense mechanisms, protein catabolism, and protein synthesis metabolism (Jozefowicz *et al.*, 2017). A proteomic draft of *Arabidopsis* has provided a thorough and quantitative atlas on the accumulation, localization, modification, and regulation of its proteins (Mergner *et al.*, 2020), along with their thermal stability (Jarzab *et al.*, 2020).

8.6.4 Metabolomics

Metabolomics has emerged as an excellent tool to obtain data for functional genomics, systems biology, metabolite profiling, and identification and quantification of metabolites (Zampieri *et al.*, 2017). The metabolomic profiling of plants is very complicated because of the complexity of the large number of metabolites produced by plants, which have diversified structures and abundance. The output of metabolomics depends mainly on the methodology and instrumentation to comprehensively localize, quantify, and identify every metabolite (Raza, 2022). Using metabolomic profiling in the pollen of *Solanum lycopersicum*, three major sets of secondary metabolites (alkaloids, polyamines, and flavonoids) that are produced in response to heat stress were reported (Paupière *et al.*, 2017). Similarly, in *Zea mays*, some essential compounds such as isoleucine, valine, malate, glucose, glycine, serine, sucrose, and proline were reported, which were effectively produced in response to heat stress and CO_2 (Qu *et al.*, 2018). In response to *Fusarium graminearum* infection, accumulation of phenolic compounds and phenylpropanoid has been reported in wheat (Gunnaiah *et al.*, 2012).

8.6.5 Phenomics

Phenomics focuses on high-throughput analysis of phenotypic variation, which is a complex network of interactions among phenotype, genotype, and environment (Chaudhary *et al.*, 2019). The data collection for phenomics is expensive, time-consuming, and complicated, and therefore integrated technical advances and collaboration between researchers can increase phenomics throughput. The complexity can also be managed by vertically integrated research teams, informatics infrastructure, and novel analytic strategies. For simultaneous evaluation of a large number of phenotypic parameters, sensors, automation, computational resources, and advanced imaging systems can be used (Tardieu *et al.*, 2017). To investigate physiological parameters in *Solanum lycopersicum*, a number of phenomics platforms such as Scanalyzer 3D are available (Laxman *et al.*, 2018). Using phenomics in Chinese wild *Vitis amurensis*, the transcription factor VaERD15 associated with cold tolerance was reported (Yu *et al.*, 2017).

8.6.6 Ionomics

Ionomics is the study of metals, non-metals, and metalloids in living organisms. It is aimed at determining how metabolism, development, genetics, and environment affect the elemental composition of target tissues, organs, and cell types (Pita-Barbosa *et al.*, 2019). Shakoor *et al.* (2016) performed ionomic analyses of staple food crops and revealed how ionomics can deal with malnutrition and improve food safety. Sun *et al.* (2020) used calcium-deficient plants and reported that the activity of antioxidant enzymes that scavenge reactive oxygen species increases with elevation in the level of Ca^{2+} ions. It has been reported that supplementation of nitrogen on foliage and soil enhanced the growth and salinity tolerance in cotton (Luo *et al.*, 2015).

8.7 Bioinformatic Tools for Metabolomic Data Analysis

Data analysis is the process of inspecting, cleansing, transforming, and modeling data to uncover meaningful insights, patterns, and trends that inform decision making and solve problems (Saggi and Jain, 2018). Each analytical technique produces a significant volume of plant metabolome data that needs to be handled differently, necessitating the use of informatics tools that include workflows, databases for annotation, and software for data processing and analysis (Zhang *et al.*, 2012). Bioinformatics tools for metabolomic data analysis continually evolve alongside advancements in high-throughput technologies, providing crucial support in unraveling the intricacies of metabolomics data sets, thereby offering numerous advantages in interpreting biological complexities and facilitating deeper insights into metabolic processes (Salem *et al.*, 2020). These are software or algorithms that are designed to analyze biological data, particularly from genomics, proteomics, or metabolomics studies. They aid in tasks such as sequence alignment, genome assembly, structure prediction, and data mining, facilitating biological research and an understanding of complex biological systems (Gangotia *et al.*, 2021). These tools streamline the processing and analysis of vast amounts of complex data generated by techniques such as MS and NMR spectroscopy, enabling researchers to efficiently handle and interpret intricate data sets (Letertre *et al.*, 2020). Moreover, they facilitate the identification of metabolites, their quantification, and the exploration of metabolic pathways, shedding light on the underlying biological mechanisms and potential biomarkers of various physiological and pathological processes. Additionally, bioinformatic tools provide robust statistical analyses, aiding in the detection of significant metabolomic changes and the identification of key metabolites associated with specific conditions or treatments. These tools collectively empower researchers to derive meaningful insights into biological systems and disease mechanisms from metabolomic data sets (Sugimoto *et al.*, 2012).

Spicer *et al.* (2017) have released a list of free and open-source metabolomic analysis software packages. The majority of software tools are written in various scripting languages, including Python, R, Java, and MATLAB, and are platform specific (O'Shea and Misra, 2020). Among the commonly utilized tools is eXtensible Computational Mass Spectrometry (e-XCMS), which efficiently detects features, aligns data, and corrects retention times in LC-MS data (Mahieu *et al.*, 2016). It offers functionalities for feature detection, alignment, retention time correction, and statistical analysis. MetaboAnalyst, a web-based e-tool platform, provides a comprehensive suite of statistical and pathway analysis tools. It includes modules for statistical analysis, pathway analysis, metabolite set enrichment analysis, and visualization (Xia and Wishart, 2011). IDEOM (Integrated Discovery of Event-Related Metabolites) is a software tool designed for identifying metabolomic biomarkers associated with biological events. It integrates statistical analysis methods with pathway analysis to identify significant metabolites and their biological relevance. MetExplore is a web server tool that provides enriched pathway analysis and metabolic network visualization. MS-DIAL is proficient in processing high-resolution MS data (Cottret *et al.*, 2010). It includes modules for peak detection, alignment, deconvolution, and identification of metabolites from spectral databases. Lastly, MAVEN (Metabolomic Analysis and Visualization Engine) caters specifically for targeted metabolomics analysis (Agrawal *et al.*, 2019). Table 8.1 lists a number of databases and tools that are accessible for metabolomics identification.

These tools offer user-friendly interfaces and visualization capabilities, making them accessible for researchers with varying levels of computational expertise to derive meaningful insights from their metabolomic data (Misra and Mohapatra, 2019). Overall, bioinformatic tools for metabolomic data analysis play a crucial role in advancing our understanding of metabolism, disease mechanisms, and personalized medicine, ultimately contributing to improved diagnostics, therapeutics, and healthcare outcomes.

Table 8.1. List of bioinformatics tools used for metabolomics.

Tool/software	Characteristics	Link[a]	Reference
XCMS	Designed for processing and analysis of high-throughput, untargeted metabolomics data	https://xcmsonline.scripps.edu/landing_page.php?pgcontent=mainPage	Mahieu *et al.* (2016)
MetaboAnalyst	A statistical, pathway analysis, enrichment, and visualization tools suite	www.metaboanalyst.ca/	Xia and Wishart (2011)
MetExplore	A web server tool that provides enriched pathway analysis and metabolic network visualization	https://metexplore.toulouse.inra.fr/index.html/	Cottret *et al.* (2010)
MS-DIAL	A proficient high-resolution MS data-processing tool with peak detection, alignment, deconvolution, and metabolite identification	http://prime.psc.riken.jp/compms/msdial/main.html	Tsugawa *et al.* (2015)
MAVEN (Metabolomic Analysis and Visualization Engine)	Designed to cater specifically for targeted metabolomics analysis	http://genomics-pubs.princeton.edu/mzroll/contents/	Agrawal *et al.* (2019)
OpenMS (Open Mass Spectrometry)	Open-source software for comprehensive analysis of MS data	https://www.openms.de/	Röst *et al.* (2016)
ADAP (Automated Data Analysis Pipeline)	Workflow for preprocessing GC-MS data that involves data acquisition, peak detection, alignment, deconvolution, and metabolite identification	www.du-lab.org/software.html	Du *et al.* (2020)
apLCMS (adaptive processing of LC-MS)	R programming package tailored for preprocessing LC-MS data	https://bio.tools/aplcms	Yu *et al.* (2009)
mzMatch	Data processing of LC-MS, which entails handling and analyzing metabolic information	http://mzmatch.sourceforge.net/	Chokkathukalam *et al.* (2013)
eRAH	Searching and distributing reference GC-MS spectra of biologically active metabolites	https://ecosystem.hubspot.com/marketplace/solutions/erah	Domingo-Almenara *et al.* (2016)
METLIN (a metabolite mass spectral database)	Searching and annotating mass spectra	https://metlin.scripps.edu/landing_page.php?pgcontent=mainPage	Domingo-Almenara *et al.* (2019)
MassBank	Designed for searching and annotating mass spectra	www.massbank.jp/	Blaženović *et al.* (2018)

VANTED (Visualization and Analysis of Networks containing Experimental Data)	Designed to analyze and visualize biochemical data within metabolic pathways or networks	https://mybiosoftware.com/vanted-2-01-visualization-analysis-networks-experimental-data.html	Rohn *et al*. (2012)
mzCloud	Designed for searching and annotating mass spectra	www.mzcloud.org/	Garcia-Aloy *et al*. (2020)
GNPS (Global Natural Products Social Molecular Networking)	Designed for searching and annotating mass spectra	http://gnps.ucsd.edu/	Schmid *et al*. (2021)
HMDB (Human Metabolome database)	Designed to compile human and plant metabolite information extracted from NMR and MS spectra	https://hmdb.ca/	Wishart *et al*. (2022)
CAMERA	A bioconductor package for extracting compound spectra and annotating LC-MS data	https://bioconductor.org/packages/release/bioc/html/CAMERA.html	Kuhl *et al*. (2012)
xMSannotator	R programming package tailored for preprocessing LC-MS	https://sourceforge.net/projects/xmsannotator/	Uppal *et al*. (2017)
RAMClust	Spectral-matching-based annotation for metabolomics data	https://github.com/cbroeckl/RAMClustR	Broeckling *et al*. (2014)
MetAssign (Metabolomics assign)	Designed for searching and annotating mass spectra	http://mzmatch.sourceforge.net/	Daly *et al*. (2014)
KEGG (Kyoto Encyclopedia of Genes and Genomes) Pathway	Designed to understand molecular interactions, reactions, and relational networks	www.genome.jp/kegg/pathway.html	Kanehisa (2016)
KNApSAcK	Designed for extensive database detailing relationships between species and metabolites	www.genome.jp/db/pcidb	Afendi *et al*. (2012)
TagFinder	Designed for quantitative assessment of metabolome profiles using GC-MS	www.expasy.org/	Luedemann *et al*. (2012)
MotoDB (Metabolome Tomato Database)	Designed for annotation of tomato using the metabolic pathway database and chemical search	www.sciex.com	Grennan (2009)
KOMICS (Kazusa Metabolomics Database)	Web-based database designed for metabolite annotation in tomato	https://sakura-kagaku.net/komics/software/	Sakurai *et al*. (2014)
MetaMap (Metabolomic map)	Designed for mapping and visualizing metabolomics data based on chemical and biochemical relationships	http://metamapp.fiehnlab.ucdavis.edu/	van Ravenzwaay *et al*. (2012)

Continued

Table 8.1. Continued.

Tool/software	Characteristics	Link[a]	Reference
Plant Reactome	Designed to curate pathway database of the Gramineae family	https://plantreactome.gramene.org /	Naithani *et al.* (2016)
MAGma	Designed for automation annotation of precise multi-stage MS spectral data	https://research-software-directory.org/software/magma	Verdegem *et al.* (2016)
SoyMetDB (Soyabean Metabolomic Database)	Designed to organize soybean metabolomics data for storage, integration, and mining in a structured manner	http://soymetdb.org/	Joshi *et al.* (2010)
COVAIN	Designed to examine statistics in both univariate and multivariate contexts, analyze time-series data, and explore correlation networks	https://bio.tools/covain	Sun and Weckwerth (2013)
Omu	Fast analysis of metabolomics data set	https://cran.r-project.org/web/packages/omu/vignettes/Omu_vignette.html	Tiffany and Bäumler (2019)
Pathvisio	Designed to create, modify, and examine biological pathways	www.pathvisio.org/	Kutmon *et al.* (2015)

GC-MS, gas chromatography–mass spectrometry; LC-MS, liquid chromatography–mass spectrometry; MS, mass spectrometry, NMR, nuclear magnetic resonance.
[a]All accessed 21 January 2025.

8.8 Multi-omics Data and Bioinformatics Tool Integration

In order to efficiently handle all types of genome-scale data sets and extract important insights, bioinformatics has proven essential in every facet of omics-based research. Incorporating the growing body of omics results will improve and expand our comprehension and enable information sharing with other model species (Mochida and Shinozaki, 2011). Following the completion of several plant genome sequences, we have been able to find conserved and/or distinctive characteristics among plant species using genome-scale evolutionary and comparative analysis. Resources for comparative genomics combine genome sequences from several organisms and correlate the data with evolutionary conclusions. Integrating omics approaches with bioinformatics may enable us to find novel pathways and biomarkers, produce better phenotypes, predict plant health and disease, and develop stress tolerance (Fig. 8.3). For instance, functional and structural annotations of published plant genomes and analytical tools are available through the online platform PLAZA for plant comparative genomics (van Bel *et al.*, 2012). The links between the various data types and the PLAZA platform capabilities are schematically represented, which provides an illustration of the level of data integration. A database for the functional and evolutionary categorization of protein-coding genes from all domains of life is PANTHER (Protein Analysis Through Evolutionary Relationships). Gene Ontology words and pathways are included in the functional classifications, which are based on an extensive library of over 15,000 phylogenetic trees for evolutionary categorization (Mi *et al.*, 2021). Another database, GreenPhylDB (www.greenphyl.org, accessed 21 January 2025) was created to make it easier to investigate gene families and homologous links across plant genomes, particularly those of staple crops that are vital to the world's food security. Following the publication of the genomes of *Oryza sativa* and *Arabidopsis thaliana* in 2007, GreenPhylDB became accessible and has subsequently undergone further releases. The enormous abundance of plant genomes makes it difficult to choose a single reference for comparative genomics examinations, but databases that use several genomes per species for orthology identification are still few in number. GreenPhylDBv5, which uses several genome sequences per species, presents the idea of comparative pangenomics (Guignon *et al.*, 2021). Founded in 2012, CottonGen is a community-based online resource that offers integrated, peer-reviewed data on breeding and genomics of cotton along with analytical tools. Utilized by researchers studying cotton all around the world and overseen by specialists knowledgeable about the crop, it has consistently been proven to be the most sensible option for incorporating fresh data and offering the required interfaces for data retrieval. Co-worker, gene, genome, genotype, germplasm, map, marker, metabolite, phenotype, publishing, quantitative trait loci, species, transcriptome, and trait data that have been carefully selected by the CottonGen

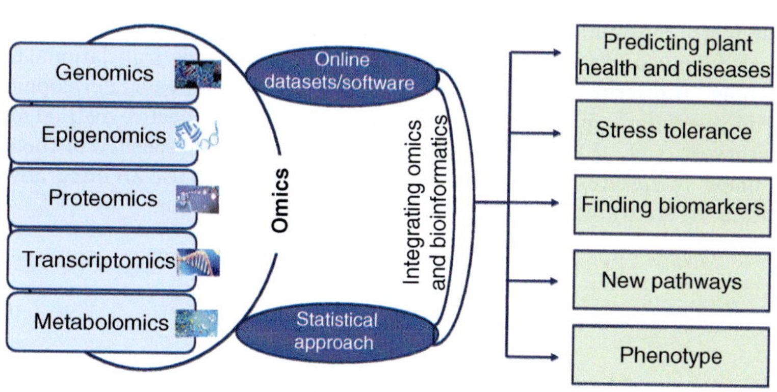

Fig. 8.3. Omics and bioinformatics integration.

team are all available in the CottonGen repository. The volume of data entries stored in CottonGen has skyrocketed. For instance, since 2014, the number of genes and mRNAs, entire genomes, and genotyping data has expanded by 18-fold, 23-fold, and 372-fold, respectively (Yu *et al.*, 2021).

Another concept named Gene Ontology (GO) serves as a vital resource for gene/protein investigations and understanding of their significance in processes and networks (Wood *et al.*, 2020). Conducting enrichment analysis on a collection of genes, transcripts, and proteins is one of the primary applications of GO. An enrichment study, for instance, uses annotations for the gene set to determine which GO words are over- or under-represented in relation to a reference list (e.g. the entire transcriptome), given a collection of genes that are up- or down-regulated under certain circumstances. Probably the most popular tool for exploring and searching the Gene Ontology (GO) database is AmiGO. A search in AmiGO for "abiotic stress" produced 12,800 genes or gene products linked to GO keywords, five GO concepts, synonyms, or definitions (Ambrosino *et al.*, 2020). Specifically, out of the five GO terms that were found, two— "abiotic stress sensitivity" and "abiotic stress sensitivity value"—are no longer in use, while three—"response to stress," "cellular response to abiotic stimulus," and "response to abiotic stimulus"—are all subsets of the other three. These hierarchical annotations frequently result in duplication and non-specificity, which might cause the enrichment findings to be misinterpreted (Carbon *et al.*, 2009). The MapMan software has greatly aided in the advancement of a better understanding of the multi-level variations of the biological systems under study by enabling users to "map" transcripts, proteins, and metabolites on cellular pathways and processes (Chaturvedi *et al.*, 2024). Users can create these maps themselves or download them directly from the software website. One of MapMan's primary benefits is its adaptability and openness, which allow any scientist to create customized maps based on their research interests by uploading pathway images and input data or by altering the standard input file that contains all the data needed to plot genes, proteins, or metabolites. MapMan was created historically for transcriptomics research, but more recently, because of its versatility, it was transformed into a tool that could look at changes in proteins and metabolites (Usadel *et al.*, 2009). The most well-known program for studying correlation networks is called Cytoscape (Muetze *et al.*, 2016). A study used correlation networks to analyze proteomic data on soybean leaves exposed to heat stress and drought. This work highlighted several protein interactions related to RuBisCO (ribulose-1,5-bisphosphate carboxylase/oxygenase) activity, electron transport, and carbon fixation, as well as a set of EF-Tu proteins that may be directly connected to mechanisms for heat stress tolerance (Das *et al.*, 2016).

8.9 Conclusion

The increasing use of high-throughput technologies and associated analytical tools has significantly accelerated the development of omics research. Metabolome analysis has been combined with other resources to provide several significant advances in metabolic system gene function. Various techniques such as GC-MS, LC-MS, and NMR for plant metabolomic analysis, and bioinformatic tools such as Python, Java, and MATLAB are available. Therefore, by combining a variety of omics data sets from various plant species with bioinformatic tools and other analytical tools, translational research has opened the door for rapid development in plant science. Thus, an effective method for determining the mechanistic specifics of the information flow in plants is to use an integrative approach with multi-omics data.

8.10 References

Afendi, F.M., Okada, T., Yamazaki, M., Hirai-Morita, A., Nakamura, Y., *et al.* (2012) KNApSAcK family databases: integrated metabolite–plant species databases for multifaceted plant research. *Plant and Cell Physiology* 53: e1.

Agrawal, S., Kumar, S., Sehgal, R., George, S., Gupta, R., *et al.* (2019) El-MAVEN: a fast, robust, and user-friendly mass spectrometry data processing engine for metabolomics. *Methods in Molecular Biology* 1978, 301–321.

Alseekh, S. and Fernie, A.R. (2018) Metabolomics 20 years on: what have we learned and what hurdles remain? *Plant Journal* 94, 933–942.

Ambrosino, L., Colantuono, C., Monticolo, F., and Chiusano, M.L. (2018) Bioinformatics resources for plant genomics: opportunities and bottlenecks in the-omics era. *Current Issues in Molecular Biology* 27, 71–88.

Ambrosino, L., Colantuono, C., Diretto, G., Fiore, A., and Chiusano, M. L. (2020) Bioinformatics resources for plant abiotic stress responses: state of the art and opportunities in the fast evolving-omics era. *Plants* 9: 591.

Ancillotti, C., Bogani, P., Biricolti, S., Calistri, E., Checchini, L., *et al.* (2015) Changes in polyphenol and sugar concentrations in wild type and genetically modified *Nicotiana langsdorffii* Weinmann in response to water and heat stress. *Plant Physiology and Biochemistry* 97, 52–61.

Anzano, A., Bonanomi, G., Mazzoleni, S., and Lanzotti, V. (2022) Plant metabolomics in biotic and abiotic stress: a critical overview. *Phytochemistry Reviews* 21, 503–524.

Ashapkin, V.V., Kutueva, L.I., Aleksandrushkina, N.I., and Vanyushin, B.F. (2020) Epigenetic mechanisms of plant adaptation to biotic and abiotic stresses. *International Journal of Molecular Sciences* 21: 7457.

Avni, R., Nave, M., Barad, O., Baruch, K., Twardziok, S.O., *et al.* (2017) Wild emmer genome architecture and diversity elucidate wheat evolution and domestication. *Science* 357, 93–97.

Bains, S., Thakur, V., Kaur, J., Singh, K., and Kaur, R. (2019) Elucidating genes involved in sesquiterpenoid and flavonoid biosynthetic pathways in *Saussurea lappa* by *de novo* leaf transcriptome analysis. *Genomics* 111, 1474–1482.

Barupal, D.K., Fan, S., and Fiehn, O. (2018) Integrating bioinformatics approaches for a comprehensive interpretation of metabolomics datasets. *Current Opinion in Biotechnology* 54, 1–9.

Benkeblia, N. (2023) Gas chromatography-mass spectrometry and liquid chromatography-mass spectrometry metabolomics platforms: tools for plant oligosaccharides analysis. *Carbohydrate Polymer Technologies and Applications* 5: 100304.

Blaženović, I., Kind, T., Ji, J., and Fiehn, O. (2018) Software tools and approaches for compound identification of LC-MS/MS data in metabolomics. *Metabolites* 8: 31.

Boness, H.V., de Sá, H.C., Dos Santos, E.K., and Canuto, G.A. (2023) Sample preparation in microbial metabolomics: advances and challenges. In: Fill, T.P. (ed.) *Advances in Experimental Medicine and Biology* 1439, 149–183.

Broeckling, C.D., Afsar, F.A., Neumann, S., Ben-Hur, A., and Prenni, J.E. (2014) RAMClust: a novel feature clustering method enables spectral-matching-based annotation for metabolomics data. *Analytical Chemistry* 86, 6812–6817.

Carbon, S., Ireland, A., Mungall, C.J., Shu, S., Marshall, B., *et al.* (2009) AmiGO: online access to ontology and annotation data. *Bioinformatics* 25, 288–289.

Chaturvedi, P., Pierides, I., Zhang, S., Schwarzerova, J., Ghatak, A., and Weckwerth, W. (2024) Multiomics for crop improvement. In: Pandey, M.K., Bently, A., Desmae, H., Roorkiwal, M., and Varshney, R.K. (eds) *Frontier Technologies for Crop Improvement*. Springer Nature, Singapore, pp. 107–141.

Chaudhary, J., Khatri, P., Singla, P., Kumawat, S., Kumari, A., *et al.* (2019) Advances in omics approaches for abiotic stress tolerance in tomato. *Biology* 8: 90.

Chen, J., Han, G., Li, J., Zhang, H., Liu, F., and Wang, J. (2015) Proteomic analyses reveal differences in cold acclimation mechanisms in freezing-tolerant and freezing-sensitive cultivars of alfalfa. *Frontiers in Plant Science* 6: 132113.

Chokkathukalam, A., Jankevics, A., Creek, D.J., Achcar, F., Barrett, M.P., and Breitling, R., (2013) mzMatch–ISO: an R tool for the annotation and relative quantification of isotope-labelled mass spectrometry data. *Bioinformatics* 29, 281–283.

Cohen, H., Szymanski, J., and Aharoni, A. (2017) Assimilation of 'omics' strategies to study the cuticle layer and suberin lamellae in plants. *Journal of Experimental Botany* 68, 5389–5400.

Cottret, L., Wildridge, D., Vinson, F., Barrett, M.P., Charles, H., *et al.* (2010) MetExplore: a web server to link metabolomic experiments and genome-scale metabolic networks. *Nucleic Acids Research* 38 (Suppl. 2), 132–137.

Daly, R., Rogers, S., Wandy, J., Jankevics, A., Burgess, K.E., and Breitling, R. (2014) MetAssign: probabilistic annotation of metabolites from LC–MS data using a Bayesian clustering approach. *Bioinformatics* 30, 2764–2771.

Das, A., Eldakak, M., Paudel, B., Kim, D.W., Hemmati, H., *et al.* (2016) Leaf proteome analysis reveals prospective drought and heat stress response mechanisms in soybean. *BioMed Research International* 2016: 6021047.

de Souza, H.M., Pereira, T.T., de Sá, H.C., Alves, M.A., Garrett, R., and Canuto, G.A. (2024) Critical factors in sample collection and preparation for clinical metabolomics of underexplored biological specimens. *Metabolites* 14: 36.

Díaz-Cruz, G.A. and Cassone, B.J. (2018) A tale of survival: molecular defense mechanisms of soybean to overcome soybean mosaic virus infection. *Physiological and Molecular Plant Pathology* 102, 79–87.

Dikilitas, M., Simsek, E., and Roychoudhury, A. (2021) Role of proline and glycine betaine in overcoming abiotic stresses. In: Roychoudhury, A. and Tripathi, D.K. (eds) *Protective Chemical Agents in the Amelioration of Plant Abiotic Stress: Biochemical and Molecular Perspective*. Wiley, Chichester, UK, pp. 1–23.

Domingo-Almenara, X., Brezmes, J., Vinaixa, M., Samino, S., Ramirez, N., *et al.* (2016) eRah: a computational tool integrating spectral deconvolution and alignment with quantification and identification of metabolites in GC/MS-based metabolomics. *Analytical Chemistry* 88, 9821–9829.

Domingo-Almenara, X., Montenegro-Burke, J.R., Guijas, C., Majumder, E.L.W., Benton, H.P., and Siuzdak, G. (2019) Autonomous METLIN-guided in-source fragment annotation for untargeted metabolomics. *Analytical Chemistry* 91, 3246–3253.

Dominguez, P.G., Cataldi, T.R., Abreu, I.N., Moritz, T., and Budzinski, I.G.F. (2024) Sample preparation for forest tree metabolomics. In: Antonio, C. (ed.) *Monitoring Forest Damage with Metabolomics Methods*. Wiley, Hoboken, New Jersey, pp. 35–70.

Du, X., Smirnov, A., Pluskal, T., Jia, W., and Sumner, S. (2020) Metabolomics data preprocessing using ADAP and MZmine 2. *Methods in Molecular Biology* 2104, 25–48.

Duportet, X., Aggio, R.B.M., Carneiro, S., and Villas-Bôas, S.G. (2012) The biological interpretation of metabolomic data can be misled by the extraction method used. *Metabolomics* 8, 410–421.

Eshawu, A.B. and Ghalsasi, V.V. (2024) Metabolomics of natural samples: a tutorial review on the latest technologies. *Journal of Separation Science* 47: 2300588.

Feng, Z., Ding, C., Li, W., Wang, D., and Cui, D. (2020) Applications of metabolomics in the research of soybean plant under abiotic stress. *Food Chemistry* 310: 125914.

Fernie, A.R., Trethewey, R.N., Krotzky, A.J., and Willmitzer, L. (2004) Metabolite profiling: from diagnostics to systems biology. *Nature Reviews Molecular Cell Biology* 5, 763–769.

Fortier, M., Lemyre, J., Ancelin, E., Oulyadi, H., Driouich, A., *et al.* (2023) Development of a root exudate collection protocol for metabolomics analysis using nuclear magnetic resonance. *Plant Science* 331: 111694.

Gangotia, D., Gupta, A., and Mani, I. (2021) Role of bioinformatics in biological sciences. In: Singh, V. and Kumar, A. (eds) *Advances in Bioinformatics*. Springer, Singapore, pp. 37–57.

Garcia-Aloy, M., Groff, N., Masuero, D., Nisi, M., Franco, A., *et al.* (2020) Exploratory analysis of commercial olive-based dietary supplements using untargeted and targeted metabolomics. *Metabolites* 10: 516.

Gorrochategui, E., Jaumot, J., Lacorte, S., and Tauler, R. (2016) Data analysis strategies for targeted and untargeted LC-MS metabolomic studies: overview and workflow. *Trends in Analytical Chemistry* 82, 425–442.

Grennan, A.K. (2009) MoTo DB: a metabolic database for tomato. *Plant Physiology* 151, 1701–1702.

Guignon, V., Toure, A., Droc, G., Dufayard, J.F., Conte, M., and Rouard, M. (2021) GreenPhylDB v5: a comparative pangenomic database for plant genomes. *Nucleic Acids Research* 49, D1464–D1471.

Gunnaiah, R., Kushalappa, A.C., Duggavathi, R., Fox, S., and Somers, D.J. (2012) Integrated metabolo-proteomic approach to decipher the mechanisms by which wheat QTL (*Fhb1*) contributes to resistance against *Fusarium graminearum*. *PLOS ONE* 7: e40695.

Guo, J., Huang, Z., Sun, J., Cui, X., and Liu, Y. (2021) Research progress and future development trends in medicinal plant transcriptomics. *Frontiers in Plant Science* 12: 691838.

Gutiérrez-Albanchez, E., Gradillas, A., García, A., García-Villaraco, A., Gutierrez-Mañero, F.J., and Ramos-Solano, B. (2020) Elicitation with *Bacillus* QV15 reveals a pivotal role of F3H on flavonoid metabolism improving adaptation to biotic stress in blackberry. *PLOS ONE* 15: e0232626.

Hamid, S.S., Wakayama, M., Ichihara, K., Sakurai, K., Ashino, Y., *et al.* (2019) Metabolome profiling of various seaweed species discriminates between brown, red, and green algae. *Planta* 249, 1921–1947.

Heyman, H.M. and Dubery, I.A. (2016) The potential of mass spectrometry imaging in plant metabolomics: a review. *Phytochemistry Reviews* 15, 297–316.

Hou, Q., Ufer, G., and Bartels, D. (2016) Lipid signalling in plant responses to abiotic stress. *Plant, Cell and Environment* 39, 1029–1048.

Hüdig, M., Schmitz, J., Engqvist, M.K.M., and Maurino, V.G. (2018) Biochemical control systems for small molecule damage in plants. *Plant Signaling and Behavior* 13: e1477906.

Jacyna, J., Kordalewska, M., and Markuszewski, M.J. (2019) Design of experiments in metabolomics-related studies: an overview. *Journal of Pharmaceutical and Biomedical Analysis* 164, 598–606.

Jarzab, A., Kurzawa, N., Hopf, T., Moerch, M., Zecha, J., *et al.* (2020) Meltome atlas—thermal proteome stability across the tree of life. *Nature Methods* 17, 495–503.

Jeppesen, M.J. and Powers, R. (2023) Multiplatform untargeted metabolomics. *Magnetic Resonance in Chemistry* 61, 628–653.

Joshi, T., Yao, Q., Levi, D.F., Brechenmacher, L., Valliyodan, B., *et al.* (2010) SoyMetDB: the soybean metabolome database. In: *2010 IEEE International Conference on Bioinformatics and Biomedicine (BIBM).* IEEE, New York, pp. 203–208.

Joshi, R., Sharma, S., and Kumar, D. (2024) Advances of ion mobility platform for plant metabolomics. *Critical Reviews in Analytical Chemistry* 54, 175–191.

Jozefowicz, A.M., Hartmann, A., Matros, A., Schum, A., and Mock, H.P. (2017) Nitrogen deficiency induced alterations in the root proteome of a pair of potato (*Solanum tuberosum* L.) varieties contrasting for their response to low N. *Proteomics* 17: 1700231.

Kanehisa, M. (2016) KEGG bioinformatics resource for plant genomics and metabolomics. *Methods in Molecular Biology* 1374, 55–70.

Kebert, M., Vuksanović, V., Stefels, J., Bojović, M., Horák, R., *et al.* (2022) Species-level differences in osmoprotectants and antioxidants contribute to stress tolerance of *Quercus robur* L., and *Q. cerris* L. seedlings under water deficit and high temperatures. *Plants* 11: 1744.

Kim, H.K., Choi, Y.H., and Verpoorte, R. (2010) NMR-based metabolomic analysis of plants. *Nature Protocols* 5, 536–549.

Kim, S.B., van den Broeck, L., Karre, S., Choi, H., Christensen, S.A., *et al.* (2021) Analysis of the transcriptomic, metabolomic, and gene regulatory responses to *Puccinia sorghi* in maize. *Molecular Plant Pathology* 22, 465–479.

Kogovsek, P., Pompe-Novak, M., Petek, M., Fragner, L., Weckwerth, W., and Gruden, K. (2016) Primary metabolism, phenylpropanoids and antioxidant pathways are regulated in potato as a response to *Potato virus Y* infection. *PLOS ONE* 11: e0146135.

Kosová, K., Vítámvás, P., Urban, M.O., Prášil, I.T., and Renaut, J. (2018) Plant abiotic stress proteomics: the major factors determining alterations in cellular proteome. *Frontiers in Plant Science* 9: 325823.

Kuhl, C., Tautenhahn, R., Bottcher, C., Larson, T.R., and Neumann, S. (2012) CAMERA: an integrated strategy for compound spectra extraction and annotation of liquid chromatography/mass spectrometry data sets. *Analytical Chemistry* 84, 283–289.

Kumar, D. (2016) Nuclear magnetic resonance (NMR) spectroscopy for metabolic profiling of medicinal plants and their products. *Critical Reviews in Analytical Chemistry* 46, 400–412.

Kutmon, M., van Iersel, M.P., Bohler, A., Kelder, T., Nunes, N., *et al.* (2015) PathVisio 3: an extendable pathway analysis toolbox. *PLOS Computational Biology* 11: e1004085.

Last, R.L., Jones, A.D., and Shachar-Hill, Y. (2007) Towards the plant metabolome and beyond. *Nature Reviews Molecular Cell Biology* 8, 167–174.

Laxman, R.H., Hemamalini, P., Bhatt, R.M., and Sadashiva, A.T. (2018) Non-invasive quantification of tomato (*Solanum lycopersicum* L.) plant biomass through digital imaging using phenomics platform. *Indian Journal of Plant Physiology* 23, 369–375.

Letertre, M.P., Dervilly, G., and Giraudeau, P. (2020) Combined nuclear magnetic resonance spectroscopy and mass spectrometry approaches for metabolomics. *Analytical Chemistry* 93, 500–518.

Li, B., Neumann, E.K., Ge, J., Gao, W., Yang, H., *et al.* (2018) Interrogation of spatial metabolome of *Ginkgo biloba* with high-resolution matrix-assisted laser desorption/ionization and laser desorption/ionization mass spectrometry imaging. *Plant, Cell and Environment* 41, 2693–2703.

Li, X., Zhang, L.P., Zhang, L., Yan, P., Ahammed, G.J., and Han, W.Y. (2019) Methyl salicylate enhances flavonoid biosynthesis in tea leaves by stimulating the phenylpropanoid pathway. *Molecules* 24: 362.

Liao, Y., Zhang, X., Li, B., Liu, T., Chen, J., *et al.* (2018) Comparison of *Oryza sativa* and *Oryza brachyantha* genomes reveals selection-driven gene escape from the centromeric regions. *Plant Cell* 30, 1729–1744.

Luedemann, A., von Malotky, L., Erban, A., and Kopka, J. (2012) TagFinder: preprocessing software for the fingerprinting and the profiling of gas chromatography–mass spectrometry based metabolome analyses. *Methods in Molecular Biology* 860, 255–286.

Luo, Z., Kong, X., Dai, J., and Dong, H. (2015) Soil plus foliar nitrogen application increases cotton growth and salinity tolerance. *Journal of Plant Nutrition* 38, 443–455.

Mahieu, N.G., Genenbacher, J.L., and Patti, G.J. (2016) A roadmap for the XCMS family of software solutions in metabolomics. *Current Opinion in Chemical Biology* 30, 87–93.

Mardoc, E., Sow, M.D., Déjean, S., and Salse, J. (2024) Genomic data integration tutorial, a plant case study. *BMC Genomics* 25: 66.

Mead, R. (2017) *Statistical Methods in Agriculture and Experimental Biology*, 3rd edn. Chapman & Hall/CRC, New York.

Meena, K.K., Sorty, A.M., Bitla, U.M., Choudhary, K., Gupta, P., *et al.* (2017) Abiotic stress responses and microbe-mediated mitigation in plants: the omics strategies. *Frontiers in Plant Science* 8: 229571.

Meena, M., Divyanshu, K., Kumar, S., Swapnil, P., Zehra, A., *et al.* (2019) Regulation of L-proline biosynthesis, signal transduction, transport, accumulation and its vital role in plants during variable environmental conditions. *Heliyon* 5: e02952.

Mergner, J., Frejno, M., List, M., Papacek, M., Chen, X., *et al.* (2020) Mass-spectrometry-based draft of the *Arabidopsis* proteome. *Nature* 579, 409–414.

Mi, H., Ebert, D., Muruganujan, A., Mills, C., Albou, L.P., *et al.* (2021) PANTHER version 16: a revised family classification, tree-based classification tool, enhancer regions and extensive API. *Nucleic Acids Research* 49, D394–D403.

Miotto-Vilanova, L., Courteaux, B., Padilla, R., Rabenoelina, F., Jacquard, C., *et al.* (2019) Impact of *Paraburkholderia phytofirmans* PsJN on grapevine phenolic metabolism. *International Journal of Molecular Sciences* 20: 5775.

Misra, B.B. and Mohapatra, S. (2019) Tools and resources for metabolomics research community: a 2017–2018 update. *Electrophoresis* 40, 227–246.

Mochida, K. and Shinozaki, K. (2011) Advances in omics and bioinformatics tools for systems analyses of plant functions. *Plant and Cell Physiology* 52, 2017–2038.

Muetze, T., Goenawan, I.H., Wiencko, H.L., Bernal-Llinares, M., Bryan, K., and Lynn, D.J. (2016) Contextual Hub Analysis Tool (CHAT): a Cytoscape app for identifying contextually relevant hubs in biological networks. *F1000Research* 5; 1745.

Mykles, D.L., Burnett, K.G., Durica, D.S., Joyce, B.L., McCarthy, F.M., *et al.* (2016) Resources and recommendations for using transcriptomics to address grand challenges in comparative biology. *Integrative and Comparative Biology* 56, 1183–1191.

Nägele, T., Fragner, L., Chaturvedi, P., Ghatak, A., and Weckwerth, W. (2017) Pollen metabolome dynamics: biochemistry, regulation and analysis. In: Obermeyer, G. and Feijó, J. (eds) *Pollen Tip Growth*. Springer, Cham. Switzerland, pp. 319–336.

Naithani, S., Preece, J., d'Eustachio, P., Gupta, P., Amarasinghe, V., *et al.* (2016) Plant Reactome: a resource for plant pathways and comparative analysis. *Nucleic Acids Research* 48, D1029–D1039.

Nakabayashi, R., Yonekura-Sakakibara, K., Urano, K., Suzuki, M., Yamada, Y., *et al.* (2014) Enhancement of oxidative and drought tolerance in *Arabidopsis* by overaccumulation of antioxidant flavonoids. *Plant Journal* 77, 367–379.

O'Shea, K. and Misra, B.B. (2020) Software tools, databases and resources in metabolomics: updates from 2018 to 2019. *Metabolomics* 16: 36.

Oh, S.W., Imran, M., Kim, E.H., and Park, S.Y. (2023) Approach strategies and application of metabolomics to biotechnology in plants. *Frontiers in Plant Science* 14: 1192235.

Patel, M.K., Pandey, S., Kumar, M., Haque, M.I., Pal, S., and Yadav, N.S. (2021) Plants metabolome study: emerging tools and techniques. *Plants* 10: 2409.

Paupière, M.J., van Haperen, P., Rieu, I., Visser, R.G., Tikunov, Y.M., and Bovy, A.G. (2017) Screening for pollen tolerance to high temperatures in tomato. *Euphytica* 213: 130.

Pei, L., Gao, Y., Feng, L., Zhang, Z., Liu, N., *et al.* (2023) Phenolic acids and flavonoids play important roles in flower bud differentiation in *Mikania micrantha*: transcriptomics and metabolomics. *International Journal of Molecular Sciences* 24: 16550.

Piasecka, A., Kachlicki, P., and Stobiecki, M. (2019) Analytical methods for detection of plant metabolomes changes in response to biotic and abiotic stresses. *International Journal of Molecular Sciences* 20: 379.

Pita-Barbosa, A., Ricachenevsky, F.K., Wilson, M., Dottorini, T., and Salt, D.E. (2019) Transcriptional plasticity buffers genetic variation in zinc homeostasis. *Scientific Reports* 9: 19482.

Qu, M., Chen, G., Bunce, J.A., Zhu, X., and Sicher, R.C. (2018) Systematic biology analysis on photosynthetic carbon metabolism of maize leaf following sudden heat shock under elevated CO_2. *Scientific Reports* 8: 7849.

Raymond, O., Gouzy, J., Just, J., Badouin, H., Verdenaud, M., *et al.* (2018) The *Rosa* genome provides new insights into the domestication of modern roses. *Nature Genetics* 50, 772–777.

Raza, A. (2022) Metabolomics: a systems biology approach for enhancing heat stress tolerance in plants. *Plant Cell Reports* 41, 741–763.

Razzaq, A., Sadia, B., Raza, A., Khalid Hameed, M., and Saleem, F. (2019) Metabolomics: a way forward for crop improvement. *Metabolites* 9: 303.

Ribbenstedt, A., Ziarrusta, H., and Benskin, J.P. (2018) Development, characterization and comparisons of targeted and non-targeted metabolomics methods. *PLOS ONE* 13: e0207082.

Riboni, N., Bianchi, F., Mattarozzi, M., Caldara, M., Gullì, M., *et al.* (2023) Ultra-high performance liquid chromatography–ion mobility–high-resolution mass spectrometry to evaluate the metabolomic response of durum wheat to sustainable treatments. *Journal of Agricultural and Food Chemistry* 71, 15407–15416.

Rodrigues, A.M., Ribeiro-Barros, A.I., and António, C. (2019) Experimental design and sample preparation in forest tree metabolomics. *Metabolites* 9: 285.

Rodziewicz, P., Swarcewicz, B., Chmielewska, K., Wojakowska, A., and Stobiecki, M. (2014) Influence of abiotic stresses on plant proteome and metabolome changes. *Acta Physiologiae Plantarum* 36, 1–19.

Rohn, H., Junker, A., Hartmann, A., Grafahrend-Belau, E., Treutler, H., *et al.* (2012) VANTED v2: a framework for systems biology applications. *BMC Systems Biology* 6: 139.

Roldan, M.V.G., Engel, B., de Vos, R.C., Vereijken, P., Astola, L., *et al.* (2014) Metabolomics reveals organ-specific metabolic rearrangements during early tomato seedling development. *Metabolomics* 10, 958–974.

Rosa, M., Prado, C., Podazza, G., Interdonato, R., González, J.A., *et al.* (2009) Soluble sugars: metabolism, sensing and abiotic stress: a complex network in the life of plants. *Plant Signaling and Behavior* 4, 388–393.

Röst, H.L., Sachsenberg, T., Aiche, S., Bielow, C., Weisser, H., *et al.* (2016) OpenMS: a flexible open-source software platform for mass spectrometry data analysis. *Nature Methods* 13, 741–748.

Saggi, M.K. and Jain, S. (2018) A survey towards an integration of big data analytics to big insights for value-creation. *Information Processing and Management* 54, 758–790.

Sakurai, N., Ara, T., Enomoto, M., Motegi, T., Morishita, Y., *et al.* (2014) Tools and databases of the KOMICS web portal for preprocessing, mining, and dissemination of metabolomics data. *BioMed Research International* 2014: 194812.

Salam, U., Ullah, S., Tang, Z.H., Elateeq, A.A., Khan, Y., *et al.* (2023) Plant metabolomics: an overview of the role of primary and secondary metabolites against different environmental stress factors. *Life* 13: 706.

Salem, M.A., Perez de Souza, L., Serag, A., Fernie, A.R., Farag, M.A., *et al.* (2020) Metabolomics in the context of plant natural products research: from sample preparation to metabolite analysis. *Metabolites* 10: 37.

Schauer, N., Semel, Y., Roessner, U., Gur, A., Balbo, I., *et al.* (2006) Comprehensive metabolic profiling and phenotyping of interspecific introgression lines for tomato improvement. *Nature Biotechnology* 24, 447–454.

Schmid, R., Petras, D., Nothias, L.F., Wang, M., Aron, A.T., *et al.* (2021) Ion identity molecular networking for mass spectrometry-based metabolomics in the GNPS environment. *Nature Communications* 12: 3832.

Shah, M., Alharby, H.F., Hakeem, K.R., Ali, N., Rahman, I.U., *et al.* (2020) *De novo* transcriptome analysis of *Lantana camara* L. revealed candidate genes involved in phenylpropanoid biosysnthesis pathway. *Scientific Reports* 10: 13726.

Shakoor, N., Ziegler, G., Dilkes, B.P., Brenton, Z., Boyles, R., *et al.* (2016) Integration of experiments across diverse environments identifies the genetic determinants of variation in *Sorghum bicolor* seed element composition. *Plant Physiology* 170, 1989–1998.

Sharma, A., Shahzad, B., Rehman, A., Bhardwaj, R., Landi, M., and Zheng, B. (2019) Response of phenylpropanoid pathway and the role of polyphenols in plants under abiotic stress. *Molecules* 24: 2452.

Shi, Q., Wang, R., Lu, W., Zhu, J., Zhang, H., *et al.* (2024) Metabolomics analysis of variation in grain quality of high-quality japonica rice. *Agronomy* 14: 430.

Soga, T. (2023) Advances in capillary electrophoresis mass spectrometry for metabolomics. *Trends in Analytical Chemistry* 158: 116883.

Spicer, R., Salek, R.M., Moreno, P., Cañueto, D., and Steinbeck, C. (2017) Navigating freely-available software tools for metabolomics analysis. *Metabolomics* 13: 106.

Sugimoto, M., Kawakami, M., Robert, M., Soga, T., and Tomita, M. (2012) Bioinformatics tools for mass spectroscopy-based metabolomic data processing and analysis. *Current Bioinformatics* 7, 96–108.

Sun, X. and Weckwerth, W. (2013) Using COVAIN to analyze metabolomics data. In: Weckwert, W. and Kahl, G. (eds) *Handbook of Plant Metabolomics*. Wiley, Weinheim, Germany, pp. 305–320.

Sun, X., Pan, B., Wang, Y., Xu, W., and Zhang, S. (2020) Exogenous calcium improved resistance to *Botryosphaeria dothidea* by increasing autophagy activity and salicylic acid level in pear. *Molecular Plant–Microbe Interactions* 33, 1150–1160.

Taha, R.S., Seleiman, M.F., Alhammad, B.A., Alkahtani, J., Alwahibi, M.S., and Mahdi, A.H. (2020) Activated yeast extract enhances growth, anatomical structure, and productivity of *Lupinus termis* L. plants under actual salinity conditions. *Agronomy* 11: 74.

Tardieu, F., Cabrera-Bosquet, L., Pridmore, T., and Bennett, M. (2017) Plant phenomics, from sensors to knowledge. *Current Biology* 27, R770–R783.

Teklić, T., Paradiković, N., Špoljarević, M., Zeljković, S., Lončarić, Z., and Lisjak, M. (2021) Linking abiotic stress, plant metabolites, biostimulants and functional food. *Annals of Applied Biology* 178, 169–191.

Templer, S.E., Ammon, A., Pscheidt, D., Ciobotea, O., Schuy, C., *et al.* (2017) Metabolite profiling of barley flag leaves under drought and combined heat and drought stress reveals metabolic QTLs for metabolites associated with antioxidant defense. *Journal of Experimental Botany* 68, 1697–1713.

Tiffany, C.R. and Bäumler, A.J. (2019) omu, a metabolomics count data analysis tool for intuitive figures and convenient metadata collection. *Microbiology Resource Announcements* 8, 10–1128.

Tsugawa, H., Cajka, T., Kind, T., Ma, Y., Higgins, B., *et al.* (2015) MS-DIAL: data-independent MS/MS deconvolution for comprehensive metabolome analysis. *Nature Methods* 12, 523–526.

Tyagi, P., Singh, D., Mathur, S., Singh, A., and Ranjan, R. (2022) Upcoming progress of transcriptomics studies on plants: an overview. *Frontiers in Plant Science* 13: 1030890.

Uppal, K., Walker, D.I., and Jones, D.P. (2017) xMSannotator: an R package for network-based annotation of high-resolution metabolomics data. *Analytical Chemistry* 89, 1063–1067.

Usadel, B., Poree, F., Nagel, A., Lohse, M., Czedik-Eysenberg, A., and Stitt, M. (2009) A guide to using MapMan to visualize and compare omics data in plants: a case study in the crop species, maize. *Plant, Cell and Environment* 32, 1211–1229.

Valifard, M., Mohsenzadeh, S., Niazi, A., and Moghadam, A. (2015) Phenylalanine ammonia lyase isolation and functional analysis of phenylpropanoid pathway under salinity stress in *Salvia* species. *Australian Journal of Crop Science* 9, 656–665.

van Bel, M., Proost, S., Wischnitzki, E., Movahedi, S., Scheerlinck, C., Van de Peer, Y., and Vandepoele, K. (2012) Dissecting plant genomes with the PLAZA comparative genomics platform. *Plant Physiology* 158, 590–600.

van Ravenzwaay, B., Herold, M., Kamp, H., Kapp, M.D., Fabian, E., *et al.* (2012) Metabolomics: a tool for early detection of toxicological effects and an opportunity for biology-based grouping of chemicals—from QSAR to QBAR. *Mutation Research/Genetic Toxicology and Environmental Mutagenesis* 746, 144–150.

van Dam, N.M., and van der Meijden, E. (2011) A role for metabolomics in plant ecology. *Annual Plant Reviews Volume 43: Biology of Plant Metabolomics* 43, 87–107.

Verdegem, D., Lambrechts, D., Carmeliet, P., and Ghesquiere, B. (2016) Improved metabolite identification with MIDAS and MAGMa through MS/MS spectral dataset-driven parameter optimization. *Metabolomics* 12: 98.

Vinay, C.M., Udayamanoharan, S.K., Prabhu Basrur, N., Paul, B., and Rai, P.S. (2021) Current analytical technologies and bioinformatic resources for plant metabolomics data. *Plant Biotechnology Reports* 15, 561–572.

Wakayama, M., Hirayama, A., and Soga, T. (2015) Capillary electrophoresis-mass spectrometry. *Methods in Molecular Biology* 1277, 113–122.

Wang, K., Liao, X., Xia, J., Xiao, C., Deng, J., and Xu, Z. (2023) Metabolomics: a promising technique for uncovering quality-attribute of fresh and processed fruits and vegetables. *Trends in Food Science and Technology* 142: 104213.

Waris, M., Kocak, E., Gonulalan, E.M., Demirezer, L.O., Kır, S., and Nemutlu, E. (2022) Metabolomics analysis insight into medicinal plant science. *Trends in Analytical Chemistry* 157: 116795.

Weckwerth, W. (2011) Green systems biology—from single genomes, proteomes and metabolomes to ecosystems research and biotechnology. *Journal of Proteomics* 75, 284–305.

Wishart, D.S., Guo, A., Oler, E., Wang, F., Anjum, A., *et al.* (2022) HMDB 5.0: the human metabolome database for 2022. *Nucleic Acids Research* 50, D622–D631.

Wood, V., Carbon, S., Harris, M.A., Lock, A., Engel, S.R., *et al.* (2020) Term matrix: a novel gene ontology annotation quality control system based on ontology term co-annotation patterns. *Open Biology* 10: 200149.

Xia, J. and Wishart, D.S. (2011) Metabolomic data processing, analysis, and interpretation using MetaboAnalyst. *Current Protocols in Bioinformatics* 34, 14–10.

Xiang, M., Ding, W., Wu, C., Wang, W., Ye, S., *et al.* (2021) Production of purple Ma bamboo (*Dendrocalamus latiflorus* Munro) with enhanced drought and cold stress tolerance by engineering anthocyanin biosynthesis. *Planta* 254: 50.

Yang, W., Su, Y., Dong, G., Qian, G., Shi, Y., *et al.* (2020) Liquid chromatography–mass spectrometry-based metabolomics analysis of flavonoids and anthraquinones in *Fagopyrum tataricum* L. Gaertn (tartary buckwheat) seeds to trace morphological variations. *Food Chemistry* 331: 127354.

Yang, Y., Saand, M.A., Huang, L., Abdelaal, W.B., and Sirohi, M.H. (2021) Applications of multi-omics technologies for crop improvement. *Frontiers in Plant Science* 12: 563953.

Yu, D., Zhang, L., Zhao, K., Niu, R., Zhai, H., and Zhang, J. (2017) VaERD15, a transcription factor gene associated with cold-tolerance in Chinese wild *Vitis amurensis*. *Frontiers in Plant Science* 8: 218625.

Yu, J., Jung, S., Cheng, C.H., Lee, T., Zheng, P., *et al.* (2021) CottonGen: the community database for cotton genomics, genetics and breeding research. *Plants* 10: 2805.

Yu, T., Park, Y., Johnson, J.M., and Jones, D.P. (2009) apLCMS—adaptive processing of high-resolution LC/MS data. *Bioinformatics* 25, 1930–1936.

Zampieri, M., Sekar, K., Zamboni, N., and Sauer, U. (2017) Frontiers of high-throughput metabolomics. *Current Opinion in Chemical Biology* 36, 15–23.

Zhang, A., Sun, H., Wang, P., Han, Y., and Wang, X. (2012) Modern analytical techniques in metabolomics analysis. *Analyst* 137, 293–300.

Zhang, X., Quinn, K., Cruickshank-Quinn, C., Reisdorph, R., and Reisdorph, N. (2018) The application of ion mobility mass spectrometry to metabolomics. *Current Opinion in Chemical Biology* 42, 60–66.

9 Applied Metabolomics for Plant Biotechnology

Ilham Dehbi[1,2], Khadija Benamar[3], Rachid Ezzouggari[1,4], Mouna Janati[2], Slimane Khayi[5], Rachid Mentag[5], Zineb Belabess[6], Yunfei Jiang[7], and Rachid Lahlali[1*]

[1]Phytopathology Unit, Department of Plant Protection, National School of Agriculture of Meknès, Meknès, Morocco; [2]Laboratory of Biotechnologies and Bioresources Valorization, Moulay Ismail University of Meknès, Meknès, Morocco; [3]Department of Biology, Laboratory of Microbial Biotechnology and Bioactive Molecules, Sciences and Technologies Faculty, Sidi Mohamed Ben Abdellah University, Fez, Morocco; [4]Laboratory of Biotechnology, Conservation, and Valorization of Natural Resources (LBCVNR), Faculty of Sciences Dhar El Mehraz, Sidi Mohamed Ben Abdallah University, Fez, Morocco; [5]Biotechnology Unit, Regional Center of Agricultural Research of Rabat, National Institute of Agricultural Research (INRA), Rabat, Morocco; [6]Plant Protection Laboratory, Regional Center of Agricultural Research of Oujda, National Institute of Agricultural Research, Oujda, Morocco; [7]Department of Plant, Food, and Environmental Sciences, Dalhousie University, Agricultural Campus, Truro, Nova Scotia, Canada

Abstract

This chapter explores the dynamic field of applied metabolomics in the context of plant biotechnology. Metabolomics, the systematic study of small molecules, offers valuable insights into the intricate biochemical processes within plant cells. The chapter begins by elucidating the fundamental principles of metabolomics and its applications, emphasizing its pivotal role in unraveling the complex metabolic networks governing plant growth, development, and the response to environmental stimuli. A comprehensive overview of cutting-edge analytical techniques, including mass spectrometry and nuclear magnetic resonance spectroscopy, is provided. The chapter further delves into case studies and success stories, showcasing how metabolomics has been instrumental in addressing key challenges in plant biotechnology, such as enhancing crop yield, stress tolerance, and the production of bioactive compounds. Moreover, it discusses the integration of metabolomics with other omics approaches, fostering a holistic understanding of plant systems. As plant biotechnology continues to evolve, the role of metabolomics as a powerful tool for precision agriculture and sustainable crop improvement will become increasingly evident. This chapter serves as a valuable resource for researchers, students, and practitioners seeking to harness the potential of metabolomics in advancing plant biotechnology.

Keywords: metabolomics, environmental stressors, plant biotechnology, NMR spectroscopy, LC-MS , GC-MS

* E-mail: rlahlali@enameknes.ac.ma

© CAB International 2025. *High-Throughput Plant Metabolomics* (Jen-Tsung Chen ed.)
DOI: 10.1079/9781800626911.0009

9.1 Overview of Metabolomics and its Significance in Plant Biotechnology

Plant biotechnology covers a wide range of techniques for modifying plant traits and developing novel crops. Metabolomics, the comprehensive analysis of the small-molecule metabolites of cells, tissues, or organisms, has become an effective methodology in this field (Wang *et al.*, 2010). To provide a detailed metabolic profile for metabolic activity, metabolomics offers unique insights into plant responses to genetic modifications, environmental perturbations, and biotic stresses (Razzaq *et al.*, 2019). It enables these small molecules to be systematically identified and measured, playing a role in the interactions between organisms' biochemistry, and helping to understand complex biological responses, interactions, and functions (Ziegler *et al.*, 2013; Hu *et al.*, 2020). Advances in biotechnology, particularly synthetic biology, are introducing new biological functions by linking functional gene expression through transcriptomes and proteomes to metabolomes (Moses *et al.*, 2013). Comparable to genomics, which describes all of an organism's genetic information, the metabolome is the term for the by-products of metabolism, which can range from hundreds to thousands of metabolites, depending on the size of the organism or biological system in question (van der Hooft *et al.*, 2020). Numerous instrumental analyses can verify the quantitative or qualitative changes in metabolites in living things, and these changes can be utilized to interpret different physiological pathways in cells or organisms (Oh *et al.*, 2023).

Metabolomics plays an important role in evaluating the metabolic consequences of genetic modifications in plants (Sharma *et al.*, 2021). Following plant transformation events, metabolomic profiling can reveal unintended metabolic disruptions or identify unexpected metabolic by-products (Zeiss *et al.*, 2023). This information is vital for optimizing transformation protocols and ensuring the metabolic stability of genetically modified plants (Singh *et al.*, 2021). Furthermore, metabolomics can be used to screen for plants with desired metabolic phenotypes, such as increased production of secondary metabolites or improved resistance to abiotic stressors (Hamany Djande *et al.*, 2020; Choudhury *et al.*, 2021). The metabolism of biological systems produces and modifies metabolites, which include lipids, peptides, nucleic acids, amino acids, organic acids, alkaloids, polyphenols, carbohydrates, vitamins, minerals, and chemical combinations created and absorbed by cells or organisms (Nikhitha Surendran *et al.*, 2020). The metabolic levels of these compounds change as a result of various environmental stimuli. Analysis of these environmental factors helps to characterize the organism and predict metabolic responses to specific factors related to metabolic pathways (Koelmel *et al.*, 2020). Plant metabolomics strives to decipher metabolic processes and mathematically formalize global metabolic flux within the framework of systems biology (Volkova *et al.*, 2020). To determine which genes contribute to the synthesis or degradation of certain metabolites, data for the metabolome may be combined with genomic and transcriptomic data via quantitative trait loci and metabolic genome-wide association studies in plant breeding (Razzaq *et al.*, 2022). As more plant genomes have been sequenced, the demand for functional genomics has increased. Furthermore, the development of multi-omics-based strategies, in which metabolomics plays a crucial role, has been promoted by the demand for functional genomics approaches (H.W. Zhang *et al.*, 2021; Yadav *et al.*, 2024). One of the most important fields is plant metabolomics for its impact on crop improvement, the identification of bioactive substances, plant development, and stress responses (Razzaq *et al.*, 2022). A more thorough understanding of the metabolic network and cellular functions is made possible by the integration of metabolomics with other omics data, such as genomic, transcriptomics, and proteomics, using systems biology techniques (Veenstra, 2021). Metabolites in plants are small, low-molecular-mass compounds found in both hydrophilic and lipophilic regions of the cell wall and membrane. These compounds can include acidic and basic ions, stable and oxidized structures, and polar compounds (Salam *et al.*, 2023). The levels of metabolites in plants can change in response to external factors such as drought, high temperatures, or circadian rhythms, as well as internal factors such as homeostasis, disease, growth, and ageing

(Bhattacharya, 2022). The proteome (the products of protein metabolism) and the genome (the genetic information) are closely related to the metabolome (the by-products of a cell's metabolic processes). This relationship is crucial in determining a cell's function (Nalbantoglu and Karadag, 2021). Moreover, it has led to a better understanding of the composition of food, the development of new nutritional markers, and the development of metabolic engineering techniques and the use of plants as bioreactor organisms (Marsafari *et al.*, 2020). Plant metabolites are becoming increasingly important for human health because they have been proven to protect against diseases such as high blood pressure, diabetes, and cancer, among others (Rana *et al.*, 2022). However, there is no separation method or single analytical platform that can define the entire metabolome (Jeppesen and Powers, 2023). Nevertheless, more sophisticated bioinformatics tools and extremely sensitive analytical methods are continuously being developed, and there are calls to use metabolomics to analyze the composition of genetically modified plants (Bedair and Glenn, 2020).

Plant metabolite research holds great value, and establishing experimental models is crucial for obtaining dependable metabolic profiling results (Wolfender *et al.*, 2015). When creating an experimental model, it is essential to carefully consider the selection and control of variables to maintain the integrity of statistical analysis (Martins *et al.*, 2018). The next stage after creating the experimental model is to prepare high-quality samples. For metabolomics based on mass spectrometry (MS), sample quality, including homogeneity, is crucial (Salem *et al.*, 2020a). To ensure the quality and dependability of sample preparation, it is necessary to choose samples that represent diverse plant-growing environments (regions), and the number of replicates must be taken into account to enhance the reliability of statistical analyses (Martins *et al.*, 2018). In the realm of plant metabolomics, it is essential to follow a systematic and progressive approach, ranging from cultivation to analysis, in order to provide an accurate and reliable interpretation of results (Raza, 2022). The detection of compounds in metabolomics can be optimized based on the choice of extraction methods and instrumental analysis techniques used for compound identification and quantification (López-Yerena *et al.*, 2021). It is critical to select appropriate extraction solvents and analytical equipment that allow for the properties of the sample (Płotka-Wasylka *et al.*, 2017). To achieve efficient and systematic metabolomics, the specifics of data processing, statistical analysis, analyte extraction techniques, and instrumental analysis must be taken into account (Yi *et al.*, 2016). To address a broad spectrum of metabolite classes, a combination of various extraction conditions and instrumentation settings might be needed (Wolfender *et al.*, 2015). Understanding the interplay between genes and metabolites is essential for the targeted manipulation of plant metabolism for the production of beneficial compounds (Lynch *et al.*, 2021). Metabolomics provides an understanding of metabolic fluxes and allows researchers to identify bottlenecks or limitations in a particular pathway (Vemuri and Aristidou, 2005; Volkova *et al.*, 2020). This information can then be used to strategically introduce new genes or modify existing ones to optimize the production of desired metabolites (Keasling, 2010). For instance, metabolomic profiling can be employed to guide the engineering of plants for the enhanced production of biofuels, pharmaceuticals, or nutraceuticals (Tatsis and O'Connor, 2016; Marchev *et al.*, 2021). Combining metabolomics with additional omics technologies, such as genomics and transcriptomics, holds immense promise for advancing plant biotechnology. This holistic approach, termed "multi-omics," allows researchers to build a comprehensive picture of gene–metabolite relationships and their regulation in response to various stimuli (W. Zhang *et al.*, 2021). By leveraging the power of metabolomics, plant biotechnologists can continue to develop novel crops with enhanced productivity and stress tolerance, and improved nutritional value (Kumar *et al.*, 2021).

The importance of metabolites and how they affect the growth of plants have been highlighted in numerous research studies, particularly concerning environmental factors. Metabolites carry out a variety of functions, including acting as fuel, providing structural support, serving as signals, enhancing or inhibiting enzyme activity, catalyzing chemical reactions (using cofactors), defending against pathogens, and facilitating nutrient uptake (Jha and Mohamed, 2022). These metabolites are divided into two groups: (i) primary metabolites including carbohydrates, vitamins, lactate, and hormones; and (ii) secondary metabolites, which are created during biosynthesis and comprise a wide range of physiologically active compounds (Bedair and Glenn, 2020). To attain comprehensive profiling of all metabolites and enhance the performance of qualitative detection for all metabolite classes, it is recommended to employ two types of analytical instruments in conjunction to supplement each other. Specifically, for secondary plant metabolite detection, liquid chromatography-mass spectrometry (LC-MS) is generally preferred as it provides a wide spectral range, enabling the identification of a diverse plant metabolite array (Halder *et al.*, 2024). Gas chromatography-mass spectrometry (GC-MS) is favored for detecting fatty acid-based compounds, as LC-MS has limitations for this (Yang *et al.*, 2024). GC-MS analysis entails a derivatization process, which allows the analysis of non-volatile compounds and ensures high data stability and reproducibility compared with high-performance liquid chromatography (HPLC) (Bubli *et al.*, 2021). Nevertheless, the chemical and structural characteristics of the raw material may change as a result of the derivatization process, making it more difficult to perform full metabolite profiling because only the fragments created by derivatization can be identified; other, non-derivatized compounds cannot (Zeki *et al.*, 2020). Metabolic profiling is done with analytical spectroscopic tools such as nuclear magnetic resonance (NMR). To minimize sample loss and identify metabolites during the extraction process, NMR is a crucial tool for directly assessing metabolites from plant material. According to Deborde *et al.* (2017), the methodology is used for the identification of unknown substances as it furnishes comprehensive details about metabolite structure. Details about analytical techniques in metabolomics can be found in section 9.3.

9.2 Applications of Metabolomics in Plant Biotechnology

9.2.1 Understanding plant metabolism

It was estimated that plants produce between 100,000 and 1 million different metabolites (Fang *et al.*, 2019). These substances are thought to be vital in supporting plants' ability to withstand and adapt to biotic and abiotic stresses (Anzano *et al.*, 2022). Only a portion of these metabolites are produced by a single plant species; estimates of the number of metabolites produced range from 5000 to tens of thousands (Fang *et al.*, 2019). Additionally, it has been demonstrated by recent studies that there is far more qualitative and quantitative metabolic variation within a species than previously believed (Carreno-Quintero *et al.*, 2013).

Plants are subject to various abiotic and biotic stresses, such as drought, cold, salinity, heat, heavy metals, diseases caused by pathogens, and insect infestations, which can impact their physiology and growth (Eid *et al.*, 2019; Anzano *et al.*, 2022). Fortunately, the emergence of metabolomics and other omics technologies has made it possible to conduct comprehensive analyses of the reactive processes that occur in plants under stress, which are triggered by changes in metabolites and gene expression (Fig. 9.1) (Anzano *et al.*, 2022). Recently, research into the variety of plant metabolism and the underlying molecular mechanisms by which the plant cell regulates its chemical composition has been made easier by the introduction of broad profiling techniques such as transcriptomics, metabolomics, and genomics (Fang *et al.*, 2019; Sharma *et al.*, 2021; Katam *et al.*, 2022). The first omics study to be used in plant sciences was genomics, which revealed the genome sequences of multiple plants, including rice and *Arabidopsis* (Okazaki and Saito, 2012). Several studies have elucidated the mechanisms involved in the role of various plant metabolites such as phytohormones, phytoalexins, and defensins in plant resistance. For instance, when beneficial microbes colonize host plants, a variety of beneficial metabolites (e.g. coumestrol, catalpol, gallic acid, daidzein, tomatidin, and myricetin) accumulate in the plant (Adeniji *et al.*, 2020). This leads to a number of benefits that increase resistance to drought, salt and the proliferation of pathogens. For example, jasmonic acid stimulates defense and metabolic processes to increase resilience in *Arabidopsis jaz* mutants and is linked to light regulation processes, HY5 activity, and photomorphogenesis (Nguyen *et al.*, 2022). Basic metabolism, endogenous phytohormone regulation, organogenesis, cell compartmentalization, senescence, fruit and flower development, and

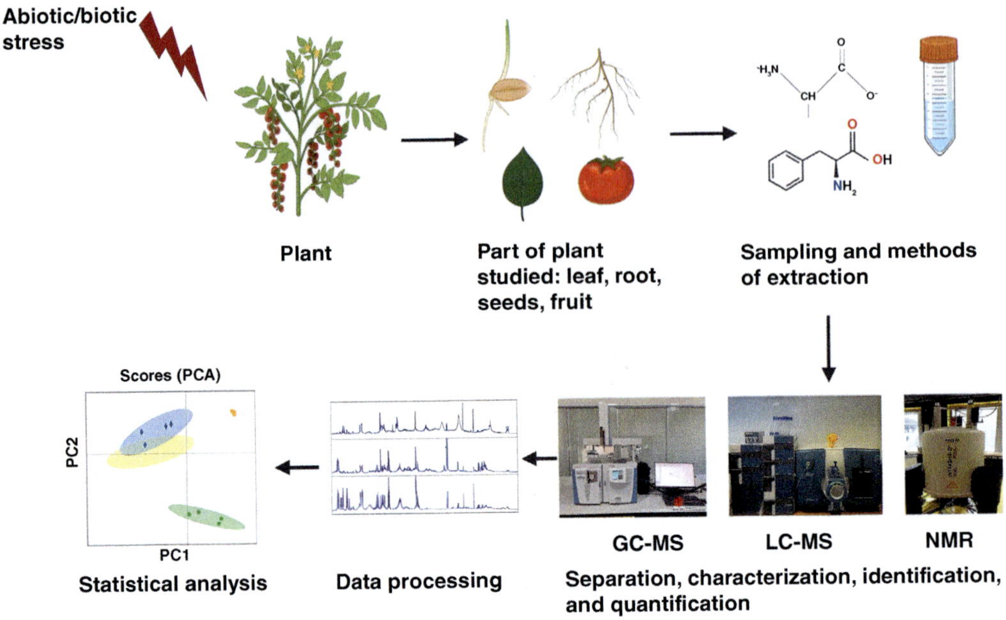

Fig. 9.1. Detection of metabolites released in response to stressors using various omics techniques. GC-MS, gas chromatography–mass spectrometry; LC-MS, liquid chromatography–mass spectrometry; NMR, nuclear magnetic resonance.

the generation of secondary metabolites as a defense mechanism are just some of the numerous biological processes in which polyamines are involved (Nandy *et al.*, 2022). Induced systemic resistance and systemic acquired resistance, the two traditional plant defense mechanisms against plant pathogen attack, have been discussed elsewhere (Reglinski *et al.*, 2023). Phytotoxins, which are produced as virulent factors by certain phytopathogens (e.g. *Rhizobium* spp. and *Pseudomonas syringae*), reduce plant nodulation and biomass by increasing ethylene production, which upsets the metabolism of the host plant (Lenard *et al.*, 2020). Furthermore, rhizobitoxin and 1-aminocyclopropane-1-carboxylic acid (ACC) deaminases, which block the synthesis and activity of the ethylene precursor ACC and the ACC synthase enzyme, respectively, are produced by a variety of advantageous microbes and represent a direct biocontrol mechanism (Jha *et al.*, 2021). Many beneficial microorganisms possess the ability to produce lytic metabolites, such as chitinases, cellulases, proteases, and 1,3-glucanases, which can break down the cell walls of pathogenic fungi (Jadhav *et al.*, 2017). In contrast, some microbes produce siderophores, which inhibit the growth and colonization of phytopathogens by preventing them from acquiring iron (Gu *et al.*, 2020). This indirect biocontrol mechanism can be an effective way to control plant pathogens (Adeniji *et al.*, 2020).

9.2.2 Investigating responses to environmental stresses

Metabolomics represents a promising avenue for deciphering the mechanisms underlying plant tolerance to both abiotic and biotic stresses (Table 9.1). Recent studies based on metabolomics and its relationship with environmental stresses have been carried out on several plants (such as *Arabidopsis*, wheat, and tomatoes) characterizing different metabolites belonging to different parts of the plant using spectrometric methods (Table 9.1) (Guo *et al.*, 2020; Sudiro *et al.*, 2022; Wang *et al.*, 2024). By analyzing metabolic profiles, researchers have identified diverse metabolites in plants, comprising amino acids, carbohydrates, phenols, polyamines, and terpenes. Among these, primary metabolites

Table 9.1. Current techniques for metabolite profiling in conventional or genetically modified crop varieties under various stressors.

Plant	Abiotic/biotic stress	Analytical technique	Specific tissue	Principle metabolites produced/ mechanism of action	Data analysis	Reference
Wheat	Drought stress	UPLC-MS	Wheat seedlings	Phenolics, thymine, a pyrimidine	Compound Discoverer 3.0, PCA, PLS-DA, R program	Guo *et al*. (2020)
Morus alba L.	Drought stress	LC-MS	Leaves	Prenol lipids	Progenesis QI v2.3 software, PCA, PLS-DA	Ackah *et al*. (2021)
Solanum lycopersicum	Drought stress	UHPLC/QTOF-MS	Tomato plant	Xanthines, chlorophylls	Mass Profiler Professional B.12.06, ANOVA, XLSTAT	Sudiro *et al*. (2022)
Wheat	Drought stress	GC-MS	Wheat plant	Palmitic acid, stearic acid, oleic acid, linoleic acid, linolenic acid	One-way ANOVA, SPSS	Ullah *et al*. (2022)
Medicago sativa L.	Drought stress	LC-MS	Roots	Sucrose, gibberellin A4, 6-gingerol, salicylic acid, abscisic acid, L-phenylalanine, *trans*-cinnamic acid, L-tyrosine, indole-3- acetic acid, nicotinic acid, succinic acid	MetaX, PCA, PLS-DA	Wang *et al*. (2024)
Tomato	Salt stress	GC-MS	Leaves	Soluble sugars, amino acids	XLStat software v.9.0, PCA,	Savy *et al*. (2022)
Lonicerae japonicae Flos	Salt stress	HPLC-triple TOF-MS/MS	Whole plant	Phenylpropanoid, monoterpenoid, alkaloids	MarkerView 1.2.1 software, PCA, OPLS-DA	Cai *et al*. (2020)
Camellia sinensis var. *sinensis*	Cold stress	UPLC-MS and GC-MS	Whole plant	Ascorbic acid, amino acids, intermediates of the TCA cycle, carbohydrates, quercetin, kaempferol, catechin, myricetin		Wang *et al*. (2022)
Kiwifruit	Cold stress	UPLC-ESI-MS/MS	Shoots	Lipids, phenolic acids, flavonoids, amino acids, organic acids, nucleotides, alkaloids, coumarins, terpenoids, tannins, lignans	PCA, OPLS-DA, MetaboAnalystR	Sun *et al*. (2021)

Continued

Table 9.1. Continued.

Plant	Abiotic/biotic stress	Analytical technique	Specific tissue	Principle metabolites produced/ mechanism of action	Data analysis	Reference
Arabidopsis thaliana	Heat stress	GC-MS	Leaves	The glycolysis pathway was enhanced, and hydrogen peroxide and antioxidant enzymes activities were increased	PCA, OPLS-DA,	Wang *et al.* (2020)
A. thaliana	Heat stress	UPLC-QTOF/MS	Whole plant	Organic acids/derivates, amines, carbohydrates, TCA cycle, indole 3-acetate, peptides, amino acids, phenylpropanoid	MetDNA2, LipidView Software, R analysis	Qian *et al.* (2023)
Wheat	Cadmium	UPLC-MS	Root	Phenylalanine, alanine, glutamate, aspartate, isoquinoline alkaloids, proline, arginine, arginine, glyoxylate, dicarboxylate, phytochelatins	SPSS 20.0, Origin 8.5, ANOVA	Lu *et al.* (2021)
Maize	*Puccinia sorghi*	UHPLC-HRMS	Maize plant	Phenylpropanoid, terpenoids, salicylic acids	MetaboAnalyst 4.0	Kim *et al.* (2021)
Soybean	*Phakopsora pachyrhizi*	UHPLC-MS/MS	Leaves	Lipids, flavonoids, isoflavonoids, fatty acyls, carboxylic acids terpenes, and others	MetaboAnalyst 3.0, PLS-DA analysis	Silva *et al.* (2021)

ANOVA, analysis of variance; GC-MS, gas chromatography–mass spectrometry; HPLC-triple TOF-MS/MS, high-performance liquid chromatography–triple time-of-flight–tandem mass spectrometry; LC-MS, liquid chromatography–mass spectrometry; OPLS-DA, orthogonal partial least squares discriminant analysis; PCA, principal component analysis; PLS-DA, partial least squares discriminant analysis; TCA, tricarboxylic acid; UHPLC/QTOF-MS, ultra-high-performance liquid chromatography–quadrupole time-of-flight–mass spectrometry; UHPLC-HRMS, ultra-high-performance liquid chromatography–high-resolution mass spectrometry; UHPLC-MS/MS, ultra-high-performance liquid chromatography–tandem mass spectrometry; UPLC-ESI-MS/MS, ultra-performance liquid chromatography–electrospray ionization–tandem mass spectrometry; UPLC-MS, ultra-performance liquid chromatography–mass spectrometry; UPLC-QTOF-MS, ultra-performance liquid chromatography–quadrupole time-of-flight–mass spectrometry.

such as amino acids, carbohydrates, and polyamine lipids play crucial roles as osmolytes and osmoprotectors, safeguarding plants against environmental stressors. In contrast, secondary metabolites derived from plants, such as phenolics, nitrogen compounds (alkaloids), and terpenoids, do not directly support development and growth. However, these secondary metabolites may serve as essential defenses, protecting plants from biotic threats posed by insects, herbivores, and pathogens (Salam *et al.*, 2023).

Environmental stress disrupts plant homeostasis, triggering a cascade of metabolic responses to ensure survival and adaptation (Singh *et al.*, 2020). The thorough examination of a plant's small-molecule composition offers a powerful tool to investigate these metabolic dynamics (Deborde *et al.*, 2017). Researchers can determine important metabolites and pathways involved in stress signaling, defense mechanisms, and metabolic reprogramming by comparing the metabolic profiles of stressed and non-stressed plants. With this knowledge, crops that can withstand stress can be developed, and agricultural practices can be optimized for a more sustainable future (Meena *et al.*, 2017).

Environmental stressors often lead to redox adjustments or changes, which are closely linked to increased synthesis and accumulation of reactive oxygen species (ROS), which can cause oxidative damage to critical organelles and structures within cells (Allwood *et al.*, 2021). ROS-triggered signaling events occur in various developmental processes, especially at low-stress intensity or in the early stages of development (Ali *et al.*, 2023). However, with increasing stress intensity and duration, oxidative stress significantly alters the overall cellular redox state, potentially affecting essential metabolic processes that regulate normal plant growth (García-Caparrós *et al.*, 2021; Borbély *et al.*, 2022). These conditions generally occur under conditions of extreme temperatures (heat or cold), salinity, water stress, metal accumulation, and attack by pathogens or microbes. Therefore, to counteract these negative effects and maintain proper cellular functions, plants accumulate and synthesize a range of metabolites that directly or indirectly contribute to maintaining protection and growth under extreme environmental conditions (García-Caparrós *et al.*, 2021). Cells constantly produce ROS in compartments such as mitochondria (particularly complexes I and III), chloroplasts, and peroxisomes (Corpas *et al.*, 2015). These ROS include singlet oxygen, hydroxyl radical, superoxide, and hydrogen peroxide (H_2O_2). They are formed mainly by the conversion of molecular oxygen. Importantly, ROS can significantly disrupt the normal functioning of many cellular pathways (Bhattacharjee, 2019). Within the chloroplast, photosystem II is the main contributor to the generation of singlet oxygen (Dmitrieva *et al.*, 2020). Superoxide and H_2O_2 are formed differently (Duong *et al.*, 2021). They can be formed during the electron transport chain in chloroplasts and mitochondria or through enzyme activity in peroxisomes (Phua *et al.*, 2021). For example, glycolate oxidase in peroxisomes and NADPH oxidases and peroxidases at the cell surface/apoplast contribute to their production (Noctor *et al.*, 2015). The highly reactive hydroxyl radical, even more harmful than the others, is formed by the breakdown of H_2O_2 by iron or similar metals (Liu *et al.*, 2021). Interestingly, research suggests that controlled ROS production can be exploited for signaling purposes, potentially leading to strategies to improve stress tolerance and normal plant growth processes such as cell expansion (Khoshru *et al.*, 2020).

Primary metabolites, a class of compounds with demonstrably vital physiological roles in organisms, show an impressive level of conservation among different plant species. This ubiquity reflects their direct involvement in fundamental processes such as growth, development, and reproduction (Salam *et al.*, 2023). In contrast, secondary metabolites display a high degree of functional specificity (Jacoby *et al.*, 2021). Unlike their primary counterparts, they are not directly engaged in core plant growth, development, or reproduction. This functional specialization often translates to a more restricted taxonomic distribution, with some secondary metabolites being unique to specific plant lineages or even individual species (Mérillon and Ramawat, 2020). Secondary metabolites have a remarkable degree of functional diversification through plant lineages (Y. Zhang *et al.*, 2021). This diversification is driven by selective pressures, with these compounds often playing a well-defined ecological role. Examples include defense against pathogen infection, enhanced tolerance to abiotic stresses, and mitigation of ultraviolet (UV) radiation damage (Yeshi *et al.*, 2022). Notably, secondary metabolites contribute to a diverse array of plant cells and tissues, related to biochemical and biophysical processes (Wink, 2010). However, unlike primary metabolites with their conserved roles in fundamental physiological processes such as photosynthesis, respiration, and nutrient transport, secondary metabolites lack such a universal function across the plant kingdom (Salam *et al.*, 2023).

9.3 Analytical Techniques in Metabolomics

Plants represent a major challenge for metabolomics because they exhibit high chemical diversity and include metabolites in a wide range of concentrations. It is estimated that plants contain over 200,000 different metabolites (Alseekh *et al.*, 2021). Currently, the most comprehensive methods are capable of identifying metabolites in various species, including approximately 700 of the 3700 metabolites in *Escherichia coli* (Guo et *al.*, 2012; Sajed *et al.*, 2016), 500 of the 2680 expected metabolites in yeast (Ramirez-Gaona *et al.*, 2017; Hautbergue *et al.*, 2018), 8000 of the 114,100 expected metabolites in humans (Wishart *et al.*, 2018), and only 14,000 of over 400,000 anticipated plant-related metabolites (Saito and Matsuda, 2010). To provide a complete metabolic profile and improve qualitative and quantitative detection, a combination of complementary analytical tools is essential (Yin *et al.*, 2024).

9.3.1 Mass spectrometry

MS is highly sensitive and capable of detecting, quantifying, and elucidating the structures of many metabolites in a single analysis. The accuracy and sensitivity of MS depend on experimental conditions and instrument settings and are influenced by factors such as metabolite extraction, separation, ionization (including ion suppression), and detection methods (Allwood and Goodacre, 2010; Alseekh *et al.*, 2021). Due to the intricacies of biological samples, it is important to separate metabolites before MS analysis. Of the three separation techniques, GC and LC methods are widely used. GC-MS provides better metabolite separation than LC and generally avoids ion suppression due to the use of the gas phase and ionization properties of MS. Therefore, unlike LC, GC often requires chemical derivatization of the metabolites prior to GC-MS analysis (González-Domínguez *et al.*, 2020; Benkeblia, 2023).

Due to its ability to identify a wider variety of intact metabolites without requiring chemical modification, LC-MS is commonly used in metabolomics. Traditional reversed-phase chromatography is used to separate non-polar or slightly polar molecules (Roberts *et al.*, 2012), while hydrophilic interaction liquid chromatography (HILIC) has emerged as the preferred method for strongly or moderately polar metabolites (Bajad *et al.*, 2006). One of the most important phases in MS-based metabolite measurements is ionization. Mass spectrometers ionize molecules by converting neutral particles into ions and then analyzing their mass:charge (m/z) ratio. The ionization process takes place inside the instrument, where the molecules are excited and then exposed to electric and magnetic fields (Chen and Lu, 2021). The microchannel plate detector then separates these ions according to their m/z value and creates a mass spectrum diagram that reveals specific element properties. The degree of ionization has a direct impact on the ability to detect and quantify metabolites. In metabolomics, the two most commonly used ionization methods are electrospray ionization (ESI) and electron impact (EI) ionization (Zhu *et al.*, 2021). ESI is preferred in HPLC-MS due to its ability to ionize molecules in the liquid phase and is suitable for a wide range of molecular sizes, from small molecules to peptides and proteins. ESI is considered a gentle ionization method that minimizes the fragmentation of molecular ions. However, its efficiency can be affected by the presence of salts, which limits chromatographic methods to volatile buffers such as formate or ammonium acetate, and can lead to ion suppression when metabolites compete for ionization. In contrast, EI is the method of choice for GC-MS analysis as it is known for its hard ionization, which induces fragmentation of metabolites (Ranjan Maji *et al.*, 2023). This fragmentation facilitates detection with minimal interference from co-produced compounds. Recently, atmospheric pressure chemical ionization and atmospheric pressure photoionization have gained popularity (Kamal *et al.*, 2016; Sedláčková *et al.*, 2022). Typically, the detection of metabolites with optimal resolution and sensitivity is preferred. However, balancing these attributes within a single metabolite detection mode is challenging because higher sensitivity often comes with lower resolution and vice versa. There are different options, such as single (MS) or tandem (MS/MS) mass analyzers, each offering different performance in terms of sensitivity and resolution (Agarwal and Goyal, 2017). Many make use of lower-resolution tools such as single quadrupoles and ion traps (both linear and three-dimensional quadrupoles). Every mass analyzer has a unique set of benefits and drawbacks. The objective of metabolomics projects, throughput, and instrument costs all play a role in the choice of a particular mass spectrometer platform (Gowda and Djukovic, 2014). Metabolomics relies heavily on accurate and sensitive detection methods, which are challenging due to the inherent trade-off between sensitivity and resolution. MS comes in a variety of configurations, each with special performance qualities in terms of sensitivity and resolution. Quadrupole (Q), Fourier-transform–ion cyclotron resonance (FT-ICR), quadrupole ion trap, time-of-flight (TOF), linear ion trap, and Orbitrap are examples of mass analyzers with a single configuration. While the TOF, FT-ICR, and Orbitrap instruments feature high mass resolution, the quadrupole and ion trap analyzers offer limited resolution but are more sensitive. Mass analyzers arranged in a tandem configuration include triple quadrupole ion trap (QTrap), quadrupole time-of-flight (Q-TOF), triple quadrupole (TQ), and linear-quadrupole ion trap-Orbitrap (LTQ-Orbitrap) (Bijukumar *et al.*, 2018). While LTQ-Orbitrap, Q-TOF, and FT-ICR analyzers are better suited for metabolite identification and global profiling (including isotopomer analysis) due to their increased mass resolving capacity, TQ and QTrap analyzers are the most frequently hyphenated to LC and used in targeted metabolic studies because of their high selectivity and sensitivity (Shan *et al.*, 2022). Single quadrupoles or TOFs are the most common mass analyzers used with GC; however, some more modern GC-MS instruments are now fitted with TQ or Q-TOF mass spectrometers. Table 9.2 outlines the various types of mass analyzers used in metabolomics, highlighting their sensitivity, resolution abilities, and typical applications in both GC-MS and LC-MS contexts (Gowda and Djukovic, 2014).

Plants use defense mechanisms to thrive under various stresses in their environment. The synthesis of specific metabolites is a well-known defense response that is crucial for their adaptation (Jing *et al.*, 2023). By studying these defense

Table 9.2. Various mass analyzers employed in metabolomics: their sensitivity, resolution abilities, and typical applications.

Mass analyzer type	Sensitivity	Resolution	Typical applications
Quadrupole	High	Limited	LC-MS for targeted metabolic studies
Linear ion trap	High	Limited	LC-MS for targeted metabolic studies
Quadrupole ion trap	High	Limited	LC-MS for targeted metabolic studies
Time-of-flight	Moderate to high	High	GC-MS, and LC-MS for metabolite identification and global profiling
Fourier-transform ion cyclotron resonance	High	Very high	LC-MS for metabolite identification and global profiling
Orbitrap	High	Very high	LC-MS for metabolite identification and global profiling
Triple quadrupole ion trap	High	Moderate to high	LC-MS for targeted metabolic studies
Triple quadrupole	High	Moderate to high	LC-MS for targeted metabolic studies
Quadrupole time-of-flight	High	High	LC-MS for metabolite identification and global profiling
Linear-quadrupole ion trap-Orbitrap	High	Very high	LC-MS for metabolite identification and global profiling

GC-MS, gas chromatography–mass spectrometry; LC-MS, liquid chromatography–mass spectrometry.

responses, metabolomics offers a holistic understanding of plant strategies against stress. While sunlight is vital for most plants, UV light poses a danger because it causes oxidative damage. For example, metabolomic analysis of wild-type *Arabidopsis* and mutants under UV-B stress revealed early metabolic responses that mainly affected primary metabolites, priming cells for later production of UV-B-absorbing secondary metabolites such as sinapoyl malate and flavonoids (Kusano *et al.*, 2011). Recent research has extensively investigated metabolic changes in response to environmental stress using metabolomics (He *et al.*, 2023; Ibrahim *et al.*, 2023; Yemelyanov *et al.*, 2023). Biomarker discovery through metabolomics is critical to understanding plant defense mechanisms and facilitating disease diagnosis, drug development, early disease detection, and improved pest control (Weitzel, 2011).

The integration of chromatographic separation techniques with MS has transformed analytical chemistry and driven advances in metabolomics, proteomics, and environmental analysis. Chromatography separates complex mixtures into individual components based on their chemical properties such as polarity, size, and affinity, and MS complements this process by precisely identifying and quantifying these components through measurements of their relative abundance and m/z ratio. By simplifying the mass spectra and the matrix effect, chromatographic separation in conjunction with MS increases the number of compounds that can be detected. Many chromatographic methods can be coupled with MS. For example, GC is employed in metabolite profiling, and GC-MS is the method most frequently used in plant metabolomics studies to date. GC is used to separate volatile polar metabolites after they have undergone derivatization. Using a coated fused-silica capillary and a carrier gas, capillary GC transfers analytes. To proceed through the capillary in GC-MS, the analyte needs to be evaporated. Thus, analytes need to be volatile or susceptible to chemical derivatization. Particularly well suited for GC-MS analysis are two sample types: essential oils and terpenoids. More polar metabolites or metabolites with highly polar functional groups can also be examined with the right derivatization, such as amines, sugars, fatty acids, organic acids, and amino acids (Zhang and Li, 2007).

LC separates samples based on mobile and stationary phase interactions. This can be based on polarities, meaning that a component of the sample that has a different polarity from the mobile phase will travel through the chromatography column more quickly. The sample is broken down into bands and individual components that can be further analyzed by MS (Rappold, 2022). Untargeted LC-MS-based metabolomics techniques have been helpful in understanding the intricacies of plant metabolism. These techniques enable identification of the metabolic components that contribute primarily to significant phenotypic and physiological traits, as well as how biosynthesis of these components is regulated (Hamany Djande *et al.*, 2020).

In contrast to GC-MS, HPLC-MS is a non-destructive method and allows analysis of thermosensitive oligosaccharides, while the use of ^{13}C-labeled substrates allows better discrimination of fragmentation patterns (Harrison *et al.*, 2011). LC-MS is characterized by the detection of secondary plant metabolites in a wide spectral range. Conversely, GC-MS is ideal for detecting fatty acid-based compounds and overcomes the limitations of LC-MS. GC-MS uses derivatization to analyze non-volatile compounds, ensuring robust reproducibility and data stability compared with HPLC methods (Oh *et al.*, 2023).

Some of the methods that are currently used to profile metabolites in conventional or genetically modified crop varieties under various environmental conditions are given in Table 9.3.

9.3.2 NMR spectroscopy

Another technology that is frequently used in metabolic studies is NMR spectroscopy, which allows identification of metabolites that would be impossible to identify by MS analysis and can also be used to determine the atomic state of compounds (Allwood and Goodacre, 2010). When metabolites are present in extracts or *in vivo*, NMR spectroscopy can give precise information about their identities and amounts (Calvo, 2015). The potential of NMR spectroscopy in metabolomics is demonstrated by the fact that most metabolomic analyses using NMR do not require any pretreatment, such as column chromatography and derivatization. However, as chemical shifts can vary depending on

Table 9.3. Some of the methods currently used to profile metabolites in conventional or genetically modified crop varieties under various environmental conditions.

Analytical tool	Test study	Crop	Detected metabolites	Reference
LC-TQ-MS	Drought	*Medicago sativa*; *M. truncatula*	Flavonoid, carbohydrate, abscisic acid, and proline	Echeverria *et al.* (2021)
GC-MS	Cold, salt, high	*Physcomitrella patens*	Sugar, amino acid, and organic acid	Arif *et al.* (2018)
GC-TOF-MS; LC-MS/MS	Cold, high light	*Arabidopsis thaliana*	Sugar, gluconic acid, organic acid hexose, amino acid, and citrate	Küstner *et al.* (2019)
LC-MS/MS	Pathogen resistance	Maize	Flavonoid and benzoxazinoid	Zhou *et al.* (2019)
LC-MS	Salt tolerance	Maize	Benzoxazinoids, amino acids, lipids terpenoids, and flavonoids	Liang *et al.* (2021)
LC-MS	Seed oil-related traits	Soybean	Asparagine, arginine, aspartic acid, alanine, and daidzein	J. Y. Liu *et al.* (2020)
LC/UPLC-MS/MS; GC-MS	Drought	*A. thaliana*	TAG, SQDG, DGDG, PC, PS, MGDG, PE, and PI	Salem *et al.* (2020b)
GC-MS; LC	Drought	*Oryza sativa*	Citric, aconitic acids, carbohydrates, benzoic acid, norvaline, GABA, benzoic acid, and proline	Auler *et al.* (2021)
LC-MS/MS	Flavonoid pathways	Wheat	Flavonoids	J. Chen *et al.* (2020)

GC-MS, gas chromatography–mass spectrometry; GC-TOF-MS, gas chromatography–time-of-flight–mass spectrometry; LC, liquid chromatography; LC-MS/MS, liquid chromatography–tandem mass spectrometry; LC-TQ-MS, liquid chromatography–triple quadrupole–mass spectrometry; UPLC-MS/MS, ultra-performance liquid chromatography–tandem mass spectrometry.

sample pH, special attention should be paid to this issue. The pH of sample solutions is frequently stabilized using buffered solutions for metabolic analysis using NMR. In addition to calibration accuracy, it is equally important to recognize the non-destructive nature of NMR, which contributes significantly to the reproducibility of experiments (Hong *et al.*, 2009; Cui *et al.*, 2021). NMR samples can be measured, stored, and tested repeatedly to ensure that previous results can be verified or corroborated. Researchers can also use NMR to answer questions left unanswered by previous methods on the same sample. This versatility is facilitated by the fact that prolonged storage of biological samples at −80°C and freeze/thaw cycles generally have little or no impact on the NMR measurements obtained (Saude and Sykes, 2007; Pinto *et al.*, 2014). In contrast to many other metabolomics technologies, NMR is not restricted to analyzing biofluids or tissue extracts. It is exceptionally well suited for examining intact tissues, organs, and other solid or semi-solid samples, facilitated by techniques such as solid-state NMR and magic-angle sample rotation (Hong *et al.*, 2009). One-dimensional ^{1}H-NMR spectra are particularly valuable in metabolomic studies due to their high degree of automation, reliability, and speed; the collection time for a single spectrum can be as short as a few minutes. Modern NMR instruments allow samples to be loaded and removed continuously using robotized sample changers, which can operate without interruption for extended periods ranging from several days to several weeks (Singh *et al.*, 2023). Moreover, researchers can acquire NMR spectra for various nuclei (e.g. ^{1}H, ^{13}C, ^{15}N, and ^{31}P) individually or concurrently, enabling the examination of diverse classes of metabolites (such as nitrogen- or phosphorus-containing compounds). Additionally, multi-dimensional NMR techniques allow the measurement of correlations between two or even three different nuclei (Mahrous and Farag, 2015).

9.3.3 Comparison of analytical techniques

Currently, LC-MS, GC-MS, and NMR spectroscopy are the three main methods of analysis in metabolomics. Although LC-MS and GC-MS techniques dominate the field, accounting for over 80% of published metabolomics research, there continues to be considerable interest in the use of NMR-based approaches. The growing adoption of NMR in metabolomics can be attributed to the distinct advantages it offers over LC-MS and GC-MS platforms (Emwas *et al.*, 2019). Specifically, NMR spectroscopy is non-destructive, unbiased, easily quantifiable, and generally requires minimal or no chromatographic separation, sample preparation, or chemical derivatization. It facilitates routine discovery of new compounds and allows high levels of automation and reproducibility, enabling effective large-scale metabolomics research (Wang *et al.*, 2011; van Erven *et al.*, 2019) that is often difficult to achieve with LC-MS or GC-MS. In addition, NMR excels in the detection and characterization of compounds such as sugars, organic acids, alcohols, polyols, and highly polar substances that can pose difficulties for LC-MS analysis. NMR also supports metabolite imaging and metabolic analysis of live samples through magnetic resonance spectroscopy and magnetic resonance imaging (Calvo, 2015; Yoon *et al.*, 2016). In contrast, LC-MS and GC-MS methods, due to their inherently destructive nature, are unsuitable for analyzing live samples. Therefore, NMR spectroscopy is ideal for real-time metabolite profiling of living cells, enabling real-time metabolic flux analysis (Motta *et al.*, 2010; Jeong *et al.*, 2017). Moreover, NMR provides unparalleled chemical insight, allowing examination of intact molecules at an atomic level and offering visibility of not only ^{1}H atoms but also various other types of atoms (^{13}C, ^{15}N) or biologically significant groups such as phosphate atoms (^{31}P) (Batool *et al.*, 2017). Additionally, NMR spectroscopy enables the exploration of distinctive metabolite classes, particularly protein-bound metabolites such as lipoprotein particles, and the measurement of specific inorganic metabolites or ions (e.g. metal ions and H^+ ions via pH) that are not accessible with LC-MS or GC-MS (Holmes *et al.*, 2018). However, NMR spectroscopy also has several drawbacks, the main one being sensitivity. Unlike LC-MS and GC-MS, NMR is generally 10–100 times less sensitive. Consequently, a metabolomics study using NMR typically identifies 50–200 metabolites with concentrations exceeding 1 μM, while LC-MS studies often identify over 1000 metabolites with concentrations ranging from 10 to 100 nM or more. Table 9.4 outlines the strengths and weaknesses of NMR spectroscopy versus MS, particularly in the context of metabolomics.

9.4 Case Studies and Success Stories

9.4.1 Crop yield enhancement

Metabolomics has emerged as a promising approach to understanding the tolerance of plant species to abiotic stress. Recently, it has been utilized to identify unique metabolites throughout plant life cycles. Both biotic and abiotic stresses play significant roles in reducing crop yields. Plants respond to these stresses through similar mechanisms, yet they induce variations in biochemical and physiological processes. During abiotic stress initiation, plants synthesize phytohormones to enhance stress resilience (Zaid *et al.*, 2021). Oxidative stress disrupts stomatal conductance and activates multiple signaling pathways (Razi and Muneer, 2021). Overall, the gene expression profile reveals the precise composition of metabolites in a given plant species. Activation of specific metabolic networks leads to the synthesis of novel bioactive agents (Rinschen *et al.*, 2019). For example, durum wheat (*Triticum durum*) accumulated antioxidant and drought tolerance-related sugars in response to water stress, especially in the ears. Furthermore, it was observed that specific metabolites related to cell walls, respiration, and protection were linked to both genotypic performance and yield stability. Specifically, the metabolomes of the leaf and ear bract independently predicted grain yield (Vergara-Diaz *et al.*, 2020). This work emphasizes the ear's critical role in wheat grain filling, particularly under stressful conditions, and offers insightful data to investigate novel approaches

Table 9.4. Comparative analysis of advantages and limitations of nuclear magnetic resonance (NMR) spectroscopy versus mass spectrometry (MS) in metabolomics.

Characteristic	NMR	MS
Reproducibility	NMR spectroscopy offers high reproducibility	MS data are generally less reproducible compared with NMR
Sensitivity	Inherently low; can be enhanced with multiple scans, higher field strengths, cryo-cooling, microprobes, and hyperpolarization	High sensitivity; allows detection of metabolites at nanomolar concentrations
Selectivity	Generally non-selective; peak overlaps pose challenges	Selective, particularly when coupled with chromatography methods (e.g. liquid- or gas-phase separation)
Sample measurement	Rapid measurement using one-dimensional ^{1}H-NMR spectroscopy; detects all metabolites above detectable levels in one scan	Different ionization methods required for maximizing metabolite detection
Sample preparation	Minimal preparation; transfer to NMR tube with deuterated locking solvent; can be automated	More rigorous; involves chromatography and sometimes sample derivatization for gas chromatography-MS
Sample recovery	Non-destructive; multiple analyses possible on the same sample; sample can be stored long term	Destructive; sample cannot be recovered, but only requires small amounts
Quantitative analysis	Inherently quantitative; signal intensity correlates directly with metabolite concentrations and number of nuclei	MS line intensity not always proportional to metabolite concentrations due to ionization efficiency variations

to enhancing wheat productivity, such as prospective biomarkers for yield prediction. Nitrogen application before drought may play a crucial role in enhancing rice yield and stress resistance, while also presenting a novel ecological perspective on nitrogen regulation in rice (Du *et al.*, 2020). In a recent study, pretreatment with chitosan of wheat susceptible to *Fusarium graminearum* resulted in increases in average kernel weight per ear, number of kernels per ear, and 1000-kernel weight, as well as activation of the jasmonic acid pathway, compared with the control group, which was sprayed with water (Deshaies *et al.*, 2022).

9.4.2 Stress tolerance improvement

Metabolomics has emerged as a valuable tool for exploring plant stress responses. This approach facilitates the investigation of stress-induced metabolite biosynthesis, providing insights into plant adaptation mechanisms under adverse environmental conditions (Raza, 2022). Metabolomic studies can not only reveal novel stress-responsive compounds and metabolic pathways but also enhance our understanding of existing pathways (Ahanger *et al.*, 2018). In the last decade, numerous studies have used metabolomics to analyze changes in metabolite concentration under various biotic and abiotic stresses.

Metabolomic approaches have also been used to study the drought-stress response in cereal crops, including sorghum, rice, wheat, and maize (Kaur *et al.*, 2021), and in vegetable crops such as tomato (Sudiro *et al.*, 2022). There have been reports of variations in other metabolites and phytohormones in the leaves of tomato plants stressed by salinity (Savy *et al.*, 2022). *Arabidopsis thaliana* is often affected by heat stress, the metabolic response to which may involve several pathways such as glycolysis, antioxidant enzymes, ROS formation, and others (Wang *et al.*, 2020). Under cold stress, kiwi fruit shoots underwent remarkable variation by activating several primary and secondary metabolite synthesis pathways (Sun *et al.*, 2021). The identification of metabolites associated with specific loci also offers a powerful tool for conservation efforts. By linking metabolite profiles to genetic variation (metabolomic quantitative trait loci (mQTL) analysis), researchers can gain a deeper understanding of the evolutionary strategies employed by different plant lineages (Omena-Garcia *et al.*, 2019). This approach allows the identification of conserved metabolic pathways and genes that contribute to plant adaptation and survival in diverse environments. Furthermore, mQTL analysis can reveal novel connections between these conserved pathways and agronomically important traits (Sethi *et al.*, 2023). Comparative mQTL studies, such as one conducted between rice and maize, highlight the potential for this approach to identify shared genetic mechanisms underlying adaptation across plant species (Raj and Nadarajah, 2022). Ultimately, this information can be harnessed to develop robust conservation strategies and improve breeding programs aimed at preserving and enhancing plant diversity (Kaur *et al.*, 2021).

9.4.3 Production of bioactive compounds

The investigation of bioactive compound production in metabolomic plants represents a burgeoning field at the interface of plant biology, chemistry, and health sciences (Marchev *et al.*, 2021; Mishra *et al.*, 2022; Oh *et al.*, 2023). Metabolomic plants are those specifically chosen for the comprehensive analysis of their metabolic profile, encompassing the entire spectrum of small molecules they generate (Mattoli *et al.*, 2023). This intricate analysis offers valuable insights into the plant's biochemical processes and its potential for the synthesis of bioactive compounds (da Silva *et al.*, 2022). Bioactive compounds are naturally occurring plant-derived molecules with demonstrably significant effects on human health (Marchev *et al.*, 2021). This diverse array of molecules includes phenolic compounds, terpenes, alkaloids, and glycosides (Desmet *et al.*, 2021). For example, *Weinmannia trichosperma* Cav. is a medicinal plant that is used to treat wound healing, inflammation, and chronic diarrhea. Three flavonols were isolated from the bark portion through metabolite profiling using ultra-HPLC-ESI-Orbitrap-MS: isoastilbin, neoastilbin, and neoisoastilbin (Barrientos *et al.*, 2020). *Ardisia elliptica* Thunb. is a plant that has been used traditionally to treat various health conditions such as fever, diarrhea, measles, and herpes. Its metabolome includes flavonoids, benzoquinones, triterpenes, and phenolic derivatives (Wong *et al.*, 2021). LC-ESI-LTQ-Orbitrap-MS has been used to study the profiles of various parts of fenugreek waste, including the leaves, stems, and bulbs. Metabolomics analysis has allowed the identification of various metabolites, most of which are derivatives of flavonoid aglycones, flavonoid glycosides, phenolic acids, hydroxycinnamic acid, iridoid derivatives, and lignanes (Crescenzi *et al.*, 2021).

9.5 Future Directions and Challenges

To ensure that metabolomics continues to advance, metabolomics technology need to be integrated with additional cutting-edge technologies. MS-based metabolomics will offer more precise results, and various combinations of techniques, including LC-MS and GC-MS, offer a scientific foundation for resolving metabolite research issues and clarifying metabolic pathways. In the field of metabolomics, MS remains a crucial analytical tool with a leading role (Tsugawa *et al.*, 2021). Developments in this area have made it possible to identify numerous potential disease biomarkers and have shed light on the etiology of numerous illnesses (Crandall *et al.*, 2020; Gonzalez-Covarrubias *et al.*, 2022). Although there are opportunities and challenges associated with the rapidly expanding applications that use evolving separation methods and protocols, one of the main obstacles is the lack of ability to contrast and analyze study findings from independent research groups using identical or comparable samples (Gowda and Djukovic, 2014). It is clear that the ability to separate metabolites and increase the number of metabolites detected has been made possible by advancements in chromatographic methods. This represents the primary obstacle impeding the advancement of this field. Data variability can also be caused by other factors, including residual effects, sample matrix, and sample preparation. The current relative metabolite concentration measures must be replaced with the more reliable absolute concentration determination, which will be independent of the analytic platform, method, and protocol used to address these challenges. Although this approach is not simple for MS, it is crucial (Yang *et al.*, 2019).

Metabolomics may be used in the future to identify metabolic markers for the study of plant metabolism and to forecast the type and degree of biotic and abiotic stress. Methods such as metabolomics-assisted breeding have the potential to significantly increase crop yields by developing crop types that are resilient to stress. To assess the risks and regulatory issues associated with genetically modified crops, metabolomics techniques can also be used for metabolic profiling of genetically modified crops using state-of-the-art genome-editing tools such as the clustered regularly interspaced short palindromic repeats (CRISPR)/Cas9 system. Finally, metabolomics has the potential to have a major impact on plant development in the field of rapid breeding (Razzaq *et al.*, 2019).

Exploring synergies between metabolomics and other omics fields (genomics, transcriptomics, and proteomics) in plant biotechnology offers numerous promising avenues for research and applications: a biosystem or organism is defined by multiple layers of an omics cascade: genome, metabolome, proteome, and transcriptome, each representing distinct levels of biomolecular structure (Yu *et al.*, 2023). The term "multi-omics" is commonly used to describe the diversity of biomolecular profiles and the complexity of the omics layers they aim to measure. Integrating various omics profiles enables the extraction of insightful information and provides more comprehensive snapshots of biological systems and molecular processes (Fig. 9.2).

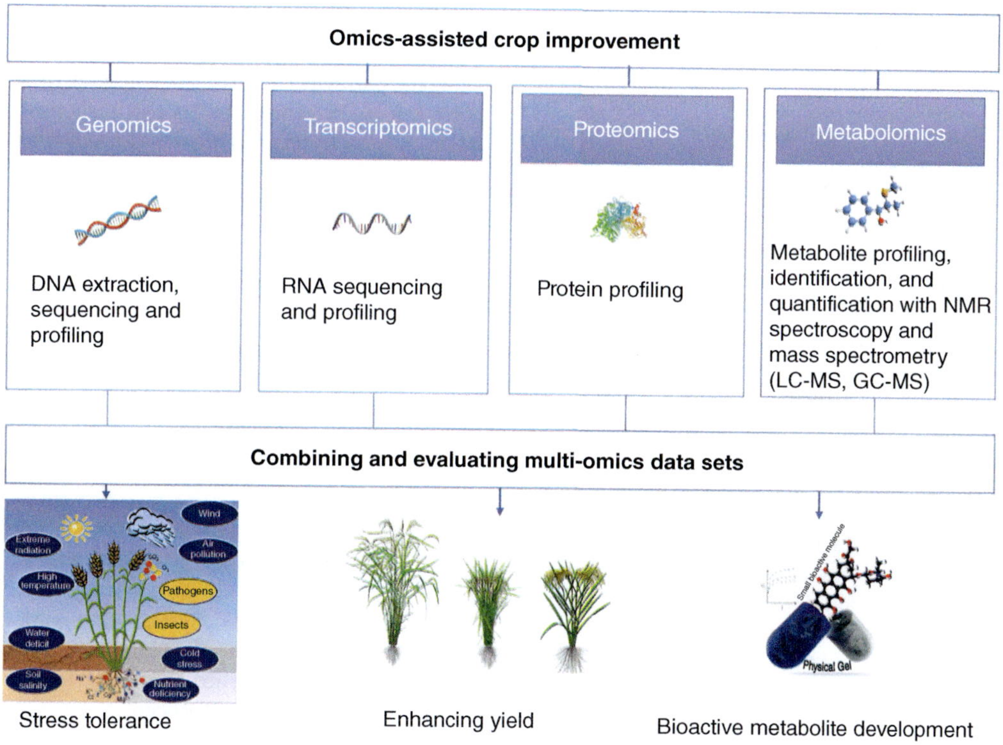

Fig. 9.2. Exploring synergies between metabolomics and other omics areas (genomics, transcriptomics, and proteomics) in plant biotechnology. GC-MS, gas chromatography–mass spectrometry; LC-MS, liquid chromatography–mass spectrometry.

Omics approaches enable detailed characterization of individual cells, facilitating the detection of subtle changes in gene and protein expression, metabolite accumulation, and metabolic pathways (Giacomello, 2021; Tripathi and Wilkins, 2021). These methods offer valuable insights into identifying key molecular players in plant biotechnology. Molecular regulators, encompassing genes, RNAs, metabolites, and proteins, along with processes such as replication, transcription, translation, and various modifications, are crucial for essential plant functions and responses to environmental stressors (Yang *et al.*, 2023). Understanding these regulatory mechanisms at the molecular level is vital for navigating the challenges posed by climate change and enhancing stress-related signaling pathways. Integrating multiple omics approaches such as transcriptomics, proteomics, and metabolomics in a single study provides novel insights into the intricate molecular mechanisms controlling various physiological responses in plants (Liu *et al.*, 2024). This holistic approach enables the discovery of key genes, proteins, and metabolites, along with their associated pathways implicated in responses to diverse environmental plant stress and tolerance. For example, recent research combining transcriptomics and metabolomics elucidated the molecular mechanisms underlying melatonin-mediated detoxification of cadmium (Li *et al.*, 2022). This study identified differentially expressed genes and differentially accumulated metabolites related to metabolic pathways such as valine, leucine, and isoleucine degradation, ABC transporters, and α-linolenic acid metabolism. The detoxification process involved three major mechanisms: increased antioxidant capacity, accumulation of secondary metabolites, and regulated ion transportation (Li *et al.*, 2022; Liu *et al.*, 2023). In omics research, biological databases are continually updated with predicted candidate genes, proteins, and metabolites implicated in responses to metal(loid) stress. A survey of these databases reveals a diversity of focuses: some are species specific or tailored to model plants, while others target specific stressors, with few encompassing multiple plant species and stress types comprehensively. However, many omics databases currently lack features addressing crop scheduling, selection panes for agricultural or visualization purposes, and correlations between metal(loid) concentrations in soil, air, and water across laboratory and greenhouse studies. Therefore, integrating omics data with artificial intelligence and high-throughput phenotyping offers a promising strategy to accelerate and improve crop breeding. These insights are crucial for identifying key genes, proteins, and metabolic pathways involved in tolerance (Song *et al.*, 2020; Jung *et al.*, 2021; Lakshmi *et al.*, 2021).

9.6 Conclusions

Applied metabolomics has emerged as a powerful tool in plant biotechnology, offering a comprehensive understanding of the complex interplay between metabolites, genes, and the environment. This chapter has highlighted the significant contributions of metabolomics in various aspects of plant improvement. These include identifying novel traits, understanding metabolic pathways for targeted engineering, optimizing plant growth and stress responses, and developing functional markers for breeding programs. In conclusion, applied metabolomics offers a powerful and versatile approach to advancing plant biotechnology. By providing a holistic view of the plant metabolome, this technology holds immense potential for developing new and improved varieties with enhanced productivity, stress tolerance, and nutritional value. Continued advances in analytical techniques and bioinformatics tools will further realize the full potential of metabolomics in shaping the future of plant biotechnology.

9.7 Acknowledgments

This research was funded by the Phytopathology Unit of the Department of Plant Protection and Environment at the Ecole Nationale d'Agriculture de Meknes, Morocco.

9.8 References

Ackah, M., Shi, Y., Wu, M., Wang, L., Guo, P., *et al.* (2021) Metabolomics response to drought stress in *Morus alba* L. variety Yu-711. *Plants* 10: 1636.

Adeniji, A.A., Babalola, O.O., and Loots, D.T. (2020) Metabolomic applications for understanding complex tripartite plant-microbes interactions: strategies and perspectives. *Biotechnology Reports* 25: e00425.

Agarwal, P. and Goyal, A. (2017) A review on analyzers for mass spectrometry. *International Journal of Pharma and Bio Sciences* 8, 140–152.

Ahanger, M.A., Gul, F., Ahmad, P., and Akram, N.A. (2018) Environmental stresses and metabolomics—deciphering the role of stress responsive metabolites. In: Ahmad, P., Ahanger, M.A., Singh, V.P., Tripathi, D.K., Alam, P., and Alyemeni, M.N. (eds) *Plant Metabolites and Regulation under Environmental Stress*. Academic Press, London.

Ali, S., Tyagi, A., and Bae, H. (2023) ROS interplay between plant growth and stress biology: challenges and future perspectives. *Plant Physiology and Biochemistry* 203: 108032.

Allwood, J.W. and Goodacre, R. (2010) An introduction to liquid chromatography–mass spectrometry instrumentation applied in plant metabolomic analyses. *Phytochemical Analysis* 21, 33–47.

Allwood, J.W., Williams, A., Uthe, H., van Dam, N.M., Mur, L.A.J., *et al.* (2021) Unravelling plant responses to stress—the importance of targeted and untargeted metabolomics. *Metabolites* 11: 558.

Alseekh, S., Aharoni, A., Brotman, Y., Contrepois, K., d'Auria, J., *et al.* (2021) Mass spectrometry-based metabolomics: a guide for annotation, quantification and best reporting practices. *Nature Methods* 18, 747–756.

Anzano, A., Bonanomi, G., Mazzoleni, S., and Lanzotti, V. (2022) Plant metabolomics in biotic and abiotic stress: a critical overview. *Phytochemistry Reviews* 21, 503–524.

Arif, M.A., Alseekh, S., Harb, J., Fernie, A., and Frank, W. (2018) Abscisic acid, cold and salt stimulate conserved metabolic regulation in the moss *Physcomitrella patens*. *Plant Biology* 20, 1014–1022.

Auler, P.A., Souza, G.M., da Silva Engela, M.R.G., do Amaral, M.N., Rossatto, T., *et al.* (2021) Stress memory of physiological, biochemical and metabolomic responses in two different rice genotypes under drought stress: the scale matters. *Plant Science* 311: 110994.

Bajad, S.U., Lu, W., Kimball, E.H., Yuan, J., Peterson, C., and Rabinowitz, J.D. (2006) Separation and quantitation of water soluble cellular metabolites by hydrophilic interaction chromatography-tandem mass spectrometry. *Journal of Chromatography A* 1125, 76–88.

Barrientos, R., Fernández-Galleguillos, C., Pastene, E., Simirgiotis, M., Romero-Parra, J., *et al.* (2020) Metabolomic analysis, fast isolation of phenolic compounds, and evaluation of biological activities of the bark from *Weinmannia trichosperma* Cav. (Cunoniaceae). *Frontiers in Pharmacology* 11: 780.

Batool, F., Emwas, A.H., Gao, X., Munawar, M.A., and Chotana, G.A. (2017) Synthesis and suzuki cross-coupling reactions of 2,6-bis(trifluoromethyl)pyridine-4-boronic acid pinacol ester. *Synthesis* 49, 1327–1334.

Bedair, M. and Glenn, K.C. (2020) Evaluation of the use of untargeted metabolomics in the safety assessment of genetically modified crops. *Metabolomics* 16: 111.

Benkeblia, N. (2023) Gas chromatography-mass spectrometry and liquid chromatography-mass spectrometry metabolomics platforms: tools for plant oligosaccharides analysis. *Carbohydrate Polymer Technologies and Applications* 5: 100304.

Bhattacharjee, S. (2019) ROS and oxidative modification of cellular components. In: *Reactive Oxygen Species in Plant Biology*. Springer, New Delhi, India, pp. 81–105.

Bhattacharya, A. (2022) Plant growth hormones in plants under low-temperature stress: a review. In: *Physiological Processes in Plants Under Low Temperature Stress*. Springer, Singapore, pp. 517–627.

Bijukumar, D.R., McGeehan, C., and Mathew, M.T. (2018) Regenerative medicine strategies in biomedical implants. *Current Osteoporosis Reports* 16, 236–245.

Borbély, P., Gasperl, A., Pálmai, T., Ahres, M., Asghar, M.A., *et al.* (2022) Light intensity- and spectrum-dependent redox regulation of plant metabolism. *Antioxidants* 11: 1311.

Bubli, S.Y., Haque, F., and Khan, M.S. (2021) Gas chromatography and mass spectroscopy (GC-MS) technique for food analysis. In: Khan, M.S. and Rahman, M.S. (eds) *Techniques to Measure Food Safety and Quality*. Springer, Cham, Switzerland, pp. 195–217.

Cai, Z., Chen, H., Chen, J., Yang, R., Zou, L., *et al.* (2020) Metabolomics characterizes the metabolic changes of *Lonicerae japonicae* Flos under different salt stresses. *PLOS ONE* 15: e0243111.

Calvo, N. (2015) Liver fat deposition and mitochondrial dysfunction in morbid obesity: an approach combining metabolomics with liver imaging and histology. *World Journal of Gastroenterology* 21: 7529.

Carreno-Quintero, N., Bouwmeester, H.J., and Keurentjes, J.J.B. (2013) Genetic analysis of metabolome–phenotype interactions: from model to crop species. *Trends in Genetics* 29, 41–50.

Chen, J., Hu, X., Shi, T., Yin, H., Sun, D., *et al.* (2020) Metabolite-based genome-wide association study enables dissection of the flavonoid decoration pathway of wheat kernels. *Plant Biotechnology Journal* 18, 1722–1735.

Chen, Y.H. and Lu, I.C. (2021) Novel ion source for a portable mass spectrometer. *Rapid Communications in Mass Spectrometry* 35 (Suppl. 1): e8503.

Choudhury, S., Sharma, P., Moulick, D., and Mazumder, M.K. (2021) Unrevealing metabolomics for abiotic stress adaptation and tolerance in plants. *Journal of Crop Science and Biotechnology* 24, 479–493.

Corpas, F.J., Gupta, D.K., and Palma, J.M. (2015) Production sites of reactive oxygen species (ROS) in organelles from plant cells. In: Gupta, D., Palma, J.M., and Corpas, F.J. (eds) *Reactive Oxygen Species and Oxidative Damage in Plants Under Stress*. Springer, Cham, Switzerland, pp. 1–22.

Crandall, S.G., Gold, K.M., Jiménez-Gasco, M., Filgueiras, C.C., and Willett, D.S. (2020) A multi-omics approach to solving problems in plant disease ecology. *PLOS ONE* 15: e0237975.

Crescenzi, M.A., d'Urso, G., Piacente, S., and Montoro, P. (2021) LC-ESI/LTQOrbitrap/MS metabolomic analysis of fennel waste (*Foeniculum vulgare* Mill.) as a byproduct rich in bioactive compounds. *Foods* 10: 1893.

Cui, T., Yuan, B., Guo, H., Tian, H., Wang, W., *et al.* (2021) Enhanced lignin biodegradation by consortium of white rot fungi: microbial synergistic effects and product mapping. *Biotechnology for Biofuels* 14: 162.

da Silva, R.F., Carneiro, C.N., de Sousa, C.B.C., Gomez, F.J.V., Espino, M., *et al.* (2022) Sustainable extraction bioactive compounds procedures in medicinal plants based on the principles of green analytical chemistry: a review. *Microchemical Journal* 175: 107184.

Deborde, C., Moing, A., Roch, L., Jacob, D., Rolin, D., and Giraudeau, P. (2017) Plant metabolism as studied by NMR spectroscopy. *Progress in Nuclear Magnetic Resonance Spectroscopy* 102–103, 61–97.

Deshaies, M., Lamari, N., Ng, C.K.Y., Ward, P., and Doohan, F.M. (2022) The impact of chitosan on the early metabolomic response of wheat to infection by *Fusarium graminearum*. *BMC Plant Biology* 22: 73.

Desmet, S., Morreel, K., and Dauwe, R. (2021) Origin and function of structural diversity in the plant specialized metabolome. *Plants* 10: 2393.

Dmitrieva, V.A., Tyutereva, E.V., and Voitsekhovskaja, O.V. (2020) Singlet oxygen in plants: generation, detection, and signaling roles. *International Journal of Molecular Sciences* 21: 3237.

Du, J., Shen, T., Xiong, Q., Zhu, C., Peng, X., *et al.* (2020) Combined proteomics, metabolomics and physiological analyses of rice growth and grain yield with heavy nitrogen application before and after drought. *BMC Plant Biology* 20: 556.

Duong, Q.V., Levitsky, Y., Dessinger, M.J., Strubbe-Rivera, J.O., and Bazil, J.N. (2021) Identifying site-specific superoxide and hydrogen peroxide production rates from the mitochondrial electron transport system using a computational strategy. *Function* 2: zqab050.

Echeverria, A., Larrainzar, E., Li, W., Watanabe, Y., Sato, M., *et al.* (2021) *Medicago sativa* and *Medicago truncatula* show contrasting root metabolic responses to drought. *Frontiers in Plant Science* 12: 652143.

Eid, A.M., Salim, S.S., Hassan, S.E.-D., Ismail, M.A., and Fouda, A. (2019) Role of endophytes in plant health and abiotic stress management. In: Kumar, V., Prasad, R., Kumar, M., and Choudhary, D.K. (eds) *Microbiome in Plant Health and Disease*. Springer, Singapore, pp. 119–144.

Emwas, A.-H., Roy, R., McKay, R.T., Tenori, L., Saccenti, E., *et al.* (2019) NMR spectroscopy for metabolomics research. *Metabolites* 9: 123.

Fang, C., Fernie, A.R., and Luo, J. (2019) Exploring the diversity of plant metabolism. *Trends in Plant Science* 24, 83–98.

García-Caparrós, P., de Filippis, L., Gul, A., Hasanuzzaman, M., Ozturk, M., *et al.* (2021) Oxidative stress and antioxidant metabolism under adverse environmental conditions: a review. *Botanical Review* 87, 421–466.

Giacomello, S. (2021) A new era for plant science: spatial single-cell transcriptomics. *Current Opinion in Plant Biology* 60: 102041.

Gonzalez-Covarrubias, V., Martínez-Martínez, E., and del Bosque-Plata, L. (2022) The potential of metabolomics in biomedical applications. *Metabolites* 12: 194.

González-Domínguez, R., González-Domínguez, Á., Sayago, A., and Fernández-Recamales, Á. (2020) Recommendations and best practices for standardizing the pre-analytical processing of blood and urine samples in metabolomics. *Metabolites* 10: 229.

Gowda, G.A.N. and Djukovic, D. (2014) Overview of mass spectrometry-based metabolomics: opportunities and challenges. *Methods in Molecular Biology* 1198, 3–12.

Gu, S., Wei, Z., Shao, Z., Friman, V.-P., Cao, K., *et al.* (2020) Competition for iron drives phytopathogen control by natural rhizosphere microbiomes. *Nature Microbiology* 5, 1002–1010.

Guo, A.C., Jewison, T., Wilson, M., Liu, Y., Knox, C., *et al.* (2012) ECMDB: the *E. coli* metabolome database. *Nucleic Acids Research* 41, D625–D630.

Guo, X., Xin, Z., Yang, T., Ma, X., Zhang, Y., *et al.* (2020) Metabolomics response for drought stress tolerance. *Plants* 9: 520.

Halder, M., Kundu, A., and Jha, S. (2024) Secondary metabolites identification techniques of the current era. In: Mérillon, J.M. and Ramawat, K.G. (eds) *Plant Specialized Metabolites*. Springer, Cham. Switzerland, pp. 1–41.

Hamany Djande, C.Y., Pretorius, C., Tugizimana, F., Piater, L.A., and Dubery, I.A. (2020) Metabolomics: a tool for cultivar phenotyping and investigation of grain crops. *Agronomy* 10: 831.

Harrison, S., Fraser, K., Lane, G., Hughes, D., Villas-Boas, S., and Rasmussen, S. (2011) Analysis of high-molecular-weight fructan polymers in crude plant extracts by high-resolution LC-MS. *Analytical and Bioanalytical Chemistry* 401, 2955–2963.

Hautbergue, T., Jamin, E.L., Debrauwer, L., Puel, O., and Oswald, I.P. (2018) From genomics to metabolomics, moving toward an integrated strategy for the discovery of fungal secondary metabolites. *Natural Product Reports* 35, 147–173.

He, Y., Zhang, K., Li, S., Lu, X., Zhao, N., *et al.* (2023) Multiomics analysis reveals the molecular mechanisms underlying virulence in *Rhizoctonia* and jasmonic acid-mediated resistance in Tartary buckwheat (*Fagopyrum tataricum*). *Plant Cell* 35, 2773–2798.

Holmes, M.V., Millwood, I.Y., Kartsonaki, C., Hill, M.R., Bennett, D.A., *et al.* (2018) Lipids, lipoproteins, and metabolites and risk of myocardial infarction and stroke. *Journal of the American College of Cardiology* 71, 620–632.

Hong, Y., Coen, M., Rhode, C.M., Reily, M.D., Robertson, D.G., *et al.* (2009) Chemical shift calibration of ^{1}H MAS NMR liver tissue spectra exemplified using a study of glycine protection of galactosamine toxicity. *Magnetic Resonance in Chemistry* 47 (Suppl. 1), S47–S53.

Hu, L., Liu, J., Zhang, W., Wang, T., Zhang, N., *et al.* (2020) Functional metabolomics decipher biochemical functions and associated mechanisms underlie small-molecule metabolism. *Mass Spectrometry Reviews* 39, 417–433.

Ibrahim, S., Ahmad, N., Kuang, L., Li, K., Tian, Z., *et al.* (2023) Transcriptome analysis reveals key regulatory genes for root growth related to potassium utilization efficiency in rapeseed (*Brassica napus* L.). *Frontiers in Plant Science* 14: 1194914.

Jacoby, R.P., Koprivova, A., and Kopriva, S. (2021) Pinpointing secondary metabolites that shape the composition and function of the plant microbiome. *Journal of Experimental Botany* 72, 57–69.

Jadhav, H.P., Shaikh, S.S., and Sayyed, R.Z. (2017) Role of hydrolytic enzymes of rhizoflora in biocontrol of fungal phytopathogens: an overview. In: Mehnaz, S. (ed.) *Rhizotrophs: Plant Growth Promotion to Bioremediation*. Springer, Singapore, pp. 183–203.

Jeong, S., Eskandari, R., Park, S.M., Alvarez, J., Tee, S.S., *et al.* (2017) Real-time quantitative analysis of metabolic flux in live cells using a hyperpolarized micromagnetic resonance spectrometer. *Science Advances* 3: e1700341.

Jeppesen, M.J. and Powers, R. (2023) Multiplatform untargeted metabolomics. *Magnetic Resonance in Chemistry* 61, 628–653.

Jha, C.K., Sharma, P., Shukla, A., Parmar, P., Patel, R., *et al.* (2021) Microbial enzyme, 1-aminocyclopropane-1-carboxylic acid (ACC) deaminase: an elixir for plant under stress. *Physiological and Molecular Plant Pathology* 115: 101664.

Jha, Y. and Mohamed, H.I. (2022) Plant secondary metabolites as a tool to investigate biotic stress tolerance in plants: a review. *Gesunde Pflanzen* 74, 771–790.

Jing, Y., Watanabe, M., Aarabi, F., Fernie, A.R., Borghi, M., and Tohge, T. (2023) Cross-species metabolomic analyses in the Brassicaceae reveals common responses to ultraviolet-B exposure. *Plant And Cell Physiology* 64, 1523–1533.

Jung, J., Maeda, M., Chang, A., Bhandari, M., Ashapure, A., and Landivar-Bowles, J. (2021) The potential of remote sensing and artificial intelligence as tools to improve the resilience of agriculture production systems. *Current Opinion in Biotechnology* 70, 15–22.

Kamal, M.S., Razzak, S.A., and Hossain, M.M. (2016) Catalytic oxidation of volatile organic compounds (VOCs)—a review. *Atmospheric Environment* 140, 117–134.

Katam, R., Lin, C., Grant, K., Katam, C.S., and Chen, S. (2022) Advances in plant metabolomics and its applications in stress and single-cell biology. *International Journal of Molecular Sciences* 23: 6985.

Kaur, B., Sandhu, K.S., Kamal, R., Kaur, K., Singh, J., *et al.* (2021) Omics for the improvement of abiotic, biotic, and agronomic traits in major cereal crops: applications, challenges, and prospects. *Plants* 10: 1989.

Keasling, J.D. (2010) Manufacturing molecules through metabolic engineering. *Science* 330, 1355–1358.

Khoshru, B., Mitra, D., Khoshmanzar, E., Myo, E.M., Uniyal, N., *et al.* (2020) Current scenario and future prospects of plant growth-promoting rhizobacteria: an economic valuable resource for the agriculture revival under stressful conditions. *Journal of Plant Nutrition* 43, 3062–3092.

Kim, S., van den Broeck, L., Karre, S., Choi, H., Christensen, S.A., *et al.* (2021) Analysis of the transcriptomic, metabolomic, and gene regulatory responses to *Puccinia sorghi* in maize. *Molecular Plant Pathology* 22, 465–479.

Koelmel, J.P., Napolitano, M.P., Ulmer, C.Z., Vasiliou, V., Garrett, T.J., *et al.* (2020) Environmental lipidomics: understanding the response of organisms and ecosystems to a changing world. *Metabolomics* 16: 56.

Kumar, A., Kumar, R., Shukla, P., and Pandey, M.K. (eds) (2021) *Omics Technologies for Sustainable Agriculture and Global Food Security, Vol. 1*. Springer, Singapore.

Kusano, M., Tohge, T., Fukushima, A., Kobayashi, M., Hayashi, N., *et al.* (2011) Metabolomics reveals comprehensive reprogramming involving two independent metabolic responses of *Arabidopsis* to UV-B light. *Plant Journal* 67, 354–369.

Küstner, L., Fürtauer, L., Weckwerth, W., Nägele, T., and Heyer, A.G. (2019) Subcellular dynamics of proteins and metabolites under abiotic stress reveal deferred response of the *Arabidopsis thaliana* hexokinase-1 mutant *gin2-1* to high light. *Plant Journal* 100, 456–472.

Lakshmi, D., Akhil, D., Kartik, A., Gopinath, K.P., Arun, J., *et al.* (2021) Artificial intelligence (AI) applications in adsorption of heavy metals using modified biochar. *Science of the Total Environment* 801: 149623.

Lenard, G.M., Williams, O.A., and Rasche, F. (2020) Tripartite interaction between *Striga* spp., cereals, and plant root-associated microorganisms: a review. *CABI Reviews* 15: 005.

Li, L., Yan, X., Li, J., Wu, X., and Wang, X. (2022) Metabolome and transcriptome association analysis revealed key factors involved in melatonin mediated cadmium-stress tolerance in cotton. *Frontiers in Plant Science* 13: 995205.

Liang, X., Liu, S., Wang, T., Li, F., Cheng, J., *et al.* (2021) Metabolomics-driven gene mining and genetic improvement of tolerance to salt-induced osmotic stress in maize. *New Phytologist* 230, 2355–2370.

Liu, C., Zhou, G., Qin, H., Guan, Y., Wang, T., *et al.* (2024) Metabolomics combined with physiology and transcriptomics reveal key metabolic pathway responses in apple plants exposure to different selenium concentrations. *Journal of Hazardous Materials* 464: 132953.

Liu, H., Jiao, Q., Fan, L., Jiang, Y., Alyemeni, M.N., *et al.* (2023) Integrated physio-biochemical and transcriptomic analysis revealed mechanism underlying of Si-mediated alleviation to cadmium toxicity in wheat. *Journal of Hazardous Materials* 452: 131366.

Liu, J.Y., Li, P., Zhang, Y.W., Zuo, J.F., Li, G., *et al.* (2020) Three-dimensional genetic networks among seed oil-related traits, metabolites and genes reveal the genetic foundations of oil synthesis in soybean. *Plant Journal* 103, 1103–1124.

Liu, Y., Zhao, Y., and Wang, J. (2021) Fenton/Fenton-like processes with in-situ production of hydrogen peroxide/hydroxyl radical for degradation of emerging contaminants: advances and prospects. *Journal of Hazardous Materials* 404: 124191.

López-Yerena, A., Domínguez-López, I., Vallverdú-Queralt, A., Pérez, M., Jáuregui, O., *et al.* (2021) Metabolomics technologies for the identification and quantification of dietary phenolic compound metabolites: an overview. *Antioxidants* 10: 846.

Lu, M., Yu, S., Lian, J., Wang, Q., He, Z., *et al.* (2021) Physiological and metabolomics responses of two wheat (*Triticum aestivum L.*) genotypes differing in grain cadmium accumulation. *Science of the Total Environment* 769: 145345.

Lynch, J.H., Huang, X.-Q., and Dudareva, N. (2021) Silent constraints: the hidden challenges faced in plant metabolic engineering. *Current Opinion in Biotechnology* 69, 112–117.

Mahrous, E.A. and Farag, M.A. (2015) Two dimensional NMR spectroscopic approaches for exploring plant metabolome: a review. *Journal of Advanced Research* 6, 3–15.

Marchev, A.S., Vasileva, L.V., Amirova, K.M., Savova, M.S., Balcheva-Sivenova, Z.P., and Georgiev, M.I. (2021) Metabolomics and health: from nutritional crops and plant-based pharmaceuticals to profiling of human biofluids. *Cellular and Molecular Life Sciences* 78, 6487–6503.

Marsafari, M., Samizadeh, H., Rabiei, B., Mehrabi, A., Koffas, M., and Xu, P. (2020) Biotechnological production of flavonoids: an update on plant metabolic engineering, microbial host selection, and genetically encoded biosensors. *Biotechnology Journal* 15: e1900432.

Martins, M.C.M., Caldana, C., Wolf, L.D., and de Abreu, L.G.F. (2018) the importance of experimental design, quality assurance, and control in plant metabolomics experiments. *Methods in Molecular Biology* 1778, 3–17.

Mattoli, L., Gianni, M., and Burico, M. (2023) Mass spectrometry-based metabolomic analysis as a tool for quality control of natural complex products. *Mass Spectrometry Reviews* 42, 1358–1396.

Meena, K.K., Sorty, A.M., Bitla, U.M., Choudhary, K., Gupta, P., *et al.* (2017) Abiotic stress responses and microbe-mediated mitigation in plants: the omics strategies. *Frontiers in Plant Science* 8: 172.

Mérillon, J.-M. and Ramawat, K.G. (eds) (2020) *Co-Evolution of Secondary Metabolites*. Springer, Cham, Switzerland.

Mishra, A.K., Sudalaimuthuasari, N., Hazzouri, K.M., Saeed, E.E., Shah, I., and Amiri, K.M.A. (2022) Tapping into plant–microbiome interactions through the lens of multi-omics techniques. *Cells* 11: 3254.

Moses, T., Pollier, J., Thevelein, J.M., and Goossens, A. (2013) Bioengineering of plant (tri)terpenoids: from metabolic engineering of plants to synthetic biology *in vivo* and *in vitro*. *New Phytologist* 200, 27–43.

Motta, A., Paris, D., and Melck, D. (2010) Monitoring real-time metabolism of living cells by fast two-dimensional NMR spectroscopy. *Analytical Chemistry* 82, 2405–2411.

Nalbantoglu, S. and Karadag, A. (2021) Metabolomics bridging proteomics along metabolites/oncometabolites and protein modifications: paving the way toward integrative multiomics. *Journal of Pharmaceutical and Biomedical Analysis* 199: 114031.

Nandy, S., Das, T., Tudu, C.K., Mishra, T., Ghorai, M., *et al.* (2022) Unravelling the multi-faceted regulatory role of polyamines in plant biotechnology, transgenics and secondary metabolomics. *Applied Microbiology and Biotechnology* 106, 905–929.

Nguyen, T.H., Goossens, A., and Lacchini, E. (2022) Jasmonate: a hormone of primary importance for plant metabolism. *Current Opinion in Plant Biology* 67: 102197.

Nikhitha Surendran, O., Haridas, M., Szakacs, G., and Sabu, A. (2020) Modified plant metabolites as nutraceuticals. In: Sukumaran, S.T., Sugathan, S., and Abdulhameed, S. (eds) *Plant Metabolites: Methods, Applications and Prospects*. Springer, Singapore, pp. 167–180.

Noctor, G., Lelarge-Trouverie, C., and Mhamdi, A. (2015) The metabolomics of oxidative stress. *Phytochemistry* 112, 33–53.

Oh, S.W., Imran, M., Kim, E.H., Park, S.Y., Lee, S.G., *et al.* (2023) Approach strategies and application of metabolomics to biotechnology in plants. *Frontiers in Plant Science* 14, 1–15.

Okazaki, Y. and Saito, K. (2012) Recent advances of metabolomics in plant biotechnology. *Plant Biotechnology Reports* 6, 1–15.

Omena-Garcia, R.P., de Ávila Silva, L., Vital, C.E., Araújo, W.L., Ramos, H.J.O., and Nunes-Nesi, A. (2019) Identification of metabolite traits from the current metabolomic approaches. *Theoretical and Experimental Plant Physiology* 31, 1–19.

Phua, S.Y., de Smet, B., Remacle, C., Chan, K.X., and van Breusegem, F. (2021) Reactive oxygen species and organellar signaling. *Journal of Experimental Botany* 72, 5807–5824.

Pinto, D., Delaby, E., Merico, D., Barbosa, M., Merikangas, A., *et al.* (2014) Convergence of genes and cellular pathways dysregulated in autism spectrum disorders. *American Journal of Human Genetics* 94, 677–694.

Płotka-Wasylka, J., Rutkowska, M., Owczarek, K., Tobiszewski, M., and Namieśnik, J. (2017) Extraction with environmentally friendly solvents. *Trends in Analytical Chemistry* 91, 12–25.

Qian, W., Zhu, Y., Chen, Q., Wang, S., Chen, L., *et al.* (2023) Comprehensive metabolomic and lipidomic alterations in response to heat stress during seed germination and seedling growth of *Arabidopsis*. *Frontiers in Plant Science* 14: 1132881.

Raj, S.R.G. and Nadarajah, K. (2022) QTL and candidate genes: techniques and advancement in abiotic stress resistance breeding of major cereals. *International Journal of Molecular Sciences* 24: 6.

Ramirez-Gaona, M., Marcu, A., Pon, A., Guo, A.C., Sajed, T., *et al.* (2017) YMDB 2.0: a significantly expanded version of the yeast metabolome database. *Nucleic Acids Research* 45, D440–D445.

Rana, A., Samtiya, M., Dhewa, T., Mishra, V., and Aluko, R.E. (2022) Health benefits of polyphenols: a concise review. *Journal of Food Biochemistry* 46: e14264.

Ranjan Maji, S., Roy, C., and Sinha, S.K. (2023) Gas chromatography-mass spectrometry (GC-MS): a comprehensive review of synergistic combinations and their applications in the past two decades. *Journal of Analytical Sciences and Applied Biotechnology* 5, 72–85.

Rappold, B.A. (2022) Review of the use of liquid chromatography-tandem mass spectrometry in clinical laboratories: part II—operations. *Annals of Laboratory Medicine* 42, 531–557.

Raza, A. (2022) Metabolomics: a systems biology approach for enhancing heat stress tolerance in plants. *Plant Cell Reports* 41, 741–763.

Razi, K. and Muneer, S. (2021) Drought stress-induced physiological mechanisms, signaling pathways and molecular response of chloroplasts in common vegetable crops. *Critical Reviews in Biotechnology* 41, 669–691.

Razzaq, A., Sadia, B., Raza, A., Hameed, M.K., and Saleem, F. (2019) Metabolomics: a way forward for crop improvement. *Metabolites* 9: 303.

Razzaq, A., Wishart, D.S., Wani, S.H., Hameed, M.K., Mubin, M., and Saleem, F. (2022) Advances in metabolomics-driven diagnostic breeding and crop improvement. *Metabolites* 12: 511.

Reglinski, T., Havis, N., Rees, H.J., and de Jong, H. (2023) The practical role of induced resistance for crop protection. *Phytopathology* 113, 719–731.

Rinschen, M.M., Ivanisevic, J., Giera, M., and Siuzdak, G. (2019) Identification of bioactive metabolites using activity metabolomics. *Nature Reviews Molecular Cell Biology* 20, 353–367.

Roberts, L.D., Souza, A.L., Gerszten, R.E., and Clish, C.B. (2012) Targeted metabolomics. *Current Protocols in Molecular Biology* 98, 30.2.1–30.2.24.

Saito, K. and Matsuda, F. (2010) Metabolomics for functional genomics, systems biology, and biotechnology. *Annual Review of Plant Biology* 61, 463–489.

Sajed, T., Marcu, A., Ramirez, M., Pon, A., Guo, A.C., *et al.* (2016) ECMDB 2.0: a richer resource for understanding the biochemistry of *E. coli*. *Nucleic Acids Research* 44, D495–D501.

Salam, U., Ullah, S., Tang, Z.H., Elateeq, A.A., Khan, Y., *et al.* (2023) Plant metabolomics: an overview of the role of primary and secondary metabolites against different environmental stress factors. *Life* 13: 706.

Salem, M.A., Perez de Souza, L., Serag, A., Fernie, A.R., Farag, M.A., *et al.* (2020a) Metabolomics in the context of plant natural products research: from sample preparation to metabolite analysis. *Metabolites* 10: 37.

Salem, M.A., Yoshida, T., Perez de Souza, L., Alseekh, S., Bajdzienko, K., *et al.* (2020b) An improved extraction method enables the comprehensive analysis of lipids, proteins, metabolites and phytohormones from a single sample of leaf tissue under water-deficit stress. *Plant Journal* 103, 1614–1632.

Saude, E.J. and Sykes, B.D. (2007) Urine stability for metabolomic studies: effects of preparation and storage. *Metabolomics* 3, 19–27.

Savy, D., Cozzolino, V., Vinci, G., Verrillo, M., Aliberti, A., *et al.* (2022) Fertilisation with compost mitigates salt stress in tomato by affecting plant metabolomics and nutritional profiles. *Chemical and Biological Technologies in Agriculture* 9: 104.

Sedláčková, S., Hubálek, M., Vrkoslav, V., Blechová, M., and Cvačka, J. (2022) Utility of atmospheric-pressure chemical ionization and photoionization mass spectrometry in bottom-up proteomics. *Separations* 9: 42.

Sethi, M., Saini, D.K., Devi, V., Kaur, C., Singh, M.P., *et al.* (2023) Unravelling the genetic framework associated with grain quality and yield-related traits in maize (*Zea mays* L.). *Frontiers in Genetics* 14: 1248697.

Shan, L., Gao, H., Zhang, J., Li, W., Su, Y., and Guo, Y. (2022) Plasma and serum exosome markers analyzed by matrix-assisted laser desorption/ionization time-of-flight mass spectrometry coupled with electron multiplier. *Talanta* 247: 123560.

Sharma, V., Gupta, P., Priscilla, K., SharanKumar, S., Hangargi, B., *et al.* (2021) Metabolomics intervention towards better understanding of plant traits. *Cells* 10: 346.

Silva, E., Perez da Graça, J., Porto, C., Martin do Prado, R., Nunes, E., *et al.* (2021) Untargeted metabolomics analysis by UHPLC-MS/MS of soybean plant in a compatible response to *Phakopsora pachyrhizi* infection. *Metabolites* 11: 179.

Singh, A., Rajput, V., Singh, A.K., Sengar, R.S., Singh, R.K., and Minkina, T. (2021) Transformation techniques and their role in crop improvements: a global scenario of GM crops. In: Singh, P., Borthakur, A., Singh, A.A., Kumar, A., and Singh, K.K. (eds) *Policy Issues in Genetically Modified Crops*. Elsevier, London, pp. 515–542.

Singh, A.K., Dhanapal, S., and Yadav, B.S. (2020) The dynamic responses of plant physiology and metabolism during environmental stress progression. *Molecular Biology Reports* 47, 1459–1470.

Singh, U., Alsuhaymi, S., Al-Nemi, R., Emwas, A.-H., and Jaremko, M. (2023) Compound-specific 1D ^{1}H NMR pulse sequence selection for metabolomics analyses. *ACS Omega* 8, 23651–23663.

Song, X., Nie, F., Chen, W., Ma, X., Gong, K., *et al.* (2020) Coriander genomics database: a genomic, transcriptomic, and metabolic database for coriander. *Horticulture Research* 7: 55.

Sudiro, C., Guglielmi, F., Hochart, M., Senizza, B., Zhang, L., *et al.* (2022) A phenomics and metabolomics investigation on the modulation of drought stress by a biostimulant plant extract in tomato (*Solanum lycopersicum*). *Agronomy* 12: 764.

Sun, S., Fang, J., Lin, M., Hu, C., Qi, X., *et al.* (2021) Comparative metabolomic and transcriptomic studies reveal key metabolism pathways contributing to freezing tolerance under cold stress in kiwifruit. *Frontiers in Plant Science* 12: 628969.

Tatsis, E.C. and O'Connor, S.E. (2016) New developments in engineering plant metabolic pathways. *Current Opinion in Biotechnology* 42, 126–132.

Tripathi, R.K. and Wilkins, O. (2021) Single cell gene regulatory networks in plants: opportunities for enhancing climate change stress resilience. *Plant, Cell & Environment* 44, 2006–2017.

Tsugawa, H., Rai, A., Saito, K., and Nakabayashi, R. (2021) Metabolomics and complementary techniques to investigate the plant phytochemical cosmos. *Natural Product Reports* 38, 1729–1759.

Ullah, S., Khan, M.N., Lodhi, S.S., Ahmed, I., Tayyab, M., *et al.* (2022) Targeted metabolomics reveals fatty acid abundance adjustments as playing a crucial role in drought-stress response and post-drought recovery in wheat. *Frontiers in Genetics* 13: 972696.

van der Hooft, J.J.J., Mohimani, H., Bauermeister, A., Dorrestein, P.C., Duncan, K.R., and Medema, M.H. (2020) Linking genomics and metabolomics to chart specialized metabolic diversity. *Chemical Society Reviews* 49, 3297–3314.

van Erven, G., Wang, J., Sun, P., de Waard, P., van der Putten, J., *et al.* (2019) Structural motifs of wheat straw lignin differ in susceptibility to degradation by the white-rot fungus *Ceriporiopsis subvermispora*. *ACS Sustainable Chemistry & Engineering* 7, 20032–20042.

Veenstra, T.D. (2021) Omics in systems biology: current progress and future outlook. *Proteomics* 21: e2000235.

Vemuri, G.N. and Aristidou, A.A. (2005) Metabolic engineering in the -omics era: elucidating and modulating regulatory networks. *Microbiology and Molecular Biology Reviews* 69, 197–216.

Vergara-Diaz, O., Vatter, T., Vicente, R., Obata, T., Nieto-Taladriz, M.T., *et al.* (2020) Metabolome profiling supports the key role of the spike in wheat yield performance. *Cells* 9: 1025.

Volkova, S., Matos, M.R.A., Mattanovich, M., and de Mas, I.M. (2020) Metabolic modelling as a framework for metabolomics data integration and analysis. *Metabolites* 10: 303.

Wang, J., Ogata, M., Hirai, H., and Kawagishi, H. (2011) Detoxification of aflatoxin B_1 by manganese peroxidase from the white-rot fungus *Phanerochaete sordida* YK-624. *FEMS Microbiology Letters* 314, 164–169.

Wang, J.H., Byun, J., and Pennathur, S. (2010) Analytical approaches to metabolomics and applications to systems biology. *Seminars in Nephrology* 30, 500–511.

Wang, K., Nan, L.L., Xia, J., Wu, S.W., and Yang, L.L. (2024) Metabolomics reveal root differential metabolites of different root-type alfalfa under drought stress. *Frontiers in Plant Science* 15: 1341826.

Wang, L., Ma, K.B., Lu, Z.G., Ren, S.X., Jiang, H.R., *et al.* (2020) Differential physiological, transcriptomic and metabolomic responses of *Arabidopsis* leaves under prolonged warming and heat shock. *BMC Plant Biology* 20: 86.

Wang, L., Di, T., Peng, J., Li, Y., Li, N., *et al.* (2022) Comparative metabolomic analysis reveals the involvement of catechins in adaptation mechanism to cold stress in tea plant (*Camellia sinensis* var. *sinensis*). *Environmental and Experimental Botany* 201: 104978.

Weitzel, K. (2011) Bond-dissociation energies of cations—pushing the limits to quantum state resolution. *Mass Spectrometry Reviews* 30, 221–235.

Wink, M. (2010) Introduction: biochemistry, physiology and ecological functions of secondary metabolites. *Annual Plant Reviews* 40, 1–19.

Wishart, D.S., Feunang, Y.D., Marcu, A., Guo, A.C., Liang, K., *et al.* (2018) HMDB 4.0: the human metabolome database for 2018. *Nucleic Acids Research* 46, D608–D617.

Wolfender, J.-L., Marti, G., Thomas, A., and Bertrand, S. (2015) Current approaches and challenges for the metabolite profiling of complex natural extracts. *Journal of Chromatography A* 1382, 136–164.

Wong, P.L., Ramli, N.S., Tan, C.P., Azlan, A., and Abas, F. (2021) Metabolomic analysis reveals the valuable bioactive compounds of *Ardisia elliptica*. *Phytochemical Analysis* 32, 685–697.

Yadav, V., Zhong, H., Patel, M.K., Zhang, S., Zhou, X., *et al.* (2024) Integrated omics-based exploration for temperature stress resilience: an approach to smart grape breeding strategies. *Plant Stress* 11: 100356.

Yang, L., Yang, Y., Huang, L., Cui, X., and Liu, Y. (2023) From single- to multi-omics: future research trends in medicinal plants. *Briefings in Bioinformatics* 24: bbac485.

Yang, L., Yuan, J., Yu, B., Hu, S., and Bai, Y. (2024) Sample preparation for fatty acid analysis in biological samples with mass spectrometry-based strategies. *Analytical and Bioanalytical Chemistry* 416, 2371–2387.

Yang, Q., Zhang, A.H., Miao, J.H., Sun, H., Han, Y., *et al.* (2019) Metabolomics biotechnology, applications, and future trends: a systematic review. *RSC Advances* 9, 37245–37257.

Yemelyanov, V.V., Puzanskiy, R.K., and Shishova, M.F. (2023) Plant life with and without oxygen: a metabolomics approach. *International Journal of Molecular Sciences* 24: 16222.

Yeshi, K., Crayn, D., Ritmejerytė, E., and Wangchuk, P. (2022) Plant secondary metabolites produced in response to abiotic stresses has potential application in pharmaceutical product development. *Molecules* 27: 313.

Yi, L., Dong, N., Yun, Y., Deng, B., Ren, D., *et al.* (2016) Chemometric methods in data processing of mass spectrometry-based metabolomics: a review. *Analytica Chimica Acta* 914, 17–34.

Yin, X., Xue, Z., Feng, L., Tan, S., Li, K., *et al.* (2024) Quantitative analysis of bio-molecules in single cells by mass spectrometry. *Trends in Analytical Chemistry* 171: 117503.

Yoon, H., Yoon, D., Yun, M., Choi, J.S., Park, V.Y., *et al.* (2016) Metabolomics of breast cancer using high-resolution magic angle spinning magnetic resonance spectroscopy: correlations with ^{18}F-FDG positron emission tomography-computed tomography, dynamic contrast-enhanced and diffusion-weighted imaging MRI. *PLOS ONE* 11: e0159949.

Yu, X., Liu, Z., and Sun, X. (2023) Single-cell and spatial multi-omics in the plant sciences: technical advances, applications, and perspectives. *Plant Communications* 4: 100508.

Zaid, A., Mushtaq, M., and Wani, S.H. (2021) Interactions of phytohormones with abiotic stress factors under changing climate. In: Aftab, T., and Rehman, K. (eds) *Frontiers in Plant–Soil Interaction*. Elsevier, London, pp. 221–236.

Zeiss, D.R., Molinaro, A., Steenkamp, P.A., Silipo, A., Piater, L.A., *et al.* (2023) Lipopolysaccharides from *Ralstonia solanacearum* induce a broad metabolomic response in *Solanum lycopersicum*. *Frontiers in Molecular Biosciences* 10: 1232233.

Zeki, Ö.C., Eylem, C.C., Reçber, T., Kır, S., and Nemutlu, E. (2020) Integration of GC–MS and LC–MS for untargeted metabolomics profiling. *Journal of Pharmaceutical and Biomedical Analysis* 190: 113509.

Zhang, H.-W., Lv, C., Zhang, L.-J., Guo, X., Shen, Y.-W., *et al.* (2021a) Application of omics- and multi-omics-based techniques for natural product target discovery. *Biomedicine & Pharmacotherapy* 141: 111833.

Zhang, W., Jiang, H., Cao, J., and Jiang, W. (2021b) Advances in biochemical mechanisms and control technologies to treat chilling injury in postharvest fruits and vegetables. *Trends in Food Science and Technology* 113, 355–365.

Zhang, Y., Deng, T., Sun, L., Landis, J.B., Moore, M.J., *et al.* (2021c) Phylogenetic patterns suggest frequent multiple origins of secondary metabolites across the seed-plant 'tree of life'. *National Science Review* 8: nwaa105.

Zhang, Z.-M. and Li, G.-K. (2007) A preliminary study of plant aroma profile characteristics by a combination sampling method coupled with GC–MS. *Microchemical Journal* 86, 29–36.

Zhou, S., Kremling, K.A., Bandillo, N., Richter, A., Zhang, Y.K., *et al.* (2019) Metabolome-scale genome-wide association studies reveal chemical diversity and genetic control of maize specialized metabolites. *Plant Cell* 31, 937–955.

Zhu, G., Shao, Y., Liu, Y., Pei, T., Li, L., *et al.* (2021) Single-cell metabolite analysis by electrospray ionization mass spectrometry. *Trends in Analytical Chemistry* 143: 116351.

Ziegler, S., Pries, V., Hedberg, C., and Waldmann, H. (2013) Target identification for small bioactive molecules: finding the needle in the haystack. *Angewandte Chemie International Edition* 52, 2744–2792.

10 Metabolomics and High-throughput Approaches to Assessing Drought-stress Tolerance in Crops

Suman Gusain[1,2], Akshita Patial[1,3], Rakhi Chandoliya[1,4], Vikas Sharma[3], and Rohit Joshi[1,2*]

[1]Biotechnology Division, CSIR-Institute of Himalayan Bioresource Technology (CSIR-IHBT), Palampur, Himachal Pradesh, India; [2]Academy of Scientific and Innovative Research (AcSIR), CSIR-HRDC Campus, Ghaziabad, Uttar Pradesh, India; [3]Department of Molecular Biology and Genetic Engineering, School of Bioengineering and Biosciences, Lovely Professional University, Phagwara, India; [4]Department of Biotechnology, School of Bioengineering and Biosciences, Lovely Professional University, Phagwara, India

Abstract

Water deficit is a serious problem that constrains crop yield. Drought, as a multi-faceted stress, activates physiological and molecular responses in plants. Alongside morphophysiological changes, plants activate various metabolic pathways to alleviate the deleterious effects of drought stress on plant development and production. Plants modulate metabolite concentrations inside the cell to maintain cellular homeostasis under drought stress. Several primary and secondary metabolites, phytohormones, and genes collectively regulate drought tolerance in plants by regulating various metabolic pathways. Therefore, metabolomics can be an essential tool to investigate plant metabolite responses under drought stress. This chapter gives an overview of techniques used in analyzing plant metabolites and also sheds light on how primary and secondary metabolites are regulated during drought. We also discuss the crosstalk of genes linked to metabolic responses during drought tolerance in plants and propose strategies to develop climate-resilient crops.

Keywords: analytical tools, metabolic analysis, metabolic pathways, phytohormone, transcription factors, water deficit

10.1 Introduction

Global climate change leads to numerous environmental fluctuations that impact crop productivity (Farooq *et al.*, 2022). Among abiotic and biotic factors, drought stress is one of the major threats to food security (Varshney *et al.*, 2021). Drought stress adversely affects all stages of crop growth and development, from germination to maturity (Dietz *et al.*, 2021), impeding the UN's goal of "zero hunger." The current situation of crop development is not sufficient to feed the expanding human population. Therefore, a progressive and sustainable approach to

* E-mail: rohitjoshi@ihbt.res.in

© CAB International 2025. *High-Throughput Plant Metabolomics* (Jen-Tsung Chen ed.)
DOI: 10.1079/9781800626911.0010

crop production and improvement is needed (Varshney *et al.*, 2021). The plant response to drought stress depends on a number of factors such as duration, intensity, and growth stage (Varshney *et al.*, 2021) and differs from species to species. Plants have developed various cellular and molecular processes to ameliorate drought stress. Under short-term drought stress, plants undergo osmotic adjustment, decreased carbon assimilation, stomatal closure, growth inhibition, and hydraulic changes, and activate drought signaling pathways (Joshi *et al.*, 2020). Plants are not severely harmed by these reactions as long as normal conditions are restored quickly. Severe drought stress halts photosynthetic machinery and reduces basic metabolic activities, resulting in cell death (Kaur and Asthir, 2017). Therefore, to cope with the adverse conditions, plants undergo metabolic alterations (Joshi and Karan, 2014). Among the metabolic responses, changes in primary metabolism, such as alterations in sugars, amino acids, and small peptide levels, as well as signaling and phytohormone signaling, are the most prominent responses under drought stress (F. Zhang *et al.*, 2024). These metabolic alterations result in the accumulation of osmolytes inside the cell. Some common osmoregulatory compounds are sugars and sugar alcohol (glucose, sucrose, and trehalose), amino acids (proline), and polyamines (spermine, spermidine, putrescine). These osmolytes inside the plant cell maintain cell turgor and stabilize the enzymes involved in maintaining the level of reactive oxygen species (ROS) in the cell under stress.

The latest advances in plant metabolomics studies provide a detailed overview of plant metabolic responses to different stresses and the intricate regulation of the underlying metabolic processes (Raza, 2022). Various metabolic profiling techniques have been utilized extensively for the molecular characterization of plants during drought stress and to assess the levels of metabolites within specific metabolic categories or pathways (Kumar *et al.*, 2021). Metabolic profiling characterizing the molecular traits associated with the stress response offers valuable insights for breeding initiatives. In this chapter, we discuss metabolomics as a key approach to investigating the effect of water-deficit stress in plants, the tools and techniques used, and the implications of drought stress on metabolic pathways (primary and secondary metabolites). We also discuss various phytohormones and their regulating genes associated with metabolic pathways under stress. This chapter provides an understanding of how plants adapt their metabolic pathways to maintain cellular balance, ensuring their survival in challenging environments such as drought stress.

10.2 Metabolic Approaches to Study the Effects of Drought Stress

Several metabolic approaches are utilized to analyze drought-mediated changes in the plant metabolome (the complete set of low-molecular-mass compounds in cells). Drought adversely affects plant metabolism by alteration of metabolites involved in different metabolic pathways. Plant metabolomics involves four major steps: sample collection, storage, extraction in suitable solvent, and sample analysis. Sample collection should be done carefully as metabolites are prone to enzymatic degradation. Harvested samples are immediately frozen in liquid nitrogen to halt any metabolic changes. Other storage techniques such as oven drying, air-drying, and freeze-drying are also used for metabolite processing. Sample preparation is a critical step and includes metabolic extraction using methods such as mechanical, quenching, and ultrasonic extraction (Ahmad *et al.*, 2018). Solvents are also used for metabolic extraction, such as methanol, methanol:water, and methanol:methyl-tert-butyl-ether:water. A solvent system containing chloroform:methanol:water is commonly utilized for a diverse range of metabolites including polar, non-polar, and water-soluble metabolites (Oh *et al.*, 2023). Extracted samples are then analyzed on a mass spectrometry (MS)-based system for data acquisition. The analytical techniques used include liquid chromatography–mass spectrometry (LC-MS), gas chromatography–mass spectrometry (GC-MS), matrix-assisted laser desorption/ionization (MALDI), ion–molecule reaction (IMR), Fourier-transform–ion cyclotron resonance–mass spectrometry (FT-ICR-MS), capillary electrophoresis–mass spectrometry (CE-MS), high-performance liquid chromatography (HPLC), and nuclear magnetic resonance (NMR) for qualitative and quantitative estimation of metabolites.

10.2.1 Gas chromatography–mass spectrometry

GC-MS is a powerful analytical method for the estimation of small metabolites (50–600 Da) such as sugars, alcohols, amino acids, amines, fatty acids, and hydroxyl acids. In GC-MS, these metabolites undergo chemical derivatization to transform them into a volatile state, enabling their analysis by GC (Fiehn, 2016). GS-MS utilizes two methods: electron ionization and chemical ionization of neutral molecules, which can further be separated based on mass, by a mass analyzer. Electron ionization is most commonly used for GC-MS. Sample screening is done through gas chromatography–time-of-flight–mass spectrometry (GC-TOF-MS). This method is preferred for metabolomic analysis because of faster data acquisition, higher mass accuracy, and its ability to separate complex mixtures (Kumar *et al.*, 2016). GC × GC-TOF-MS or triple quadrupole mass spectrometry has also been used for metabolic profiling with a high acquisition rate of 500 Hz. These techniques offer greater resolution and higher sensitivity. However, these analytical tools are confined to thermostable volatiles (500–800 Da), usually providing a range of primary metabolites but excluding secondary metabolites (Kumar *et al.*, 2016). Among the various techniques, GC-MS stands out as one of the most efficient and productive methods employed in plant metabolomics.

10.2.2 Liquid chromatography–mass spectrometry

LC-MS is another analytical method used to quantify a wide range of complex metabolites. It is suitable for higher-molecular-mass metabolites (>500 Da) with chemically unstable structures, thermolabile functional groups, and high boiling points. LC-MS does not involve sample volatilization. This technique is effective for profiling various secondary metabolites (phenolics, flavonoids, alkaloids), complex lipids (sphingolipids, phospholipids, glycerolipids), and sterols and steroids (Okazaki and Saito, 2011). LC-MS utilizes various separating columns (reversed phase, hydrophobic interaction, ion exchange) to separate metabolites based on their distinct chemical properties (Obata and Fernie, 2012). Reversed-phase columns (C8 or C18) are commonly used. Various ionization tools can be used in LC-MS, such as atmospheric pressure ionization (API) and electrospray ionization (ESI) (Okazaki and Saito, 2011). Sample ionization via API provides limited structural information, but further fragmentation of ions using collision-induced dissociation offers more comprehensive structural details. Ion fragmentation is performed using a tandem MS instrument, which enables two or more sequential stages of MS analysis (Jorge *et al.*, 2016). Tandem MS instruments come with two different configurations: tandem-in-time and tandem-in-space for LC-MS. Tandem-in-time instruments include ion trap MS such as Orbitrap, FT-ICR-MS, and quadrupole ion trap-MS. Tandem-in-space uses TOF, quadrupole TOF (Q-TOF), and triple quadrupole. Despite its numerous advantages, a significant limitation of LC-MS is its inability to analyze secondary metabolites in cases where there is no standard reference available.

10.2.3 Capillary electrophoresis–mass spectrometry

CE is a potent analytical technique used for the estimation of ionic metabolites. CE-MS separates ionic metabolite on the basis of mass:charge (m/z) ratio (Obata and Fernie, 2012). It requires small injection volumes and provides rapid and high resolution of charged compounds. CE-MS provides mass information and performs metabolic characterization on the basis of mass fragmentation (Salem *et al.*, 2020). Metabolic coverage by CE-MS largely coincides with that of GC-MS but does not require derivatization. It uses API as an ionization source. CE is conducted using a silica tube, with its extremities immersed in a buffer solution, across which a high voltage of 20–30 kV is applied. The limitations of CE-MS include low sensitivity, low reproducibility, and lack of reference libraries.

10.2.4 Fourier-transform–ion cyclotron resonance–mass spectrometry

FT-ICR-MS offers the highest resolution and mass accuracy among other mass spectrometers

(Nikolaev *et al.*, 2016). FT-ICR-MS is particularly used for metabolic and proteomic analysis. It is also compatible with multi-stage MS analyzers. However, due to the higher magnetic field, complex ionic interactions, and higher cost of this technique, it is not widely applicable to metabolomic studies (Tanna and Mishra, 2018).

10.2.5 MALDI–MS

MALDI is an ionization-based MS imaging technique that enables the spatial metabolomic analysis of biological samples. MALDI-MS imaging is used to assess the spatial distribution of various compounds such as lipids, proteins, sugars, and various other metabolites (Baker *et al.*, 2017). Samples are coated with a thin film of a matrix composed of α-cyano-4-hydroxycinnamic acid, sinapinic acid, and 2,5-dihydroxybenzoic acid, after which a laser beam is passed through the coated sample to obtain a mass spectrum. MALDI is coupled with a variety of analyzers, such as Q-TOF, TOF, TOF-TOF, and FT-ICR, for MS imaging. Other techniques also used for MS imaging are laser ablation ESI, desorption ESI, and secondary ion MS (Baker *et al.*, 2017).

10.2.6 Ion mobility spectrometry

Ion mobility spectrometry (IMS) differentiates gas ions according to their m/z ratio. IMS-MS enables the multi-dimensional characterization of detected compounds. IMS separation occurs within milliseconds and therefore can be integrated into conventional GC-MS and LC-MS processes. IMS-MS integrated with front-end LC can increase separation capabilities and peak capacity (Lapthorn *et al.*, 2013). The essential component of metabolite identification in IMS-MS is the careful maintenance and refinement of the collision cross-section (CCS) database. Various CCS databases such as IsiCLE, MetCCS, and LipidCCS are maintained, containing more than 1 million CCS values. Zhou *et al.* (2020) created a new CCS atlas (AllCCS) along with a strategy for annotating metabolites, encompassing both known and unknown compounds. AllCCS comprises an extensive array of chemical structures, including more than 5000 experimental CCS records and approximately 12 million predicted CCS values for more than 1.6 million chemical molecules. In a separate study, Schroeder *et al.* (2019) detected 146 plant-derived natural compounds, annotated 343 CCS values, and identified 29 isomers (including various isoflavonoids and flavonoids) in *Medicago truncatula* based on mass, CCS, accurate mass, chemical formula, and retention time. The integration of an extensive CCS database with diverse MS/MS spectra will facilitate the exploration of novel metabolites.

10.2.7 Nuclear magnetic resonance

NMR is another widely used analytical technique for studying the diverse metabolome within plants, encompassing examination of molecular structure, composition, and purity of molecules in the sample. Metabolic identification via NMR relies on the spectral characteristics of atomic nuclei with either an odd mass number (^{13}C) or an odd atomic number (^{1}H) when subjected to a strong magnetic field. This technique does not require the derivatization of samples and is a non-destructive, highly repeatable procedure for high-throughput and rapid metabolic analysis (Deborde *et al.*, 2017). The process is pH sensitive and requires a buffered solution for stable pH. Ionic liquids such as 1-butyl-3-methylimidazolium chloride or a mixture of phosphate buffer and methanol in a 1:1 ratio (pH 6.0) have mostly been used for evaluation of primary and secondary metabolites (Kim *et al.*, 2011). ^{1}H-NMR is a rapid and simple process for metabolic profiling but is not suitable for complex mixtures due to signal overlapping. Other advanced two-dimensional (2D) NMR techniques including 2D J-resolved NMR, heteronuclear single quantum coherence spectroscopy, heteronuclear multiple quantum coherence, total correlation spectroscopy, and nuclear overhauser effect spectroscopy are also used for quantification of primary and secondary metabolites (Kim *et al.*, 2011). High-resolution magic angle spinning NMR is generally used for lyophilized solid tissue. It does not require chemical extraction and can be utilized for MS and liquid-state NMR (Kumar *et al.*, 2016). Data acquisition time for 2D-NMR is approximately 20 min, while one-dimensional NMR takes 1 min. In maize seeds, ^{1}H-NMR was used to detect various compounds such as glycine betaine, trehalose, ethanol, and citric acid levels (Piccioni *et al.*, 2009), while Pandey *et al.* (2019) quantified the external absorption of glucose in chickpeas during seed germination when nitric oxide was present using this technique.

Alongside the analytical methods used to quantify different metabolites, there are numerous software programs and databases designed for the processing, analysis, and interpretation of metabolic data. Tools such as XCMS and mzMINE are utilized for metabolite identification, while others such as MetaboAnalyst and SIMCA are used for the analysis of metabolic data. Metabolic pathways can also be investigated using platforms such as MetaboAnalyst and MetScape.

10.3 Impact of Drought Stress on Metabolic Pathways

Plants encounter a range of environmental variations (abiotic and biotic factors) in their surroundings that lead to disruption of metabolic processes and ultimately hinder plant development and yield (Shiade *et al.*, 2024). Drought is a major factor that significantly impacts agricultural output worldwide (Vadez *et al.*, 2024). Plants undergo various adaptations involving changes in metabolic dynamics and gene expression under drought stress. A plant's stress tolerance is typically measured by how well it can maintain key metabolic processes and its defense mechanisms. Therefore, metabolic analysis can be an essential tool to examine and identify key differences between tolerant and sensitive plant species or genotypes and establish links between genetic and observable changes in plants (Carrera *et al.*, 2021). Two types of approaches—non-targeted and targeted—are used to determine metabolic reprogramming in plants under stress. The non-targeted approach gives an overview of most of the metabolites in plants, while the targeted approach measures and analyzes selected metabolites based on known chemistry and biology (Demiwal *et al.*, 2024). Therefore, metabolomics can reveal the role of metabolic reprogramming such as accumulation and regulation of primary and secondary metabolites in plants under drought stress.

10.3.1 Drought-mediated changes in primary metabolism

Drought stress regulates plant metabolism by altering biosynthesis and transport of primary and secondary metabolites (Table 10.1). Primary metabolites essentially regulate various mechanisms in plants such as growth and development, photosynthesis and respiration (Shiade *et al.*, 2024). Primary metabolites such as lipids, amino acids, and carbohydrates increase the plant's resilience and mitigate the consequence of water deficit.

Carbohydrates such as sucrose, glucose, and fructose are essential primary metabolites. They act as a primary energy source, supplying the necessary energy for various metabolic processes, including respiration and photosynthesis (Shiade *et al.*, 2024) and also play a crucial role in osmotic regulation under drought stress (Bistgani *et al.*, 2024). For example, Harsha *et al.* (2024) noted an increase in sugar concentration (sucrose, fructose, glucose, mannose, xylose, and ribose) in eggplants under drought stress. Elevated sugar levels may positively influence energy metabolism, osmotic adjustment, and signaling, thereby enhancing the response to drought stress. Sugars also play a significant role during grain filling under drought stress. In wheat, the level of soluble sugars and sucrose synthase activity increased in endosperm under drought stress (Li *et al.*, 2023), while overexpression of the *GhTPPS2* gene in tobacco resulted in enhanced drought tolerance (Wei *et al.*, 2024). *GhTPPS2* was involved in regulating sugar metabolism and enhancing the accumulation of soluble sugars in the plants under drought stress.

Amino acids are vital metabolic components that regulate protein biosynthesis and cellular activities and also act as osmolytes and ROS scavengers inside the cell under stress. Amino acids such as proline, mannitol, inositol, sorbitol, polyamines, and glycine betaine accumulate to protect the cytoplasmic compartment, maintain cell turgor, and facilitate the water uptake gradient. In faba bean, drought stress led to an increased accumulation of proline, which had an osmoprotective effect (Abid *et al.*, 2024), and exogenous application of proline increased growth, bulb yield, and photosynthetic efficiency of onion under drought stress (Semida *et al.*,

Table 10.1. Alterations in the metabolism of plants under drought stress.

Metabolites	Plant species	Response of metabolites	Effect on plant	Reference
Soluble sugars	*Solanum melongena*	Increase in glucose, fructose, sucrose, mannose, xylose, and ribose	Increased energy metabolism, osmotic adjustment	Harsha *et al.* (2024)
Soluble sugars	*Triticum aestivum*	Increase in glucose and sucrose	Energy and material accumulation	Li *et al.* (2023)
Proline	*Allium cepa*	Increase in sugar content	Increased bulb yield, photosynthetic efficiency	Semida *et al.* 2020)
Polyamines	*Triticum aestivum*	Increase in proline content	Maintains photosystem II activity	Doneva *et al.* (2021)
Polyamines	*Triticum aestivum*	Increased putrescene and spermine	Reduces chlorophyll degradation and RuBisCO level	Hassan *et al.* (2020)
Fatty acids	*Triticum aestivum*	Increase in palmitic acid and oleic acid	Maintains membrane integrity	Ullah *et al.* (2022)
Fatty acids	*Zea mays*	Increase in C18:3, C16:2, and C16:3 fatty acids	Enhances membrane fluidity and reduces membrane damage	Yin *et al.* (2024)
Phenolics	*Citrus latifolia* and *Citrus aurantifolia*	Increase in phenols and flavonoids	Reduces oxidative damage in the cell	Jafari and Shahsavar (2021)
Nitrogen- and sulfur-containing compounds	*Zea mays*	Increase in zealexins and kauralexins	Facilitates increased root:shoot ratio	Vaughan *et al.* (2015)
Cyanogenic glycoside	*Morus alba*	Increase dhurrin level	Membrane protection	Ackah *et al.* (2021)
Polyphenols	*Camellia sinensis*	Lignin accumulation	Maintains structural integrity of tracheary elements and prevents drought-induced xylem collapse	Li *et al.* (2024)

RuBisCO, ribulose-1,5-bisphosphate carboxylase/oxygenase.

2020). Polyamines also play an osmoregulatory role under drought stress, with putrescene, spermidine, and spermine having critical roles in plant development and stress tolerance. A study on wheat indicated that endogenous synthesis of polyamines increased drought tolerance during the grain-filling stage (Zhao *et al.*, 2024), while exogenous application of putrescene increased proline levels in wheat leaves and maintained photosystem II activity under drought stress (Doneva *et al.*, 2021). Similarly, exogenous application of putrescene and spermine in wheat reduced chlorophyll degradation and increased RuBisCO (ribulose-1,5-bisphosphate carboxylase/oxygenase) levels in the leaves, leading to enhanced growth under drought stress (Hassan *et al.*, 2020).

Besides carbohydrates and amino acids, fatty acids also improve drought tolerance in plants. A study on maize demonstrated an increase in C18:3 fatty acid in the plasma membrane, responsible for drought tolerance, while an increase in C16:2 and C16:3 was responsible for drought recovery (Yin *et al.*, 2024). Such changes in lipid composition may enhance membrane fluidity and reduce membrane damage, thus retaining photosynthetic capacity under drought. GC-MS analysis of wheat revealed an increase in fatty acids such as palmitic acid and oleic acid under drought stress to maintain membrane integrity (Ullah *et al.*, 2022), while an integrated study of transcriptomics and metabolomics in soybeans revealed upregulation of intrinsic membrane proteins, linoleic acid metabolism, and sugar metabolism pathways, including galactose, glucuronate, and pentose metabolism (Singh *et al.*, 2015). Additionally, there was an increase in thaumatin protein gene expression. These changes contribute to cell-wall remodeling, thereby enhancing water absorption and improving drought tolerance (Gao *et al.*, 2024).

10.3.2 Changes in secondary metabolism under drought conditions

In plants, the tricarboxylic acid (TCA) cycle and glycolysis are essential metabolic pathways that interconnect protein, carbohydrate, and lipid metabolism, facilitating growth and development. However, during adverse conditions, primary metabolism shifts to secondary metabolism, leading to the synthesis and accumulation of various small organic molecules, known as secondary metabolites (Yadav *et al.*, 2021). Based on their chemical properties and biosynthetic pathways, secondary metabolites are categorized into three groups: phenolics (lignans, flavonoids, tannins, phenolic acids, and stilbenes), terpenes (glycosides, carotenoid, sterols, and volatile compounds), and nitrogen- and sulfur-containing compounds (alkaloids and glucosinolates). Under drought conditions, various secondary metabolites play a critical role in improving survival of plants. Drought stress causes oxidative damage in plants by producing ROS. Various studies have demonstrated an increase in secondary metabolites, particularly from the phenylpropanoid pathway, that facilitates ROS scavenging to protect plant cells from lipid peroxidation-mediated oxidative stress (Sharma *et al.*, 2019; Xu *et al.*, 2021). During water deficit, citrus species reduce ROS-mediated cellular damage by increasing the level of antioxidant substances such as phenols and flavonoids to increase their survivability (Jafari and Shahsavar, 2021).

Secondary metabolites also play an important role in strengthening cell-wall components under stress. Several studies have shown an increase in total lignin content under drought stress (Sharma *et al.*, 2020; Liu *et al.*, 2021). The reduction in soil water potential leads to a discrepancy between water supply in the roots and demand from the leaves (Gusain *et al.*, 2024). This imbalance increases tension within the tracheary elements, causing their collapse. Consequently, the accumulation of lignin enhances the structural integrity of the tracheary elements, thereby improving resistance to drought-induced xylem collapse (Li *et al.*, 2024). Mazloom *et al.* (2020) demonstrated that the application of lignin-hydrogel decreased leaf electrolyte leakage and enhanced proline accumulation and biomass in maize. Terpenoids also regulate drought stress tolerance in plants, and various studies have demonstrated an increase in volatile terpene and resinous acid in conifer during drought stress (Lüpke *et al.*, 2016; Quan *et al.*, 2020). In Masson pine, it was observed that needles release more volatile terpenoids during drought stress (Quan *et al.*, 2020). Similarly, a greenhouse study on *Pinus sylvestris* seedlings showed an increase in biogenic volatile compounds under drought stress (Lüpke *et al.*, 2016). Nitrogen- and sulfur-containing compounds including phytoalexins, alkaloids, glucosinolates, and cyanogenic glucosides were also found to be altered under drought stress. Vaughan *et al.* (2015) reported an increase in zealexins and kauralexins in the roots of maize under drought stress, which facilitates an increased root:shoot ratio. The level of dhurrin, a cyanogenic glucoside, was also altered in the mulberry plant, suggesting its role in membrane protection under drought stress (Ackah *et al.*, 2021).

10.4 Role of Phytohormones and Transcription Factors in Regulating Plant Metabolism Under Water-Deficit Stress

Phytohormones are key signaling molecule that regulate various processes in plants, such as growth, productivity, and stress response. Under stress, plants acquire various strategies such as transcriptional regulation, post-transcriptional modification, and metabolic adjustment facilitated by phytohormones. Among the different phytohormones (abscisic acid (ABA), jasmonic acid, salicylic acid, auxin, and cytokinin), ABA is essential for plants experiencing drought stress. Drought stress disrupts the solute balance in the cytoplasm, prompting the production of various metabolites, and secondary messengers such as ROS in plants. ABA has been researched extensively as a crucial signaling molecule involved in drought-stress responses. Transcriptomic and metabolic studies in *Arabidopsis* have revealed a reduction in proline and branched-chain amino acid levels and an increase in oligosaccharides such as raffinose in ABA-biosynthesis mutants (Urano *et al.*, 2009), while Thalmann *et al.* (2016) showed a degradation in leaf starch under osmotic stress that could provide free soluble sugars for cellular osmotic balance in *Arabidopsis*. Various studies have shown the involvement of tryptophan and cysteine in ABA synthesis and ABA-associated processes (Liu *et al.*, 2022; Soda *et al.*, 2022).

Sato *et al.* (2018) demonstrated that sulfate and its derivative cysteine increased ABA synthesis in guard cells and facilitated stomatal closure. The early stages of ABA biosynthesis occur in plastids. SULTR, a chloroplast-localized sulfate transporter, has been implicated in this process (Chen *et al.*, 2019). Mutant analysis of the *sultr3* mutant revealed that sulfate application failed to induce stomatal closure, suggesting that sulfate uptake into chloroplasts is essential for activating ABA synthesis in guard cells. The breakdown of cysteine, which generates hydrogen sulfide, also plays a key role in ABA synthesis and signal transduction. Scuffi *et al.* (2014) showed that expression of a cysteine-catabolizing enzyme in guard cells, which produces hydrogen sulfide, resulted in ABA-mediated stomatal closure. Further research has shown that hydrogen sulfide produced by DES1 persulfidation activates signaling components such as OST1 (OPEN STOMATA 1) and RBOHD (RESPIRATORY BURST OXIDASE HOMOLOG D) involved in stomatal closure (Chen *et al.*, 2020; Shen *et al.*, 2020).

Under drought stress, auxin can influence carbon flow by regulating primary metabolic pathways. Shi *et al.* (2014) observed auxin-mediated modulation of carbon metabolites such as amino acids, sugar and sugar alcohols, aromatic amine, and organic acids in *Arabidopsis* under water-deficit stress. Exogenous treatment with indole-3-acetic acid (IAA) on white clover enhanced relative water content and chlorophyll content under polyethylene glycol-induced drought stress (Zhang *et al.*, 2020). Additionally, exogenous IAA increased the ABA and jasmonic acid (JA) levels in the leaves under water-deficit stress (Zhang *et al.*, 2020). In *Arabidopsis*, auxins also regulate glucosinolate levels under water-deficit stress. Salehin *et al.* (2019) observed auxin/IAA-mediated regulation of aliphatic glucosinolates in *Arabidopsis*, which has a major role in stomatal regulation under drought stress. The phytohormone ethylene can act as a positive or negative regulator in abiotic stress. HPLC analysis of soybean leaves under ethylene treatment showed an increase in isoflavones such as genistin, genistein, and daidzein. In squash, it was observed that ethylene receptor mutant *etr2b* lines showed accumulation of fatty acids (unsaturated and polyunsaturated), phenolic compounds, soluble sugars, and trigonelline under drought stress to prevent oxidative damage in the cell (Iglesias-Moya *et al.*, 2023).

Cytokinins act as negative regulators under drought stress (Gusain *et al.*, 2024). Therefore, cytokinin levels can be altered by genetic modification to increase yield and stress tolerance (Joshi *et al.*, 2018). In barley, overexpression of the cytokinin dehydrogenase gene from *Arabidopsis* caused an increase in lignin biosynthesis in roots under water-deficit stress (Pospíšilová *et al.*, 2016). Conversely, *Arabidopsis* seedlings with decreased levels of cytokinin displayed enhanced tolerance to drought and salinity stress (Nishiyama *et al.*, 2011). This tolerant phenotype may be linked to improved membrane integrity and heightened ABA sensitivity, although ABA levels under normal conditions were reduced in these cytokinin-deficient seedlings. Nishiyama *et al.* (2013) observed that *ahk2*, *ahk3*, *ahk4* multiple mutants in *Arabidopsis* resulted in a drought-tolerant phenotype, suggesting that cytokinin biosynthesis and signaling negatively regulate ABA signaling during drought-stress conditions.

Brassinosteroids (BRs) regulate several developmental processes in plants. Apart from their essential role in plant development, studies have indicated the involvement of BRs in drought-stress responses and ABA signaling. BRs are perceived by a receptor complex containing BRI1 (BRASSINOSTEROID INSENSITIVE 1) and BAK1

(BRI1-ASSOCIATED RECEPTOR KINASE1) (Planas-Riverola *et al.*, 2019). This receptor complex activates a kinase signaling cascade, stabilizing and activating transcription factors such as BES1 (BRI1-EMS-SUPPRESSOR), which induces BR-regulated genes. Caño-Delgado *et al.* (2004) showed that, unlike the BRI1 receptor, its homologs *BRL1* and *BRL3* (*BRI1-LIKE*) were found in the innermost part of vascular tissues, indicating their role in vascular development. Fàbregas *et al.* (2018) observed that the loss of function in *BRI1* conferred drought tolerance but severely impaired plant growth, whereas the overexpression of *BRL3* enhanced drought tolerance without compromising plant growth. Multi-omics studies have demonstrated an increase in proline and sugar accumulation, which can contribute to enhanced drought tolerance by maintaining cellular turgor. Genes such as *PDH1* (*PROLINE DEHYDROGENASE 1*), *TPP* (*TREHALOSE-6-PHOSPHATE PHOSPHATASE*), *GolS2* (*GALACTINOL SYNTHASE 2*), sucrose synthases, and hexokinases are differentially regulated in response to *BRL3* overexpression.

BR signaling has been shown to interact with ABA signaling in *Arabidopsis* during seed germination (Steber and McCourt, 2001). BIN2, a glycogen synthase kinase 3-like enzyme, was found to activate SnRK2, an important signaling regulator of the ABA signaling pathway (Fujita *et al.*, 2009; Cai *et al.*, 2014). In the absence of BRs, BIN2 phosphorylates and inactivates transcription factors such as BES1, and ABA was found to inhibit BIN2-dependent signaling (Zhang *et al.*, 2009). Cai *et al.* (2014) suggested that BIN2 activates the ABA signaling pathway by regulating SnRK2 kinases. BIN2 is regulated by BRs, as well as by ABA signaling. Studies also show transcription factor-mediated antagonistic interaction between BR and ABA signaling pathways. Fujita *et al.* (2004) observed that upregulation of the *RD26* (*RESPONSIVE TO DESICCATION 26*) transcription factor gene increased the expression of ABA-inducible genes in response to drought stress in *Arabidopsis*, while Yu *et al.* (2011) found that *BES1* negatively regulates *RD26* expression (Yu *et al.*, 2011). However, Ye *et al.* (2017) showed an antagonistic interaction between *RD26* and *BES1* via transcriptional repression and protein–protein interaction, thus turning on and off the ABA- and BR-regulated drought and growth signaling pathway. Moreover, it is yet to be determined whether the interaction between *RD26* and *BES1* is specific to their homologs. Thus, overall, BR and ABA signaling are proposed to interact through transcription factors and kinases, potentially regulating genes involved in proline synthesis and sugar metabolism (Fàbregas *et al.*, 2018), and these interactions might regulate growth and stress responses.

JA is another phytohormone that regulates drought-stress response in plants. A transcriptomic study on *Arabidopsis* showed that genes involved in JA biosynthesis and signaling did not show any change in expression at the initial stage of mild drought stress but were reduced during later stages. It is suggested that JA signaling is decreased under prolonged drought stress to reduce its inhibitory effects on growth. Kim *et al.* (2017) showed that exogenous application of acetate increased drought-stress resistance in *Arabidopsis*. Acetate accumulation has been found to enhance drought resistance in various species, most likely due to the initiation of JA biosynthesis and signaling.

10.5 Metabolomics: An Essential Tool for Understanding Drought-stress Adaptation in Plants

Metabolomics has surfaced as a powerful tool in plant biology, offering a view of the complex biochemical responses that plants undergo to withstand adverse conditions such as drought. Understanding how plants adapt metabolically is essential for creating strategies to improve drought resistance in crops. Metabolomics enables comprehensive profiling of small molecules (metabolites) within plant tissues, providing a snapshot of the metabolic state under different conditions, including drought stress. By identifying and quantifying the metabolites involved in primary and secondary metabolic pathways, metabolomics sheds light on how plants allocate resources, manage energy, and synthesize protective compounds during drought (Fig. 10.1). Recently, metabolomics has been used for the identification of metabolic

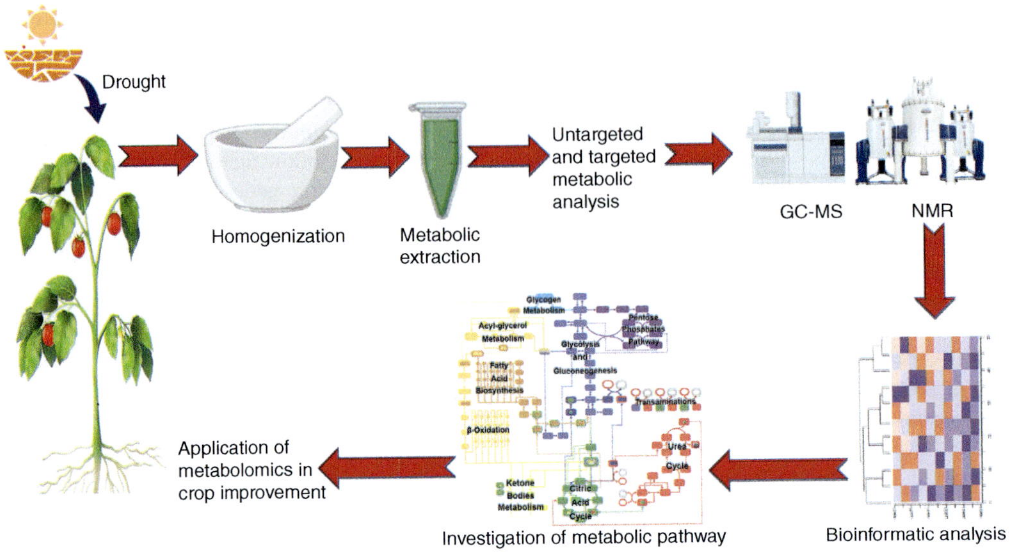

Fig. 10.1. Schematic representation of metabolic analysis of plants exposed to drought stress. The figure illustrates various steps involved in metabolic analysis such as extract preparation, metabolic analysis via analytical methods such as gas chromatography–mass spectrometry (GC-MS) and nuclear magnetic resonance (NMR) spectroscopy, and bioinformatic analysis using a software program and databases designed for the processing, analysis, and interpretation of metabolic data and utilization of metabolic information for crop improvement.

markers for drought tolerance because of their close association with yield phenotype. Metabolomic studies in *Quercus suber* have revealed the involvement of secondary metabolites such as terpenes (tarennoside and leonuridine), phenols, and flavonoids in drought-stress tolerance.

Accumulation of flavonoids such as epigallocatechin 3-gallate and 3,5-digalloylepicatechin in *Q. suber* helps to curb ROS accumulation during drought stress (Almeida *et al.*, 2020). Ultra-performance liquid chromatography tandem mass spectrometry (UPLC-MS/MS) analysis in licorice root extract revealed differential expression of 18 metabolites comprising three phenols, five flavonoids, three amino acids, three sugars and alcohols, one nucleotide, two lipids, and three organic acids under drought stress (Zhang *et al.*, 2023). Wei *et al.* (2020) performed LC-MS analysis to determine drought-regulated metabolites in two safflower genotypes (PI560169 and PI401477) under water-deficit and well-watered conditions. Principal component analysis of the metabolic data showed 13 common differential metabolites in both genotypes and 56 common differential metabolites in well-watered and drought-stress conditions in both genotypes.

Metabolomics highlights the activation of pathways leading to the biosynthesis of secondary metabolites such as phenolics, terpenoids, and alkaloids under drought conditions (Zhang *et al.*, 2023). For instance, phenylalanine ammonia-lyase and cinnamate-4-hydroxylase activate the phenylpropanoid pathway, leading to the accumulation of phenolic and flavonoid compounds (Yang *et al.*, 2021). These compounds play diverse roles such as strengthening of cell walls and acting as antioxidants to prevent cellular damage in cells under drought stress. Similarly, enzymes such as 1-deoxy-D-xylulose-5-phosphate synthase lead to terpenoid biosynthesis, which further contributes to the plant's defense mechanisms and can also help mitigate the oxidative stress associated with drought. Metabolomics elucidates the biosynthesis, accumulation, and signaling pathways of various plant hormones, revealing their roles in the regulation of stomatal closing, water-use efficiency, and stress-responsive gene expression (Shailani *et al.*, 2021). For instance, ABA biosynthesis is induced by 9-*cis*-epoxycarotenoid dioxygenase under drought stress, initiating cascades that enhance drought tolerance through osmotic adjustment and antioxidant defense.

Combining metabolomics with other omics disciplines such as genomics, transcriptomics, and proteomics can be helpful for the comprehensive assessment of plant reactions to drought stress. Multi-omics approaches facilitate the identification of key regulatory genes, transcription factors, and metabolic networks that govern adaptive responses (Joshi *et al.*, 2016). These insights pave the way for targeted breeding strategies and genetic engineering aimed at developing drought-tolerant crop varieties. A comparative study using transcriptomics and metabolomics on two asparagus varieties, Pacific Early (drought tolerant) and Jilv3 (drought sensitive), revealed that 2567 and 7187 genes were differentially expressed, respectively, related to carbon metabolism, amino acid biosynthesis, and plant hormone signaling under drought stress (X. Zhang *et al.*, 2024). Additionally, Pacific Early and Jilv3 showed accumulation of 112 and 254 differentially accumulated metabolites under drought conditions. The results showed that Jilv3 demonstrated drought tolerance due to enhanced TCA cycle activity and increased flavonoid biosynthesis.

10.6 Conclusion and Future Perspectives

Drought stress presents a significant challenge to plant growth and productivity, necessitating a comprehensive understanding of its effects on plant metabolism. Metabolic approaches have proven invaluable in unraveling the complex biochemical responses that plants exhibit under water-deficit conditions. These approaches, facilitated by advanced tools and techniques for metabolic analysis, have highlighted critical changes in both primary and secondary metabolic pathways. Primary metabolism, responsible for essential processes such as photosynthesis and respiration, undergoes notable alterations to conserve resources and maintain cellular homeostasis. Similarly, secondary metabolism, which contributes to plant defense and stress tolerance, is also profoundly impacted, resulting in the accumulation of protective metabolites. Phytohormones and transcription factors play crucial roles in regulating these metabolic adjustments, orchestrating a coordinated response to drought stress. The intricate interplay between these signaling molecules ensures that plants can adapt to varying degrees of water availability, enhancing their resilience. Metabolomics has come out as an important tool in this field, providing comprehensive insights into the metabolic networks associated with drought stress tolerance. By utilizing metabolomics, researchers can identify key metabolites and biomarkers associated with drought tolerance, paving the way for the development of drought-resistant crop varieties. Overall, the integration of metabolic approaches, phytohormonal regulation, and advanced metabolomic techniques offers the possibility of a comprehensive understanding of plant responses to drought stress. This knowledge will be important for developing innovative strategies to alleviate the adverse effects of drought on agriculture, ensuring sustainable food production in the face of increasing global water scarcity. Future research should focus on refining analytical techniques, expanding metabolite databases, and elucidating metabolic interactions to accelerate the development of resilient crop varieties capable of withstanding drought and other environmental stresses.

10.7 Acknowledgments

The authors thank the Director, CSIR-Institute of Himalayan Bioresource Technology, Palampur, India, for infrastructure and support. S.G. thanks the University Grants Commission, Govt. of India (UGC Ref. No.: 574/CSIR-UGC NET JUNE 2019), for providing a Senior Research Fellowship. The authors acknowledge the financial support provided by CSIR (MLP-201) and DBT (BT/PR45280/NER/95/1918/2022). This manuscript represents CSIR-IHBT Communication No. 5625.

10.9 References

Abid, G., Jebara, M., Debode, F., Vertommen, D., dit Ruys, S.P., *et al.* (2024) Comparative physiological, biochemical and proteomic analyses reveal key proteins and crucial regulatory pathways related to drought stress tolerance in faba bean (*Vicia faba* L.) leaves. *Current Plant Biology* 37: 100320.

Ackah, M., Shi, Y., Wu, M., Wang, L., Guo, P., *et al.* (2021) Metabolomics response to drought stress in *Morus alba* L. variety Yu-711. *Plants* 10: 1636.

Ahmad, P., Ahanger, M.A., Singh, V.P., Tripathi, D.K., Alam, P., and Alyemeni, M.N. (eds) (2018) *Plant Metabolites and Regulation under Environmental Stress*. Academic Press, London.

Almeida, T., Pinto, G., Correia, B., Gonçalves, S., Meijon, M., and Escandón, M. (2020) In-depth analysis of the *Quercus suber* metabolome under drought stress and recovery reveals potential key metabolic players. *Plant Science* 299: 110606.

Baker, T.C., Han, J., and Borchers, C.H. (2017) Recent advancements in matrix-assisted laser desorption/ionization mass spectrometry imaging. *Current Opinion in Biotechnology* 43, 62–69.

Bistgani, Z.E., Barker, A.V., and Hashemi, M. (2024) Physiology of medicinal and aromatic plants under drought stress. *Crop Journal* 12, 330–339.

Cai, Z., Liu, J., Wang, H., Yang, C., Chen, Y., *et al.* (2014) GSK3-like kinases positively modulate abscisic acid signaling through phosphorylating subgroup III SnRK2s in *Arabidopsis*. *Proceedings of the National Academy of Sciences USA* 111, 9651–9656.

Caño-Delgado, A., Yin, Y., Yu, C., Vafeados, D., Mora-García, S., *et al.* (2004) BRL1 and BRL3 are novel brassinosteroid receptors that function in vascular differentiation in *Arabidopsis*. *Development* 131, 5341–5351.

Carrera, F.P., Noceda, C., Maridueña-Zavala, M.G., and Cevallos-Cevallos, J.M. (2021) Metabolomics, a powerful tool for understanding plant abiotic stress. *Agronomy* 11: 824.

Chen, S., Jia, H., Wang, X., Shi, C., Wang, X., *et al.* (2020) Hydrogen sulfide positively regulates abscisic acid signaling through persulfidation of SnRK2.6 in guard cells. *Molecular Plant* 13, 732–744.

Chen, Z., Zhao, P.X., Miao, Z.Q., Qi, G.F., Wang, Z., *et al.* (2019) SULTR3s function in chloroplast sulfate uptake and affect ABA biosynthesis and the stress response. *Plant Physiology* 180, 593–604.

Deborde, C., Moing, A., Roch, L., Jacob, D., Rolin, D., and Giraudeau, P. (2017) Plant metabolism as studied by NMR spectroscopy. *Progress in Nuclear Magnetic Resonance Spectroscopy* 102, 61–97.

Demiwal, P., Tayade, S., Yadav, S.R., and Sircar, D. (2024) A metabolomics perspective on root-derived plant immunity and phytohormone interaction. *Physiologia Plantarum* 176: e14150.

Dietz, K.J., Zörb, C., and Geilfus, C.M. (2021) Drought and crop yield. *Plant Biology* 23, 881–893.

Doneva, D., Pál, M., Brankova, L., Szalai, G., Tajti, J., *et al.* (2021) The effects of putrescine pre-treatment on osmotic stress responses in drought-tolerant and drought-sensitive wheat seedlings. *Physiologia Plantarum* 171, 200–216.

Fàbregas, N., Lozano-Elena, F., Blasco-Escámez, D., Tohge, T., Martínez-Andújar, C., *et al.* (2018) Overexpression of the vascular brassinosteroid receptor BRL3 confers drought resistance without penalizing plant growth. *Nature Communications* 9: 4680.

Farooq, M.S., Uzair, M., Raza, A., Habib, M., Xu, Y., *et al.* (2022) Uncovering the research gaps to alleviate the negative impacts of climate change on food security: a review. *Frontiers in Plant Science* 13: 927535.

Fiehn, O. (2016) Metabolomics by gas chromatography–mass spectrometry: combined targeted and untargeted profiling. *Current Protocols in Molecular Biology* 114, 30–34.

Fujita, M., Fujita, Y., Maruyama, K., Seki, M., Hiratsu, K., *et al.* (2004) A dehydration-induced NAC protein, RD26, is involved in a novel ABA-dependent stress-signaling pathway. *Plant Journal* 39, 863–876.

Fujita, Y., Nakashima, K., Yoshida, T., Katagiri, T., Kidokoro, S., *et al.* (2009) Three SnRK2 protein kinases are the main positive regulators of abscisic acid signaling in response to water stress in *Arabidopsis*. *Plant and Cell Physiology* 50, 2123–2132.

Gao, S., Li, M., Hu, Y., Zhang, T., Guo, J., *et al.* (2024) Comparative differences in maintaining membrane fluidity and remodeling cell wall between *Glycine soja* and *Glycine max* leaves under drought. *Plant Physiology and Biochemistry* 209: 108545.

Gusain, S., Kumari, K., and Joshi, R. (2024) Physiological, hormonal and molecular dynamics of root system architectural response to drought stress signaling in crops. *Rhizosphere* 31: 100922.

Harsha, S.G., Girish, B., Dinsha, M., Patil, M.I., Singh, T.H., and Prathibha, M.D. (2024) Comparative study on physiological intricacies and sugar accumulation dynamics in brinjal (*Solanum melongena* L.) under drought stress. *Scientia Horticulturae* 325: 112633.

Hassan, N., Ebeed, H., and Aljaarany, A. (2020) Exogenous application of spermine and putrescine mitigate adversities of drought stress in wheat by protecting membranes and chloroplast ultra-structure. *Physiology and Molecular Biology of Plants* 26, 233–245.

Iglesias-Moya, J., Abreu, A.C., Alonso, S., Torres-García, M.T., Martínez, C., *et al.* (2023) Physiological and metabolomic responses of the ethylene insensitive squash mutant *etr2b* to drought. *Plant Science* 336: 111853.

Jafari, M. and Shahsavar, A. (2021) The effect of foliar application of melatonin on changes in secondary metabolite contents in two citrus species under drought stress conditions. *Frontiers in Plant Science* 12: 692735.

Jorge, T.F., Rodrigues, J.A., Caldana, C., Schmidt, R., van Dongen, J.T., *et al.* (2016) Mass spectrometry-based plant metabolomics: metabolite responses to abiotic stress. *Mass Spectrometry Reviews* 35, 620–649.

Joshi, R. and Karan, R. (2014) Physiological, biochemical and molecular mechanisms of drought tolerance in plants. In: Gaur, R.K. and Sharma P. (eds) *Molecular Approaches in Plant Abiotic Stress*. CRC Press, Boca Raton, Florida, pp. 209–231.

Joshi, R., Wani, S.H., Singh, B., Bohra, A., Dar, Z.A., *et al.* (2016) Transcription factors and plants response to drought stress: current understanding and future directions. *Frontiers in Plant Science* 7: 1029.

Joshi, R., Sahoo, K.K., Tripathi, A.K., Kumar, R., Gupta, B.K., *et al.* (2018) Knockdown of an inflorescence meristem-specific cytokinin oxidase—OsCKX2 in rice reduces yield penalty under salinity stress condition. *Plant, Cell & Environment* 41, 936–946.

Joshi, R., Sahoo, K.K., Singh, A.K., Anwar, K., Pundir, P., *et al.* (2020) Enhancing trehalose biosynthesis improves yield potential in marker-free transgenic rice under drought, saline, and sodic conditions. *Journal of Experimental Botany* 71, 653–668.

Kaur, G. and Asthir, B. (2017) Molecular responses to drought stress in plants. *Biologia Plantarum* 61, 201–209.

Kim, H.K., Choi, Y.H., and Verpoorte, R. (2011) NMR-based plant metabolomics: where do we stand, where do we go? *Trends in Biotechnology* 29, 267–275.

Kim, J.M., To, T.K., Matsui, A., Tanoi, K., Kobayashi, N.I., *et al.* (2017) Acetate-mediated novel survival strategy against drought in plants. *Nature Plants* 3: 17097.

Kumar, M., Kuzhiumparambil, U., Pernice, M., Jiang, Z., and Ralph, P.J. (2016) Metabolomics: an emerging frontier of systems biology in marine macrophytes. *Algal Research* 16, 76–92.

Kumar, M., Kumar Patel, M., Kumar, N., Bajpai, A.B., and Siddique, K.H. (2021) Metabolomics and molecular approaches reveal drought stress tolerance in plants. *International Journal of Molecular Sciences* 22: 9108.

Lapthorn, C., Pullen, F., and Chowdhry, B.Z. (2013) Ion mobility spectrometry-mass spectrometry (IMS-MS) of small molecules: separating and assigning structures to ions. *Mass Spectrometry Reviews* 32, 43–71.

Li, C., Fu, K., Guo, W., Zhang, X., Li, C., and Li, C. (2023) Starch and sugar metabolism response to post-anthesis drought stress during critical periods of elite wheat (*Triticum aestivum* L.) endosperm development. *Journal of Plant Growth Regulation* 42, 5476–5494.

Li, J., Ren, J., Lei, X., Fan, W., Tang, L., *et al.* (2024) CsREV-CsTCP4-CsVND7 module shapes xylem patterns differentially between stem and leaf to enhance tea plant tolerance to drought. *Cell Reports* 43: 113987.

Liu, W., Jiang, Y., Jin, Y., Wang, C., Yang, J., and Qi, H. (2021) Drought-induced ABA, H_2O_2 and JA positively regulate *CmCAD* genes and lignin synthesis in melon stems. *BMC Plant Biology* 21: 83.

Liu, W.C., Song, R.F., Zheng, S.Q., Li, T.T., Zhang, B.L., *et al.* (2022) Coordination of plant growth and abiotic stress responses by tryptophan synthase β subunit 1 through modulation of tryptophan and ABA homeostasis in *Arabidopsis*. *Molecular Plant* 15, 973–990.

Lüpke, M., Leuchner, M., Steinbrecher, R., and Menzel, A. (2016) Impact of summer drought on isoprenoid emissions and carbon sink of three Scots pine provenances. *Tree Physiology* 36, 1382–1399.

Mazloom, N., Khorassani, R., Zohury, G.H., Emami, H., and Whalen, J. (2020) Lignin-based hydrogel alleviates drought stress in maize. *Environmental and Experimental Botany* 175: 104055.

Nikolaev, E.N., Kostyukevich, Y.I., and Vladimirov, G.N. (2016) Fourier transform ion cyclotron resonance (FT ICR) mass spectrometry: theory and simulations. *Mass Spectrometry Reviews* 35, 219–258.

Nishiyama, R., Watanabe, Y., Fujita, Y., Le, D.T., Kojima. M., *et al.* (2011) Analysis of cytokinin mutants and regulation of cytokinin metabolic genes reveals important regulatory roles of cytokinins in drought, salt and abscisic acid responses, and abscisic acid biosynthesis. *Plant Cell* 23, 2169–2183.

Nishiyama, R., Watanabe, Y., Leyva-Gonzalez, M.A., van Ha, C., Fujita, Y., *et al.* (2013) *Arabidopsis* AHP2, AHP3, and AHP5 histidine phosphotransfer proteins function as redundant negative regulators of drought stress response. *Proceedings of the National Academy of Sciences USA* 110, 4840–4845.

Obata, T. and Fernie, A.R. (2012) The use of metabolomics to dissect plant responses to abiotic stresses. *Cellular and Molecular Life Sciences* 69, 3225–3243.

Oh, S.W., Imran, M., Kim, E.H., Park, S.Y., Lee, S.G., *et al.* (2023) Approach strategies and application of metabolomics to biotechnology in plants. *Frontiers in Plant Science* 14: 1192235.

Okazaki, Y. and Saito, K. (2011) Recent advances of metabolomics in plant biotechnology. *Plant Biotechnology Reports* 6, 1–15.

Pandey, S., Kumari, A., Shree, M., Kumar, V., Singh, P., *et al.* (2019) Nitric oxide accelerates germination via the regulation of respiration in chickpea. *Journal of Experimental Botany* 70, 4539–4555.

Piccioni, F., Capitani, D., Zolla, L., and Mannina, L. (2009) NMR metabolic profiling of transgenic maize with the *Cry1A(b)* gene. *Journal of Agricultural and Food Chemistry* 57, 6041–6049.

Planas-Riverola, A., Gupta, A., Betegón-Putze, I., Bosch, N., Ibañes, M., and Caño-Delgado, A.I. (2019) Brassinosteroid signaling in plant development and adaptation to stress. *Development* 146: dev151894.

Pospíšilová, H., Jiskrova, E., Vojta, P., Mrizova, K., Kokáš, F., *et al.* (2016) Transgenic barley overexpressing a cytokinin dehydrogenase gene shows greater tolerance to drought stress. *New Biotechnology* 33, 692–705.

Quan, W., Zhao, C., and Ding, G. (2020) Dynamics of volatile organic compounds in Masson pine (*Pinus massoniana* lamb.) seedling needles under drought stress. *Pakistan Journal of Botany* 52, 375–382.

Raza, A. (2022) Metabolomics: a systems biology approach for enhancing heat stress tolerance in plants. *Plant Cell Reports* 41, 741–763.

Salehin, M., Li, B., Tang, M., Katz, E., Song, L., *et al.* (2019) Auxin-sensitive Aux/IAA proteins mediate drought tolerance in *Arabidopsis* by regulating glucosinolate levels. *Nature Communications* 10: 4021.

Salem, M.A., Perez de Souza, L., Serag, A., Fernie, A.R., Farag, M.A., *et al.* (2020) Metabolomics in the context of plant natural products research: from sample preparation to metabolite analysis. *Metabolites* 10: 37.

Sato, H., Takasaki, H., Takahashi, F., Suzuki, T., Iuchi, S., *et al.* (2018) *Arabidopsis thaliana* NGATHA1 transcription factor induces ABA biosynthesis by activating *NCED3* gene during dehydration stress. *Proceedings of the National Academy of Sciences USA* 115, E11178–E11187.

Schroeder, M., Meyer, S.W., Heyman, H.M., Barsch, A., and Sumner, L.W. (2019) Generation of a collision cross section library for multi-dimensional plant metabolomics using UHPLC-trapped ion mobility-MS/MS. *Metabolites* 10: 13.

Scuffi, D., Álvarez, C., Laspina, N., Gotor, C., Lamattina, L., and García-Mata, C. (2014) Hydrogen sulfide generated by L-cysteine desulfhydrase acts upstream of nitric oxide to modulate abscisic acid-dependent stomatal closure. *Plant Physiology* 166, 2065–2076.

Semida, W.M., Abdelkhalik, A., Rady, M.O., Marey, R.A., and Abd El-Mageed, T.A. (2020) Exogenously applied proline enhances growth and productivity of drought stressed onion by improving photosynthetic efficiency, water use efficiency and up-regulating osmoprotectants. *Scientia Horticulturae* 272: 109580.

Shailani, A., Joshi, R., Singla-Pareek, S.L., and Pareek, A. (2021) Stacking for future: pyramiding genes to improve drought and salinity tolerance in rice. *Physiologia Plantarum* 172, 1352–1362.

Sharma, A., Shahzad, B., Rehman, A., Bhardwaj, R., Landi, M., and Zheng, B. (2019) Response of phenylpropanoid pathway and the role of polyphenols in plants under abiotic stress. *Molecules* 24: 2452.

Sharma, N.K., Gupta, S.K., Dwivedi, V., and Chattopadhyay, D. (2020). Lignin deposition in chickpea root xylem under drought. *Plant Signaling & Behavior* 15: 1754621.

Shen, J., Zhang, J., Zhou, M., Zhou, H., Cui, B., *et al.* (2020) Persulfidation-based modification of cysteine desulfhydrase and the NADPH oxidase RBOHD controls guard cell abscisic acid signaling. *Plant Cell* 32, 1000–1017.

Shi, H., Chen, L., Ye, T., Liu, X., Ding, K., and Chan, Z. (2014) Modulation of auxin content in *Arabidopsis* confers improved drought stress resistance. *Plant Physiology and Biochemistry* 82, 209–217.

Shiade, S.R.G., Zand-Silakhoor, A., Fathi, A., Rahimi, R., Minkina, T., *et al.* (2024) Plant metabolites and signaling pathways in response to biotic and abiotic stresses: exploring bio stimulant applications. *Plant* Stress 12: 100454.

Singh, B., Bohra, A., Mishra, S., Joshi, R., and Pandey, S. (2015) Embracing new-generation 'omics' tools to improve drought tolerance in cereal and food-legume crops. *Biologia Plantarum* 59, 413–428.

Soda, M.N., Hayashi, Y., Takahashi, K., and Kinoshita, T. (2022) Tryptophan synthase β subunit 1 affects stomatal phenotypes in *Arabidopsis thaliana*. *Frontiers in Plant Science* 13: 1011360.

Steber, C.M. and McCourt, P. (2001) A role for brassinosteroids in germination in *Arabidopsis*. *Plant Physiology* 125, 763–739.

Tanna, B. and Mishra, A. (2018) Metabolomics of seaweeds: tools and techniques. In: Ahmad, P., Ahanger, M.A., Singh, V.P., Tripathi, D.K., Alam, P., and Alyemeni, M.N. (eds) *Plant Metabolites and Regulation under Environmental Stress*. Academic Press, London, pp. 37–52.

Thalmann, M., Pazmino, D., Seung, D., Horrer, D., Nigro, A., *et al.* (2016) Regulation of leaf starch degradation by abscisic acid is important for osmotic stress tolerance in plants. *Plant Cell* 28, 1860–1878.

Ullah, S., Khan, M.N., Lodhi, S.S., Ahmed, I., Tayyab, M., *et al.* (2022) Targeted metabolomics reveals fatty acid abundance adjustments as playing a crucial role in drought-stress response and post-drought recovery in wheat. *Frontiers in Genetics* 13: 972696.

Urano, K., Maruyama, K., Ogata, Y., Morishita, Y., Takeda, M., *et al.* (2009) Characterization of the ABA-regulated global responses to dehydration in *Arabidopsis* by metabolomics. *Plant Journal* 57, 1065–1078.

Vadez, V., Grondin, A., Chenu, K., Henry, A., Laplaze, L., *et al.* (2024) Crop traits and production under drought. *Nature Reviews Earth & Environment* 5, 211–225.

Varshney, R.K., Barmukh, R., Roorkiwal, M., Qi, Y., Kholova, J., *et al.* (2021) Breeding custom-designed crops for improved drought adaptation. *Advanced Genetics* 2: e202100017.

Vaughan, M.M., Christensen, S., Schmelz, E.A., Huffaker, A., Mcauslane, H.J., *et al.* (2015) Accumulation of terpenoid phytoalexins in maize roots is associated with drought tolerance. *Plant, Cell & Environment* 38, 2195–2207.

Wei, B., Hou, K., Zhang, H., Wang, X., and Wu, W. (2020) Integrating transcriptomics and metabolomics to studies key metabolism, pathways and candidate genes associated with drought-tolerance in *Carthamus tinctorius* L. under drought stress. *Industrial Crops and Products* 151: 112465.

Wei, Y., Song, Y., Khan, M.A., Liang, C., Meng, Z., *et al.* (2024) GhTPPA_2 enhancement of tobacco sugar accumulation and drought tolerance. *Gene* 894: 147969.

Xu, C., Wei, L., Huang, S., Yang, C., Wang, Y., *et al.* (2021) Drought resistance in qingke involves a reprogramming of the phenylpropanoid pathway and UDP-glucosyltransferase regulation of abiotic stress tolerance targeting flavonoid biosynthesis. *Journal of Agricultural and Food Chemistry* 69, 3992–4005.

Yang, Y.H., Yang, H., Li, R.F., Li, C.X., Zeng, L., *et al.* (2021) A *Rehmannia glutinosa* cinnamate 4-hydroxylase promotes phenolic accumulation and enhances tolerance to oxidative stress. *Plant Cell Reports* 40, 375–391.

Ye, H., Liu, S., Tang, B., Chen, J., Xie, Z., *et al.* (2017) RD26 mediates crosstalk between drought and brassinosteroid signalling pathways. *Nature Communications* 8: 14573.

Yin, L., Xu, J., Zhang, L., Liu, D., Zhang, C., *et al.* (2024) Altered fatty acid composition confers improved drought acclimation in maize. *Plant Physiology and Biochemistry* 206: 108274.

Yu, X., Li, L., Zola, J., Aluru, M., Ye, H., *et al.* (2011) A brassinosteroid transcriptional network revealed by genome-wide identification of BESI target genes in *Arabidopsis thaliana*. *Plant Journal* 65, 634–646.

Zhang, D., Liu, Y., Yang, Z., Song, X., Ma, Y., *et al.* (2023) Widely target metabolomics analysis of the differences in metabolites of licorice under drought stress. *Industrial Crops and Products* 202: 117071.

Zhang, F., Rosental, L., Ji, B., Brotman, Y., and Dai, M. (2024) Metabolite-mediated adaptation of crops to drought and the acquisition of tolerance. *Plant Journal* 118, 626–644.

Zhang, S., Cai, Z., and Wang, X. (2009) The primary signaling outputs of brassinosteroids are regulated by abscisic acid signaling. *Proceedings of the National Academy of Sciences USA* 106, 4543–4548.

Zhang, X., Han, C., Wang, Y., Liu, T., Liang, Y., and Cao, Y. (2024) Integrated analysis of transcriptomics and metabolomics of garden asparagus (*Asparagus officinalis* L.) under drought stress. *BMC Plant Biology* 24: 563.

Zhang, Y., Li, Y., Hassan, M.J., Li, Z., and Peng, Y. (2020) Indole-3-acetic acid improves drought tolerance of white clover via activating auxin, abscisic acid and jasmonic acid related genes and inhibiting senescence genes. *BMC Plant Biology* 20: 150.

Zhao, P., Chen, X., Xue, X., Wang, Y., Wang, Y., *et al.* (2024) Improvement of polyamine synthesis maintains photosynthetic function in wheat during drought stress and rewatering at the grain filling stage. *Plant Growth Regulation* 102, 497–513.

Zhou, Z., Luo, M., Chen, X., Yin, Y., Xiong, X., *et al.* (2020) Ion mobility collision cross-section atlas for known and unknown metabolite annotation in untargeted metabolomics. *Nature Communications* 11: 4334.

11 Applied Metabolomics for Crop Improvement and Plant Breeding

Saika Nabi[1]*, M. Altaf Wani[1], Asif. B. Shikari[1], Zahoor A. Dar[2], P.A. Sofi[1], Khazin Hussain[1], Asma Shakeel[1], Mehnaz Shakeel[1], Liset George[1], and Seelam Giri Uma Sankar Rao[1]

[1]*Division of Genetics and Plant Breeding, Faculty of Agriculture, Sher-e-Kashmir University of Agricultural Sciences and Technology (SKUAST), Kashmir, India;* [2]*Dryland Agriculture Research Station Rangreth, Sher-e-Kashmir University of Agricultural Sciences and Technology (SKUAST), Kashmir, India*

Abstract

A new and expanding branch of "omics" that identifies and quantifies cellular pathway metabolites in different biological species is known as metabolomics. It is important to discover gene–environment interactions, characterize mutants, and identify metabolic markers. Metabolomics effectively decodes metabolic networks linked to stress tolerance and crop yield. Advanced analytical tools such as mass spectrometry, nuclear magnetic resonance, and high-performance thin-layer chromatography have been developed to speed up metabolic profiling. Integrative analysis of genomes, transcriptomes, and metabolomes in multi-omics is a tool of metabolic engineering that can be used for crop improvement.

Keywords: crop yield, metabolic engineering, metabolic markers, multi-omics, stress tolerance

11.1 Introduction

The production of economically significant yields by crops relies on various internal and external factors. Internally, factors such as plant structure and cellular processes play vital roles, while externally, both biotic and abiotic stresses, along with climate changes, impact yield levels. Researchers have concentrated primarily on enhancing plant structure and introducing genes or quantitative trait loci (QTLs) to enhance resilience to stresses, with the aim of increasing productivity. Despite advancements in these methods benefiting crop enhancement, they have not entirely met yield expectations. One key reason for this shortfall is the pivotal role of metabolites in enabling crops to achieve higher yields and tackle challenges. Metabolites—organic compounds vital for biochemical reactions within cells—are distinguished into primary and secondary metabolites (Fig. 11.1). Primary metabolites, such as carbohydrates and proteins, directly support plant growth and development, while secondary metabolites, such as essential oils and pigments, perform protective and competitive functions and are produced in response to plant needs. Unique secondary metabolites are crucial markers for evaluating the performance of crops under stress in crop improvement initiatives. The study of the metabolome, encompassing

*E-mail: saikanabi@skuastkashmir.ac.in

© CAB International 2025. *High-Throughput Plant Metabolomics* (Jen-Tsung Chen ed.)
DOI: 10.1079/9781800626911.0011

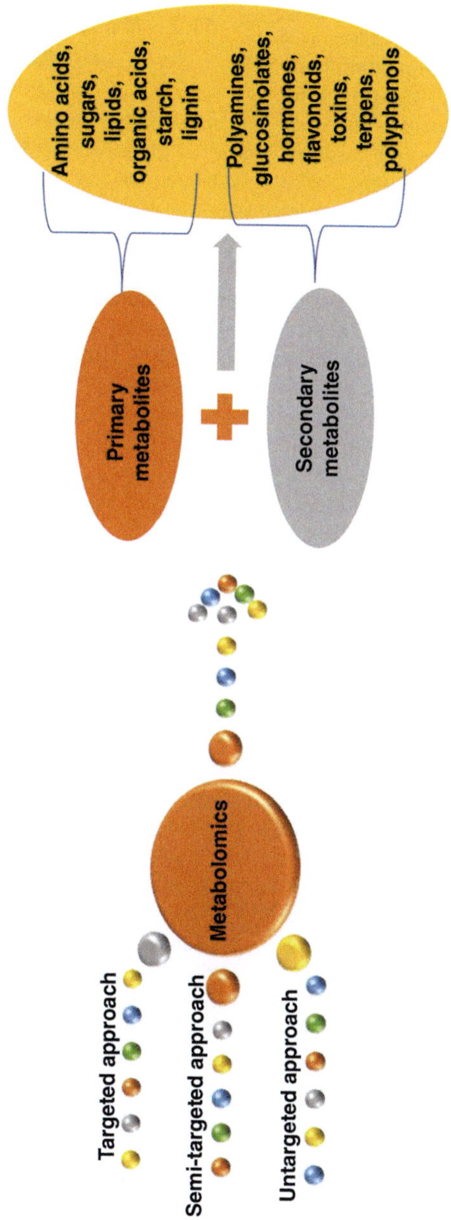

Fig. 11.1. Primary and secondary metabolites.

all primary and secondary metabolites within a particular biological unit, is known as metabolomics. Metabolomics has been utilized in plant science since the mid-1990s, starting with *Arabidopsis thaliana* as a model plant because of its genetic information and later expanding to agriculture for crop enhancement. It serves as a vital tool in understanding plant metabolic pathways by identifying and estimating low-molecular-mass (<1500 Da) molecules, and their chemical composition, and assessing multiple interactions within an organism. Metabolomics offers a comprehensive insight into cellular metabolites that reflect the cell's physiological state, aiding in the identification of genes involved in cellular metabolism and their impact on connected pathways. Furthermore, it aids in understanding the dynamic between plants and their environment, as plant development and growth under varying conditions rely heavily on metabolic processes. Integrating metabolomics into crop enhancement programs is crucial as it provides a deeper understanding of the interactions between a plant's genotype and phenotype and its environment at the metabolite level. Plant metabolites play a pivotal role in such interactions, especially the challenges posed by pests, diseases, and environmental factors impacting crop yields. By incorporating metabolomics, researchers can seamlessly navigate the complexities associated with breeding programs, offering opportunities for phenotype prediction even without genomic data. An integrated approach encompassing transcriptomics, proteomics, genomics, and metabolomics simplifies the identification and prioritization of genes/QTLs capable of mitigating biotic and abiotic stresses effectively. While traditional breeding goals have focused on yield improvement, recent years have emphasized the significance of enhancing crop nutrient content to meet nutritional needs. Understanding the nutrient accumulation potential of crops is essential, and metabolomics is emerging as a key player in elucidating the metabolic pathways governing crop quality parameters in terms of nutrient composition. Therefore, integrating various omics technologies holds promise for achieving food security by enhancing both yield quantity and nutritional quality. This chapter delves into the harmonized approach of diverse omics methodologies in terms of their relevance to metabolomics-guided breeding in crop enhancement and their usefulness. It also discusses the potential for metabolite manipulation and successful integration of biofortification in crop cultivation.

11.2 Metabolomics: Techniques and Approaches

In early studies, various methods were standardized to identify and detect specific metabolites (Wishart, 2011). However, profiling the entire range of metabolites in a cell can pose challenges because of varied chemical composition, their abundance in organelles, and the complex structures of plant cells. Metabolomics plays a pivotal role in exploring the relationship between genes, metabolites, and traits in plants (Yang *et al.*, 2018). Plants, distinct from most other organisms, can produce a wide array of metabolites varying in characteristics such as size, adaptability, volatility, stability, polarity, and solubility (Hein *et al.*, 2016; Ma *et al.*, 2018; Kang *et al.*, 2019).

Plant metabolome studies have progressed significantly in recent decades, allowing researchers to analyze numerous identified and unidentified metabolites in individual experiments and to visualize metabolites within organisms at fine spatial resolution (Bjarnholt *et al.*, 2014). However, a single technique is not sufficient to scrutinize all metabolites present in a metabolome; thus, a combination of technologies is required for comprehensive coverage. The instrumental setup for plant metabolite assessment is typically spectroscopy, chromatography, or a combination of both. Various metabolomics approaches have been developed for different experimental aims and sample types such as mass spectrometry (MS), nuclear magnetic resonance (NMR) spectroscopy, capillary electrophoresis–mass spectrometry (CE-MS), ultra-performance liquid chromatography (UPLC), gas chromatography–mass spectrometry (GC-MS), liquid chromatography–mass spectrometry (LC-MS), and other techniques such as Fourier-transform–ion cyclotron resonance–mass spectrometry (FT-ICR-MS), direct infusion mass spectrometry (DIMS), high-resolution mass spectrometry (HRMS), and high-performance thin-layer chromatography (Fig. 11.2) (Razzaq *et al.*, 2019). The non-destructive NMR method, although

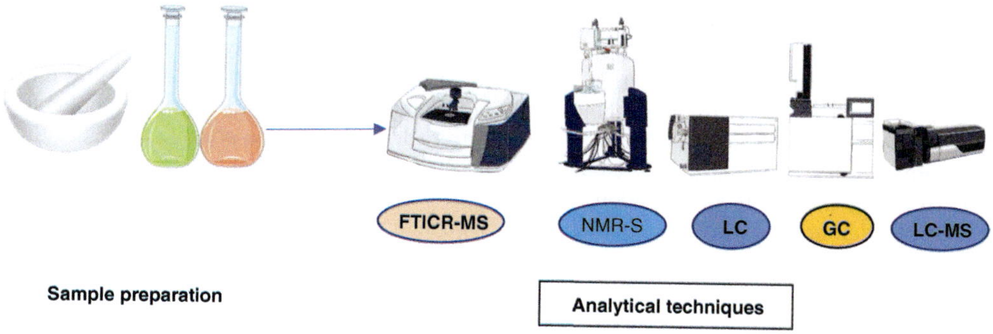

Fig. 11.2. Analytical techniques used in metabolomic analysis. FT-ICR-MS, Fourier-transform–ion cyclotron resonance–mass spectrometry; GC, gas chromatography; LC, liquid chromatography; LC-MS, liquid chromatography–mass spectrometry; NMR-S, nuclear magnetic resonance spectroscopy.

effective in detecting lower-molecular-mass metabolites (<50 kDa), faces challenges in identifying low-abundance metabolites. MS, particularly in combination with chromatography, enhances metabolite coverage and aids in the discovery of critical metabolites essential for reconstructing metabolic pathways. For metabolomics research, the choice of analytical tool is largely influenced by the necessary speed, accuracy, sensitivity, and selectivity.

NMR spectroscopy is frequently chosen for an in-depth comprehension of metabolites in plants because of its non-destructive, accurate, and highly repeatable methodology (Boiteau *et al.*, 2018). Important limitations of this technique have been overcome by the recent creation of more effective instruments such as multi-dimensional NMR, multi-coil NMR, and isotope-labeled NMR.

Throughput is greatly increased by combining techniques, such as MS and chromatography. The GC-MS approach is especially useful for non-targeted assessment for metabolite profiling due to its high sensitivity (Dutta *et al.*, 2012). However, this method is only applicable to the identification of volatile and thermally unstable substances (Jorge *et al.*, 2016). Aliferis *et al.* (2014) used this technique in soybean in conjunction with DIMS to determine the metabolic control during *Rhizoctonia solani* infection and to evaluate the biochemical profile of several cultivars in drought stress. GC-MS was used by Ogbaga *et al.* (2016) in sorghum in combination with FT-ICR-MS and by Obata *et al.* (2015) in maize to determine the association between grain yield and metabolism by metabolite profiling under conditions of heat, drought, and a combination of these. LC-MS is used for analysis of high-molecular-mass polar and thermolabile molecules (Turner *et al.*, 2016). Although it can be used for both primary and secondary metabolites in both untargeted and targeted analysis, it is mainly applied to study secondary metabolites. For example, Yang *et al.* (2014) used this technique in conjunction with NMR to classify and assess the molecular makeup of 36 distinct metabolites in rice. In order to understand the composition of metabolites and resistance mechanisms underlying molybdenum toxicity in soybean, Xu *et al.* (2018) used a UPLC-MS approach, while Farahbakhsh *et al.* (2019) conducted a comparative metabolomics study of temperature-sensitive resistance to wheat stripe mosaic virus in susceptible and resistant cultivars of wheat. In addition to the methods described above, targeted and untargeted procedures can be used with CE-MS to achieve high-resolution partitioning of various chemical groups, such as neutral, charged, polar, and hydrophobic (Ramautar and de Jong, 2014).

Metabolomics experiments face a number of challenges that determine the reliability of the data. Special hurdles arise for known and unknown metabolites (Weckwerth, 2011). Non-targeted approaches produce metabolites the majority of which are either putatively identified or unidentified, which poses a significant obstacle in metabolomics studies. However, technological enhancements in metabolomics tools have simplified and enhanced the proficiency of metabolomics research and addressed many of the problems. Metabolomics generates a massive amount of raw data that has to be handled (Kuhn *et al.*, 2008). Processing such data necessitates a highly adaptable and effectual data-processing platform. Widely used metabolomics data-processing tools include MET-COFEA, XCMS, Met-Align, MAVEN, and ChromaTOF. Following data processing, metabolite identification is carried out for additional research using databases such as METLIN, CFM-ID, MMCD, and GOLM. Next, using statistical software such as MetaboAnalyst, COVAIN, and Babelomics, the detected molecules are statistically analyzed.

11.3 The Metabolomics Analysis Workflow

The core procedures of experimental metabolomics include sample preparation, data collection using analytical techniques, and the use of suitable chemometric methods for data mining (Fig. 11.3) (Kim and Verpoorte, 2010).

Although high-resolution magic angle spinning is commonly used in plant metabolomics investigations, it is not ideal for extracting plant secondary metabolites, which play an essential role in plant self-defense mechanisms (Li *et al.*, 2016). The initial phase in the metabolomics workflow is sample preparation, which consists of four steps: harvesting plant material, quenching, extraction, and analysis of the sample. The extraction and cryo stages are optional, depending on the type of metabolites and the analytical method utilized. Screening must be done meticulously as the plant metabolome is extremely susceptible to enzyme processes that destroy different metabolites. To prevent metabolic changes, the sample is typically quenched soon after harvest with liquid nitrogen (Kim and Verpoorte, 2010).

Similarly, it may be necessary to select mature plants, as the metabolic makeup of young leaves differs greatly from that of mature leaves. Many extraction techniques have been developed in recent years for metabolomics research (Vuckovic, 2012; Wu *et al.*, 2014), and the selection of extraction solvent is equally critical. Microwave-assisted extraction is a fast and effective method that can be used in sample extraction (Wu *et al.*, 2014), while Pourmortazavi and Hajimirsadeghi (2007) identified an efficient approach for volatile metabolites known as supercritical fluid extraction. There are numerous sample preparation methods, including ultrasound-assisted extraction (Altemimi *et al.*, 2016), enzyme-assisted extraction (Zuorro *et al.*, 2013), solid-phase microextraction, and the Swiss rolling technique (Veličković *et al.*, 2019). With new advances in metabolomics, new extraction strategies are emerging, based on the analytical technique and molecule type. Because of the high number of metabolites found in

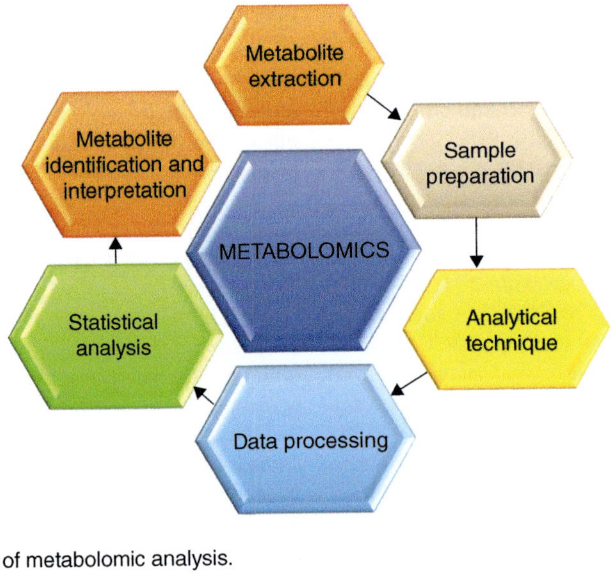

Fig. 11.3. Workflow of metabolomic analysis.

different regions of plant cells and tissues, analyzing the entire metabolome has resulted in massive amounts of data. The intricacy of the nature and composition of the metabolites in various plant samples complicates data analysis. The ultimate objective of whole-metabolome assessment is to classify many metabolites in various plant samples induced by a wide range of variables (Aoki-Kinoshita, 2006). The fundamental problem in plant metabolomics is extracting precise information regarding specific metabolites from the massive data sets created with modern methodologies. The fundamental steps of data mining include data pretreatment, preprocessing, and statistical analysis (Liland, 2011; Sun and Weckwerth, 2012). As a result, complex statistical procedures are necessary to rapidly identify and quantify each target in a sample. A number of statistical analyses are performed on the raw data obtained from the sample analysis in order to create a numerical data matrix and prepare it for additional processing. Metabolomics can measure metabolite abundance as a prognostic indicator for medical diagnosis. Data analysis utilizing a variety of statistical techniques is necessary for biomarker identification. High-throughput metabolomics data are processed using statistical approaches, most of which are based on existing omics technology. For example, statistical tools originally designed for transcriptome analysis are effective for metabolome research. Typically, for metabolomics data, there are two categories of analysis approaches: (i) multivariate analysis, which is used to screen plant varieties and ecotypes, detect diseases, and identify metabolic markers; and (ii) univariate analysis, which is frequently used for biomarker discovery at initial stages in systems biology and for examining one variable at a specific point in time (Foito and Stewart, 2018).

11.4 Crop Improvement Through Metabolomics

Metabolomics has proven to be the best tool for deciphering abiotic stress tolerance among plant species. Recently, metabolomics has been used to explore unique metabolites throughout the plant life cycle. Biotic and abiotic stresses play an important role in reducing crop yield by inducing changes in biochemical and physiological processes in the plant (Hein *et al.*, 2016). Plants react to these two types of stress using almost identical processes. Identification of particular molecular entities is necessary for the discovery of invasive species. Plant defense mechanisms such as effector-triggered immunity, pattern-triggered immunity, and pattern recognition receptors are activated by invaders (Boutrout and Zipfel, 2017). When abiotic stress begins, plants synthesize phytohormones to confer stress tolerance (Yang *et al.*, 2012). García-Cristobal *et al.* (2015) concluded that oxidative stress disrupts stomatal conductance and stimulates multiple signaling mechanisms. Collectively, a particular gene expression report indicates the precise conformation of metabolites in a specific plant species. By activating specific metabolic networks, novel bioactive substances are synthesized (Yang *et al.*, 2012). The flowchart in Fig. 11.4 illustrates how metabolomics is used to evaluate biotic and abiotic stress tolerance and metabolically aided breeding for improved crop yields.

11.4.1 Metabolic profiling/breeding for trait selection

The metabolome is an indicator of a plant's metabolic phenotype that aids in understanding the genetic connections that exist between an organism and its surroundings. The systematic screening of crop plants in order to ascertain the metabolic and chemical signature of regulatory systems in plants is known as metabolome-assisted breeding, with the outcomes being higher yield and quicker stress tolerance. Combined with other omics methods such as transcriptomics, proteomics, and genomics, metabolomics can be used to examine plant resistance to infections, resilience to abiotic stresses, ecotype wealth, and enhancement of crops. It is crucial in establishing the connection between genes and their functions. The goal of agricultural research is to identify desirable crop features from huge populations. The initial screening procedure used to be labor-intensive and dependent on the appearance of the plant. However, with the advent of marker-based crop selection and screening, the time required by previous methods has been reduced. Genome-wide association studies

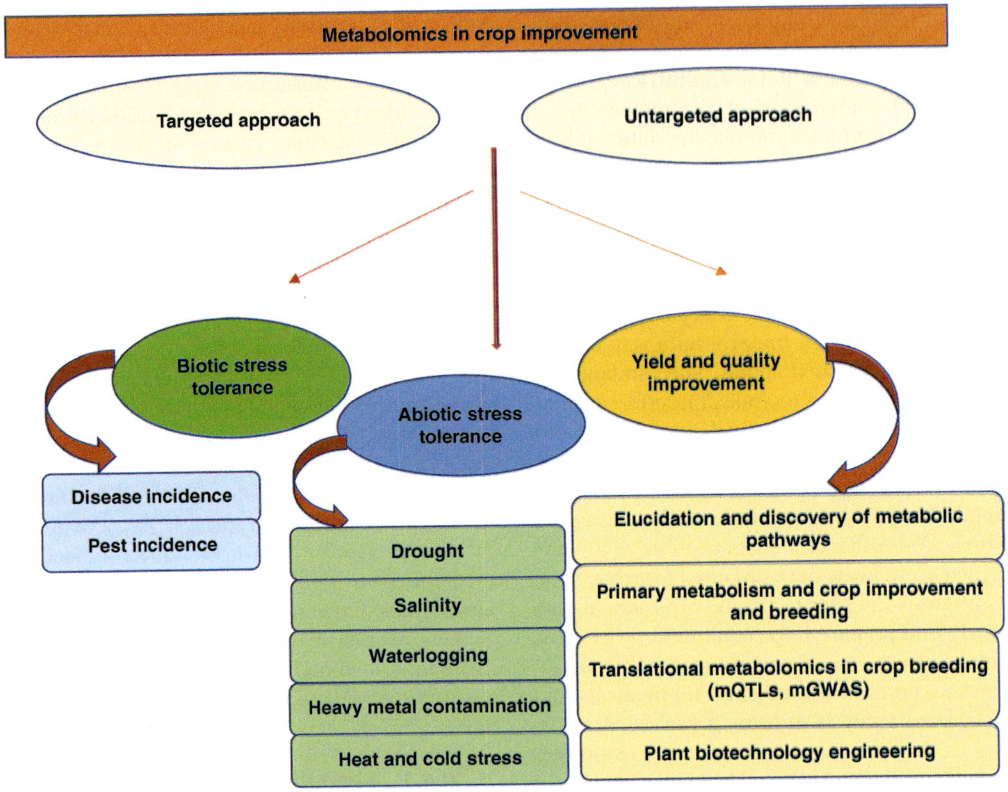

Fig. 11.4. Use of metabolomics in crop improvement. mGWAS, metabolic genome-wide association studies; mQTLs, metabolic quantitative trait loci.

(GWAS) using metabolic QTLs and metabolomics approaches (metabolic GWAS) have been used to expedite the selection and development of superior traits and breeding materials (Fernie and Schauer, 2009). By combining metabolomics with other omics techniques, researchers can identify trait-specific markers, functional candidate genes, and metabolites to boost growth and yield in commercially relevant crops. Furthermore, a plant's metabolic composition affects the crop's nutritional status, which directly affects human health. Thus, investigation of biochemical mechanisms resulting in unique phenotypes, recognition of agronomic characteristics, and the identification of genetic variants related to metabolic characteristics are important. In a study on rice varieties, 840 metabolites from 524 varieties were analyzed to determine the potential of these metabolites in crops (Chen *et al.*, 2014). By combining transcriptomics, metabolomics, and proteomics data, a metabolic map was created to improve grain quality and maize yield (Rao *et al.*, 2014). The study of mechanisms governing biomass buildup in a variety of agriculturally significant plants, leading to enhanced photosynthesis and production, is ideally suited for metabolic approaches (Hu *et al.*, 2014). Rhizobial nodulation factors in *Medicago sativa* are mostly caused by elevated oxylipin concentrations, and this pathway regulates rhizobial signaling throughout the initial phases of legume–rhizobia symbiosis in plants (Zhang *et al.*, 2012). According to Turner *et al.* (2016), shikimate may be connected to the enhancement of plant development and ultimate biomass production in sorghum, determined by metabolic examination of dichromatic leaf tissue, while research in maize identified the role of chlorogenic acid in enhanced grain yield quality using a combination of different approaches (biochemical, fluxomic, and metabolic) (Cañas *et al.*, 2017). Thus, metabolomics is considered one of the breakthrough advances in plant science, paving the way for better plant growth, improvement, and higher yields.

11.4.2 Plant responses to environmental stresses

Around the world, a variety of abiotic and biotic stresses are the primary cause of annual crop output losses. These plants react to the acting pressures in different ways, but metabolism plays a role in how these plants react biochemically. Plants use innate immune responses including effector-regulated immunity and pattern-regulated immunity to fight invasive pathogens. In response to abiotic stresses, plants also manufacture a range of phytohormones, metabolites, and oxidative stress responses. Metabolomics has advanced to the point that many metabolic networks linked to abiotic and biotic stresses in plants may now be translated.

Biotic stress

Plant productivity is constantly threatened by biotic stress caused by numerous pathogens such as bacteria, fungi, viruses, and pests. These pathogens influence host plant metabolism and cause multiple diseases and disorders that lead to significant yield losses and economic problems. Host plants and pathogens are engaged in an evolutionary arms race, with pathogens exploiting host metabolism to improve survival, and plants activating defense responses to overcome biochemical changes in pathogens. In addition to prevailing defense mechanisms such as cell walls, plants have highly proficient biochemical and metabolic defense mechanisms. Therefore, metabolic profiling during host–pathogen interactions helps understand plant defense and physiology, with the aim of achieving viable crop improvement. Metabolomic profiling can help identify important variations in primary and secondary metabolites when plants are attacked by pathogens (Allwood *et al*., 2008). It is difficult to fully understand the plant metabolome because plants activate different defense pathways against pathogens (Tenenboim and Brotman, 2016). However, plants under biotic stress accumulate certain metabolites that serve as biomarkers of tolerance to this type of stress. For example, accumulation of benzoxazinoids has been identified in grasses as important secondary metabolites for defense against biotic stress, and high concentrations of benzoxazinoids have been found in maize (Niculaes *et al*., 2018). Plant–microbe associations can also lead to pathogen resistance by triggering molecular mechanisms such as complex metabolite biosynthesis and cell signaling. Metabolites isolated from *A. thaliana* have been shown to increase disease resistance in barley (Lenk *et al*., 2018). A complex network associated with plant–pathogen interactions has been identified by metabolic profiling of infected and healthy plants. Phenylpropanoids are essential for cell-wall lignin and have a crucial role in the reaction of plants to pathogens by altering cell-wall conformation and root architecture.

Bollina *et al*. (2010) conducted a study in barley infected with *Fusarium* head blight, and kaempferol was identified along with 496 metabolites, highlighting their important role in enhancing plant defense in resistant barley cultivars. Different metabolic pathways have also been identified in wheat such as phenylpropanoid, terpenoid, and fatty acid metabolic pathways contributing to the development of resistance to *Fusarium graminearum* (Gunnaiah *et al*., 2012). Ahmad *et al*. (2011) validated the role of benzoxazinones such as 2,4-dihydroxy-7-methoxy-2H-1,4-benzoxazine-3(4H) (DIMBOA), a powerful biocidal metabolite, against *Setosphaeria trusica* in maize, as well as for resistance to aphids, while Erb *et al*. (2009) concluded that DIMBOA within maize roots protected against the attack by the rootworm *Diabrotica virgifera*, as well as against *Spodoptera littoralis*. Jones *et al*. (2011) assessed the response of rice plants to *Magnaporthe grisea* infection at multiple time points using MS- and NMR-based metabolomics, and the results suggested that overproduction of alanine was induced in plants to prevent fungal invasion. Oat (*Avena sativa*) induces the *O*-methyl-apigenin-C-deoxyhexoside-*O*-hexoside metabolic pathway during infection by the nematode *Pratylenchus* and *Heterodera* spp., which efficiently protects plants (Soriano *et al*., 2004), while in maize infected with *Ustilago maydis*, tissue-specific metabolomic analysis revealed that the shikimate and flavonoid signaling pathways are induced in response to infection caused by fungal infestation (Balmer *et al*.,

2013). During the assessment of elite barley plants resistant to Gibberellaceae, a metabolomics approach was used to successfully identify potential biomarkers for the fungus (Kumaraswamy *et al.*, 2011). Perkowski *et al.* (2012) used a GC-MS-based metabolomics approach to discriminate uninfected grains from naturally mycotoxin-infected grains such as barley, oats, and rye. Accumulation of momilactone A, an important phytoalexin in rice, was found to have a repressive effect on *Xanthomonas oryzae* and *M. grisea* infection (Sawada *et al.*, 2004).

Further enhancements in metabolomics tools combined with other omics may provide a deeper knowledge of the function of metabolites in biotic stress. For example, Seybold *et al.* (2019) recently used FT-ICR-MS to investigate the chemical response of wheat metabolites to *Zymoseptoria tritici* and found immunological and defense-related metabolites in resistant and susceptible wheat cultivars. Another NMR-based metabolic analysis of wheat resistance to *F. graminearum* showed increased production of galactose, lactic acid, spermine, putrescine, γ-aminobutyric acid (GABA), and inositol (Cuperlovic-Culf *et al.*, 2016).

Abiotic stress

Plants will suffer stunted growth and reduced metabolic rates while growing under abiotic stress conditions. Abiotic stresses such as waterlogging, drought, mineral deficiency, soil contamination with heavy metals, extreme temperatures, high salinity, or a mixture of multiple stresses negatively affect plant development and physiology. Metabolomics is an evolving tool that helps decipher plant abiotic stress tolerance. The main objective of metabolomics is to reconstruct plant homeostasis and normalize metabolic pathways. For example, Wu *et al.* (2013) conducted a study on barley in high-salinity conditions and concluded that an osmotic imbalance is created due to an extreme influx of sodium ions in the cells, disrupting numerous physiological conditions of the plant. Kim *et al.* (2007) conducted a temporal metabolic profile of *Arabidopsis*, which suggested that salt stress stimulates multiple metabolic pathways for lignin production such as the methylation cycle, the phenylpropanoid pathway, and glycine-betaine biosynthesis. Against this background, plants engage in a variety of metabolic pathways to promote stress tolerance by inducing cellular abnormalities. When plants are cold stressed, metabolic mechanisms such as levels of antioxidants rise to ensure they can tolerate the cold conditions (Zhang *et al.*, 2016). Similarly, abscisic acid, a phytohormone, confers drought resistance to plants (Yamaguchi-Shinozaki and Shinozaki, 2006). Hence, metabolomics has emerged as a choice of approach to detect and quantify stress responses in plants, which will contribute to a deeper understanding of plant physiology.

DROUGHT STRESS. Drought significantly reduces agricultural productivity around the world. When plants suffer from drought stress, they exhibit a variety of physiological responses, including leaf abscission, reductions in leaf size, and increased nutrient uptake through the roots. Stomatal closure reduces transpiration and contributes to water conservation, but also limits carbon dioxide uptake, thereby reducing photosynthesis. As a defense mechanism, plants produce polyamines such as spermine, spermidine, and putrescine (Bitrian *et al.*, 2012). Metabolic profiling of six wheat genotypes in drought conditions showed that drought-tolerant cultivars were able to produce important amino acids such as proline, GABA, threonine, myoinositol, oxalate, malate, glucose, fructose, and sucrose (Marcek *et al.*, 2019), while comparative metabolic analysis showed increased levels of methionine, lysine, proline, and arginine in drought-tolerant wheat varieties (Michaletti *et al.*, 2018). Similarly, metabolomics revealed increased concentrations of different metabolites in wheat during drought stress such as pipecolate, proline, linamarin, and tryptophan, and these metabolites can serve as biomarkers for screening desiccation-tolerant genotypes (Rahman *et al.*, 2017). In wheat, recent GC-MS-based profiling in eight cultivars showed increased levels of methionine, lysine, glutamine, asparagine, and serine under drought stress. Yang *et al.* (2018) used reversed-phase UPLC-MS for metabolic profiling of maize under drought and demonstrated upregulation of lipid and carbohydrate metabolism and improved glutathione cycling. LC-MS and GC-MS analysis by Skirycz and Inze (2010) and Nam *et al.* (2019) showed the accumulation of different metabolites in young and mature plant leaves, while another study of ten maize hybrids using GC-MS found lower levels of sugars such as erythritol and maltose, whereas xylitol and raffinose remained unchanged during drought conditions. Hyperaccumulation of GABA, leucine, glycine, serine, alanine, and tryptophan was found in drought-tolerant maize varieties (Obata *et al.*, 2015), while GC-MS was used to demonstrate that production of 4-hydroxycinnamic acid, ferulic acid, stearic acid, and xylitol was increased in rice under drought (Ma *et al.*, 2016). Moreover, a combination of GC-TOF-MS strategies in rice highlighted the coordinated control of polyamine biosynthesis, thereby revealing that it increases spermine synthesis and controls drought tolerance (Do *et al.*, 2013). Metabolomic profiling of wheat, barley, rice, and soybean has identified key metabolites that confer drought tolerance (Silvente *et al.*, 2012; Witt *et al.*, 2012; Degenkolbe *et al.*, 2013; Hein *et al.*, 2016; Kang *et al.*, 2019).

REGULATION OF SALINITY STRESS. High soil salinity can cause ion toxicity and ion uptake disorders, leading to metabolic disorders and osmotic imbalance. This can impede growth and affect physiological activity (Guo *et al.*, 2015). Elevated levels of sodium ion concentrations impede nutrient and water absorption under hypersaline conditions by causing ion toxicity (Gayen *et al.*, 2019). The plant's response to salt stress is mainly the production of primary and secondary metabolites. For example, when rice seedlings are subjected to salt stress, the levels of leucine, isoleucine, valine, and proline increase. Increased production of xylose, maltose, organic acids, proline, and sucrose was observed in barley roots under salt stress, indicating a salt tolerance mechanism (Shelden *et al.*, 2016). In rice, salinity stress decreases the levels of quinate and shikimate while increasing the concentrations of sucrose and mannitol, which may contribute to salt tolerance (Chang *et al.*, 2019), and changes in the abscisic acid pathway may affect different hormone pathways related to salt stress, such as the jasmonic acid pathway, which reduces salt damage (Kurotani *et al.*, 2015). Various metabolomic tools have been used to study changes in the metabolic profile of crops such as tomato, maize, barley, and wheat under salt stress (Zorb *et al.*, 2013; Borrelli *et al.*, 2018; Rouphael *et al.*, 2018).

WATERLOGGING STRESS CONTROL. Waterlogging affects plant growth and yield by limiting carbon dioxide and oxygen supply and inhibiting photosynthesis. Prolonged waterlogging causes hypoxic conditions, which directly affect roots and prevent carbon dioxide assimilation. Plants adapt to waterlogging through signal transduction, metabolic changes, and morphological changes (Barding *et al.*, 2012). Studies on soybean have shown a negative relationship between waterlogging stress and metabolite synthesis (Coutinho *et al.*, 2018). In one experiment, 73 metabolites were identified in the root tips of flooded soybean, including phosphoenolpyruvate, $NADH_2$, glycine, and GABA (Komatsu *et al.*, 2014). A study on wheat varieties identified methionine, tryptophan, lysine, and proline as important biomarkers for waterlogging tolerance (Herzog *et al.*, 2018). Metabolic profiling of different rice varieties under flood stress revealed unique metabolites such as 6-phosphogluconate, phenylalanine, and lactic acid, which facilitated survival of the plants during flooding (Locke *et al.*, 2018).

REGULATION OF TEMPERATURE STRESS. Plants require optimal temperatures for their development and growth. Aberrant variations in temperature seriously damage plants and halt their developmental processes, while increased temperatures disrupt homeostasis and physiological formations (Luengwilai *et al.*, 2012; Thomason *et al.*, 2018). Under heat stress, plants synthesize secondary metabolites such as 2-ketoisocaproic acid, arachidic acid, allantoin, rhamnose, putrescine, myo-inositol, and alanine (Luengwilai *et al.*, 2012). An LC-HRMS study on wheat under heat stress showed increased synthesis of pipecolate and L-tryptophan, which affects secondary metabolite biosynthesis and aminoacyl-tRNA biosynthesis (Thomason *et al.*, 2018). Metabolic profiles of different grains using LC-MS/MS-high-performance LC showed that sucrose and HisG1p concentrations were higher under heat stress (Wang *et al.*, 2018), while comparative metabolomics analysis of heat-resistant and heat-sensitive soybean varieties revealed that the heat-resistant genotype synthesizes elevated levels of carbohydrates such as 1,3-dihydroxyacetone, ribose, and glycolic acid (Chebrolu *et al.*, 2016). Metabolomics analyses have been performed on crops such as tomato, maize, and wheat to investigate the effects of heat stress (Sun *et al.*, 2016; Paupière *et al.*, 2017; Qi *et al.*, 2017).

METAL-INDUCED STRESS MANAGEMENT. Heavy-metal stress has a substantial impact on crop productivity, causing complicated biochemical, molecular, and physiological responses. Trace elements such as cobalt, tungsten, vanadium, chromium, zinc, nickel, and copper can be fatal in high doses (Xu *et al.*, 2018). Chromium, nickel, zinc, lead, cesium, and cadmium are the primary contaminants causing plant stress. Metal stress inhibits metabolism, cellular oxidation, and enzyme activity, resulting in stunted growth and cell death (Ibarra *et al.*, 2019). Iron, manganese, and copper are vital for plant growth but heavy metals can affect reproduction, metabolism, and growth. Plants require multi-dimensional biochemical and physiological adjustments, protein structure stabilization, and metabolomic adaptations to develop tolerance to metal stress. Metabolic profiling of beans revealed that carbohydrate concentration increased under high zinc and copper treatment (Manivasagaperumal *et al.*, 2011). In *Brassica rapa*, heavy-metal stress caused enhanced accumulation of metabolites as shown by ^{1}H-NMR and two-dimensional *J*-based metabolomics analysis (Jahangir *et al.*, 2008). Similarly, chromium stress in sunflower roots showed increased production of specific fatty acids using capillary high-performance LC-ESI-Q-TOF-MS-mediated profiling of leaves (Ibarra *et al.*, 2019).

NUTRIENT DEFICIENCIES. Plants require a specific balance of critical nutrients to thrive and grow effectively. Plant key organic compounds and metabolites are composed of elements such as nitrogen, carbon, phosphorus, and sulfur. When plants lack these nutrients, it can significantly impact their growth, metabolism, and overall health. Nitrogen, in particular, plays a crucial role in forming important cellular components such as nucleic acids, amino acids, proteins, and other secondary metabolites (Foito *et al.*, 2013). When plants experience nitrogen deficiency, they produce a number of metabolites such as threonate, glycerate, inositol, and soluble carbohydrates as a response to stress (Comadira *et al.*, 2015). Studies on wheat metabolism using GC-MS and LC-MS showed that, under nitrogen stress, the production of metabolites such as tyrosine, lysine, allo-inositol, and L-ascorbic acid increased (Khan, 2019). Similarly, metabolic profiling revealed that organic acids and amino acids decreased in tomato plants under nitrogen and phosphorus deficiency, while soluble sugars increased under nitrogen stress (Sung *et al.*, 2015). Sulfur is essential for the production of various key metabolites and sulfur-containing amino acids (Hoefgen and Nikiforova, 2008). A study on barley roots and leaves under sulfur stress found increased synthesis of various amino acids, organic acids, and sulfur-sensitive metabolites (Ghosson *et al.*, 2018). Phosphorus is crucial for making DNA, RNA, and ATP, as well as for various energy-related metabolites (Theodorou and Plaxton, 1993). Metabolic profiling of barley under phosphorus deficiency showed increased production of certain sugars and decreased organic acids and phosphorus-containing intermediates (Huang *et al.*, 2008). A study on kidney bean tubers and roots under phosphorus stress and on wheat under nitrogen deficiency also revealed important metabolic changes (Zhang *et al.*, 2017).

11.4.3 Metabolic engineering and synthetic biology

Metabolic engineering is the conscious modification of cellular metabolism and its components by the manipulation of numerous metabolic pathways using tools such as recombinant DNA technology and gene and RNA editing (Fu *et al.*, 2018). In its natural form, the plant is composed of over 200,000 metabolites (Fiehn, 2011). Metabolomics to identify these molecules plays an important role in the study of biotic resistance, abiotic resistance, active ecotypes, and plant breeding (Razzaq *et al.*, 2019). Over the past 20 years, research in plant metabolic engineering has made significant progress. Complex biological pathways are being investigated, using both modern metabolic engineering and classic breeding strategies, to uncover new regulatory networks, and enhance plant health and productivity (Dangl *et al.*, 2013). Genetic engineering is often used when traditional breeding approaches are insufficient. This approach uses genes from different species and has become more successful due to the availability of the whole-genome sequences. Advancements in gene editing, including tools such as zinc-finger nucleases, transcription activator-like effector nucleases (TALENs), and clustered regularly interspaced short palindromic repeats (CRISPR)-Cas9, have allowed precise adjustments to targeted genes (Jaganathan *et al.*, 2018). However, despite extensive research in this field, the commercial use of genetically modified organisms is limited by political and social opposition in many countries (Watanabe *et al.*, 2005).

11.4.4 Application of multi-omics in crop improvement

Different omics techniques—genomics, transcriptomics, proteomics, and metabolomics—are important to the fields of molecular biology and biotechnology. Structural and functional genomics are subsets of genomics. Whereas functional genomics seeks to understand the large-scale functions of different genes, structural genomics is concerned with figuring an organism's entire genome sequence (Leister, 2005). Researchers use transcriptomics, proteomics, and metabolomics to explore genome function in detail. Large-scale, high-throughput data collection has been transformed by advances in integrated MS, chromatography systems, and next-generation sequencing (Fondi and Liò, 2015). However, data from numerous omics research must be connected in order to understand biological systems. To fully comprehend biological systems, integrated multi-omics techniques are crucial due to the intricacy of omics data sets and biological systems themselves (Hughes, 2015). The complicated biology, various metabolomes, and huge, incompletely annotated genomes of plants limit the adoption of multi-omics techniques, despite their widespread use in studies involving humans, animals, and microbes. Large biological data sets are difficult to handle, and precise results and interpretations need highly qualified researchers. To effectively and accurately analyze enormous volumes of data, these researchers must be skilled in choosing the right combination of software applications from the many platforms available (Pinu *et al.*, 2019).

Previous integration methods were not particularly effective, but current approaches have been reformulated into three distinct categories: element based, pathway based, and mathematics based (Jamil *et al.*, 2020).

Techniques based on elements

Three subclasses of element-based omics data-set analysis techniques exist: multivariate analysis, clustering, and correlation analysis. The link between various omics data sets is measured by correlation analysis. The Spearman (Myers and Sirois, 2006) and Pearson (Benesty *et al.*, 2009) correlation coefficients are frequently used to evaluate ranked and linear correlations, respectively. For skewed data sets, the Fisher transformation is sometimes applied to achieve a normal distribution before evaluating correlations (Mata *et al.*, 2018). In order to find underlying relationships between omics data sets, clustering analysis groups them according to shared features such as expression levels or concentrations. Both hierarchical and non-hierarchical clustering are possible. To integrate data, non-hierarchical approaches frequently use machine learning algorithms such as random forests and *k*-means clustering (Silva *et al.*, 2019). According to Rai *et al.* (2017), multivariate analysis provides great versatility in experimental design and metadata analysis, and can handle increasingly complicated omics data sets. With this method, scientists can forecast several characteristics of a data set, investigate correlations and dynamic networks across transcript, protein, and metabolite components, and evaluate variance or covariance links (Weckwerth, 2011). Common multivariate analysis techniques include principal component analysis, partial least squares, and orthogonal projective discriminant analysis of latent structures. Success depends on choosing the correct techniques, ideal parameters, and model validation.

In a study to integrate the maize proteome and metabolomic data sets, an advanced multivariate technique called multi-coin inertial analysis was used to determine metabolic variations in sugar metabolism and polyamine biosynthesis pathways among two strains (Mesnage *et al.*, 2016). By combining transcriptome and metabolome data, a different variation called GFLASSO (graph-guided fusion least absolute shrinkage and selection operator) was used on maize to interpret the pathways involved in lipid production (de Abreu E. Lima *et al.*, 2018).

Techniques based on pathways

There are two approaches for analyzing omics data sets. The first approach is pathway mapping, which involves mapping data sets to existing metabolic pathway databases. The Kyoto Encyclopedia of Genes and Genomes (KEGG) is a commonly used database, but there are also organism-specific databases such as AraCyc for *Arabidopsis*, SolCyc for Solanaceae species, and CitrusCyc for citrus fruits. Biochemical pathways can also be manually reconstructed without the need for sophisticated computing tools. For instance, the transcriptome and metabolome of soybeans infested with the fungus *Phytophthora sojae* (Zhu *et al.*, 2018) and cyst nematode (Kang *et al.*, 2018) were examined using the KEGG database revealing transcriptional and metabolic regulation of isoflavonoid and phenylpropanoid biosynthetic pathways. The second approach is coexpression analysis, which calculates the similarity between compounds based on statistical correlations between different omics data sets. These associations are converted into weighted networks that can be visualized using tools such as WGCNA (weighted gene coexpression network analysis) in R and Cytoscape. This method has been widely used in plant studies to identify modules, clusters, and hubs that provide deeper insights into specific signaling pathways and regulatory molecules. For example, MapMan functional annotation has been used to understand the enriched pathways of highly connected coexpression hubs (Walley *et al.*, 2016). Jiang *et al.*, (2019) used the same data set to reveal the involvement of several transcription factors in the molecular alteration of maize by fusing the differently scaled networks obtained from every omics study into a consensus network.

Mathematics-based approaches

Mathematics-based approaches for analyzing omics data sets can be divided into two subgroups: differential and genome-level analysis. These approaches are highly detailed and require comprehensive data coverage and well-documented databases.

Differential analysis includes defining the topology and regulation of the system, identifying its constituent parts, creating the necessary mathematical formulas, and choosing and maximizing its parameters (Voit, 2017). Pathway techniques for this analysis can be further divided into targeted and non-targeted categories. The efficacy of targeted pathway analysis depends on the comprehensiveness of metabolic pathway annotation. It models certain metabolic pathways and dynamics. Differential analysis is especially helpful when merging omics data with genome-level analysis (Cavill *et al.*, 2016). Using a top-down method, it develops stoichiometric equations for particular objectives, including calculating translation rates or metabolic fluxes (Voit, 2017).

In contrast, genome-scale analysis follows a bottom-up methodology. It involves building models after extensive curation and then validating them (Goh, 2018). This method provides a comprehensive view of the system and helps in understanding the broader aspects of genome function and regulation. This approach aims to abstract metabolic pathways at the biological and cellular levels so that each reaction can be comprehensively evaluated mathematically (de Oliveira Dal'Molin *et al.*, 2018). The process of genome-scale analysis can be broken down into four key steps: (i) mathematically building the network; (ii) utilizing experimental data to refine the path; (iii) retrieving the design based on the annotated genome; and (iv) verifying and repeating the model to assure fidelity (Thiele and Palsson, 2010). Ten thoroughly annotated plant genomes can be found in the genome-scale modeling database PlantSEED (Seaver *et al.*, 2018). Novel plant metabolic pathways can be annotated using the data in this database; however, it works best for reconstructing primary metabolism as human curation is required for secondary metabolism (Pinu *et al.*, 2019). New methods are being developed to integrate omics data from C4 plants, as the majority of genome-wide metabolic studies on plants are focused on C3 species such as *Arabidopsis* (de Oliveira Dal'Molin *et al.*, 2016). Genome-scale modeling utilizes techniques such as the Plant/Eukaryotic and Microbial Metabolomics System Resource and MetNetDB to integrate transcriptomics and metabolomics data sets. This method has been applied to operationalize metabolic models for plants such as soybean and rapeseed (Li *et al.*, 2015). Similarly, genome-scale modeling in maize has been updated to confirm leaf development through transcriptomic and biochemical assays (Bogart and Myers, 2016).

11.5 Conclusion and Future Outlook

By enabling research into plant responses to diverse stresses, understanding important metabolic pathways, determining gene functions, and enhancing plant quality, metabolomics has greatly enhanced the field of agricultural sciences. Furthermore, it offers deeper understanding of the connections among genotype, metabolome, and phenotype. The efficiency of analyzing the complicated metabolome in plants has increased with the introduction of high-throughput technologies. Researchers can now use metabolic engineering to more accurately target specific metabolites by producing full profiles of many metabolites in a single assay, rather than evaluating single metabolites. Many metabolic QTLs and metabolite markers have been discovered, and these findings are currently being utilized in breeding initiatives.

The metabolome can also be better understood through the use of low-cost, next-generation sequencing technology, which increases the impact and accessibility of metabolomics. By providing an improved understanding of plants as living systems, the integration of metabolomics into other omics technologies is increasing our knowledge in this field.

Metabolomics research still has room to grow. More metabolite markers and metabolic QTLs will be found and developed to help increase crop quality, yield, and stress tolerance. Future applications of cutting-edge technologies such as CRISPR-Cas9 could include altering genes that regulate metabolic pathways and metabolite profiling of plants with genome editing. Future developments in the integration of metabolomics in agriculture could greatly expand its ability to successfully handle difficulties in the field.

11.6 References

Ahmad, S., Veyrat, N., Gordon-Weeks, R., Zhang, Y., Martin, J., *et al.* (2011) Benzoxazinoid metabolites regulate innate immunity against aphids and fungi in maize. *Plant Physiology* 157, 317–327.

Aliferis, K.A., Faubert, D., and Jabaji, S. (2014) A metabolic profiling strategy for the dissection of plant defense against fungal pathogens. *PLOS ONE* 9: e111930.

Allwood, J.W., Ellis, D.I., and Goodacre, R. (2008) Metabolomic technologies and their application to the study of plants and plant–host interactions. *Physiologia Plantarum* 132, 117–135.

Altemimi, A., Watson, D.G., Choudhary, R., and Dasari, M.R. (2016) Ultrasound assisted extraction of phenolic compounds from peaches and pumpkins. *PLOS ONE* 11: e0148758.

Aoki-Kinoshita, K.F. (2006) Overview of KEGG applications to omics-related research. *Journal of Pesticide Science* 31, 296–299.

Balmer, D., Flors, V., Glauser, G., and Mauch-Mani, B. (2013) Metabolomics of cereals under biotic stress: current knowledge and techniques. *Frontiers in Plant Science* 4: 82.

Barding, G.A., Jr., Béni, S., Fukao, T., Bailey-Serres, J., and Larive, C.K. (2012) Comparison of GC-MS and NMR for metabolite profiling of rice subjected to submergence stress. *Journal of Proteome Research* 12, 898–909.

Benesty, J., Chen, J., Huang, Y., and Cohen, I. (2009) *Pearson Correlation Coefficient*. In: Benesty, J., Chen, J., and Huang, Y. (eds) *Noise Reduction in Speech Processing. Springer Topics in Signal Processing, Vol. 2*. Springer, Berlin, Heidelberg, Germany, pp. 37–40.

Bitrián, M., Zarza, X., Altabella, T., and Tiburcio, A.F. (2012) Polyamines under abiotic stress: metabolic crossroads and hormonal crosstalks in plants. *Metabolites* 2, 516–528.

Bjarnholt, N., Li, B., d'Alvise, J., and Janfelt, C. (2014) Mass spectrometry imaging of plant metabolites principles and possibilities. *Natural Product Reports* 31, 818–837.

Bogart, E. and Myers, C.R. (2016) Multiscale metabolic modeling of C4 plants: connecting nonlinear genome-scale models to leaf-scale metabolism in developing maize leaves. *PLOS ONE* 11: e0151722.

Boiteau, R.M., Hoyt, D.W., Nicora, C.D., Kinmonth-Schultz, H.A., Ward, J.K., *et al.* (2018) Structure elucidation of unknown metabolites in metabolomics by combined NMR and MS/MS prediction. *Metabolites* 8: 8.

Bollina, V., Kumaraswamy, G.K., Kushalappa, A.C., Choo, T.M., Dion, Y., *et al.* (2010) Mass spectrometry-based metabolomics application to identify quantitative resistance-related metabolites in barley against *Fusarium* head blight. *Molecular Plant Pathology* 11, 769–782.

Borrelli, G.M., Fragasso, M., Nigro, F., Platani, C., Papa, R., and Trono, D. (2018) Analysis of metabolic and mineral changes in response to salt stress in durum wheat (*Triticum turgidum* ssp. *durum*) genotypes, which differ in salinity tolerance. *Plant Physiology and Biochemistry* 133, 57–70.

Boutrot, F. and Zipfel, C. (2017) Function, discovery, and exploitation of plant pattern recognition receptors for broad-spectrum disease resistance. *Annual Review of Phytopathology* 55, 257–286.

Cañas, R.A., Yesbergenova-Cuny, Z., Simons, M., Chardon, F., Armengaud, P., *et al.* (2017) Exploiting the genetic diversity of maize using a combined metabolomic, enzyme activity profiling, and metabolic modeling approach to link leaf physiology to kernel yield. *Plant Cell* 29, 919–943.

Cavill, R., Jennen, D., Kleinjans, J., and Briedé, J.J. (2016) Transcriptomic and metabolomic data integration. *Briefings in Bioinformatics* 17, 891–901.

Chang, J., Cheong, B.E., Natera, S., and Roessner, U. (2019) Morphological and metabolic responses to salt stress of rice (*Oryza sativa* L.) cultivars which differ in salinity tolerance. *Plant Physiology and Biochemistry* 144, 427–435.

Chebrolu, K.K., Fritschi, F.B., Ye, S., Krishnan, H.B., Smith, J.R., and Gillman, J.D. (2016) Impact of heat stress during seed development on soybean seed metabolome. *Metabolomics* 12: 28

Chen, W., Gao, Y., Xie, W., Gong, L., Lu, K., *et al.* (2014) Genome-wide association analyses provide genetic and biochemical insights into natural variation in rice metabolism. *Nature Genetics* 46, 714–721.

Comadira, G., Rasool, B., Karpinska, B., Morris, J., Verrall, S.R., *et al.* (2015) Nitrogen deficiency in barley (*Hordeum vulgare*) seedlings induces molecular and metabolic adjustments that trigger aphid resistance. *Journal of Experimental Botany* 66, 3639–3655.

Coutinho, I.D., Henning, L.M.M., Döpp, S.A., Nepomuceno, A., Moraes, L.A.C., *et al.* (2018) Flooded soybean metabolomic analysis reveals important primary and secondary metabolites involved in the hypoxia stress response and tolerance. *Environmental and Experimental Botany* 153, 176–187.

Cuperlovic-Culf, M., Wang, L., Forseille, L., *et al.* (2016) Metabolic biomarker panels of response to fusarium head blight infection in different wheat varieties. *PLOS ONE* 11: e0153642.

Dangl, J.L., Horvath, D.M., and Staskawicz, B.J. (2013) Pivoting the plant immune system from dissection to deployment. *Science* 341, 746–751.

de Abreu E. Lima, F., Li, K., Wen, W., Yan, J., *et al.* (2018) Unraveling lipid metabolism in maize with time resolved multi-omics data. *Plant Journal* 93, 1102–1115.

de Oliveira Dal'Molin, C.G., Orellana, C., Gebbie, L., Steen, J., Hodson, M.P., *et al.* (2016) Metabolic reconstruction of *Setaria italica*: a systems biology approach for integrating tissue-specific omics and pathway analysis of bioenergy grasses. *Frontiers in Plant Science* 7: 1138.

de Oliveira Dal'Molin, C.G., Quek, L.E., Saa, P.A., Palfreyman, R., and Nielsen, L.K. (2018) From reconstruction to C_4 metabolic engineering: a case study for overproduction of polyhydroxybutyrate in bioenergy grasses. *Plant Science* 273, 50–60.

Degenkolbe, T., Do, P.T., Kopka, J., Zuther, E., Hincha, D.K., and Köhl, K.I. (2013) Identification of drought tolerance markers in a diverse population of rice cultivars by expression and metabolite profiling. *PLOS ONE* 8: e63637.

Do, P.T., Degenkolbe, T., Erban, A., *et al.* (2013) Dissecting rice polyamine metabolism under controlled long-term drought stress. *PLOS ONE* 8: e60325.

Dutta, A., Shetty, P., Bhat, S.,Ramachandra, Y., and Hegde, S. (2012) A mass spectrometric study for comparative analysis and evaluation of metabolite recovery from plasma by various solvent systems. *Journal of Biomolecular Techniques* 23, 128–135.

Erb, M., Flors, V., Karlen, D., de Lange, E., Planchamp, C., *et al.* (2009) Signal signature of aboveground-induced resistance upon belowground herbivory in maize. *Plant Journal* 59, 292–302.

Farahbakhsh, F., Hamzehzarghani, H., Massah, A., Tortosa, M., Yassaie, M., and Rodriguez, V.M. (2019) Comparative metabolomics of temperature sensitive resistance to wheat streak mosaic virus (WSMV) in resistant and susceptible wheat cultivars. *Journal of Plant Physiology* 237, 30–42.

Fernie, A.R. and Schauer, N. (2009) Metabolomics-assisted breeding: a viable option for crop improvement? *Trends in Genetics* 25, 39–48.

Fiehn, O., Barupal, D.K., and Kind, T. (2011) Extending biochemical databases by metabolomic surveys. *Journal of Biological Chemistry* 286, 23637–23643.

Foito, A. and Stewart, D. (2018) Metabolomics: a high-throughput screen for biochemical and bioactivity diversity in plants and crops. *Current Pharmaceutical Design* 24, 2043–2054.

Foito, A., Byrne, S.L., Hackett, C.A., Stewart, D. and Barth, S. (2013) Short-term response in leaf metabolism of perennial ryegrass (*Lolium perenne*) to alterations in nitrogen supply. *Metabolomics* 9, 145–156.

Fondi, M. and Liò, P. (2015) Multi-omics and metabolic modelling pipelines: challenges and tools for systems microbiology. *Microbiological Research* 171, 52–64.

Fu, R., Martin, C., and Zhang, Y. (2018) Next-generation plant metabolic engineering, inspired by an ancient Chinese irrigation system. *Molecular Plant* 11, 47–57.

García-Cristobal, J., García-Villaraco, A., Ramos, B., Gutierrez-Mañero, J., and Lucas, J.A. (2015) Priming of pathogenesis related-proteins and enzymes related to oxidative stress by plant growth promoting rhizobacteria on rice plants upon abiotic and biotic stress challenge. *Journal of Plant Physiology* 188, 72–79.

Gayen, D., Barua, P., Lande, N.V., Varshney, S., Sengupta, S., *et al.* (2019) Dehydration-responsive alterations in the chloroplast proteome and cell metabolomic profile of rice reveals key stress adaptation responses. *Environmental and Experimental Botany* 160, 12–24.

Ghosson, H., Schwarzenberg, A., Jamois, F., and Yvin, J.C. (2018) Simultaneous untargeted and targeted metabolomics profiling of underivatized primary metabolites in sulfur-deficient barley by ultra-high performance liquid chromatography-quadrupole/time-of-flight mass spectrometry. *Plant Methods* 14: 62.

Goh, H.H. (2018) Integrative multi-omics through bioinformatics. *Advances in Experimental Medicine and Biology* 1102, 69–80.

Gunnaiah, R., Kushalappa, A.C., Duggavathi, R., Fox, S., and Somers, D.J. (2012) Integrated metabolo-proteomic approach to decipher the mechanisms by which wheat QTL (*Fhb1*) contributes to resistance against *Fusarium graminearum*. *PLOS ONE* 7: e40695.

Guo, R., Yang, Z., Li, F., Yan, C., Zhong, X., *et al.* (2015) Comparative metabolic responses and adaptive strategies of wheat (*Triticum aestivum*) to salt and alkali stress. *BMC Plant Biology* 15: 170.

Hein, J.A., Sherrard, M.E., Manfredi, K.P., and Abebe, T. (2016) The fifth leaf and spike organs of barley (*Hordeum vulgare* L.) display different physiological and metabolic responses to drought stress. *BMC Plant Biology* 16: 248.

Herzog, M., Fukao, T., Winkel, A., Konnerup, D., Lamichhane, S., *et al.* (2018) Physiology, gene expression, and metabolome of two wheat cultivars with contrasting submergence tolerance. *Plant, Cell and Environment* 41, 1632–1644.

Hoefgen, R. and Nikiforova, V.J. (2008) Metabolomics integrated with transcriptomics: assessing systems response to sulfur-deficiency stress. *Physiologia Plantarum* 132, 190–198.

Hu, C., Shi, J., Quan, S., Cui, B., Kleessen, S., *et al.* (2014) Metabolic variation between japonica and indica rice cultivars as revealed by non-targeted metabolomics. *Scientific Reports* 4: 5067.

Huang, C.Y., Roessner, U., Eickmeier, I., Genc, Y., Callahan, D.L., *et al.* (2008) Metabolite profiling reveals distinct changes in carbon and nitrogen metabolism in phosphate-deficient barley plants (*Hordeum vulgare* L.). *Plant, Cell and Environment* 49, 691–703.

Hughes, M. (2015) Systems biology tools for integrated omics analysis. *Genetic Engineering and Biotechnology News* 35, 18–19.

Ibarra, A.A.G., Wrobel, K., Barrientos, E.Y., Escobosa, A.R.C., Corona, J.F.G., *et al.* (2019) Impact of Cr (VI) on the oxidation of polyunsaturated fatty acids in *Helianthus annuus* roots studied by metabolomic tools. *Chemosphere* 220, 442–451.

Jaganathan, D., Ramasamy, K., Sellamuthu, G., Jayabalan, S., and Venkataraman, G. (2018) CRISPR for crop improvement: an update review. *Frontiers in Plant Science* 9: 985.

Jahangir, M., Abdel-Farid, I.B., Choi, Y.H., and Verpoorte, R. (2008) Metal ion-inducing metabolite accumulation in *Brassica rapa*. *Journal of Plant Physiology* 165, 1429–1437.

Jamil, I.N., Remali, J., Azizan, K.A., Nor Muhammad, N.A., Arita, M., *et al.* (2020) Systematic multi-omics integration (MOI) approach in plant systems biology. *Frontiers in Plant Science* 11: 944.

Jiang, J., Xing, F., Wang, C., Zeng, X., and Zou, Q. (2019) Investigation and development of maize fused network analysis with multi-omics. *Plant Physiology and Biochemistry* 141, 380–387.

Jones, O.A., Maguire, M.L., Griffin, J.L., Jung, Y.-H., Shibato, J., *et al.* (2011) Using metabolic profiling to assess plant-pathogen interactions: an example using rice (*Oryza sativa*) and the blast pathogen *Magnaporthe grisea*. *European Journal of Plant Pathology* 129, 539–554.

Jorge, T.F., Rodrigues, J.A., Caldana, C., Schmidt, R., van Dongen, J.T., *et al.* (2016) Mass spectrometry-based plant metabolomics: metabolite responses to abiotic stress. *Mass Spectrometry Reviews* 35, 620–649.

Kang, W., Zhu, X., Wang, Y., Chen, L., and Duan, Y. (2018) Transcriptomic and metabolomic analyses reveal that bacteria promote plant defense during infection of soybean cyst nematode in soybean. *BMC Plant Biology* 18: 86.

Kang, Z., Babar, M.A., Khan, N., Guo, J., Khan, J., *et al.* (2019) Comparative metabolomic profiling in the roots and leaves in contrasting genotypes reveals complex mechanisms involved in post-anthesis drought tolerance in wheat. *PLOS ONE* 14: e0213502.

Khan, F.U. (2019) Comparative metabolite profiling of two wheat genotypes as affected by nitrogen stress at seedling stage. *Journal of Animal and Plant Sciences* 29, 260–268.

Kim, H.K. and Verpoorte, R. (2010) Sample preparation for plant metabolomics. *Phytochemical Analysis* 21, 4–13.

Kim, J.K., Bamba, T., Harada, K., *et al.* (2007) Time-course metabolic profiling in *Arabidopsis thaliana* cell cultures after salt stress treatment. *Journal of Experimental Botany* 58, 415–424.

Komatsu, S., Nakamura, T., Sugimoto, Y., and Sakamoto, K. (2014) Proteomic and metabolomic analyses of soybean root tips under flooding stress. *Protein and Peptide Letters* 21, 865–884.

Kuhn, S., Egert, B., Neumann, S., and Steinbeck, C. (2008) Building blocks for automated elucidation of metabolites: machine learning methods for NMR prediction. *BMC Bioinformatics* 9: 400.

Kumaraswamy, K.G., Kushalappa, A.C., Choo, T.M., Dion, Y., and Rioux S. (2011) Mass spectrometry based metabolomics to identify potential biomarkers for resistance in barley against fusarium head blight (*Fusarium graminearum*). *Journal of Chemical Ecology* 37, 846–856.

Kurotani, K.-I., Hayashi, K., Hatanaka, S., Toda, Y., Ogawa, D., *et al.* (2015) Elevated levels of CYP94 family gene expression alleviate the jasmonate response and enhance salt tolerance in rice. *Plant Cell Physiology* 56, 779–789.

Leister, D. (2005) *Plant Functional Genomics*. Food Products Press, New York.

Lenk, M., Wenig, M., Mengel, F., Häußler, F., and Vlot, A. (2018) *Arabidopsis thaliana* immunity-related compounds modulate disease susceptibility in barley. *Agronomy* 8: 142.

Li, L., Hur, M., Lee, J.Y., Zhou, W., Song, Z., *et al.* (2015) A systems biology approach toward understanding seed composition in soybean. *BMC Genomics* 16, S3–S9.

Li, N., Peng, Song, Y., Tang, H., and Wang, Y. (2016) Recent developments in sample preparation and data pre-treatment in metabonomics research. *Archives of Biochemistry and Biophysics* 589, 4–9.

Liland, K.H. (2011) Multivariate methods in metabolomics—from pre-processing to dimension reduction and statistical analysis. *Trends in Analytical Chemistry* 30, 827–841.

Locke, A.M., Barding, G.A., Sathnur, S., Larive, C.K., and Bailey-Serres, J. (2018) Rice *SUB1A* constrains remodelling of the transcriptome and metabolome during submergence to facilitate post-submergence recovery. *Plant, Cell and Environment* 41, 721–736.

Luengwilai, K., Saltveit, M., and Beckles, D.M. (2012) Metabolite content of harvested Micro-Tom tomato (*Solanum lycopersicum* L.) fruit is altered by chilling and protective heat-shock treatments as shown by GC–MS metabolic profiling. *Postharvest Biology and Technology* 63, 116–122.

Ma, N.L., Che Lah, W.A., Kadir, N.A., Mustaqim, M., Rahmat, Z., *et al.* (2018) Susceptibility and tolerance of rice crop to salt threat: physiological and metabolic inspections. *PLOS ONE* 13: e0192732.

Ma, X., Xia, H., Liu, Y., Wei, H., Zheng, X., *et al.* (2016) Transcriptomic and metabolomic studies disclose key metabolism pathways contributing to well-maintained photosynthesis under the drought and the consequent drought-tolerance in rice. *Frontiers in Plant Science* 7: 1886.

Manivasagaperumal, R., Balamurugan, S., Thiyagarajan, G., and Sekar, J. (2011) Effect of zinc on germination, seedling growth and biochemical content of cluster bean (*Cyamopsis tetragonoloba* (L.) Taub). *Current Botany* 2, 11–15.

Marček, T., Hamow, K.Á., Végh, B., Janda, T., and Darko, E. (2019) Metabolic response to drought in six winter wheat genotypes. *PLOS ONE* 14: e0212411.

Mata, C.I., Fabre, B., Parsons, H.T., Hertog, M.L.A.T.M., van Raemdonck, G., *et al.* (2018) Ethylene receptors, CTRs and EIN2 target protein identification and quantification through parallel reaction monitoring during tomato fruit ripening. *Frontiers in Plant Science* 9: 1626.

Mesnage, R., Agapito-Tenfen, S.Z., Vilperte, V., Renney, G., Ward, M., *et al.* (2016) An integrated multi-omics analysis of the NK603 roundup-tolerant GM maize reveals metabolism disturbances caused by the transformation process. *Scientific Reports* 6: 37855.

Michaletti, A., Naghavi, M.R., Toorchi, M., Zolla, L., and Rinalducci, S. (2018) Metabolomics and proteomics reveal drought-stress responses of leaf tissues from spring-wheat. *Scientific Reports* 8: 5710.

Myers, L. and Sirois, M.J. (2006) Spearman correlation coefficients, differences between. In: *Encyclopedia of Statistical Sciences*. Wiley, Hoboken, New Jersey.

Nam, K.H., Kim, H.J., Pack, I.S., Kim, H.J., Chung, Y.S., *et al.* (2019) Global metabolite profiling based on GC–MS and LC–MS/MS analyses in ABF3-overexpressing soybean with enhanced drought tolerance. *Applied Biological Chemistry* 62: 15.

Niculaes, C., Abramov, A., Hannemann, L., and Frey, M. (2018) Plant protection by benzoxazinoids—recent insights into biosynthesis and function. *Agronomy* 8: 143.

Obata, T., Witt, S., Lisec, J., Palacios-Rojas, N., Florez-Sarasa, I., *et al.* (2015) Metabolite profiles of maize leaves in drought, heat, and combined stress field trials reveal the relationship between metabolism and grain yield. *Plant Physiology* 169, 2665–2683.

Ogbaga, C.C., Stepien, P., Dyson, B.C., Rattray, N.J.W., Ellis, D.I., *et al.* (2016) Biochemical analyses of *Sorghum* varieties reveal differential responses to drought. *PLOS ONE* 11: e0154423.

Paupière, M.J., Müller, F., Li, H., Rieu, I., Tikunov, Y.M., *et al.* (2017) Untargeted metabolomic analysis of tomato pollen development and heat stress response. *Plant Reproduction* 30, 81–94.

Perkowski, J., Stuper, K., Buśko, M., Góral, T., Kaczmarek, A., and Jeleń, H. (2012) Differences in metabolomic profiles of the naturally contaminated grain of barley, oats and rye. *Journal of Cereal Science* 56, 544–551.

Pinu, F.R., Beale, D.J., Paten, A.M., Kouremenos, K., Swarup, S., *et al.* (2019) Systems biology and multi-omics integration: viewpoints from the metabolomics research community. *Metabolites* 9: 76.

Pourmortazavi, S.M. and Hajimirsadeghi, S.S. (2007) Supercritical fluid extraction in plant essential and volatile oil analysis. *Journal of Chromatography A* 1163, 2–24.

Qi, X., Xu, W., Zhang, J., Guo, R., Zhao, M., *et al.* (2017) Physiological characteristics and metabolomics of transgenic wheat containing the maize C_4 phosphoenolpyruvate carboxylase (PEPC) gene under high temperature stress. *Protoplasma* 254, 1017–1030.

Rahman, M.A., Akond, M., Babar, M.A., Beecher, C., Erickson, J., *et al.* (2017) LC-HRMS based non-targeted metabolomic profiling of wheat (*Triticum aestivum* L.) under post-anthesis drought stress. *American Journal of Plant Sciences* 8: 3024.

Rai, A., Saito, K., and Yamazaki, M. (2017) Integrated omics analysis of specialized metabolism in medicinal plants. *Plant Journal* 90, 764–787.

Ramautar, R. and de Jong, G.J. (2014) Recent developments in liquid-phase separation techniques for metabolomics. *Bioanalysis* 6, 1011–1026.

Rao, J., Cheng, F., Hu, C., Quan, S., Lin, H., *et al.* (2014) Metabolic map of mature maize kernels. *Metabolomics* 10, 775–787.

Razzaq, A., Sadia, B., Raza, A., Khalid Hameed, M., and Saleem, F. (2019) Metabolomics: a way forward for crop improvement. *Metabolites* 9: 303.

Rouphael, Y., Raimondi, G., Lucini, L., Carillo, P., Kyriacou, M.C., *et al.* (2018) Physiological and metabolic responses triggered by omeprazole improve tomato plant tolerance to NaCl stress. *Frontiers in Plant Science* 9: 249.

Sawada, K., Hasegawa, M., Tokuda, L., Kameyama, J., Kodama, O., *et al.* (2004) Enhanced resistance to blast fungus and bacterial blight in transgenic rice constitutively expressing *OsSBP*, a rice homologue of mammalian selenium-binding proteins. *Bioscience, Biotechnology, and Biochemistry* 68, 873–880.

Seaver, S.M.D., Lerma-Ortiz, C., Conrad, N., Mikaili, A., Sreedasyam, A., *et al.* (2018) PlantSEED enables automated annotation and reconstruction of plant primary metabolism with improved compartmentalization and comparative consistency. *Plant Journal* 95, 1102–1113.

Seybold, H., Demetrowitsch, T., Hassani, M.A., Szymczak, S., Reim, E., *et al.* (2019) A fungal pathogen induces systemic susceptibility and systemic shifts in wheat metabolome and microbiome composition. *Nature Communications* 11: 1910.

Shelden, M.C., Dias, D.A., Jayasinghe, N.S., Bacic, A., and Roessner, U. (2016) Root spatial metabolite profiling of two genotypes of barley (*Hordeum vulgare* L.) reveals differences in response to short-term salt stress. *Journal of Experimental Botany* 67, 3731–3745.

Silva, J.C.F., Teixeira, R.M., and Silva, F.F., *et al.* (2019) Machine learning approaches and their current application in plant molecular biology: a systematic review. *Plant Science* 284, 37–47.

Silvente, S., Sobolev, A.P., and Lara, M. (2012) Metabolite adjustments in drought tolerant and sensitive soybean genotypes in response to water stress. *PLOS ONE* 7: e38554.

Skirycz, A. and Inzé, D. (2010) More from less: plant growth under limited water. *Current Opinion in Biotechnology* 21, 197–203.

Soriano, I.R., Asenstorfer, R.E., Schmidt, O., and Riley, I.T. (2004) Inducible flavone in oats (*Avena sativa*) is a novel defense against plant-parasitic nematodes. *Phytopathology* 94, 1207–1214.

Sun, C., Gao, X., Li, M., Fu, J., and Zhang, Y. (2016) Plastic responses in the metabolome and functional traits of maize plants to temperature variations. *Plant Biology* 18, 249–261.

Sun, X. and Weckwerth, W. (2012) COVAIN: a toolbox for uni-and multivariate statistics, time-series and correlation network analysis and inverse estimation of the differential Jacobian from metabolomics covariance data. *Metabolomics* 8, 81–93.

Sung, J., Sonn, Y., Lee, Y., Kang, S., Ha, S., *et al.* (2015) Compositional changes of selected amino acids, organic acids, and soluble sugars in the xylem sap of N, P, or K-deficient tomato plants. *Journal of Plant Nutrition and Soil Science* 178, 792–797.

Tenenboim, H. and Brotman, Y. (2016) Omic relief for the biotically stressed: metabolomics of plant biotic interactions. *Trends in Plant Science* 21, 781–791.

Theodorou, M.E. and Plaxton, W.C. (1993) Metabolic adaptations of plant respiration to nutritional phosphate deprivation. *Plant Physiology* 101, 339–344.

Thiele, I. and Palsson, B. (2010) A protocol for generating a high-quality genome-scale metabolic reconstruction. *Nature Protocols* 5, 93–121.

Thomason, K., Babar, M.A., Erickson, J.E., Mulvaney, M., Beecher, C., and MacDonald, G., (2018) Comparative physiological and metabolomics analysis of wheat (*Triticum aestivum* L.) following post-anthesis heat stress. *PLOS ONE* 13: e0197919.

Turner, M.F., Heuberger, A.L., Kirkwood, J.S., Collins, C.C., Wolfrum, E.J., *et al.* (2016) Non-targeted metabolomics in diverse *Sorghum* breeding lines indicates primary and secondary metabolite profiles are associated with plant biomass accumulation and photosynthesis. *Frontiers in Plant Science* 7: 953.

Veličković, D., Chu, R.K., Myers, G.L., Ahkami, A.H., and Anderton, C.R. (2019) An approach for visualizing the spatial metabolome of an entire plant root system inspired by the Swiss-rolling technique. *Journal of Mass Spectrometry* 55: e4363.

Voit, E.O. (2017) The best models of metabolism. *Wiley Interdisciplinary Reviews: Systems Biology and Medicine* 9: e1391.

Vuckovic, D. (2012) Current trends and challenges in sample preparation for global metabolomics using liquid chromatography–mass spectrometry. *Analytical and Bioanalytical Chemistry* 403, 1523–1548.

Walley, J.W., Sartor, R.C., Shen, Z., Schmitz, R.J., Wu, K.J., *et al.* (2016) Integration of omic networks in a developmental atlas of maize. *Science* 353, 814–818.

Wang, X., Hou, L., Lu, Y., Wu, B., Gong, X., *et al.* (2018) Metabolic adaptation of wheat grain contributes to a stable filling rate under heat stress. *Journal of Experimental Botany* 69, 5531–5545.

Watanabe, K.N., Sassa, Y., Suda, E., Chen, C.-H., Inaba, M., and Kikuchi, A. (2005) Global political, economic, social and technological issues on transgenic crops. *Plant Biotechnology* 22, 515–522.

Weckwerth, W. (2011) Unpredictability of metabolism—the key role of metabolomics science in combination with next-generation genome sequencing. *Analytical and Bioanalytical Chemistry* 400, 1967–1978.

Wishart, D.S. (2011) Advances in metabolite identification. *Bioanalysis* 3, 1769–1782.

Witt, S., Galicia, L., Lisec, J., Cairns, J., Tiessen, A., *et al.* (2012) Metabolic and phenotypic responses of greenhouse-grown maize hybrids to experimentally controlled drought stress. *Molecular Plant* 5, 401–417.

Wu, D., Cai, S., Chen, M., Ye, L., Chen, Z., *et al.* (2013) Tissue metabolic responses to salt stress in wild and cultivated barley. *PLOS ONE* 8: e55431.

Wu, X., Li, N., Li, H., and Tang, H. (2014) An optimized method for NMR-based plant seed metabolomic analysis with maximized polar metabolite extraction efficiency, signal-to-noise ratio, and chemical shift consistency. *Analyst* 139, 1769–1778.

Xu, S., Hu, C., Hussain, S., Tan, Q., Wu, S., and Sun, X. (2018) Metabolomics analysis reveals potential mechanisms of tolerance to excess molybdenum in soybean seedlings. *Ecotoxicology and Environmental Safety* 164, 589–596.

Yamaguchi-Shinozaki, K. and Shinozaki, K. (2006) Transcriptional regulatory networks in cellular responses and tolerance to dehydration and cold stresses. *Annual Review of Plant Biology* 57, 781–803.

Yang, H., Li, X.A., Li, F., Zhao, M.R., and Wang, W. (2012) Characterization of a wheat (*Triticum aestivum* L.) expansin gene, *TaEXPB23*, involved in the abiotic stress response and phytohormone regulation. *Plant Physiology and Biochemistry* 54, 49–58.

Yang, L., Fountain, J.C., Ji, P., Ni, X., Chen, S., *et al.* (2018) Deciphering drought-induced metabolic responses and regulation in developing maize kernels. *Plant Biotechnology Journal* 16, 1616–1628.

Yang, Z., Nakabayashi, R., Okazaki, Y., Mori, T., Takamatsu, S., *et al.* (2014) Toward better annotation in plant metabolomics: isolation and structure elucidation of 36 specialized metabolites from *Oryza sativa* (rice) by using MS/MS and NMR analyses. *Metabolomics* 10, 543–555.

Zhang, J., Luo, W., Zhao, Y., Xu, Y., Song, S., and Chong, K. (2016) Comparative metabolomic analysis reveals a reactive oxygen species-dominated dynamic model underlying chilling environment adaptation and tolerance in rice. *New Phytologist* 211, 1295–1310.

Zhang, N., Venkateshwaran, M., Boersma, M., Harms, A., Howes-Podoll, M, *et al.* (2012) Metabolomic profiling reveals suppression of oxylipin biosynthesis during the early stages of legume-rhizobia symbiosis. *FEBS Letters* 586, 3150–3158.

Zhang, Y., Ma, X.M., Wang, X.C., Liu, J.H., Huang, B.Y., *et al.* (2017) UPLC-QTOF analysis reveals metabolomic changes in the flag leaf of wheat (*Triticum aestivum* L.) under low-nitrogen stress. *Plant Physiology and Biochemistry* 111, 30–38.

Zhu, L., Zhou, Y., Li, X., Zhao, J., Guo, N., and Xing, H. (2018) Metabolomics analysis of soybean hypocotyls in response to *Phytophthora sojae* infection. *Frontiers in Plant Science* 9: 1530.

Zörb, C., Geilfus, C.M., Mühling, K.H., and Ludwig-Müller, J. (2013) The influence of salt stress on ABA and auxin concentrations in two maize cultivars differing in salt resistance. *Journal of Plant Physiology* 170, 220–224.

Zuorro, A., Lavecchia, R., Medici, F., and Piga, L. (2013) Enzyme-assisted production of tomato seed oil enriched with lycopene from tomato pomace. *Food and Bioprocess Technology* 6, 3499–3509.

12 High-throughput Metabolomics for Plant Phenotyping

Sarfraz Ahmad[1]*, Maksud Hasan Shah[2], Saniya[3], Mohammad Ghani Raghib[4], Vikas Belwal[5], Shaukat Ali[6], and Rubina Khan[7]

[1]Department of Genetics and Plant Breeding, College of Agriculture Mandawa, Swami Keshwanand Rajasthan Agriculture University, Bikaner, Rajasthan, India; [2]Department of Agronomy, College of Agriculture Chandgothi, Swami Keshwanand Rajasthan Agriculture University, Bikaner, Rajasthan, India; [3]Department of Horticulture, ICAR-National Research Centre for Grapes, Pune, Maharashtra, India; [4]Department of Horticulture, Faculty of Agriculture, Nangarhar University, Jalalabad, Afghanistan; [5]Department of Tree Improvement, College of Forestry Ranichauri, VCSG Uttarakhand University of Horticulture and Forestry, Bharsar, Uttarakhand, India; [6]Department of Agronomy, Krishi Vigyan Kendra Chandgothi, Swami Keshwanand Rajasthan Agriculture University, Bikaner, Rajasthan, India; [7]Department of Genetics and Plant Breeding, College of Agriculture Jobner, Sri Karan Narendra Agriculture University, Jobner, Rajasthan, India

Abstract

High-throughput metabolomics has emerged as a powerful tool in plant phenotyping, enabling the comprehensive analysis of small molecules (metabolites) that serve as intermediates and products of metabolism. This chapter explores the application of high-throughput metabolomics in plant phenotyping, emphasizing its potential to accelerate the identification of biochemical markers associated with desirable traits and enhance our understanding of plant metabolism for developing resilient and high-yielding crops. The chapter begins by providing an overview of metabolomics workflow, discussing the techniques and technologies used for high-throughput analysis, including mass spectrometry and nuclear magnetic resonance spectroscopy. These methods allow the simultaneous measurement of thousands of metabolites across diverse plant tissues, providing a snapshot of the plant's metabolic state. The integration of these techniques with advanced bioinformatics tools for data processing, analysis, and interpretation is also discussed, highlighting the challenges and solutions in handling large-scale metabolomic data sets. The chapter addresses the integration of metabolomics with other omics approaches, such as genomics and transcriptomics, to provide a more comprehensive understanding of plant phenotypes. Furthermore, the use of high-throughput metabolomics in plant phenotyping is highlighted, particularly its potential to expedite the selection of superior genotypes by linking metabolite profiles to phenotypic traits such as biotic and abiotic tolerance, metabolomics-assisted breeding, and nutritional quality. This systems biology approach is essential for unraveling the complex interactions between genes, proteins, and metabolites that underpin plant traits. High-throughput metabolomics is a transformative approach to plant phenotyping, offering insights that can drive the next generation of crop-improvement strategies. As technology advances, the potential of metabolomics to contribute to sustainable agriculture and food security will become increasingly significant.

Keywords: high-throughput metabolomics, mass spectrometry, metabolomics, mGWAS, mQTLs, multi-omics, phenotyping

* E-mail: sarfraz_firoz@rediffmail.com

© CAB International 2025. *High-Throughput Plant Metabolomics* (Jen-Tsung Chen ed.)
DOI: 10.1079/9781800626911.0012

12.1 Introduction

Plant phenotyping, the process of measuring and analyzing the observable traits of plants, has always been a cornerstone of plant biology and agriculture. These traits, known as phenotypes, encompass various characteristics, including morphology, physiology, development, biochemical properties, and environmental interactions. Traditionally, plant phenotyping has relied on manual measurements and visual assessments, which, while effective, are often time-consuming, labor-intensive, and prone to human error (Basavaraj and Rane, 2020). This conventional approach limits the scale at which phenotyping can be conducted, thereby constraining the speed of breeding programs and the discovery of important trait variations. As the global population continues to rise and climate change poses new challenges to agricultural systems, there is an urgent and compelling need to accelerate the development of crop varieties that are more resilient, productive, and sustainable. In response to this need, the field of plant phenotyping is transforming. Technological advances have enabled the transition from traditional phenotyping methods to high-throughput approaches, which allow the rapid and automated measurement of plant traits across large populations. Among these advanced techniques, high-throughput metabolomics stands out as a powerful tool for phenotyping, offering deep insights into the biochemical and physiological processes underlying plant traits.

Metabolomics is the comprehensive study of metabolites, the small molecules that are the intermediates and end products of cellular metabolism. These metabolites include primary compounds such as sugars, amino acids, organic acids, and lipids, as well as secondary metabolites such as flavonoids, alkaloids, and terpenoids. In plants, metabolites are not only involved in basic cellular functions but also play crucial roles in growth, development, defense, and adaptation to environmental changes (Okazaki and Saito, 2012). The plant metabolome, which encompasses the complete set of metabolites present within a cell or plant, is highly complex and dynamic. It reflects the plant's physiological state at any given time, as well as its interactions with the environment. Because metabolites are the direct products of enzymatic reactions and biochemical pathways, they are closely linked to the phenotype (Rinschen *et al.*, 2019). This makes metabolomics a powerful tool for connecting genotype to phenotype, providing insights into the functional consequences of genetic variation. Furthermore, metabolomics has several advantages over other omics approaches, such as genomics (DNA based), transcriptomics (RNA based), and proteomics (protein based) in plant phenotyping. While genomics and transcriptomics provide information about the potential and actual expression of genes, metabolomics reflects the biochemical activities taking place within the plant (Pott *et al.*, 2021). This allows observation of how genetic and environmental factors translate into specific metabolic profiles influencing plant traits.

The field of metabolomics has evolved rapidly over the past two decades, driven by advances in analytical technologies, such as mass spectrometry (MS) and nuclear magnetic resonance (NMR) spectroscopy, capillary electrophoresis–mass spectrometry (CE-MS), and Fourier-transform–ion cyclotron resonance–mass spectrometry (FT-ICR-MS). These technologies have made it possible to detect and quantify thousands of metabolites simultaneously, providing a comprehensive view of the plant metabolome (Khakimov *et al.*, 2014; Letertre *et al.*, 2020). However, the complexity and diversity of plant metabolites present significant challenges, particularly when analyzing large numbers of samples. High-throughput metabolomics addresses these challenges by integrating advanced analytical techniques with automation, data processing, and bioinformatics (Salem *et al.*, 2020; Hall *et al.*, 2022). This approach enables the rapid and large-scale analysis of metabolic profiles, making it feasible to conduct metabolomics studies on hundreds or even thousands of plant samples. High-throughput metabolomics is particularly valuable in plant phenotyping, where it allows screening of large populations of plants to identify those with desirable metabolic profiles associated with specific traits. Furthermore, high-throughput metabolomics has been made possible by automation, robotics, and MS and chromatography improvements, as well as data processing and bioinformatics, which have streamlined sample preparation, increased sensitivity, accuracy, and resolution, and enabled efficient management and interpretation of vast amounts of data (Elpa *et al.*, 2020; Bisht *et al.*, 2023).

High-throughput metabolomics has many applications in plant phenotyping, particularly in crop breeding and precision agriculture. By linking metabolic profiles to specific phenotypic traits, metabolomics can identify biomarkers associated with desirable characteristics, such as stress tolerance, disease resistance, nutritional quality, and yield (Alseekh *et al.*, 2018; Razzaq *et al.*, 2022). These biomarkers can then guide the selection of superior plant varieties in breeding programs, accelerating the development of improved crops. In order to generate more resilient crops, high-throughput metabolomics is essential for investigating how plants respond to various abiotic (e.g. drought, salinity, temperature extremes) and biotic (e.g. pathogens, pests) stresses. High-throughput metabolomics allows researchers to monitor the dynamic changes in the plant metabolome in response to stress, providing insights into the underlying stress-tolerance mechanisms (Raza, 2022). Moreover, metabolomics can profile the nutritional content of different plant varieties, identify metabolic pathways that contribute to nutritional quality, and guide the selection of varieties with enhanced nutritional properties (Hall *et al.*, 2008).

High-throughput metabolomics offers potential for plant phenotyping but also presents challenges due to the complexity of the plant metabolome, large data volumes, and lack of standardization of methods and data formats (Peters *et al.*, 2018; Hall *et al.*, 2022). Despite these challenges, advancements in analytical technologies, data processing, and bioinformatics are expected to enhance the sensitivity, accuracy, and throughput of metabolomic analyses. Emerging technologies such as single-cell metabolomics and MS imaging will provide deeper insights into plant metabolism.

This chapter comprehensively reviews the significant advancements in high-throughput metabolomics for plant phenotyping, delving into the latest developments in metabolomics techniques. The significance of inherent genetic variables influencing natural metabolic variance among plants is underlined, and the application of plant metabolomics in crop improvement and its future prospects are also discussed.

12.2 The Metabolomics Workflow

Applying high-throughput metabolomics in plant phenotyping involves a structured workflow integrating sample collection, metabolite extraction, data acquisition, data processing, and biological interpretation (Fig. 12.1).

12.2.1 Sample collection and preparation

Selection of plant material

Plant samples are collected from different tissues (e.g. leaves, roots, seeds) and at various developmental stages or under different environmental conditions to capture the dynamic nature of the metabolome.

Sample harvesting

Once the plant material is selected, it is crucial to harvest the samples carefully to preserve the plant's metabolic state. Metabolites are highly sensitive to changes in environmental conditions, and their levels can fluctuate rapidly after harvest (Li *et al.*, 2020). Plant samples are often harvested quickly and sometimes immediately frozen in liquid nitrogen to minimize metabolic changes. This process halts metabolic activity and stabilizes the metabolites, preserving the metabolic profile during harvest.

Sample preparation and extraction

The plant tissue is typically ground into a fine powder, often frozen, to ensure that metabolites are evenly distributed and to prevent degradation. Various extraction methods can be used, depending on the types of metabolites of interest. Common extraction solvents include methanol, water, chloroform, and acetone, which can be used alone or in combinations to extract a wide range of metabolites (Roopashree and Naik, 2019). For instance, a methanol: chloroform:water extraction method is commonly used to separate polar metabolites (in the methanol:water phase) from non-polar metabolites (in the chloroform phase) (Mushtaq *et al.*, 2014).

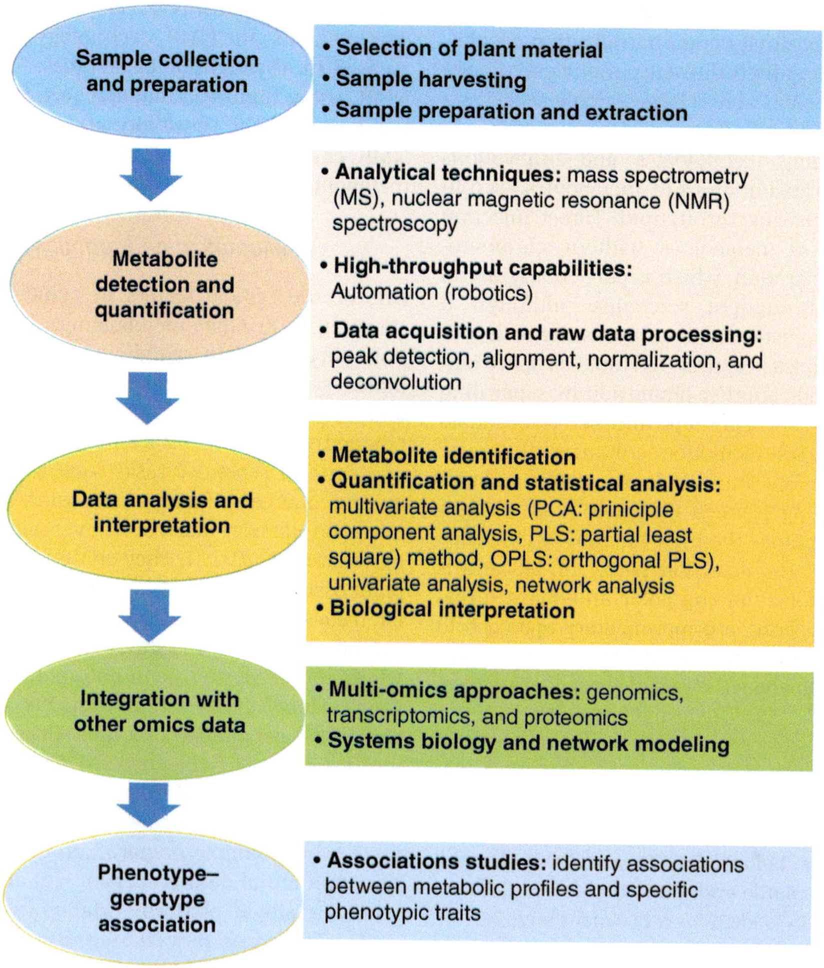

Fig. 12.1. A typical metabolomics workflow for high-throughput plant phenotyping.

12.2.2 Metabolite detection and quantification

Analytical techniques and high-throughput capabilities

The core of the high-throughput metabolomics workflow involves detecting and quantifying metabolites, with the goal being to analyze a large number of samples in a relatively short time period. The most common analytical techniques used in high-throughput metabolomics are as follows.

MS. This is the cornerstone of high-throughput metabolomics due to its sensitivity and ability to analyze a wide range of metabolites. MS identifies and quantifies metabolites based on their mass:charge (m/z) ratio. Recent advancements in MS technology have significantly enhanced its high-throughput capabilities. Quadrupole-time-of-flight-mass spectrometry (Q-TOF-MS) involves a mass spectrometer with high mass accuracy and fast scanning speeds, ideal for high-throughput applications. It can quickly acquire data from large samples while maintaining high resolution, which is essential for distinguishing between closely related metabolites (Yang *et al.*, 2023). Orbitrap MS offers ultra-high resolution and mass accuracy, allowing precise identification of metabolites in complex mixtures

(Yan *et al.*, 2022). Triple quadrupole MS offers excellent sensitivity and quantification capabilities, enabling high-throughput analysis of specific metabolites across large sample sets (Yang *et al.*, 2024).

Emerging technologies and innovations allow the rapid analysis of metabolites, significantly increasing throughput. Direct injection MS analyses metabolites without chromatographic separation, which is particularly useful for high-throughput screening, although it requires sophisticated data processing to deconvolute complex mixtures. Ion mobility spectrometry MS adds another dimension by separating ions based on their shape and size before mass analysis. This technique enhances the resolution of complex metabolomic profiles and aids in identifying isomers (Miggiels *et al.*, 2019). CE-MS combines the high separation efficiency of CE with the detection power of MS and is beneficial for analyzing polar and charged metabolites, offering a complementary approach to traditional liquid chromatography–mass spectrometry (LC-MS) and gas chromatography–mass spectrometry (GC-MS) (Zhang and Ramautar, 2022).

CHROMATOGRAPHIC TECHNIQUES. MS is often coupled with chromatographic techniques (GC-MS and LC-MS) to separate complex mixtures of metabolites before detection. GC-MS is well suited for volatile and semi-volatile compounds, while LC-MS is ideal for a broader range of metabolites, including polar and non-volatile compounds (Lopes *et al.*, 2017). Advances in chromatography have contributed to the high-throughput capabilities of metabolomics. Ultra-performance LC operates at higher pressures than conventional high-performance LC, resulting in faster separation and improved resolution. This allows shorter run times and higher throughput, making it possible to analyze more samples in a given time frame (Chawla and Ranjan, 2016). CE-MS is less common, although it offers high efficiency and resolution for separating polar and charged metabolites. Its quick processing of small sample volumes makes it a valuable tool for high-throughput metabolomics (Ramautar *et al.*, 2017).

NMR SPECTROSCOPY. While NMR spectroscopy is traditionally less sensitive than MS, it has distinct advantages in terms of reproducibility and quantification (Emwas *et al.*, 2019). Recent developments in NMR technology have increased its throughput capabilities. High-field NMR, using high-field magnets (e.g. 900 MHz) has improved the sensitivity and resolution of NMR, enabling the detection of a wider range of metabolites in less time (Wishart *et al.*, 2022).

Automation and robotics

One of the critical factors in achieving high throughput in plant metabolomics is automation, particularly in sample preparation. Automated systems offer robotic sample handling and extraction systems, reducing preparation time and effort for large plant samples. These systems can perform repetitive tasks such as pipetting, mixing, and centrifugation, ensuring consistent extraction efficiency and reducing variability (Malinowska *et al.*, 2022). They can handle various plant tissues and solvents, making them versatile for metabolomics applications. Automation also extends to high-throughput data acquisition, where robotic systems can automatically inject samples into MS or NMR instruments. Some MS platforms have multiplexing capabilities, allowing multiple samples to be analyzed simultaneously (Bertić *et al.*, 2023). Moreover, high-throughput metabolomics often involves integrated workflows where sample preparation, data acquisition, and initial data processing are automated within a single platform (Malinowska *et al.*, 2022). These workflows minimize manual intervention and streamline the metabolomics process from start to finish.

Data acquisition and raw data processing

The vast amount of data generated in high-throughput metabolomics necessitates robust and efficient data-processing pipelines. These pipelines are designed to handle the complexities of large-scale metabolomics data sets, including peak detection, alignment, normalization, and deconvolution. Peak detection involves identifying signals in spectra corresponding to individual metabolites, while peak alignment corrects for variations in retention times or chemical shifts across samples. Normalization adjusts for sample concentration or instrument sensitivity. Deconvolution separates overlapping peaks when multiple metabolites have similar ratios or shifts (Vinay *et al.*, 2021). Once the raw data have been processed, they are ready for further analysis to identify and quantify the metabolites present in each sample.

12.2.3 Data analysis and interpretation

Metabolite identification

This is typically done by comparing the processed spectra against reference databases that contain information on known metabolites, including their mass spectra (for MS) or chemical shifts (for NMR). Accurate identification links metabolic changes to specific biochemical pathways and phenotypic traits.

Advanced bioinformatics tools and software platforms are often used to automate identification. These tools can match the spectral data to large metabolite libraries, considering factors such as isotopic patterns, fragmentation patterns, and retention times. However, metabolite identification remains a challenging aspect of metabolomics, particularly for novel or low-abundance compounds that may not be well represented in existing databases.

Quantification and statistical analysis

After metabolites have been identified, the next step is quantification, which determines the concentration of each metabolite in the samples. In MS-based metabolomics, quantification is typically relative, comparing the signal intensity of each metabolite across different samples. Absolute quantification, which provides exact concentrations, can be achieved using calibration curves from known standards (Alseekh *et al.*, 2021).

The quantified metabolite data are then subjected to statistical analysis to identify patterns, correlations, and differences between experimental groups. Common statistical approaches are multivariate analysis, univariate analysis, and network analysis. Multivariate analysis techniques, such as principal component analysis (PCA) and partial least squares-discriminant analysis (PLS-DA), are used to reduce the dimensionality of the data and identify the primary sources of variation. Databases such as SMPDB 2.0, KEGG, MetaboAnalyst 4.0, and Plant Metabolic Network 15 allow statistical analysis of specific metabolite pathways (Shen *et al.*, 2023). A list of online databases/tools and software for analyzing, processing, statistical analysis, and functional interpretation of metabolomics data suitable for identifying metabolites is presented in Table 12.1. PCA and other discriminant analyses have been widely used since the onset of plant metabolomics analysis for comparative evaluations.

Univariate analysis uses statistical tests such as *t*-tests or analysis of variance (ANOVA) to find specific metabolites that differ significantly between groups. At the same time, metabolic networks are constructed to visualize the relationships between metabolites and identify critical pathways involved in the observed phenotypic traits.

Biological interpretation

This involves integrating metabolomic data with other biological information, such as genomics, transcriptomics, and proteomics data, to provide a comprehensive view of the underlying biological processes. By linking changes in metabolite levels to specific genetic variations, environmental conditions, or developmental stages, researchers can gain insights into the mechanisms that drive plant phenotypes.

Pathway analysis tools are often used to map the identified metabolites to known biochemical pathways, allowing us to see how metabolic changes correspond to different biological functions. This can help identify critical metabolites or pathways associated with important traits, such as stress tolerance, yield, or nutritional quality.

12.2.4 Integration with other omics data

Multi-omics approaches

One of the strengths of high-throughput metabolomics is its ability to be integrated with other omics approaches, such as genomics, transcriptomics, and proteomics (Table 12.1). This multi-omics approach provides a more holistic view of plant biology, allowing researchers to link genetic variation to metabolic and phenotypic traits. Here we provide an overview of how

Table 12.1. Online databases/tools and software for analyzing, processing, statistical analysis, and integrating omics of data sets (Hall *et al.*, 2022).

Databases/tools	Functionalities	Omics approach	Remarks	Reference
mixOmics, ggmixOmics	Multivariant-based framework (PCA, CCA, PLS-DA)	Transcriptomics, proteomics, metabolomics, regulatory omics	The process involves reducing dimensions, extracting variable subgroups linked to traits, and creating visualizations	Rohart *et al.* (2017)
SOMbrero	Tools for self-organizing maps	Transcriptomics, proteomics, metabolomics	Identifying profiles linked to specific traits	Villa-Vialaneix (2017)
IntLIM	Linear models	Transcriptomics, metabolomics	Matching gene–metabolite pairs to phenotype	Siddiqui *et al.* (2018)
RaMP	Enrichment analysis for genes and metabolites	Transcriptomics, metabolomics	The process involves clustering over-represented pathways based on their similarity	Zhang *et al.* (2018)
mixKernel	A machine learning approach for data integration	Transcriptomics, proteomics, metabolomics	Pattern identification and phenotypic prediction	Mariette and Villa-Vialaneix (2018)
MarVis	Self-organizing maps and pathway enrichment analysis	Transcriptomics, metabolomics	The process involves identifying profiles and markers linked to experimental conditions and displaying an output visualization	Feussner and Feussner (2019)
MetaGeneAlyse	Concatenation and multivariant-based tools	Transcriptomics, metabolomics	Statistical approach for normalization, clustering, and PCA/ICA	Wenke *et al.* (2019)
COBRA	A 'constraint-based reconstruction and analysis' environment for flux balance analysis	Transcriptomics, proteomics, metabolomics	The process involves creating, managing, calculating, and visualizing flux models for analysis and management	Heirendt *et al.* (2019)
multiGSEA	Enrichment analysis	Transcriptomics, proteomics, metabolomics	Numerous pathway databases support the multi-omics enrichment test	Canzler and Hackermüller (2020)
GAIT-GM	KEGG pathways can be mapped using a galaxy tool that maps genes and metabolites	Transcriptomics, metabolomics	The text-mining algorithm has enhanced annotation revealing metabolite abundance influenced by gene expression	McIntyre *et al.* (2020)
MetaboAnalyst	A pathway-based approach to analyze metabolites and genes using enrichment analysis, either through hypergeometric or Fisher's exact tests	Transcriptomics, metabolomics	The support is restricted to particular plant species, specifically *Arabidopsis* and rice	Pang *et al.* (2021)
OmicsAnalyst	Analysis and visualization of partial, univariate, and multivariate data	Transcriptomics, proteomics, metabolomics	Data processing, differential analysis, correlation network, and reduction of network dimensionality, and data visualization	Zhou *et al.* (2021)
MiBiOmics	A correlation-based tool for generating, reducing dimensionality, and exploring networks	Transcriptomics, proteomics, metabolomics	The task involves providing tools for data processing, including filtration, normalization, and transformation	Zoppi *et al.* (2021)

CCA, canonical correlation analysis; ICA, independent component analysis; PCA, principal component analysis; PLS-DA, partial least squares discriminant analysis.

metabolomics is integrated with other omics in plant research.

GENOMICS AND METABOLOMICS INTEGRATION. Understanding gene–metabolite relationships is crucial for understanding how genetic variations control plant metabolic pathways and phenotypes. Quantitative trait loci (QTLs) mapping identifies genomic regions associated with specific metabolic traits, allowing the identification of genes controlling important processes such as secondary metabolite production (Bilbrey *et al.*, 2021). Genome-wide association studies (GWAS) associate genetic variants across the plant genome with variations in metabolite levels, identifying genes and pathways involved in metabolic traits of interest (Luo, 2015).

TRANSCRIPTOMICS AND METABOLOMICS INTEGRATION. Transcriptomics and metabolomics data can be used to analyze gene expression and metabolic changes, providing insights into plant responses to environmental stressors. This helps build regulatory networks, explaining how transcription factors and other elements control metabolic pathways. Studying gene expression and metabolite levels over time, such as during development or stress, also provides insights into plants' dynamic regulation of metabolism (Zheng *et al.*, 2022).

PROTEOMICS AND METABOLOMICS INTEGRATION. Proteomics data help link protein abundance to metabolic activity, allowing enzyme activity and metabolism to be studied. It also helps understand post-translational modifications affecting metabolic pathways and overall plant metabolism. Furthermore, proteomics data can functionally annotate unknown proteins, enhancing our understanding of plant genomes and their role in metabolic pathways (Zheng *et al.*, 2022).

EPIGENOMICS AND METABOLOMICS INTEGRATION. Epigenomic data, such as DNA methylation and histone modification profiles, can be integrated with metabolomics to understand how epigenetic changes influence plant metabolic pathways (Hayashi *et al.*, 2023). Plants often retain a "memory" of stress events, which epigenetic changes can mediate. Integrating epigenomics with metabolomics can help unravel how these memories affect plant metabolism and stress resilience (Lephatsi *et al.*, 2021). Moreover, epigenomic modifications can be inherited and may influence the metabolism of subsequent generations. Integrating metabolomics with epigenomics can shed light on these transgenerational metabolic effects.

PHENOMICS AND METABOLOMICS INTEGRATION. Phenomics involves the high-throughput measurement of plant phenotypes, such as growth rates, yield, and stress tolerance. Integrating metabolomics with phenomics allows specific metabolites to be linked with observable plant traits. Additionally, metabolomics data can be used to predict complex traits, such as yield or stress resistance, through machine learning models that integrate phenotypic and metabolomic data (Cembrowska-Lech *et al.*, 2023). Furthermore, understanding how metabolic changes contribute to phenotypic plasticity (the ability of plants to adapt their phenotype in response to environmental fluctuations) is critical to improving crop resilience.

Systems biology and network modeling

In systems biology, metabolomics data can be used to construct metabolic networks and models that describe the flow of metabolites through different pathways. These models can predict the effects of genetic or environmental changes on plant metabolism and phenotypes, providing a powerful tool for understanding complex biological systems.

Network modeling and systems biology approaches are beneficial for identifying critical regulatory nodes in metabolic pathways, which can be targeted for crop improvement. By manipulating these nodes, it may be possible to enhance desirable traits, such as increased yield or improved stress tolerance.

12.2.5 Phenotype–genotype association

The ultimate goal of high-throughput metabolomics in plant phenotyping is to identify associations between metabolic profiles and specific phenotypic traits. These associations can be used to identify biomarkers for traits of interest, which can be used in breeding programs or precision agriculture.

12.3 Applications of High-throughput Metabolomics in Plant Phenotyping

High-throughput metabolomics is increasingly applied in plant phenotyping to advance our understanding of plant biology and improve agricultural practices. Some recent applications of high-throughput metabolomics in crop development are presented in Table 12.2. Some key applications of high-throughput metabolomics in plant phenotyping are discussed below.

12.3.1 Trait discovery and characterization

High-throughput metabolomics allows identification of specific metabolites that serve as biomarkers for various plant traits, such as drought tolerance, disease resistance, or nutrient content (Table 12.2). These biomarkers can be used to characterize and select plants with desirable traits. For example, metabolites such as glucose, fructose, tyrosine, glutamine, valine, and serine have been identified as diagnostic biomarkers for potato tuber quality trait identification and were used to select transgressive segregant progeny from a breeding population (Steinfath *et al.*, 2010). Recent research has focused on analyzing parental metabolomes to predict heterosis power for biomarker selection, with predictive biomarkers identified for biomass heterosis in rice, demonstrating the efficiency of genetic data (Dan *et al.*, 2021). Furthermore, non-targeted metabolomics can reveal unexpected metabolites or metabolic pathways associated with unique traits. This can lead to the discovery of novel traits that were previously unrecognized in plant breeding programs.

12.3.2 Crop improvement and breeding

By correlating metabolomic profiles with phenotypic traits, breeders can select plants with optimal metabolic profiles for yield, nutritional value, or stress resilience. Metabolite-assisted breeding can be used alongside traditional breeding techniques to improve the efficiency of selecting desirable traits. For instance, plants with specific metabolite profiles linked to high yield or stress tolerance can be prioritized in breeding programs.

High-throughput metabolome analysis has significantly improved software tool design and instrumentation innovation in fields such as genomics, disease diagnostics, and precision breeding. Recent advancements have speeded up screening processes, shortened the time for developing elite crop varieties, and increased the potential for metabolite diagnosis and plant phenotyping. Traditional QTL and GWAS methods are limited to identifying macroscale phenotype–genotype relationships due to unclear effects of genes, proteins, and metabolites. By focusing on molecular-scale phenotypes such as metabolite concentrations, metabolomics-based QTL (mQTLs) studies and metabolomics-based GWAS (mGWAS) have emerged as powerful tools for dissecting the relationship between genotype and phenotype and deciphering metabolic traits associated with genetic variants.

Advances in functional genomics and metabolomics technologies have enabled mQTL mapping to identify diagnostic markers and plant performance studies. Metabolomics and other omics techniques are combined to study plant phenology and genetic diversity under stress conditions. LC-MS-based profiling of barely inbred lines has revealed stress-responsive metabolites, potentially aiding in breeding of elite barley genotypes under drought and heat stress (Piasecka *et al.*, 2017). Furthermore, mGWAS analysis is crucial for understanding rice's genetic and metabolic architecture, identifying genes associated with natural metabolism variations and candidate genes modulating metabolite levels from different strains (Chen *et al.*, 2014). Comparative analysis using a combination of pGWAS and mGWAS on a diverse rice accession collection revealed over 830 metabolic features with high heritabilities for amino acids and flavonoids, indicating a relationship among phenotype–metabolites for complex trait mapping (Chen *et al.*, 2016).

12.3.3 Stress response analysis

Metabolomics is crucial in understanding how plants respond to various stressors, both biotic (e.g. pathogens, pests) and abiotic (e.g. drought, salinity, temperature extremes). It can reveal how plants alter their metabolism in response to abiotic stresses such as drought, salinity, or temperature extremes. For instance, eight wheat cultivars underwent GC-MS metabolic profiling to understand drought tolerance mechanisms. Results showed elevated amino acid concentrations under drought stress (Yadav *et al.*, 2019). Similarly, maize and rice exhibited modified levels of metabolites under stress conditions (Yang *et al.*, 2018; Ma *et al.*, 2016). Further, metabolomics helps in understanding the underlying stress adaptation mechanisms in plants. For example, key pathways that enhance plant resilience can be identified by analyzing the accumulation of osmoprotectants or antioxidants in response to drought (Anzano *et al.*, 2022).

Metabolomics analysis detects significant deviations in plant primary and secondary metabolites in response to biotic stressors such as disease and insect pests. Plants have developed many metabolites in response to biotic stressors, which serve as indicators for controlling biotic stress tolerance in various plants. Maize has significant levels of benzoxazinoids (BXs), which are crucial secondary metabolites that operate as a defense mechanism against biotic stress (Niculaes *et al.*, 2018). These insights can lead to the development of disease-resistant crop varieties.

12.3.4 Nutritional quality improvement

High-throughput metabolomics can identify and quantify metabolites that contribute to the nutritional quality of crops, such as vitamins, minerals, and phytochemicals. Plants need essential nutrients for growth and development, as carbon, nitrogen, phosphorus, and sulfur metabolites are formed in plant cells. Nitrogen is crucial for cellular metabolites such as proteins and nucleic acids. A study using non-targeted GC-MS/MS identified medicinally important metabolites in traditional Mappillai Samba rice (Rajagopalan *et al.*, 2022), while a wheat metabolomic profiling study found increased synthesis of lysine, tyrosine, L-ascorbic acid, and allo-inositol in wheat under nitrogen stress (Khan *et al.*, 2019). Moreover, increased quantities of ribulose, fructose, and lyxose were found in a wheat metabolism investigation using an integrated GC-TOF-MS method (Heyneke *et al.*, 2017). These findings highlight the importance of nitrogen in metabolizing various nutrients. Metabolomics also plays a role in biofortification efforts by identifying metabolic pathways that help synthesize essential nutrients. This knowledge can be applied to increase the content of nutrients such as iron, zinc, and provitamin A in staple crops (Gupta *et al.*, 2023).

12.3.5 Metabolic pathway elucidation

Metabolomics provides detailed insights into the metabolic pathways active in plants to map out complex metabolic networks. This is particularly useful for understanding the biosynthesis of secondary metabolites such as flavonoids, alkaloids, and terpenes (Stitt *et al.*, 2010). By elucidating the metabolic pathways involved in producing valuable compounds, metabolomics can guide efforts to genetically engineer plants for enhanced production of these compounds, such as pharmaceuticals, fragrances, and biofuels (Abbas *et al.*, 2023).

12.3.6 Phenotyping for precision agriculture

Metabolomics can be integrated into high-throughput phenotyping platforms to screen large numbers of genotypes for desirable traits. This approach enables rapid assessment of genetic diversity and the identification of superior genotypes. Moreover, advances in metabolomics technology enable the real-time monitoring of plant metabolic changes, which can be used to make informed decisions in precision agriculture. For example, metabolomic data can inform irrigation schedules or nutrient application based on the metabolic status of crops (Hall *et al.*, 2022).

12.3.7 Climate change adaptation

Metabolomics aids in understanding plant responses to climate change, enabling the breeding of climate-resilient crops. Researchers can model future plant responses by integrating metabolomics data with other omics and environmental data, guiding breeding programs for climate-adapted crops. This knowledge is crucial

Table 12.2. Some recent applications of high-throughput metabolomics in crop improvement (Raza, 2022; Razzaq *et al.*, 2022).

Crop	Analytical tools/platforms	Metabolite/trait detected	Reference
Arabidopsis	GC-TOF-MS analysis	Upregulation of basic leucine zippers, DREB, and HSFA	Wang *et al.* (2020)
Rice	LC-MS/MS	The nutritional value of leucine, L-alanine, threonine, histidine, chlorogenic acid, and syringenone was found to be significant	Chen *et al.* (2014)
	LC-MS/MS	Exploration of the relationship between flavonoids and amino acids, focusing on their impact on grain size, weight, and color	Chen *et al.* (2016)
	LC-MS/MS	Feruloyl serotonin, L-asparagine	Li *et al.* (2019)
	LC-MS, GC-MS, CE-MS,	Amino acids, sugar phosphates, nucleotides, organic acids, triacylglycerols, sterols, phenolic acids	Zeković *et al.* (2012)
	GC-MS	Phenylalanine, threonine, raffinose	Lawas *et al.* (2019)
	picoPPESI-MS	Phosphatidylinositol, organic acids, carbohydrates, amino acids	Wada *et al.* (2020)
Wheat	LC-HRMS	L-tryptophan, pipecolate, drummondol, anthranilate	Thomason *et al.* (2018)
	GC-MS, NMR	Amino acids, sugars, organic acids	Tangkhavanich *et al.* (2012)
	GC-MS	Sugars, fatty acids, phenolic acids, organic acids, sterols	Beleggia *et al.* (2013)
	LC-MS/MS	Deoxyinosine-5′-monophosphate, betaine	Shi *et al.* (2020)
Maize	LC-MS/MS	Benzoxazinoid, flavonoid	Zhou *et al.* (2019)
	LC-MS	Amino acids, flavonoids, lipids, terpenoids, benzoxazinoids	Liang *et al.* (2021)
	GC-MS, LC-MS, CE-MS, NMR, electronic spectroscopy	Amino acids, sugars, phenolic acids, organic acids, flavonoids, benzoxazinones, mycotoxins, sugar alcohols, fatty acids, sterols, phytic acid, nucleotides, amines, fluorescence	Khakimov *et al.* (2014)
Pearl millet	FIE-HRMS	Vitamins, antioxidants, dietary starch	Yadav *et al.* (2021)
Soybean	GC/LC-MS	Phytochemicals: genistin, daidzein, formononetin, syringic acid, genistein	Das *et al.* (2017)
	LC-MS	Aspartic acid, alanine, arginine, daidzein, asparagine	Liu *et al.* (2020)
Barley	LC-MS	Acylated flavonoid glycosides, phenylpropenoic acid, glycosides	Piasecka *et al.* (2017)
Tomato	ESI-Qq-TOF-MS/MS	Amino acids, alkaloids, vitamins, polyamines, polyphenols	Zhu *et al.* (2018)
Capsicum	LC-MS/MS	Amino and organic acids, flavonoids, sugars	Wang *et al.* (2019)

CE-MS, capillary electrophoresis–mass spectrometry; ESI-Qq-TOF-MS/MS, electrospray ionization–double quadrupole time-of-flight–tandem mass spectrometry; FIE-HRMS, flow infusion electrospray–high-resolution mass spectrometry; GC-MS, gas chromatography–mass spectrometry; GC-TOF-MS, gas chromatography–time-of-flight–mass spectrometry; LC-HRMS, liquid chromatography–high-resolution mass spectrometry; LC-MS, liquid chromatography–mass spectrometry; LC-MS/MS, liquid chromatography–tandem mass spectrometry; NMR,; picoPPESI-MS, picolitre pressure probe electrospray ionization–mass spectrometry for breeding crops that can withstand changing climate conditions.

12.3.8 Sustainable agriculture and environmental impact

Metabolomics can be used to study how plants use water, nutrients, and other resources, leading to the development of more efficient crops. This contributes to sustainable agriculture practices. By understanding how plants metabolically respond to environmental stresses, metabolomics can guide strategies to mitigate the impact of such stresses on crop production, contributing to more sustainable agricultural systems.

12.4 Challenges and Future Directions

While high-throughput metabolomics offers tremendous potential for plant phenotyping, it also presents several challenges. The complexity of the plant metabolome, with its vast diversity of metabolites and the influence of environmental factors, makes data interpretation challenging. Additionally, the high-throughput nature of the approach generates large volumes of data, requiring sophisticated data-processing and bioinformatics tools to manage and analyze the data effectively.

Another challenge is the standardization of metabolomics methods and data formats. The lack of standardized protocols can lead to variability in results across different studies, making comparing and replicating findings difficult. Efforts are under way to develop standardized methods and data formats for metabolomics, ensuring reproducibility and comparability in high-throughput phenotyping studies.

Despite these challenges, the future of high-throughput metabolomics in plant phenotyping is bright. Advances in analytical technologies, data processing, and bioinformatics will continue enhancing the sensitivity, accuracy, and throughput of metabolomic analyses. Emerging technologies, such as single-cell metabolomics and MS imaging, are poised to provide even deeper insights into plant metabolism at unprecedented spatial and temporal resolution. Moreover, as high-throughput metabolomics becomes more integrated with other omics approaches and machine learning, it will unlock new possibilities for understanding plant biology and improving crop traits. These advancements will play a critical role in addressing the global challenges of food security, climate change, and sustainable agriculture.

12.5 Conclusion

High-throughput metabolomics has become a pivotal tool in plant phenotyping, offering detailed insights into the metabolic processes that drive plant traits. This approach allows researchers to connect metabolic profiles with phenotypic characteristics such as stress tolerance, yield, and nutritional content by enabling the simultaneous analysis of thousands of metabolites. Integrating metabolomics with other omics technologies, such as genomics and transcriptomics, further enhances our understanding of the complex biological networks underlying plant phenotypes. The application of high-throughput metabolomics in breeding programs holds significant potential to accelerate the development of more resilient crops and crops better adapted to environmental challenges. However, the field faces ongoing challenges, including the need for standardized methodologies and advanced data analysis tools. Addressing these challenges will be essential to fully realizing the potential of metabolomics in plant science. As technological advancements continue to refine this approach, it is poised to play an increasingly critical role in crop improvement and sustainable agriculture, contributing to global food security in the face of climate change and population growth.

12.6 References

Abbas, F., Zhou, Y., O'Neill Rothenberg, D., Alam, I., Ke, Y., and Wang, H. C. (2023) Aroma components in horticultural crops: chemical diversity and usage of metabolic engineering for industrial applications. *Plants* 12: 1748.

Alseekh, S., Bermudez, L., de Haro, L.A., Fernie, A.R., and Carrari, F. (2018) Crop metabolomics: from diagnostics to assisted breeding. *Metabolomics* 14: 148.

Alseekh, S., Aharoni, A., Brotman, Y., Contrepois, K., d'Auria, J., *et al.* (2021) Mass spectrometry-based metabolomics: a guide for annotation, quantification and best reporting practices. *Nature Methods* 18, 747–756.

Anzano, A., Bonanomi, G., Mazzoleni, S., and Lanzotti, V. (2022) Plant metabolomics in biotic and abiotic stress: a critical overview. *Phytochemistry Reviews* 21, 503–524.

Basavaraj, P.S. and Rane, J. (2020) Avenues to realize potential of phenomics to accelerate crop breeding for heat tolerance. *Plant Physiology Reports* 25, 594–610.

Beleggia, R., Platani, C., Nigro, F., de Vita, P., Cattivelli, L., and Papa, R. (2013) Effect of genotype, environment and genotype-by-environment interaction on metabolite profiling in durum wheat (*Triticum durum* Desf.) grain. *Journal of Cereal Science* 57, 183–192.

Bertić, M., Zimmer, I., Andrés-Montaner, D., Rosenkranz, M., Kangasjärvi, J., *et al.* (2023) Automatization of metabolite extraction for high-throughput metabolomics: case study on transgenic isoprene-emitting birch. *Tree Physiology* 43, 1855–1869.

Bilbrey, E.A., Williamson, K., Hatzakis, E., Miller, D.D., Fresnedo-Ramírez, J., and Cooperstone, J.L. (2021) Integrating genomics and multiplatform metabolomics enables metabolite quantitative trait loci detection in breeding-relevant apple germplasm. *New Phytologist* 232, 1944–1958.

Bisht, A., Saini, D.K., Kaur, B., Batra, R., Kaur, S., *et al.* (2023) Multi-omics assisted breeding for biotic stress resistance in soybean. *Molecular Biology Reports* 50, 3787–3814.

Canzler, S. and Hackermüller, J. (2020) multiGSEA: a GSEA-based pathway enrichment analysis for multi-omics data. *BMC Bioinformatics* 21: 561.

Cembrowska-Lech, D., Krzemińska, A., Miller, T., Nowakowska, A., Adamski, C., *et al.* (2023) An integrated multi-omics and artificial intelligence framework for advance plant phenotyping in horticulture. *Biology* 12: 1298.

Chawla, G. and Ranjan, C. (2016) Principle, instrumentation, and applications of UPLC: a novel technique of liquid chromatography. *Open Chemistry Journal* 3, 1–16.

Chen, W., Gao, Y., Xie, W., Gong, L., Lu, K., *et al.* (2014) Genome-wide association analyses provide genetic and biochemical insights into natural variation in rice metabolism. *Nature Genetics* 46, 714–721.

Chen, W., Wang, W., Peng, M., Gong, L., Gao, Y., *et al.* (2016) Comparative and parallel genome-wide association studies for metabolic and agronomic traits in cereals. *Nature Communications* 7: 12767.

Dan, Z., Chen, Y., Li, H., Zeng, Y., Xu, W., *et al.* (2021) The metabolomic landscape of rice heterosis highlights pathway biomarkers for predicting complex phenotypes. *Plant Physiology* 187, 1011–1025.

Das, A., Rushton, P.J., and Rohila, J.S. (2017) Metabolomic profiling of soybeans (*Glycine max* L.) reveals the importance of sugar and nitrogen metabolism under drought and heat stress. *Plants* 6: 21.

Elpa, D.P., Prabhu, G.R.D., Wu, S.P., Tay, K.S., and Urban, P.L. (2020) Automation of mass spectrometric detection of analytes and related workflows: a review. *Talanta* 208: 120304.

Emwas, A.H., Roy, R., McKay, R.T., Tenori, L., Saccenti, E., *et al.* (2019) NMR spectroscopy for metabolomics research. *Metabolites* 9: 123.

Feussner, K. and Feussner, I. (2019) Comprehensive LC-MS-based metabolite fingerprinting approach for plant and fungal-derived samples. *Methods in Molecular Biology* 1978, 167–185.

Gupta, A., Sharma, T., Singh, S.P., Bhardwaj, A., Srivastava, D., and Kumar, R. (2023) Prospects of microgreens as budding living functional food: breeding and biofortification through OMICS and other approaches for nutritional security. *Frontiers in Genetics* 14: 1053810.

Hall, R.D., Brouwer, I.D., and Fitzgerald, M.A. (2008) Plant metabolomics and its potential application for human nutrition. *Physiologia Plantarum* 132, 162–175.

Hall, R.D., d'Auria, J.C., Ferreira, A.C.S., Gibon, Y., Kruszka, D., *et al.* (2022) High-throughput plant phenotyping: a role for metabolomics? *Trends in Plant Science* 27, 549–563.

Hayashi, K., Alseekh, S., and Fernie, A.R. (2023) Genetic and epigenetic control of the plant metabolome. *Proteomics* 23: 2200104.

Heirendt, L., Arreckx, S., Pfau, T., Mendoza, S.N., Richelle, A., *et al.* (2019) Creation and analysis of biochemical constraint-based models using the COBRA Toolbox v. 3.0. *Nature Protocols* 14, 639–702.

Heyneke, E., Watanabe, M., Erban, A., Duan, G., Buchner, P., *et al.* (2017) Characterization of the wheat leaf metabolome during grain filling and under varied N-supply. *Frontiers in Plant Science* 8: 2048.

Khakimov, B., Bak, S., and Engelsen, S.B. (2014) High-throughput cereal metabolomics: current analytical technologies, challenges and perspectives. *Journal of Cereal Science* 59, 393–418.

Khan, F.U., Fuentes, D., Threthowan, R., Mohammad, F., and Ahmad, M. (2019) Comparative metabolite profiling of two wheat genotypes as affected by nitrogen stress at seedling stage. *Journal of Animal and Plant Sciences* 29, 260–268.

Lawas, L.M.F., Li, X., Erban, A., Kopka, J., Jagadish, S.K., *et al.* (2019) Metabolic responses of rice cultivars with different tolerance to combined drought and heat stress under field conditions. *GigaScience* 8: giz050.

Lephatsi, M.M., Meyer, V., Piater, L.A., Dubery, I.A., and Tugizimana, F. (2021) Plant responses to abiotic stresses and rhizobacterial biostimulants: metabolomics and epigenetics perspectives. *Metabolites* 11: 457.

Letertre, M.P., Dervilly, G., and Giraudeau, P. (2020) Combined nuclear magnetic resonance spectroscopy and mass spectrometry approaches for metabolomics. *Analytical Chemistry* 93, 500–518.

Li, K., Wang, D., Gong, L., Lyu, Y., Guo, H., *et al.* (2019) Comparative analysis of metabolome of rice seeds at three developmental stages using a recombinant inbred line population. *Plant Journal* 100, 908–922.

Li, Y., Kong, D., Fu, Y., Sussman, M.R., and Wu, H. (2020) The effect of developmental and environmental factors on secondary metabolites in medicinal plants. *Plant Physiology and Biochemistry* 148, 80–89.

Liang, X., Liu, S., Wang, T., Li, F., Cheng, J., *et al.* (2021) Metabolomics-driven gene mining and genetic improvement of tolerance to salt-induced osmotic stress in maize. *New Phytologist* 230, 2355–2370.

Liu, J.Y., Li, P., Zhang, Y.W., Zuo, J.F., Li, G., *et al.* (2020) Three-dimensional genetic networks among seed oil-related traits, metabolites and genes reveal the genetic foundations of oil synthesis in soybean. *Plant Journal* 103, 1103–1124.

Lopes, A.S., Cruz, E.C.S., Sussulini, A. and Klassen, A. (2017) Metabolomic strategies involving mass spectrometry combined with liquid and gas chromatography. In: Sussulini, A. (ed.) *Metabolomics: from Fundamentals to Clinical Applications*, Springer, Cham, Switzerland, pp. 77–98.

Luo, J. (2015) Metabolite-based genome-wide association studies in plants. *Current Opinion in Plant Biology* 24, 31–38.

Ma, X., Xia, H., Liu, Y., Wei, H., Zheng, X., *et al.* (2016) Transcriptomic and metabolomic studies disclose key metabolism pathways contributing to well-maintained photosynthesis under the drought and the consequent drought-tolerance in rice. *Frontiers in Plant Science* 7: 1886.

Malinowska, J.M., Palosaari, T., Sund, J., Carpi, D., Lloyd, G.R., *et al.* (2022) Automated sample preparation and data collection workflow for high-throughput in vitro metabolomics. *Metabolites* 12: 52.

Mariette, J. and Villa-Vialaneix, N. (2018) Unsupervised multiple kernel learning for heterogeneous data integration. *Bioinformatics* 34, 1009–1015.

McIntyre, L.M., Huertas, F., Moskalenko, O., Llansola, M., Felipo, V., *et al.* (2020) GAIT-GM: Galaxy tools for modeling metabolite changes as a function of gene expression. *bioRxiv* (preprint). Available at: www.biorxiv.org/content/10.1101/2020.12.25.424407v1 (accessed 29 January 2025).

Miggiels, P., Wouters, B., van Westen, G.J., Dubbelman, A.C., and Hankemeier, T. (2019) Novel technologies for metabolomics: more for less. *Trends in Analytical Chemistry* 120: 115323.

Mushtaq, M.Y., Choi, Y.H., Verpoorte, R., and Wilson, E.G. (2014) Extraction for metabolomics: access to the metabolome. *Phytochemical Analysis* 25, 291–306.

Niculaes, C., Abramov, A., Hannemann, L., and Frey, M. (2018) Plant protection by benzoxazinoids—recent insights into biosynthesis and function. *Agronomy* 8: 143.

Okazaki, Y. and Saito, K. (2012) Recent advances of metabolomics in plant biotechnology. *Plant Biotechnology Reports* 6, 1–15.

Pang, Z., Chong, J., Zhou, G., de Lima Morais, D.A., Chang, L., *et al.* (2021) MetaboAnalyst 5.0: narrowing the gap between raw spectra and functional insights. *Nucleic Acids Research* 49, W388–W396.

Peters, K., Worrich, A., Weinhold, A., Alka, O., Balcke, G., *et al.* (2018) Current challenges in plant eco-metabolomics. *International Journal of Molecular Sciences* 19: 1385.

Piasecka, A., Sawikowska, A., Kuczyńska, A., Ogrodowicz, P., Mikołajczak, K., *et al.* (2017) Drought-related secondary metabolites of barley (*Hordeum vulgare* L.) leaves and their metabolomic quantitative trait loci. *Plant Journal* 89, 898–913.

Pott, D.M., Durán-Soria, S., Osorio, S., and Vallarino, J.G. (2021) Combining metabolomic and transcriptomic approaches to assess and improve crop quality traits. *CABI Agriculture and Bioscience* 2: 1.

Rajagopalan, V.R., Manickam, S., and Muthurajan, R. (2022) A comparative metabolomic analysis reveals the nutritional and therapeutic potential of grains of the traditional rice variety Mappillai Samba. *Plants* 11; 543.

Ramautar, R., Somsen, G.W., and de Jong, G.J. (2017) CE–MS for metabolomics: developments and applications in the period 2014–2016. *Electrophoresis* 38, 190–202.

Raza, A. (2022) Metabolomics: a systems biology approach for enhancing heat stress tolerance in plants. *Plant Cell Reports* 41, 741–763.

Razzaq, A., Wishart, D.S., Wani, S.H., Hameed, M.K., Mubin, M., and Saleem, F. (2022) Advances in metabolomics-driven diagnostic breeding and crop improvement. *Metabolites* 12: 511.

Rinschen, M.M., Ivanisevic, J., Giera, M., and Siuzdak, G. (2019) Identification of bioactive metabolites using activity metabolomics. *Nature Reviews Molecular Cell Biology* 20, 353–367.

Rohart, F., Gautier, B., Singh, A., and Lê Cao, K.A. (2017) mixOmics: an R package for 'omics feature selection and multiple data integration. *PLOS Computational Biology* 13: e1005752.

Roopashree, K.M. and Naik, D. (2019) Advanced method of secondary metabolite extraction and quality analysis. *Journal of Pharmacognosy and Phytochemistry* 8, 1829–1842.

Salem, M.A., Perez de Souza, L., Serag, A., Fernie, A.R., Farag, M.A., *et al.* (2020) Metabolomics in the context of plant natural products research: from sample preparation to metabolite analysis. *Metabolites* 10: 37.

Shen, S., Zhan, C., Yang, C., Fernie, A.R., and Luo, J. (2023) Metabolomics-centered mining of plant metabolic diversity and function: past decade and future perspectives. *Molecular Plant* 16, 43–63.

Shi, T., Zhu, A., Jia, J., Hu, X., Chen, J., *et al.* (2020) Metabolomics analysis and metabolite-agronomic trait associations using kernels of wheat (*Triticum aestivum*) recombinant inbred lines. *Plant Journal* 103, 279–292.

Siddiqui, J.K., Baskin, E., Liu, M., Cantemir-Stone, C.Z., Zhang, B., *et al.* (2018) IntLIM: integration using linear models of metabolomics and gene expression data. *BMC Bioinformatics* 19: 81.

Steinfath, M., Strehmel, N., Peters, R., Schauer, N., Groth, D., *et al.* (2010) Discovering plant metabolic biomarkers for phenotype prediction using an untargeted approach. *Plant Biotechnology Journal* 8, 900–911.

Stitt, M., Sulpice, R., and Keurentjes, J. (2010) Metabolic networks: how to identify key components in the regulation of metabolism and growth. *Plant Physiology* 152, 428–444.

Tangkhavanich, B., Kobayashi, T., and Adachi, S. (2012) Properties of rice straw extract after subcritical water treatment. *Bioscience, Biotechnology and Biochemistry* 76, 1146–1149.

Thomason, K., Babar, M.A., Erickson, J.E., Mulvaney, M., Beecher, C., and MacDonald, G. (2018) Comparative physiological and metabolomics analysis of wheat (*Triticum aestivum* L.) following post-anthesis heat stress. *PLOS ONE* 13: e0197919.

Villa-Vialaneix, N. (2017) Stochastic self-organizing map variants with the R package SOMbrero. In: *2017 12th International Workshop on Self-Organizing Maps and Learning Vector Quantization, Clustering and Data Visualization (WSOM)*. IEEE, New York, pp. 1–7.

Vinay, C.M., Udayamanoharan, S.K., Prabhu Basrur, N., Paul, B., and Rai, P.S. (2021) Current analytical technologies and bioinformatic resources for plant metabolomics data. *Plant Biotechnology Reports* 15, 561–572.

Wada, H., Hatakeyama, Y., Nakashima, T., Nonami, H., Erra-Balsells, R., *et al.* (2020) On-site single pollen metabolomics reveals varietal differences in phosphatidylinositol synthesis under heat-stress conditions in rice. *Scientific Reports* 10: 2013.

Wang, J., Lv, J., Liu, Z., Liu, Y., Song, J., *et al.* (2019) Integration of transcriptomics and metabolomics for pepper (*Capsicum annuum* L.) in response to heat stress. *International Journal of Molecular Sciences* 20: 5042.

Wang, L., Ma, K.B., Lu, Z.G., Ren, S.X., Jiang, H.R., *et al.* (2020) Differential physiological, transcriptomic and metabolomic responses of *Arabidopsis* leaves under prolonged warming and heat shock. *BMC Plant Biology* 20: 86.

Wenke, K., Kopka, J., Schwachtje, J., van Dongen, J.T., and Piechulla, B. (2019) Volatiles of rhizobacteria *Serratia* and *Stenotrophomonas* alter growth and metabolite composition of *Arabidopsis thaliana*. *Plant Biology* 21, 109–119.

Wishart, D.S., Cheng, L.L., Copié, V., Edison, A.S., Eghbalnia, H.R., *et al.* (2022) NMR and metabolomics—a roadmap for the future. *Metabolites* 12: 678.

Yadav, A.K., Carroll, A.J., Estavillo, G.M., Rebetzke, G.J., and Pogson, B.J. (2019) Wheat drought tolerance in the field is predicted by amino acid responses to glasshouse-imposed drought. *Journal of Experimental Botany* 70, 4931–4948.

Yadav, C.B., Srivastava, R.K., Gangashetty, P.I., Yadav, R., Mur, L.A., and Yadav, R.S. (2021) Metabolite diversity and metabolic genome-wide marker association studies (MGWAS) for health benefiting nutritional traits in pearl millet grains. *Cells* 10: 3076.

Yan, X.T., Zhang, Y., Zhou, Y., Li, G.H., and Feng, X.S. (2022) Technical overview of orbitrap high resolution mass spectrometry and its application to the detection of small molecules in food (update since 2012). *Critical Reviews in Analytical Chemistry* 52, 593–626.

Yang, L., Fountain, J.C., Ji, P., Ni, X., Chen, S., *et al.* (2018) Deciphering drought-induced metabolic responses and regulation in developing maize kernels. *Plant Biotechnology Journal* 16, 1616–1628.

Yang, M., Li, J., Zhao, C., Xiao, H., Fang, X., and Zheng, J. (2023) LC-Q-TOF-MS/MS detection of food flavonoids: principle, methodology, and applications. *Critical Reviews in Food Science and Nutrition* 63, 3750–3770.

Yang, Q., Zhou, T., Liu, Y., Weng, J., Gao, L., *et al.* (2024) Analysis of 78 trace liquid crystal monomers in air by gas chromatography coupled with triple quadrupole mass spectrometry. *Science of the Total Environment* 928: 172459.

Zeković, I., Lenhardt, L., Dramićanin, T., and Dramićanin, M.D. (2012) Classification of intact cereal flours by front-face synchronous fluorescence spectroscopy. *Food Analytical Methods* 5, 1205–1213.

Zhang, B., Hu, S., Baskin, E., Patt, A., Siddiqui, J.K., and Mathé, E.A. (2018) RaMP: a comprehensive relational database of metabolomics pathways for pathway enrichment analysis of genes and metabolites. *Metabolites* 8: 16.

Zhang, W. and Ramautar, R. (2022) Utility and advances of capillary electrophoresis-mass spectrometry for metabolomics. In: Ramautar, R. and Chen, D.D.Y. (eds) *Capillary Electrophoresis-Mass Spectrometry for Proteomics and Metabolomics: Principles and Applications*. Wiley, Weinheim, Germany, pp. 147–177.

Zheng, L., Wu, W., Chen, Q., Zhang, G., Gao, F., and Zhou, Y. (2022) Integrated transcriptomics, proteomics, and metabolomics identified biological processes and metabolic pathways involved in heat stress response in jojoba. *Industrial Crops and Products* 183: 114946.

Zhou, G., Ewald, J., and Xia, J. (2021) OmicsAnalyst: a comprehensive web-based platform for visual analytics of multi-omics data. *Nucleic Acids Research* 49, W476-–W482.

Zhou, S., Kremling, K.A., Bandillo, N., Richter, A., Zhang, Y.K., *et al.* (2019) Metabolome-scale genome-wide association studies reveal chemical diversity and genetic control of maize specialized metabolites. *Plant Cell* 31, 937–955.

Zhu, G., Wang, S., Huang, Z., Zhang, S., Liao, Q., *et al.* (2018) Rewiring of the fruit metabolome in tomato breeding. *Cell* 172, 249–261.

Zoppi, J., Guillaume, J.F., Neunlist, M., and Chaffron, S. (2021) MiBiOmics: an interactive web application for multi-omics data exploration and integration. *BMC Bioinformatics* 22: 6.

13 High-throughput Metabolomics for the Dissection of Plant Biotic and Abiotic Stress Responses

Berivan Ozlem Gumus, Alberto Ribes-Navarro, Eva Cañizares, Juan Manuel Acién, Miguel González-Guzmán*, and Vicent Arbona*

Departament de Biologia, Bioquímica i Ciències Naturals, Universitat Jaume I, Castelló de la Plana, Spain

Abstract

More than 20 years after the term was coined, metabolomics has become one of the most popular omics technologies due to its transversality, and being species independent and relatively easy to implement in different analytical platforms. However, despite the growing interest in this technology and the increasing number of workflows, databases, and applications, a gold standard is still lacking for data processing and annotation. One of the leading fields in the development and implementation of metabolomics is plant science in a number of different areas including plant physiology and biochemistry, plant biotechnology, and plant breeding. This chapter discusses the fundamentals of plant metabolomics, from data acquisition and the different platforms available to processing and annotation of metabolites, emphasizing the important role of metabolomics as a functional evaluation tool as well as in the discovery and validation of selection metabolic markers for crops under constraints imposed by climate change.

Keywords: disease resistance, metabolic engineering, metabolite annotation, plant breeding, stress tolerance

13.1 Introduction

13.1.1 Metabolomics analysis: fundamentals and available platforms

In the intricate realm of biological systems, low-molecular-mass compounds (<1500 Da), also known as metabolites, are essential molecules that regulate various biochemical pathways vital to life in organs, tissues, and cells (Kumar *et al.*, 2017). Metabolomics, the study of these metabolites within organisms, provides a deeper understanding of cellular processes and their implications for health and disease status at the molecular level (Brunetti *et al.*, 2013). The rapidly progressing field of plant metabolomics offers the prospect of studying the metabolite profiles of transgenic lines or mutants of potentially important genes, which can help decipher the metabolic homeostasis underlying environmental interactions, such as abiotic and biotic stresses in plants (Sharma *et al.*, 2021). A myriad of these metabolites are involved in plant responses and tolerance to environmental conditions (Fernie and Tohge, 2017). Comparative analysis of metabolomes extracted from stressed and control plants

* E-mail: mguzman@uji.es or arbona@uji.es

© CAB International 2025. *High-Throughput Plant Metabolomics* (Jen-Tsung Chen ed.)
DOI: 10.1079/9781800626911.0013

provides a window into the broader regulatory mechanisms operating within the plant system, offers insights into transcriptional responses as well as translational and post-translational regulation, and enables a comprehensive view of how plants adapt to environmental challenges at different levels of gene expression (Dinakar and Bartels, 2013). Through this integrated approach, researchers can elucidate the dynamic interplay between genetic composition and environmental cues, paving the way for enhancing plant resilience and agricultural productivity in the face of climate change and the emergence of new pathogens and pests. The application of metabolomics techniques in plant research represents a powerful tool for unraveling the complex network of interactions governing plant metabolism and responses to environmental stimuli.

Metabolomics usually employs two distinct approaches: targeted and untargeted. Both approaches subject biological samples to the same analytical techniques, such as gas chromatography–mass spectrometry (GC-MS) or liquid chromatography–mass spectrometry (LC-MS), to profile the metabolites present. Notably, GC-MS requires derivatization of soluble metabolites for analysis, whereas LC-MS is more versatile and sensitive for a wide range of metabolites and does not require previous derivatization of metabolites (Jorge *et al.*, 2016). Targeted metabolomics focuses on the quantification of a predefined set of metabolites using an instrumental setup that is especially suited for specific compounds. This approach is particularly useful for unbiased analysis and quantification of a set of metabolites with physiological and/or ecological implications in a large number of samples (Roberts *et al.*, 2012). MS-based targeted metabolomics techniques can be categorized into selected reaction monitoring and multiple reaction monitoring to monitor single or multiple precursor-to-product transitions, respectively. These analyses are typically conducted using a triple quadrupole mass spectrometer, which offers sensitive and efficient quantification of specific metabolites (Choudhury *et al.*, 2022), an ion trap-type mass spectrometer or, more recently, an Orbitrap, which combines sensitivity and multiple-stage ion fragmentation within the trap (Chang *et al.*, 2020). In general, these approaches involve monitoring specific product ions, which, after fragmentation in the collision cell, render one or more predominant product ions. These techniques have been applied to a range of plant biological studies, including the analysis of plant hormones (Chiwocha *et al.*, 2003; Durgbanshi *et al.*, 2005; Manzi *et al.*, 2015) in different plant matrices or redox-related metabolites involved in systemic acquired resistance in the model plant *Arabidopsis thaliana* infected with the biotrophic pathogen *Pseudomonas syringae* pv. *tomato* DC3000 (David *et al.*, 2020), using targeted metabolomics. Additionally, targeted metabolomics has provided valuable insights into the metabolic profiles of crop species, enabling the identification of specific metabolites and pathways that are crucial for plant stress responses (Razzaq *et al.*, 2019; Litvinov *et al.*, 2021). For instance, the response of two wheat varieties, sensitive Wyalkatchem and tolerant Young, to cold stress was investigated by Cheong *et al.* (2019), who found that Wyalkatchem showed reduced levels of amino acids after the first cold night, followed by the accumulation of osmolytes over 4 days, indicating acclimation to water stress. The Young wheat cultivar displayed a higher unsaturated:-saturated lipid ratio than the sensitive one after cold treatment, suggesting superior membrane fluidity maintenance compared with Wyalkatchem. Hence, targeted metabolomics research provides detailed metabolic profiles and the identification of key metabolites associated with various biological functions and stress responses, allowing metabolomics-assisted breeding and the development of improved crop varieties (Allwood *et al.*, 2021). However, while targeted metabolomics offers high sensitivity and specificity, it is sometimes limited by the reduced coverage of metabolites compared with untargeted approaches (Schrimpe-Rutledge *et al.*, 2016).

Untargeted metabolomics has become an increasingly valuable tool in plant research, enabling the identification of a wide array of metabolites and providing insight into metabolic pathways (Perez de Souza *et al.*, 2019). This approach is particularly beneficial when there is no predefined hypothesis regarding which metabolites may be involved, allowing the discovery of novel biomarkers and metabolite accumulation patterns (Gertsman and Barshop, 2018). Untargeted metabolite analysis involves several key steps, starting with collection, preparation of biological samples, and metabolomic profiling using one or several analytical techniques. As there is no single technique that exhaustively analyzes the whole plant metabolome, it is becoming common to combine GC-MS and LC-MS to attain a more complete metabolite coverage. Usually, LC separation is better suited for soluble metabolites ranging from polar to semi-polar to relatively non-polar, while GC has traditionally been used for volatile compound analysis and central carbon metabolites after derivatization (Jorge *et al.*, 2016). Interestingly, while GC-MS and LC-MS are common and widespread analytical platforms, other methods, such as nuclear magnetic resonance (NMR) and infrared spectroscopy, which are also used for metabolite fingerprinting without prior separation (Schuhmacher *et al.*, 2013), are less prevalent in the scientific literature. Another technique is flow injection analysis–mass spectrometry (FIA-MS), situated between chromatography MS and ambient MS, that enables high-throughput screening of liquid samples without the need for chromatographic separation (Nanita and Kaldon *et al.*, 2016). Plant samples are extracted in a mixture of polar and non-polar solvents (usually water and methanol) and injected into the mass spectrometer through an electrospray ionization (ESI) or an atmospheric pressure chemical ionization source at a very low flow rate (usually 25 ml/min). Acquisition and grouping of mass chromatographic features allow the identification of metabolites, but mostly have been used as a fingerprinting technique for classification purposes, using mass features as classification variables. FIA-tandem MS (FIA-MS/MS) has been used in the analysis of different plant extracts, such as its application in *Cannabis sativa* L. for detection of cannabinoids (Büttenbender *et al.*, 2022). The utility of the technique in plant analysis has been further demonstrated in tea authentication, where FIA-MS fingerprinting differentiated some tea varieties and detected adulteration with chicory extracts (Vilà *et al.*, 2022). Another interesting group of techniques collectively known as MS imaging enables visualization of the spatial distribution of biomolecules within plant tissues using a label-free approach (Boughton *et al.*, 2016). Different ionization sources are typically used for MS imaging that allow access to internal metabolites by using tissue sections of plant material. The direct analysis in real time (DART) ion source operates by using a heated stream of inert gas, such as helium or nitrogen, to directly ionize samples. This process involves excited-state gas molecules interacting with the analyte, allowing immediate MS analysis without the need for extensive sample preparation (Gross, 2014). Pan *et al.* (2014) combined DART and LC-ESI-MS, to identify compounds in 'Anliu' sweet orange and its bud mutant 'Hong Anliu'. A total of 54 compounds were identified through DART-MS, while 105 were identified with LC-ESI-MS, indicating that matrix effects are a significant factor hindering DART measurements. Bai *et al.* (2012) proposed a time-efficient method using a single-drop liquid–liquid–liquid microextraction technique combined with DART-MS to analyze phytohormones in juices of pineapple, pear, and watermelon. Matrix-assisted laser desorption/ionization (MALDI) is acknowledged for its ability to induce soft ionization of high-molecular-mass molecules using an ultraviolet or infrared laser source and requires minimal sample preparation (Donnarumma *et al.*, 2010). Klein *et al.* (2015) visualized internal metabolites using MALDI-MS imaging and developed novel sample preparation methods to investigate the chemical interface in soybean–aphid and rice–bacteria interactions. Another study by Shiono and Taira, 2020 showed that nanoparticle-assisted laser desorption/ionization MS can detect and visualize multiple plant hormones in rice root sections, overcoming the limitations of MALDI-MS such as self-ionization of the organic matrix (typically picolinic acid or similar compounds). The desorption electrospray ionization (DESI) method involves directing a stream of charged droplets on to the surface of a sample, where the droplets extract and ionize the target analytes and the resulting ions are then introduced into the mass spectrometer for examination and analysis (Venter *et al.*, 2006). This method has been used in plant research for various applications, including the imaging of metabolite distributions in plant materials such as leaves and petals (Janfelt, 2015), and the identification and spatial quantification of chlorophyll degradation products in senescent plant tissues (Müller *et al.*, 2011). Further research into these mechanisms, especially in the context of plant analysis, could enhance the capabilities of these ionization techniques. However, most of the studies discussed in this chapter will be centered on MS-based technologies preceded by a chromatographic separative technique and on untargeted metabolomics studies.

As in any other omics technology, the most important aspect is to maintain the unbiased character of the technique. Therefore, during the setup of a metabolomics procedure, it is important that, after careful optimization and homogenization, each step is carried out equally for all samples in order to provide accurate metabolite quantification and identification. After obtaining data from an experiment, the post-analytical steps, including data processing, normalization, and identification of metabolites, often use databases such as METLIN (Montenegro-Burke *et al.*, 2020), Massbank (Horai *et al.*, 2010) or mzCloud (Vinaixa *et al.*, 2016). The integration of various data dimensions (MS and MS/MS spectra and ion intensity, retention time, and, when available, ion mobility data), is of the utmost importance in the construction of public or in-house libraries, which are among the most popular strategies used for metabolite annotation.

To process and visualize MS-based untargeted metabolomics data, several cloud-based and stand-alone software platforms have been developed. One of the most popular is XCMS (Smith *et al.*, 2006; Tautenhahn *et al.*, 2012). Several steps are invloved, including data preprocessing, peak detection and alignment, and retention time correction. It constructs extracted ion chromatograms (EICs), detects the peaks from these EICs, and aligns them across multiple samples (Myers *et al.*, 2017). XCMS is available in different forms, such as an R package (XCMS/R) that allows manual coding and customization, and a user-friendly web interface known as XCMS Online, which automates most tasks and integrates statistical analysis and metabolic pathway mapping (Domingo-Almenara and Siuzdak, 2020). However, there are limitations to the current methods used by XCMS, as they have been noted to detect false-positive EIC peaks and fail to detect real EIC peaks, which can lead to incorrect compound identification (Myers *et al.*, 2017). The optimization of data-processing parameters is crucial for reliable results. Moreover, ion mobility spectrometry (IMS) has been utilized to separate ions, based on their size, shape, and charge, which adds an additional dimension to the data and can improve peak resolution and identification accuracy (Du *et al.*, 2018). Integration of IMS with LC-MS can significantly enhance the specificity and resolution of metabolomic analyses, allowing better separation and identification of complex mixtures (Gil-Solsona *et al.*, 2021). Another popular tool is MzMine (Katajamaa *et al.*, 2006; Pluskal *et al.*, 2010; Heuckeroth *et al.*, 2024). MzMine is based on Java and can be run on any machine encompassing a graphical user interface. Essentially, it performs the same tasks as XCMS but also performs streamlined database searches and statistical analysis. In addition, the toolbox includes different data visualization modes. Other processing tools include: (i) msInspect (May *et al.*, 2009), also based on Java, which stopped being updated back in 2011: (ii) OpenMS, an open source and cross-platform software (written in C++), which can be operated from a graphical user interface to perform basic mass chromatogram processing (Röst *et al.*, 2016); and (iii) MET-XAlign, which was released in 2015 and performs cross-alignment of mass chromatographic features in LC-MS samples from a set previously extracted by MET-COFEA (W. Zhang *et al.*, 2015). There are other proprietary tools that have been developed and distributed as software suites for metabolomics, such as MassHunter from Agilent Ltd, which performs data acquisition and qualitative and quantitative data analyses, as well as metabolite library search. Thermo Fisher Scientific also developed several tools such as MassFrontier for small-molecule identification, and Lipidsearch focused on the automated annotation and identification of cell lipids. Mass Analytica MARS from Waters Ltd performs identification of multiple adducts, automated in-source fragmentation detection, and *in silico* MS/MS spectrum validation. Finally, Progenesis QI, also distributed by Waters Ltd, encompasses several tools from raw mass chromatogram processing, and alignment of mass chromatographic features and retention time correction, to the annotation of mass chromatographic features and metabolite identification and quantitation. Other complete user-friendly tools for full metabolomics analysis include MS-Dial (Tsugawa *et al.*, 2015), which performs independent MS/MS deconvolution for mass data with different acquisition functions (i.e. one for profiling at low collision energy and another to assess fragmentation at high collision energy, acquired independently) or data-independent acquisition, providing important information for database searches of mass spectra. MS-Dial processes native files from different instruments performing peak picking, alignment, annotation of compounds, and statistical analysis. Another interesting tool is MetaboAnalyst, a web-based tool that can also be used as an R package, first released in 2009. The software includes several methods for metabolomics data processing, normalization, multivariate statistical analysis, and data annotation (Xia and Wishart, 2016).

13.1.2 Main bottleneck: metabolite annotation

Metabolite annotation in metabolomics, particularly for MS-based metabolomics, remains the main bottleneck in this technique. Three different levels of metabolite annotation are widely acknowledged. Level 1 is the definitive annotation, which implies running a standard sample of the identified metabolite on the same instrument or instruments, linking at least two orthogonal characters. This matches the mass spectra and retention time of a putatively annotated metabolite in a plant sample to those of an authenticated purified standard (Aharoni *et al.*, 2023). This is currently not possible for a single laboratory, as most naturally occurring metabolites are not available as pure standards and there are also economic reasons behind purchasing large compound libraries. Therefore, level 2 refers to putatively annotated compounds, relying on spectral libraries and computational tools to match mass spectra against databases such as the Human Metabolome Database (HMDB), METLIN, and others (Table 13.1). While the retention time data for level 2 annotations may not be definitive, they still provide valuable insights and are commonly used in metabolomics studies. Finally, level 3 annotations derive from molecular formulas based on accurate mass measurements and isotope patterns, suggesting potential compound classes or substructures without identifying exact compounds.

Advances in bioinformatics and cheminformatics tools enhance metabolite annotation by processing and analyzing MS data. Initiatives include Compound Discoverer, MS-Finder, and SIRIUS, which integrate fragmentation trees and *in silico* MS/MS spectra prediction to improve annotation accuracy. For clustering similar fragmentation spectra (MS/MS) into molecular networking, the Global Natural Products Social Molecular Networking (GNPS) is utilized extensively; by assessing the relationships among MS/MS spectra based on spectral similarity, this process categorizes and organizes metabolites into molecular families (Aron *et al.*, 2020). Molecular networking has evolved with the integration of feature detection methods, resulting in the development of feature-based molecular networking. It enhances the resolution of isomeric compounds and integrates quantitative data into molecular networks, which is particularly useful in the analysis of complex biological samples. The use of alternative fragmentation methods such as electron-activated dissociation in conjunction with feature-based molecular networking has recently been used to improve compound annotation in the freshwater sentinel organism *Gammarus fossarum* (Calabrese *et al.*, 2024). Moreover, various techniques have been developed to group related features and minimize redundant data during the metabolite annotation process. For instance, RAMClustR was designed to perform a similar role but employing a different strategy (Broeckling *et al.*, 2014). It primarily uses feature clustering based on coelution and similar MS fragmentation patterns without relying on predefined rules for ion species annotation. Hence, RAMClustR integrates spectral similarity into the clustering process, allowing it to identify and group features that coelute and share similar MS/MS spectra, which is particularly useful for identifying unknown compounds and reducing data redundancy. This method enhances the annotation process by providing a holistic view of the metabolite fragmentation pattern.

The CAMERA package (coupled with automated molecular feature extraction) (Kuhl *et al.*, 2012), is a powerful tool for refining XCMS feature lists by aggregating related features into single compound spectra. The package groups mass chromatographic features based on the correlation of retention time, mass:charge (m/z) ratio difference, and peak intensity across samples to identify those from the same analyte. By combining isotopologs, adducts, and fragments, this approach simplifies data complexity and enhances the accuracy of metabolite identification in downstream analysis from LC-MS data. To automate the identification of metabolites

Table 13.1. List of data repositories and databases for metabolomics.

Resource name	URL[a]	Content	Reference
Databases with spectral libraries available			
Metabolights	https://www.ebi.ac.uk/metabolights	Experiment repository (MS and NMR), metabolite information, and different online resources for training in metabolomics	Yurekten *et al.* (2024)
Metabolomics workbench	https://metabolomicsworkbench.org/	Data repository, metabolite database with access to submitted analyses data, metabolomics tools, and protocols	Sud *et al.* (2016)
Massbank	https://helmholtz.software/software/massbank	Mass spectral library of high-resolution MS data of small molecules	Horai *et al.* (2010)
Human Metabolome Database	https://hmdb.ca/	Library of metabolites including a complete description (metabolite class and chemical taxonomy), organisms in which the metabolite has been identified, and MS and NMR spectral data (*in silico* and experimental)	Wishart *et al.* (2022)
METLIN	https://metlin.scripps.edu/	Metabolite repository of MS data including tandem mass data at multiple collision energies in positive and negative ionization modes, and links to further information for each metabolite	Montenegro-Burke *et al.* (2020)
NIST 14	https://www.nist.gov/srd/nist-special-database-14	Spectral libraries of compound (derivatized and non-derivatized), including high and low resolution (unit mass difference) and retention time indices (for matching purposes)	Vinaixa *et al.* (2016)
Golm Metabolome Database	http://gmd.mpimp-golm.mpg.de/	Metabolite database, particularly for derivatized compounds, including mass spectra and retention time indices, useful for matching purposes using TargetSearch (Cuadros-Inostroza *et al.*, 2009) or TagFinder (Luedemann *et al.*, 2008)	Hummel *et al.* (2013)
mzCloud	https://www.mzcloud.org/	High-resolution tandem MS library arranged into spectral trees from authentic standard samples	Hummel *et al.* (2013)
Databases with no spectral libraries available			
Knapsack	http://www.knapsackfamily.com/KNApSAcK/	Constitutes a comprehensive species–metabolite relationship database with no mass spectra available	Afendi *et al.* (2012)

Continued

Table 13.1. Continued.

Resource name	URL[a]	Content	Reference
LipidMaps	https://www.lipidmaps.org/	A comprehensive database on lipid structures and related metabolic pathways and biochemical reactions involving lipids; currently encompasses MS and ion mobility-based searches	Conroy *et al.* (2024)
KEGG	https://www.genome.jp/kegg/	The Kyoto Encyclopedia of Genes and Genomes (KEGG) is a database encompassing genetic, genomic, and metabolite data of different organisms. It is especially suited for the integration of massive omics data sets into comprehensive pathway-based models. It also includes several tools, such as the recently added MWsearch, which allows matching of MS metabolomics data with KEGG identification	Kanehisa and Subramaniam (2002)

MS, mass spectrometry; NMR, nuclear magnetic resonance.
[a]All accessed 27 January 2025.

from MS data by leveraging *in silico* fragmentation and spectral matching techniques, MetFrag, an advanced *in silico* metabolite annotation tool, was developed by Wolf *et al.* (2010). Furthermore, to enhance the functionality of the CAMERA package's feature grouping capabilities, Gaquerel *et al.* (2013) proposed a new integrative approach that merges high-resolution LC-MS profiles with in-source fragmentation patterns and co-regulation patterns (relying on deconvoluted MS compound spectra and excluding isotopic and multi-charged ions), enabling rapid annotation of metabolites using MetFrag on metabolic profiles of *Nicotiana attenuata* leaves subjected to mechanical wounding and exposure to oral secretions of *Manduca sexta* larvae. Techniques such as MS/MS fragmentation patterns and database searches are used to infer metabolite identities, but the absence of reference standards remains a significant bottleneck. Ruttkies *et al.* (2015) proposed a workflow that incorporated *in silico* databases, including the Kyoto Encyclopedia of Genes and Genomes (KEGG), PubChem, and the Golm Metabolome Database (GMD), with MetFrag for annotation of GC-atmospheric pressure chemical ionization-MS/MS spectra of *A. thaliana* and *Solanum tuberosum* samples. The authors emphasized the significance of *in silico* derivatization and the metabolite-likeness score in enhancing identification performance, particularly when reference spectra are unavailable.

Despite these advancements in the field of metabolomics, it still has to confront considerable challenges such as the complexity of plant metabolomes, with their vast diversity of chemical structures, or the dynamic character of metabolite profiles under various environmental conditions, which add layers of complexity to metabolite annotation. Future directions in metabolomic research will emphasize the integration of multi-omics approaches combining metabolomics with genomics, transcriptomics, and proteomics to gain comprehensive insights into biological systems (Fig. 13.1).

13.1.3 Metabolomics as a functional analysis and breeding tool in plant science

Metabolites are the downstream product of gene expression and hence might reflect the endpoint of various plant strategies to cope with different environmental or biotic stressors (Arbona *et al.*, 2015). Therefore, metabolomics can be used as a functional analysis tool in plant science similar to transcriptomics and proteomics. Metabolites can be grouped into metabolic pathways (as gene ontologies) and the relative enrichment is used as a proxy of the different mechanisms set in motion in response to the different biotic or abiotic factors. However, as already mentioned, there is no single metabolomics technique that can cover all metabolite classes and it is unlikely that an integrated methodology covering all metabolite classes will arise in the future. Unlike transcripts or, to a lesser extent, proteins, the plant metabolome encompasses compounds with different chemical and physical properties that require different extraction and analytical techniques. For this reason, it is common to combine different metabolite profiling techniques to determine as much as possible of the plant metabolome. For instance, by combining GC-MS and LC-MS, it is possible to collect information from the central carbon metabolism and specialized metabolism (both semi-polar and relatively non-polar metabolites and volatiles) at a quantitative or semi-quantitative level (Zandalinas *et al.*, 2017; de Ollas *et al.*, 2021). In addition, to capture the plant functional profile in response to stress or genetic modification, it is becoming clear that several omics need to be combined and, generally, the most widely used combination is transcript and metabolite profiling. These data can be also correlated with physiological traits, such as gas exchange and chlorophyll fluorescence parameters or stress-induced damage of any kind, to investigate stress tolerance or important breeding traits such as yield, fruit quality, or nutritional values, and to better understand the biochemical or molecular mechanisms and extract useful information for breeders. The idea behind using metabolite profiling as a tool to steer plant breeding was first verbalized by Fernie and Schauer (2009). It then became clear that metabolomics, aside from being cheaper than sequencing-based technologies, has the advantage of not requiring any pre-existing biological information (genetic or other) to achieve full performance. Moreover, it is possible to screen multiple metabolites at the same time and detect unanticipated consequences of breeding such as the accumulation of unwanted compounds and

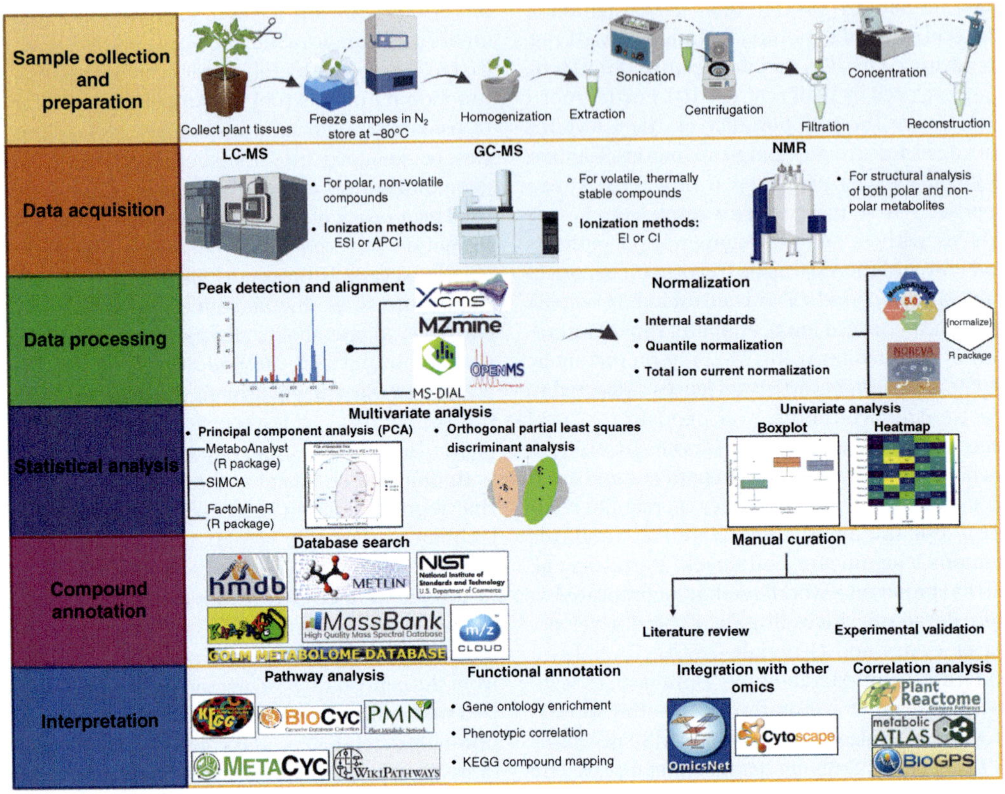

Fig. 13.1. Comprehensive workflow for plant metabolomics analysis. APCI, atmospheric pressure chemical ionization; CI, chemical ionization; EI, electron ionization; ESI, electrospray ionization; GC-MS, gas chromatography–mass spectrometry; LC-MS, liquid chromatography–mass spectrometry; NMR, nuclear magnetic resonance; The example graphs for the statistical analysis part are taken from Modesto *et al.* (2021) for the boxplot and from Arbona *et al.* (2020) for the PCA and heatmap. This figure was modified with BioRender.com.

also enable a more profound understanding of the complex metabolite network and its interaction with stress or developmental processes. One potential drawback, still true now and then, is the occurrence of many unannotated signals representing unknown compounds. Nevertheless, it is only a matter of time before these handicaps progressively disappear, not only due to the improvement in profiling technologies but also to the increasing participation of metabolomics researchers in the buildup of comprehensive metabolite databases (Table 13.1). In this respect, metabolite profiling data have already been used to predict stress tolerance and to investigate hybrid vigor or heterosis in the model plant *A. thaliana* (Meyer *et al.*, 2007; Lisec *et al.*, 2009; Korn *et al.*, 2010). This was achieved by employing a recombinant inbred line population obtained from backcrossing C24 × Col-0 F_1 lines with either Col-0 or C24. Subsequently, using canonical correlations, several metabolites showed a significant correlation with seedling mass. Despite this approach achieving limited to moderate success in model plant species, there are several examples of metabolite × agronomic trait correlations, although metabolomics, as a genotype selection tool, has not been efficiently implemented in breeding programs. Hence, to date, genomic selection has been the preferred choice for marker-assisted selection to introgress alleles of interesting agronomic traits in elite cultivars through backcrosses. This is also true when the desired trait is governed by the action of several genes, but the procedure requires complex crossing schemes or pyramiding and dealing with large numbers of progeny (Fernandez *et al.*,

2021). In this case, the genomic selection approach is applied to capture the most important part of the genetic variance enabling a decrease in the number of breeding cycles, increasing the precision of the genetic values, and allowing prediction of the performance of breeding lines in situations when the environment interacts with the genotype (Fernandez *et al.*, 2021). Therefore, genomic selection models can predict the additive genetic values and could potentially predict the statistical epistasis and genotype–environment interactions. However, a combination of genetic and phenotypic information is not enough to understand the biological processes leading to a particular phenotype in crops, particularly under strong environmental pressure. Metabolomics can provide insight into processes associated with development, tolerance to stress, and interactions with other organisms, as well as those obviously related to plant yield and quality, integrating both the genetic and the environmental contribution. Therefore, by exploiting all the available information from mutants, near-inbred lines, introgression lines, backcross-inbred lines, etc., metabolic genome-wide association studies have been used to investigate the genetics of metabolic networks and enabled the identification of new regulatory genes (Fernandez *et al.*, 2021). In wheat, the metabolome of flag leaves and spike bracts correlated well with genotypic outperformance and yield stability, predicting grain yield with high accuracy (Vergara-Diaz *et al.*, 2020). Moreover, in hybrid maize, the prediction accuracy of metabolites paralleled that of single-nucleotide polymorphisms (Riedelsheimer *et al.*, 2012), and under drought conditions, levels of the metabolites myo-inositol and glycine, a photorespiratory amino acid, correlated well with grain yield (Obata *et al.*, 2015).

A typical application of plant metabolomics is the identification of metabolites with nutritional or health properties as markers of crop improvement, but also toxic metabolites to steer *de novo* domestication initiatives (Cárdenas *et al.*, 2023). In this respect, all plants have evolved mechanisms to defend against other organisms that try to eat them. Even the most widespread and consumed fleshy fruit in the world, the tomato, has defense metabolites (Cárdenas *et al.*, 2019). All solanaceous crops (tomato, eggplant, and potato) have anti-nutritional steroidal glycoalkaloids, which are terpenoid-derived molecules that harbor several glycosidic moieties. These molecules play an important role as deterrents of pest feeding, particularly against lepidopterans, but can also exert important anti-nutritional effects on humans upon consumption. For instance, α-solanine and α-chaconine in potato are highly toxic compounds that cause cell-membrane permeabilization leading to the occurrence of gastrointestinal and neurological disorders, and the ingestion of solanine even at moderate levels can cause death (Dalvi and Bowie, 1983; Gao *et al.*, 2006). In tomato, α-tomatine is the main steroidal glycoalkaloid that accumulates in fruits and leaves. In ripening fruits, this metabolite is converted into esculeoside A, a non-bitter steroidal alkaloid. This transformation involves several steps including hydroxylation, acetylation, and glycosylation (Cárdenas *et al.*, 2019), which have been acquired during the domestication process to remove bitterness and toxicity in ripe tomato fruits. The recapitulation of the domestication process as well as elucidation of the biosynthetic pathways of toxic metabolites have been possible due to the availability of tomato wild relatives as well as the technological advances in high-throughput metabolite profiling techniques (Szymański *et al.*, 2020). Nevertheless, it must be noted that the removal of toxic metabolites also has a negative impact on the competitive potential of crops, as toxic compounds have an ecological role in protecting plants from harmful microbes and phytophagous arthropods ensuring that they reach the reproductive stage. In the Cruciferae family, there are lots of species that serve humans as food including cabbage, rapeseed, Brussels sprouts, turnip, cauliflower, broccoli, mustard, and kale; they all belong to the genus *Brassica* and are distantly related to the model plant *A. thaliana*. Their wild relatives within this family produce high amounts of indolic phytoalexins and indole and aliphatic glucosinolates as phytoanticipins (Pedras *et al.*, 2000, 2006). Glucosinolates in cruciferous plants are constitutively present molecules that are hydrolyzed upon pathogen attack or arthropod feeding by the action of a particular type of thiolase known as myrosinase, which is an inducible response. Hence, while intact glucosinolates are relatively stable and non-toxic molecules, their breakdown products are highly toxic and labile metabolites (Grubb and Abel, 2006). Conversely, phytoalexins are synthesized only in response to pathogen attack or upon imposition of an abiotic stress (Pedras *et al.*, 2000). Cultivated brassicas (e.g. *Brassica oleracea* or cabbage) share 17–98% of their genome with *A. thaliana*, with defense and specialized metabolism, suberin biosynthesis, and wax metabolism being the most conserved processes (Mabry *et al.*, 2021). However, as in tomato, large differences between cultivated and wild brassicas exist in terms of glucosinolate content, with wild relatives having higher glucosinolate content than cultivated species, which is directly related to resistance to fungal pathogens (Mithen *et al.*, 1987). More recently, the levels of glucosinolates and, more importantly, the ability to hydrolyze them have been associated with enhanced resistance to the spider mite *Tetranychus urticae* (Zhurov *et al.*, 2014; Widemann *et al.*, 2021). Moreover, the hydrolyzed glucosinolate aglycones have been described as elicitors of a new leaf-to-leaf molecular signaling mechanism to generate an electrical signal that promotes defense gene activation in distal leaves against wounding and insects (Gao *et al.*, 2023). Legumes are also important in human nutrition. The domestication history of beans was scrutinized by performing metabolite profiling analyses in wild and domesticated bean accessions, which showed a shift from specialized (usually toxic) metabolites to metabolites belonging to the central carbon metabolism (Perez de Souza *et al.*, 2023). Therefore, as already mentioned, domestication generally produces new genotypes with lower ecological competitive capacities. As an exception, using a metabolomics approach, Ben-Abu and Itsko (2022) found that during the primary and secondary domestication of wheat, the groups of plant metabolites involved in pathogen and antioxidant defense as well as those annotated as plant hormones tended to accumulate in the embryo, whereas they were mostly conserved in the endosperm, with fold changes in amino acid content up to four times higher with respect to wild relatives (wild emmer, domesticated emmer, landrace durum, and modern durum wheat). Using a similar approach, Beleggia *et al.* (2016) found that a reduction in the number of fatty acids was associated with wild emmer domestication (primary domestication), whereas durum wheat domestication (secondary domestication) was characterized by alterations in amino acid content. These apparently unrelated changes in wheat kernel metabolomes were partially found to be the result of an extensive metabolome rearrangement due to the complex domestication of wheat. In general, crop domestication leads to a dramatic loss of diversity leading to homogeneous edible plant species, primarily evidenced in the loss or silencing of genes involved in specialized metabolism, involved in plant defense against phytophagous arthropods and pathogens (González Guzmán *et al.*, 2022). This loss of genetic diversity is mirrored by a loss in metabolite diversity, showing a reduction in "exotic" metabolites and an accumulation of nutritional metabolites (e.g. carbohydrates, lipids, amino acids). Therefore, domestication has always had an impact on the ecological relationships of plants with their biotic environment, by reducing plant toxicity and increasing yield and nutritional values.

To sum up, the ability to screen hundreds of metabolites at the same time provided by modern metabolomics techniques constitutes an advantage not only in breeding but also as an essential tool to steer *de novo* domestication of emerging crops. Overall, large-scale screening of breeding populations by combining different omics technologies with metabolomics in non-model species is now possible due to the enormous amount of metabolite and metabolic pathway data available, including co-expression networks databases (e.g. STRING, https://string-db.org/, accessed 27 January 2025).

13.2 Plants in a Challenging Environment: Abiotic and Biotic Constraints

In natural environments, plants face a range of abiotic and biotic constraints that affect their growth and survival. These constraints necessitate complex metabolic adaptations to mitigate the adverse effect. Major metabolic pathways involved in plant responses to abiotic stressors can be categorized such as drought, salinity, and high temperatures, as well as biotic stressors such as pathogens and herbivores, which are crucial for their adaptation.

Under normal circumstances, plants have access to an adequate amount of water, which is necessary for their survival and growth. However, when water supply is restricted or transpiration increases, plants experience a significant reduction in water availability, leading to drought stress (Basu *et al.*, 2016). To cope with this, plants accumulate osmoprotectants such as proline, glycine betaine, and sugars (e.g. raffinose) (Saxena *et al.*, 2013). These compounds help to maintain cell turgor, stabilize proteins and membranes, and protect cellular structures from damage caused by dehydration. The key metabolic pathways involved include proline biosynthesis, where glutamate is converted to proline via the intermediate pyrroline-5-carboxylate, which is regulated by the enzyme pyrroline-5-carboxylate synthetase. It has been shown that under water-deficit conditions, proline levels increase (Yoshiba *et al.*, 1997). The glycine betaine pathway involves the oxidation of choline to betaine aldehyde, which is then converted to glycine betaine by betaine aldehyde dehydrogenase. Glycine betaine accumulates under drought stress to protect biomolecules (Chen and Murata, 2011) by ensuring the stability of membranes, proteins, and DNA structure. In *Arabidopsis*, the presence of the sugar raffinose has been observed in the shoots and roots of plants during later stages of drought (Fàbregas and Fernie, 2019). These pathways are known to be essential for osmotic adjustment and cellular protection during drought conditions.

Salinity stress in plants leads to a cascade of physiological, biochemical, and molecular changes, which include alterations in the metabolomic profile (Dubey and Verma, 2019). These changes are part of the adaptive mechanisms of plants to cope with the detrimental effects of high salt concentrations. Metabolomic studies have identified specific metabolites that are modulated in response to salinity, such as osmolytes and osmoprotectants, which help in osmotic adjustment and the protection of cellular structures (Llanes *et al.*, 2018). Interestingly, not all plant species respond to salinity stress by synthesizing the same metabolites in the same quantities. Some species may exhibit low levels of certain metabolites or may not produce them at all, suggesting that the flux through metabolic pathways rather than the accumulation of specific metabolites might be crucial for stress tolerance (Llanes *et al.*, 2018). In addition, salinity stress induces the production of reactive oxygen species (ROS), which can cause oxidative damage. To cope with this, plants enhance the activity of antioxidant enzymes such as superoxide dismutase, catalase, and peroxidases. These enzymes scavenge ROS, protect cells from oxidative damage, and maintain cellular integrity under high-salinity conditions. Related to ROS regulation during salt stress, it has been found that the salt stress-resistant rice cultivar Patnai exhibited a greater utilization of cinnamic acid, leading to increased production of gentisic and syringic acids. Conversely, the salt-sensitive cultivar IR29 accumulated more cinnamic and benzoic acids because of their lower utilization in the synthesis of syringic acid, salicylic acid, and gentisic acid (Banik and Bhattacharjee, 2020).

The metabolomic changes in plants during high-temperature stress are complex and involve alterations in a wide array of metabolites and pathways. In *Pinus radiata*, an integrative omics approach revealed significant shifts in the metabolome in response to high-temperature stress, with a notable turning point on day 3 of heat exposure. Key pathways such as cytokinin and fatty acid metabolism as well as flavonoid and terpenoid biosynthesis were affected. Among these, specific metabolites such as L-phenylalanine, hexadecanoic acid, and dihydromyricetin were identified as important traits for thermotolerance (Escandón *et al.*, 2018). Similarly, a study on the warm-season forage grass *Paspalum wettsteinii* showed changes in metabolites such as amino acids, organic acids, flavonoids, and sugars under heat stress, associated with growth promotion at moderate high temperatures and a reduction in photosynthesis at extreme temperatures (Zhao *et al.*, 2022). High-temperature stress induces a wide range of metabolomic changes in plants, affecting various metabolic pathways and leading to both up- and downregulation of metabolites. These changes are part of a complex response mechanism that includes both immediate stress responses and longer-term acclimation strategies. Following a similar approach, Zandalinas *et al.* (2017) performed metabolite profiling in two contrasting citrus genotypes in response to combined heat and drought conditions and found that sensitivity to the stress combination was associated with activation of energy metabolism and, particularly, the part of the specialized metabolism leading to the accumulation of photoprotective and antioxidant metabolites, to cope with the damaging effects of stress. Anzano *et al.* (2022) performed a review of 151 publications and showed that the most studied abiotic stresses were salinity and drought and focused mainly on amino acids, sugars, and organic acids belonging to the tricarboxylic acid cycle. In this respect, in response to drought or drought-simulated conditions (adding an osmotic agent such as polyethylene glycol), plants overacumulate amino acids (alanine, leucine, methionine, proline, ornithine, tryptophan, and valine), as well as carbohydrates and sugar derivatives (sucrose and mannitol). In response to salinity, the amino acid proline was again found as a main metabolic response, but other primary metabolites also consistently accumulated in different plant species, such as galactose and sucrose and other amino acids such as serine and alanine. Heat stress again consistently induced proline accumulation, as well as sugar derivatives such as galactinol and myo-inositol, along with other amino acids such as threonine and valine. However, when different combinations such as heat stress/salinity, heat stress/drought, and drought/salinity were studied, different responses were observed compared with those under single stresses. Although the typical proline accumulation response remained, it cannot be considered a general trend as it highly depends on the plant species (Anzano *et al.*, 2022). This was the case for a heat and stress combination in tomato (Rivero *et al.*, 2014), in which the accumulation of proline could only be observed in response to salinity, whereas its levels reduced below control values on combining heat and salinity, or with heat alone. However, glycine betaine accumulated in response to heat and heat/salinity stress, suggesting a specific role of this metabolite in osmotic adjustment under these conditions in tomato. Moreover, it was found that the accumulation of trehalose in tomato leaves constituted a specific response to heat and salinity combination.

In addition to abiotic stressors, plants must also contend with biotic stressors such as pathogens and herbivores. To defend against pathogens, plants have evolved complex immune responses involving both constitutive and inducible defenses. Pattern recognition receptors recognize pathogen-associated molecular patterns (PAMPs), activating PAMP-triggered immunity (Rathore and Ghosh, 2018). This response involves the generation of ROS and antimicrobial compounds, as well as the reinforcement of cell walls to prevent pathogen entry. NADPH oxidases and various biosynthetic pathways that produce specialized metabolites with antimicrobial properties are key enzymes in this process (Hunyadi, 2019). Effector-triggered immunity increases when plant resistance proteins recognize pathogen effectors, leading to a strong defense response that often includes localized cell death (hypersensitive response) to limit pathogen spread (Nguyen *et al.*, 2021).

Plant defenses against herbivores involve both physical and chemical defenses. One of the key chemical defense mechanisms is the jasmonic acid (JA) pathway, which plays a crucial role in plant responses to herbivory. When plants are exposed to herbivores, mechanical damage or specific herbivore-associated molecular patterns trigger the synthesis of JA from α-linolenic acid through the octadecanoid (LOX3) pathway (Howe and Jander, 2008). Activation of the JA pathway leads to the expression of defense-related genes that encode proteinase inhibitors, which hinder herbivore digestive enzymes, and the production of secondary metabolites such as alkaloids and terpenoids (Erb *et al.*, 2012). These compounds act as deterrents to herbivores, reducing their feeding efficiency and minimizing damage to the plant. Additionally, the salicylic acid (SA) pathway, which is known primarily for its role in plant defense against pathogens, also contributes to defense responses against herbivores. SA signaling induces the expression of pathogenesis-related proteins, which have antimicrobial properties and also play a role in plant resistance to herbivores (dos Santos and Franco, 2023). The interaction between the JA and SA pathways in response to herbivory involves a complex signaling crosstalk, where SA may antagonize JA-dependent defenses against herbivores in some plant species (Wei *et al.*, 2014). Combining gene expression analysis using quantitative polymerase chain reaction and metabolite profiling in tomato, Arbona *et al.* (2020) identified metabolic and genetic responses to combined drought and *Tetranychus evansi* spider mite attack in tomato that were associated with abscisic acid (ABA) and SA signaling. In this respect, it has been suggested earlier that *T. evansi* specialized in tomato feeding through modulation of JA-dependent defenses using protein effectors (Villarroel *et al.*, 2016; Schimmel *et al.*, 2018). In this study, a set of metabolites related to lipid metabolism such as glycerophosphocholine and phosphatidylethanolamine were found to be induced by the concurring effect of *T. evansi* and drought at early time points. The disaccharide trehalose was also found to be induced exclusively by the combined action of drought and *T. evansi*. Nevertheless, amino acids such as L-aspartate were depleted in response to the concurrence of drought and *T. evansi* infestation. These results highlight that metabolism reprogramming responds differently to single and combined stresses (Arbona *et al.*, 2020). Coordinated activation of JA and SA pathways demonstrates the sophisticated defense strategies that plants employ to deter herbivores and enhance their resilience in the face of biotic stressors such as herbivory, although some other signaling molecules, such as ABA, might play an important role in balancing both JA and SA signaling pathways. This highlights the enormous complexity of the plant metabolome and its regulation, and the existing difficulties in identifying stress and/or tolerance markers. Therefore, the combination of metabolomics along with other omics strategies might help explain phenotypes and provide interesting tolerance markers useful in breeding and crop improvement. A summary of the major metabolite pathways involved in abiotic and biotic stress responses is shown in Fig. 13.2.

13.3 Metabolic Engineering for Crop Improvement

The number of compounds that plants can synthesize has been estimated at more than 200,000 (Saito and Matsuda, 2010), some of which have key roles in plant adaptation to or defense against different environmental conditions (Marone *et al.*, 2022). In fact, using the classical biotechnological strategy that exploits genetic variability to improve the plant stress tolerance by increasing specific metabolite content (i.e. myo-inositol and glycine betaine), it has been possible to link changes in the metabolome to maize response tolerance to drought, providing important selection markers for marker-assisted selection programs (Obata *et al.*, 2015). Moreover, the exogenous application of metabolites such as some amino acids and sugars, and other specialized metabolites, has effectively increased stress tolerance in various crop plants (Godoy *et al.*, 2021; Marone *et al.*, 2022), supporting the role of certain compounds in plant stress tolerance. In *A. thaliana*, a natural ecotype screening was conducted to identify accessions with enhanced resistance *to T. urticae* spider mite attack. The accession with the highest resistance showed enrichment in JA and SA as well as specialized metabolism-related transcripts such as *MYB29* and *CYP79B2* involved in aliphatic and indolic glucosinolate biosynthesis, respectively (Zhurov *et al.*, 2014). Indeed, the impossibility of synthesizing indolics (glucosinolates and the phytoalexin camalexin) in the mutant *cyp79b2/cyp79b3* was clearly associated with increased spider mite-induced damage. As well as the ability to synthesize indole glucosinolates, their biotransformation to breakdown products (i.e. nitriles and isothiocyanates; see Grubb and Abel, 2006b, for an extensive review) through the activity of myrosinases AtTGG1 and AtTGG2 (Widemann *et al.*, 2021) was also important. Therefore, not only is the biosynthesis of precursors an important trait, but their endogenous biotransformation products also result in short-lived, highly reactive molecules that are the true defensive metabolites.

Hence, targeted plant metabolic engineering (TPME) has become a promising biotechnological strategy to increase plant tolerance to environmental stresses and generate plant biofactories to synthesize huge amounts of specialized metabolites of nutritional pharmaceutical interest (Verpoorte *et al.*, 2002). TPME could be defined as modification of plant metabolism by optimizing genetic and/or regulatory processes with the aim of increasing the production of the desired compound by modifying its biosynthetic pathway (Fig. 13.3) (Farré *et al.*, 2014). To achieve this goal, researchers must combine cutting-edge genome-editing technologies to modify plant genetic information with fundamental knowledge of the biosynthetic pathways and regulatory factors (Selma *et al.*, 2023). This last factor is crucial to boost accumulation of the target product as researchers must modify plant metabolic fluxes by increasing the flux of the compound biosynthesis pathway and, very often, reducing the flux of competing or catabolic pathways (Farré *et al.*, 2014). However, metabolic flux modification could be very complex for some metabolite biosynthetic pathways whose activity is modulated by feedback regulation exerted by-product metabolites (Caspi *et al.*, 2013).

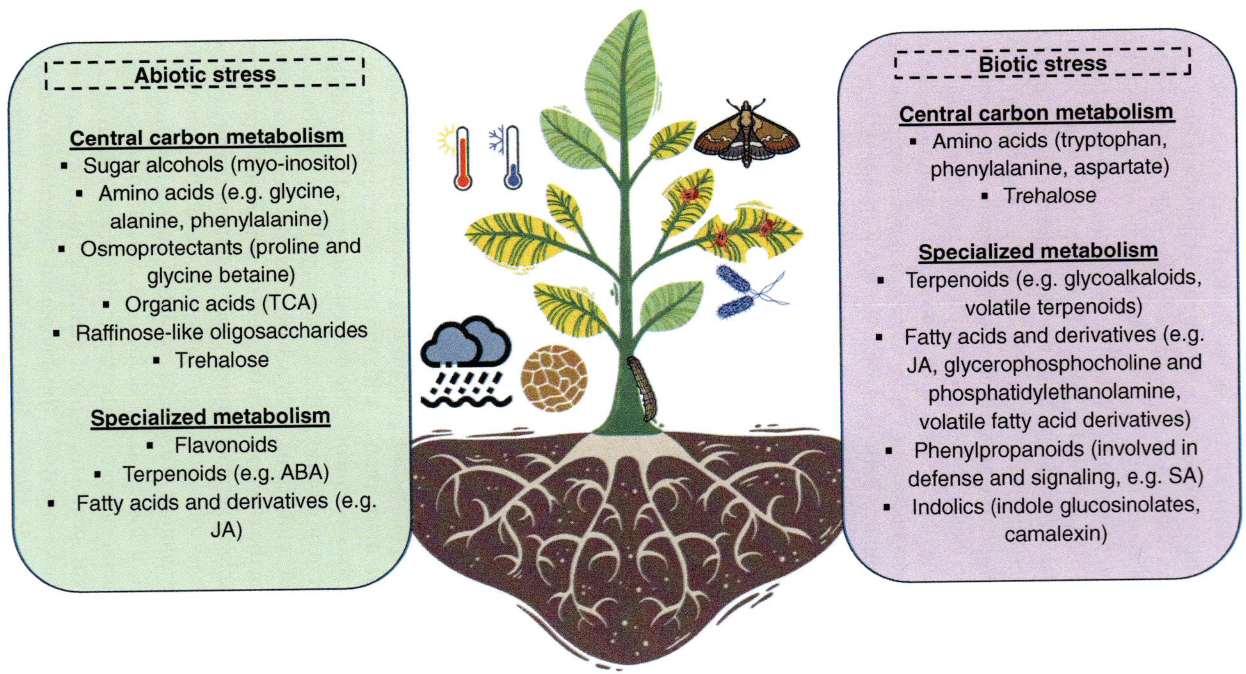

Fig. 13.2. Diagram of the major pathways and metabolites involved in the response of plants to abiotic (green) and biotic (purple) stressors. A diverse range of abiotic stressors such as temperature, flooding, and drought are represented. Regarding biotic stressors, some examples of pathogens and herbivores are illustrated. ABA, abscisic acid; JA, jasmonic acid; SA, salicylic acid, TCA, tricarboxylic acid.

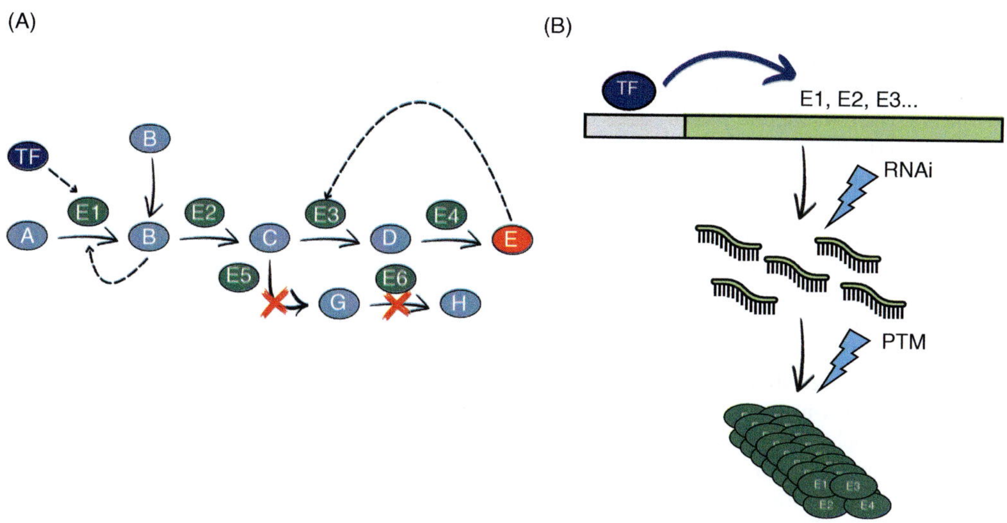

Fig. 13.3. Basics strategies for targeted plant metabolic engineering. (A) At the genetic level, the biosynthesis pathway of the desired metabolite E could be engineered to increase the biosynthesis metabolic flux by supplying one metabolite precursor (metabolite B), by increasing the expression of structural genes mediated by transcriptional regulators (TFs), and/or by blocking the gene expression of side pathways that uses metabolite intermediates (E5 and E6) by new genome-editing technologies. Two feedback regulation loops regulating biosynthetic metabolic flux are also indicated, one exerted by the desired metabolite E and the other by an intermediate metabolite. (B) Regulation of the biosynthetic metabolic flux may be exerted by regulating mRNA and protein levels of structural and regulatory genes by RNA interference (RNAi) and post-translational modification (PTM) methodology, respectively.

Basic TPME strategies generally involve enhancement, suppression, and/or sequestration of metabolic fluxes by supplying metabolite precursors or by overexpressing, silencing, or reducing the expression of structural or regulatory metabolic genes (Fig. 13.3). Many different studies of TPME have been carried out by overexpressing endogenous or orthologous transcription factors to produce protective compounds induced by environmental stresses, such as phenylpropanoids and flavonoids, carotenoids, and alkaloids (Farré *et al.*, 2014; Y. Zhang *et al.*, 2015; Stanley and Yuan, 2019; Yamada *et al.*, 2021; Selma *et al.*, 2023). For instance, transgenic flavonoid-deficient *Arabidopsis* lines overexpressing MYB12 or MYB75 transcriptions factors showed flavonoid overaccumulation that correlated with enhancing tolerance to oxidative and drought stress (Nakabayashi *et al.*, 2014). In addition, the enhancement of lignin content in rice by overexpressing the HD-Zip transcription factor OsTF1L showed improved tolerance to drought stress (Bang *et al.*, 2019).

Importantly, plant metabolites are objects of increasing scientific and industrial interest because of their possible beneficial effects against environmental stresses not only on crops but also on human health. For instance, flavonoids, polyphenols, and carotenoids participate as antioxidants in the scavenging of ROS, and when accumulated in plants and eaten, they can protect humans against cancer, diabetes, and coronary and cardiovascular diseases (Cory *et al.*, 2018). In fact, a huge number of plant extracts containing a myriad of metabolites are sold as nutraceuticals and functional foods that increase consumer health but with the potential to produce significant toxicity (Ronis *et al.*, 2018; Chopra *et al.*, 2022).

Another popular approach consists of modification of the abundance of structural biosynthetic activities, by up- or downregulating expression of key genes involved in metabolite biosynthesis pathways. As an example, and using RNA interference methodology, the levels of cottonseed oil in *Gossypium hirsutum* were increased by downregulating the phosphoenolpyruvate carboxylase 1 gene (*GhPEPC1*), and the content of gossypol, a toxic polyphenolic bisesquiterpene present in the cotton plant, was reduced by downregulating the D-cadinene synthase-encoding gene (*GhCAD1*) (Rathore *et al.*, 2012; Xu *et al.*, 2016). Therefore, there is no direct correlation between the specific genetic modification and the metabolic outcome, due to the enormous complexity of the plant metabolism.

The emergence of genome-editing tools such as zinc-finger nucleases (ZFNs), transcription activator-like effectors (TALEs), and clustered regularly interspaced short palindromic repeats (CRISPR)/CRISPR-associated (Cas) systems enables more sophisticated approaches that modify metabolic fluxes by targeting specific regulatory or biosynthetic activities, thereby avoiding the limitations of the classical gene loss-of-function, downregulation, or overexpression strategies (Selma *et al.*, 2023). Briefly, ZFNs comprise a zinc-finger DNA-binding domain fused to a non-specific DNA cleavage domain from the *Fok*I nuclease that produces double-strand breaks. ZFNs can be produced that target specific DNA sequences with complex genomes (Kim *et al.*, 1996). The TALEs system is derived from *Xanthomonas* TALE proteins and contains a DNA-binding motif with multiple amino acid repeats that differ only in the repeat variable di-residues, which recognize specific nucleotides and provide the DNA specificity recognition (Boch *et al.*, 2009). Finally, the CRISPR/Cas system is derived from prokaryotic immunity systems and comprises a nuclease (Cas) and a guide RNA that drives Cas to a specific DNA target site and produces double-strand breaks (Mojica *et al.*, 2000).

Using TALEs, a synthetic transcription factor that controls overexpression of the tobacco MYB transcription factor AtPAP1 (PRODUCTION OF ANTHOCYANIN PIGMENT 1) was constructed, leading to an increase in anthocyanin accumulation in the plants (Liu *et al.*, 2014). However, ZFNs and TALENs show lower versatility than the CRISPR/Cas system, which has emerged as an easy, specific, and versatile alternative. Specific CRISPR knockouts have been generated to reroute metabolic pathways and obtain biofortified or more stress-tolerant crops. CRISPR-based editing of the *LCYε* gene increased the production of β-carotene in banana fruits and decreased the content of the α-carotene and lutein metabolites at the same time (Kaur *et al.*, 2020), while simultaneous soybean genome edition of the *GmF3H1*, *GmF3H2*, and *GmFNSII-1* genes led to an increase in isoflavone content and thus improve its resistance to the soybean mosaic virus (Zhang *et al.*, 2020). In addition, CRISPR/Cas-based transcription factors have been designed to avoid the endonuclease activity of the Cas protein while maintaining CRISPR/Cas binding capacity, allowing the CRISPR/Cas tool to regulate gene expression of specific enzymatic steps that provoke the accumulation of flavonoids, polyphenols, or carotenoids metabolites (Qi *et al.*, 2013; Pan *et al.*, 2021; Selma *et al.*, 2022).

Finally, metabolic fluxes could be regulated by controlling the protein levels of regulatory or structural protein activities. In fact, levels of biosynthetic enzymes, regulators, or transcription factors are constantly used, renewed, and replaced at the cellular level by post-translational modification systems that add or removal chemical groups from these proteins. Examples of enzymatic post-translational modifications include phosphorylation, acetylation, and ubiquitination (Friso and van Wijk, 2015). For instance, by studying the JA-dependent transcriptional reprogramming of *Medicago truncatula* specialized metabolism, Pollier *et al.* (2013) identified an ERAD-type RING membrane-anchor E3 ubiquitin ligase controlling 3-hydroxy-3-methylglutaryl-CoA reductase (HMGR) protein levels. HMGR is the rate-limiting enzyme in the supply of the ubiquitous terpene precursor isopentenyl diphosphate, steering the central carbon and specialized metabolism biosynthetic fluxes. In fact, gene downregulation of this E3 ligase provoked a change in the regulation of biosynthetic fluxes and overaccumulation of xenotoxic triterpene saponins.

13.4 Conclusions and Future Prospects

Metabolites are pivotal in defining the plant phenotype with respect to its surrounding environment, as they determine odors, flavors, and nutritional aspects as well as semiotic and defensive traits in both cultivated plants and wild relatives. Therefore, metabolites integrate environmental (abiotic and biotic) as well as developmental cues as the downstream products of signal perception, transduction, and response elaboration. The dynamic nature of biological processes makes it difficult to extract conclusions, even from multidimensional single-omics data. Hence, it is becoming more usual to integrate different omics to provide an overview of the processes connecting endogenous signaling and regulatory processes with the phenotype. While genetic strategies, genome sequencing, and whole transcript-profiling are well-established methodologies that have undergone a number of improvements over the years, it must be noted that a single molecule class is extracted and analyzed in these methodologies (DNA and RNA), whereas several metabolite classes spanning different polarities, masses, and chemical and physical properties are considered in metabolomics. The term was coined 26 years ago by Oliver *et al.* (1998) and the same bottlenecks remain, which are the unbiased identification of metabolites and the reproducibility of data between different instruments and analysis platforms. Nevertheless, several improvements in terms of instrument sensitivity and accuracy have been made, as well as the availability of other sampling strategies, such as MALDI, DART, and DESI, which enable the direct analysis of metabolites at the tissue and cell scale. The integration of metabolite fluxes through tissues and cell types over time with gene expression is one of the challenges that this group of techniques will face in the future (Aharoni *et al.*, 2023). Moreover, the advent of artificial-intelligence approaches including analytical detection, data preprocessing, biomarker discovery, predictive modeling, and multi-omics data integration (Chi *et al.*, 2024) provides a new whole understanding of biological processes.

13.5 Acknowledgments

Financial support was provided through projects OPTIMUS PRIME, grant PCI2021-121920 funded by MCIN/AEI/10.13039/501100011033 and by the "European Union NextGenerationEU/PRTR," EXTREMO grant PID2020-118126RB-I00 funded by MCIN/AEI/10.13039/501100011033, LAYERS grant CNS2023-143759 funded by MICIU/AEI /10.13039/501100011033, and by the "European Union NextGenerationEU/PRTR," UJI-B2019-24, and UJI-B2022-23 funded by Universitat Jaume I. Funding from a "Santiago Grisolia" fellowship grant from Conselleria d'Innovació, Universitats, Ciència i Societat Digital (Generalitat Valenciana, CIGRIS/2021/014) to B.O.G. and a predoctoral fellowship from Conselleria d'Educació, Universitats i Ocupació (Generalitat Valenciana, CIACIF/2022/241) to J.M.A. are gratefully acknowledged.

13.6 References

Afendi, F.M., Okada, T., Yamazaki, M., Hirai-Morita, A., Nakamura, Y., *et al.* (2012) KNApSAcK family databases: integrated metabolite–plant species databases for multifaceted plant research. *Plant and Cell Physiology* 53: e1.

Aharoni, A., Goodacre, R., and Fernie, A.R. (2023) Plant and microbial sciences as key drivers in the development of metabolomics research. *Proceedings of the National Academy of Sciences USA* 120: e2217383120.

Allwood, J.W., Williams, A., Uthe, H., van Dam, N.M., Luis, L.A.J., *et al.* (2021) Unravelling plant responses to stress—the importance of targeted and untargeted metabolomics. *Metabolites* 11: 558.

Anzano, A., Bonanomi, G., Mazzoleni, S., and Lanzotti, V. (2022) Plant metabolomics in biotic and abiotic stress: a critical overview. *Phytochemistry Reviews* 21, 503–524.

Arbona, V., Iglesias, D.J., and Gómez-Cadenas, A. (2015) Non-targeted metabolite profiling of citrus juices as a tool for variety discrimination and metabolite flow analysis. *BMC Plant Biology* 15: 38.

Arbona, V., Ximénez-Embún, M.G., Echevarri-Muñoz, A., Martin-Sanchez, M., Gómez-Cadenas, A., *et al.* (2020) Early molecular responses of tomato to combined moderate water stress and tomato red spider mite attack. *Plants* 9: 1131.

Aron, A.T., Gentry, E.C., McPhail, K.L., Nothias, L.F., Nothias-Esposito, M., *et al.* (2020) Reproducible molecular networking of untargeted mass spectrometry data using GNPS. *Nature Protocols* 15, 1954–1991.

Bai, Y., Zhang, J., Bai, Y., and Liu, H. (2012) Direct analysis in real time mass spectrometry combined with single-drop liquid-liquid-liquid microextraction for the rapid analysis of multiple phytohormones in fruit juice. *Analytical and Bioanalytical Chemistry* 403, 2307–2314.

Bang, S.W., Lee, D.K., Jung, H., Chung, P.J., Kim, Y.S., *et al.* (2019) Overexpression of *OsTF1L*, a rice HD-Zip transcription factor, promotes lignin biosynthesis and stomatal closure that improves drought tolerance. *Plant Biotechnology Journal* 17, 118–131.

Banik, N. and Bhattacharjee, S. (2020) Complementation of ROS scavenging secondary metabolites with enzymatic antioxidant defense system augments redox-regulation property under salinity stress in rice. *Physiology and Molecular Biology of Plants* 26: 1623.

Basu, S., Ramegowda, V., Kumar, A., and Pereira, A. (2016) Plant adaptation to drought stress. *F1000Res* 5: F1000.

Beleggia, R., Rau, D., Laid, G., Platani, C., Nigro, F., *et al.* (2016) Evolutionary metabolomics reveals domestication-associated changes in tetraploid wheat kernels. *Molecular Biology and Evolution* 33, 1740–1753.

Ben-Abu, Y. and Itsko, M. (2022) Metabolome dynamics during wheat domestication. *Scientific Reports* 12: 8532.

Boch, J., Scholze, H., Schornack, S., Landgraf, A., Hahn, S., *et al.* (2009) Breaking the code of DNA binding specificity of TAL-type III effectors. *Science* 326, 1509–1512.

Boughton, B.A., Thinagaran, D., Sarabia, D., Bacic, A., and Roessner, U. (2016) Mass spectrometry imaging for plant biology: a review. *Phytochemistry Reviews* 15, 445–488.

Broeckling, C.D., Afsar, F.A., Neumann, S., Ben-Hur, A., and Prenni, J.E. (2014) RAMClust: a novel feature clustering method enables spectral-matching-based annotation for metabolomics data. *Analytical Chemistry* 86, 6812–6817.

Brunetti, C., George, R.M., Tattini, M., Field, K., and Davey, M.P. (2013) Metabolomics in plant environmental physiology. *Journal of Experimental Botany* 64, 4011–4020.

Büttenbender, S.L., Carvalho, Â.R., de Souza Barbosa, F., Scorsatto Ortiz, R., Limberger, R.P., and Mendez, A.S.L. (2022) Fragmentation of cannabinoids by flow injection analysis tandem mass spectrometry (FIA–MS/MS). *Journal of AOAC International* 105, 915–927.

Calabrese, V., Brunet, T.A., Degli-Esposti, D., Chaumot, A., Geffard, O., *et al.* (2024) Electron-activated dissociation (EAD) for the complementary annotation of metabolites and lipids through data-dependent acquisition analysis and feature-based molecular networking, applied to the sentinel amphipod *Gammarus fossarum*. *Analytical and Bioanalytical Chemistry* 416, 2893–2911.

Cárdenas, P.D., Sonawane, P.D., Heinig, U., Jozwiak, A., Panda, S., *et al.* (2019) Pathways to defense metabolites and evading fruit bitterness in genus *Solanum* evolved through 2-oxoglutarate-dependent dioxygenases. *Nature Communications* 10: 5169.

Cárdenas, P.D., Landtved, J.P., Larsen, S.H., Lindegaard, N., Wøhlk, S., *et al.* (2023) Phytoalexins of the crucifer *Barbarea vulgaris*: structural profile and correlation with glucosinolate turnover. *Phytochemistry* 213: 113742.

Caspi, R., Dreher, K., and Karp, P.D. (2013) The challenge of constructing, classifying, and representing metabolic pathways. *FEMS Microbiology Letters* 345, 85–93.

Chang, W.C.W., Chen, Y.T., Chen, H.J., Hsieh, C.W., and Liao, P.C. (2020) Comparative UHPLC-Q-Orbitrap HRMS-based metabolomics unveils biochemical changes of black garlic during aging process. *Journal of Agricultural and Food Chemistry* 68, 14049–14058.

Chen, T.H.H. and Murata, N. (2011) Glycinebetaine protects plants against abiotic stress: mechanisms and biotechnological applications. *Plant, Cell and Environment* 34, 1–20.

Cheong, B.E., Ho, W.W.H., Biddulph, B., Wallace, X., Rathjen, T., *et al.* (2019) Phenotyping reproductive stage chilling and frost tolerance in wheat using targeted metabolome and lipidome profiling. *Metabolomics* 15: 144.

Chi, J., Shu, J., Li, M., Mudappathi, R., Jin, Y., *et al.* (2024) Artificial intelligence in metabolomics: a current review. *Trends in Analytical Chemistry* 178: 117852.

Chiwocha, S.D.S., Abrams, S.R., Ambrose, S.J., Cutler, A.J., Loewen, M., *et al.* (2003) A method for profiling classes of plant hormones and their metabolites using liquid chromatography-electrospray ionization tandem mass spectrometry: an analysis of hormone regulation of thermodormancy of lettuce (*Lactuca sativa* L.) seeds. *Plant Journal* 35, 405–417.

Chopra, A.S., Lordan, R., Horbańczuk, O.K., Atanasov, A.G., Chopra, I., *et al.* (2022) The current use and evolving landscape of nutraceuticals. *Pharmacological Research* 175: 106001.

Choudhury, F.K., Pandey, P., Meitei, R., Cardona, D., Gujar, A.C., and Shulaev, V. (2022) GC-MS/MS profiling of plant metabolites. *Methods in Molecular Biology* 2396, 101–115.

Conroy, M.J., Andrews, R.M., Andrews, S., Cockayne, L., Dennis, E.A., *et al.* (2024) LIPID MAPS: update to databases and tools for the lipidomics community. *Nucleic Acids Research* 52, D1677–D1682.

Cory, H., Passarelli, S., Szeto, J., Tamez, M., and Mattei, J. (2018) The role of polyphenols in human health and food systems: a mini-review. *Frontiers in Nutrition* 5: 87.

Dalvi, R.R. and Bowie, W.C. (1983) Toxicology of solanine: an overview. *Veterinary and Human Toxicology* 25, 13–15.

David, L., Kang, J., and Chen, S. (2020) Targeted metabolomics of plant hormones and redox metabolites in stomatal immunity. *Methods in Molecular Biology* 2085, 79–92.

de Ollas, C., Gonzalez-Guzman, M., Pitarch, Z., Matus, J.T., Candela, H., *et al.* (2021) Identification of ABA-mediated genetic and metabolic responses to soil flooding in tomato (*Solanum lycopersicum* L. Mill). *Frontiers in Plant Science* 12: 613059.

Dinakar, C. and Bartels, D. (2013) Desiccation tolerance in resurrection plants: new insights from transcriptome, proteome, and metabolome analysis. *Frontiers in Plant Science* 4: 65937.

Domingo-Almenara, X. and Siuzdak, G. (2020) Metabolomics data processing using XCMS. *Methods in Molecular Biology* 2104, 11–24.

Donnarumma, F., Murray, K.K., and Hanley, L. (2021). Fundamentals of laser desorption ionization. In: Zimmermann, R. and Hanley, L. (eds) *Photoionization and Photo-Induced Processes in Mass Spectrometry: Fundamentals and Applications*. Wiley, Weinheim, Germany, pp. 305–325.

dos Santos, C. and Franco, O.L. (2023) Pathogenesis-related proteins (PRs) with enzyme activity activating plant defense responses. *Plants* 12: 2226.

Du, Z., Sun, T., Zhao, J., Wang, D., Zhang, Z., and Yu, W. (2018) Development of a plug-type IMS-MS instrument and its applications in resolving problems existing in in-situ detection of illicit drugs and explosives by IMS. *Talanta* 184, 65–72.

Dubey, A. and Verma, A.K. (2019) Salinity stress-dependent coordination of metabolic networks in relation to salt tolerance in plants. *Microorganisms in Saline Environments: Strategies and Functions* 2019, 401–422.

Durgbanshi, A., Arbona, V., Pozo, O., Miersch, O., Sancho, J.V., and Gómez-Cadenas, A. (2005) Simultaneous determination of multiple phytohormones in plant extracts by liquid chromatography-electrospray tandem mass spectrometry. *Journal of Agricultural and Food Chemistry* 53, 8437–42.

Erb, M., Meldau, S., and Howe, G.A. (2012) Role of phytohormones in insect-specific plant reactions. *Trends in Plant Science* 17, 250–259.

Escandón, M., Meijón, M., Valledor, L., Pascual, J., Pinto, G., and Cañal, M.J. (2018) Metabolome integrated analysis of high-temperature response in *Pinus radiata*. *Frontiers in Plant Science* 9: 328576.

Fàbregas, N. and Fernie, A.R. (2019) The metabolic response to drought. *Journal of Experimental Botany* 70, 1077–1085.

Farré, G., Blancquaert, D., Capell, T., van der Straeten, D., Christou, P., and Zhu, C. (2014) Engineering complex metabolic pathways in plants. *Annual Review of Plant Biology* 65, 187–223.

Fernandez, O., Millet, E.J., Rincent, R., Prigent, S., Pétriacq, P., and Gibon, Y. (2021) Plant metabolomics and breeding. *Advances in Botanical Research* 98, 207–235.

Fernie, A.R. and Schauer, N. (2009) Metabolomics-assisted breeding: a viable option for crop improvement? *Trends in Genetics* 25, 39–48.

Fernie, A.R. and Tohge, T. (2017) The genetics of plant metabolism. *Annual Review of Genetics* 51, 287–310.

Friso, G. and van Wijk, K.J. (2015) Posttranslational protein modifications in plant metabolism. *Plant Physiology* 169, 1469–87.

Gao, S.-Y., Wang, Q.-J., and Ji, Y.B. (2006) Effect of solanine on the membrane potential of mitochondria in HepG2 cells and $[Ca^{2+}]_i$ in the cells. *World Journal of Gastroenterology* 12, 3359–3367.

Gao, Y.-Q., Jimenez-Sandoval, P., Tiwari, S., Stolz, S., Wang, J., *et al.* (2023) Ricca's factors as mobile proteinaceous effectors of electrical signaling. *Cells* 186, 1337–1351.e20.

Gaquerel, E., Kuhl, C., and Neumann, S. (2013) Computational annotation of plant metabolomics profiles via a novel network-assisted approach. *Metabolomics* 9, 904–918.

Gertsman, I. and Barshop, B.A. (2018) Promises and pitfalls of untargeted metabolomics. *Journal of Inherited Metabolic Disease* 41, 355–366.

Gil-Solsona, R., Sancho, J. V., Gassner, A.L., Weyermann, C., Hernández, F., *et al.* (2021) Use of ion mobility-high resolution mass spectrometry in metabolomics studies to provide near MS/MS quality data in a single injection. *Journal of Mass Spectrometry* 56: e4718.

Godoy, F., Olivos-Hernández, K., Stange, C., and Handford, M. (2021) Abiotic stress in crop species: improving tolerance by applying plant metabolites. *Plants (Basel)* 10: 186.

González Guzmán, M., Cellini, F., Fotopoulos, V., Balestrini, R., and Arbona, V. (2022) New approaches to improve crop tolerance to biotic and abiotic stresses. *Physiologia Plantarum* 174: e13547.

Gross, J.H. (2014) Direct analysis in real time-a critical review on DART-MS. *Analytical and Bioanalytical Chemistry* 406, 63–80.

Grubb, C.D. and Abel, S. (2006) Glucosinolate metabolism and its control. *Trends in Plant Science* 11, 89–100.

Heuckeroth, S., Damiani, T., Smirnov, A., Mokshyna, O., Brungs, C., *et al.* (2024) Reproducible mass spectrometry data processing and compound annotation in MZmine 3. *Nature Protocols* 19, 2597–2641.

Horai, H., Arita, M., Kanaya, S., Nihei, Y., Ikeda, T., *et al.* (2010) MassBank: a public repository for sharing mass spectral data for life sciences. *Journal of Mass Spectrometry* 45, 703–714.

Howe, G.A. and Jander, G. (2008) Plant immunity to insect herbivores. *Annual Review of Plant Biology* 59, 41–66.

Hummel, J., Strehmel, N., Bölling, C., Schmidt, S., Walther, D., and Kopka, J. (2013) Mass spectral search and analysis using the Golm Metabolome Database. In: Weckwerth, W. and Kahl, G. (eds) *Handbook of Plant Metabolomics*. Wiley, Weinheim, Germany, pp. 321–343.

Hunyadi, A. (2019) The mechanism(s) of action of antioxidants: from scavenging reactive oxygen/nitrogen species to redox signaling and the generation of bioactive secondary metabolites. *Medicinal Research Reviews* 39, 2505–2533.

Janfelt, C. (2015) Imaging of plant materials using indirect desorption electrospray ionization mass spectrometry. *Methods in Molecular Biology* 1203, 91–97.

Jorge, T.F., Rodrigues, J.A., Caldana, C., Schmidt, R., van Dongen, J.T., *et al.* (2016) Mass spectrometry-based plant metabolomics: metabolite responses to abiotic stress. *Mass Spectrometry Reviews* 35, 620–649.

Kanehisa, M., and Subramaniam (2002) The KEGG Database. *Novartis Foundation Symposium* 247, 91–103.

Katajamaa, M., Miettinen, J., and Oresic, M. (2006) MZmine: toolbox for processing and visualization of mass spectrometry based molecular profile data. *Bioinformatics* 22, 634–636.

Kaur, N., Alok, A., Shivani, K.P., Kaur, N., Awasthi, P., *et al.* (2020) CRISPR/Cas9 directed editing of *lycopene epsilon-cyclase* modulates metabolic flux for β-carotene biosynthesis in banana fruit. *Metabolic Engineering* 59, 76–86.

Kim, Y.G., Cha, J., and Chandrasegaran, S. (1996) Hybrid restriction enzymes: zinc finger fusions to Fok I cleavage domain. *Proceedings of the National Academy of Sciences USA* 93, 1156–1160.

Klein, A.T., Yagnik, G.B., Hohenstein, J.D., Ji, Z., Zi, J., *et al.* (2015) Investigation of the chemical interface in the soybean–aphid and rice–bacteria interactions using MALDI-mass spectrometry imaging. *Analytical Chemistry* 87, 5294–5301.

Korn, M., Gärtner, T., Erban, A., Kopka, J., Selbig, J., and Hincha, D.K. (2010) Predicting *Arabidopsis* freezing tolerance and heterosis in freezing tolerance from metabolite composition. *Molecular Plant* 3, 224–35.

Kuhl, C., Tautenhahn, R., Böttcher, C., Larson, T.R., and Neumann, S. (2012) CAMERA: an integrated strategy for compound spectra extraction and annotation of liquid chromatography/mass spectrometry data sets. *Analytical Chemistry* 84, 283–289.

Kumar, R., Bohra, A., Pandey, A.K., Pandey, M.K., and Kumar, A. (2017) Metabolomics for plant improvement: status and prospects. *Frontiers in Plant Science* 8: 271676.

Lisec, J., Steinfath, M., Meyer, R.C., Selbig, J., Melchinger, A.E., *et al.* (2009) Identification of heterotic metabolite QTL in *Arabidopsis thaliana* RIL and IL populations. *Plant Journal* 59, 777–788.

Litvinov, D.Y., Karlov, G.I., and Divashuk, M.G. (2021) Metabolomics for crop breeding: general considerations. *Genes* 12: 1602.

Liu, W., Rudis, M.R., Peng, Y., Mazarei, M., Millwood, R.J., *et al.* (2014) Synthetic TAL effectors for targeted enhancement of transgene expression in plants. *Plant Biotechnology Journal* 12, 436–446.

Llanes, A., Andrade, A., Alemano, S., and Luna, V. (2018) Metabolomic approach to understand plant adaptations to water and salt stress. In: Ahmad, P., Ahanger, M.A., Singh, V.P., Tripathi, D.K., and Alyemeni, M.N. (eds) *Plant Metabolites and Regulation under Environmental Stress*. Elsevier, London, pp. 133–144.

Mabry, M.E., Turner-Hissong, S.D., Gallagher, E.Y., McAlvay, A.C., An, H., *et al.* (2021) The evolutionary history of wild, domesticated, and feral *Brassica oleracea* (Brassicaceae). *Molecular Biology and Evolution* 38, 4419–4434.

Manzi, M., Gómez-Cadenas, A., and Arbona, V. (2015) Rapid and reproducible determination of active gibberellins in citrus tissues by UPLC/ESI-MS/MS. *Plant Physiology and Biochemistry* 94, 1–9.

Marone, D., Mastrangelo, A.M., Borrelli, G.M., Mores, A., Laidò, G., *et al.* (2022) Specialized metabolites: physiological and biochemical role in stress resistance, strategies to improve their accumulation, and new applications in crop breeding and management. *Plant Physiology and Biochemistry* 172, 48–55.

May, D., Law, W., Fitzgibbon, M., Fang, Q., and McIntosh, M. (2009) Software platform for rapidly creating computational tools for mass spectrometry-based proteomics. *Journal of Proteome Research* 8, 3212–3217.

Meyer, R.C., Steinfath, M., Lisec, J., Becher, M., Witucka-Wall, H., *et al.* (2007) The metabolic signature related to high plant growth rate in *Arabidopsis thaliana*. *Proceedings of the National Academy of Sciences USA* 104, 4759–764.

Mithen, R.F., Lewis, B.G., Heaney, R.K., and Fenwick, G.R. (1987) Glucosinolates of wild and cultivated *Brassica* species. *Phytochemistry* 26, 1969–1973.

Modesto, I., Sterck, L., Arbona, V., Gómez-Cadenas, A., Carrasquinho, I., *et al.* (2021) Insights into the mechanisms implicated in *Pinus pinaster* resistance to pinewood nematode. *Frontiers in Plant Science* 12: 690857.

Mojica, F.J.M., Díez-Villaseñor, C., Soria, E., and Juez, G. (2000) Biological significance of a family of regularly spaced repeats in the genomes of Archaea, Bacteria and mitochondria. *Molecular Microbiology* 36, 244–246.

Montenegro-Burke, J.R., Guijas, C., and Siuzdak, G. (2020) METLIN: a tandem mass spectral library of standards. *Methods in Molecular Biology* 2104, 149–163.

Müller, T., Oradu, S., Ifa, D.R., Cooks, R.G., and Kräutler, B. (2011) Direct plant tissue analysis and imprint imaging by desorption electrospray ionization mass spectrometry. *Analytical Chemistry* 83, 5754–5761.

Myers, O.D., Sumner, S.J., Li, S., Barnes, S., and Du, X. (2017) Detailed investigation and comparison of the XCMS and MZmine 2 chromatogram construction and chromatographic peak detection methods for preprocessing mass spectrometry metabolomics data. *Analytical Chemistry* 89, 8689–8695.

Nakabayashi, R., Yonekura-Sakakibara, K., Urano, K., Suzuki, M., Yamada, Y., *et al.* (2014) Enhancement of oxidative and drought tolerance in *Arabidopsis* by overaccumulation of antioxidant flavonoids. *Plant Journal* 77, 367–379.

Nanita, S.C. and Kaldon, L.G. (2016) Emerging flow injection mass spectrometry methods for high-throughput quantitative analysis. *Analytical and Bioanalytical Chemistry* 408, 23–33.

Nguyen, Q.M., Iswanto, A.B.B., Son, G.H., and Kim, S.H. (2021) Recent advances in effector-triggered immunity in plants: new pieces in the puzzle create a different paradigm. *International Journal of Molecular Sciences* 22: 4709.

Obata, T., Witt, S., Lisec, J., Palacios-Rojas, N., Florez-Sarasa, I., *et al.* (2015) Metabolite profiles of maize leaves in drought, heat, and combined stress field trials reveal the relationship between metabolism and grain yield 1. *Plant Physiology* 169, 2665–2683.

Oliver, S.G., Winson, M.K., Kell, D.B., and Baganz, F. (1998) Systematic functional analysis of the yeast genome. *Trends in Biotechnology* 16, 373–378.

Pan, Z., Li, Y., Deng, X., and Xiao, S. (2014) Non-targeted metabolomic analysis of orange (*Citrus sinensis* [L.] Osbeck) wild type and bud mutant fruits by direct analysis in real-time and HPLC-electrospray mass spectrometry. *Metabolomics* 10, 508–523.

Pan, C., Wu, X., Markel, K., Malzahn, A.A., Kundagrami, N., *et al.* (2021) CRISPR–Act3.0 for highly efficient multiplexed gene activation in plants. *Nature Plants* 7, 942–953.

Pedras, M.S.C., Okanga, F.I., Zaharia, I.L., and Khan, A.Q. (2000) Phytoalexins from crucifers: synthesis, biosynthesis, and biotransformation. *Phytochemistry* 53, 161–176.

Pedras, M.S.C., Adio, A.M., Suchy, M., Okinyo, D.P.O., Zheng, Q.-A., *et al.* (2006) Detection, characterization and identification of crucifer phytoalexins using high-performance liquid chromatography with diode array detection and electrospray ionization mass spectrometry. *Journal of Chromatography A* 1133, 172–83.

Perez de Souza, L., Alseekh, S., Naake, T., Fernie, A. (2019) Mass spectrometry-based untargeted plant metabolomics. *Current Protocols in Plant Biology* 4: e20100.

Perez de Souza, L., Bitocchi, E., Papa, R., and Tohge, T. (2023) Decreased metabolic diversity in common beans associated with domestication revealed by untargeted metabolomics, information theory, and molecular networking. *Plant Journal* 115, 1021–1036.

Pluskal, T., Castillo, S., Villar-Briones, A., and Oresic, M. (2010) MZmine 2: modular framework for processing, visualizing, and analyzing mass spectrometry-based molecular profile data. *BMC Bioinformatics* 11: 395.

Pollier, J., Moses, T., González-Guzmán, M., de Geyter, N., Lippens, S., *et al.* (2013) The protein quality control system manages plant defence compound synthesis. *Nature* 504, 148–152.

Qi, L.S., Larson, M.H., Gilbert, L.A., Doudna, J.A., Weissman, J.S., *et al.* (2013) Repurposing CRISPR as an RNA-guided platform for sequence-specific control of gene expression. *Cells* 152, 1173–1183.

Rathore, J.S. and Ghosh, C. (2018) Pathogen-associated molecular patterns and their perception in plants. In: Singh, A. and Singh, I. (eds) *Molecular Aspects of Plant–Pathogen Interaction*. Springer, Singapore, pp. 79–113.

Rathore, K.S., Sundaram, S., Sunilkumar, G., Campbell, L.M., Puckhaber, L., *et al.* (2012) Ultra-low gossypol cottonseed: generational stability of the seed-specific, RNAi-mediated phenotype and resumption of terpenoid profile following seed germination. *Plant Biotechnology Journal* 10, 174–183.

Razzaq, A., Sadia, B., Raza, A., Khalid Hameed, M., and Saleem, F. (2019) Metabolomics: a way forward for crop improvement. *Metabolites* 9: 303.

Riedelsheimer, C., Czedik-Eysenberg, A., Grieder, C., Lisec, J., Technow, F., *et al.* (2012) Genomic and metabolic prediction of complex heterotic traits in hybrid maize. *Nature Genetics* 44, 217–220.

Rivero, R.M., Mestre, T.C., Mittler, R., Rubio, F., Garcia-Sanchez, F., and Martinez, V. (2014) The combined effect of salinity and heat reveals a specific physiological, biochemical and molecular response in tomato plants. *Plant, Cell and Environment* 37, 1059–1073.

Roberts, L.D., Souza, A.L., Gerszten, R.E., and Clish, C.B. (2012) Targeted metabolomics. *Current Protocols in Molecular Biology* 30.2.1–30.2.24.

Ronis, M.J.J., Pedersen, K.B., and Watt, J. (2018) Adverse effects of nutraceuticals and dietary supplements. *Annual Review of Pharmacology and Toxicology* 58, 583–601.

Röst, H.L., Sachsenberg, T., Aiche, S., Bielow, C., Weisser, H., *et al.* (2016) OpenMS: a flexible open-source software platform for mass spectrometry data analysis. *Nature Methods* 13, 741–748.

Ruttkies, C., Strehmel, N., Scheel, D., and Neumann, S. (2015) Annotation of metabolites from gas chromatography/atmospheric pressure chemical ionization tandem mass spectrometry data using an *in silico* generated compound database and MetFrag. *Rapid Communications in Mass Spectrometry* 29, 1521–1529.

Saito, K. and Matsuda, F. (2010) Metabolomics for functional genomics, systems biology, and biotechnology. *Annual Review of Plant Biology* 61, 463–489.

Saxena, S.C., Kaur, H., Verma, P., Petla, B.P., Andugula, V.R., and Majee, M. (2013) Osmoprotectants: potential for crop improvement under adverse conditions. In: Tuteja, N. and Singh Gill, S. (eds) *Plant Acclimation to Environmental Stress*. Springer, New York, pp. 197–232.

Schimmel, B.C.J., Alba, J.M., Wybouw, N., Glas, J.J., Meijer, T.T., *et al.* (2018) Distinct signatures of host defense suppression by plant-feeding mites. *International Journal of Molecular Sciences* 19: 3265.

Schrimpe-Rutledge, A.C., Codreanu, S.G., Sherrod, S.D., and McLean, J.A. (2016) Untargeted metabolomics strategies—challenges and emerging directions. *Journal of the American Society for Mass Spectrometry* 27, 1897–1905.

Schuhmacher, R., Krska, R., Weckwerth, W., and Goodacre, R. (2013) Metabolomics and metabolite profiling. *Analytical and Bioanalytical Chemistry* 405, 5003–5004.

Selma, S., Sanmartín, N., Espinosa-Ruiz, A., Gianoglio, S., Lopez-Gresa, M.P., *et al.* (2022) Custom-made design of metabolite composition in *N. benthamiana* leaves using CRISPR activators. *Plant Biotechnology Journal* 20, 1578–1590.

Selma, S., Ntelkis, N., Nguyen, T.H., and Goossens, A. (2023) Engineering the plant metabolic system by exploiting metabolic regulation. *Plant Journal* 114, 1149–1163.

Sharma, V., Gupta, P., Priscilla, K., Sharankumar, Hangargi, B., *et al.* (2021) Metabolomics intervention towards better understanding of plant traits. *Cells* 10: 346.

Shiono, K. and Taira, S. (2020) Imaging of multiple plant hormones in roots of rice (*Oryza sativa*) using nanoparticle-assisted laser desorption/ionization mass spectrometry. *Journal of Agricultural and Food Chemistry* 68, 6770–6775.

Smith, C.A., Want, E.J., O'Maille, G., Abagyan, R., and Siuzdak, G. (2006) XCMS: processing mass spectrometry data for metabolite profiling using nonlinear peak alignment, matching and identification. *Analytical Chemistry* 78, 779–787.

Stanley, L. and Yuan, Y.W. (2019) Transcriptional regulation of carotenoid biosynthesis in plants: so many regulators, so little consensus. *Frontiers in Plant Science* 10: 473012.

Sud, M., Fahy, E., Cotter, D., Azam, K., Vadivelu, I., *et al.* (2016) Metabolomics Workbench: an international repository for metabolomics data and metadata, metabolite standards, protocols, tutorials and training, and analysis tools. *Nucleic Acids Research* 44, D463–D470.

Szymański, J., Bocobza, S., Panda, S., Sonawane, P., Cárdenas, P.D., *et al.* (2020) Analysis of wild tomato introgression lines elucidates the genetic basis of transcriptome and metabolome variation underlying fruit traits and pathogen response. *Nature Genetics* 52, 1111–1121.

Tautenhahn, R., Patti, G., Rinehart, D., and Siuzdak, G. (2012) XCMS Online: a web-based platform to process untargeted metabolomic data. *Analytical Chemistry* 84, 5035–5039.

Tsugawa, H., Cajka, T., Kind, T., Ma, Y., Higgins, B., *et al.* (2015) MS-DIAL: data-independent MS/MS deconvolution for comprehensive metabolome analysis. *Nature Methods* 12, 523–526.

Venter, A., Sojka, P.E., and Cooks, R.G. (2006) Droplet dynamics and ionization mechanisms in desorption electrospray ionization mass spectrometry. *Analytical Chemistry* 78, 8549–8555.

Vergara-Diaz, O., Vatter, T., Vicente, R., Obata, T., Nieto-Taladriz, M.T., *et al.* (2020) Metabolome profiling supports the key role of the spike in wheat yield performance. *Cells* 9: 1025.

Verpoorte, R., Contin, A., and Memelink, J. (2002) Biotechnology for the production of plant secondary metabolites. *Phytochemistry Reviews* 1, 13–25.

Vilà, M., Bedmar, A., Saurina, J., Núñez, O., and Sentellas, S. (2022) High-throughput flow injection analysis–mass spectrometry (FIA-MS) fingerprinting for the authentication of tea application to the detection of teas adulterated with chicory. *Food* 11: 2153.

Villarroel, C.A., Jonckheere, W., Alba, J.M., Glas, J.J., Dermauw, W., *et al.* (2016) Salivary proteins of spider mites suppress defenses in *Nicotiana benthamiana* and promote mite reproduction. *Plant Journal* 86, 119–131.

Vinaixa, M., Schymanski, E. L., Neumann, S., Navarro, M., Salek, R. M., and Yanes, O. (2016) Mass spectral databases for LC/MS- and GC/MS-based metabolomics: state of the field and future prospects. *Trends in Analytical Chemistry* 78, 23–35.

Wei, J., van Loon, J.J.A., Rieta, G., Menzel, T.R., Li, N., *et al.* (2014) Reciprocal crosstalk between jasmonate and salicylate defence-signalling pathways modulates plant volatile emission and herbivore host-selection behaviour. *Journal of Experimental Botany* 65, 3289–3298.

Widemann, E., Bruinsma, K., Walshe-Roussel, B., Rioja, C., Arbona, V., *et al.* (2021) Multiple indole glucosinolates and myrosinases defend *Arabidopsis* against *Tetranychus urticae* herbivory. *Plant Physiology* 187, 116–132.

Wishart, D.S., Guo, A.C., Oler, E., Wang, F., Anjum, A., *et al.* (2022) HMDB 5.0: the human metabolome database for 2022. *Nucleic Acids Research* 50, D622–D631.

Wolf, S., Schmidt, S., Müller-Hannemann, M., and Neumann, S. (2010) *In silico* fragmentation for computer assisted identification of metabolite mass spectra. *BMC Bioinformatics* 11: 148.

Xia, J. and Wishart, D.S. (2016) Using metaboanalyst 3.0 for comprehensive metabolomics data analysis. *Current Protocols in Bioinformatics* 2016, 14.10.1–14.10.91.

Xu, Z., Li, J., Guo, X., Jin, S., and Zhang, X. (2016) Metabolic engineering of cottonseed oil biosynthesis pathway via RNA interference. *Scientific Reports* 6: 33342.

Yamada, Y., Sato, F., Back, K., Bru-Martinez, R., Malacarne, G., and Matus, J.T. (2021) Transcription factors in alkaloid engineering. *Biomolecules* 11: 1719.

Yoshiba, Y., Kiyosue, T., Nakashima, K., Yamaguchi-Shinozaki, K., and Shinozaki, K. (1997) Regulation of levels of proline as an osmolyte in plants under water stress. *Plant and Cell Physiology* 38, 1095–1102.

Yurekten, O., Payne, T., Tejera, N., Amaladoss, F.X., Martin, C., *et al.* (2024) MetaboLights: open data repository for metabolomics. *Nucleic Acids Research* 52, D640–D646.

Zandalinas, S.I., Sales, C., Beltrán, J., Gómez-Cadenas, A., and Arbona, V. (2017) Activation of secondary metabolism in citrus plants is associated to sensitivity to combined drought and high temperatures. *Frontiers in Plant Science* 7: 1954.

Zhang, W., Lei, Z., Huhman, D., Sumner, L.W., and Zhao, P.X. (2015a) MET-XAlign: a metabolite cross-alignment tool for LC/MS-based comparative metabolomics. *Analytical Chemistry* 87, 9114–9119.

Zhang, Y., Butelli, E., Alseekh, S., Tohge, T., and Rallapalli, G. (2015b) Multi-level engineering facilitates the production of phenylpropanoid compounds in tomato. *Nature Communications* 6: 8635.

Zhang, P., Du, H., Wang, J., Pu, Y., Yang, C., *et al.* (2020) Multiplex CRISPR/Cas9-mediated metabolic engineering increases soya bean isoflavone content and resistance to soya bean mosaic virus. *Plant Biotechnology Journal* 18, 1384–1395.

Zhao, X., Huang, L.J., Sun, X.F., Zhao, L.L., and Wang, P.C. (2022) Differential physiological, transcriptomic, and metabolomic responses of *Paspalum wettsteinii* under high-temperature stress. *Frontiers in Plant Science* 13: 865608.

Zhurov, V., Navarro, M., Bruinsma, K.A., Arbona, V., Santamaria, M.E., *et al.* (2014) Reciprocal responses in the interaction between *Arabidopsis* and the cell-content-feeding chelicerate herbivore spider mite. *Plant Physiology* 164, 384–399.

14 High-throughput Metabolomics for Assessing CRISPR-edited Plants

Ajay Kumar Keot[1,2], Yogita N. Sarki[1,2], Riwandahun Marwein[1,2], Channakeshavaiah Chikkaputtaiah[1,2*], and Dhanawantari L. Singha[3*]

[1]Biological Sciences and Technology Division, CSIR-North East Institute of Science and Technology (CSIR- NEIST), Jorhat 785 006, Assam, India; [2]Academy of Scientific and Innovative Research (AcSIR), Ghaziabad 201 002, India; [3]Department of Botany, Rabindranath Tagore University, Hojai, Assam, India

Abstract

Clustered regularly interspaced short palindromic repeats (CRISPR)-edited plants are being generated globally by precisely manipulating plant genomes utilizing Cas nuclease and single guide RNA for biotic and abiotic stress resistance, yield improvement, and nutritional enhancement. As well as genomics, proteomics, and phenotypic screening of CRISPR-edited plants, assessment of the differential metabolome is crucial for understanding the comprehensive metabolite crosstalk that determines various aspects of plant growth development, defense mechanisms, fruit taste, etc. indicative of gene editing. Editing events (e.g. single-base mutations) and unknown off-target effects are challenging to differentiate between edited and non-edited genotypes, and therefore high-throughput technologies such as liquid chromatography–mass spectrometry, gas chromatography–mass spectrometry, and nuclear magnetic resonance spectroscopy have been used to study the plant metabolomics. This chapter highlights their application for studying holistic metabolomics in CRISPR-edited plants. Metabolite screening can be used to ascertain the novelty and authenticity of CRISPR-edited plants for market release and consumption.

Keywords: abiotic stress, biotic stress, metabolomics, CRISPR/Cas9, GC-MS, LC-MS, NMR

14.1 Introduction

Environmental conditions have been rapidly changing due to global warming and climate change. These conditions pose biotic and abiotic stresses for plants and alter morphological and physiological aspects, often resulting in yield reduction in crop plants. However, the growing population demands higher yields for global food security. To cope with this, the DNA of plants could be manipulated by the clustered regularly interspaced short palindromic repeats (CRISPR)/Cas gene-editing method to make them more climate resilient. CRISPR/Cas gene editing has transformed gene editing in plants, especially crops. CRISPR-edited plants (plants generated by the CRISPR/Cas gene-editing method) are characterized for gene editing by DNA sequencing. The integration of metabolomics could advance our understanding of the effect of gene

* E-mail: channakeshav@neist.res.in or dhanawantarisingha@gmail.com

© CAB International 2025. *High-Throughput Plant Metabolomics* (Jen-Tsung Chen ed.)
DOI: 10.1079/9781800626911.0014

editing, because CRISPR provides the means for precise genetic modifications, while metabolomics offers a detailed view of the resulting metabolic changes. Together, these technologies can drive innovations in crop improvement, and metabolic engineering, contributing to produce sustainable and resilient agricultural systems. The plant metabolome comprises two types of metabolites, primary and secondary, and their metabolic profiling offers comprehensive insights into the biochemical processes involved in plant metabolism. Metabolomics is studied using techniques such as liquid chromatography–mass spectrometry (LC-MS), gas chromatography–mass spectrometry (GC-MS), nuclear magnetic resonance (NMR) spectroscopy, and other derived versions, for identification and quantification of plant metabolites. For example, GC-MS analysis of seeds reveals an altered profile of aromatic compounds in CRISPR-edited rice plants compared with wild-type plants, and the T_1 seeds of edited plants were found to have several unique aromatic compounds (Ashokkumar *et al.*, 2020).

Primary metabolites are essential for basic metabolic processes necessary for plant growth, development, and reproduction, and are typically found in all plant cells. Primary metabolites are produced by plants during the active growth phase known as the trophophase (Zaynab *et al.*, 2019). As well as their role in growth and development, there have been reports showing that the components of primary metabolic pathways are directly or indirectly involved in defense responses to prevent or halt the expansion of potential pathogens (Rojas *et al.*, 2014). Examples include amino acids, which serve as protein building blocks and precursors for other metabolites; carbohydrates, which provide energy, and structural components; nucleotides (adenine, guanine, cytosine, thymine, and uracil), which are the building blocks of DNA and RNA; and fats, storage lipids, polyamines, and glycine betaine (Pagare *et al.*, 2015).

Secondary metabolites are crucial for a plant's interaction with its environment, often serving ecological functions including defense responses against pathogens, resistance to abiotic stress, attraction of pollinators, and competition with other plants. The secondary metabolome in plants is represented by approximately 200,000–1,000,000 compounds, which are grouped into terpenes/terpenoids, phenolics, nitrogen-containing compounds, sulfur-containing compounds, and lectins (Al-Khayri *et al.*, 2023; Huang and Dudareva, 2023). These compounds are derivatives of primary metabolites produced by plants for their growth and development functions (Jogawat *et al.*, 2021). There are some metabolites (e.g. terpenoids) that can function as both primary and secondary metabolites, making it difficult to categorize them. Secondary metabolite production in plants increases during stress situations or specific plant functions. For example, to attract pollinators, plants synthesize flavonoids and carotenoids, which are responsible for flower color, thereby playing a vital role in plant reproduction (Pagare *et al.*, 2015).

CRISPR/Cas is a revolutionary technology for gene editing in plants. Its development has significantly advanced our ability to modify the DNA of living organisms with unprecedented precision, efficiency, and flexibility. It originated from a natural defense machinery of bacteria or archaea. These microorganisms use CRISPR sequences and Cas (CRISPR-associated) nuclease to fend off viral infections. When a virus attacks, the bacteria capture snippets of the virus's DNA and insert them into their own genome within the CRISPR sequences. These sequences are then transcribed into RNA, which guides specialized Cas proteins to recognize and cut the viral DNA if the virus attacks again (Wang and Doudna, 2023). CRISPR/Cas tools induce targeted nicks or double-stranded breaks in DNA, activating two key repair pathways: non-homologous end-joining and homology-directed repair. Homology-directed repair is considered an accurate DNA-repair mechanism that utilizes a homologous donor DNA template to repair double-stranded breaks, allowing precise gene insertions or corrections. In contrast, non-homologous end-joining is an error-prone mechanism that directly ligates the broken DNA ends and does not require a template, often leading to insertions or deletions (indels) and sometimes substitutions as well (Miyaoka *et al.*, 2016).

The most commonly used CRISPR system in gene editing is CRISPR/Cas9, originating from *Streptococcus pyogenes*. The Cas9 protein consists of six domains: recognition lobe REC I, REC II, protospacer adjacent motif (PAM)-interacting domain, arginine-rich bridge helix, and the nuclease domains RuvC and HNH. REC I binds primarily to the guide RNA (gRNA; also known as single guide RNA (sgRNA)), while the function of REC II is not well understood. The arginine-rich bridge helix is important for cleavage activity when it binds to target sequences. The PAM-interacting domain enables binding to the target sequence, thus providing specificity. The nuclease domain cleaves the target DNA and generates double-stranded breaks resulting in blunt ends (Hillary and Ceasar, 2023). The other component of CRISPR/Cas9 is gRNA, which directs Cas9 to target DNA sites. The gRNA is a simple combined version of CRISPR RNA (crRNA) and transactivating crRNA (tracrRNA). When gRNA binds to Cas9, it induces conformational changes that activate Cas9. Subsequently, the complex attaches to the DNA and searches for the gRNA complementary target sequence adjacent to PAM (5′-NGG-3′). Cas9 then cuts the double-strand of DNA at three nucleotides upstream of the PAM motif using the nuclease domains, resulting in blunt ends. The gRNA complementary strand is cleaved by the HNH domain, while the other DNA strand is targeted by the RuvC domain (Jinek *et al.*, 2012).

Another key player in gene editing is the Cas12a protein (formerly known as Cpf1, CRISPR from *Prevotella* and *Francisella* 1). It requires only crRNAs to efficiently cleave single or double-stranded DNA. It features the RuvC and nuclease lobe (NUC) domains, essential for its nuclease activity. Cas12a identifies a target site downstream from a PAM sequence (5′-TTTN-3′, 5′-TTTV-3′, etc.). Upon encountering the target DNA, Cas12 initiates complementary base pairing between the crRNA and DNA strand. This is known as R-loop formation. Cas12a matches approximately 17 base pairs of the target DNA sequence during this process. Subsequently, Cas12a cleaves the non-target DNA strand by its active RuvC domain. In contrast to Cas9, Cas12 lacks the HNH domain. However, the exact role of the RuvC domain in cleaving the target strand is not fully understood (Hillary and Ceasar, 2023).

14.2 Principles of CRISPR/Cas Gene Editing in Plants

A basic understanding of the pipeline of gene editing in plants using CRISPR/Cas is crucial and is outlined below.

14.2.1 Selection of plant

The first step in editing plant genes using CRISPR/Cas is selection of an appropriate plant species, based on criteria such as the availability of its whole-genome sequence, tissue-culture response, life cycle, and economic importance. It is generally preferred to edit genes in plants with publicly available genome sequences, as these facilitate gRNA design and off-target searches. During gRNA design, researchers check for off-target binding sites in the plant genome, opting for gRNAs with minimal off-target effects.

14.2.2 Selection of target gene

For basic research, scientists often target genes with unknown functions to generate knockouts to understand the gene function. In applied research, well-characterized genes are targeted for crop improvement. The choice of target gene depends on the experiment's goals, such as optimizing CRISPR protocols using model genes such as phytoene desaturase (*PDS*) (Kaur *et al.*, 2018) or editing genes based on literature and experimental data. Base editing is used to alter specific nucleotides for desirable traits, such as herbicide resistance from single-nucleotide polymorphisms (SNPs) in the acetolactate synthase (*ALS*) gene (Li *et al.*, 2020). Transcriptomics and proteomics also help in identifying target genes for editing, aiming to enhance yield, stress tolerance, and disease resistance in crops.

14.2.3 Selection of a suitable CRISPR/Cas system

After selecting a target gene in a plant, choosing the appropriate CRISPR/Cas system is crucial. CRISPR/Cas9 is ideal for G-rich sequences due to its 5′-NGG-3′ PAM requirement, while CRISPR/Cas12 is better for T-rich sequences with its 5′-TTTV-3′ PAM. For experiments involving delivery of the ribonucleoprotein complex, using a smaller Cas protein can enhance delivery efficiency into cells.

14.2.4 gRNA design

A good gRNA should have two features: efficacy and specificity. This is crucial for successful gene editing. To design gRNA, many algorithms and computational tools have been developed such as CHOPCHOP, E-CRISP, Cas-OFFinder, Benchling, CRISP-P, CCTop, CRISPR.mit, sgRNA designer, CRISPRseek, CRISPRdirect, CRISPRscan, PROTOSPACER, and CRISPOR. It is important to consider off-target sites when designing gRNAs; those with no off-target binding sites and maximum efficiency at the target are preferred (Cui *et al.*, 2018).

14.2.5 Design of CRISPR/Cas gene-editing construct

Designing a CRISPR/Cas gene-editing construct is crucial for successful gene editing and depends on several factors, including the type of plant (monocot or dicot). The choice of promoters varies between these plant types. Monocots and dicots require specific promoters to express Cas proteins and gRNA effectively. Choosing plant-selectable markers (e.g. hygromycin) and reporter genes (e.g. *GFP*, *GUS*, *RUBY*) is also vital when designing the construct. High expression levels of Cas and gRNA are essential for efficient editing. Tang *et al.* (2017) generated constructs by driving LbCpf1 and AsCpf1 expression using a promoter from the maize ubiquitin gene (*ZmUbi*), while the expression of gRNA (crRNA) was controlled by the rice *U6* promoter to target four rice genes for stable transformation.

14.2.6 Transfer of the CRISPR/Cas tool into cells and plantlet regeneration

The most common approach is *Agrobacterium*-mediated transformation, which involves the integration of CRISPR/Cas DNA into the plant genome via the bacterium's transfer DNA (T-DNA) (Altpeter *et al.*, 2016). Particle bombardment by means of a gene gun, can also be used. Both methods result in random DNA integration into the plant genome. In another method, protoplasts can be transfected with CRISPR plasmids or ribonucleoproteins. However, regenerating plants from protoplasts involves extensive tissue-culture procedures, which often lead to somaclonal variations. Several methods for plant genetic transformation that are independent of *in vitro* tissue culture have been established such as the floral dip method, *in planta* transformation, *de novo* meristem induction, the cut–dip–budding method, and nanoparticle-mediated gene delivery (Verma *et al.*, 2023).

14.2.7 Screening for gene-editing events

The regenerants are screened for positive transformants by the polymerase chain reaction (PCR) using Cas-specific primers. Next, DNA Sanger sequencing is commonly used to detect editing events. The sequence data can be analyzed manually or with an online computational tool to identify mutations and their types. Depending on the gene being edited, some edits may result in observable phenotypic changes, such as the formation of an albino phenotype when the *PDS* gene is edited (Komatsu *et al.*, 2020)

14.3 Characterization of CRISPR/Cas-edited Plants Using Sequencing Technology

Alterations in target genes due to CRISPR/Cas gene editing in plants are confirmed using DNA sequencing technologies. The change in sequence occurs as a result of deletion, insertion, or both (indels), or substitution, and the types of mutations generated are biallelic, homozygous, heterozygous, or chimeric (Shillito *et al.*, 2021). The detection of Cas-mediated mutations in both target and off-target sites is achieved using the Sanger sequencing method. This is done by sequencing the gRNA target site in the gene using a specific PCR primer set. A suitable online or offline tool is then used to examine the sequences to detect editing. Various computational tools are available for analyzing the Sanger sequencing output files such as TIDE (Tracking of Indels by Decomposition), DECODR (Deconvolution of Complex DNA Repair), DSDecode (Degenerate Sequence Decode), ICE (Inference of CRISPR Edits), and EditR (for base editing analysis). These tools also calculate the number of indels, and their size and frequency. However, Sanger sequencing has its limitations when used for polyploid plants because in such cases it results in confusing superimposed chromatograms. Moreover, it is laborious and expensive when a large number of samples need to be sequenced (Cardi *et al.*, 2023; Aoki *et al.*, 2024). In the case of confusing sequencing chromatograms, the PCR product of the sample can be cloned and sequenced to avoid superimposing chromatograms. Most importantly, successful decoding of mutations relies on the purity of the PCR-amplified product and high-quality sequencing reads in the wild type and samples. Off-target mutations are also validated by Sanger sequencing (Zischewski *et al.*, 2017).

Whole-genome sequencing (WGS) of a CRISPR/Cas-edited plant provides a comprehensive detection method for off-target effects but is expensive and time-consuming, and needs expertise. The method can scan for small (SNPs) or large genomic changes (structural variations caused by large deletions, duplications, rearrangements, and inversions) that occur at the target or unknown regions as a result of DNA double-stranded break repair mechanisms (Atkins *et al.*, 2021). For example, whole-genome comparative analysis between CRISPR/Cas9-edited and wild-type plants in grapevine (*Vitis vinifera*) identified around 200,000 SNPs and approximately 55,000 indels. Most of these mutations were shared between edited and wild-type plants, and induced variations were concluded to be a result of *in vitro* tissue-culture somaclonal variations (Wang *et al.*, 2021). In another study in rice, a large-scale WGS was conducted to analyze off-target mutations using the Cas9 and Cas12a editing systems (Tang *et al.*, 2018).

14.4 Importance of Metabolomics for Characterization of CRISPR/Cas-edited Plants

A considerable effort is needed to screen CRISPR/Cas-mediated genome-edited plant products for their required characteristics and to avoid undesirable changes caused by Cas before approval for public release. Genome screening using next-generation sequencing has been widely adopted to characterize variations in genomic sequences between edited and non-edited plants; however, the bottleneck lies with the difficulty in analyzing small changes or single-base variations that differ from the normal phenotype (Fraser *et al.*, 2020). Therefore, a more prominent assessment can be achieved by comparing associated metabolites, which are the end products of the gene. It is already established that gene editing can lead to unintended modifications at the genomic level and downstream processes in plants. For example, gene editing of a few nucleotide bases could disrupt alternative splicing sites in genes, and this might affect transcriptome, proteome, and metabolome levels differently, leading to undesirable or unpredicted outcomes in plant–environment interactions (Kawall, 2021). For instance, microbe–plant interactions play an important role in the growth and development of plants by controlling immunity and inhibiting colonization by pathogens. A mutation in a particular gene may affect the colonization of bacteria around the rhizosphere if the root exudate composition is changed as a result of the mutation. Metabolomics profiling could be used to estimate this effect in gene-edited plants. In another example to elucidate the robustness of the metabolomics approach, Drapal *et al.* (2023) performed comparative metabolomic profiling using LC-MS in tobacco (non-food) and tomato (food) crops to analyze whether the existence of Cas9 protein with or without the target gene and/or with or without mutation affected the plant metabolome in primary and subsequent generations. Principal component analysis in tobacco found significant variations within and between the control groups (*Agrobacterium* without binary plasmid and tissue-culture-regenerated plants only) and Cas9-transformed groups. These variations were assumed to be a result of *in vitro* tissue-culture regeneration and the *Agrobacterium*-mediated stress response. Significant compounds isolated from Cas9-positive plants included glucose, fructose, malic acid, chlorogenic acid, mannose, and uncharacterized glucosides, whereas linolenic acid, disaccharides, and lutein were isolated from the *Agrobacterium*-infected control. No distinct biological variations were observed in the T_1 progeny plants between and within the groups. Interestingly, no significant alterations of metabolites occurred between tobacco T_1 genotypes of edited *DET1* plants, tissue-culture-generated, and wild-type plants. Principal component analysis in the leaf and fruits of tomato showed no significant segregation of metabolic profiling among the plants harboring Cas9, null segregants, and control plants in T_1 and T_2 plants. However, metabolite segregation was observed between Cas9-positive and -negative plants when using orthogonal partial least squares-discriminant analysis showing a higher level of glucose and rutin in Cas9-harboring plants and disaccharides in Cas9-negative plants. However, this variation was neglected because of insignificant predictive ability and no significant differences observed using analysis of variance.

Metabolic pathways are interconnected and the use of genome editing to prevent activation of a gene for intended purposes could also disrupt the functioning of another important metabolic pathway. This can shift the metabolite production of unintended targets from its normal concentration. As plants interact with numerous environmental factors in a given ecosystem, unintended changes in the metabolic product could negatively impact a plant's ability to withstand changes caused by environmental stress conditions. A review article by Kawall (2021) highlighted the possible ecological risks resulting from unintended changes in the genome-edited crop plant *Camelina sativa*. The genome of this plant was edited by targeting a fatty acid desaturase gene (*OsFAD2*) in an attempt to decrease polyunsaturated fatty acids (PUFAs) and linoleic and linolenic acid to increase oleic acid content in the seeds to extend their shelf life. However, these PUFAs aid in maintaining cellular membrane fluidity, which is essential for the plant in regulating physiological stresses. *FAD2* mutants in a closely related species, *Arabidopsis thaliana*, were reported to have low survivability under cold and salt stress with impaired root formation and seed germination (Matos *et al.*, 2007; Zhang *et al.*, 2012). A possible alteration is the impairment of jasmonic acid biosynthesis, as linolenic acid is a metabolic precursor. Low levels of jasmonic acid in *fad2* mutants have been reported with increasing susceptibility to aphid insects in soybean (Kanobe *et al.*, 2015). Additionally, reduced accumulation of PUFAs in plants can negatively affect the whole food chain as animals are unable to synthesize PUFAs and thus depend on plants for their diet supply.

In conclusion, metabolomics can be used to ascertain whether any undesirable or unanticipated metabolite has accumulated in plants as a consequence of gene editing. This information will be critical before the release of gene-edited plant products to the market.

14.5 Techniques for Metabolic Profiling in CRISPR/Cas-edited Plants

Understanding the metabolic landscape of plants, especially those modified through CRISPR/Cas editing, is crucial because manipulating a gene affects related metabolic compounds. In plants, the metabolome plays a critical role in stress adaptation and defense responses, and affects various agronomic phenotypes. Generally, characterizing the metabolome involves major steps such as separating and identifying the metabolites. However, separating mixtures and detecting metabolites is often challenging due to the lack of standardized protocols and sample complexity. No single technique can analyze all metabolites within a metabolome, necessitating multiple technologies for comprehensive coverage (Mirnaghi and Caudy, 2014; Segers *et al.*, 2019; Collins *et al.*, 2021). Key analytical techniques include high-performance liquid chromatography (HPLC) and ultra-high-performance liquid chromatography (UHPLC), GC-MS, and LC-MS. After acquiring high-resolution metabolite profiles, identifying the eluted compounds is important. Mass spectrometry (MS), high-resolution mass spectrometry (HRMS), and NMR spectroscopy are mainly used for dereplication (identifying known compounds), while direct infusion mass spectrometry (DIMS) and Fourier-transform–ion cyclotron resonance–mass spectrometry (FT-ICR-MS) can be used without prior chromatographic separation. A generalized comparison of these analytical techniques for metabolomics profiling is presented in Table 14.1.

14.6 Chromatographic Techniques for Metabolic Profiling in Plants

14.6.1 High-performance liquid chromatography

HPLC works by separating a mixture of plant metabolites based on their polarity as they travel through a column. Imagine a race track where

Table 14.1. A generalized comparison of analytical techniques used for metabolomics profiling.

Technique	Sensitivity	Reproducibility	Resolution	Quantification	Sample volume for injection	Range of metabolites	Identification of metabolites
HPLC	Moderate	High	Moderate	High	10–100 µl	Broad, mainly non-volatile	Moderate, based on retention time and standards
UHPLC	High	High	High	High	1–10 µl	Broad, mainly non-volatile	Moderate, based on retention time and standards
LC-MS	Very high	High	Very high	High	1–10 µl	Broad, including polar and non-polar	High, based on accurate mass and fragmentation patterns
GC	Moderate	High	High	High	1–2 µl	Polar and non-polar but volatile	Moderate, based on retention time and standards
GC-MS	Very high	High	Very high	High	1–2 µl	Polar and non-polar but volatile	High, based on accurate mass and fragmentation patterns
HRMS	Very high	High	Very high	Very high	1–10 µl	Broad	Very high, based on very accurate mass and fragmentation patterns
NMR	Low	Very high	Low to moderate	Moderate	200–1000 µl	Limited, mainly small molecules but both polar and non-polar	Moderate, based on chemical shifts and coupling patterns
FT-ICR-MS	Very high	High	Very high	Very high	1–10 µl	Broad	Very high, based on very accurate mass and fragmentation patterns
DIMS	High	High	High	Moderate	1–10 µl	Broad	High, based on accurate mass and fragmentation patterns

DIMS, direct infusion mass spectrometry; FT-ICR-MS, Fourier-transform–ion cyclotron resonance–mass spectrometry; GC, gas chromatography; GC-MS, gas chromatography–mass spectrometry; HPLC, high-performance liquid chromatography; HRMS, high-resolution mass spectrometry; LC-MS, liquid chromatography–mass spectrometry; NMR, nuclear magnetic resonance; UHPLC, ultra-high-performance liquid chromatography.

each metabolite competes based on its interaction with the track (the stationary phase). Different metabolites interact with the stationary phase for varying lengths of time, resulting in distinct times to cross the finish line (retention times). Typically, reversed-phase chromatography on C18 material is employed, using a gradient of acetonitrile–water or methanol–water with 0.1% formic acid. Separation efficiency can be further enhanced by adding various modifiers or buffers to the mobile phase (Segers *et al.*, 2019). Detection generally involves methods such as ultraviolet light absorbance, which identifies metabolites based on their unique light absorption properties. While HPLC offers high sensitivity and resolution for specific metabolites, it can struggle to identify unknown compounds. Additionally, some metabolite classes might require derivatization, a process that modifies the molecule for better analysis (Wolfender, 2009). Since its introduction, HPLC has seen significant improvements, becoming more convenient, faster, and more applicable to a wider range of samples. The development of columns with reversed-phase chemistries has been a game changer, allowing separation of nearly any type of natural product (Halder *et al.*, 2024). For detailed information about the separated compounds, detectors such as photodiode array, MS, HRMS, or NMR can be combined with the chromatography methods.

14.6.2 Ultra-high-performance liquid chromatography

This powerful technique builds upon HPLC principles but offers significant advantages in terms of speed, resolution, and sensitivity. UHPLC achieves this by using even smaller particles in the separation column. These tiny particles allow much sharper separation but also generate higher backpressure (the resistance encountered by the mobile phase as it passes through the chromatographic system). Specialized UHPLC instruments handle these pressures, enabling faster analysis times, better peak capacity, and improved overall performance compared with conventional HPLC (Becker *et al.*, 2013; Nahar *et al.*, 2020; Collins *et al.*, 2021). UHPLC has become a popular choice for metabolite profiling, dereplication, fast fingerprinting, and metabolomics. By adjusting column length, UHPLC can theoretically achieve analysis times nine times faster than HPLC. Additionally, UHPLC can achieve high-resolution separations with longer columns, offering a powerful tool for metabolite profiling. UHPLC methods often rely on electrospray ionization (ESI) mass spectrometry for detection. This combination results in narrow peaks and high detection sensitivity (Becker *et al.*, 2013; Nahar *et al.*, 2020). When UHPLC is coupled with HRMS, the data generate detailed ion maps that are invaluable for various analyses in plant research. These ion maps allow precise localization and identification of metabolites, minimizing issues caused by overlapping peaks.

14.6.3 Liquid chromatography–mass spectrometry

LC-MS is another potent hyphenated method for plant metabolome research. It excels at analyzing a broad range of complex plant metabolites, including those with high molecular mass, heat-labile structures, or high vapor points. LC-MS is highly effective in profiling various metabolite classes, including secondary metabolites such as alkaloids and phenolics, lipids such as phospholipids, and sterols (Elshafie *et al.*, 2023). Depending on the specific metabolites of interest, LC-MS can be adapted with various ionization methods and separation columns. However, the requirement for standard reference compounds for identification and quantification of some metabolites limits the analysis of entirely new discoveries.

14.6.4 Gas chromatography

GC is a powerful separation technique used to separate and analyze compounds that can be vaporized without decomposition. However, nonvolatile constituents can be analyzed with GC following sample enrichment and derivatization steps. Derivatization methods such as alkylation, silylation, and acylation enhance the volatility and thermal stability of molecules, crucial for detailed metabolomics studies using GC (Beale *et al.*, 2018; Segers *et al.*, 2019). The physical components of a GC system include the injection port (inlet), column, oven, detector, and data system. The carrier gas, normally an inert gas such as helium or nitrogen, is vital for transporting the sample through the column, with selection and flow rates (usually in the range of 1–3 ml/min) impacting separation efficiency and analysis time (Beale *et al.*, 2018; Segers *et al.*, 2019; Turner *et al.*, 2019; Collins *et al.*, 2021). The stationary phase, coated inside the column, must be chosen based on the chemical properties of the target compounds to achieve optimal separation. Inlet types, such as split/splitless or on-column, and their corresponding flow rates are selected based on the sample's complexity and concentration. Sample size and injection technique are critical for maintaining the integrity of the analysis, with typical injection volumes ranging from 0.1 to 5 µl. Column selection involves choosing the appropriate length, diameter, and stationary-phase material to match the analytical needs. The column temperature and temperature program, often starting at a lower initial temperature and ramping up gradually, are meticulously controlled to optimize the separation of compounds based on their volatilities and affinities to the stationary phase. Together, these parameters must be precisely managed to achieve accurate and reproducible results in GC (Beale *et al.*, 2018; Turner *et al.*, 2019).

14.7 Non-chromatographic Identification Techniques for Plant Metabolites

Various spectroscopic techniques are utilized to identify metabolites from crude natural extracts, providing valuable spectroscopic fingerprints. Direct MS, NMR, and infrared (IR) approaches have proven successful in metabolomics studies. These methods facilitate the discrimination of complex metabolite mixtures through multivariate data analysis. Techniques such as MS, with tandem MS (MS/MS) spectra, and NMR (through two-dimensional NMR) offer crucial insights for the direct identification of metabolites in crude mixtures. NMR and direct MS analyses not only generate fingerprints but also provide structural information for compounds.

14.7.1 Mass spectrometry

MS is a powerful analytical technique that calculate the mass:charge (*m*/*z*) ratio of ions, offering detailed insights into the molecular mass and structure of compounds. The process begins with ion creation methods, which are crucial for converting sample molecules into ions. ESI and matrix-assisted laser desorption/ionization (MALDI) are common ionization methods. ESI is often used for large biomolecules, where a high voltage is applied to generate a liquid aerosol, while MALDI is suitable for analyzing large biomolecules and polymers by embedding the sample in a matrix and using a laser to ionize it (Segers *et al.*, 2019). After ionization, the ions are directed to the mass spectrometer for mass selection. This is done in the mass analyzer, which splits the ions depending on their *m*/*z* value. Depending on the intended resolution and application, several mass analyzer types are employed. Time-of-flight (TOF) analyzers calculate the time for ions to travel a fixed distance: light ions reach the detector earlier than heavier ones. Quadrupole analyzers use oscillating electric fields to filter ions by their *m*/*z* value, allowing only ions with a specific *m*/*z* value to pass through at any given time (Segers *et al.*, 2019). Ion trap analyzers capture ions in a three-dimensional electric field and sequentially eject them to the detector based on their *m*/*z* value. The separated ions are then detected and quantified by detectors, which convert the ion impacts into an electrical signal that can be measured. Common detectors include electron multipliers, which amplify the signal produced by ion impacts, and photomultiplier tubes, which are highly sensitive and capable of detecting low levels of ions (Collins *et al.*, 2021). To further enhance the analysis, separation techniques such as LC or GC are often coupled with MS. These techniques separate complex mixtures into individual components before they enter the mass spectrometer, improving the ability to analyze complex samples. LC-MS is particularly useful for analyzing non-volatile and thermally labile compounds, while GC-MS is ideal for volatile and semi-volatile compounds (Segers *et al.*, 2019; Collins *et al.*, 2021).

HRMS differs from traditional MS in several technical and theoretical aspects, making HRMS a more precise analytical tool (Collins *et al.*, 2021). Technically, HRMS uses advanced mass analyzers such as Orbitrap, TOF, and FT-ICR, which offer significantly higher resolving power and mass accuracy compared with the quadrupole or ion trap analyzers typically used in conventional

MS (Zarrouk *et al.*, 2022; Lai and Wang, 2023). Theoretically, HRMS can distinguish between ions with very small differences in their *m*/*z* ratios, enabling the resolution of isobars and isotopic variants that standard MS cannot separate. This higher resolution and accuracy stem from the ability of HRMS to measure mass with an accuracy of less than 1 ppm, compared with the 10–100 ppm accuracy typical of MS (Feng, 2022; Zarrouk *et al.*, 2022). Furthermore, HRMS provides detailed structural information through high-resolution fragmentation patterns in MS/MS experiments. This precision in mass determination and structural elucidation makes HRMS particularly valuable for identifying and quantifying complex mixtures and detecting low-abundance compounds that would be challenging or impossible to analyze using traditional MS.

14.7.2 Fourier-transform–ion cyclotron resonance–mass spectrometry

FT-ICR-MS is a cutting-edge technique renowned for its high-resolution and high-precision mass-analysis abilities. The working principle of FT-ICR-MS involves trapping ions in a magnetic field (Penning trap: a magnetic field with electric trapping plates) and detecting their cyclotron motion frequencies, which are directly proportional to their *m*/*z* ratios (Nikolaev *et al.*, 2019). The instrumentation includes a superconducting magnet to create a strong magnetic field, ion traps, or cells where ions are confined, and an electrode system that facilitates ion detection. The ion-trapping cells are crucial, with two common types being quadrupole and hexapole cells. Toggle instrumentation is used to switch between different ion storage and detection modes, enhancing the versatility of the system. Additionally, the stored waveform inverse Fourier transform (SWIFT) method is employed to manipulate ion trajectories and improve the signal:noise ratio, enabling the detection of ions with extremely low abundances. For sample preparation of plant metabolites, the process typically involves extracting metabolites from plant tissues using solvents such as methanol or ethanol, followed by filtration and concentration. The extract is then introduced into the FT-ICR-MS system, where further steps such as desalting, spiking with internal standards, and pH adjustment may be necessary to enhance the quality and reliability of the mass spectra. This approach ensures that even complex mixtures of plant metabolites can be analyzed effectively, providing detailed structural and compositional insights (Floris and O'Connor, 2019).

14.7.3 Nuclear magnetic resonance

NMR, unlike many other techniques, provides both qualitative and quantitative data from biological extracts without requiring chromatographic separation or sample derivatization. This method detects radio frequency electromagnetic radiation emitted by atomic nuclei with odd atomic numbers (e.g. ^{1}H) or odd mass numbers (e.g. ^{13}C) when placed in a strong magnetic field. One of the key advantages of NMR is its ease of sample preparation (Segers *et al.*, 2019a; Collins *et al.*, 2021). First, the plant material is ground and extracted with a suitable solvent such as methanol or chloroform. The extract is then concentrated and redissolved in a deuterated solvent such as deuterated methanol or deuterated chloroform for NMR analysis. After filtration, an internal standard might be added for quantification. Finally, the sample is transferred to an NMR tube and analyzed (Kohler *et al.*, 2016; Deborde *et al.*, 2019; Segers *et al.*, 2019). The technique is highly reproducible and non-destructive, allowing rapid analysis and making it particularly suitable for high-throughput studies. Additionally, NMR does not need specific sample preparation techniques, which simplifies the process and reduces potential sources of error. In typical ^{1}H-NMR acquisition, spectra can be obtained in just a few minutes, depending on the instrument's sensitivity and sample amount (Kohler *et al.*, 2016; Segers *et al.*, 2019).

NMR's versatility is further demonstrated by its ability to provide crucial structural information for compound identification through the combination of chemical shifts, spin–spin coupling, and peak intensity. The proton signal intensity in NMR is directly proportional to the molar concentration of metabolites, allowing direct quantitative analysis. Furthermore, advanced two-dimensional NMR techniques such as heteronuclear single quantum correlation, heteronuclear multiple bond correlation, total correlation spectroscopy, and nuclear Overhauser effect spectroscopy offer more detailed analysis by reducing signal overlap and providing insights into atom connectivities within complex mixtures (Pelantová *et al.*, 2014; Kohler *et al.*, 2016). High-resolution magic angle spinning NMR is particularly useful for analyzing solid samples without chemical extraction (Pelantová *et al.*, 2014).

The general approach for identifying natural products or annotating LC peaks involves database searches where MS, HRMS, MS/MS, or NMR spectra are compared against extensive databases. For NMR, these searches are constrained by solvent composition, as chemical shifts can vary significantly between different solvents. Despite the existence of large structural databases containing over 200,000 natural compounds, obtaining strong evidence for annotation remains complex (Pelantová *et al.*, 2014; Kohler *et al.*, 2016; Wolfender *et al.*, 2019). Consequently, alternative strategies are needed to identify unknown metabolites. NMR is less sensitive than MS and is pH sensitive, necessitating the use of buffered solutions (Kohler *et al.*, 2016; Segers *et al.*, 2019). Effective solvents for comprehensive metabolite profiling include methanol, aqueous phosphate buffer, and ionic liquids such as 1-butyl-3-methylimidazolium chloride. Despite its lower sensitivity, the method's reproducibility of chemical shifts under specific solvent or buffer conditions allows unambiguous metabolite identification when compared with database entries.

14.7.4 Direct infusion mass spectrometry

DIMS is another powerful analytical technique for characterizing the composition of mixed samples. This method involves the direct infusion of biological samples into the mass spectrometer without prior chromatographic separation, significantly increasing throughput compared with LC-MS (Koulman *et al.*, 2007). For optimal results, DIMS is typically performed on HRMS instruments, which include TOF, Orbitrap, and FT-ICR. These instruments enable feature deconvolution and metabolite annotation in a single run through precise molecular formula determination. TOF provides a resolution of 50,000 and accuracy of 1 ppm, Orbitrap offers a resolution of 100,000 and accuracy of less than 1 ppm, and FT-ICR delivers ultra-high resolution of over 500,000 and accuracy of less than 1 ppm (Wolfender *et al.*, 2015; Donato *et al.*, 2020). Combining HRMS fingerprints with ^{13}C isotope labeling has improved the detection and assignment of true metabolites. Additionally, DIMS is compatible with atmospheric pressure ionization sources, such as ESI or atmospheric pressure chemical ionization, and samples can be infused using syringe pumps, short LC runs, or nanoinfusion devices such as the chip-based Nanomate™. One of the main challenges in DIMS is ionization suppression, where competitive ionization from other matrix components can obscure signals from minor compounds. Solutions include dilution assays to mitigate ion suppression and periodically infusing quality-control samples representative of the biological matrix to measure and correct variations within and between batches (Wolfender *et al.*, 2015; Donato *et al.*, 2020).

DIMS is used primarily for recording metabolic fingerprints for differential studies rather than comprehensive compound identification. However, HRMS/MS experiments provide additional structural information, making DIMS suitable for various applications, including quality control, authentication, and metabolite identification. For example, multivariate data analysis of DIMS fingerprints has been used to classify wine by grape variety and ageing process, certify the origin and detect adulteration in complex natural products, and identify biomarkers in grapevine leaves infested with downy mildew (Becker *et al.*, 2013).

14.8 Assessing Gene Editing in Plants by a Combined Approach of Sequencing and Metabolite Characterization

Gene editing by CRISPR/Cas is achieved in plants mainly for agricultural trait improvement and gene-function studies. Successful editing of the DNA sequence is confirmed by sequencing the genomic region targeted by the gene-editing complex (Cas and gRNA). In plant gene-editing experiments, the gene selected for editing (by knockout or modification by base editing) may participate in metabolic pathways at the protein or metabolite level, or may be involved in protein–protein interactions or cross talk between metabolites, thus demonstrating the functions

Table 14.2. Examples of CRISPR/Cas gene editing where metabolomics along with sequencing has been used for the characterization of target traits in gene-edited plants.

Plant	Trait(s)	Target gene/locus	Metabolomics	Reference
Soybean	Improvement in isoflavone content	*GmF3H1*, *GmF3H2* and *GmFNSII-1*	LC-ESI-MS/MS to characterize isoflavone in gene-edited plants	Zhang *et al.* (2020)
Tomato	Lycopene improvement	*SGR1*, *LCY-E*, *Blc*, *LCY-B1*, and *LCY-B2*	RT-HPLC and HPLC-MS were employed for quantification of lycopene in gene-edited plants	Li *et al.* (2018)
Rice	Improvement of aroma	*BADH2*	Untargeted HS-SPME-GC-MS analysis in CRISPR/Cas-edited and wild-type seeds of rice was performed to validate the formation of 2-acetyl-1-pyrroline in edited rice seed	Hui *et al.* (2022)
Brassica napus L.	Improvement of phytic acid quality	*BnITPK*	HPLC was used to analyze phytic acid quality following CRISPR/Cas9-mediated editing of the inositol tetrakisphosphate kinase (*BnITPK*) gene	Sashidhar *et al.* (2020)
Solanum melongena L.	Reduction of flesh browning	Polyphenol oxidases (PPOs)	PPO activity in gene-edited lines was determined colorimetrically to confirm a reduction of targeted protein activity	Maioli *et al.* (2020)
Rice	Enhancement of rice grain quality	*OsAAP11*	CRISPR/Cas9-edited mutants of *OsAAP11* homozygous lines were used to analyze the protein, amino acid, and starch content of seeds	Yang *et al.* (2023)
Tomato	Improvement of γ-aminobutyric acid (GABA)	Glutamate decarboxylase	CRISPR/Cas9-mediated mutation in the SIGAD2 and SIGAD3 proteins improved GABA content, which was determined by a biochemical method	Nonaka *et al.* (2017)
Rice	Cold response	*OsHPL1*	LC-MS/MS metabolomics was performed in wild-type and CRISPR/Cas9-mediated *OsHPL1* mutant plants to investigate its role in cold stress	Wu *et al.* (2023)
Potato	Lowering the content of acrylamide and reducing browning	Vacuolar invertase and asparagine synthetase 1	Biochemical methods were used to estimate the content of glucose, sucrose, and fructose to reveal the final effect of gene editing	Ly *et al.* (2023)

CRISPR, clustered regularly interspaced short palindromic repeats; HPLC-MS, high-performance liquid chromatography–mass spectrometry; HS-SPME-GC-MS, headspace solid-phase microextraction–gas chromatography–mass spectrometry; LC-ESI-MS/MS, liquid chromatography–electrospray ionization–tandem mass spectrometry; LC-MS/MS, liquid chromatography–tandem mass spectrometry; RT-HPLC, reversed-phase high-performance liquid chromatography.

of individual metabolites in multiple pathways. It is also important to bear in mind that knockout of a gene might not completely abolish the final metabolic pathway due to the presence of paralogs or isoforms (Kawall, 2021). This necessitates checking metabolite alterations in CRISPR/Cas-edited plants by adopting comprehensive metabolomics or targeted metabolite estimations after successful editing confirmation by sequencing. Such a combined approach will increase the clarity of gene-editing events and their related function. There is also scope for the metabolite profile of gene-edited plants to increase their authenticity and allow risk assessment before commercialization. Fraser *et al.* (2020) discussed the challenges and opportunities of gene editing in plants and suggested adding metabolomics techniques for characterization of CRISPR/Cas-edited plants. Table 14.2 summarizes CRISPR/Cas gene-editing experiments where metabolomics was used as a method to annotate gene-edited plants, while a generalized pipeline involving metabolomic characterization of CRISPR/Cas-edited plant is summarized in Fig. 14.1.

14.9 Conclusion and Future Directions

Plant genetic improvement using CRISPR/Cas gene-editing technology is implemented worldwide. Genetic enhancement of crop plants for better agricultural traits is essentially for global food security. The world population is rising rapidly and is projected to reach approximately 10 billion by 2050, demanding the need to enhance agricultural yield. The situation is especially worrying due to the impact of climate change on crop productivity. It is predicted that major food crops including staple cereals and legumes may suffer yield losses as a result of biotic and abiotic stress due to altered environmental conditions as the outcome of climate change. CRISPR/Cas gene editing in plants can help improve disease tolerance, stress resistance, and nutritional and yield enhancement, as well as be used for functional studies of genes, and is possible in all major crop plants. Unlike conventional genetically modified plants, a CRISPR/Cas gene-edited plant can be made transgene-free by genetic segregation, which is important for biosafety, regulation, commercialization, and public acceptance.

To characterize gene-editing events in plants Sanger sequencing is often the chosen method whereas confirmation of foreign gene (Cas) elimination due to segregation is achieved by PCR. The consequential effects of plant gene editing are studied by implementing specific methods; for example, drought stress tolerance can be assessed by performing drought stress assays on gene-edited plants. Similarly, disease resistance can be checked by infection with pathogens. In addition to these methods, transcriptomics is a significant addition to

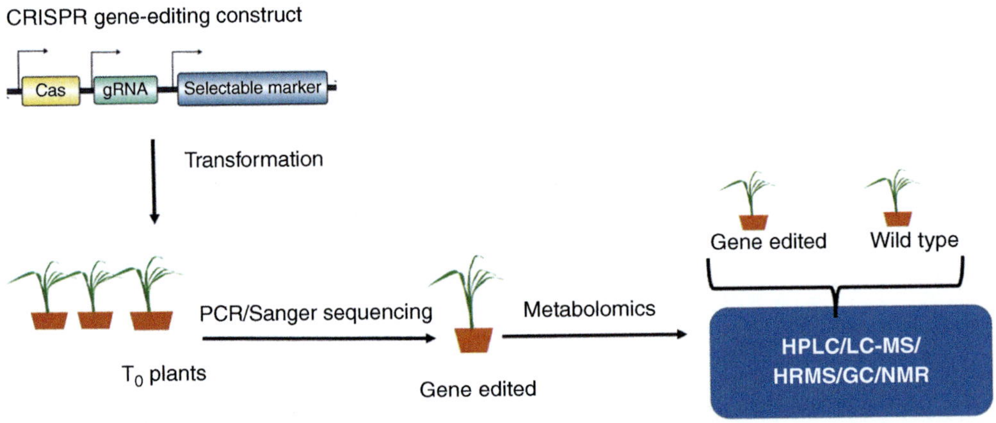

Fig. 14.1. Pipeline for high-throughput metabolomics for assessing clustered regularly interspaced short palindromic repeats (CRISPR)-edited plants. GC, gas chromatography; gRNA, guide RNA; HPLC, high-performance liquid chromatography; HRMS, high-resolution mass spectrometry; LC-MS, liquid chromatography–mass spectrometry; NMR, nuclear magnetic resonance.

the characterization methods preferred for gene-edited plants. Comprehensive transcriptome profiling on gene-edited plants could play a critical role because metabolites are frequently linked with phenomena such as stress adaptation or the taste of fruits. Moreover, in plants, often a single metabolite may participate in more than one pathway, thus affecting directly or indirectly other metabolites and their related processes. The metabolic profiling of wild-type and gene-edited plants could unveil whether any alteration has occurred due to metabolic crosstalk, as well as any off-target effects of gene editing at the metabolic level. This will help in the selection of a suitable gene or gene-edited plant to take forward for commercialization.

CRISPR/Cas gene editing in plants and validation involving target DNA sequencing, experimental evaluation of the effect of gene editing, and finally comprehensive metabolite profiling will establish a more specific pipeline for gene editing. It could be argued that removing or modulating the concentration of a particular metabolite may result in physiological changes in plants. These challenges could be overcome by metabolite estimation, which also benefits validating a crop plant's suitability for consumption and commercialization. It is also important to mention that gene editing in medicinal and aromatic plants for improving particular metabolites is crucial, and in such experiments, performing metabolomics is very important to link gene editing with the final effect. In this chapter, we discussed the importance and benefits of incorporating metabolomics in CRISPR/Cas gene editing in plants.

14.10 References

Al-Khayri, J.M., Rashmi, R., Toppo, V., Chole, P.B., Banadka, A., *et al.* (2023) Plant secondary metabolites: the weapons for biotic stress management. *Metabolites* 13: 716.

Altpeter, F., Springer, N.M., Bartley, L.E., Blechl, A., Brutnell, T.P., *et al.* (2016) Advancing crop transformation in the era of genome editing. *Plant Cell* 28, 1510–1520.

Aoki, K., Yamasaki, M., Umezono, R., Hamamoto, T., and Kamachi, Y. (2024) Systematic comparison of computational tools for Sanger sequencing-based genome editing analysis. *Cells* 13: 261.

Ashokkumar, S., Jaganathan, D., Ramanathan, V., Rahman, H., Palaniswamy, R., *et al.* (2020) Creation of novel alleles of fragrance gene *OsBADH2* in rice through CRISPR/Cas9 mediated gene editing. *PLOS ONE* 15: e0237018.

Atkins, A., Chung, C.-H., Allen, A.G., Dampier, W., Gurrola, T.E., *et al.* (2021) Off-target analysis in gene editing and applications for clinical translation of CRISPR/Cas9 in HIV-1 therapy. *Frontiers in Genome Editing* 3: 673022.

Beale, D.J., Pinu, F.R., Kouremenos, K.A., Poojary, M.M., Narayana, V.K., *et al.* (2018) Review of recent developments in GC–MS approaches to metabolomics-based research. *Metabolomics* 14: 152.

Becker, L., Poutaraud, A., Hamm, G., Muller, J.-F., Merdinoglu, D., *et al.* (2013) Metabolic study of grapevine leaves infected by downy mildew using negative ion electrospray–Fourier transform ion cyclotron resonance mass spectrometry. *Analytica Chimica Acta* 795, 44–51.

Cardi, T., Murovec, J., Bakhsh, A., Boniecka, J., Bruegmann, T., *et al.* (2023) CRISPR/Cas-mediated plant genome editing: outstanding challenges a decade after implementation. *Trends in Plant Science* 28, 1144–1165.

Collins, S.L., Koo, I., Peters, J.M., Smith, P.B., and Patterson, A.D. (2021) Current challenges and recent developments in mass spectrometry-based metabolomics. *Annual Review of Analytical Chemistry* 14, 467–487.

Cui, Y., Xu, J., Cheng, M., Liao, X., and Peng, S. (2018) Review of CRISPR/Cas9 sgRNA design tools. *Interdisciplinary Sciences* 10, 455–465.

Deborde, C., Fontaine, J.-X., Jacob, D., Botana, A., Nicaise, V., *et al.* (2019) Optimizing 1D ^{1}H-NMR profiling of plant samples for high throughput analysis: extract preparation, standardization, automation and spectra processing. *Metabolomics* 15: 28.

Donato, P., Arena, P., and Mondello, L. (2020) Theoretical and practical aspects of LC-MS analysis. In: Tranchida, P.Q. and Mondello, L. (eds) *Hyphenations of Capillary Chromatography with Mass Spectrometry*. Elsevier, Amsterdam, the Netherlands, pp. 297–317.

Drapal, M., Enfissi, E.M.A., Almeida, J., Rapacz, E., Nogueira, M., and Fraser, P.D. (2023) The potential of metabolomics in assessing global compositional changes resulting from the application of CRISPR/Cas9 technologies. *Transgenic Research* 32, 265–278.

Elshafie, H.S., Camele, I., and Mohamed, A.A. (2023) A comprehensive review on the biological, agricultural and pharmaceutical properties of secondary metabolites based-plant origin. *International Journal of Molecular Sciences* 24: 3266.

Feng, Y.-L. (2022) Chromatography high-resolution mass spectrometry in food and environmental chemistry. In: Picó, Y. and Campo, J. (eds) *Mass Spectrometry in Food and Environmental Chemistry. The Handbook of Environmental Chemistry, Vol. 119*. Springer, Cham, Switzerland, pp. 149–185.

Floris, F. and O'Connor, P.B. (2019) Fundamentals of two dimensional Fourier transform mass spectrometry. In: Kanawati, B. and Schmitt-Kopplin, P. (eds) *Fundamentals and Applications of Fourier Transform Mass Spectrometry*. Elsevier, Amsterdam, the Netherlands, pp. 187–232.

Fraser, P.D., Aharoni, A., Hall, R.D., Huang, S., Giovannoni, J.J., *et al.* (2020) Metabolomics should be deployed in the identification and characterization of gene-edited crops. *Plant Journal* 102, 897–902.

Halder, M., Kundu, A., and Jha, S. (2024) Secondary metabolites identification techniques of the current era. In: Mérillon, J.M., and Ramawat, K.G. (eds) *Plant Specialized Metabolites.* Springer, Cham, Switzerland, pp. 1–41.

Hillary, V.E. and Ceasar, S.A. (2023) A review on the mechanism and applications of CRISPR/Cas9/Cas12/Cas13/Cas14 proteins utilized for genome engineering. *Molecular Biotechnology* 65, 311–325.

Huang, X.-Q. and Dudareva, N. (2023) Plant specialized metabolism. *Current Biology* 33, R473–R478.

Hui, S., Li, H., Mawia, A.M., Zhou, L., Cai, J., *et al.* (2022) Production of aromatic three-line hybrid rice using novel alleles of *BADH2*. *Plant Biotechnology Journal* 20, 59–74.

Jinek, M., Chylinski, K., Fonfara, I., Hauer, M., Doudna, J.A., and Charpentier, E. (2012) A programmable dual-RNA-guided DNA endonuclease in adaptive bacterial immunity. *Science* 337, 816–821.

Jogawat, A., Yadav, B., Chhaya, Lakra, N., Singh, A.K., and Narayan, O.P. (2021) Crosstalk between phytohormones and secondary metabolites in the drought stress tolerance of crop plants: a review. *Physiologia Plantarum* 172, 1106–1132.

Kanobe, C., McCarville, M.T., O'Neal, M.E., Tylka, G.L., and MacIntosh, G.C. (2015) Soybean aphid infestation induces changes in fatty acid metabolism in soybean. *PLOS ONE* 10: e0145660.

Kaur, N., Alok, A., Shivani, Kaur, N., Pandey, P., *et al.* (2018) CRISPR/Cas9-mediated efficient editing in phytoene desaturase (*PDS*) demonstrates precise manipulation in banana cv. Rasthali genome. *Functional & Integrative Genomics* 18, 89–99.

Kawall, K. (2021) Genome-edited *Camelina sativa* with a unique fatty acid content and its potential impact on ecosystems. *Environmental Sciences Europe* 33: 38.

Kohler, I., Verhoeven, A., Derks, R.J., and Giera, M. (2016) Analytical pitfalls and challenges in clinical metabolomics. *Bioanalysis* 8, 1509–1532.

Komatsu, H., Abdellatif, I.M.Y., Yuan, S., Ono, M., Nonaka, S., *et al.* (2020) Genome editing in *PDS* genes of tomatoes by non-selection method and of *Nicotiana benthamiana* by one single guide RNA to edit two orthologs. *Plant Biotechnology* 37, 213–221.

Koulman, A., Tapper, B.A., Fraser, K., Cao, M., Lane, G.A., and Rasmussen, S. (2007) High-throughput direct-infusion ion trap mass spectrometry: a new method for metabolomics. *Rapid Communications in Mass Spectrometry* 21, 421–428.

Lai, Y. and Wang, Y. (2023) Advances in high-resolution mass spectrometry techniques for analysis of high mass-to-charge ions. *Mass Spectrometry Reviews* 42, 2426–2445.

Li, X., Wang, Y., Chen, S., Tian, H., Fu, D., *et al.* (2018) Lycopene is enriched in tomato fruit by CRISPR/Cas9-mediated multiplex genome editing. *Frontiers in Plant Science* 9: 00559.

Li, Y., Zhu, J., Wu, H., Liu, C., Huang, C., *et al.* (2020) Precise base editing of non-allelic acetolactate synthase genes confers sulfonylurea herbicide resistance in maize. *Crop Journal* 8, 449–456.

Ly, D.N.P., Iqbal, S., Fosu-Nyarko, J., Milroy, S., and Jones, M.G.K. (2023) Multiplex CRISPR-Cas9 gene-editing can deliver potato cultivars with reduced browning and acrylamide. *Plants* 12: 379.

Maioli, A., Gianoglio, S., Moglia, A., Acquadro, A., Valentino, D., *et al.* (2020) Simultaneous CRISPR/Cas9 editing of three *PPO* genes reduces fruit flesh browning in *Solanum melongena* L. *Frontiers in Plant Science* 11: 607161.

Matos, A.R., Hourton-Cabassa, C., Ciçek, D., Rezé, N., Arrabaça, J.D., *et al.* (2007) Alternative oxidase involvement in cold stress response of *Arabidopsis thaliana fad2* and *FAD3*+ cell suspensions altered in membrane lipid composition. *Plant and Cell Physiology* 48, 856–865.

Mirnaghi, F.S. and Caudy, A.A. (2014) Challenges of analyzing different classes of metabolites by a single analytical method. *Bioanalysis* 6, 3393–3416.

Miyaoka, Y., Berman, J.R., Cooper, S.B., Mayerl, S.J., Chan, A.H., *et al.* (2016) Systematic quantification of HDR and NHEJ reveals effects of locus, nuclease, and cell type on genome-editing. *Scientific Reports* 6: 23549.

Nahar, L., Onder, A., and Sarker, S.D. (2020) A review on the recent advances in HPLC, UHPLC and UPLC analyses of naturally occurring cannabinoids (2010–2019). *Phytochemical Analysis* 31, 413–457.
Nikolaev, E.N., Kostyukevich, Y.I., and Vladimirov, G. (2019) Fundamentals and simulations in FT-ICR-MS. In: Kanawati, B. and Schmitt-Kopplin, P. (eds) *Fundamentals and Applications of Fourier Transform Mass Spectrometry*. Elsevier, Amsterdam, the Netherlands, pp. 89–111.
Nonaka, S., Arai, C., Takayama, M., Matsukura, C., and Ezura, H. (2017) Efficient increase of γ-aminobutyric acid (GABA) content in tomato fruits by targeted mutagenesis. *Scientific Reports* 7: 7057.
Pagare, S., Bhatia, M., Tripathi, N., Pagare, S., and Bansal, Y.K. (2015) Secondary metabolites of plants and their role: overview. *Current Trends in Biotechnology and Pharmacy* 9, 293–304.
Pelantová, H., Bártová, S., and Kuzma, M. (2014) Advanced techniques for NMR analysis of complex biological mixtures—from simple NMR to hyphenated techniques. In: Havlíček, V. and Spížek, J. (eds) *Natural Products Analysis: Instrumentation, Methods, and Applications*. Wiley, Hoboken, New Jersey, pp. 239–284.
Rojas, C.M., Senthil-Kumar, M., Tzin, V., and Mysore, K.S. (2014) Regulation of primary plant metabolism during plant–pathogen interactions and its contribution to plant defense. *Frontiers in Plant Science* 5: 17.
Sashidhar, N., Harloff, H.J., Potgieter, L., and Jung, C. (2020) Gene editing of three *BnITPK* genes in tetraploid oilseed rape leads to significant reduction of phytic acid in seeds. *Plant Biotechnology Journal* 18, 2241–2250.
Segers, K., Declerck, S., Mangelings, D., Vander Heyden, Y. and van Eeckhaut, A. (2019) Analytical techniques for metabolomic studies: a review. *Bioanalysis* 11, 2297–2318.
Shillito, R.D., Whitt, S., Ross, M., Ghavami, F., de Vleesschauwer, D., *et al.* (2021) Detection of genome edits in plants—from editing to seed. *In Vitro Cellular & Developmental Biology – Plant* 57, 595–608.
Tang, X., Lowder, L.G., Zhang, T., Malzahn, A.A., Zheng, X., *et al.* (2017) A CRISPR-Cpf1 system for efficient genome editing and transcriptional repression in plants. *Nature Plants* 3: 17018.
Tang, X., Liu, G., Zhou, J., Ren, Q., You, Q., *et al.* (2018) A large-scale whole-genome sequencing analysis reveals highly specific genome editing by both Cas9 and Cpf1 (Cas12a) nucleases in rice. *Genome Biology* 19: 84.
Turner, D.C., Schäfer, M., Lancaster, S., Janmohamed, I., Gachanja, A., and Creasey, J. (2019) *Gas Chromatography-Mass Spectrometry: How Do I Get the Best Results?* Royal Society of Chemistry, London.
Verma, V., Kumar, A., Partap, M., Thakur, M., and Bhargava, B. (2023) CRISPR-Cas: a robust technology for enhancing consumer-preferred commercial traits in crops. *Frontiers in Plant Science* 14: 1122940.
Wang, J.Y. and Doudna, J.A. (2023) CRISPR technology: a decade of genome editing is only the beginning. *Science* 379: eadd8643.
Wang, X., Tu, M., Wang, Y., Yin, W., Zhang, Y., *et al.* (2021) Whole-genome sequencing reveals rare off-target mutations in CRISPR/Cas9-edited grapevine. *Horticulture Research* 8: 114.
Wolfender, J.-L. (2009) HPLC in natural product analysis: the detection issue. *Planta Medica* 75, 719–734.
Wolfender, J.-L., Marti, G., Thomas, A., and Bertrand, S. (2015) Current approaches and challenges for the metabolite profiling of complex natural extracts. *Journal of Chromatography A* 1382, 136–164.
Wolfender, J.-L., Nuzillard, J.-M., van der Hooft, J.J.J., Renault, J.-H., and Bertrand, S. (2019) Accelerating metabolite identification in natural product research: toward an ideal combination of liquid chromatography–high-resolution tandem mass spectrometry and NMR profiling, *in silico* databases, and chemometrics. *Analytical Chemistry* 91, 704–742.
Wu, Z., Guo, Z., Wang, K., Wang, R., and Fang, C. (2023) Comparative metabolomic analysis reveals the role of *OsHPL1* in the cold-induced metabolic changes in rice. *Plants (Basel)* 12: 2032.
Yang, Y., Zhang, Y., Sun, Z., Shen, Z., Li, Y., *et al.* (2023) Knocking out *OsAAP11* to improve rice grain quality using CRISPR/Cas9 system. *International Journal of Molecular Sciences* 24: 14360.
Zarrouk, E., Lenski, M., Bruno, C., Thibert, V., Contreras, P., *et al.* (2022) High-resolution mass spectrometry: theoretical and technological aspects. *Toxicologie Analytique et Clinique* 34, 3–18.
Zaynab, M., Fatima, M., Sharif, Y., Zafar, M.H., Ali, H., and Khan, K.A. (2019) Role of primary metabolites in plant defense against pathogens. *Microbial Pathogenesis* 137: 103728.
Zhang, J., Liu, H., Sun, J., Li, B., Zhu, Q., *et al.* (2012) *Arabidopsis* fatty acid desaturase FAD2 is required for salt tolerance during seed germination and early seedling growth. *PLOS ONE* 7: e30355.
Zhang, P., Du, H., Wang, J., Pu, Y., Yang, C., *et al.* (2020) Multiplex CRISPR/Cas9-mediated metabolic engineering increases soya bean isoflavone content and resistance to soya bean mosaic virus. *Plant Biotechnology Journal* 18, 1384–1395.
Zischewski, J., Fischer, R., and Bortesi, L. (2017) Detection of on-target and off-target mutations generated by CRISPR/Cas9 and other sequence-specific nucleases. *Biotechnology Advances* 35, 95–104.

15 Metabolomics-assisted Breeding: A New Omics-integrated Avenue for Improved Stress Response in Plants

Rajib Roychowdhury[1,2*], Archana Prasad[3], Soumya Prakash Das[4], Priya Shah[2], Ajay Kumar[5], Umakanta Sarker[6], Sunil S. Gangurde[2], and Manish K. Pandey[2*]

[1]*Agricultural Research Organization (ARO) – The Volcani Institute, Rishon Lezion, Israel;* [2]*Center of Excellence in Genomics & Systems Biology (CEGSB), and Center for Pre-breeding Research (CPBR), International Crops Research Institute for the Semi-Arid Tropics (ICRISAT), Hyderabad, Telangana, India;* [3]*Grape Physiology and Breeding Lab, Department of Fruit Tree Sciences, Institute of Plant Sciences, Agricultural Research Organization (ARO) – Volcani Institute, Rishon Lezion, Israel;* [4]*School of Life Sciences, Seacom Skills University, Bolpur, West Bengal, India;* [5]*Amity Institute of Biotechnology, Amity University, Noida, India;* [6]*Department of Genetics and Plant Breeding, Faculty of Agriculture, Gazipur Agricultural University, Gazipur, Bangladesh*

Abstract

In the face of increasing environmental challenges such as climate change, enhancing plant resilience to stress has become a critical goal in agriculture. Metabolomics-assisted breeding (MetAB) has emerged as a promising approach that integrates metabolomics with traditional breeding strategies to develop stress-tolerant plant varieties. This chapter explores the principles, methodologies, and potential applications of MetAB in improving plant stress responses. Metabolomics, a powerful tool in systems biology, enables comprehensive analysis of metabolites present in plant tissues. By profiling the metabolome under stress conditions, MetAB provides valuable insights into the biochemical pathways and metabolic networks involved in stress tolerance mechanisms. Through the identification of stress-responsive metabolites and metabolic signatures associated with stress adaptation, MetAB facilitates the selection of superior plant genotypes with enhanced stress resilience. Key components of MetAB include metabolite profiling, biomarker discovery, and integration with other omics tools. Metabolite profiling techniques such as mass spectrometry and nuclear magnetic resonance spectroscopy allow the high-throughput analysis of diverse metabolites, while advanced statistical and bioinformatics tools aid in biomarker identification and selection. Integration of metabolomics data with genomic information enables the identification of candidate genes and quantitative trait loci underlying stress tolerance traits, facilitating targeted breeding efforts. The application of MetAB holds immense potential for crop-improvement programs aimed at developing stress-tolerant varieties with increased yield stability and resilience to biotic and abiotic stressors. By harnessing the synergistic effects of metabolomics and breeding, MetAB offers a novel approach to accelerate the development of climate-resilient crops, thereby contributing to global food security and sustainable agriculture.

Keywords: abiotic stress tolerance, crop improvement, disease resistance, metabolomics, omics

*E-mail: rajibroychowdhury86@yahoo.com or manish.pandey@icrisat.org

© CAB International 2025. *High-Throughput Plant Metabolomics* (Jen-Tsung Chen ed.)
DOI: 10.1079/9781800626911.0015

15.1 Introduction

In the face of escalating environmental challenges such as climate change, soil degradation, and emerging plant pathogens, ensuring global food security has become a paramount concern for agricultural scientists and policy makers worldwide. Agriculture, the backbone of human civilization, is intricately tied to the ability of crops to withstand and adapt to various stressors, including drought, salinity, heat, and pests. However, traditional breeding methods, while successful in enhancing crop productivity over the past century, are often limited by the lengthy breeding cycles and the complexity of stress tolerance traits. In recent years, the advent of omics technologies has revolutionized our understanding of plant biology and opened new avenues for crop improvement (Varshney *et al.*, 2014). Among these, metabolomics, the comprehensive study of small-molecule metabolites within biological systems, has emerged as a powerful tool for dissecting the complex metabolic networks underlying plant stress responses (Roychowdhury *et al.*, 2023a). Metabolomics offers a holistic view of plant metabolism by simultaneously profiling a wide range of metabolites, including sugars, amino acids, organic acids, secondary metabolites, and phytohormones. The dynamic nature of the metabolome makes it a sensitive readout of plant physiological status, reflecting both genetic and environmental influences. In the context of plant stress responses, metabolomics provides valuable insights into the biochemical pathways and metabolic adjustments that plants undergo to cope with adverse conditions (Wu *et al.*, 2023). By identifying stress-responsive metabolites and metabolic signatures associated with stress adaptation, metabolomics holds promise for elucidating the molecular mechanisms underlying stress tolerance traits in plants. In parallel with the advancements in metabolomics, breeding strategies have also evolved to incorporate molecular and genomic approaches for accelerating crop improvement. Marker-assisted selection, a technique that utilizes molecular markers linked to desired traits for efficient selection of superior genotypes, has revolutionized traditional breeding programs (Roychowdhury *et al.*, 2020, 2023b). However, the identification of reliable molecular markers for complex traits such as stress tolerance remains a challenge due to the polygenic nature and environmental sensitivity of these traits. Integrating metabolomics with breeding approaches offers a novel solution to this challenge by providing a direct link between genotype and phenotype through the analysis of metabolic phenotypes (Anzano *et al.*, 2022). Metabolomics-assisted breeding (MetAB) represents an innovative approach that leverages metabolomics data to enhance the efficiency and precision of breeding for stress tolerance in plants. By integrating metabolomics with genomics, phenomics, and traditional breeding methods, MetAB aims to expedite the development of stress-tolerant crop varieties with improved yield, quality, and resilience to environmental fluctuations. In this chapter, we provide an overview of the principles, methodologies, and applications of MetAB in plant breeding, with a focus on its potential to revolutionize the field of stress resilience breeding. The foundation of MetAB lies in the comprehensive analysis of plant metabolomes under stress conditions to identify metabolic signatures associated with stress tolerance. Metabolomics technologies, including mass spectrometry (MS), nuclear magnetic resonance (NMR) spectroscopy, and chromatography, enable the high-throughput profiling of metabolites in plant tissues. These techniques offer unparalleled sensitivity and resolution, allowing the detection and quantification of hundreds to thousands of metabolites in a single experiment (Salam *et al.*, 2023). By comparing the metabolite profiles of stress-tolerant and susceptible genotypes, MetAB facilitates the identification of stress-responsive metabolites and metabolic pathways that are crucial for stress adaptation.

15.2 Plant Metabolites in Response to Abiotic and Biotic Stresses

15.2.1 Plant metabolites in response to abiotic stresses

Glycine betaine acts as an osmoprotectant and mitigates abiotic stress-mediated oxidative stress (Hasanuzzaman *et al.*, 2015). Accumulation of soluble sugars and sugar alcohols was reported with abiotic stresses such as drought, temperature, and salinity stress. Sugar molecules act as signaling molecules, energy reservoirs, and structural and storage components during adaptation to adverse environmental conditions (Salam *et al.*, 2023). Among polyamines, putrescine, spermidine, and spermine enhanced abiotic stress tolerance in plants. Both exogenous application of polyamines and transgenic plants with altered polyamine biosynthesis showed increased physiological responses against adverse environmental conditions. Abiotic stress-induced cellular responses lead to the generation of various reactive oxygen species (ROS), such as superoxide radicals (O_2^-), hydrogen peroxide (H_2O_2) and hydroxyl radicals (•OH) (Sarker and Oba, 2018a, 2020a). In plants, both enzymatic and non-enzymatic antioxidants were used to combat ROS-induced oxidative stress (Xu and Fu, 2022). Among non-enzymatic antioxidants, ascorbic acid and glutathione were reported to play a significant role. Glutathione detoxifies ROS, methylglyoxal, toxic metals, metalloids, and xenobiotics, while ascorbic acid is associated with the activation of various antioxidant enzymes. Tocopherols or vitamin E are liposoluble antioxidants and effectively scavenge lipid peroxides, oxygen radicals, or singlet oxygen in higher plants. Among enzymatic antioxidants, different isoforms of superoxide dismutase quench ROS by catalyzing the dismutation of superoxide into molecular H_2O_2 and oxygen (O_2) (Sarker and Oba, 2018b). Similarly, another key antioxidant enzyme, catalase, efficiently reduces ROS by converting H_2O_2 into water (H_2O) and O_2. Among phenolic compounds, flavonoids have high antioxidant activity and participate in H_2O_2 quenching. Unfavorable environmental conditions led to elevated concentrations of phenolic compounds and supported defensive effects (Šamec *et al.*, 2021).

Secondary metabolites (colorant pigments), such as betalains (Sarker and Oba, 2019a) and polyphenolic compounds (Sarker *et al.*, 2018) play an important role in the salinity tolerance of the vegetable amaranth. Betalains are synthesized from the amino acid tyrosine through a pathway involving a number of enzymatic reactions (Sarker *et al.*, 2022a). Plant polyphenols can act as natural antioxidants (Sarker *et al.*, 2022b,c, 2024). These metabolites are grouped into simple phenols (Jahan *et al.*, 2023), coumarin and phenolic acids, including a group of hydroxybenzoic acids and hydroxycinnamic acids (Sarker and Oba, 2020b), flavonoids, including flavonols (Sarker and Oba, 2019b), flavones, flavanols, flavanones (Sarker and Oba, 2020c,d), isoflavones, anthocyanins, and chalcones, and non-flavonoids, including tannins, lignans, and stilbenes, which are accumulated by shikimate and/or phenylpropanoid pathways from phenylalanine. Various studies have shown that drought and salinity stress increase secondary metabolites in the leaves and protect amaranth plants from the adverse effects of drought (Sarker and Oba, 2018c,d) and salinity stress (Sarker and Oba, 2018e; Sarker and Ercisli, 2022; Hossain *et al.*, 2022; Sarker *et al.*, 2023). Potassium (K) amendment increases protein content and drought tolerance in soybean (Mamun *et al.*, 2022), and the nutrient content and drought tolerance in mung bean (Islam *et al.*, 2024). Nano-iron oxide accelerates protein content and drought tolerance, while reducing the chlorophyll content of soybean (Dola *et al.*, 2022), while a biostimulant improved protein, carbohydrate, and drought tolerance in soybean (Mannan *et al.*, 2023).

A number of studies have reported the accumulation of different amino acids while plants were exposed to adverse environmental conditions. Among these amino acids, a predominance of proline was observed in various adverse environmental conditions. Phytohormones also participate in plants' abiotic stress response by controlling and regulating internal as well as external signaling pathways. Abscisic acid (ABA) acts as a key stress hormone by enhancing the tolerance of plants to abiotic stimuli. Under drought stress, ABA is associated with the biosynthesis of osmoprotectants, dehydrins, and other protective proteins. Auxin (indole-3-acetic acid) was also reported to control plant growth under salt and heavy-metal stress (Mathur *et al.*, 2022). Similarly, cytokinins act as ABA antagonists and play an important role under water-deficit conditions (Wani *et al.*, 2016). Another multi-functional phytohormone, jasmonic acid (JA), participates in salt- and water-stress conditions (Wang *et al.*, 2020). JA was also reported to reduce stresses induced by heavy metals such as copper and cadmium (Cd). The lipid composition of plants significantly alters with varying environmental conditions due to changes in lipid biosynthesis and lipid signaling pathways (Sharma *et al.*, 2023). Volatile terpenes scavenge abiotic stress-induced oxidative stresses and play an important role in abiotic stress tolerance in plants (Boncan *et al.*, 2020).

15.2.2 Plant metabolites in response to biotic stresses

Phytoalexins exhibit antimicrobial activities, and synthesis and activation of numerous phytoalexins are triggered by pathogenic invasion. *Arabidopsis* synthesizes an indole phytoalexin, 3′-thiazol-2′-yl-indole, under pathogen invasion (Kumar *et al.*, 2023). Resveratrol is a phenylpropanoid synthesized in higher plants such as crops as a result of pathogen infection and provides pathogen immunity. Pathogen-inducible resveratrol production has been reported in pulses, grapevine, and peanuts (Al-Khayri *et al.*, 2023). Similarly, there are other phenylalanine-derived phytoalexins such as pisatin in peas, sakuranetin in rice, medicarpin in alfalfa, and glyceollins in soybean. Pathogen infection also induces the release of several volatiles, as well as non-volatile terpenoids. Solanaceous vegetables upon pathogen attack emit various sesquiterpenoid phytoalexins including gossypol and capsidiol (Jan *et al.*, 2021). Crop plants also produce various terpenoid phytoalexins that provide defense against biotic stresses. In maize, several diterpenoid phytoalexins such as kauralexins and sesquiterpenoid phytoalexins such as zealexin protect against fungal infection. Pyrethrins are monoterpenes synthesized in *Chrysanthemum indicum* and act as a neurotoxin for insects and parasites (Saijo and Loo, 2020). The sesquiterpene hemigossypol and diterpene gossypol both exhibit antifungal activity and are obtained from *Gossypium hirsutum*. Cucurbitacins are triterpenoid molecules secreted by *Cucurbita* spp. and are toxic to spider mites (Dresselhaus and Hückelhoven, 2018). Leptin is a cyanogenic glycoside that provides defense against *Leptinotarsa decemlineata* in *Solanum chacoense* (Banothu and Uma, 2022). α-Chaconine and α-solanine are two glycoalkaloids with antifungal activity, providing protection against *Ascobolus crenulatus* and *Rhizoctonia solani* in *Solanum tuberosum* (Li et al., 2023). 2/3-Methylbutyronitrile is another cyanogenic glycoside and protects *Populus trichocarpa* plants from *Lymantria dispar* (Irmisch *et al.*, 2014). Dhurrin is a cyanogenic glucoside that acts as a defensive chemical synthesized in *Sorghum bicolor* to protect against *Phyllotreta nemorum* (Yadav *et al.*, 2023). Nicotine is a pyridine alkaloid synthesized by *Nicotiana attenuata* and provides protection against *Spodoptera exigua* and *Manduca sexta* (Sun *et al.*, 2024). Other alkaloids such as atropine, caffeine, morphine, and quinine prevent pathogen infection and protect various plant families against herbivores.

Among phenolic compounds, flavonoids play an important role in the biotic stress tolerance of plants. For example, *Cajanus platycarpus* synthesizes a flavonoid that prevents herbivory by *Helicoverpa armigera* (D. Chen *et al.*, 2022). *Hevea brasiliensis* synthesizes scopoletin, a naturally occurring antifungal coumarin, whose concentration increased upon *Microcyclus ulei* invasion (Ferreira-Neto *et al.*, 2022). Lignin synthesized by *Pinus nigra* prevents blight disease caused by *Sphaeropsis sapinea* (Yadav and Chattopadhyay, 2023), whereas increased synthesis of lignin in *Triticum* spp. inhibited stem rust by *Puccinia graminis* (Menden *et al.*, 2007). *Zea mays* synthesizes a polyphenol, caffeic acid, which prevents leaf blight induced by *Glomerella graminicola* (Hajam *et al.*, 2024). Xanthone is synthesized by both *Hypericum perforatum* suspension cultures and hairy roots of *Gentiana dinarica* and increases biotic stress tolerance (Khattab and Farag, 2020). Among phytohormones, salicylic acid (SA) and JA exert a significant role in the perception and tolerance of biotic stress tolerance. SA has a specialized role in plant–pathogen interactions and regulates the expression of plant stress signaling genes to control the expression of pathogenesis-related proteins (Song *et al.*, 2023; Mishra *et al.*, 2024a; Roychowdhury *et al.*, 2024a). JA and its derivatives are intrinsic chemical compounds that play a significant role in plant signaling pathways induced by biotic stress (Roychowdhury *et al.*, 2024b). Under biotic stress conditions, JA acts as an endogenous signaling molecule and regulates stress-responsive gene expression (Rehman *et al.*, 2023).

15.3 Tools and Techniques Used in Plant Metabolomics

Metabolomics is an emerging field of science that comprehensively and quantitatively analyzes vast arrays of biological samples for the detection and quantification of endogenous metabolites. Tracking of physical and chemical characteristics changes in metabolites under different conditions provides information on their upstream molecular species (Aderemi *et al.*, 2021). Due to the non-volatility, high polarity, and poor detectability of metabolites, a number of different analytical methods have been developed and utilized for metabolomics study. Currently, large-scale metabolomics analysis has been achieved by combining analytical platforms such as MS, NMR, various chromatographic techniques, and a series of bioinformatic and statistical tools.

Traditionally, NMR spectroscopy and MS are commonly used for the direct analysis of metabolic processes in biological samples. NMR spectroscopy utilizes hydrogen (^{1}H-NMR), carbon (^{13}C-NMR), and phosphorus (^{31}P-NMR) spectra for the detection and quantification of diverse metabolites (Gowda and Raftery, 2019). NMR-based metabolomics proficiently identifies and elucidates the structure of the low-molecular-mass molecules present in biological samples. By using stable isotope tracers, NMR technologies can determine the metabolic fate of metabolites, monitor metabolite–protein interactions, and reveal new metabolic pathways. Although NMR is much less sensitive in comparison with MS-based techniques, NMR spectroscopy offers some advantages such as simpler sample preparation, more convenient, high-throughput fingerprinting, and the non-discriminating and non-destructive nature of the technique. Subsequently, more and more NMR-based new technologies such as differential off-line liquid chromatography (LC)-NMR have been utilized for the identification and characterization of metabolites.

MS is a powerful analytical technique that analyzes the mass:charge (m/z) ratio of metabolites and identifies and quantifies them (Alseekh *et al.*, 2021). MS-based analytical methods combined with separation techniques provide in-depth information about the physicochemical properties of metabolites. In MS studies, molecules are ionized and the m/z ratios of the produced ions are calculated. Therefore, for successful MS, it is essential to select an appropriate ionization technique and type of analyzer. A high-resolution mass analyzer estimates the elemental composition and molecular mass of the studied molecules presented in the mass spectra. The ions present in the mass spectra also provide valuable information about the structure of the analyzed compounds as well as their degree of unambiguity. All MS system consists of an ionization chamber, a mass analyzer, and a detector. While an ionization chamber ionizes the analyzed metabolite present in the given sample, the mass analyzer uses an electromagnetic field for separation of the ions based on their m/z ratio, and finally, the detector records m/z. With the advancement of high-resolution MS techniques, matrix-assisted laser desorption (MALDI) or desorption electrospray (DESI) ionization procedures have been coupled with MS. Overall, every MS method comprises the following five basic steps; sampling, sample preparation, sample analysis, data export, and data analysis. Consequently, MS is as an important analytical instrument in metabolomic analysis.

MS-based techniques require prior separation of the analyzed mixture for precise results. Therefore, MS is usually coupled with chromatographic techniques such as liquid chromatography (LC-MS) and gas chromatography (GC-MS), or with capillary electrophoresis (CE-MS), followed by analysis using various statistical and bioinformatics tools for detection and quantification of molecules in biological samples. Due to high specificity and higher detection sensitivity, LC-MS is used as a universal metabolic profiling technique (Wang *et al.*, 2023). However, due to the ion suppression (or enhancement) effect, LC-MS lacks some quantification accuracy. LC-MS efficiently detects polar compounds such as organic acids, polyamines, nucleotides, fatty acids, alcohols, phenols, vitamins, polyphenols, terpenes, flavonoids, lipids, and other compounds. In contrast, GC-MS shows high resolution and sensitivity and effectively profiles volatile metabolites but is not suitable for non-volatile, thermolabile, and polar compounds. A derivatization process is required for analysis of water-soluble substances in GC-MS. Therefore, compounds that are not derivatized cannot be analyzed by GC-MS. Moreover, GC-MS is less sensitive than LC-MS. CE-MS is currently considered a powerful tool for the comprehensive analysis of highly polar and charged metabolites (Soga, 2023). CE-MS-based metabolomics approaches provide an advantage over GC-MS and LC-MS methods due to their requirement for low sample throughput. As volume-limited biological samples are easily separated and analyzed by CE-MS technology, its usefulness is gradually increasing in metabolic studies. Development of MS imaging technologies such as MALDI or DESI ionization procedures coupled with high-resolution mass spectrometers is enabling *in situ* metabolomics studies.

To properly understand any biological phenomena, metabolomics requires an integrated analysis of transcriptomics, proteomics, and other omics approaches. Such integration of different omics studies is achieved by combining specialized statistical and bioinformatics tools and software. Because of the complexity of the metabolomics data, numerous univariate and multivariate statistical analysis methods are used extensively in metabolomics studies. Currently, several powerful bioinformatical tools are available for large-scale metabolomics data analysis such as MS-DIAL, MZmine 3, El-MAVEN, LipidMatch, MetFlow, LION, 3Omics, IMPaLA, MassTRIX, and OmicsNet (Y. Chen *et al.*, 2022). However, the field of metabolomics still faces a number of challenges for which new and improved bioinformatical tools and software are continuously being developed. Advanced metabolomic approaches such as MS or NMR use stable isotope-labeled compounds to attain quantitative information on many metabolites. Numerous studies using isotope-labeled nuclei such as ^{13}C, ^{15}N, ^{2}H, and ^{31}P have been reported (Moco, 2022).

15.4 Metabolomics for Plant Abiotic Stress Tolerance

Plant encounters various environmental stresses such as abiotic and biotic factors that significantly influence their morphological and biochemical as well as physiological signatures, resulting in a drop in crop yield by altering their growth and development. The leading abiotic stresses such as salinity, drought, heavy metals, oxidative stress, ultraviolet radiation, and cold are known for their destructive nature and lowering of crop productivity (Tinte *et al.*, 2021; Salam *et al.*, 2023). In plants, these abiotic stresses lead to activation of various complex cellular responses, resulting in the stimulation of different biochemical and metabolic pathways. These events are multi-layered and involve the interactions of a cascade of genes and the regulation/activation of different molecular and metabolic pathways that lead to the synthesis and accumulation of primary and secondary metabolites. The major group of primary metabolites includes sugars, organic acids, carbohydrates, polyamines, lipids, glycine betaine, and amino acids and their derivatives that are produced as osmoprotectants in response to various environmental stresses (Michaletti *et al.*, 2018). In addition to this, levels of secondary metabolites, which include terpenes, flavonoids, phenolics, and alkaloids, are directly influenced by various abiotic factors and play a defensive role under different stress conditions (Zandalinas *et al.*, 2022). Comprehensive knowledge of metabolic signatures/markers associated with stress tolerance is crucial for developing stress-tolerant plants for sustainable agricultural practices, and this comprehensive analysis of metabolites also leads to the identification of bioactive phytochemicals and their associated biochemical and physiological stress regulatory mechanisms in plants. Table 15.1 lists some metabolomics studies performed on various plant species under different abiotic stress conditions. Plant species adopt various mechanisms to tolerate different types of stress. Identification of the variation of different metabolites synthesized in response to stress will unveil the molecular events governed by the plant in response to the defense mechanism for stress tolerance. Some of the recent metabolomics studies dealing with overcoming various abiotic factors/stresses are discussed below.

15.4.1 Drought/water-deficit stress

Drought stress is one of the major threats that leads to various metabolic changes resulting in a reduction of plant growth and productivity. Under water-deficit conditions, plants regulate many physiological and metabolic processes to compensate for water loss and damage. Various physiological parameters such as reduced photosynthesis, cell damage, reduced turgor pressure, and depleted nutrient uptake significantly affect crop quality and hence yield (Kumar *et al.*, 2021). A reduction in photosynthesis is associated with the decline in turgor pressure of the leaf, which leads to the closure of stomata, resulting in reduced assimilation of carbon dioxide (CO_2). Interruption of the diffusion of CO_2 via the mesophyll leads to reduced quantum yield of photosystem II and increases the production of

Table 15.1. Potential metabolites and metabolomic tools used for plant abiotic stress tolerance responses.

Types of abiotic stresses	Plant species	Metabolites/metabolic pathway	Tools used for metabolomics	Reference
Salinity	*Medicago sativa*, *Medicago arborea*	Saponins and hydroxycinnamic acids	UHPLC-HRMS/MS	Sarri *et al.* (2021)
	Lycium chinense, *Lycium ruthenicum*	Flavone and flavonoid contents	LC-MS/MS	Qin *et al.* (2022)
	Sorghum bicolor L. Moench	Salicylic acid and betaine	LC-MS/MS	Zhang *et al.* (2022)
	Gliricidia sepium	Steroid and lysine metabolism	UHPLC-ESI-Q-TOF-HRMS	Braga *et al.* (2022)
	Salicornia europaea	Sucrose, glucose, *p*-proline, quercetin, kaempferol derivatives, coumaric, caffeic, and ferulic acids	UPLC-ESI-MS/MS	Duan *et al.* (2023)
	Spirodela polyrhiza	Metabolite profiling identified 524 metabolites	LC-MS	Hui *et al.* (2023)
	Avena sativa	Citric acid, serotonin, pyroglutamic acid, L-citrulline, ferulic acid, and caffeic acid	UPLC-MS/MS	Bai *et al.* (2023)
Drought	*Illicium difengpi*	Metabolite profiling	UPLC-MS/MS	Zhang *et al.* (2024)
	Brassica napus L.	Galactose, tricarboxylic acid cycle, aminoacyl-tRNA, nicotinate and nicotinamide metabolisms	LC-MS and GC-MS	Zhi *et al.* (2024)
	Elaeis guineensis	Starch, sucrose glyoxylate, alanine, aspartate, glutamate arginine, proline, glycine, serine, and threonine metabolism	UHPLC-ESI-Q-TOF-HRMS	Neto *et al.* (2021)
	Arabidopsis thaliana	Proline, trehalose, raffinose, and sucrose	GC-MS	Lozano-Elena *et al.* (2022)
	Arachis hypogaea L.	Carbohydrates: glucose D-ribose, D-mannitol, D-xylopyranose, xylonic acid, D-glucopyranose, 2-deoxyribose, L-mannopyranose, myo-inositol, galactose oxime, D-fructose, D-turanose, malic acid, succinic acid, 2-butenedioic acid, 2-deoxyribose, and myo-inositol Fatty acids: stearic acid, pentadecanoic acid, 8,11-octadecadienoic acid, palmitic acid, and pentadecanoic acid	GC-MS	Gundaraniya *et al.* (2020)
	Portulaca oleracea	Palmitic acid, linolenic acid, linoleic acid, oleic acid, stearic acid, arachidic acid, and behenic acid	GC	Saheri *et al.* (2020)
	Nigella sativa	Stearic acid, palmitic acid, oleic acid, linoleic acid, linolenic acid, myristic acid, and arachidic acid	GC	Bayati *et al.* (2020)

Heavy metals	*Oryza sativa* L.	Carboxylic acids, benzene derivatives, carbonyl compounds, cinnamic acids, fatty acyls, and nitrogen compounds	LC-MS/MS	Cao *et al.* (2023)
		L-iduronate	UPLC-MS	Ma *et al.* (2023)
	Triticum aestivum L.	Cinnamic acid, caffeic acid, *p*-coumaraldehyde, coniferyl aldehyde, and ferulic acid	LC-MS/MS	Qian *et al.* (2024)
	Malus spp.	Isoquinoline, flavonol, flavone, leucine, valine, isoleucine, ascorbate and aldarate metabolism; pentose, glucuronate, phenylalanine, tryptophan, mannose, fructose, galactose, proline, arginine, and citrate biosynthesis	LC-MS	Liu *et al.* (2024)
	Nicotiana tabacum	Amino acids, nicotinamide and nicotinate, proline, arginine, flavone and flavonol biosynthesis	UPLC	Zou *et al.* (2022)
Temperature	*Camellia sinensis*	Molecular chaperones, or chaperone-like proteins, flavonols accumulation	LC-MS	Huang *et al.* (2024)
	Actinidia arguta	Nucleotides and phenolic acids metabolism	UPLC-ESI-MS/MS	Sun *et al.* (2021)
	Medicago sativa	Phenylpropanoids, polyketides, and organic acids	LC-MS	Zhang *et al.* (2023)
	Nepenthes sp.	Flavonols, flavonoid, amino acids, lignins, and coniferin metabolism	UHPLC-Q-TOF–MS	Wong *et al.* (2020)
	Triticum aestivum cv. Norin 61	β-alanine, serotonin, tryptophan, proline, and putrescine	LC-MS/MS	Matsunaga *et al.* (2021)

GC, gas chromatography; GC-MS, gas chromatography–mass spectrometry; LC-MS/MS, liquid chromatography–tandem mass spectrometry; UHPLC-HRMS/MS, ultra-performance liquid chromatography–high-resolution tandem mass spectrometry; UHPLC-ESI-Q-TOF-MS, ultra-high performance liquid chromatography–electrospray ionization–quadrupole time-of-flight–mass spectrometry; UPLC, ultra-performance liquid chromatography; UPLC-ESI-MS/MS, ultra-performance liquid chromatography–electrospray ionization–tandem mass spectrometry; UPLC-MS, ultra-performance liquid chromatography–mass spectrometry; UPLC-MS/MS, ultra-performance liquid chromatography–tandem mass spectrometry.

H_2O_2, resulting in elevated levels of ROS accompanied by severe cell damage (Djoukeng *et al.*, 2008). Plants produce a number of metabolites, both primary and secondary, to overcome the adverse effect of this ROS-produced damage. Major antioxidant substances such as polyphenol, glutathione, and amino acids (leucine, methionine, valine, tryptophan, and isoleucine) in combination with carbohydrates and proline vary in their levels during this stress (Ghatak *et al.*, 2018). Proline plays a crucial role as an osmoprotectant (Anzano *et al.*, 2022), and several metabolomics studies have shown its accumulation under drought stress conditions (Filippou *et al.*, 2016). Other metabolites, including tryptophan, valine, β-alanine, mannitol, and sucrose, also help to combat osmotic pressure under water-deficit-related stress conditions (Filippou *et al.*, 2016; Jespersen *et al.*, 2017).

ABA is a key phytohormone involved in stress mitigation, with increased levels observed under drought conditions (Ali *et al.*, 2020). ABA regulates stomatal conductance and genes associated with water scarcity in plants. Studies have shown that ABA in the xylem is associated with stomatal closure and drought resistance (Dodd, 2013). However, every plant species adapts different defense mechanisms against drought stress (Jia and Zhang, 2008). In *Arabidopsis* and tomato, ABA-regulated genes such as ascorbate peroxidase 1 and multiprotein bridging factor are crucial for acclimatization under drought stress (Suzuki *et al.*, 2016; Xue *et al.*, 2021). JA is another vital signaling molecule involved in stress regulation including salinity, drought, and temperature stress (Roychowdhury *et al.*, 2024a), while methyl jasmonate upregulates genes related to the production of osmolytes such as sugars, polyamines, and sugar alcohols involved in combating stress in some plants (Feng *et al.*, 2020).

15.4.2 Salinity stress

Salinity stress is another important limiting factor that affects the growth and development of plants. Increased levels of sodium ions (Na^+) resulted in an interruption of ionic balance that affected plant metabolic processes (Anumalla *et al.*, 2016). A high concentration of salt results in reduced water absorption, nutrient uptake, lipid metabolism, and protein assimilation (Anzano *et al.*, 2022). Different plant genotypes of the same species may respond diversely against salinity stress. For example, Borrelli *et al.* (2018) observed that five different genotypes of *Triticum durum* Desf. treated with varied concentrations of sodium chloride, accumulated higher levels of proline and lower levels of organic acids, with varying levels of metabolites such as γ-aminobutyric acid, glutamic acid, leucine, glycine, mannose, threonine, and fructose. Proline accumulation helps enhance water uptake and protect proteins, enzymes, and cell membranes under stress. Polyamines also play the role of osmoprotectants by regulating cellular turgor pressure (Arif *et al.*, 2020).

Additionally, plants synthesize sugar alcohols and sugars such as sucrose, trehalose, and galactose to stabilize cell membranes (Li *et al.*, 2017). SA is a key phytohormone that helps to mitigate Na^+ toxicity by maintaining cell membrane permeability and ionic balance, with increased SA levels reported in *Poplar* × *canescens* roots under salinity stress (Mishra *et al.*, 2024; Roychowdhury *et al.*, 2024a). SA aids in improving osmolytes such as glycine betaine and proline, which regulate Na^+/K^+ flux to diminish salinity toxicity (Chen and Murata, 2011). Additionally, ABA and JA are involved in stomatal closure and ion flux regulation, hence increasing tolerance against salinity stress (Kang *et al.*, 2005). Metabolomics studies will be helpful for the identification of a wide range of responses attributed to stress-tolerant plants.

15.4.3 Heavy-metal stress

Heavy metals include the group of selected metals that are required by plants in small amounts to perform their biological functions. These metals are used in the relevant quantities for the functional regulation of many key enzymes involved in metabolic events such as photosynthesis, homeostasis, synthesis of DNA, and redox metabolism (Riyazuddin *et al.*, 2021). However, excess amounts result in various deleterious effects such as lower biomass, inadequate growth, deficient osmoregulation, chlorosis of leaves, altered nutrient absorption, reduced photosynthesis, or eventually cell death (Mashabela *et al.*, 2022). Plants activate various defense mechanisms to develop tolerance against the toxicity of heavy metals. One of these is the production of chelating agents such as malonic acid, oxalic acid, tartaric acid, and malic acid, which can chelate positively charged metal ions and further restrict the intake of ions from the roots. Components of the cell wall such as pectin and cellulose also play a role in protecting against heavy-metal toxicity in plants (Kosakivska *et al.*, 2021), while chaperones and metallothioneins are another group of proteins involved in sequestering heavy metals. The osmoprotectant activity of proline is also elevated under heavy-metal stress. Proline combats heavy-metal toxicity by neutralizing ROS and other oxidative molecules induced by stress (Ghori *et al.*, 2019). The role of proline in improving photosynthetic efficiency and uptake of nutrients has been studied in *Cajanus cajan* L. (pigeon pea) under the influence of Cd stress (Hayat *et al.*, 2021).

ABA is another crucial molecule produced by plants for preventing heavy-metal stress. ABA in combination with glutathione decreases the conductance of heavy metals from roots to the aerial parts of the plant. It increases the levels of antioxidants that help in reducing the toxic effect conferred by heavy metals (Roychowdhury *et al.*, 2019). Many researchers have also documented the involvement of SA in the production of glutathione and the activation of antioxidant defense mechanisms mitigating the effect of ROS caused by free metals (Mishra *et al.*, 2024). Metabolomics studies using ultra-performance LC-electrospray ionization-quadrupole-time-of-flight-MS in *Oryza sativa* L. revealed overexpression of fatty acids and lipids in the TY816 low Cd-accumulating cultivar compared with the JY841 high Cd-accumulating cultivar (Liu *et al.*, 2021).

15.4.4 Oxidative stress

One of the major consequences of abiotic stresses that influences plant growth and development is the occurrence of oxidative stress (Mittler, 2006). Different stresses are regulated by a cascade of metabolic fluxes that lead to the production of ROS, causing oxidative damage at cellular levels by altering proteins, lipids, DNA, etc. (Møller *et al.*, 2007). Glutathione is one of the important antioxidant signatures that helps in protecting cells against ROS damage (Noctor *et al.*, 2012). Fukushima *et al.* (2017) reported that glutathione-deficient mutants underwent differential metabolic alterations under oxidative stress in *Arabidopsis thaliana*. Another study performed by Pascual (2017) in *Pinus radiata* showed that ultraviolet radiation led to the production of ROS and alterations in other biochemical changes and played a significant role in protecting the plant from oxidation as well as radiation. Recently, metabolomics studies have led to the assessment of the defense mechanisms of plants against various toxicities such as minerals, toxins, and other harmful chemicals, further helping in the identification of tolerant species. For example, *Brassica napus* L. is recognized as a tolerant species against Cd toxicity and accumulates high concentrations of inositol-derived metabolites as well as unsaturated fatty acids in its aerial parts (Mwamba *et al.*, 2020), while exposure to silver nanoparticles resulted in increased production of ROS and other antioxidants to overcome the deleterious effects caused by oxidative stress in *Cucumis sativus* (Zhang *et al.*, 2018).

15.5 Metabolomics for Plant Disease Resistance

Plants are the major source of food for the human population. Food limitations and insecurity are always major global alarming threats. Plant disease is one of the main issues that severely decreases the crop yield, and it has been reported that viral, bacterial, fungal, and nematode parasites led to the loss of US$26 billion in soybean crops in the USA in just 4 years from 2010 to 2014 (Allen *et al.*, 2017). Many fungi such as *Colletotrichum graminicola* (Ces.) Wils. and *Fusarium* spp. are causative agents for plant diseases that lead to major economic losses in crops such as maize, wheat, and barley in various countries (Mielniczuk and Skwaryło-Bednarz, 2020). Therefore, it is imperative to have comprehensive knowledge of plant disease management as well as different protection approaches to prevent major outbreaks and economic losses. Plant pathologists performed phenotypic, metabolic, and molecular-based studies to explore gene expression, development, and protein interactions to gain in-depth knowledge associated with these plant–pathogen interactions.

Plants have innate defense mechanisms to fight against pathogens. Based on expression patterns, there are two major categories of defense mechanisms possessed by plants (Fig. 15.1). The primary mechanism comprises a constitutive defense system that includes a continuous expression system involved in the protection of plants from pathogens. It includes physical barriers such as cuticles, cell walls, and bark. In addition to these constitutive defense systems, plants undergo various metabolic adaptations resulting in the production of an array of secondary metabolites conferring resistance against pathogenic attack (Choudhary *et al.*, 2021). Recently, metabolomics has contributed greatly to our understanding of the mechanisms used by plants against pathogenic outbreaks by determining the coordination of genes, proteins, and transcription factors involved in the various biological processes that eventually lead to the production of metabolites. Hence, the utilization of metabolomics along with other omics tools will pave the path for a better understanding of the defense mechanisms against pathogens in plants. To improve our understanding of the metabolites involved in the crosstalk of plant pathogens, these molecules are categorized based on their localization, function, biosynthesis, and/or structure (Castro-Moretti *et al.*, 2020). Below, we focus on the molecules intervening in important categories involved in the regulation of plant-pathogen defense systems using metabolomics practices.

15.5.1 Phytoanticipins

Plants naturally produced some pre-formed specialized defense chemicals/molecules in a constitutive manner that act as the first line of defense against pathogenic attack. These secondary metabolites are termed phytoanticipins (Muñoz-Hoyos and Stam, 2023). The major classes of phytoanticipins include saponins, cyanogenic glycosides, and glucosinolates, which are synthesized via the shikimate, polyketoid, isoprenoid, and phenylpropanoid pathways (Andersen *et al.*, 2018).

15.5.2 Saponins

Saponins are one of the major, widely distributed, glycosylated plant secondary metabolites that are responsible for antifungal activities in plants. On the basis of the aglycone backbone, saponins are categorized into the major classes of triterpenoidal and steroidal/steroidal alkaloid compounds. Grzywaczyk *et al.* (2024) showed that saponins possess antifungal activity by creating a complex with sterols present in the fungal membranes, leading to the loss of integrity of the membrane. An extensive study performed in oat (*Avena* spp.) demonstrated accumulation of non-toxic steroidal avenacosides A and B in leaves and triterpenoid avenacins in roots that converted into antifungal saponin steroidal 26-desglucoavenacosides in response to pathogen attack and tissue damage (Moses *et al.*, 2014). Saponins are also major secondary metabolites of members of the Solanaceae family. Tomato (*Solanum lycopersicum*) contains the steroidal glycoalkaloid α-tomatine contributing to protection against various fungi including *Botrytis cinerea*, *Septoria lycopersici*, *Fusarium oxysporum* f. sp. *lycopersici*, *Alternaria solani*, and

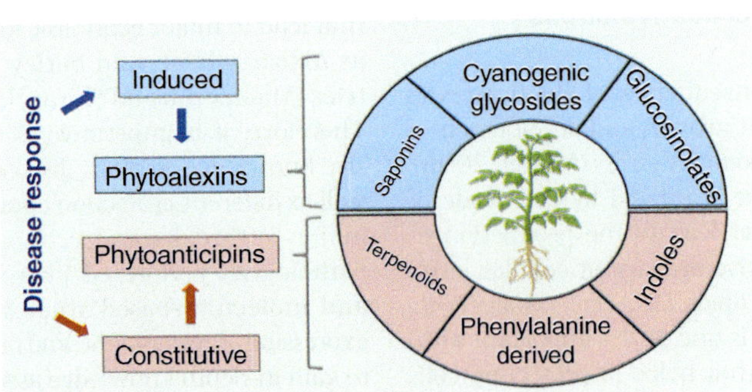

Fig. 15.1. An overview of the specialized metabolic compounds produced by plants in response to defense against pathogenic attacks.

Verticillium alboatrum (Osbourn, 1996). Potato (*Solanum tuberosum*) another member of the Solanacae family, contains a wide range of antifungal compounds such as the steroidal glycoalkaloids α-solanine and α-chaconine. Both α-chaconine and α-solanine are degraded by the fungal pathogen *Gibberella pulicaris*, which causes storage rot disease in potato tubers (Weltring *et al.*, 1997). Recently, many studies have led to the identification of new steroidal saponins that are actively associated with the defense regulation against oomycetes as well as fungi (Baur *et al.*, 2022).

15.5.3 Glucosinolates

Glucosinolates are sulfur-containing glycosides found predominantly in members of the Brassicaceae family. Glucosinolates consist of β-thioglucose and *N*-hydroxyimmunosulfate and are attached with aglycone side chains (Ishida *et al.*, 2014). Glucosinolates are consequentially synthesized from different amino acids such as tryptophan, phenylalanine, leucine, methionine, and glutamic acids. The biological activity of specific glucosinolate-derived products depends on the amino acid from which it is derived, as well as the decorative proteins. Different *Brassica* spp. and *Arabidopsis* are known to contain variable hydrolysis products of glucosinolates (Burrow and Halkier, 2017). The major interest in glucosinolates is due to their hydrolyzed products such as aglycone-based isothiocyanates, which have insect dissuasion properties (Hopkins *et al.*, 2009). In *Brassica* spp., glucosinolates have been reported to aid in resistance against *Peronospora parasitica* and *Leptosphaeria maculans* (Lv *et al.*, 2020). In *Arabidopsis*, they have been shown to inhibit the growth of *Pseudomonas synringe* (Fan *et al.*, 2011). In cauliflower and cabbage, the role of glucosinolates has been documented in protecting the leaves against *Mycosphaerella brassicae* (Yi *et al.*, 2015).

15.5.4 Cyanogenic glycosides

Cyanogenic glycosides are amino acid-derived α-hydroxynitriles that are stored in cellular vacuoles. These compounds are inherently non-toxic and biologically inactive, but when hydrolyzed into their respective cyanogenic glycosides they reveal their biological utility (Gleadow and Møller, 2014). Some metabolomics studies support antimicrobial efficiency in plants. For example, Nielsen *et al.* (2002) showed that the accumulation of cyanogenic glycosides in leaves of barley (*Hordeum vulgare*) supported the fight against many pathogens.

15.5.5 Phytoalexins

Phytoalexins are compounds that are synthesized by plants in response to the defense system induced by pathogen attack (Ahuja *et al.*, 2012). Pathogen attack induces signals that lead to activation of enzymes involved in the biosynthesis of phytoalexins (Thakur and Sohal, 2013). The biosynthesis of phytoalexins is known to be directed from primary metabolites, derivatives of coenzyme A, amino acids, and mevalonic acids (Iriti and Faoro, 2009). Plants accumulate a wide range of phytoalexins and their diversity is highly species and family specific (Schmelz *et al.*, 2014). The production of isoflavones is related to the Fabaceae family, while diterpenes are associated with the Poaceae family: sesquiterpene is commonly produced in members of the Solanaceae family, and Brassicaceae family plants are known to be associated with indole phytoalexins (Tiku, 2018). Bizuneh (2021) reported that the mode of action of phytoalexins is associated with metabolism disruption, delayed maturation, and restriction of the reproductive mechanisms of pathogens.

One of the major classes of phytoalexins is indole phytoalexins, such as camalexin, which is synthesized by *A. thaliana* under pathogenic attack (Kettles *et al.*, 2013). The biosynthesis of this compound is governed by *WRKY* transcription factors and mitogen-activated protein kinases (Pandey *et al.*, 2010). It has also been reported that chitosan, oligogalacturonides, and bacterial peptides are associated with upregulation of the camalexin biosynthetic pathway gene (Povero *et al.*, 2011). A wide range of pathogens including the fungi *Alternaria brassicicola*, *Botrytis cinerea*, and *Plectosphaerella cucumerina*, bacteria such as *P. syringae*, oomycetes such as *Phytophthora brassicae*, and viruses are associated with camalexin accumulation (Jeandet *et al.*, 2014).

Some plant families such as grape (*Vitis vinifera*), peanut (*Arachis hypogaea*), and tobacco (*Nicotiana tabacum*) are known to produce phenylalanine-derived phytoalexins. Resveratrol is one of the major compounds of this class and is produced in response to fungus attack and catalyzed by the stilbene synthase enzyme (Veitch, 2007). Another phytoalexin, scopoletin, is produced in the leaves of tobacco as a defense response against *Alternaria alternata* (Sun *et al.*, 2014). Synthesis of various flavonoids is also associated with the defense response against pathogen attacks. Rice blast disease caused by *Pyricularia oryzae* has been associated with sakuranetin production providing resistance against this disease (González-Lamothe *et al.*, 2009).

Terpenes are one of the largest classes of secondary metabolites and their role is well defined in the plant defense system. These compounds are known to be synthesized against various stimuli such as wounds, herbivores, and pathogen attacks. These compounds disrupt the cell walls of bacteria and fungi by influencing their integrity and permeability. For example, rice plants have been shown to accumulate sesquiterpenes such as (E)-nerolidol and monoterpenes in response to *Xanthomonas oryza* infection (Upadhyay *et al.*, 2024), while a study performed in *Allium sativum* identified terpenes (nerolidol, terpinolene, and α-pinene) associated with antifungal activity against *Sclerotium cepivorum* (Pointin *et al.*, 2015).

15.6 Integration of Metabolomics with Other Multi-omics Pipelines

The integration of metabolomics with other multi-omics pipelines holds great promise for advancing our understanding of plant stress tolerance mechanisms and accelerating crop-improvement efforts. Metabolomics can be effectively integrated with genomics, transcriptomics, and proteomics to unravel the complex molecular networks underlying plant stress responses for both abiotic and biotic stress tolerance. Integration of metabolomics with genomics enables the identification of metabolic quantitative trait loci and candidate genes underlying stress-responsive metabolic pathways. Genome-wide association studies and quantitative trait locus mapping are used to identify genetic loci associated with metabolic phenotypes and stress tolerance traits. Genomic information is integrated with metabolomics data to elucidate the genetic regulation of metabolic pathways and facilitate targeted breeding efforts (Wen *et al.*, 2014). In addition, transcriptomic studies, such as RNA sequencing, capture changes in gene expression in response to stress. Integrating transcriptomics with metabolomics allows the identification of genes involved in stress-responsive metabolic pathways and the dynamic regulation of metabolic gene expression. Correlating changes in transcript levels with alterations in metabolite abundance provides a deeper understanding of the transcriptional regulation of metabolic pathways under stress conditions. In addition, proteomic analyses, including MS-based proteomics, characterize changes in protein abundance and post-translational modifications in response to stress (Olanrewaju *et al.*, 2023). Proteomics–metabolomics integration allows the identification of stress-responsive proteins and their downstream metabolic effects. This integration facilitates the elucidation of protein–metabolite interactions and metabolic enzyme activities, shedding light on the regulation of metabolic pathways under stress conditions. Phenomics technologies, such as high-throughput phenotyping platforms and imaging techniques, are utilized to quantify plant growth, morphology, and physiological traits under stress conditions (Tardieu *et al.*, 2017). Integration of metabolomics with phenomics enables the correlation of metabolic changes with phenotypic responses, providing insights into the physiological mechanisms underlying stress tolerance. Finally, integrating metabolomics with other omics data sets within a systems biology framework allows the construction of comprehensive metabolic networks and regulatory models (Mergner *et al.*, 2020). Systems biology approaches, such as metabolic modeling and network analysis, enable the prediction of metabolic phenotypes and identification of key regulatory nodes in stress-responsive pathways. This integration provides a holistic view of plant stress responses, from the genome to the metabolome, and facilitates the discovery of novel targets for crop improvement through metabolic engineering and breeding.

15.7 Metabolomics-assisted Breeding: Tools and Techniques

MetAB integrates metabolomics technologies with traditional breeding methods to enhance the efficiency and precision of crop improvement for stress tolerance. Metabolite profiling techniques, including MS, NMR spectroscopy, and chromatography (e.g. GC-MS, LC-MS), enable the comprehensive analysis of metabolites in plant tissues. These techniques allow the simultaneous detection and quantification of a wide range of metabolites, including sugars, amino acids, organic acids, secondary metabolites, and phytohormones, providing a holistic view of plant metabolism. For example, in a study on maize, LC-MS was used to identify metabolites associated with drought tolerance, which led to the identification of specific osmoprotectants with a role in stress adaptation (Katam *et al.*, 2022). In a metabolomics study on peanut, it was observed that a drought-tolerant cultivar efficiently accumulated protective metabolites such as trehalose, proline, and γ-aminobutyric acid in its nodules during biological nitrogen fixation. Following re-irrigation, the drought-stressed nodules in the tolerant cultivar were able to revert to a similar metabolic state to that of the unstressed plant. In contrast, the drought-sensitive cultivar failed to restore the metabolic balance after rehydration, leading to inhibited growth (Furlan *et al.*, 2017). Similarly, in rice, NMR-based metabolomics revealed metabolites associated with salinity tolerance, helping breeders to select genotypes with enhanced salt stress resistance (Wu *et al.*, 2023).

Advanced statistical and bioinformatics tools are essential for the analysis and interpretation of metabolomics data in MetAB. Multivariate statistical methods, such as principal component analysis and partial least squares-discriminant analysis, are used for data exploration, visualization, and pattern recognition. In wheat, these bioinformatics tools have been employed for the metabolic profiling of drought-tolerant and -susceptible genotypes (Katam *et al.*, 2022). Metabolite identification and annotation are facilitated by databases, spectral libraries, and computational tools for metabolite matching and annotation. In addition, biomarkers are specific metabolites or metabolic signatures associated with stress tolerance traits in plants. MetAB employs statistical approaches, such as univariate and multivariate analysis, to identify biomarkers that discriminate between stress-tolerant and -susceptible genotypes. In tomatoes, the metabolites linked to oxidative stress tolerance have been identified, which allows the selection of resilient genotypes (Jorge and António, 2018). Machine learning algorithms, including random forests and support vector machines, are utilized for feature selection and classification of stress-responsive metabolites. In this current aspect of MetAB, marker-assisted selection could be useful as a breeding strategy by utilizing molecular markers linked to desired traits for the efficient selection of superior genotypes.

In MetAB, metabolite-based markers, known as metabolic markers, are identified from metabolomics data and integrated with genomic markers for marker-assisted selection. Metabolic markers can be used to phenotype plants early in development, allowing the selection of stress-tolerant genotypes at the seedling stage. This approach has been applied successfully in several crops such as barley, where the metabolic markers have been used for the selection of genotypes with improved resistance to several abiotic stresses (Carrera *et al.*, 2021). By leveraging these tools and techniques, MetAB offers a powerful approach to accelerate the development of stress-tolerant crop varieties with improved yield, quality, and resilience to environmental fluctuations (Fig. 15.2).

15.8 Conclusion and Future Prospects

MetAB represents a transformative approach to plant breeding, offering a holistic view of plant metabolism and stress responses. By integrating metabolomics with traditional breeding methods, MetAB facilitates the identification of metabolic markers and pathways associated with stress tolerance, enabling the targeted selection of superior genotypes for crop improvement. The synergy between metabolomics and breeding holds immense potential for addressing the challenges posed by climate change, resource limitations, and emerging pests and diseases in agriculture. Looking ahead, the future of MetAB holds exciting prospects for enhancing the resilience

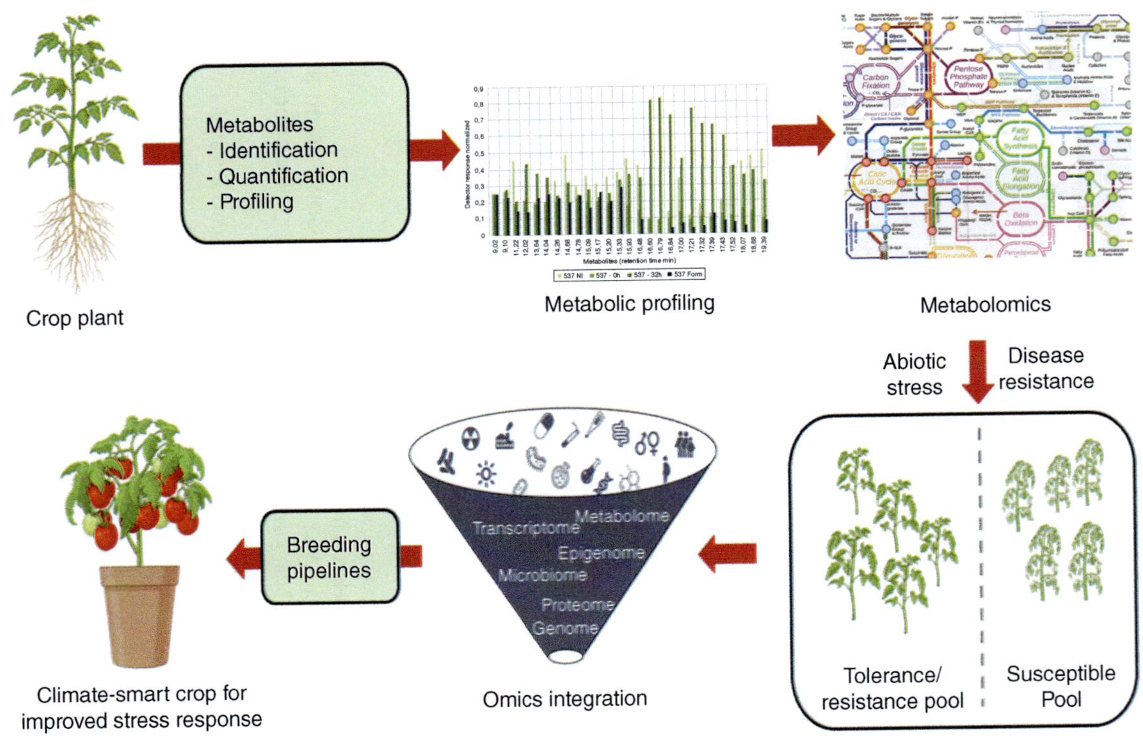

Fig. 15.2. Schematic diagram representing the pipeline for metabolomics-assisted breeding to create climate-smart crops with improved stress tolerance.

and sustainability of agricultural systems. Continued advancements in metabolomics technologies, including improvements in analytical sensitivity, throughput, and data-analysis pipelines, will further accelerate the discovery of stress-responsive metabolites and metabolic networks. Additionally, integrating multi-omics approaches, such as genomics, transcriptomics, and proteomics, will provide a more comprehensive understanding of the molecular mechanisms underlying stress tolerance traits in plants. The application of MetAB in diverse crops and agroecosystems will enable the development of tailored solutions to address specific stress challenges faced by different regions and cropping systems. Overall, MetAB stands poised to revolutionize plant breeding practices, paving the way for the development of climate-resilient crops and ensuring global food security in the face of changing environmental conditions.

15.9 References

Aderemi, A.V., Ayeleso, A.O., Oyedapo, O.O., and Mukwevho, E. (2021) Metabolomics: a scoping review of its role as a tool for disease biomarker discovery in selected non-communicable diseases. *Metabolites* 11: 418.

Ahuja, I., Kissen, R., and Bones, A.M. (2012) Phytoalexins in defense against pathogens. *Trends in Plant Science* 17, 73–90.

Ali, S., Hayat, K., Iqbal, A., and Xie, L. (2020) Implications of abscisic acid in the drought stress tolerance of plants. *Agronomy* 10: 1323.

Al-Khayri, J.M., Rashmi, R., Toppo, V., Chole, P.B., Banadka, A., *et al.* (2023) Plant secondary metabolites: the weapons for biotic stress management. *Metabolites* 13: 716.

Allen, T.W., Bradley, C.A., Sisson, A.J., Byamukama, E., Chilvers, M.I., *et al.* (2017) Soybean yield loss estimates due to diseases in the United States and Ontario, Canada, from 2010 to 2014. *Plant Health Progress* 18, 19–27.

Alseekh, S., Aharoni, A., Brotman, Y., Contrepois, K., d'Auria, J., *et al.* (2021) Mass spectrometry-based metabolomics: a guide for annotation, quantification and best reporting practices. *Nature Methods* 18, 747–756.

Andersen, E.J., Ali, S., Byamukama, E., Yen, Y., and Nepal, M.P. (2018) Disease resistance mechanisms in plants. *Genes* 9: 339.

Anumalla, M., Roychowdhury, R., Geda, C.K., Bharathkumar, S., Goutam, K.D., and Mohandev, T.S.S. (2016) Mechanism of stress signal transduction and involvement of stress inducible transcription factors and genes in response to abiotic stresses in plants. *International Journal of Recent Scientific Research* 7, 12754–12771.

Anzano, A., Bonanomi, G., Mazzoleni, S., and Lanzotti, V. (2022) Plant metabolomics in biotic and abiotic stress: a critical overview. *Phytochemistry Reviews* 21, 503–524.

Arif, Y., Singh, P., Siddiqui, H., Bajguz, A., and Hayat, S. (2020) Salinity induced physiological and biochemical changes in plants: an omic approach towards salt stress tolerance. *Plant Physiology and Biochemistry* 156, 64–77.

Bai, J., Lu, P., Li, F., Li, L., and Yin, Q. (2023) Metabolome and transcriptome analyses reveal the differences in the molecular mechanisms of oat leaves responding to salt and alkali stress conditions. *Agronomy* 13: 1441.

Banothu, V. and Uma, A. (2022) Effect of biotic and abiotic stresses on plant metabolic pathways. In: Badria, F.A. and Surguchov, A. (eds) *Phenolic Compounds—Chemistry, Synthesis, Diversity, Non-Conventional Industrial, Pharmaceutical and Therapeutic Applications*. IntechOpen, London. Available at: www.intechopen.com/chapters/78421 (accessed 3 February 2025).

Baur, S., Bellé, N., Frank, O., Wurzer, S., Pieczonka, S.A., *et al.* (2022) Steroidal saponins—new sources to develop potato (*Solanum tuberosum* L.) genotypes resistant against certain *Phytophthora infestans* strains. *Journal of Agricultural and Food Chemistry* 70, 7447–7459.

Bayati, P., Karimmojeni, H., and Razmjoo, J. (2020) Changes in essential oil yield and fatty acid contents in black cumin (*Nigella sativa* L.) genotypes in response to drought stress. *Industrial Crops and Products* 155: 112764.

Bizuneh, G.K. (2021) The chemical diversity and biological activities of phytoalexins. *Advances in Traditional Medicine* 21, 31–43.

Boncan, D.A.T., Tsang, S.S., Li, C., Lee, I.H., Lam, H.M., *et al.* (2020) Terpenes and terpenoids in plants: interactions with environment and insects. *International Journal of Molecular Sciences* 21: 7382.

Borrelli, G.M., Fragasso, M., Nigro, F., Platani, C., Papa, R., *et al.* (2018) Analysis of metabolic and mineral changes in response to salt stress in durum wheat (*Triticum turgidum* ssp. *durum*) genotypes, which differ in salinity tolerance. *Plant Physiology and Biochemistry* 133, 57–70.

Braga, Í.D.O., Carvalho da Silva, T.L., Belo Silva, V.N., Rodrigues Neto, J.C., Ribeiro, J.A.D.A., *et al.* (2022) Deep untargeted metabolomics analysis to further characterize the adaptation response of *Gliricidia sepium* (Jacq.) Walp. to very high salinity stress. *Frontiers in Plant Science* 13: 869105.

Burow, M. and Halkier, B.A. (2017) How does a plant orchestrate defense in time and space? Using glucosinolates in *Arabidopsis* as case study. *Current Opinion in Plant Biology* 38, 142–147.

Cao, H., Chen, D., Kuang, L., Yan, T., Gao, F., and Wu, D. (2023) Metabolomic analysis reveals the molecular responses to copper toxicity in rice (*Oryza sativa*). *Plant Physiology and Biochemistry* 199: 107727.

Carrera, F.P., Noceda, C., Maridueña-Zavala, M.G., and Cevallos-Cevallos, J.M. (2021) Metabolomics, a powerful tool for understanding plant abiotic stress. *Agronomy* 11: 824.

Castro-Moretti, F.R., Gentzel, I.N., Mackey, D., and Alonso, A.P. (2020) Metabolomics as an emerging tool for the study of plant–pathogen interactions. *Metabolites* 10: 52.

Chen, D., Mubeen, B., Hasnain, A., Rizwan, M., Adrees, M., *et al.* (2022) Role of promising secondary metabolites to confer resistance against environmental stresses in crop plants: current scenario and future perspectives. *Frontiers in Plant Science* 13: 881032.

Chen, T.H. and Murata, N. (2011) Glycinebetaine protects plants against abiotic stress: mechanisms and biotechnological applications. *Plant, Cell & Environment* 34, 1–20.

Chen, Y., Li, E.M., and Xu, L.Y. (2022) Guide to metabolomics analysis: a bioinformatics workflow. *Metabolites* 12: 357.

Choudhary, P., Aggarwal, P.R., Rana, S., Nagarathnam, R., and Muthamilarasan, M. (2021) Molecular and metabolomic interventions for identifying potential bioactive molecules to mitigate diseases and their impacts on crop plants. *Physiological and Molecular Plant Pathology* 114: 101624.

Djoukeng, J.D., Arbona, V., Argamasilla, R., and Gomez-Cadenas, A. (2008) Flavonoid profiling in leaves of citrus genotypes under different environmental situations. *Journal of Agricultural and Food Chemistry* 56, 11087–11097.

Dodd, I.C. (2013) Abscisic acid and stomatal closure: a hydraulic conductance conundrum? *New Phytologist* 197, 6–8.

Dola, D.B., Mannan, M.A., Sarker, U., Mamun, M.A.A., Islam, T., *et al.* (2022) Nano-iron oxide accelerates growth, yield, and quality of *Glycine max* seed in water deficits. *Frontiers in Plant Science* 13: 992535.

Dresselhaus, T. and Hückelhoven, R. (2018) Biotic and abiotic stress responses in crop plants. *Agronomy* 8: 267.

Duan, H., Tiika, R.J., Tian, F., Lu, Y., Zhang, Q., *et al.* (2023) Metabolomics analysis unveils important changes involved in the salt tolerance of *Salicornia europaea*. *Frontiers in Plant Science* 13: 1097076.

Fan, J., Crooks, C., Creissen, G., Hill, L., Fairhurst, S., *et al.* (2011) *Pseudomonas sax* genes overcome aliphatic isothiocyanate-mediated non-host resistance in *Arabidopsis*. *Science* 331, 1185–1188.

Feng, X., Zhang, L., Wei, X., Zhou, Y., Dai, Y., and Zhu, Z. (2020) *OsJAZ13* negatively regulates jasmonate signaling and activates hypersensitive cell death response in rice. *International Journal of Molecular Sciences* 21: 4379.

Ferreira-Neto, J.R., Kido, É.A., Aburjaile, F.F., da Silva, M.D., de Souza, M.C., and Benko-Iseppon, A.M. (2022) The journey through the gene: a focus on plant anti-pathogenic agents mining in the omics era. In: Rai, M. and feitosa, C.M. (eds) *Eco-Friendly Biobased Products Used in Microbial Diseases*. CRC Press, Boca Raton, Florida, pp. 3–22.

Filippou, P., Antoniou, C., Obata, T., van der Kelen, K., Harokopos, V., *et al.* (2016) Kresoxim-methyl primes *Medicago truncatula* plants against abiotic stress factors via altered reactive oxygen and nitrogen species signalling leading to downstream transcriptional and metabolic readjustment. *Journal of Experimental Botany* 67, 1259–1274.

Fukushima, A., Iwasa, M., Nakabayashi, R., Kobayashi, M., Nishizawa, T., *et al.* (2017) Effects of combined low glutathione with mild oxidative and low phosphorus stress on the metabolism of *Arabidopsis thaliana*. *Frontiers in Plant Science* 8: 1464.

Furlan, A.L., Bianucci, E., Castro, S., and Dietz, K.J. (2017) Metabolic features involved in drought stress tolerance mechanisms in peanut nodules and their contribution to biological nitrogen fixation. *Plant Science* 263, 12–22.

Ghatak, A., Chaturvedi, P., and Weckwerth, W. (2018) Metabolomics in plant stress physiology. *Plant Genetics and Molecular Biology* 164, 187–236.

Ghori, N.H., Ghori, T., Hayat, M.Q., Imadi, S.R., Gul, A., *et al.* (2019) Heavy metal stress and responses in plants. *International Journal of Environmental Science and Technology* 16, 1807–1828.

Gleadow, R.M. and Møller, B.L. (2014) Cyanogenic glycosides: synthesis, physiology, and phenotypic plasticity. *Annual Review of Plant Biology* 65, 155–185.

González-Lamothe, R., Mitchell, G., Gattuso, M., Diarra, M.S., Malouin, F., and Bouarab, K. (2009) Plant antimicrobial agents and their effects on plant and human pathogens. *International Journal of Molecular Sciences* 10, 3400–3419.

Gowda, G.N. and Raftery, D. (eds) (2019) *NMR-Based Metabolomics: Methods and Protocols*. Humana Press, Totowa, New Jersey.

Grzywaczyk, A., Smułek, W., and Kaczorek, E. (2024) *Saponaria officinalis* saponins as a factor increasing permeability of *Candida* yeasts' biomembrane. *World Journal of Microbiology and Biotechnology* 40: 152.

Gundaraniya, S.A., Ambalam, P.S., and Tomar, R.S. (2020) Metabolomic profiling of drought-tolerant and susceptible peanut (*Arachis hypogaea* L.) genotypes in response to drought stress. *ACS Omega* 5, 31209–31219.

Hajam, Y.A., Diksha, Kumar, R., and Lone, R. (2024) Plant phenolics and their versatile promising role in the management of nematode stress. In: Lone, R., Khan, S., and Al-Sadi, A.M. (eds) *Plant Phenolics in Biotic Stress Management*. Springer Nature, Singapore, pp. 389–416.

Hasanuzzaman, M., Roychowdhury, R., Karmakar, J., Dey, N., Nahar, K., and Fujita, F. (2015) Recent advances in biotechnology and genomic approaches for abiotic stress tolerance in crop plants. In: Thangadurai, D. and Sangeetha, J. (eds) *Genomics and Proteomics: Concepts, Technologies and Applications*. Apple Academic Press, Burlington, Vermont, pp. 333–366.

Hayat, K., Khan, J., Khan, A., Ullah, S., Ali, S. (2021) Ameliorative effects of exogenous proline on photosynthetic attributes, nutrients uptake, and oxidative stresses under cadmium in pigeon pea (*Cajanus cajan* L.). *Plants* 10: 796.

Hopkins, R.J., van Dam, N.M., and van Loon, J.J. (2009) Role of glucosinolates in insect–plant relationships and multitrophic interactions. *Annual Review of Entomology* 54, 57–83.

Hossain, M.N., Sarker, U., Raihan, M.S., Al-Huqail, A.A., Siddiqui, M.H., and Oba, S. (2022) Influence of salinity stress on color parameters, leaf pigmentation, polyphenol and flavonoid contents, and antioxidant activity of *Amaranthus lividus* leafy vegetables. *Molecules* 27: 1821.

Huang, F., Lei, Y., Duan, J., Kang, Y., Luo, Y., *et al.* (2024) Investigation of heat stress responses and adaptation mechanisms by integrative metabolome and transcriptome analysis in tea plants (*Camellia sinensis*). *Scientific Reports* 14: 10023.

Hui, T., Zhang, Y., Jia, R., Hu, Y., Wang, W., *et al.* (2023) Metabolomic analysis reveals responses of *Spirodela polyrhiza* L. to salt stress. *Journal of Plant Interactions* 18: 2210163.

Irmisch, S., Clavijo McCormick, A., Günther, J., Schmidt, A., Boeckler, G.A., *et al.* (2014) Herbivore-induced poplar cytochrome P450 enzymes of the CYP71 family convert aldoximes to nitriles which repel a generalist caterpillar. *Plant Journal* 80, 1095–1107.

Iriti, M. and Faoro, F. (2009) Chemical diversity and defence metabolism: how plants cope with pathogens and ozone pollution. *International Journal of Molecular Sciences* 10, 3371–3399.

Ishida, M., Hara, M., Fukino, N., Kakizaki, T., and Morimitsu, Y. (2014) Glucosinolate metabolism, functionality and breeding for the improvement of Brassicaceae vegetables. *Breeding Science* 64, 48–59.

Islam, M.R., Sarker, U., Azam, M.G., Hossain, J., Alam, M.A., *et al.* (2024) Potassium augments growth, yield, nutrient content, and drought tolerance in mung bean (*Vigna radiata* L. *Wilczek.). Scientific Reports* 14: 9378.

Jahan, N., Sarker, U., Saikat, M.M.H., Hossain, M.M., Azam, M.G., *et al.* (2023) Evaluation of yield attributes and bioactive phytochemicals of twenty amaranth genotypes of Bengal floodplain. *Heliyon* 9: e19644.

Jan, R., Asaf, S., Numan, M., Lubna, and Kim, K.-M., (2021) Plant secondary metabolite biosynthesis and transcriptional regulation in response to biotic and abiotic stress conditions. *Agronomy* 11: 968.

Jeandet, P., Hébrard, C., Deville, M.A., Cordelier, S., Dorey, S., *et al.* (2014) Deciphering the role of phytoalexins in plant-microorganism interactions and human health. *Molecules* 19, 18033–18056.

Jespersen, D., Yu, J., and Huang, B. (2017) Metabolic effects of acibenzolar-*S*-methyl for improving heat or drought stress in creeping bentgrass. *Frontiers in Plant Science* 8: 1224.

Jia, W. and Zhang, J. (2008) Stomatal movements and long-distance signaling in plants. *Plant Signaling & Behavior* 3, 772–777.

Jorge, T.F. and António, C. (2018) Plant metabolomics in a changing world: metabolite responses to abiotic stress combinations. In: Andjelkovic, V. (ed.) *Plant Abiotic Stress Responses to Climate Change*. IntechOpen, London. Available at: www.intechopen.com/chapters/57908 (accessed 3 February 2025).

Kang, D.J., Seo, Y.J., Lee, J.D., Ishii, R., Kim, K.U., *et al.* (2005) Jasmonic acid differentially affects growth, ion uptake and abscisic acid concentration in salt-tolerant and salt-sensitive rice cultivars. *Journal of Agronomy and Crop Science* 191, 273–282.

Katam, R., Lin, C., Grant, K., Katam, C.S., and Chen, S. (2022) Advances in plant metabolomics and its applications in stress and single-cell biology. *International Journal of Molecular Sciences* 23: 6985.

Kettles, G.J., Drurey, C., Schoonbeek, H.J., Maule, A.J., and Hogenhout, S.A. (2013) Resistance of *Arabidopsis thaliana* to the green peach aphid, *Myzus persicae*, involves camalexin and is regulated by micro RNAs. *New Phytologist* 198, 1178–1190.

Khattab, A.R. and Farag, M.A. (2020) Current status and perspectives of xanthones production using cultured plant biocatalyst models aided by *in-silico* tools for its optimization. *Critical Reviews in Biotechnology* 40, 415–431.

Kosakivska, I.V., Babenko, L.M., Romanenko, K.O., Korotka, I.Y., and Potters, G. (2021) Molecular mechanisms of plant adaptive responses to heavy metals stress. *Cell Biology International* 45, 258–272.

Kumar, M., Kumar Patel, M., Kumar, N., Bajpai, A.B., and Siddique, K.H. (2021) Metabolomics and molecular approaches reveal drought stress tolerance in plants. *International Journal of Molecular Sciences* 22: 9108.

Kumar, S., Korra, T., Thakur, R., Arutselvan, R., Kashyap, A.S., *et al.* (2023) Role of plant secondary metabolites in defence and transcriptional regulation in response to biotic stress. *Plant Stress* 8: 100154.

Li, H., Li, M., Fan, Y., Liu, Y., and Qin, S. (2023) Antifungal activity of potato glycoalkaloids and its potential to control severity of dry rot caused by F*usarium sulphureum*. *Crop Science* 63, 801–811.

Li, M., Guo, R., Jiao, Y., Jin, X., Zhang, H., and Shi, L. (2017) Comparison of salt tolerance in *Soja* based on metabolomics of seedling roots. *Frontiers in Plant Science* 8: 1101.

Liu, C., Lan, M.M., He, E.K., Yao, A.J., Wang, G.B., *et al.* (2021) Phenomic and metabolomic responses of roots to cadmium reveal contrasting resistance strategies in two rice cultivars (*Oryza sativa* L.). *Soil Ecology Letters* 3, 220–229.

Liu, C., Zhou, G., Qin, H., Guan, Y., Wang, T., *et al.* (2024) Metabolomics combined with physiology and transcriptomics reveal key metabolic pathway responses in apple plants exposure to different selenium concentrations. *Journal of Hazardous Materials* 464: 132953.

Lozano-Elena, F., Fàbregas, N., Coleto-Alcudia, V., and Caño-Delgado, A.I. (2022) Analysis of metabolic dynamics during drought stress in *Arabidopsis* plants. *Scientific Data* 9: 90.

Lv, H., Fang, Z., Yang, L., Zhang, Y., and Wang, Y. (2020) An update on the arsenal: mining resistance genes for disease management of *Brassica* crops in the genomic era. *Horticulture Research* 7: 34.

Ma, L., Zeng, J., Zhang, R.Q, Wang, L., Zhang, F., *et al.* (2023) Integrated transcriptomic and metabolomic analysis the variation of rice cultivars response to arsenite stress. *Environmental Technology & Innovation* 31: 103207.

Mamun, M.A.A., Sarker, U., Mannan, M.A., Rahman, M.M., Karim, M.A., *et al.* (2022) Application of potassium after waterlogging improves quality and productivity of soybean seeds. *Life* 12: 1816.

Mannan, M.A., Yasmin, A., Sarker, U., Bari, N., Dola, D.B., *et al.* (2023) Biostimulant red seaweed (*Gracilaria tenuistipitata* var. liui) extracts spray improves yield and drought tolerance in soybean. *PeerJ* 11: 15588.

Mashabela, M.D., Piater, L.A., Steenkamp, P.A., Dubery, I.A., Tugizimana, F., and Mhlongo, M.I. (2022) Comparative metabolite profiling of wheat cultivars (*Triticum aestivum*) reveals signatory markers for resistance and susceptibility to stripe rust and aluminium (Al^{3+}) toxicity. *Metabolites* 12: 98.

Mathur, P., Tripathi, D.K., Baluška, F., and Mukherjee, S. (2022) Auxin-mediated molecular mechanisms of heavy metal and metalloid stress regulation in plants. *Environmental and Experimental Botany* 196: 104796.

Matsunaga, S., Yamasaki, Y., Mega, R., Toda, Y., Akashi, K., and Tsujimoto, H. (2021) Metabolome profiling of heat priming effects, senescence, and acclimation of bread wheat induced by high temperatures at different growth stages. *International Journal of Molecular Sciences* 22: 13139.

Menden, B., Kohlhoff, M., and Moerschbacher, B.M. (2007) Wheat cells accumulate a syringyl-rich lignin during the hypersensitive resistance response. *Phytochemistry* 68, 513–520.

Mergner, J., Frejno, M., List, M., Papacek, M., Chen, X., *et al.* (2020) Mass-spectrometry-based draft of the *Arabidopsis* proteome. *Nature* 579, 409–414.

Michaletti, A., Naghavi, M.R., Toorchi, M., Zolla, L., and Rinalducci, S. (2018) Metabolomics and proteomics reveal drought-stress responses of leaf tissues from spring-wheat. *Scientific Reports* 8: 5710.

Mielniczuk, E. and Skwaryło-Bednarz, B. (2020) *Fusarium* head blight, mycotoxins and strategies for their reduction. *Agronomy* 10: 509.

Mishra, S., Roychowdhury, R., Ray, S., Hada, A., Kumar, A., *et al.* (2024) Salicylic acid (SA)-mediated plant immunity against biotic stresses: an insight on molecular components and signaling mechanism. *Plant Stress* 11: 100427.

Mittler, R. (2006) Abiotic stress, the field environment and stress combination. *Trends in Plant Science* 11, 15–19.

Moco, S. (2022) Studying metabolism by NMR-based metabolomics. *Frontiers in Molecular Biosciences* 9: 882487.

Møller, I.M., Jensen, P.E., and Hansson, A. (2007) Oxidative modifications to cellular components in plants. *Annual Review Plant Biology* 58, 459–481.

Moses, T., Papadopoulou, K.K., and Osbourn, A. (2014) Metabolic and functional diversity of saponins, biosynthetic intermediates and semi-synthetic derivatives. *Critical Reviews in Biochemistry and Molecular Biology* 49, 439–462.

Muñoz-Hoyos, L. and Stam, R. (2023) Metabolomics in plant pathogen defense: from single molecules to large-scale analysis. *Phytopathology* 113, 760–770.

Mwamba, T.M., Islam, F., Ali, B., Lwalaba, J.L.W., Gill, R.A., *et al.* (2020) Comparative metabolomic responses of low- and high-cadmium accumulating genotypes reveal the cadmium adaptive mechanism in *Brassica napus*. *Chemosphere* 250: 126308.

Neto, J.C.R., Vieira, L.R., de Aquino Ribeiro, J.A., de Sousa, C.A.F., Júnior, M.T.S., and Abdelnur, P.V. (2021) Metabolic effect of drought stress on the leaves of young oil palm (*Elaeis guineensis*) plants using UHPLC-MS and multivariate analysis. *Scientific Reports* 11: 18271.

Nielsen, K.A., Olsen, C.E., Pontoppidan, K., and Møller, B.L. (2002) Leucine-derived cyano glucosides in barley. *Plant Physiology* 129, 1066–1075.

Noctor, G., Mhamdi, A., Chaouch, S., Han, Y.I., Neukermans, J., *et al.* (2012) Glutathione in plants: an integrated overview. *Plant, Cell & Environment* 35, 454–484.

Olanrewaju, G.O., Haveman, N.J., Naldrett, M.J., Paul, A.L., Ferl, R.J., and Wyatt, S.E. (2023) Integrative transcriptomics and proteomics profiling of *Arabidopsis thaliana* elucidates novel mechanisms underlying spaceflight adaptation. *Frontiers in Plant Science* 14: 1260429.

Osbourn, A. (1996) Saponins and plant defence—a soap story. *Trends in Plant Science* 1, 4–9.

Pandey, S.P., Roccaro, M., Schön, M., Logemann, E., and Somssich, I.E. (2010) Transcriptional reprogramming regulated by WRKY18 and WRKY40 facilitates powdery mildew infection of *Arabidopsis*. *Plant Journal* 64, 912–923.

Pascual, J., Cañal, M.J., Escandon, M., Meijon, M., Weckwerth, W., and Valledor, L. (2017) Integrated physiological, proteomic, and metabolomic analysis of ultra violet (UV) stress responses and adaptation mechanisms in *Pinus radiata*. *Molecular & Cellular Proteomics* 16, 485–501.

Pontin, M., Bottini, R., Burba, J.L., and Piccoli, P. (2015) *Allium sativum* produces terpenes with fungistatic properties in response to infection with *Sclerotium cepivorum*. *Phytochemistry* 115, 152–160.

Povero, G., Loreti, E., Pucciariello, C., Santaniello, A., Di Tommaso, D., *et al.* (2011) Transcript profiling of chitosan-treated *Arabidopsis* seedlings. *Journal of Plant Research* 124, 619–629.

Qian, R., Li, Y., Liu, Y., Sun, N., Liu, L., *et al.* (2024) Integrated transcriptomic and metabolomic analysis reveals the potential mechanisms underlying indium-induced inhibition of root elongation in wheat plants. *Science of the Total Environment* 908: 168477.

Qin, X., Yin, Y., Zhao, J., An, W., Fan, Y., *et al.* (2022) Metabolomic and transcriptomic analysis of *Lycium chinese* and *L. ruthenicum* under salinity stress. *BMC Plant Biology* 22: 8.

Rehman, M., Saeed, M.S., Fan, X., Salam, A., Munir, R., *et al.* (2023) The multifaceted role of jasmonic acid in plant stress mitigation: an overview. *Plants* 12: 3982.

Riyazuddin, R., Nisha, N., Ejaz, B., Khan, M.I.R., Kumar, M., *et al.* (2021) A comprehensive review on the heavy metal toxicity and sequestration in plants. *Biomolecules* 12: 43.

Roychowdhury, R., Khan, M.H., and Choudhury, S. (2019) Physiological and molecular responses for metalloid stress in rice—a comprehensive overview. In: Hasanuzzaman, M., Fujita, M., Nahar, K., and Biswas, J.K. (eds) *Advances in Rice Research for Abiotic Stress Tolerance*. Woodhead Publishing, Duxford, UK, pp. 341–369.

Roychowdhury, R., Choudhury, S., Hasanuzzaman, M., and Srivastava, S. (eds) (2020) *Sustainable Agriculture in the Era of Climate Change*. Springer, Cham, Switzerland.

Roychowdhury, R., Das, S.P., Gupta, A., Parihar, P., Chandrasekhar, K., *et al.* (2023a) Multi-omics pipeline and omics-integration approach to decipher plant's abiotic stress tolerance responses. *Genes* 14: 1281.

Roychowdhury, R., Ballén-Taborda, C., and Chaturvedi, P. (2023b) Characterizing and improving traits for resilient crop development. *Frontiers in Plant Science* 14: 1307327.

Roychowdhury, R., Hada, A., Biswas, S., Mishra, S., Prusty, M.R., *et al.* (2024a) Jasmonic acid (JA) in plant immune response: unravelling complex molecular mechanisms and networking of defence signalling against pathogens. *Journal of Plant Growth Regulation* 44, 89–114.

Roychowdhury, R., Mishra, S., Anand, G., Dalal, D., Gupta, R., *et al.* (2024b) Decoding the molecular mechanism underlying salicylic acid (SA)-mediated plant immunity: an integrated overview from its biosynthesis to the mode of action. *Physiologia Plantarum* 176: e14399.

Saheri, F., Barzin, G., Pishkar, L., Boojar, M.M.A., and Babaeekhou, L. (2020) Foliar spray of salicylic acid induces physiological and biochemical changes in purslane (*Portulaca oleracea* L.) under drought stress. *Biologia* 75, 2189–2200.

Saijo, Y. and Loo, E.P.I. (2020) Plant immunity in signal integration between biotic and abiotic stress responses. *New Phytologist* 225, 87–104.

Salam, U., Ullah, S., Tang, Z.H., Elateeq, A.A., Khan, Y., *et al.* (2023) Plant metabolomics: an overview of the role of primary and secondary metabolites against different environmental stress factors. *Life* 13: 706.

Šamec, D., Karalija, E., Šola, I., Vujčić Bok, V., and Salopek-Sondi, B. (2021) The role of polyphenols in abiotic stress response: the influence of molecular structure. *Plants* 10: 118.

Sarker, U. and Ercisli, S. (2022) Salt eustress induction in red amaranth (*Amaranthus gangeticus*) augments nutritional, phenolic acids and antiradical potential of leaves. *Antioxidants* 11: 2434.

Sarker, U. and Oba, S. (2018a) Drought stress effects on growth, ROS markers, compatible solutes, phenolics, flavonoids, and antioxidant activity in *Amaranthus tricolor*. *Applied Biochemistry and Biotechnology* 186, 999–1016.

Sarker, U. and Oba, S. (2018b) Catalase, superoxide dismutase and ascorbate-glutathione cycle enzymes confer drought tolerance of *Amaranthus tricolor*. *Scientific Report* 8: 16496.

Sarker, U. and Oba, S. (2018c) Response of nutrients, minerals, antioxidant leaf pigments, vitamins, polyphenol, flavonoid and antioxidant activity in selected vegetable amaranth under four soil water content. *Food Chemistry* 252, 72–83.

Sarker, U. and Oba, S. (2018d) Drought stress enhances nutritional and bioactive compounds, phenolic acids and antioxidant capacity of *Amaranthus* leafy vegetable. *BMC Plant Biology* 18: 258.

Sarker, U. and Oba, S. (2018e) Augmentation of leaf color parameters, pigments, vitamins, phenolic acids, flavonoids and antioxidant activity in selected *Amaranthus tricolor* under salinity stress. *Scientific Reports* 8: e12349.

Sarker, U. and Oba, S. (2019a) Salinity stress enhances color parameters, bioactive leaf pigments, vitamins, polyphenols, flavonoids and antioxidant activity in selected *Amaranthus* leafy vegetables. *Journal of the Science of Food and Agriculture* 99, 2275–2284.

Sarker, U. and Oba, S. (2019b) Antioxidant constituents of three selected red and green color *Amaranthus* leafy vegetable. *Scientific Reports* 9: 18233.

Sarker, U. and Oba, S. (2020a) The response of salinity stress-induced *A. tricolor* to growth, anatomy, physiology, non-enzymatic and enzymatic antioxidants. *Frontiers in Plant Science* 11: 559876.

Sarker, U. and Oba, S. (2020b) Phenolic profiles and antioxidant activities in selected drought-tolerant leafy vegetable amaranth. *Scientific Reports* 10: 18287.

Sarker, U. and Oba, S. (2020c) Nutraceuticals, phytochemicals, and radical quenching ability of selected drought-tolerant advance lines of vegetable amaranth. *BMC Plant Biology* 20: 564.

Sarker, U. and Oba, S. (2020d) Polyphenol and flavonoid profiles and radical scavenging activity in selected leafy vegetable *Amaranthus gangeticus*. *BMC Plant Biology* 20: 499.

Sarker, U., Islam, M.T., and Oba, S. (2018) Salinity stress accelerates nutrients, dietary fiber, minerals, phytochemicals and antioxidant activity in *Amaranthus tricolor* leaves. *PLOS ONE* 13: e0206388.

Sarker, U., Lin, Y.P., Oba, S., Yoshioka, Y., and Ken, H. (2022a) Prospects and potentials of underutilized leafy amaranths as vegetable use for health-promotion. *Plant Physiology and Biochemistry* 182, 104–123.

Sarker, U., Iqbal, M.A., Hossain, M.N., Oba, S., Ercisli, S., *et al.* (2022b) Colorant pigments, nutrients, bioactive components, and antiradical potential of danta leaves (*Amaranthus lividus*). *Antioxidants* 11: 1206.

Sarker, U., Oba, S., Alsanie, W.F., and Gaber, A. (2022c) Characterization of phytochemicals, nutrients, and antiradical potential in slim amaranth. *Antioxidants* 11: 1089.

Sarker, U., Hossain, M.N., Oba, S., Ercisli, S., Marc, R.A., and Golokhvast, K.S. (2023) Salinity stress ameliorates pigments, minerals, polyphenolic profiles, and antiradical capacity in Lalshak. *Antioxidants* 12: 173.

Sarker, U., Oba, S., Ullah, R., Bari, A., Ercisli, S., *et al.* (2024) Nutritional and bioactive properties and antioxidant potential of *Amaranthus tricolor*, *A. lividus*, *A viridis,* and *A. spinosus* leafy vegetables. *Heliyon* 10: e30453.

Sarri, E., Termentzi, A., Abraham, E.M., Papadopoulos, G.K., Baira, E., *et al.* (2021) Salinity stress alters the secondary metabolic profile of *M. sativa*, *M. arborea* and their hybrid (Alborea). *International Journal of Molecular Sciences* 22: 4882.

Schmelz, E.A., Huffaker, A., Sims, J.W., Christensen, S.A., Lu, X., *et al.* (2014) Biosynthesis, elicitation and roles of monocot terpenoid phytoalexins. *Plant Journal* 79, 659–678.

Sharma, P., Lakra, N., Goyal, A., Ahlawat, Y.K., Zaid, A., and Siddique, K.H. (2023) Drought and heat stress mediated activation of lipid signaling in plants: a critical review. *Frontiers in Plant Science* 14: 1216835.

Soga, T. (2023) Advances in capillary electrophoresis mass spectrometry for metabolomics. *Trends in Analytical Chemistry* 158: 116883.

Song, W., Shao, H., Zheng, A., Zhao, L., and Xu, Y. (2023) Advances in roles of salicylic acid in plant tolerance responses to biotic and abiotic stresses. *Plants* 12: 3475.

Sun, G., Zhang, X., Liu, Y., Chai, L., Liu, D., *et al.* (2024) Plant chemical defenses against insect herbivores—using the wild tobacco as a model. *Phyton-International Journal of Experimental Botany* 93, 641–659.

Sun, H., Wang, L., Zhang, B., Ma, J., Hettenhausen, C., *et al.* (2014) Scopoletin is a phytoalexin against *Alternaria alternata* in wild tobacco dependent on jasmonate signalling. *Journal of Experimental Botany* 65, 4305–4315.

Sun, S., Fang, J., Lin, M., Hu, C., Qi, X., *et al.* (2021) Comparative metabolomic and transcriptomic studies reveal key metabolism pathways contributing to freezing tolerance under cold stress in kiwifruit. *Frontiers in Plant Science* 12: 628969.

Suzuki, N., Bassil, E., Hamilton, J.S., Inupakutika, M.A., Zandalinas, S.I., *et al.* (2016) ABA is required for plant acclimation to a combination of salt and heat stress. *PLOS ONE* 11: e0147625.

Tardieu, F., Cabrera-Bosquet, L., Pridmore, T., and Bennett, M. (2017) Plant phenomics, from sensors to knowledge. *Current Biology* 27, R770–R783.

Thakur, M. and Sohal, B.S. (2013) Role of elicitors in inducing resistance in plants against pathogen infection: a review. *International Scholarly Research Notices* 2013: 762412.

Tiku, A.R. (2018) Antimicrobial compounds and their role in plant defense. In: Singh, A., Singh, I. (eds) *Molecular Aspects of Plant–Pathogen Interaction*. Springer, Singapore, pp. 283–307.

Tinte, M.M., Chele, K.H., van der Hooft, J.J., and Tugizimana, F. (2021) Metabolomics-guided elucidation of plant abiotic stress responses in the 4IR era: an overview. *Metabolites* 11: 445.

Upadhyay, R., Saini, R., Shukla, P.K., and Tiwari, K.N. (2024) Role of secondary metabolites in plant defense mechanisms: a molecular and biotechnological insights. *Phytochemistry Reviews* 24(1): 953–983.

Varshney, R.K., Terauchi, R., and McCouch, S.R. (2014) Harvesting the promising fruits of genomics: applying genome sequencing technologies to crop breeding. *PLoS Biology* 12: e1001883.

Veitch, N.C. (2007) Isoflavonoids of the Leguminosae. *Natural Product Reports* 24, 417–464.

Wang, J., Song, L., Gong, X., Xu, J., and Li, M. (2020) Functions of jasmonic acid in plant regulation and response to abiotic stress. *International Journal of Molecular Sciences* 21: 1446.

Wang, Z., Zhu, H., and Xiong, W. (2023) Advances in mass spectrometry-based multi-scale metabolomic methodologies and their applications in biological and clinical investigations. *Science Bulletin* 68, 2268–2284.

Wani, S.H., Kumar, V., Shriram, V., and Sah, S.K. (2016) Phytohormones and their metabolic engineering for abiotic stress tolerance in crop plants. *Crop Journal* 4, 162–176.

Weltring, K.M., Wessels, J., and Geyer, R. (1997) Metabolism of the potato saponins α-chaconine and α-solanine by *Gibberella pilicaris*. *Phytochemistry* 46, 1005–1009.

Wen, W., Li, D., Li, X., Gao, Y., Li, W., *et al.* (2014) Metabolome-based genome-wide association study of maize kernel leads to novel biochemical insights. *Nature Communications* 5: 3438.

Wong, C., Ling, Y.S., Wee, J.L.S., Mujahid, A., and Müller, M. (2020) A comparative UHPLC-Q/TOF–MS-based eco-metabolomics approach reveals temperature adaptation of four *Nepenthes* species. *Scientific Reports* 10: 21861.

Wu, M., Northen, T.R., and Ding, Y. (2023) Stressing the importance of plant specialized metabolites: omics-based approaches for discovering specialized metabolism in plant stress responses. *Frontiers in Plant Science* 14: 1272363.

Xu, Y. and Fu, X. (2022) Reprogramming of plant central metabolism in response to abiotic stresses: a metabolomics view. *International Journal of Molecular Sciences* 23: 5716.

Xue, F., Liu, W., Cao, H., Song, L., Ji, S., *et al.* (2021) Stomatal conductance of tomato leaves is regulated by both abscisic acid and leaf water potential under combined water and salt stress. *Physiologia Plantarum* 172, 2070–2078.

Yadav, M., Singh, I.K. and Singh, A. (2023) Dhurrin: a naturally occurring phytochemical as a weapon against insect herbivores. *Phytochemistry* 205: 113483.

Yadav, S. and Chattopadhyay, D. (2023) Lignin: the building block of defense responses to stress in plants. *Journal of Plant Growth Regulation* 42, 6652–6666.

Yi, G.E., Robin, A.H.K., Yang, K., Park, J.I., Kang, J.G., *et al.* (2015) Identification and expression analysis of glucosinolate biosynthetic genes and estimation of glucosinolate contents in edible organs of *Brassica oleracea* subspecies. *Molecules* 20, 13089–13111.

Zandalinas, S.I., Balfagón, D., Gómez-Cadenas, A., and Mittler, R. (2022) Plant responses to climate change: metabolic changes under combined abiotic stresses. *Journal of Experimental Botany* 73, 3339–3354.

Zhang, F., Lu, F., Wang, Y., Zhang, Z., Wang, J., *et al.* (2022) Combined transcriptomic and physiological metabolomic analyses elucidate key biological pathways in the response of two sorghum genotypes to salinity stress. *Frontiers in Plant Science* 13: 880373.

Zhang, H., Du, W., Peralta-Videa, J.R., Gardea-Torresdey, J.L., White, J.C., *et al.* (2018) Metabolomics reveals how cucumber (*Cucumis sativus*) reprograms metabolites to cope with silver ions and silver nanoparticle-induced oxidative stress. *Environmental Science & Technology* 52, 8016–8026.

Zhang, X.J., Wu, C., Liu, B.Y., Liang, H.L., Wang, M.L., and Li, H. (2024) Transcriptomic and metabolomic profiling reveals the drought tolerance mechanism of *Illicium difengpi* (Schisandraceae). *Frontiers in Plant Science* 14: 1284135.

Zhang, Y., Li, X., and Zhao, Y. (2023) Pseudotargeted metabolomics reveals the effects of high root-zone temperature stress on leaves and roots of *Medicago sativa*. *Journal of Plant Growth Regulation* 42, 5667–5682.

Zhi, X., Bian, X., Yu, J., Xiao, X., Duan, B., *et al.* (2024) Comparative metabolomics analysis of tolerant and sensitive genotypes of rapeseed (*Brassica napus* L.) seedlings under drought stress. *Agricultural Water Management* 296: 108797.

Zou, C., Lu, T., Wang, R., Xu, P., Jing, Y., *et al.* (2022) Comparative physiological and metabolomic analyses reveal that Fe_3O_4 and ZnO nanoparticles alleviate Cd toxicity in tobacco. *Journal of Nanobiotechnology* 20: 302.

16 Plant High-throughput Metabolomics: Recent Achievements, Applications and Future Perspectives

Lumbini Devi Darapu, Shivani Tiwari, and Prateek Gupta*
Department of Biological Sciences, SRM University AP, Amaravati, Andhra Pradesh, India

Abstract
Plants produce a vast array of metabolites, far more than those produced by most other organisms. Large-scale metabolite profiling assays now provide unmatched access to global data sets of metabolites and their corresponding pathways, greatly enhancing our understanding of plant biology. High-throughput metabolome analysis platforms have accelerated the discovery of diverse biochemical metabolites and new pathways while enhancing our understanding of existing ones. Despite many metabolites remaining unidentified, metabolomics has significantly advanced our comprehension of plant physiology and biology through the study of these small molecules. As well as the latest advancements in analytical techniques, the integration of metabolomics with other omics will greatly enhance our understanding of biological systems. In this chapter, we briefly describe the latest developments in the field of analytical techniques and their application in plant metabolomics.

Keywords: GC-MS, LC-MS, mass spectrometry, metabolomics, NMR

16.1 Introduction

Advances in technology have greatly enhanced our ability to study biological systems at the molecular level by allowing us to gather and analyze multiple data sets more effectively. This progress has led to an increased focus on identifying, measuring, and understanding the key molecules that drive biochemical processes. A critical area of research in this context is metabolomics, which involves the comprehensive study of metabolites—the final products of biological activities—within cells, tissues, or entire organisms. Oliver *et al.* (1998) first used the word "metabolome" to describe the study of the quantitative and qualitative collection of all the metabolites in a particular cell or tissue. The metabolome refers to the complete set of small molecules, or metabolites, within a cell. Metabolomics, therefore, is the study of all these low-molecular-mass compounds in an organism (Oliver *et al.*, 1998; Tweeddale *et al.*, 1998). At the start of the 21st century, the field of metabolomics has become part of the suite of techniques used in system biology. It provides a quantitative and qualitative summary of all the metabolites, which are small molecules with a molecular weight of 30–3000 Da that are present in an organism and have a variety of characteristics and functions (Markley *et al.*, 2017). Since then, the

* E-mail: prateekg04@gmail.com

© CAB International 2025. *High-Throughput Plant Metabolomics* (Jen-Tsung Chen ed.)
DOI: 10.1079/9781800626911.0016

application of metabolomics techniques has gained popularity, especially in the field of plant sciences (Fiehn, 2002).

With the advancements in genomic technologies, many economically important plants and crops now have high-quality reference genome sequences, which are crucial for discovering the genes that control important biochemical pathways. Additionally, transcriptomics and proteomics studies have been made available for most of these plants, helping us better understand gene expression. However, integrated studies that link genomics, transcriptomics, proteomics, and metabolomics are still lacking, even though metabolomics is vital for understanding phenotype expression, especially in crop improvement. Interest in metabolomics has grown, particularly with the use of non-targeted approaches to analyze plant metabolites (Roessner *et al.*, 2000). These metabolites are essential to plant metabolism because they significantly influence plant growth and structure. In recent years, metabolomics has emerged as a major scientific breakthrough, enabling precise profiling of metabolites in microbes, plants, and animals (Heyman and Dubery, 2016; van Dam and Bouwmeester, 2016). This field allows the detection of a wide range of metabolites from a single sample, providing a detailed view of a cell's metabolic state. As metabolomics advances, it offers great potential for studying mutants and transgenic plants to understand metabolic networks and identify key genes. It also helps clarify the function of specific genes and how they impact metabolic pathways, revealing different layers of regulation.

The study of plant metabolism started in 1817, when Bienaimé Caventou extracted chlorophyll for the first time (Delépine, 1951). Plant metabolomics is a rapidly developing field within omics science. Since then, our knowledge of the chemical makeup and biological functions of plants has improved due to the discovery and isolation of alkaloids, phytohormones, and other metabolites (Misra and Chen., 2015). Metabolites are bioindicators of the physiological and biochemical characteristics of plants, revealing metabolic pathways, gene expression, and signaling. Primary and secondary metabolites make up the two categories of metabolites that account for the plant metabolome. A comprehensive understanding of the biochemical processes involved in plant metabolism is possible through the metabolic profiling of primary and secondary metabolites (Sung *et al.*, 2015). A few primary and secondary metabolites of plants can be interlinked with highly complex metabolic pathways.

In plants, the biosynthesis of lipids, carbohydrates, and amino acids depends on primary metabolites. They influence the growth and development of plants by mediating the tricarboxylic acid and glycolysis cycles during photosynthesis. Secondary metabolites such as flavonoids, atropine, carotenoids, and phytic acid are produced because of primary metabolism, in response to various stress situations, such as high temperature, cold, drought, salinity, and insect/pest assault, but they are not necessary for plant life. Coenzymes, antioxidants, and reactive oxygen species are examples of secondary metabolites (Dawid and Hille, 2018). Specialized secondary metabolites that provide resistance against biotic and abiotic stressors, such as phenolics (~10,000), alkaloids (~21,000), and terpenoids (>25,000), are also included in the plant metabolome. Many of these specialized chemicals are now essential parts of numerous crop-development programs because they are distinct biomarkers that evaluate plant performance in stressful settings. Plant growth, development, and reproduction are intrinsic processes that are mediated by both primary and specialized metabolites. A small number of metabolites function as bridges between primary and secondary metabolites, including alcohols, organic acids, and antioxidants. A single plant species can have anything from a few thousand to tens of thousands of primary and secondary metabolites in total, and there are already more than 200,000 known metabolites (Mazza and Miniati, 1993). Overall, plants produce a vast array of metabolites ranging from a few hundred thousand to 1 million, the majority of which are still unknown (Fang *et al.*, 2019). These compounds occur in plant tissues in various concentrations differing by several orders of magnitude: from hormones present at femto- or picomolar levels to abundant sequestered compounds/natural products existing in the plant tissue in millimolar concentrations.

16.2 High-throughput Analysis in Plant Metabolomics

High-throughput metabolomics is a powerful approach in plant science that enables the comprehensive analysis of metabolites in complex natural extracts. It provides insights into plant physiology, biochemistry, and interactions with the environment. Plant metabolomics utilizes advanced techniques and methodologies to swiftly and extensively analyze metabolites in plant tissues. This process involves the efficient handling and examination of numerous samples in a brief timeframe. High-throughput approaches are vital for gaining a comprehensive understanding of metabolic changes in plants under various conditions, including different growth stages, environmental stresses, or genetic modifications. Key steps in plant metabolomics are sample preparation, advanced analytical techniques, data processing and analysis, integration with other omics data, and high-throughput screening platforms. The success of a metabolomics study relies heavily on the methodologies and instruments used to identify and measure the metabolites comprehensively (Hong *et al.*, 2016). Given the complexity of various metabolic characteristics and molecular abundances, metabolomics remains a challenging task.

Currently, automated technologies that homogenize plant tissues, extract metabolites, and prepare samples can process many samples at once or swiftly in succession. For gathering data, a variety of sophisticated analytical methods are available, with nuclear magnetic resonance (NMR) and mass spectrometry (MS) being the two key techniques driving these platforms. Several integrated technologies and methodologies, particularly MS-based methods, are employed for large-scale exploration of highly complex plant extracts. These include liquid chromatography–mass spectrometry (LC-MS) and gas chromatography–mass spectrometry (GC-MS), matrix-assisted laser desorption/ionization (MALDI), capillary electrophoresis–mass spectrometry (CE-MS), Fourier-transform–ion cyclotron resonance–mass spectrometry (FT-ICR-MS), and ion mobility spectrometry (IMS) (Patel *et al.*, 2021). LC-MS, GC-MS, and NMR have been continuously developed, with recent advancements in IMS-MS proving useful in resolving isomeric and isobaric metabolites. The combination of MS and NMR in metabolomics has significantly enhanced the identification and quantification of metabolites in both untargeted and targeted studies. Modern metabolomics platforms now allow large-scale surveys that encompass both known and unknown metabolites.

High-throughput metabolomics techniques generate vast amounts of data, which can be challenging to manage and analyze effectively. Processing, evaluating, and interpreting large data sets in high-throughput data production requires robust bioinformatics tools. This includes software for peak detection, alignment, normalization, and statistical analysis along with databases for metabolite identification. High-throughput metabolomics commonly supports a variety of omics approaches, including transcriptomics, genomics, and proteomics. Integrative analysis facilitates comprehension of the complex linkages and regulatory systems inside the plant. High-throughput screening platforms can help with large-scale screening by processing hundreds to thousands of samples daily (Patel *et al.*, 2021).

16.3 Sample Preparation and Automation

The first and most critical step in metabolomics studies is the preparation of plant samples and the extraction of metabolites. This involves selecting appropriate tissues, homogenizing the samples, and using solvents to extract a wide range of metabolites (Weckwerth, 2003). High-throughput sample preparation and automation are critical for advancing plant metabolomics, allowing the efficient and reproducible analysis of large numbers of samples. Sample preparation is indeed critical in metabolomics, impacting the quality and reliability of the results. This process aims to isolate metabolites from other components and enrich the desired metabolites for analysis (Kim and Verpoorte, 2010). The steps in plant sample preparation involve harvesting, quenching, extraction, and sample analysis.

Harvesting is the initial step and involves carefully collecting plant tissues (e.g. seeds, leaves, fruit, roots) to ensure minimal metabolic changes due to enzymatic reactions. It is crucial to harvest plant material quickly and efficiently (Wu *et al.*, 2014), and immediately after harvesting, the plant material is quenched in liquid nitrogen to halt enzymatic activity and preserve the metabolic state. This step is crucial for maintaining the integrity of the metabolome, preventing degradation of metabolites (Joo *et al.*, 2019). Extraction involves isolating metabolites from the plant tissue using suitable solvents and methods. The choice of extraction method depends on the metabolite characteristics and the analytical techniques to be used. Some commonly used extraction methods include solid-phase extraction, liquid–liquid extraction, and solvent-based extraction (Zuorro *et al.*, 2013). After extraction, samples are prepared for analysis using various analytical techniques. The sample preparation method should ensure that the metabolites remain stable and are present in concentrations suitable for detection and quantification.

These steps highlight the significant role of automation in enhancing the efficiency and reliability of high-throughput metabolomics. By automating labor-intensive steps, researchers can process more samples in less time, with greater precision and consistency, facilitating large-scale studies and accelerating the pace of discovery in plant sciences.

16.4 Data Acquisition

Data acquisition in plant high-throughput metabolomics involves systematically collecting and recording large volumes of data from plant samples using advanced analytical techniques (Dettmer *et al.*, 2007). A wide range of integrated technologies are currently utilized, including NMR spectroscopy, MS, CE-MS, LC-MS, GC-MS, MALDI, and FT-ICR-MS (Fig. 16.1) (Kim and Buell, 2010; Kim *et al.*, 2010; Peukert *et al.*, 2016; Vinaixa *et al.*, 2016; Mitchell *et al.*, 2018). By combining these approaches, researchers can effectively analyze large numbers of plant samples, enabling more comprehensive metabolomics studies and enhancing our understanding of plant metabolism in various contexts.

In the early 2000s, the combination of high-throughput MS with GC significantly expanded the ability to identify and quantify a wider range of primary metabolites, covering amino acids, organic acids, carbohydrates, lipids, and various sugar phosphates (Fiehn *et al.*, 2000; Roessner *et al.*, 2000). LC-MS profiling was reported to detect metabolites at the femtomole level, including acyl-CoA esters and flavonoids (Larson and Graham, 2001; Hall *et al.*, 2022). During the same period, NMR started being applied in the field of plant metabolomics for structural elucidation and metabolite fingerprinting (Bailey *et al.*, 2003; Defernez and Colquhoun, 2003). CE-MS provides high-resolution separation of metabolites, which includes polar and charged compounds (Zhang and Ramautar, 2021); however, it has been less frequently applied to plant metabolomics. Additionally, MS, when combined with field asymmetric waveform ion mobility spectrometry (FAIMS), can effectively detect volatile chemicals based on ion mobility (Zrodnikov and Davis, 2012). Since then, advances in MS and NMR techniques have led to the discovery of an increasing number of metabolites with molecular masses lower than 1000 Da. The identification of metabolites using MS and NMR is based on their mass:charge (m/z) ratio in electric fields or their resonance frequency in magnetic fields. Hyphenated techniques such as GC-MS and LC-MS work by separating complex mixtures based on how molecules partition between the mobile and stationary phases within a column.

The advancement of MS imaging techniques, such as MALDI and desorption electrospray (DESI) ionization platforms combined with high-resolution MS now allows *in situ* metabolome analysis (Blanksby and Mitchell, 2010). In liquid phase separations, ultra-performance liquid chromatography (UPLC) and high-performance LC (HPLC) are robust analytical methods used for metabolite analysis in various applications, particularly when integrated with other techniques such as MS, for the discovery of plant metabolites (Theodoridis *et al.*, 2012; Khan *et al.*, 2017). A detailed list of analytical methods used in the field of metabolomics is summarized in Table 16.1.

16.4.1 Mass spectrometry

MS has become an essential tool in metabolomics due to its higher sensitivity compared with NMR, enabling researchers to cover a wide range of metabolites. This enhanced sensitivity has led to the identification of novel metabolic biomarkers and molecules that are crucial for reconstructing metabolic networks and pathways. Recent advancements in ionization methods, such as atmospheric pressure chemical ionization (APCI), electrospray ionization (ESI), and MALDI–time-of-flight (MALDI-TOF), have further improved the accuracy of MS analyses.

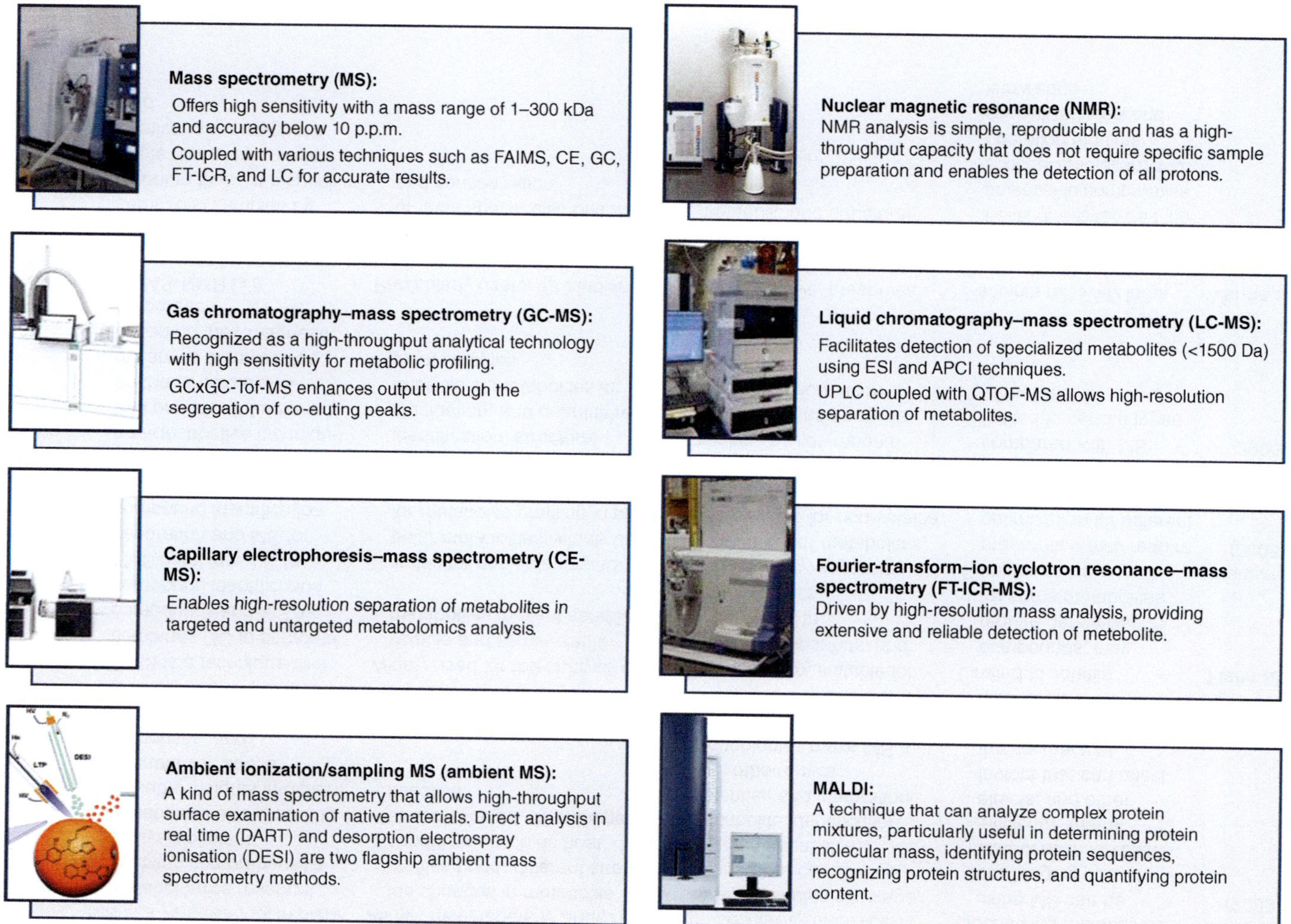

Fig. 16.1. Types of analytical techniques used in the field of plant metabolomics. APCI, atmospheric pressure chemical ionization; CE, capillary electrophoresis; ESI, electrospray ionization; FAIMS, field asymmetric waveform ion mobility spectrometry; FT-ICR, Fourier-transform–ion cyclotron resonance; GC, gas chromatography; LC, liquid chromatography; MALDI, matrix-assisted laser desorption/ionization; Q, quadrupole; TOF, time-of-flight; UPLC, ultra-performance liquid chromatography.

Table 16.1. Summary of analytical methods used in metabolomics.

Method	Description	Applications	Advantages	Disadvantages	Reference
MS	MS is a versatile tool in plant metabolomics research, enabling comprehensive analysis of plant metabolites and providing valuable insights into plant metabolism, stress responses, and biochemical pathways	Allows researchers to analyze the changes in metabolite profiles under different stress conditions, such as heat stress, drought, or pathogen infection	High sensitivity, specificity, wide metabolite coverage, quantitative capabilities, imaging capabilities, metabolite identification potential, and integration with other omics technologies make MS a powerful tool	Quantifying metabolites using MS can be challenging due to variations in ionization efficiency, matrix effects, and other factors that can affect the accuracy of quantitative measurements	Issaq *et al.* (2009)
GC-MS	GC-MS is a technique that combines GC to separate compounds and MS for compound identification	Widely used for the analysis of volatile and semi-volatile metabolites in plant samples	High resolution, suitable for volatile compounds, and widely used in metabolomics	Limited to volatile compounds; may require derivatization for some metabolites	Fiehn (2002)
LC-MS	LC-MS combines LC for separation and MS for compound identification	Suitable for non-volatile and polar metabolites; widely used for metabolite profiling in plant samples	High sensitivity, separates a wide range of metabolites, and suitable for non-volatile compounds	Complex sample preparation may require optimization for different metabolite classes	Want *et al.* (2005)
NMR spectroscopy	NMR spectroscopy is a non-destructive technique that provides information on the structure and dynamics of molecules based on their magnetic properties	Used for metabolite identification, structural elucidation, and quantitative analysis of metabolites in plant samples	Non-destructive and can analyze a wide range of metabolites and provide structural information	Lower sensitivity compared with MS; limited to certain types of metabolite	Krishnan *et al.* (2005)
HR-MAS-NMR	HR-MAS-NMR is a specialized NMR technique that allows the analysis of semi-solid samples by spinning them at a specific angle (the magic angle) relative to the magnetic field	Particularly useful for studying metabolites in intact plant tissues, providing information on their spatial distribution and concentration	Non-destructive, preserves the sample's physical state, and provides detailed metabolic and structural information	Requires relatively large sample volumes and can result in sample heating, overlap has the potential to compromise spectral resolution, and operation requires specialized tools and knowledge	Verma and Kumar (2020)

CE-MS	CE-MS combines CE for separation and MS for compound identification	High separation efficiency, suitable for charged metabolites, and complementary to LC-MS and GC-MS	High separation efficiency, suitable for charged metabolites	Limited to certain types of metabolite and requires specialized equipment	Soga and Heiger (2000)
FT-ICR-MS	FT-ICR-MS is a high-resolution MS technique that provides accurate mass measurements of metabolites	Used for the analysis of complex samples, and high-resolution metabolite profiling in plant metabolomics	High resolution and mass accuracy, suitable for complex samples	Expensive equipment and complex data analysis, requiring skilled operators	Marshall *et al.* (1998)
Ambient MS					
DART-MS	DART-MS is a well-known method for quickly analyzing mass spectra of a wide range of compounds	Used for drug analysis, allowing rapid and sensitive detection of drugs in various samples	Rapid analysis, sensitivity, and non-destructive	Faces challenges such as matrix effects, limited compound coverage, and ionization variability	Feider *et al.* (2019)
DESI-MS	DESI-MS is a technique that allows chemical information to be obtained directly from a wide range of surfaces	Used to investigate secondary metabolites *in situ*, specifically looking at both spatial and temporal distributions	Offers advantages such as high sensitivity, imaging capabilities, and non-destructiveness	Faces challenges including matrix effects, limited quantitative accuracy, and ion suppression effects	Ferreira *et al.* (2019)
MALDI	A widely employed technique for MS imaging that can be performed under vacuum or at atmospheric pressure, depending on the ionization source capabilities	MALDI has various applications, including proteomics, lipidomics, metabolomics, and MS imaging for biomolecular analysis	High sensitivity, ability to analyze complex samples, and suitability for imaging mass spectrometry	Challenges include limited mass range, potential matrix interference, and variability in ionization efficiency	Cornett *et al.* (2007)

CE-MS, capillary electrophoresis–mass spectrometry; DART-MS, direct analysis in real time–mass spectrometry; DESI-MS, desorption electrospray ionization–mass spectrometry; FT-ICR-MS, Fourier-transform–ion cyclotron resonance–mass spectrometry; GC-MS, gas chromatography–mass spectrometry; HR-MAS-NMR, high-resolution magic angle spinning–nuclear magnetic resonance; LC-MS, liquid chromatography–mass spectrometry; MALDI, Matrix-assisted laser desorption/ionization; MS, mass spectrometry; NMR, nuclear magnetic resonance.

To increase throughput and separate complex mixtures, MS is often coupled with various chromatography techniques, including GC, LC, CE, and advanced methods such as FT-ICR and FAIMS. These combined approaches are vital due to the diverse physicochemical properties of metabolites and the complexity of the metabolome.

GC-MS

GC-MS separates and identifies volatile metabolites and those amenable to derivatization. In plant metabolomics, it is used primarily for analyzing primary metabolites such as sugars, amino acids, and organic acids (Lisec *et al.*, 2006). GC-MS, especially using electron-impact ionization, is well suited for high-throughput analysis due to its reproducibility and the availability of extensive libraries for compound identification (Fiehn, 2008).

With the significant advancements in the field of GC-MS, the development of two-dimensional GC-MS (GC×GC-MS) further enhances separation by utilizing two columns with different stationary phases, improving the separation of co-eluting peaks (Ralston-Hooper *et al.*, 2008). GC-MS has been used to analyze the primary metabolite profile of tomato fruit at different stages of ripening, revealing changes in sugar and carbonic acid content, which helps to understand the biochemical changes produced during fruit development (Osorio *et al.*, 2013), and a study using GC-MS to analyze volatile organic compounds (VOCs) released by different plant species after herbivore attack found unique VOC signatures that are important for understanding plant protection methods (Dudareva *et al.*, 2006).

Overall, GC-MS remains one of the most used and versatile platforms in plant metabolomics, due to its extensive libraries of both commercial and in-house metabolite databases (Dutta *et al.*, 2012). Despite being a "gold standard" in metabolomics, it has its own limitations. It can only be utilized to analyze thermally stable and volatile metabolites or those metabolites that can be chemically altered to generate volatile derivatives.

LC-MS

LC-MS is used to separate and identify a wide range of metabolites, including non-volatile and thermally labile compounds, even at low concentrations (Dass, 2006). It is versatile and used for both primary and secondary metabolite analysis. LC-MS primarily uses soft ionization techniques such as ESI and APCI, which are well suited for different types of compounds (Turner *et al.*, 2016). One of the strengths of LC-MS is its ability to operate in both positive and negative ionization modes, allowing broader metabolite coverage compared with using a single polarity mode. Advances in ultra-HPLC coupled with MS have significantly improved throughput and resolution, allowing more efficient metabolite profiling (Cajka and Fiehn, 2016).

By utilizing LC-MS, *Arabidopsis thaliana* secondary metabolites have been investigated, leading to the identification of flavonoids and other phenolic chemicals that are crucial to plant defense systems, and examination of the cellular stress caused to the plant by variations in salt (Kim and Buell, 2010). Alkaloids in therapeutic plants, such as periwinkle, have been analyzed using LC-MS to help discover novel chemicals with potential medical uses (van der Heijden *et al.* 2004). LC-MS is recognized for its versatility and effectiveness, with both low-resolution mass spectrometers such as single quadrupoles and high-resolution instruments such as TOF-MS and Orbitraps being widely used in metabolomics research.

TOF-MS

TOF-MS measures the time it takes for ions to travel a fixed distance, allowing the determination of their m/z ratios and providing high mass accuracy and resolution. TOF-MS is often used in combination with LC or GC for high-throughput applications, providing rapid and precise metabolite identification (Zhang *et al.*, 2016).

In rice, TOF-MS was used to analyze metabolites produced in response to drought stress, and found changes in metabolites such as amino acids and sugars, demonstrating the mechanisms used by the plant in response to stress (Zuther *et al.*, 2007). TOF-MS was used to profile the metabolome of wheat leaves infected with the fungal pathogen *Blumeria graminis*, and identified significant metabolic shifts associated with the plant's defense responses (Walters and McRoberts, 2006). TOF-MS was also applied to generate metabolic fingerprints of genetically modified crops, facilitating the assessment of metabolic changes and potential unintended effects (Ricroch *et al.*, 2011).

Quadrupole (Q)-MS

Q-MS utilizes oscillating electric fields to filter ions based on their m/z ratio. It is often used with other MS techniques for targeted metabolite analysis. Q-MS is beneficial for high-throughput screening when combined with automated sample handling and data-processing systems (Kell and Oliver, 2016).

Q-MS has been used in maize research to measure the levels of growth hormones such as auxin and gibberellin, which are important for understanding growth and development processes (Saini and Keum, 2017). It has also been used for biochemical studies in soybean roots to measure isoflavone levels, which play an important role in plant–microbial interactions, especially in symbiosis and in nitrogen-fixing bacteria (Sugiyama *et al.*, 2007). Q-MS can measure flavonoids and other chemical substances in various fruits and vegetables, which helps to study their effect on diet and improve health (Manach *et al.*, 2004).

FT-ICR-MS

FT-ICR-MS is known for its ultra-high resolution and mass accuracy, making it ideal for comprehensive and detailed metabolomic studies. Although traditionally slower, recent advancements have improved its throughput, making it more suitable for extensive metabolomics projects (Marshall *et al.*, 1998).

FT-ICR-MS was used for the comprehensive characterization of grape leaves, providing a detailed description of complex metabolites, including polyphenols and tannins, important for plant protection and human health benefits (Vaclavik *et al.*, 2013). It was also used to analyze the composition of complex metabolites of tea leaves to identify and elucidate new bioactive compounds that have health benefits (Chen *et al.*, 2011). Using FT-ICR-MS, the complex structure of saponins was elucidated in ginseng providing improved understanding of their medicinal properties (Sun *et al.*, 2009).

CE-MS

CE-MS is another powerful analytical technique used in plant metabolomics, which combines the high separation efficiency of C) with the sensitive detection capabilities of MS. It is used to evaluate a wide range of ionic metabolites based on their m/z ratio. This combination is particularly useful for analyzing polar and charged metabolites, such as amino acids, peptides, and small organic acids, which can be challenging to separate using traditional chromatographic methods (Thurman and Ferrer, 2010).

CE-MS is used for the investigation of metabolites in various plant tissues, including leaves, stems, and roots. These applications demonstrate the technology's ability to handle large sample sizes and its suitability for large-scale metabolic studies (Huhn *et al.*, 2010). For the analysis of secondary metabolites in tobacco plants, CE-MS allowed the identification of alkaloids and phenolic compounds, which are important for the conservation and medicinal properties of plants. CE-MS analysis of metabolites involved in the interaction between soybean roots and nitrogen-fixing bacteria helped to find changes in amino acids and organic acids that contribute to our understanding of symbiotic relationships in plants (Okazaki and Saito, 2012). Despite its strengths, CE-MS has limitations, as it suffers from low sensitivity, poor reproducibility, and issues with migration time. Additionally, the lack of reference libraries makes it less suitable for complex sample analysis (Soga and Imaizumi, 2001).

MALDI-MS

Large biomolecules such as proteins, peptides, and secondary metabolites can be analyzed using the robust MALDI-MS technology, which is extensively employed in plant metabolomics. The capacity of MALDI-MS to evaluate complicated mixtures with minimal sample preparation and high specificity is highly valued. The sample is embedded in a matrix that absorbs laser energy and helps with the analytes' desorption and ionization. The analytes are carried into the gas phase and ionized there when the matrix vaporizes by absorbing the laser energy. The mass spectrometer then examines these ions. The method is particularly helpful for studying biomolecules that are challenging for other techniques to ionize (Hillenkamp and Peter-Katalinic, 2013).

MALDI-MS has mostly been used for proteomic studies such as understanding protein expression and post-translational changes under various physiological settings, which is made easier with the use of this application (Koller *et al.*, 2002). The spatial distribution of metabolites in plant tissues, such as leaves and roots, has been visualized using MALDI-MS imaging. Using this method, certain chemicals can be mapped throughout the tissue to reveal information about their location and functions (Kaspar *et al.*, 2011). Alkaloids, flavonoids, and terpenes are examples of secondary metabolites that have been profiled by MALDI-MS in medicinal plants such as *Ginkgo biloba* (Beck and Stengel, 2016).

Ambient ionization techniques in MS

Direct analysis in real time (DART) and DESI are two ambient ionization techniques in MS that have found significant applications in plant metabolomics due to their ability to analyze samples in their native state with minimal or no sample preparation.

DART is an ambient ionization technique where samples are ionized in open air at atmospheric pressure. A gas stream (usually helium or nitrogen) is heated and passed through a corona discharge, producing excited-state species. These species then interact with the sample, desorbing and ionizing molecules from the sample surface. The ions are subsequently introduced into the mass spectrometer for analysis. DART-MS has been used to rapidly analyze the metabolic profiles of various plant species. For instance, it was applied to study the alkaloid content in medicinal plants such as *Erythroxylum coca* leaves, providing rapid and reliable identification of cocaine and related alkaloids (Cody *et al.*, 2005).

DESI is another ambient ionization technique that combines ESI with desorption processes. A charged solvent spray is directed at the sample surface, creating a thin liquid film. The interaction between the solvent droplets and the sample surface causes desorption of analytes, which are then ionized by the charged droplets and transferred into the mass spectrometer. DESI-MS has been utilized for *in situ* metabolic profiling of plant tissues, such as tomato leaves, enabling the mapping of metabolites directly from the surface of the leaves. This application aids in understanding the spatial distribution of metabolites within plant tissues (Takats *et al.*, 2004).

16.4.2 NMR spectroscopy

NMR spectroscopy is a powerful tool for identifying and characterizing plant bioactive compounds and serves as a complementary technique to MS. It offers detailed structural information, non-destructive analysis, quantitative measurements, high-throughput capabilities, *de novo* structural elucidation, and multivariate data analysis. It involves the alignment of atomic nuclei in a magnetic field, the application of radio frequency pulses to induce resonance and energy absorption, relaxation of the nuclei back to their original state with energy release, and detection of this energy release to generate an NMR spectrum. NMR spectroscopy relies on the behavior of nuclear spins in a magnetic field to provide valuable insights into molecular structure, dynamics, and interactions (Hore, 2015). It detects metabolites, particularly those with a smaller molecular mass (<50 kDa) (Winning *et al.*, 2009).

These features make NMR essential for understanding the structure–activity relationships of bioactive compounds, assessing compound potency, and identifying unknown compounds in plant samples for nutraceutical research. NMR spectroscopy in plant metabolomics has diverse applications, including identifying bioactive compounds, quality control of nutraceuticals, monitoring plant responses to environmental factors, differentiating plant varieties, optimizing nutraceutical production, detecting adulterations, enabling quantitative analysis, and facilitating high-throughput screening. These applications showcase the versatility and significance of NMR in plant-based nutraceutical research for understanding plant metabolites and developing effective nutraceutical products (Markley *et al.*, 2017). Plant-based nutraceutical research has benefited significantly from NMR spectroscopy in identifying bioactive compounds. For instance, NMR-based metabolomics has been used to identify compounds such as gallic acid, hesperidin, kaempferol, and quercetin in various plant sources (Valentino *et al.*, 2020). Additionally, NMR analysis has helped in characterizing bioactive compounds such as lutein, zeaxanthin, caffeic acid, and chlorogenic acid in plants such as *Diospyros kaki* L., *Musa* spp., and *Bougainvillea spectabilis* (Valentino *et al.*, 2020). These examples demonstrate the diverse applications of NMR spectroscopy in plant-based nutraceutical research for identifying and characterizing bioactive compounds. NMR is advantageous for structure elucidation in plant metabolomics when identifying unknown metabolites; however, its poor sensitivity and limited detection of low-abundance biomarkers limit its applications.

16.5 Data Processing and Analysis

Data processing and analysis in plant high-throughput metabolomics involve meticulous steps from sample preparation to data visualization. Each step requires specific tools and methods to handle the complexity and volume of data generated. Advances in computational tools and bioinformatics have significantly enhanced the ability to analyze and interpret metabolomics data, leading to new insights into plant biology. Data collection and preprocessing are the initial steps, involving quality control (data cleaning, noise reduction, and peak detection), alignment, normalization, and transformation of data from various sources such as next-generation sequencing for genomics (Goodwin *et al.*, 2016), RNA sequencing for transcriptomics (Wang *et al.*, 2009), MS for proteomics (Aebersold and Mann, 2003), and LC-MS or GC-MS for metabolomics (Dettmer *et al.*, 2007).

In quality control, data cleaning involves removing or correcting errors and inconsistencies in the raw data. This can be achieved by filtering and de-noising (Smith *et al.*, 2006), wavelet transform, and Savitzky–Golay filtering, which are essential to distinguish true metabolite signals from background noise (Eilers, 2004). The centWave algorithm and local maximum searches were used to detect the peaks in XCMS and LC-MS data (Tautenhahn *et al.*, 2008). Peak alignment corrects retention time shifts and ensures that the same metabolite is correctly identified across multiple samples (Lommen, 2009), while probabilistic quotient normalization and total ion current normalizations are used to correct for systematic variations between samples that are not related to biological differences (van den Berg *et al.*, 2006). Data transformation involves converting the data to a format suitable for statistical analysis and visualization.

Data preprocessing is essential for preparing the raw data for further analysis, ensuring accuracy and consistency across data sets. Data-processing platforms such as MetAlignID, MET-COFEA, MZmine, XCMS, TagFinder, msInspect, MetaboliteDetector and MET-XAlign are commonly used (Lommen *et al.*, 2012; Zhang *et al.*, 2014, 2015; Ma *et al.*, 2016; Misra and van der Hooft, 2016; Perez de Souza *et al.*, 2017; Pluskal *et al.*, 2010; Tautenhahn *et al.*, 2012; Luedemann *et al.*, 2008; Hiller *et al.*, 2009). The list of software commonly used in preprocessing of metabolite data is given in Table 16.2. The metabolite identification process is supported by extensive databases such as METLIN, NIST, and GOLM (Johnson and Lange, 2015). Once metabolites are identified, they undergo statistical analysis using tools such as MetaboAnalyst, Cytoscape, and other statistical software. These analyses can reveal metabolic markers and reconstruct metabolic pathways, aiding in the understanding of biological processes.

16.6 Integration with Other Omics Data

Integrating omics data in plant metabolomics involves combining data from various high-throughput biological studies, such as genomics, transcriptomics, proteomics, and metabolomics, to provide a comprehensive understanding of plant biology. This integrative approach, known as multi-omics, leverages the strengths of each omics discipline to uncover complex biological processes and interactions. Data-integration techniques include statistical methods such as correlation analysis and principal component analysis to identify relationships between different omics layers (Johnson *et al.*, 2007), network analysis to visualize interactions between genes, proteins, and metabolites (Kamburov *et al.*, 2011), and pathway analysis using tools such as KEGG and Reactome to map data on to biological pathways (Kanehisa *et al.*, 2017).

Computational platforms such as Cytoscape (Shannon *et al.*, 2003) and repositories such as MetaboLights (Yurekten *et al.*, 2024) facilitate data integration and management, while machine learning and artificial intelligence are used for pattern recognition and predictive modeling (Libbrecht and Noble, 2015). Applications of multi-omics integration in plant biology are vast, including functional genomics to link genetic variations with metabolic changes (Fang *et al.*, 2019), pathway elucidation to refine metabolic networks (Krumsiek, *et al.*, 2011), and studies

Table 16.2. List of software tools commonly used for data preprocessing in metabolomics.

Software	Applications	Reference
XCMS	Open-source software for processing MS data, including peak detection and alignment	Smith *et al.* (2006)
MetAlign	Provides functionalities for baseline correction, peak detection, and alignment in GC-MS and LC-MS data	Lommen (2009)
MZmine	Offers a comprehensive platform for MS data processing with tools for peak detection, deconvolution, alignment, and identification	Pluskal *et al.* (2010)
MeltDB	An independent software for the analysis and integration of metabolomic software, which currently hosts more than 30 experiments predominantly from GC-MS measurement	Neuweger *et al.* (2008)
Metabo Analyst	Primarily used for metabolomic data analysis and interpretation, focusing on statistical and functional analysis of biological samples; supports data processing, filtering, normalization, and multi-omics integration tasks; the platform is designed to facilitate complex bioinformatics needs, particularly for LC-HRMS-based global metabolomics	Pang *et al.* (2022)
Metabolite detector	A software tool designed to efficiently and automatically analyze GC-MS-based metabolomics data. It starts by detecting and identifying potential metabolites from raw MS data and can perform comparative analysis on multiple chromatograms using either targeted or non-targeted approaches	Hiller *et al.* (2009)
Tagfinder	The ideal tool for modern metabolomics and fluxomics studies; it supports non-biased metabolomic fingerprinting, footprinting and profiling experiments and the metabolite targeted analysis of changes in both metabolite pools and flux	Luedemann *et al.* (2008)

GC-MS, gas chromatography–mass spectrometry; HRMS, high-resolution mass spectrometry; LC-MS, liquid chromatography–mass spectrometry; MS, mass spectrometry.

on stress response to environmental factors (Saito and Matsuda, 2010). This integration also aids in identifying biomarkers for desirable traits (Zivy *et al.*, 2015), enhancing plant breeding and genetic engineering, and providing a holistic view of biological systems through systems biology. However, challenges such as data heterogeneity, computational complexity, and the need for robust data-interpretation methods persist, necessitating the development of standardized data formats, advanced bioinformatics tools, and high-performance computing solutions. Despite these challenges, advances in computational tools and collaborative efforts continue to enhance the feasibility.

16.7 Application of Plant High-throughput Metabolomics

The application of metabolomics is expanding rapidly due to recent advancements in high-throughput technologies. With the help of emerging techniques and methodology, methods based on MS and NMR are commonly used for large-scale analysis of complex plant extracts, allowing the comprehensive exploration of their chemical composition. These include GC-MS, LC-MS, NMR spectroscopy, MALDI, capillary-based MS and other MS-based methods (Manickam *et al.*, 2023). In addition, metabolomics plays an important role in molecular networking to fill the gaps between genotypes and phenotypes (Katam *et al.*, 2022) through genome-editing tools, which has enabled the researcher to perform specific and effective target modification in a large variety of plant species to observe gene functions and manipulation in metabolic pathways (Manickam *et al.*, 2023). Over the last two decades, analytical techniques have been extensively used in the identification of metabolites in economically important plant species. Some examples of these applications are summarized in Table 16.3.

Table 16.3. Application of metabolomics to plant breeding.

Analytical technique	Plant	Purpose	Reference
NMR	Barley	High malting quality	Guo *et al.* (2020)
NMR	Cambuci	Nutritional value	Spricigo et.al.,2021)
NMR	Chicory (*Cichorium intybus* L.)	Biosynthesis	Testone *et al.* (2019)
NMR	Eggplant, pepper, cucumber, apple, peach, clementine, kiwifruit, grape	Difference in productivity	Roch *et al.* (2020)
NMR	Gladiolus (*Gladiolus hybridus* L.)	Insect resistance	Wahyuni *et al.* (2021)
NMR	Kiwifruit (*Actinidia chinensis*)	Pathogen response	Li *et al.* (2020)
NMR	Lentil (*Lens culinaris* L.)	Genotype effect by environment	Rochfort *et al.* (2019)
NMR	*Leucojum aestivum*	Biosynthesis	Spina *et al.* (2021)
NMR	Maize	Chilling tolerance	Urrutia *et al.* (2021)
NMR	Olive	Pathogen response	Girelli *et al.* (2019)
GC-MS	*Aegilops tauschii* Coss.	Fungal resistance	Shelenga *et al.* (2020)
GC-MS	Alphonso mango (*Mangifera indica* L.)	Shelf life	Dar *et al.* (2020)
GC-MS	Apple	Peel ripening	Karagiannis *et al.* (2020)
GC-MS	Asia lotus (*Nelumbo nucifera* Gaertn.)	Flower coloration	Zhu *et al.* (2019)
GC-MS	Asparagus bean (*Vigna unguiculata* L. Walp.)	Nutritional properties	Perchuk *et al.* (2020)
GC-MS	Barley	Cytokinin response	Berka *et al.* (2020)
GC-MS	Cabbage *(Brassica oleracea* var. *capitata)*	Antioxidant activity and taste attributes	Mabuchi *et al.* (2019)
GC-MS	Cannabis	Terpene metabolomics	Mudge *et al.* (2019)
GC-MS	*Celosia argentea*	Copper stresses	Wang *et al.* (2021)
GC-MS	*Chaetoceros tenuissimus*	Growth phase	Hano and Tomaru (2019)
LC-MS	Almond (*Prunus dulcis*)	Endodormancy release	Guillamon *et al.* (2020)
LC-MS	Apple	Peel ripening	Karagiannis *et al.* (2020)
LC-MS	*Arabidopsis*	Root exudates	Durr *et al.* (2021)
LC-MS	Bamboo (*Bambusa pervariabilis* × *Dendrocalamopsis grandis*)	Pathogen response	Li *et al.* (2019)
LC-MS	*Camelina sativa*	Antinutritionals	Amyot *et al.* (2018)
LC-MS	Common bean (*Phaseolus vulgaris*)	Pod degreening	Hu *et al.* (2020)
LC-MS	Cucumber	Fruit color	Wang *et al.* (2020)
LC-MS	Date palm (*Phoenix dactylifera*)	Sucrose metabolism	Malek *et al.* (2020)
LC-MS	Lettuce	Minerals for leaf pigmentation	Corrado *et al.* (2021)
CE-MS	Rice	Osmotic stress tolerance	Matsunami *et al.* (2020)
CE-MS	Tea	Stress status	Yamashita *et al.* (2020)
FT-ICR	*Pisum sativum*	Investigation of root exudate	Calabrese *et al.* (2023)

FT-ICR, Fourier-transform–ion cyclotron resonance; CE-MS, capillary electrophoresis–mass spectrometry; GC-MS, gas chromatography–mass spectrometry; LC-MS, liquid chromatography–mass spectroscopy; NMR, nuclear magnetic resonance.

16.8 Future Perspectives for High-throughput Metabolomics in Plants

High-throughput metabolomics in plants covers a large area for advancing our understanding of plant biology, improving crop performance, environmental sustainability, and human health. Studying small molecules or metabolites within cells, tissues, or organisms has historically been a complex and time-consuming process. However, recent technological advancements in analytical instrumentation and data-processing capabilities are fundamentally transforming this field (Hall *et al.*, 2022). Metabolite coverage in plants is still limited, with only a fraction of an estimated one million plant metabolites studied. Despite the estimation of over one million plant metabolites, the most extensive database, KNApSAcK, only records 50,048 metabolites from 20,741 plant species (Baumeister *et al.*, 2019). Similarly, other databases such as PlantCyc and KEGG contain just 4806 and 2854 metabolite records related to metabolic pathways, respectively (Katam *et al.*, 2022). Enhancing the development of comprehensive metabolite libraries and databases, alongside advancements in metabolite identification techniques, will facilitate the future discovery and detailed analysis of a broader spectrum of plant metabolites (Fang *et al.*, 2019b; Katam *et al.*, 2022).

Compared with genomics and proteomics, metabolomics is still in its early stage, with a considerable need for improved technologies that can increase metabolite coverage. Integrating high-throughput metabolomics data with other omics disciplines such as genomics, transcriptomics, and proteomics will offer a comprehensive understanding of plant systems biology. This synergy can reveal new associations between genetic traits, metabolic pathways, and observable plant characteristics, fostering the development of predictive models for plant performance (Manickam *et al.*, 2023). The integration of metabolomics with functional genomics and advanced analytical platforms is essential for providing novel insights into plant biology. Future studies will likely focus on enhancing the metabolomics platform to enable accurate identification and quantification of metabolites, as well as integrating metabolomics data with other omics platforms to better understand plant metabolism and its applications in agriculture. Advancing spatial metabolomics within plant tissues, cells, and subcellular compartments will provide valuable insight into how metabolites and signaling pathways are compartmentalized. This research may enhance our understanding of how plants adapt to their surroundings (Katam *et al.*, 2022; Shen *et al.*, 2023).

In conclusion, while plant metabolomics has advanced considerably, there is still much work to be done to fully realize its full potential in understanding plant metabolism and improving crop traits. The ongoing development of technologies and methodologies will continue to push the boundaries of this rapidly evolving field.

16.9 Acknowledgments

Dr Prateek Gupta would like to acknowledge the SRM University-AP for providing the seed grant (Sanction No. SRMAP/URG/SEED/2023-24/010) for financial assistance.

16.10 References

Aebersold, R. and Mann, M. (2003) Mass spectrometry-based proteomics. *Nature* 422, 198–207.

Amyot, L., McDowell, T., Martin, S.L., Renaud, J., Gruber, M.Y., and Hannoufa, A. (2018) Assessment of antinutritional compounds and chemotaxonomic relationships between *Camelina sativa* and its wild relatives. *Journal of Agricultural and Food Chemistry* 67, 796–806.

Bailey, N.J., Oven, M., Holmes, E., Nicholson, J.K., and Zenk, M.H. (2003) Metabolomic analysis of the consequences of cadmium exposure in *Silene cucubalus* cell cultures via ^{1}H NMR spectroscopy and chemometrics. *Phytochemistry* 62, 851–858.

Baumeister, T.U., Vallet, M., Kaftan, F., Svatoš, A., and Pohnert, G. (2019) Live single-cell metabolomics with matrix-free laser/desorption ionization mass spectrometry to address microalgal physiology. *Frontiers in Plant Science* 10: 172.

Beck, S. and Stengel, J. (2016) Mass spectrometric imaging of flavonoid glycosides and biflavonoids in *Ginkgo biloba* L. *Phytochemistry* 130, 201–206.

Berka, M., Luklová, M., Dufková, H., Berková, V., Novák, J., *et al.* (2020) Barley root proteome and metabolome in response to cytokinin and abiotic stimuli. *Frontiers in Plant Science* 11: 590337.

Blanksby, S.J. and Mitchell, T.W. (2010) Advances in mass spectrometry for lipidomics. *Annual Review of Analytical Chemistry* 3, 433–465.

Cajka, T. and Fiehn, O. (2016) Toward merging untargeted and targeted methods in mass spectrometry-based metabolomics and lipidomics. *Analytical Chemistry* 88, 524–545.

Calabrese, V., Schmitz-Afonso, I., Riah-Anglet, W., Trinsoutrot-Gattin, I., Pawlak, B., and Afonso, C. (2023) Direct introduction MALDI FTICR MS based on dried droplet deposition applied to non-targeted metabolomics on *Pisum sativum* root exudates. *Talanta* 253: 123901.

Chen, L., Xin, X., Yuan, Q., Su, D., and Liu, W. (2011) Phytochemical properties and antioxidant capacities of commercial tea products in China. *Journal of Food Science* 76, C139–C144.

Cody, R.B., Laramée, J.A., and Durst, H.D. (2005) Versatile new ion source for the analysis of materials in open air under ambient conditions. *Analytical Chemistry* 77, 2297–2302.

Cornett, D.S., Reyzer, M.L., Chaurand, P., and Caprioli, R.M. (2007) MALDI imaging mass spectrometry: molecular snapshots of biochemical systems. *Nature Methods* 4, 828–833.

Corrado, G., Vitaglione, P., Soteriou, G.A., Kyriacou, M.C., and Rouphael, Y. (2021) Configuration by osmotic eustress agents of the morphometric characteristics and the polyphenolic content of differently pigmented baby lettuce varieties in two successive harvests. *Horticulturae* 7: 264.

Dar, M.S., Dholakia, B.B., Shanmugam, H., Gupta, V.S., Subramanian, K.S., *et al.* (2020) Differential modulation in metabolites revealed with the improvement in the shelf-life of alphonso fruits. *Molecular Biotechnology* 62, 508–520.

Dass, C. (2006) Hyphenated separation techniques. In: *Fundamentals of Contemporary Mass Spectrometry*. Wiley, Hoboken, New Jersey, pp. 151–194.

Dawid, C. and Hille, K. (2018) Functional metabolomics—a useful tool to characterize stress-induced metabolome alterations opening new avenues towards tailoring food crop quality. *Agronomy* 8: 138.

Defernez, M. and Colquhoun, I.J. (2003) Factors affecting the robustness of metabolite fingerprinting using ^{1}H NMR spectra. *Phytochemistry* 62, 1009–1017.

Delépine, M. (1951) Joseph Pelletier and Joseph Caventou. *Journal of Chemical Education* 28, 454.

Dettmer, K., Aronov, P.A., and Hammock, B.D. (2007) Mass spectrometry-based metabolomics. *Mass Spectrometry Reviews* 26, 51–78.

Dudareva, N., Negre, F., Nagegowda, D.A., and Orlova, I. (2006) Plant volatiles: recent advances and future perspectives. *Critical Reviews in Plant Sciences* 25, 417–440.

Durr, J., Reyt, G., Spaepen, S., Hilton, S., Meehan, C., *et al.* (2021) A novel signaling pathway required for *Arabidopsis* endodermal root organization shapes the rhizosphere microbiome. *Plant and Cell Physiology* 62, 248–261.

Dutta, A., Shetty, P., Bhat, S., Ramachandra, Y., and Hegde, S. (2012) A mass spectrometric study for comparative analysis and evaluation of metabolite recovery from plasma by various solvent systems. *Journal of Biomolecular Techniques* 23: 128.

Eilers, P.H. (2004) Parametric time warping. *Analytical Chemistry* 76, 404–411.

Fang, C., Fernie, A.R., and Luo, J. (2019) Exploring the diversity of plant metabolism. *Trends in Plant Science* 24, 83–98.

Feider, C.L., Krieger, A., DeHoog, R.J., and Eberlin, L.S. (2019) Ambient ionization mass spectrometry: recent developments and applications. *Analytical Chemistry* 91, 4266–4290.

Ferreira, C., Alfaro, C.M., Pirro, V., and Cooks, R.G. (2019) Desorption electrospray ionization mass spectrometry imaging: recent developments and perspectives. *Journal of Biomolecular Techniques* 30 (Suppl.): S46.

Fiehn, O. (2002) Metabolomics—the link between genotypes and phenotypes. *Functional Genomics* 48, 155–171.

Fiehn, O. (2008) Extending the breadth of metabolite profiling by gas chromatography coupled to mass spectrometry. *Trends in Analytical Chemistry* 27, 261–269.

Fiehn, O., Kopka, J., Dörmann, P., Altmann, T., Trethewey, R.N., and Willmitzer, L. (2000) Metabolite profiling for plant functional genomics. *Nature Biotechnology* 18, 1157–1161.

Girelli, C.R., Angilè, F., Del Coco, L., Migoni, D., Zampella, L., *et al.* (2019) ^{1}H-NMR metabolite fingerprinting analysis reveals a disease biomarker and a field treatment response in *Xylella fastidiosa* subsp. *pauca*-infected olive trees. *Plants* 8: 115.

Goodwin, S., McPherson, J.D., and McCombie, W.R. (2016) Coming of age: ten years of next-generation sequencing technologies. *Nature Reviews Genetics* 17, 333–351.

Guillamón, J.G., Prudencio, Á.S., Yuste, J.E., Dicenta, F., and Sánchez-Pérez, R. (2020) Ascorbic acid and prunasin, two candidate biomarkers for endodormancy release in almond flower buds identified by a nontargeted metabolomic study. *Horticulture Research* 7: 203.

Guo, X., Sarup, P., Jensen, J.D., Orabi, J., Kristensen, N.H., *et al.* (2020) Genetic variance of metabolomic features and their relationship with malting quality traits in spring barley. *Frontiers in Plant Science* 11: 575467.

Hall, R.D., d'Auria, J.C., Ferreira, A.C.S., Gibon, Y., Kruszka, D. *et al.* (2022) High-throughput plant phenotyping: a role for metabolomics? *Trends in Plant Science* 27, 549–563.

Hano, T. and Tomaru, Y. (2019) Metabolomics-based approach to explore growth phase-dependent markers in cultured diatom *Chaetoceros tenuissimus*. *Journal of Chromatography B* 1128: 121779.

Heyman, H.M. and Dubery, I.A. (2016) The potential of mass spectrometry imaging in plant metabolomics: a review. *Phytochemistry Reviews* 15, 297–316.

Hillenkamp, F. and Peter-Katalinic, J. (eds.) (2013) *MALDI MS: A Practical Guide to Instrumentation, Methods and Applications*. Wiley, Weinheim, Germany.

Hiller, K., Hangebrauk, J., Jäger, C., Spura, J., Schreiber, K., and Schomburg, D. (2009) MetaboliteDetector: comprehensive analysis tool for targeted and nontargeted GC/MS based metabolome analysis. *Analytical Chemistry* 81, 3429–3439.

Hong, J., Yang, L., Zhang, D., and Shi, J. (2016) Plant metabolomics: an indispensable system biology tool for plant science. *International Journal of Molecular Sciences* 17: 767.

Hore, P.J. (2015) *Nuclear Magnetic Resonance*. Oxford University Press, Oxford, UK.

Hu, B., Zhu, J., Wu, H., Xu, K., Zhai, H., *et al.* (2020) Enhanced chlorophyll degradation triggers the pod degreening of "Golden Hook," a special ecotype in common bean (*Phaseolus vulgaris* L.). *Frontiers in Genetics* 11: 570816.

Huhn, C., Ramautar, R., Wuhrer, M., and Somsen, G.W. (2010) Relevance and use of capillary coatings in capillary electrophoresis–mass spectrometry. *Analytical and Bioanalytical Chemistry* 396, 297–314.

Issaq, H.J., Van, Q.N., Waybright, T.J., Muschik, G.M., and Veenstra, T.D. (2009) Analytical and statistical approaches to metabolomics research. *Journal of Separation Science* 32, 2183–2199.

Johnson, S.R. and Lange, B.M. (2015) Open-access metabolomics databases for natural product research: present capabilities and future potential. *Frontiers in Bioengineering and Biotechnology* 3: 22.

Johnson, W.E., Li, C., and Rabinovic, A. (2007) Adjusting batch effects in microarray expression data using empirical Bayes methods. *Biostatistics* 8, 118–127.

Joo, M., Park, J.M., Duong, V.A., Kwon, D., Jeon, J., *et al.* (2019) An automated high-throughput sample preparation method using double-filtration for serum metabolite LC-MS analysis. *Analytical Methods* 11, 4060–4065.

Kamburov, A., Pentchev, K., Galicka, H., Wierling, C., Lehrach, H., and Herwig, R. (2011) ConsensusPathDB: toward a more complete picture of cell biology. *Nucleic Acids Research* 39 (Suppl. 1), D712–D717.

Kanehisa, M., Furumichi, M., Tanabe, M., Sato, Y., and Morishima, K. (2017) KEGG: new perspectives on genomes, pathways, diseases and drugs. *Nucleic Acids Research* 45, D353–D361.

Karagiannis, E., Michailidis, M., Tanou, G., Scossa, F., Sarrou, E., *et al.* (2020) Decoding altitude-activated regulatory mechanisms occurring during apple peel ripening. *Horticulture Research* 7: 120.

Kaspar, S., Peukert, M., Svatos, A., Matros, A., and Mock, H.P. (2011) MALDI-imaging mass spectrometry—an emerging technique in plant biology. *Proteomics* 11, 1840–1850.

Katam, R., Lin, C., Grant, K., Katam, C.S., and Chen, S. (2022) Advances in plant metabolomics and its applications in stress and single-cell biology. *International Journal of Molecular Sciences* 23: 6985.

Kell, D.B. and Oliver, S.G. (2016) The metabolome 18 years on: a concept comes of age. *Metabolomics* 12: 148.

Khan, W., Chester, K., Anjum, V., Ahmad, W., Ahmad, S., *et al.* (2017) Chromatographic profiling of Pancharishta at different stages of its development using HPTLC, HPLC, GC–MS and UPLC–MS. *Phytochemistry Letters* 20, 391–400.

Kim, H.K. and Verpoorte, R. (2010) Sample preparation for plant metabolomics. *Phytochemical Analysis* 21, 4–13.

Kim, J. and Buell, C.R. (2010) A high-throughput, LC-MS/MS-based, targeted metabolomics assay for plant and microbial derived phenylpropanoids. *Plant Methods* 6: 19.

Kim, H.K., Choi, Y.H., and Verpoorte, R. (2010) NMR-based metabolomic analysis of plants. *Nature Protocols* 5, 536–549.

Koller, A., Washburn, M.P., Lange, B.M., Andon, N.L., Deciu, C., *et al.* (2002) Proteomic survey of metabolic pathways in rice. *Proceedings of the National Academy of Sciences USA* 99, 11969–11974.

Krishnan, P., Kruger, N.J., and Ratcliffe, R.G. (2005) Metabolite fingerprinting and profiling in plants using NMR. *Journal of Experimental Botany* 56, 255–265.

Krumsiek, J., Suhre, K., Illig, T., Adamski, J., and Theis, F.J. (2011) Gaussian graphical modeling reconstructs pathway reactions from high-throughput metabolomics data. *BMC Systems Biology* 5: 21.

Larson, T.R. and Graham, I.A. (2001) Technical advance: a novel technique for the sensitive quantification of acyl CoA esters from plant tissues. *Plant Journal* 25, 115–125.

Li, S., He, Q., Peng, Q., Fang, X., Zhu, T., *et al.* (2019) Metabolomics responses of *Bambusa pervariabilis* × *Dendrocalamopsis grandis* varieties to biotic (pathogenic fungus) stress. *Phytochemistry* 167: 112087.

Li, Y., Wang, X., Zeng, Y., and Liu, P. (2020) Metabolic profiling reveals local and systemic responses of kiwifruit to *Pseudomonas syringae* pv. *actinidiae*. *Plant Direct* 4: e00297.

Libbrecht, M.W. and Noble, W.S. (2015) Machine learning applications in genetics and genomics. *Nature Reviews Genetics* 16, 321–332.

Lisec, J., Schauer, N., Kopka, J., Willmitzer, L., and Fernie, A.R. (2006) Gas chromatography mass spectrometry-based metabolite profiling in plants. *Nature Protocols* 1, 387–396.

Lommen, A. (2009) MetAlign: interface-driven, versatile metabolomics tool for hyphenated full-scan mass spectrometry data preprocessing. *Analytical Chemistry* 81, 3079–3086.

Lommen, A., van der Kamp, H.J., Kools, H.J., van der Lee, M.K., van der Weg, G., and Mol, H.G. (2012) metAlignID: a high-throughput software tool set for automated detection of trace level contaminants in comprehensive LECO two-dimensional gas chromatography time-of-flight mass spectrometry data. *Journal of Chromatography A* 1263, 169–178.

Luedemann, A., Strassburg, K., Erban, A., and Kopka, J. (2008) TagFinder for the quantitative analysis of gas chromatography–mass spectrometry (GC-MS)-based metabolite profiling experiments. *Bioinformatics* 24, 732–737.

Ma, P., Zhang, Z., Zhou, X., Yun, Y., Liang, Y., and Lu, H. (2016) Feature extraction from resolution perspective for gas chromatography-mass spectrometry datasets. *RSC Advances* 6, 113997–114004.

Mabuchi, R., Tanaka, M., Nakanishi, C., Takatani, N., and Tanimoto, S. (2019) Analysis of primary metabolites in cabbage (*Brassica oleracea* var. *capitata*) varieties correlated with antioxidant activity and taste attributes by metabolic profiling. *Molecules* 24: 4282.

Malek, J.A., Mathew, S., Mathew, L.S., Younuskunju, S., Mohamoud, Y.A., and Suhre, K. (2020) Deletion of beta-fructofuranosidase (invertase) genes is associated with sucrose content in date palm fruit. *Plant Direct* 4: e00214.

Manach, C., Scalbert, A., Morand, C., Rémésy, C., and Jiménez, L. (2004) Polyphenols: food sources and bioavailability. *American Journal of Clinical Nutrition* 79, 727–747.

Manickam, S., Rajagopalan, V.R., Kambale, R., Rajasekaran, R., Kanagarajan, S., and Muthurajan, R. (2023) Plant metabolomics: current initiatives and future prospects. *Current Issues in Molecular Biology* 45, 8894–8906.

Markley, J.L., Bruschweiler, R., Edison, A.S., Eghbalnia, H.R., Powers, R., *et al.* (2017) The future of NMR-based metabolomics. *Current Opinion in Biotechnology* 43, 34–40.

Marshall, A.G., Hendrickson, C.L., and Jackson, G.S. (1998) Fourier transform ion cyclotron resonance mass spectrometry: a primer. *Mass Spectrometry Reviews* 17, 1–35.

Matsunami, M., Toyofuku, K., Kimura, N., and Ogawa, A. (2020) Osmotic stress leads to significant changes in rice root metabolic profiles between tolerant and sensitive genotypes. *Plants* 9: 1503.

Mazza, G. and Miniati, E. (1993) *Anthocyanins in Fruits, Vegetables, and Grains*. CRC Press, Boca Raton, Florida.

Misra, B.B. and Chen, S. (2015) Advances in understanding CO_2 responsive plant metabolomes in the era of climate change. *Metabolomics* 11, 1478–1491.

Misra, B.B. and van der Hooft, J.J. (2016) Updates in metabolomics tools and resources: 2014–2015. *Electrophoresis* 37, 86–110.

Mitchell, J.M., Flight, R.M., Wang, Q.J., Higashi, R.M., Fan, T.W.M., *et al.* (2018) New methods to identify high peak density artifacts in Fourier transform mass spectra and to mitigate their effects on high-throughput metabolomic data analysis. *Metabolomics* 14: 125.

Mudge, E.M., Brown, P.N., and Murch, S.J. (2019) The terroir of cannabis: terpene metabolomics as a tool to understand *Cannabis sativa* selections. *Planta Medica* 85, 781–796.

Neuweger, H., Albaum, S.P., Dondrup, M., Persicke, M., Watt, T., *et al.* (2008) MeltDB: a software platform for the analysis and integration of metabolomics experiment data. *Bioinformatics* 24, 2726–2732.

Okazaki, Y. and Saito, K. (2012) Recent advances of metabolomics in plant biotechnology. *Plant Biotechnology Reports* 6, 1–15.

Oliver, S.G., Winson, M.K., Kell, D.B., and Baganz, F. (1998) Systematic functional analysis of the yeast genome. *Trends in Biotechnology* 16, 373–378.

Osorio, S., Vallarino, J.G., Szecowka, M., Ufaz, S., Tzin, V., *et al.* (2013) Alteration of the interconversion of pyruvate and malate in the plastid or cytosol of ripening tomato fruit invokes diverse consequences on sugar but similar effects on cellular organic acid, metabolism, and transitory starch accumulation. *Plant Physiology* 161, 628–643.

Pang, Z., Zhou, G., Ewald, J., Chang, L., Hacariz, O., *et al.* (2022) Using MetaboAnalyst 5.0 for LC–HRMS spectra processing, multi-omics integration and covariate adjustment of global metabolomics data. *Nature Protocols* 17, 1735–1761.

Patel, M.K., Pandey, S., Kumar, M., Haque, M.I., Pal, S., and Yadav, N.S. (2021) Plants metabolome study: emerging tools and techniques. *Plants* 10: 2409.

Perchuk, I., Shelenga, T., Gurkina, M., Miroshnichenko, E., and Burlyaeva, M. (2020) Composition of primary and secondary metabolite compounds in seeds and pods of asparagus bean (*Vigna unguiculata* (L.) Walp.) from China. *Molecules* 25: 3778.

Perez de Souza, L., Naake, T., Tohge, T., and Fernie, A.R. (2017) From chromatogram to analyte to metabolite. How to pick horses for courses from the massive web resources for mass spectral plant metabolomics. *Gigascience* 6: gix037.

Peukert, M., Lim, W.L., Seiffert, U., and Matros, A. (2016) Mass spectrometry imaging of metabolites in barley grain tissues. *Current Protocols in Plant Biology* 1, 574–591.

Pluskal, T., Castillo, S., Villar-Briones, A., and Orešič, M. (2010) MZmine 2: modular framework for processing, visualizing, and analyzing mass spectrometry-based molecular profile data. *BMC Bioinformatics* 11: 395.

Ralston-Hooper, K., Hopf, A., Oh, C., Zhang, X., Adamec, J., and Sepúlveda, M.S. (2008) Development of GCxGC/TOF-MS metabolomics for use in ecotoxicological studies with invertebrates. *Aquatic Toxicology* 88, 48–52.

Ricroch, A.E., Bergé, J.B., and Kuntz, M. (2011) Evaluation of genetically engineered crops using transcriptomic, proteomic, and metabolomic profiling techniques. *Plant Physiology* 155, 1752–1761.

Roch, L., Prigent, S., Klose, H., Cakpo, C.B., Beauvoit, B., *et al.* (2020) Biomass composition explains fruit relative growth rate and discriminates climacteric from non-climacteric species. *Journal of Experimental Botany* 71, 5823–5836.

Rochfort, S., Vassiliadis, S., Maharjan, P., Brand, J., and Panozzo, J. (2019) NMR based metabolomic analysis of health promoting phytochemicals in lentils. *Metabolites* 9: 168.

Roessner, U., Wagner, C., Kopka, J., Trethewey, R.N., and Willmitzer, L. (2000) Simultaneous analysis of metabolites in potato tuber by gas chromatography–mass spectrometry. *Plant Journal* 23, 131–142.

Saini, R.K. and Keum, Y.S. (2017) An updated review on use of metabolomics in plant stress response studies. *Plant Ecology and Diversity* 10, 509–526.

Saito, K. and Matsuda, F. (2010) Metabolomics for functional genomics, systems biology, and biotechnology. *Annual Review of Plant Biology* 61, 463–489.

Shannon, P., Markiel, A., Ozier, O., Baliga, N.S., Wang, J.T., *et al.* (2003) Cytoscape: a software environment for integrated models of biomolecular interaction networks. *Genome Research* 13, 2498–2504.

Shelenga, T.V., Malyshev, L.L., Kerv, Y.A., Diubenko, T.V., Konarev, A.V., *et al.* (2020) Metabolomic approach to search for fungal resistant forms of *Aegilops tauschii* Coss. from the VIR collection. *Vavilovskii Journal of Genetics and Breeding* 24, 252–258.

Shen, S., Zhan, C., Yang, C., Fernie, A.R., and Luo, J. (2023) Metabolomics-centered mining of plant metabolic diversity and function: past decade and future perspectives. *Molecular Plant* 16, 43–63.

Smith, C.A., Want, E.J., O'Maille, G., Abagyan, R., and Siuzdak, G. (2006) XCMS: processing mass spectrometry data for metabolite profiling using nonlinear peak alignment, matching, and identification. *Analytical Chemistry* 78, 779–787.

Soga, T. and Heiger, D.N. (2000) Amino acid analysis by capillary electrophoresis electrospray ionization mass spectrometry. *Analytical Chemistry* 72, 1236–1241.

Soga, T. and Imaizumi, M. (2001) Capillary electrophoresis method for the analysis of inorganic anions, organic acids, amino acids, nucleotides, carbohydrates and other anionic compounds. *Electrophoresis* 22, 3418–3425.

Spina, R., Saliba, S., Dupire, F., Ptak, A., Hehn, A., *et al.* (2021) Molecular identification of endophytic bacteria in *Leucojum aestivum in vitro* culture, NMR-based metabolomics study and LC-MS analysis leading to potential Amaryllidaceae alkaloid production. *International Journal of Molecular Sciences* 22: 1773.

Spricigo, P.C., Correia, B.S.B., Borba, K.R., Taver, I.B., Machado, G.D.O., *et al.* (2021) Classical food quality attributes and the metabolic profile of cambuci, a native Brazilian Atlantic rainforest fruit. *Molecules* 26: 3613.

Sugiyama, A., Shitan, N., and Yazaki, K. (2007) Involvement of a soybean ATP-binding cassette-type transporter in the secretion of genistein, a signal flavonoid in legume–*Rhizobium* symbiosis. *Plant Physiology* 144, 2000–2008.

Sun, B.S., Gu, L.J., Fang, Z.M., Wang, C.Y., Wang, Z., and Lee, M.R. (2009) Simultaneous quantification of 14 ginsenosides in Panax ginseng with different growth ages using high-performance liquid chromatography coupled with evaporative light scattering detection. *Journal of Chromatography A* 1216, 5137–5145.

Sung, J., Lee, S., Lee, Y., Ha, S., Song, B., *et al.* (2015) Metabolomic profiling from leaves and roots of tomato (*Solanum lycopersicum* L.) plants grown under nitrogen, phosphorus or potassium-deficient conditions. *Plant Science* 241, 55–64.

Takats, Z., Wiseman, J.M., Gologan, B., and Cooks, R.G. (2004) Mass spectrometry sampling under ambient conditions with desorption electrospray ionization. *Science* 306, 471–473.

Tautenhahn, R., Böttcher, C., and Neumann, S. (2008) Highly sensitive feature detection for high resolution LC/MS. *BMC Bioinformatics* 9: 504.

Tautenhahn, R., Patti, G.J., Rinehart, D., and Siuzdak, G. (2012) XCMS Online: a web-based platform to process untargeted metabolomic data. *Analytical Chemistry* 84, 5035–5039.

Testone, G., Sobolev, A., Gonnella, M., Renna, M., Mannina, L., *et al.* (2019) Insights into sucrose pathway of chicory stems by integrative transcriptomic and metabolic analyses. *Phytochemistry* 167: 112086.

Theodoridis, G., Gika, H., Franceschi, P., Caputi, L., Arapitsas, P., *et al.* (2012) LC-MS based global metabolite profiling of grapes: Solvent extraction protocol optimisation. *Metabolomics* 8, 175–185.

Thurman, E.M. and Ferrer, I. (2010) Capillary electrophoresis–mass spectrometry: a perfect marriage of high efficiency and molecular identification. *Journal of Chromatography A* 1217, 3955–3967.

Turner, M.F., Heuberger, A.L., Kirkwood, J.S., Collins, C.C., Wolfrum, E.J., *et al.* (2016) Non-targeted metabolomics in diverse sorghum breeding lines indicates primary and secondary metabolite profiles are associated with plant biomass accumulation and photosynthesis. *Frontiers in Plant Science* 7: 953.

Tweeddale, H., Notley-McRobb, L., and Ferenci, T. (1998) Effect of slow growth on metabolism of *Escherichia coli*, as revealed by global metabolite pool ("metabolome") analysis. *Journal of Bacteriology* 180, 5109–5116.

Urrutia, M., Blein-Nicolas, M., Prigent, S., Bernillon, S., Deborde, C., *et al.* (2021) Maize metabolome and proteome responses to controlled cold stress partly mimic early-sowing effects in the field and differ from those of *Arabidopsis*. *Plant, Cell & Environment* 44, 1504–1521.

Vaclavik, L., Lacina, O., Hajslova, J., and Zweigenbaum, J. (2013) The use of high resolution mass spectrometry and accurate mass analyzers in food analysis. *Analytica Chimica Acta* 787, 27–35.

Valentino, G., Graziani, V., d'Abrosca, B., Pacifico, S., Fiorentino, A., and Scognamiglio, M. (2020) NMR-based plant metabolomics in nutraceutical research: an overview. *Molecules* 25: 1444.

van Dam, N.M. and Bouwmeester, H.J. (2016) Metabolomics in the rhizosphere: tapping into belowground chemical communication. *Trends in Plant Science* 21, 256–265.

van den Berg, R.A., Hoefsloot, H.C., Westerhuis, J.A., Smilde, A.K., and van der Werf, M.J. (2006) Centering, scaling, and transformations: improving the biological information content of metabolomics data. *BMC Genomics* 7: 142.

van der Heijden, R., Jacobs, D.I., Snoeijer, W., Hallard, D., and Verpoorte, R. (2004) The *Catharanthus* alkaloids: pharmacognosy and biotechnology. *Current Medicinal Chemistry* 11, 607–628.

Verma, A. and Kumar, N. (2020) Advances in NMR-based metabolomics of plants. *Methods in Molecular Biology* 2072, 63–84.

Vinaixa, M., Schymanski, E.L., Neumann, S., Navarro, M., Salek, R.M., and Yanes, O. (2016) Mass spectral databases for LC/MS- and GC/MS-based metabolomics: state of the field and future prospects. *Trends in Analytical Chemistry* 78, 23–35.

Wahyuni, D.S., Choi, Y.H., Leiss, K.A., and Klinkhamer, P.G. (2021) Morphological and chemical factors related to western flower thrips resistance in the ornamental gladiolus. *Plants* 10: 1384.

Walters, D.R. and McRoberts, N. (2006) Plants and biotrophs: a pivotal role for cytokinins? *Trends in Plant Science* 11, 581–586.

Wang, M., Chen, L., Liang, Z., He, X., Liu, W., *et al.* (2020) Metabolome and transcriptome analyses reveal chlorophyll and anthocyanin metabolism pathway associated with cucumber fruit skin color. *BMC Plant Biology* 20: 386.

Wang, S., He, T., Xu, F., Li, X., Yuan, L., *et al.* (2021) Analysis of physiological and metabolite response of *Celosia argentea* to copper stress. *Plant Biology* 23, 391–399.

Wang, Z., Gerstein, M., and Snyder, M. (2009) RNA-Seq: a revolutionary tool for transcriptomics. *Nature Reviews Genetics* 10, 57–63.

Want, E.J., Cravatt, B.F., and Siuzdak, G. (2005) The expanding role of mass spectrometry in metabolite profiling and characterization. *Chembiochem* 6, 1941–1951.

Weckwerth, W. (2003) Metabolomics in systems biology. *Annual Review of Plant Biology* 54, 669–689.

Winning, H., Viereck, N., Wollenweber, B., Larsen, F.H., Jacobsen, S., *et al.* (2009) Exploring abiotic stress on asynchronous protein metabolism in single kernels of wheat studied by NMR spectroscopy and chemometrics. *Journal of Experimental Botany* 60, 291–300.

Wu, X., Li, N., Li, H., and Tang, H. (2014) An optimized method for NMR-based plant seed metabolomic analysis with maximized polar metabolite extraction efficiency, signal-to-noise ratio, and chemical shift consistency. *Analyst* 139, 1769–1778.

Yamashita, H., Sonobe, R., Hirono, Y., Morita, A., and Ikka, T. (2020) Dissection of hyperspectral reflectance to estimate nitrogen and chlorophyll contents in tea leaves based on machine learning algorithms. *Scientific Reports* 10: 17360.

Yurekten, O., Payne, T., Tejera, N., Amaladoss, F.X., Martin, C., *et al.* (2024) MetaboLights: open data repository for metabolomics. *Nucleic Acids Research* 52, D640–D646.

Zhang, A., Sun, H., and Wang, X. (2016) Mass spectrometry-driven drug discovery for development of herbal medicine. *Mass Spectrometry Reviews* 37, 307–320.

Zhang, W. and Ramautar, R. (2021) CE-MS for metabolomics: developments and applications in the period 2018–2020. *Electrophoresis* 42, 381–401.

Zhang, W., Chang, J., Lei, Z., Huhman, D., Sumner, L.W., and Zhao, P.X. (2014) MET-COFEA: a liquid chromatography/mass spectrometry data processing platform for metabolite compound feature extraction and annotation. *Analytical Chemistry* 86, 6245–6253.

Zhang, W., Lei, Z., Huhman, D., Sumner, L.W., and Zhao, P.X. (2015) MET-XAlign: a metabolite cross-alignment tool for LC/MS-based comparative metabolomics. *Analytical Chemistry* 87, 9114–9119.

Zhu, H.H., Yang, J.X., Xiao, C.H., Mao, T.Y., Zhang, J., and Zhang, H.Y. (2019) Differences in flavonoid pathway metabolites and transcripts affect yellow petal colouration in the aquatic plant *Nelumbo nucifera*. *BMC Plant Biology* 19: 277.

Zivy, M., Wienkoop, S., Renaut, J., Pinheiro, C., Goulas, E., and Carpentier, S. (2015) The quest for tolerant varieties: the importance of integrating "omics" techniques to phenotyping. *Frontiers in Plant Science* 6: 448.

Zrodnikov, Y. and Davis, C.E. (2012) The highs and lows of FAIMS: predictions and future trends for high field asymmetric waveform ion mobility spectrometry. *Journal of Nanomedicine and Nanotechnology* 3: 109e.

Zuorro, A., Lavecchia, R., Medici, F., and Piga, L. (2013) Enzyme-assisted production of tomato seed oil enriched with lycopene from tomato pomace. *Food and Bioprocess Technology* 6, 3499–3509.

Zuther, E., Koehl, K., and Kopka, J. (2007) Comparative analysis of the transcriptome and metabolome of *Arabidopsis thaliana* acclimated to thermal stress and naturally grown in different seasons. *Plant Journal* 51, 271–292.

17 Integrating Advanced Metabolomics with Plant Functional Genomics

Akash Shaikh[1], Varsha Toppo[1], Jameel M. Al-Khayri[2], and Praveen Nagella[1]*
[1]Department of Life Sciences, School of Sciences, CHRIST (Deemed to be University), Hosur Road, Bengaluru, India; [2]Department of Agricultural Biotechnology, College of Agriculture and Food Sciences, King Faisal University, Al-Ahsa, Saudi Arabia

Abstract

Metabolomics encompasses the entire suite of small-molecule compounds or metabolites synthesized by an organism, whereas functional genomics refers to the gene-level functioning of an organism. The genome of a plant will dictate its metabolome, but the link between the two omics data may not always be clearly visible or properly explored. This chapter delves into the integration of advanced metabolomics with plant functional genomics, highlighting its pivotal role in advancing our understanding of plant biology and its applications in agriculture. Metabolomics provides a comprehensive analysis of small molecules, bridging the gap between genotype and phenotype by elucidating the dynamic interactions within plant systems. Key techniques such as mass spectrometry and nuclear magnetic resonance are explored, emphasizing their importance in high-throughput and high-resolution metabolite profiling. The chapter further discusses the synergy between metabolomics and other omics technologies, including genomics, transcriptomics, and proteomics, underscoring its significance in identifying gene functions and metabolic pathways linked to complex traits such as stress tolerance. Applications in plant breeding are also highlighted, showcasing how metabolomics can drive the development of crops with enhanced stress resilience, yield, and nutritional quality. The chapter concludes by emphasizing the transformative potential of this integrated approach in shaping future agricultural practices and improving food security.

Keywords: functional genomics, gene function, metabolomics, omics integration, plant breeding

17.1 Introduction

In any living system, metabolites—the small-molecule end products of metabolism—are arguably the most important functional compounds, and collectively, the entire array of primary and secondary metabolites produced by an organism is called the metabolome. The metabolome of a functional living organism is entirely responsible for the metabolic reactions that maintain normal functionality and growth of cells in the organism (Bechert, 2012). Metabolomics is the field that involves the comprehensive analysis of the metabolome within a biological system, providing insights into metabolic pathways and their regulation. As primary and secondary metabolites of plants are invaluable in terms of nutritional and medicinal value, it is important to utilize the field of metabolomics to explore more about the unique molecules that impart such nutritional and medicinal values to plants. For such analysis, this field leverages advanced

* E-mail: praveen.n@christuniversity.in

© CAB International 2025. *High-Throughput Plant Metabolomics* (Jen-Tsung Chen ed.)
DOI: 10.1079/9781800626911.0017

analytical techniques such as mass spectrometry (MS) and nuclear magnetic resonance (NMR) spectroscopy to identify and quantify metabolites (Fig. 17.1) (Hagel and Facchini, 2008). These techniques enable researchers to map out the complex biochemical networks within plants, revealing how various metabolic pathways are interconnected and regulated. However, simply identifying the important metabolites in a plant serves a singular purpose and leaves many questions unanswered as to the genetic and environmental basis of the production and monitoring of such molecules. This creates the opportunity for the integration of metabolomics data with other omics data, such as genomics and transcriptomics, which can enhance our understanding of the dynamic metabolic changes occurring in response to genetic and external environmental factors (Hagel and Facchini, 2008). Functional genomics aims to understand the roles and interactions of genes and their products, often using high-throughput techniques such as sequencing and gene editing. This field focuses on elucidating gene functions and their regulatory mechanisms by analyzing gene-expression patterns, protein interactions, and phenotypic outcomes. Integrating metabolomics with functional genomics allows a holistic understanding

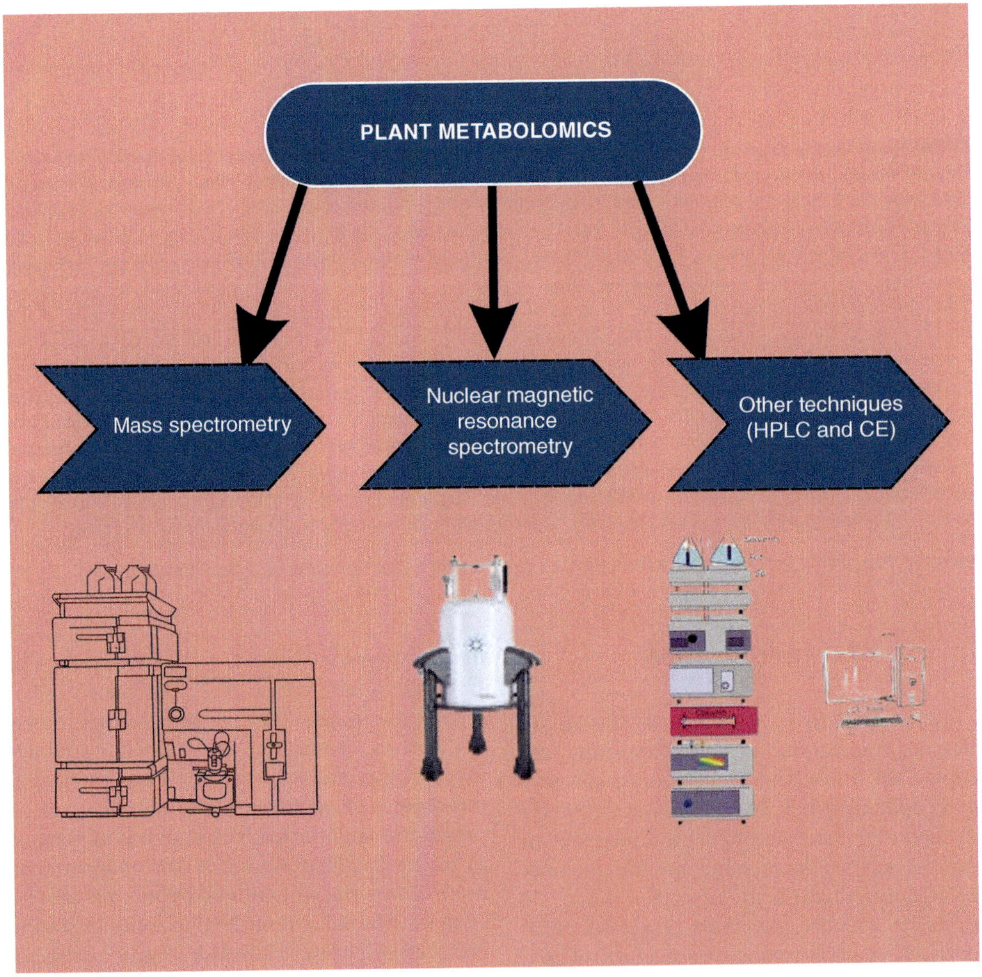

Fig 17.1. Overview of plant metabolomics techniques: mass spectrometry, nuclear magnetic resonance spectrometry, and other methods such as high-performance liquid chromatography (HPLC) and capillary electrophoresis (CE), are used to analyze plant metabolites. (Figure constructed using Canva.)

of plant biology, linking genetic variations to metabolic phenotypes. This chapter aims to provide a comprehensive overview of the methodologies and technologies employed in both metabolomics and functional genomics. It will cover advanced analytical techniques, high-throughput sequencing methods, and bioinformatics tools used to analyze and integrate data from both fields. The chapter will also discuss the benefits of integrating metabolomics with functional genomics, highlighting how this combined approach can enhance our understanding of plant biology. Furthermore, we will explore various applications, including the identification of metabolic pathways, gene-function elucidation, and the development of stress-resistant crops. In order to support the practicality of such applications, the chapter will present case studies demonstrating the successful integration of metabolomics and functional genomics in plant research. These case studies will illustrate the practical applications and potential of this integrated approach. Additionally, we will discuss future directions and emerging trends in the field, emphasizing the importance of continued research and technological advancements.

17.2 Technological Platforms in Metabolomics

17.2.1 Mass spectrometry

Metabolomics examines the makeup of metabolites in all species, including plants. Metabolomics allows us to obtain insight into plant phenotypic expression as well as explore metabolic modifications and regulation in order to better understand how plants adapt and defend themselves against environmental stress. The examination of primary and secondary metabolites starts with sample preparation, which entails extracting metabolites using a variety of procedures. Quenching, mechanical extraction, and ultrasonic extraction are all common procedures for metabolomic analysis, and they are frequently used in conjunction to optimize findings (Piasecka *et al.*, 2019).

Metabolomics research generally uses a variety of tools and approaches. Among these, MS stands out as an effective and commonly used analytical technique. MS provides a variety of sensitivity and resolution choices, which can be achieved with single or tandem mass analyzers. Various MS-based approaches are currently being used for both untargeted and targeted metabolic profiling. These techniques frequently combine MS with other separation methods such as liquid chromatography (LC-MS), gas chromatography (GC-MS), or capillary electrophoresis (CE-MS). Furthermore, advanced MS techniques such as Fourier transform–ion cyclotron resonance, matrix-assisted laser desorption/ionization, ion mobility spectrometry, and NMR are used. Notably, GC-MS outperforms LC-MS in metabolite separation and reduces ion suppression by exploiting gaseous-phase features and the MS ionization process (Shen *et al.*, 2023).

GC-MS, LC-MS, and tandem MS (MS/MS) are essential techniques in plant metabolomics, valued for their sensitivity and precision. Among these, LC-MS/MS has gained prominence as a preferred method for analyzing diverse metabolites, such as amino acids and secondary metabolites, due to its exceptional specificity and ability to effectively manage complex biological samples (Gautam *et al.*, 2023). LC-MS enables comprehensive metabolite profiling, supporting both qualitative and quantitative analyses crucial for advancing plant biology and meeting regulatory standards (Kılınc *et al.*, 2023). Although GC-MS is particularly effective for analyzing volatile compounds, LC-MS/MS offers greater versatility across a wider range of metabolites, making it the preferred method in numerous studies. Together, these techniques have significantly enhanced our analytical abilities in plant metabolomics, with the choice of method typically influenced by the specific metabolites being studied and the complexity of the sample matrix (Yi *et al.*, 2022).

LC-MS is employed in plant metabolomics for untargeted metabolite discovery, including the identification of specialized metabolites, providing a thorough understanding of plant metabolic diversity. LC-MS/MS is widely utilized in plant metabolomics for detecting phytohormones, folic acid, flavonoids, and other secondary metabolites, highlighting its crucial role in the analysis of plant secondary metabolites (Mori *et al.*, 2023). Exact-mass GC-MS is beneficial in plant metabolomics for profiling primary metabolites, resolving isotopic patterns, and localizing sulfur-containing metabolites. It also provides a curated database, enabling precise metabolic analysis (Abadie *et al.*, 2022).

17.2.2 NMR spectroscopy

NMR spectroscopy is a crucial technique in plant metabolomics, distinguished by several key features. First, NMR offers detailed molecular insights with minimal sample preparation, preserving the integrity of biological samples, which is vital for conducting longitudinal studies in plant systems. It supports comprehensive metabolic profiling by detecting a broad spectrum of metabolites simultaneously in complex matrices, such as plant tissues and extracts (Peng *et al.*, 2024). Additionally, the capacity of NMR to elucidate molecular structures through multi-dimensional techniques enhances our understanding of metabolic pathways and interactions. Recent advancements, such as hyperpolarization methods, have significantly improved sensitivity, enabling the detection of low-abundance metabolites and the real-time monitoring of metabolic activities. Despite these advancements, challenges such as sensitivity limitations persist, highlighting the need for continuous methodological refinement. Overall, NMR's non-destructive nature, structural elucidation capabilities, and extensive profiling make it an essential tool for advancing plant metabolomics research (Olorunkoya *et al.*, 2024).

NMR spectroscopy in plant metabolomics offers detailed structural insights of metabolites non-destructively, aiding in comprehensive chemical analysis and molecular identification (Kumar and Jaitak, 2024). NMR spectroscopy assesses organic molecule structures, useful in plant metabolomics for non-destructive analysis. It utilizes nuclear spin interactions with magnetic fields, spin coupling, and two-dimensional techniques (Lo Meo, 2024). Recent advancements, such as the development of the automated quantification algorithm (AQuA), have enhanced the efficiency of metabolite quantification in complex samples such as plant root exudates, addressing challenges posed by background signals (Alexandersson *et al.*, 2023). Additionally, high-resolution magic angle spinning NMR eliminates the need for metabolite extraction, simplifying the workflow and enhancing the analysis of intact plant tissues. Moreover, innovative techniques such as the pure shift yielded by chirp excitation (PSYCHE) method improve spectral resolution, making it easier to evaluate overlapping signals in complex plant extracts. These advancements collectively highlight NMR's versatility and effectiveness in deepening our understanding of plant metabolomics (Stark *et al.*, 2020).

17.2.3 Other techniques

High-performance liquid chromatography (HPLC) is vital in plant metabolomics because it enables the simultaneous quantification and identification of many compounds. For example, Araujo-León *et al.* (2024) created an HPLC ultraviolet diode array detector method to assess flavonoids and phenolic chemicals in *Amaranthus cruentus*, highlighting their potential health advantages, such as antidiabetic and antihypertensive treatments. Similarly, Hai *et al.* (2024) used HPLC to distinguish processed rhizomes of *Atractylodes macrocephala*, indicating its utility in quality assurance and metabolite profiling. Furthermore, combining HPLC with chemometric techniques such as principal component analysis improves our understanding of complicated metabolomic data, allowing the discovery of bioactive chemicals and their roles in medicinal plants. Overall, HPLC is an effective analytical method in plant metabolomics, yielding useful information about the biochemical makeup and possible applications of plant-derived compounds (Xiao *et al.*, 2022).

CE-MS detects and quantifies hundreds of charged metabolites using high-resolution separations, sensitive mass determination, and internal chemical standards in both positive and negative ionization modes. CE advances analytical chemistry from the micro- to the nanoscale, enabling investigation of individual cells or molecules. CE is ideal for analyzing highly polar and charged metabolites such as amino acids, nucleotides, tiny organic acids, and phosphoric sugars (Carrera *et al.*, 2021).

17.3 Functional Genomics Approaches

17.3.1 Genomic sequencing

Functional genomics techniques in plant metabolomics use comparative analysis and modern sequencing technology to elucidate the genetic basis for metabolic variation. Recent research has emphasized the importance of intra- and inter species comparisons in identifying metabolic polymorphisms and gene clusters that contribute to specialized metabolism in plants (Watanabe and Tohge, 2023). The combination of metabolomics, genetics, and proteomics has created new opportunities for understanding the biological importance of metabolic chemicals that are critical for plant adaptability and stress response (Saifi *et al.*, 2023). These techniques support the development of metabolic engineering solutions that meet societal objectives in addition to helping identify genes involved in the production of metabolites (Chandok *et al.*, 2022).

Techniques

Functional genomics provides a comprehensive understanding of gene expression and control through the use of contemporary techniques such as whole-genome sequencing (WGS), RNA sequencing (RNA-seq), and single-cell sequencing. The identification of genetic variations and mutations that may impact gene function and disease manifestation is made possible by WGS (Shen *et al.*, 2023). RNA-seq, a powerful transcriptome analytic technique, allows quantification of RNA levels throughout the entire genome, making it easier to analyze how gene expression changes under different conditions. Furthermore, single-cell sequencing methods provide insights into gene expression at the individual cell level, revealing cellular heterogeneity and developmental processes that are usually ignored in bulk analysis (Vo and Thorne, 2024).

Applications

Functional genomics methods such as WGS, RNA-seq, and single-cell applications are critical in improving our understanding of plant metabolomics. Comprehensive insights into genetic diversity and metabolic pathways are provided by WGS, which makes it possible to compare studies between species and cultivars to find metabolic polymorphisms (Watanabe and Tohge, 2023). Understanding plant development and metabolism requires an understanding of cellular heterogeneity and spatial gene-expression patterns, which can be found by RNA-seq, particularly single-cell RNA-seq, which offers a thorough picture of gene expression at the individual cell level (Wang *et al.*, 2024). Additionally, integrating transcriptomics with metabolomics improves our ability to correlate gene expression with metabolic profiles, shedding light on the functional roles of specific genes in specialized metabolism. Despite promising advancements, the application of these cutting-edge techniques in plant research is still developing, highlighting the need for continued exploration to fully leverage their potential in horticulture and plant sciences (Shen *et al.*, 2023).

17.3.2 Gene editing and functional studies

CRISPR-Cas9

Clustered regularly interspaced short palindromic repeats (CRISPR)-Cas9 technology has become a transformative tool for precise gene editing in plant metabolomics, enabling researchers to explore gene functions related to secondary metabolite production. This technology facilitates targeted modifications in plant genomes, allowing the investigation of previously uncharacterized gene clusters involved in the biosynthesis of valuable phytometabolites, such as terpenoids and flavonoids (Asha and Sunita, 2023). By employing knock-out and knock-in strategies to manipulate metabolic pathways, researchers can enhance the yield of these important compounds, which are critical for medicinal applications (Atia *et al.*, 2024). The straightforward operation of CRISPR-Cas9 and its broad target range make it an appealing choice for both fundamental and applied research in plant genetics. However, challenges such as off-target effects and regulatory constraints are significant considerations in its deployment. Overall, CRISPR-Cas9 not only deepens our understanding of plant metabolomics but also offers potential for improving agricultural practices and developing new medicinal compounds (Cheng, 2024).

RNA interference

RNA interference (RNAi) has emerged as an important tool in plant metabolomics, allowing the precise regulation of secondary metabolite pathways and the enhancement of crop attributes. RNAi enables post-transcriptional gene silencing, allowing researchers to target specific genes involved in metabolic pathways and increase the generation of useful secondary metabolites with important biological applications (Mitra *et al.*, 2022). Studies have demonstrated that RNAi can successfully affect gene expression related to stress tolerance and nutritional content, hence promoting sustainable agriculture methods. Comparative assessments show that RNAi-based genetic engineering in crops such as maize has no unintended consequences beyond those observed in conventional breeding, highlighting its safety and effectiveness (Kaur and Talekar, 2024). Furthermore, microRNAs inside the RNAi mechanism play important roles in regulating multiple biosynthetic pathways, broadening the scope of metabolic engineering in plants (Pandita, 2022).

17.3.3 Case studies

RNAi and CRISPR-Cas9 technologies have made substantial advances in plant metabolomics, particularly in stress tolerance through gene editing in flavonoid biosynthesis (Chongtham and Yadav, 2023). CRISPR-Cas9 has emerged as a highly accurate tool for changing metabolic pathways, allowing the controlled increase of secondary metabolites such as flavonoids, which play an important role in plant stress responses (Mohamed *et al.*, 2024). This method has been successfully applied to more than 20 crops, increasing their resistance to abiotic challenges such as drought and salinity. Furthermore, the capacity of CRISPR-Cas9 to modify phytohormones and stress-related genes facilitates the generation of stress-resistant plant varieties. While CRISPR-Cas9 has enormous potential, problems such as off-target consequences and the necessity for effective delivery systems must be considered when using it (Erdoğan *et al.*, 2023).

Global warming has exacerbated drought and temperature stress, prompting the adoption of CRISPR-Cas9 to improve drought tolerance in plants. In rice, CRISPR/Cas9 has been used to target genes such as *OsERA1*, *OsDST*, *SRL1*, *SRL2*, and *SAPK2*, leading to higher water retention, decreased stomatal density, improved reactive oxygen species scavenging, and overall drought tolerance (Yue *et al.*, 2020). Further studies using CRISPR-Cas9 to edit ERF family members, including *OsBIERF1*, *OsBIERF3*, and *OsBIERF4* in rice, *ARGOS8* in maize, and *TaERF3* in wheat, have also demonstrated enhanced drought tolerance (Nascimento *et al.*, 2023). Low temperatures cause stress in plants by reducing metabolic responses. CRISPR-Cas9 was used to create a rice variant with *pyl1/4/6* triple knockdown, leading to higher yield, improved temperature tolerance, and reduced preharvest germination. Additionally, editing genes such as *CBF1*, *OsANN3*, *SAPK2*, and *OsMYB30* in rice improved drought tolerance by nearly 63% (Zeng *et al.*, 2020).

Nilaparvata lugens, commonly known as the brown planthopper, is a highly detrimental insect pest to rice crops. By targeting the vacuolar ATP synthase component E of *N. lugens* by ingestion of double-stranded RNA, researchers were able to successfully induce RNAi (Halder *et al.*, 2022). Furthermore, a cotton plant that expresses CYP6AE14 dsRNA was created, which increased the plant's resistance to the *Helicoverpa armigera* (cotton bollworm) (Halder *et al.*, 2022). The receptor for activated C-kinase 1 (RACK-1) is a highly conserved scaffold protein crucial for plant growth and development. RNAi-mediated downregulation of the *RACK-1* gene in transgenic *Oryza sativa* (rice) plants resulted in greater tolerance to drought stress compared with non-transgenic plants. The *OsTZF1* gene, when overexpressed in transgenic plants, improved tolerance to high salt and drought stressors. In contrast, *O. sativa* plants with the *OsTZF1* gene silenced via RNAi technology showed decreased tolerance, demonstrating the critical involvement of the *OsTZF1* gene in abiotic stress tolerance (Rajput *et al.*, 2021).

17.4 Data Integration and Analysis

17.4.1 Combining omics data

The seamless conjoining of metabolomics with genomics, transcriptomics, proteomics, and other omics data is essential for a comprehensive understanding of how organisms work. This integration can be achieved through various bioinformatics tools and databases that facilitate the combination and analysis of multi-omics data. One approach involves the use of pathway-based methods, which map metabolites, genes, and proteins on to known biochemical pathways. This method allows researchers to identify how changes in one omics layer affect others within the same biological pathway (Wieder *et al.*, 2024). Such an approach was proposed by the team behind PathIntegrate, because they recognized that the usual multi-omics approach involved outputting clusters or lists of molecules associated with a particular outcome, which had to be further investigated to discern the specific pathways involved, resulting in a time-consuming process. Pathway-based integration reverses this idea, focusing on simple single-pathway systems instead, and encompassing multi-omics data from single molecules to complex biological pathways (Wieder *et al.*, 2024).

An alternative method is the use of constraint-based stoichiometric metabolic models, as demonstrated by the INTEGRATE pipeline. This approach integrates transcriptomics with metabolomics to predict metabolic fluxes and identify regulatory mechanisms at both the gene expression and metabolic levels (Di Filippo *et al.*, 2022). These stoichiometric models, which represent metabolic networks mathematically, incorporate data from genomics, transcriptomics, proteomics, and metabolomics to analyze complex biological interactions. By applying constraints based on biochemical laws, the models can predict cellular responses under various conditions. This helps to create a biologically accurate model, mimicking existing biochemical pathways, and aiding in the identification of biomarkers, understanding of disease mechanisms, and design of metabolic engineering strategies. This integrative approach enhances the analysis and interpretation of biological data, leading to more accurate insights into cellular functions and disease states (Di Filippo *et al.*, 2022). Backed by multi-omics data, such modeling does away with the need for guesswork and ensures accurate predictions.

Tools and software

There are various software and platforms available that allow seamless integration of multi-omics data. Some of these are discussed below and listed in Table 17.1.

METABOANALYST. This comprehensive platform offers a range of statistical and functional analysis tools for metabolomics data. It supports multi-omics data integration through various modules designed for pathway analysis, biomarker discovery, and data visualization (Pang *et al.*, 2024). It is an entirely web-based platform, and offers robust tools for data normalization, transformation, and scaling, ensuring that the data are ready for analysis and integration with multi-omics data.

Table 17.1. List of popular platforms and software for multi-omics data integration.

Name	Year introduced	Main use	Reference
KEGG (Kyoto Encyclopedia of Genes and Genomes)	1995	Pathway mapping and data integration	Kanehisa and Goto (2000)
Cytoscape	2003	Network visualizations and integration of molecular interaction data	Shannon *et al.* (2003)
MetaboAnalyst	2009	Comprehensive metabolomics data analysis and integration	Xia *et al.* (2009)
mixOmics	2016	Multivariate analysis and data integration in R	Rohart *et al.* (2017)
INTEGRATE	2018	Integrative analysis of multi-omics data in Python	Di Filippo *et al.* (2022)
GraphOmics	2021	Integrates transcriptomics, proteomics, and metabolomics data	Wandy and Daly (2021)
MiBiOmics	2021	Multi-omics data visualization, exploration, and integration	Zoppi *et al.* (2021)
OmicsNet 2.0	2022	Multi-omics integration and network visual analytics	Zhou *et al.* (2022)
ALASCA	2022	Analyses data from various omics	Jarmund *et al.* (2022)

CYTOSCAPE. Cytoscape was started as a network analysis software that supports the visualization and analysis of complex biological networks (Wanichthanarak *et al.*, 2015). Over time, analysis of multi-omics data on this platform has been made possible by the use of plugins such as Metscape, which enables users to interpret and visualize metabolomics data along with expression profiling, particularly in the context of human metabolism (Karnovsky *et al.*, 2012). By adapting this plugin for other extended purposes, scientists have been able to integrate metabolomics data with genomics and proteomics data, facilitating the construction and analysis of gene-to-metabolite networks (Basu *et al.*, 2017).

KYOTO ENCYCLOPEDIA OF GENES AND GENOMES (KEGG). KEGG is a database resource that provides information on genes, proteins, and metabolites, as well as their interactions within biological pathways. KEGG is widely used for pathway mapping and functional annotation of multi-omics data (Wanichthanarak *et al.*, 2015). Using KEGG's extensive database as a base platform, several useful tools have been developed in recent years for multi-omics integration, such as OmicsNet 2.0 and GraphOmics (Wandy and Daly, 2021; Zhou *et al.*, 2022). This shows the utility of consolidating multi-omics data under one umbrella, allowing scientists easy access to vast data sets and the opportunity to work simultaneously with different data sets on compatible platforms.

OMICSNET 2.0. As mentioned above, OmicsNet 2.0 builds on the availability of huge databases such as KEGG, and functions as a one-stop solution for integration and synchronization of such multiple data sets. Since its inception, OmicsNet has helped scientists correlate disease occurrence with data available across transcriptomics, metabolomics, and genomics to decipher causes and correlations previously unexplained when approached from a singular point of view such as genomics (Zhou *et al.*, 2022). This holds the potential for linking disease-resistance mechanisms in plants and the genetic pathways responsible for secondary metabolite productions that cause this resistance.

17.4.2 Statistical and computational approaches

The rapid growth of the number of platforms developed for integrating multi-omics data highlights that the field has been deemed to be extremely important for multiple disciplines, including disease management, physiological analysis, healthcare, agriculture, and drug design. All these platforms follow a few common statistical and computational approaches depending on the field they are being deployed in, and such approaches rely mainly on machine learning, network analysis, and pathway modeling.

Machine learning

Machine learning techniques are increasingly being applied to multi-omics data integration to uncover patterns and build predictive models. Such techniques can tackle the high dimensionality and complexity of multi-omics data sets, making them suitable for tasks such as classification, regression, and clustering. Such extensive analysis and grouping work would previously have been done manually, but algorithmic pattern recognition through machine learning and deep learning makes that process extremely streamlined and quick. This allows rapid transition from theoretical data obtained from multi-omics analysis to practical applications of it, such as plant breeding (Yan and Wang, 2023). Machine learning approaches can be both supervised and unsupervised. In supervised learning algorithms, the predictors and the proposed model are clearly defined, and the aim is to establish a link. Such approaches have been used to tag important sequences in the rice genome (Do, *et al.*, 2018), differentiate between the regulatory and non-regulatory regions of the maize genome (Mejía-Guerra and Buckler, 2019), and find correlations between different plant stresses and their phenotypic manifestations (Ghosal *et al.*, 2018). Unsupervised learning approaches help identify underlying structures and patterns within the data and can be immensely useful in linking metabolomics and functional genomics to determine any common genomic functions in metabolic pathways across different species of plants.

Network analysis

Network analysis is a powerful approach to understanding the complex interactions between genes, proteins, and metabolites. By constructing and analyzing biological networks, researchers can identify key regulatory nodes and pathways involved in various biological processes. Tools such as Cytoscape and Metscape, as discussed earlier, facilitate the construction and visualization of gene-to-metabolite networks (Basu *et al.*, 2017). These networks provide insights into the functional relationships between different omics layers and help in identifying potential targets for genetic modifications. In the context of linking metabolomics and plant functional genomics, network analysis can do wonders in predicting and mapping out complete metabolic pathways and determining the important and crucial ones among multi-pathway systems, such as those governed by core genomes.

Pathway modeling

Pathway modeling involves the use of computational models to simulate metabolic pathways and predict the effects of genetic modifications. These models can integrate data from multiple omics layers to provide a holistic view of metabolic regulation. For instance, constraint-based metabolic models, such as those used in the INTEGRATE pipeline, can predict how changes in gene expression or metabolite levels affect metabolic fluxes (Di Filippo *et al.*, 2022). These predictions can guide experimental designs and inform strategies for metabolic engineering. As an offshoot of pathway modeling, metabolomic quantitative trait locus (mQTL) mapping has already been used in conjunction with plant genomics to predict diagnostic markers in plants, identifying them as potential candidates for gene transplantation (Dumas and Gauguier, 2012; Fernandez *et al.*, 2016).

17.5 Applications of Integrated Metabolomics and Functional Genomics

17.5.1 Gene discovery and functional characterization

The integration of metabolomics with functional genomics has significantly advanced gene discovery and functional characterization in plants. For gene discovery and linking of metabolites to previously unmapped genes across a genome, a graph-clustering approach has already proven to be successful in finding glucosinolate genes in *Arabidopsis thaliana* ((Harun *et al.*, 2021). Another practical use of integrating metabolomics with functional genomics is to accurately link metabolites in plants with their phenotypic expressions, especially in terms of stress resistance and disease markers. One of the key techniques used in this integration is mQTL mapping (Dumas and Gauguier, 2012). This method links variations in metabolite profiles to specific genetic loci, thereby identifying candidate genes responsible for metabolic traits. For instance, mQTL mapping was successfully utilized to find a correlation between certain secondary metabolites in maize and the drought tolerance ability of the plant across the population (Zhang *et al.*, 2021). In the same vein, functional genomics, when integrated with metabolomics, enables the characterization of unknown enzymes and their roles in metabolic pathways. This approach uses both transcriptomics and proteomics data to identify candidate genes, followed by metabolomics to validate their functions. For example, in rice, the integration of data from transcriptomics and metabolomics led to the discovery of a novel biosynthetic gene cluster involved in the synthetic pathway of diterpenoid phytoalexins, which are crucial for plant defense (Liang *et al.*, 2021). By analyzing the changes in metabolite profiles in mutant lines, researchers were able to confirm that the particular gene cluster and the enzymes transcribed from the genes play a crucial role in disease resistance against rice fungal blasts. Similarly, in rice itself, another study took advantage of the integration of the transcriptome and metabolome to determine the differences in the metabolic signaling pathways between different rice varieties and their different tolerances to salinity stress (Zhou *et al.*, 2023). In the above examples and many more, linking metabolomics and functional genomics has proven to be very effective in accelerating our understanding of how plants function at a molecular and genetic level when it comes to disease- and stress-resistance mechanisms.

17.5.2 Plant breeding and crop improvement

As discussed above, the integration of metabolomics and functional genomics has helped in the identification of several genes and metabolic pathways responsible for various desirable plant traits such as better yields and higher stress resistance. This has opened up new pathways for plant breeding and crop improvement. By manipulating metabolic pathways, researchers can enhance these traits, either by upregulation of genes deemed to be useful by multi-omics-based modeling, or by splicing in genes from other plants that have been linked with desirable traits. For instance, in tomato, overexpression of a gene regulating the polyamine biosynthesis pathway led to increased tolerance to salinity and water-stress conditions (Zhao *et al.*, 2021). Metabolomics analysis revealed that the transgenic plants accumulated higher levels of osmoprotectants, which helped them withstand abiotic stresses (Nandy *et al.*, 2022). Similarly, in rice, changes in the metabolic pathways responsible for the biosynthesis of essential amino acids have resulted in improved nutritional quality (Zhao *et al.*, 2020). In another notable achievement, breeding programs that integrate metabolomics and functional genomics have successfully enhanced drought resistance in maize. By targeting specific metabolic pathways, researchers have developed maize varieties that can better withstand water scarcity. One notable example is the identification and manipulation of genes responsible for the synthesis of osmoprotectants such as proline and trehalose. Metabolomics analysis of drought-tolerant maize lines revealed higher levels of these compounds, which help maintain cellular homeostasis under water-deficit conditions (Zhang *et al.*, 2021). Another similar study revealed that lipid remodeling plays a crucial role in the drought-adapting ability of maize (Kränzlein *et al.*, 2023). These findings have been translated into breeding programs, which have successfully created drought-resistant maize varieties. Moreover, such studies also facilitate scientists to bypass the usage of genetic modifications using genes from other plants or organisms, and allows the introduction of specific models for plant breeding, which makes the crops more acceptable and accessible across markets.

17.6 Challenges and Future Directions

As with any fledgling branch of science, the integration of metabolomics with functional genomics in plant biology faces several technical and analytical challenges. One of the primary limitations is the sensitivity and resolution of current technologies. Metabolomics techniques such as MS and NMR spectroscopy often struggle to detect low-abundance metabolites, which can be crucial for understanding metabolic pathways. While known metabolites are easily identified, the difficulty of identifying unknown metabolites hinders the process of linking that data with genomics and transcriptomics, which undermines the reason for such multi-omics studies (Bingol and Brüschweiler, 2017). The sheer complexity and amount of data generated by metabolomics and functional genomics studies present another major challenge. High-throughput techniques can produce vast amounts of raw data, which is difficult to manage, analyze, and interpret. A single metabolomics experiment can generate thousands of data points, each representing a different metabolite. When combined with genomics and transcriptomics data, the complexity increases exponentially (Pinu *et al.*, 2019). This also leads to an additional challenge—that of integrating such a vast array of data. Combining data from metabolomics with genomics, transcriptomics, and proteomics requires sophisticated computational tools and algorithms. However, current methods often fall short in effectively integrating these diverse data sets, leading to incomplete or biased interpretations. For instance, the INTEGRATE pipeline, which aims to integrate metabolomics and transcriptomics data, highlights the complexity of accurately characterizing metabolic regulation due to the intricate interplay between different regulatory layers (Di Filippo *et al.*, 2022). Moreover, the heterogeneity of the data adds another layer of complexity. Different omics data sets often have varying formats, scales, and levels of noise, complicating the integration process (Flores *et al.*, 2023). Advanced computational methods and machine learning algorithms are required to handle this complexity, but these tools are still in their infancy and require further development.

Despite the challenges, the future of integrating metabolomics with functional genomics in plant biology looks promising, thanks to emerging technologies. Single-cell metabolomics is one such advancement that holds great potential (Yu *et al.*, 2023). Recent studies have shown the potential for single-cell metabolomics in plants, paving the way for more detailed and accurate metabolic profiling. Spatial metabolomics is another emerging field that combines metabolomics with imaging techniques to map the spatial distribution of metabolites within tissues (Hu *et al.*, 2023). This approach can reveal spatially resolved metabolic information, which is crucial for understanding tissue-specific metabolic processes and interactions. The combination of single-cell and partial metabolomics can be used to study the distribution of secondary metabolites in plant tissues, helping scientists to understand how they function in defense and signaling (Yu *et al.*, 2023). Integrative multi-omics approaches, which combine metabolomics with genomics, transcriptomics, and proteomics, are also gaining traction, and thanks to advances in computational methods and data integration tools, it is becoming increasingly feasible to implement these approaches in plant biology. In the near future, we can reasonably expect to see more plant breeding programs taking advantage of multi-omics approaches to try and integrate desirable features into crop plants, once the genetic and metabolomic causes responsible for such traits have been deciphered and modeled accurately.

17.7 Conclusion

The profound impact that advanced metabolomics, when integrated with plant functional genomics, has on our understanding of plant functions and agriculture cannot be overstated. The journey from genotype to transcriptome to phenotype is complex and multi-faceted, involving a variety of interconnected metabolic pathways that are intricately regulated by environmental and genetic factors. Through a detailed examination of metabolomics, particularly its role in capturing the dynamic nature of metabolites, this approach provides a unique and comprehensive view of the plant phenotype far beyond what traditional genomics alone can offer.

To achieve such results, the latest analytical and software techniques are indispensable. Cutting-edge techniques such as MS and NMR are crucial for metabolite profiling. These technologies have evolved to offer unprecedented sensitivity and accuracy, enabling the precise identification and quantification of thousands of metabolites, even those present in trace amounts. The advancements in ultra-performance LC coupled with MS have further enhanced the ability to dissect complex plant metabolomes, paving the way for more detailed functional analysis. Building on the foundation laid by such analytical techniques, the integration of metabolomics with other omics technologies, namely genomics, transcriptomics, and proteomics, is the key to understanding the complexities of plant systems, particularly in mapping and predicting the regulatory networks that control stress responses, growth, and development. By linking metabolite profiles to specific gene activities, researchers are able to identify candidate genes responsible for desirable traits, providing valuable insights into plant biology that can drive targeted breeding strategies, with a particular emphasis on the potential to enhance stress tolerance, yield, and nutritional content in crops. The identification of metabolic biomarkers linked to these traits allows more efficient selection processes in breeding programs, ultimately helping to develop crop varieties that are better suited to withstand environmental challenges while meeting the increasing global food demands due to the growing population.

In summary, the integration of advanced metabolomics with plant functional genomics is a transformative approach that enhances our ability to decode the complex interactions within plant systems. As we continue to refine these technologies and expand their applications, the potential for improving agricultural productivity and sustainability becomes increasingly attainable.

17.8 Acknowledgments

The authors express their appreciation for the support provided by the Deanship of Scientific Research, Vice Presidency for Graduate Studies and Scientific Research, King Faisal University, Saudi Arabia (grant no. KFU251719).

17.9 References

Abadie, C., Lalande, J., and Tcherkez, G. (2022) Exact mass GC-MS analysis: protocol, database, advantages and application to plant metabolic profiling. *Plant, Cell & Environment* 45, 3171–3183.

Alexandersson, E., Sandström, C., Meijer, J., Nestor, G., Broberg, A., and Röhnisch, H.E. (2023) Extended automated quantification algorithm (AQuA) for targeted ^{1}H NMR metabolomics of highly complex samples: application to plant root exudates. *Metabolomics* 20: 11.

Araujo-León, J.A., Sánchez-Del Pino, I., Ortiz-Andrade, R., Hidalgo-Figueroa, S., Carrera-Lanestosa, A., *et al.* (2024) HPLC-based metabolomic analysis and characterization of *Amaranthus cruentus* leaf and inflorescence extracts for their antidiabetic and antihypertensive potential. *Molecules* 29: 2003.

Asha, A. and Sunita, J. (2023) CRISPR-Cas-mediated optimization of plant-derived natural products of biomedical significance. In: Mérillon, J.-M. and Ramawat, K.G. (eds) *Plant Specialized Metabolites: Phytochemistry, Ecology and Biotechnology*. Springer, Cham, Switzerland, pp. 1–40.

Atia, M.A.M., Salah, A., Elsawah, A., Karam, A., and Abdelmoneim, T.K. (2024) CRISPR-based precision breeding of fruits, vegetables, and ornamental plants. In: Abd-Elsalam, K.A., Ahmad, A., and Zhang, B. (eds) *CRISPRized Horticulture Crops*. Academic Press, London, pp. 191–216.

Basu, S., Duren, W., Evans, C.R., Burant, C.F., Michailidis, G., and Karnovsky, A. (2017) Sparse network modeling and metscape-based visualization methods for the analysis of large-scale metabolomics data. *Bioinformatics* 33, 1545–1553.

Bechert, U. (2012) Noninvasive techniques to assess health and ecology of wildlife populations. In: Miller, R.E. and Fowler, M. (eds) *Fowler's Zoo and Wild Animal Medicine*. W.B. Saunders, Saint Louis, Missouri, pp. 60–70.

Bingol, K. and Brüschweiler, R. (2017) Knowns and unknowns in metabolomics identified by multidimensional NMR and hybrid MS/NMR methods. *Current Opinion in Biotechnology* 43, 17–24.

Carrera, F.P., Noceda, C., Maridueña-Zavala, M.G., and Cevallos-Cevallos, J.M. (2021) Metabolomics, a powerful tool for understanding plant abiotic stress. *Agronomy* 11: 824.

Chandok, I.K., Afreen, H., Afreen, R., Haider, S., Moharana, D.P., *et al.* (2022) Chapter 9 – Functional genomics tools for studying microbe-mediated stress tolerance in plants. In: Santoyo, G., Kumar, A., Aamir, M., and Uthandi, S. (eds) *Mitigation of Plant Abiotic Stress by Microorganisms*. Academic Press, London, pp. 175–204.

Cheng, Q. (2024) The application of CRISPR/Cas9 technology in plant gene editing. In: *Third International Conference on Biological Engineering and Medical Science (ICBioMed2023)*. SPIE, Bellingham, Washington, pp. 332–336.

Chongtham, I. and Yadav, P. (2023) CRISPR/Cas9 genome editing and plant stress management. In: *Global Climate Change and Plant Stress Management*. Wiley, Hoboken, New Jersey, pp. 335–359.

Di Filippo, M., Pescini, D., Galuzzi, B.G., Bonanomi, M., Gaglio, D., *et al.* (2022) INTEGRATE: model-based multi-omics data integration to characterize multi-level metabolic regulation. *PLOS Computational Biology* 18: e1009337.

Do, H., Than, K., and Larmande, P. (2018) Evaluating named-entity recognition approaches in plant molecular biology. In: Kaenampornpan, M., Malaka, R., Nguyen, D., and Schwind, N. (eds) *Multi-disciplinary Trends in Artificial Intelligence. MIWAI 2018. Lecture Notes in Computer Science, Vol. 11248*. Springer, Cham, Switzerland, pp. 219–225.

Dumas, M.-E. and Gauguier, D. (2012) Mapping metabolomic quantitative trait loci (mQTL): a link between metabolome-wide association studies and systems biology. In: Suhre, K. (ed.) *Genetics Meets Metabolomics: From Experiment to Systems Biology*. Springer, New York, pp. 233–254.

Erdoğan, İ., Cevher-Keskin, B., Bilir, Ö., Hong, Y., and Tör, M. (2023) Recent developments in CRISPR/Cas9 genome-editing technology related to plant disease resistance and abiotic stress tolerance. *Biology* 12: 1037.

Fernandez, O., Urrutia, M., Bernillon, S., Giauffret, C., Tardieu, F., *et al.* (2016) Fortune telling: metabolic markers of plant performance. *Metabolomics* 12: 158.

Flores, J.E., Claborne, D.M., Weller, Z.D., Webb-Robertson, B.M., Waters, K.M., and Bramer, L.M. (2023) Missing data in multi-omics integration: recent advances through artificial intelligence. *Frontiers in Artificial Intelligence* 6: 1098308.

Gautam, G., Parveen, R., and Ahmad, S. (2023) LC-MS-based metabolomics of medicinal plants. In: Ahmad, A. and Asif, A. (eds.) *Omics Studies of Medicinal Plants*. CRC Press, Boca Raton, Florida, pp. 157–180.

Ghosal, S., Blystone, D., Singh, A.K., Ganapathysubramanian, B., Singh, A., and Sarkar, S. (2018) An explainable deep machine vision framework for plant stress phenotyping. *Proceedings of the National Academy of Sciences USA* 115, 4613–4618.

Hagel, J.M. and Facchini, P.J. (2008) Plant metabolomics: analytical platforms and integration with functional genomics. *Phytochemistry Reviews* 7, 479–497.

Hai, C.T., Uyen, N.T., Giang, D.H., Minh, N.T.T., Duong, H.T., *et al.* (2024) Quantitative HPLC-based metabolomics approach for the discrimination of processed rhizomes of *Atractylodes macrocephala*. *Current Analytical Chemistry* 20, 41–51.

Halder, K., Chaudhuri, A., Abdin, M.Z., Majee, M., and Datta, A. (2022) RNA interference for improving disease resistance in plants and its relevance in this clustered regularly interspaced short palindromic repeats-dominated era in terms of dsRNA-based biopesticides. *Frontiers in Plant Science* 13: 885128.

Harun, S., Afiqah-Aleng, N., Karim, M.B., Altaf Ul Amin, M., Kanaya, S., and Mohamed-Hussein, Z.A. (2021) Potential *Arabidopsis thaliana* glucosinolate genes identified from the co-expression modules using graph clustering approach. *PeerJ* 9: e11876.

Hu, T., Allam, M., Cai, S., Henderson, W., Yueh, B., *et al.* (2023) Single-cell spatial metabolomics with cell-type specific protein profiling for tissue systems biology. *Nature Communications* 14: 8260.

Jarmund, A.H., Madssen, T.S., and Giskeødegård, G.F. (2022) ALASCA: an R package for longitudinal and cross-sectional analysis of multivariate data by ASCA-based methods. *Frontiers in Molecular Biosciences* 9: 962431.

Kanehisa, M. and Goto, S. (2000) KEGG: Kyoto encyclopedia of genes and genomes. *Nucleic Acids Research* 28, 27–30.

Karnovsky, A., Weymouth, T., Hull, T., Tarcea, V.G., Scardoni, G., *et al.* (2012) Metscape 2 bioinformatics tool for the analysis and visualization of metabolomics and gene expression data. *Bioinformatics* 28, 373–380.

Kaur, H. and Talekar, N. (2024) The function of RNA interference (RNAi) in crop enhancement—mechanism and applications: a review. *Journal of Experimental Agriculture International* 46: 20.

Kılınc, H., d'Urso, G., Paolillo, A., Alankus, O., Piacente, S., and Masullo, M. (2023) LC-MS and NMR based plant metabolomics: a comprehensive phytochemical investigation of *Symphytum anatolicum*. *Metabolites* 13: 1051.

Kränzlein, M., Schmöckel, S.M., Geilfus, C.M., Schulze, W.X., Altenbuchinger, M., *et al.* (2023) Lipid remodeling of contrasting maize (*Zea mays* L.) hybrids under repeated drought. *Frontiers in Plant Science* 14: 1050079.

Kumar, N. and Jaitak, V. (2024) Recent advancement in NMR based plant metabolomics: techniques, tools, and analytical approaches. *Critical Reviews in Analytical Chemistry* 2024, 1–25.

Liang, J., *et al.* (2021) Rice contains a biosynthetic gene cluster associated with production of the casbane-type diterpenoid phytoalexin ent-10-oxodepressin. *New Phytologist* 231, 85–93.

Lo Meo, P. (2024) Basics of NMR spectroscopy (for novices). In: Conte, P., Martino, D.F.C., and Lo meo, P. (eds) *The Environment in a Magnet: Applications of NMR Techniques to Environmental Problems*. Royal Society of Chemistry, London, pp. 434–459.

Mejía-Guerra, M.K. and Buckler, E.S. (2019) A k-mer grammar analysis to uncover maize regulatory architecture. *BMC Plant Biology* 19: 103.

Mitra, S., Ghorai, M., Kumar, V., Mandal, S., Jha, N.K., *et al.* (2022) Deciphering the potential of RNAi technology as modulator of plant secondary metabolites with biomedical significance. In: Swamy, M.K. and Kumar, A. (eds) *Phytochemical Genomics: Plant Metabolomics and Medicinal Plant Genomics*. Springer, Singapore, pp. 591–604.

Mohamed, H.I., Khan, A., and Basit, A. (2024) CRISPR-Cas9 system mediated genome editing technology: an ultimate tool to enhance abiotic stress in crop plants. *Journal of Soil Science and Plant Nutrition* 24, 1799–1822.

Mori, T., Rai, A., Tsugawa, H., Yamada, Y., and Saito, K. (2023) A liquid chromatography-mass spectrometry-based metabolomics strategy to explore plant metabolic diversity. *Methods in Enzymology* 680, 247–273.

Nandy, S., Das, T., Tudu, C.K., Mishra, T., Ghorai, M., *et al.* (2022) Unravelling the multi-faceted regulatory role of polyamines in plant biotechnology, transgenics and secondary metabolomics. *Applied Microbiology and Biotechnology* 106, 905–929.

Nascimento, F.D.S., Rocha, A.J., Soares, J.M.D.S., Mascarenhas, M.S., Ferreira, M.D.S., *et al.* (2023) Gene editing for plant resistance to abiotic factors: a systematic review. *Plants* 12: 305.

Olorunkoya, S.O., Fijabi, O., Ogbo, M.O., and Olugbola, T.S. (2024) The role of nuclear magnetic resonance (NMR) spectroscopy in cattle metabolism. *Engineering Science & Technology Journal* 5, 2110–2138.

Pandita, D. (2022) miRNA- and RNAi-mediated metabolic engineering in plants. In: Aftab, T. and Hakeem, K.R. (eds) *Metabolic Engineering in Plants*. Springer, Singapore, pp. 171–186.

Pang, Z., Lu, Y., Zhou, G., Hui, F., Xu, L., *et al.* (2024) MetaboAnalyst 6.0: towards a unified platform for metabolomics data processing, analysis and interpretation. *Nucleic Acids Research* 52, W398–W406.

Peng, Y., Zhang, Z., He, L., Li, C., and Liu, M. (2024) NMR spectroscopy for metabolomics in the living system: recent progress and future challenges. *Analytical and Bioanalytical Chemistry* 416, 2319–2334.

Piasecka, A., Kachlicki, P., and Stobiecki, M. (2019) Analytical methods for detection of plant metabolomes changes in response to biotic and abiotic stresses. *International Journal of Molecular Sciences* 20: 379.

Pinu, F.R., Beale, D.J., Paten, A.M., Kouremenos, K., Swarup, S., *et al.* (2019) Systems biology and multi-omics integration: viewpoints from the metabolomics research community. *Metabolites* 9: 76.

Rajput, M., Choudhary, K., Kumar, M., Vivekanand, V., Chawade, A, *et al.* (2021) RNA interference and CRISPR/Cas gene editing for crop improvement: paradigm shift towards sustainable agriculture. *Plants* 10: 1914.

Rohart, F., Gautier, B., Singh, A., and Lê Cao, K.A. (2017) mixOmics: an R package for 'omics feature selection and multiple data integration. *PLoS Computational Biology* 13: e1005752.

Saifi, M., Khan, M.A., and Abdin, M.Z. (2023) Functional genomics of secondary metabolite biosynthesis in medicinal plants. *Frontiers in Genetics* 14: 1223492.

Shannon, P., Markiel, A., Ozier, O., Baliga, N.S., Wang, J.T., *et al.* (2003) Cytoscape: a software environment for integrated models of biomolecular interaction networks. *Genome Research* 13, 2498–2504.

Shen, S., Zhan, C., Yang, C., Fernie, A.R., and Luo, J. (2023) Metabolomics-centered mining of plant metabolic diversity and function: past decade and future perspectives. *Molecular Plant* 16, 43–63.

Stark, P., Zab, C., Porzel, A., Franke, K., Rizzo, P., and Wessjohann, L.A. (2020) PSYCHE—a valuable experiment in plant NMR-metabolomics. *Molecules* 25: 5125.

Vo, D.H.T. and Thorne, T. (2024) Shrinkage estimation of gene interaction networks in single-cell RNA sequencing data. *BMC Bioinformatics* 25: 339.

Wandy, J. and Daly, R. (2021) GraphOmics: an interactive platform to explore and integrate multi-omics data. *BMC Bioinformatics* 22: 603.

Wang, H., Xu, Y., Zhang, Z., Zhang, G., Tan, C., and Ye, L. (2024) Development and application of transcriptomics technologies in plant science. *Crop Design* 3: 100057.

Wanichthanarak, K., Fahrmann, J.F., and Grapov, D. (2015) Genomic, proteomic, and metabolomic data integration strategies. *Biomarker Insights* 10(Suppl. 4), 1–6.

Watanabe, M. and Tohge, T. (2023) Species-specific "specialized" genomic region provides the new insights into the functional genomics characterizing metabolic polymorphisms in plants. *Current Opinion in Plant Biology* 75: 102427.

Wieder, C., Cooke, J., Frainay, C., Poupin, N., Bowler, R, *et al.* (2024) PathIntegrate: multivariate modelling approaches for pathway-based multi-omics data integration. *PLOS Computational Biology* 20: e1011814.

Xia, J., Psychogios, N., Young, N., and Wishart, D.S. (2009) MetaboAnalyst: a web server for metabolomic data analysis and interpretation. *Nucleic Acids Research* 37, W652–W660.

Xiao, Q., Mu, X., Liu, J., Li, B., Liu, H., *et al.* (2022) Plant metabolomics: a new strategy and tool for quality evaluation of Chinese medicinal materials. *Chinese Medicine* 17: 45.

Yan, J. and Wang, X. (2023) Machine learning bridges omics sciences and plant breeding. *Trends in Plant Science* 28, 199–210.

Yi, Y., Sun, Y., Wang, D., Li, X., Wu, X., *et al.* (2022) [*Advances in liquid chromatography-mass spectrometry analysis of several important secondary metabolites in plant metabolomics*.] *Chinese Journal of Biotechnology* 38, 3674–3681 (in Chinese).

Yu, X., Liu, Z., and Sun, X. (2023) Single-cell and spatial multi-omics in the plant sciences: technical advances, applications, and perspectives. *Plant Communications* 4: 100508.

Yue, E., Cao, H., and Liu, B. (2020) *OsmiR535*, a potential genetic editing target for drought and salinity stress tolerance in *Oryza sativa*. *Plants* 9: 1337.

Zeng, Y., Wen, J., Zhao, W., Wang, Q., and Huang, W. (2020) Rational improvement of rice yield and cold tolerance by editing the three genes *OsPIN5b*, *GS3*, and *OsMYB30* with the CRISPR-Cas9 system. *Frontiers in Plant Science* 10: 1663.

Zhang, F., Wu, J., Sade, N., Wu, S., Egbaria, A., *et al.* (2021) Genomic basis underlying the metabolome-mediated drought adaptation of maize. *Genome Biology* 22: 260.

Zhao, M., Lin, Y., and Chen, H. (2020) Improving nutritional quality of rice for human health. *Theoretical and Applied Genetics* 133, 1397–1413.

Zhao, T., Wu, T., Pei, T., Wang, Z., Yang, H., *et al.* (2021) Overexpression of SlGATA17 promotes drought tolerance in transgenic tomato plants by enhancing activation of the phenylpropanoid biosynthetic pathway. *Frontiers in Plant Science* 12: 634888.

Zhou, G., Pang, Z., Lu, Y., Ewald, J., and Xia, J. (2022) OmicsNet 2.0: a web-based platform for multi-omics integration and network visual analytics. *Nucleic Acids Research* 50, W527–W533.

Zhou, Z., Liu, J., Meng, W., Sun, Z., Tan, Y., *et al.* (2023) Integrated analysis of transcriptome and metabolome reveals molecular mechanisms of rice with different salinity tolerances. *Plants* 12: 3359.

Zoppi, J., Guillaume, J-F., Neunlist, M. and Chaffron, S. (2021) MiBiOmics: an interactive web application for multi-omics data exploration and integration. *BMC Bioinformatics* 22: 6.

18 Insights into Medicinal Herbs Based on Metabolomics and High-throughput Technologies

Akshita Patial[1,2], Rakhi Chandoliya[1,3], Suman Gusain[1,4], Vikas Sharma[2], and Rohit Joshi[1,4*]

[1]*Biotechnology Division, CSIR-Institute of Himalayan Bioresource Technology (CSIR-IHBT), Palampur, Himachal Pradesh, India;* [2]*Department of Molecular Biology and Genetic Engineering, School of Bioengineering and Biosciences, Lovely Professional University, Phagwara, India;* [3]*Department of Biotechnology, School of Bioengineering and Biosciences, Lovely Professional University, Phagwara, India;* [4]*Academy of Scientific and Innovative Research (AcSIR), CSIR-HRDC Campus, Ghaziabad, Uttar Pradesh, India*

Abstract

Metabolomics is a potent technique for analysis of therapeutic plants and identification of their bioactive chemicals with possible pharmaceutical uses and acts as a supplement to transcriptomics, proteomics, and genomics. With the goal of providing an exhaustive and precise evaluation of all low-molecular-mass metabolites, metabolomics studies in medicinal plants also seek to identify new active compounds, investigate adulteration, evaluate relationships between metabolite levels and their activities, and implement quality control. The distinction between primary and secondary metabolites in complex natural products presents analytical challenges, despite the significant progress made in metabolite separation and analysis technology. To address these issues, an approach integrating metabolomics with other omics data is needed, requiring specialized statistical and bioinformatics software. Plant metabolomics is of utmost importance in plant science, enabling detailed component analysis, quality assessment, nutritional and organoleptic attribute assessment, and functional and toxicological aspects. The number of metabolites in the plant kingdom is estimated to range from 200,000 to 1 million, and thus plant science has to rely heavily on the metabolomics approach, which allows comprehensive analysis of a plant's components, such as their quality, nutritional, and organoleptic attributes, as well as their functional and toxicological properties. This chapter provides an overview of different high-throughput techniques used in metabolomics with a focus on medicinal herbs, applications, and the challenges ahead.

Keywords: analytical methods, applications of metabolomics, bioactive compounds, data analysis, medicinal plants, phytochemical profiling, secondary metabolites

18.1 Introduction

Plants produce a diverse array of bioactive compounds such as polyphenols, terpenoids, alkaloids, saponins, and coumarins. These compounds play crucial roles in helping plants adapt to various environmental challenges and are highly valued in the pharmaceutical industry for their medicinal

* E-mail: rohitjoshi@ihbt.res.in

© CAB International 2025. *High-Throughput Plant Metabolomics* (Jen-Tsung Chen ed.)
DOI: 10.1079/9781800626911.0018

properties. However, despite the increasing global demand for natural medicines, large-scale identification and production of these plant-derived compounds remain limited. To overcome these challenges, metabolomics has evolved as a field that focuses on identifying and quantifying low-molecular-mass compounds or elements of the metabolome, produced by plants at different stages of their life cycle and under various environmental conditions. According to Ulrich-Merzenich *et al.* (2007), metabolomics can be applied across multiple domains, including drug development and discovery, as well as high-throughput screening for the assessment of plant-based medications. The continuous search for novel compounds derived from medicinal plants persists because the pharmaceutical industry's synthetic libraries have not yielded as many promising candidates as anticipated. The evaluation of plant secondary metabolites has been explored using various modern technologies. Among these, metabolomics stands out as a revolutionary approach with great potential for comprehensive secondary metabolite profiling.

Metabolomics has become a powerful analytical method for assessing the wide range of secondary metabolites found in herbal medicine (Tweeddale *et al.*, 1998). By detecting and analyzing these metabolites in medicinal plants, metabolomics has become a valuable tool in drug research (Gahlaut *et al.*, 2013). Metabolite profiling in herbal products requires a number of analytical approaches and procedures, including sample preparation, instrumental analysis, and data processing. With the aid of metabolomics, many metabolites and chemical markers present in herbal products can be readily characterized (Wolfender *et al.*, 2013). Significant advancements in science and technology over the past 30 years have introduced the era of multi-omics, combining various omics techniques. While metabolomics is still less developed compared with genomics, transcriptomics, and proteomics, it offers a unique advantage by focusing on the cellular components that most directly influence the final phenotype.

18.2 Metabolomics of Medicinal Plants

Metabolomics research in plants is crucial due to the vast diversity of primary and secondary metabolites they produce, many of which hold significant therapeutic potential. It is estimated that plants contain between 200,000 and 1 million secondary metabolites of which 5000 flavonoids and tannins, 22,000 terpenoids, and 21,000 alkaloids are known (Trethewey, 2004; Kroymann, 2011). These metabolites are essential not only for plant survival and adaptation but also for various activities that benefit human health and the environment. Numerous reports have highlighted the potential anticancer, antimalarial, antidiabetic, and other therapeutic effects of plant-derived metabolites. As a result, natural products and medicinal plants are being extensively explored as alternative sources for discovering novel chemical entities for drug development. For instance, paclitaxel, a compound derived from the bark of the Pacific yew tree (*Taxus brevifolia*), is widely used for its potent anticancer properties. Similarly, etoposide and teniposide, derived from podophyllotoxin, are used to treat small-cell lung cancer and testicular cancer (Gahtori *et al.*, 2023). In addition, aspirin, a semi-synthetic derivative of salicylic acid from the willow tree (*Salix alba*), has been utilized for over a century as an effective pain reliever and anti-inflammatory agent. The traditional medicinal use of plants by various ethnic groups, whether in their raw or extracted forms, remains one of the most effective approaches for discovering and isolating novel bioactive compounds. The discovery of morphine from opium in the early 19th century is a classic example of how traditional knowledge has led to the development of critical pharmaceutical agents.

In this context, metabolomics serves as a valuable tool for understanding the intricate phytochemical properties of medicinally active plant constituents (Table 18.1). By providing comprehensive profiles of metabolites, metabolomics helps in identifying, quantifying, and understanding the biological roles of these compounds within the plant and their potential therapeutic effects on human health. Two primary approaches are utilized in metabolomics research: targeted and non-targeted methods. Targeted metabolomics focuses on the characterization and quantification of a specific group of metabolites. It is particularly useful when the metabolites of interest are already known or when the study aims to compare the levels of

Table 18.1. Examples of the use of metabolomics for analysis of different plant parts used for medicinal purposes.

Species	Plant part used for analysis	Type of metabolite extracted	Method	Reference
Zanthoxylum spp.	Stem bark	Alkaloids (berberine and chelerythrine)	LC-MS	Plazas *et al.* (2019)
Carissa opaca	Fruit	Flavonoids (phenolics and triterpenoids)	LC-MS/MS	Bashir *et al.* (2021)
Prosopis cineraria	Bark	Flavonoid and phenols	FTIR	Jangid *et al.* (2023)
Gymnosporia montana	Leaf and fruit	Flavonoid	FTIR	Sutariya *et al.* (2023)
Medicago trunculata	Root	Flavonoids and isoflavonoids	IMS	Schroeder *et al.* (2019)
Momordica charantia	Fruit	Polysterols (charantin and momordicine)	NMR	Mishra *et al.* (2021)
Derris scandens	Root, leaves, stem bark, and root bark	Flavonoid (dalpanitin)	NMR	Mohotti *et al.* (2020)
Schima wallichii	Leaf	Phenol and flavonoid	GC-MS	Chanu *et al.* (2024)
Artemisia annua	Whole plant	Artemisinin, kaempferol, camphor	Raman spectroscopy	Rai *et al.* (2024)

FTIR, Fourier-transform infrared; GC-MS, gas chromatography–mass spectrometry; IMS, ion mobility spectroscopy; LC-MS, liquid chromatography–mass spectrometry; NMR, nuclear magnetic resonance.

these compounds under different conditions. Techniques such as nuclear magnetic resonance (NMR) spectroscopy and mass spectrometry (MS) are commonly employed in targeted metabolomics. For example, focused metabolomics approaches can determine the relative abundance of approximately 200 specified compounds in a sample, providing insights into the metabolic pathways and biological significance of these metabolites (Dudley *et al.*, 2010). In contrast, non-targeted metabolomics aims to capture a global profile of all metabolites present in a sample without prior knowledge of the compounds. This approach is more exploratory, allowing the discovery of new metabolites and the generation of comprehensive metabolic profiles that can reveal novel insights into plant biochemistry and potential therapeutic applications. Metabolomics, through both targeted and non-targeted approaches, is instrumental in advancing our understanding of plant metabolites and their potential applications in drug development. By bridging traditional knowledge with modern analytical techniques, metabolomics paves the way for the discovery of novel compounds that can be developed into new therapeutic agents, addressing the growing demand for natural and effective medications.

18.3 Analytical Methods and Techniques in Metabolomics for Medicinal Plants

In metabolomics research, biological samples are first collected, processed, and quantified, after which the resulting data is evaluated. The reproducibility of the metabolomics data is significantly impacted by sample preparation, making it an essential step in the process. Small adjustments made during sample extraction, storage, or collection can have a large impact on metabolite stability and can therefore cause significant changes in the metabolome being studied. Because of the rapid enzymatic turnover rate, metabolomics samples must be obtained consistently and quickly to prevent changes (Riekeberg and Powers, 2017). The goal is to reduce the changes that sample processing causes that are not relevant to the biology. The most likely cause of errors/deviations in metabolomic investigations is improper treatment of biological samples (Riekeberg and Powers, 2017).

18.3.1 Preparation of samples for metabolomic studies

In metabolomics research, the process begins with the collection, processing, and quantification of biological samples, followed by a thorough evaluation of the resulting data (Fig. 18.1). Sample preparation plays a critical role in the overall reproducibility of metabolomics data, making it an essential step in the research process. Even small adjustments during sample extraction, storage, or collection can significantly impact metabolite stability, leading to considerable changes in the metabolome under study. The stability of metabolites is highly sensitive to the conditions under which samples are handled. Due to the rapid enzymatic turnover of many metabolites, it is crucial that metabolomics samples are collected consistently and processed quickly to prevent unwanted alterations that may distort the biological relevance of the data (Riekeberg and Powers, 2017). The primary objective in sample processing is to minimize any changes that do not reflect true biological variations, thereby ensuring that the data accurately represent the metabolome of interest. Improper handling and treatment of biological samples are among the most likely sources of errors/deviations in metabolomics studies. Such errors/deviations can lead to inaccurate results and misinterpretations, which could compromise the reliability of the research findings.

To address this issue, it is recommended that metabolomics studies adhere to the guidelines established by the Metabolomics Standards Initiative (MSI). The MSI provides minimum criteria for experimental design, sample extraction, and data analysis, which helps to ensure that plant metabolomics investigations are conducted with consistency and reliability (Sumner *et al.*, 2007). Adhering to these standardized protocols is crucial for producing high-quality, reproducible metabolomics data. By following best practices in sample preparation and processing, researchers can reduce the risk of introducing non-biological variations into the data, thereby enhancing the validity and reliability of their findings. This systematic approach is essential for advancing our understanding of the complex metabolome and its implications for medicinal plant research.

18.3.2 Sample collection

Metabolomics research has published extensive data on a wide variety of biological samples (Vuckovic, 2012; Cajka and Fiehn, 2016). Due to their diverse matrices, these complicated samples call for differing sample preparation techniques. Different amounts of biological material per sample must be considered, depending on the level of detection and the natural abundance of certain metabolites. Typically, each biological replication needs 1–100 mg of tissue (Cajka and Fiehn, 2016). In metabolomic research, a minimum of three to five biological replicates is advised for each condition (Sumner *et al.*, 2007). Technical replicates, repeated measurements of the same sample, should not be prioritized over biological replicates, which involve parallel measurements of samples from various individuals (Blainey *et al.*, 2014).

18.3.3 Harvesting methods

When harvesting fresh samples, it is crucial to use sample freezing techniques such as liquid nitrogen or dry ice to prevent enzyme-induced metabolic changes (Sumner *et al.*, 2007). Additionally, removing unwanted elements such as dirt particles before collection is recommended. However, it is important to avoid storing samples for an extended period before extraction. For short-term storage, samples can be kept at −80°C in a freezer, on dry ice, or in liquid nitrogen for a few days to up to 2 weeks (Sumner *et al.*, 2007).

Depending on the biological material, the collected samples may require processing techniques such as lyophilization, cell lysis, or grinding before extraction. As metabolites are significantly influenced by these factors, it is essential to report specific details such as the type of tissue collected, cultivation conditions, seasonality, developmental stage, and timing of sample processing and harvesting for each condition (Kim and Verpoorte, 2010). Environmental factors

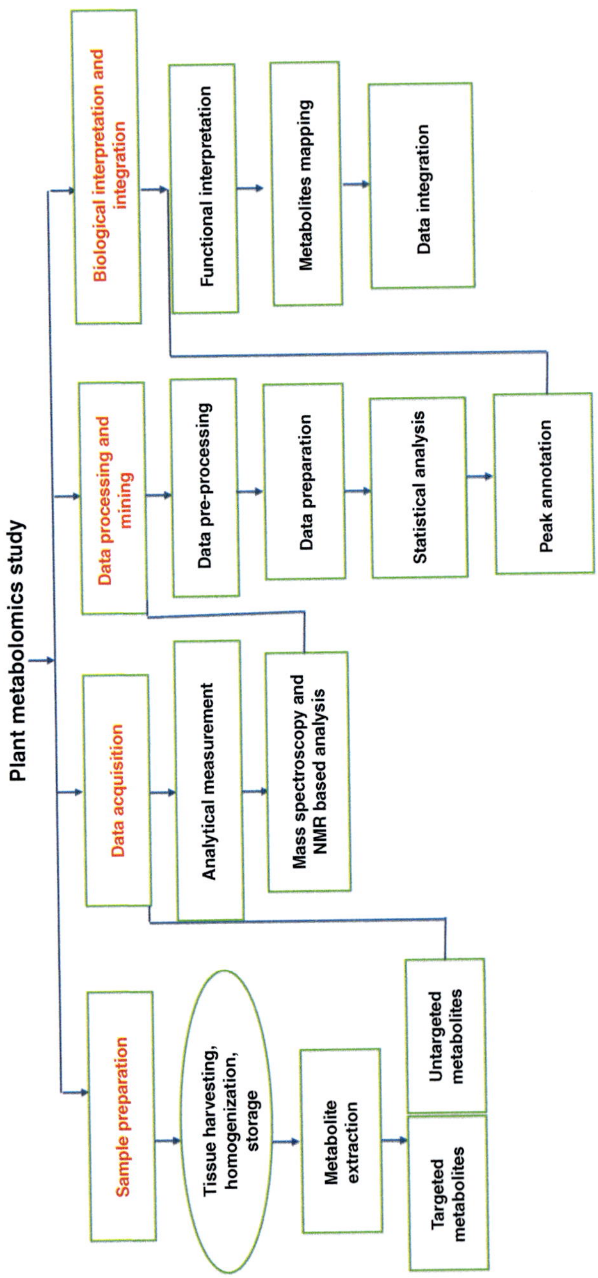

Fig. 18.1. Flow diagram representing multiple steps in plant metabolomic study. NMR, nuclear magnetic resonance.

can also cause both qualitative and quantitative variations in the composition of plant primary and secondary metabolites (Ernst *et al.*, 2014).

18.3.4 Sample extraction

The complex chemistry of metabolites poses a challenge in capturing the entire metabolome using a single solvent, unlike the proteome and genome, which can be extracted with a single technique (Fernie *et al.*, 2004). Because complex biological materials have varied matrices, as discussed previously, different extraction strategies are necessary. The principle of "like dissolves like" is typically applied to preferentially extract metabolites based on their solubility in a particular solvent. For instance, hydrophobic solvents are used to extract lipids, while polar and semi-polar metabolites are extracted using hydrophilic solvents, such as hydro-alcoholic solutions. Numerous metabolome extraction procedures have been developed and extensively evaluated (Bylda *et al.*, 2014; Mushtaq *et al.*, 2014). The chosen extraction method must efficiently and rapidly cover a broad spectrum of target metabolites while ensuring high accuracy and precision. Compared with single extraction techniques, liquid–liquid fractionation offers a more streamlined process (Salem *et al.*, 2018). This method partially purifies each fraction by removing impurities: polar metabolites are removed in hydrophobic solvents, while hydrophilic compounds are concentrated in lipophilic solvents (Salem *et al.*, 2017). Most extraction techniques include a deproteinization step, although they vary in the type and quantity of solvents used. The presence of proteins in the sample can significantly impact the accuracy, precision, and longevity of the analytical equipment (Raterink *et al.*, 2014). Extraction efficiency can often be improved by using a combination of solvents in one or two stages, rather than relying on a single solvent (Cajka and Fiehn, 2016).

Liquid–liquid extraction techniques are based on the gold standard protocols of the Folch and Bligh and Dyer methods, which used a mixture of methanol and chloroform in different ratios (Bligh and Dyer, 1959; Folch *et al.*, 1957). This method has been used to analyze both lipids and polar metabolites, employing a chloroform:methanol:water mixture (Lisec *et al.*, 2006). Methyl tert-butyl ether has also been introduced as a safer and cleaner alternative to chloroform, particularly for the recovery of lipids and metabolites from plant samples (Giavalisco *et al.*, 2011). To assess matrix effects, the use of isotope-labeled internal standards during sample preparation is recommended (Doppler *et al.*, 2016).

18.3.5 Concentration of the extract and its dilution, enrichment, and re-suspension

To partially purify samples before analytical measurement, various techniques are used in addition to liquid–liquid extraction, including organic solvent precipitation and single- or multiple-step solid-phase extraction (SPE) (Cajka and Fiehn, 2014). After sample extraction, it is important to concentrate the metabolites by evaporating the solvents. This is typically done at room temperature without heating, using a vacuum concentrator. When dealing with metabolites that are sensitive to oxidative changes, such as certain lipids, a nitrogen flow evaporator is recommended to prevent degradation. Once concentrated, the dry extracts are usually resuspended in a solvent that is appropriate for subsequent analysis (Raterink *et al.*, 2014). In some cases, such as before gas chromatography–mass spectrometry (GC-MS) analysis, the dried materials may require additional processing steps such as derivatization (Lisec *et al.*, 2006). It is generally not advisable to store extracted liquid samples for short periods, even at low temperatures such as −20°C. Therefore, if short-term storage is necessary, samples are best kept in a dry form before analysis (Fernie *et al.*, 2011).

18.3.6 Analytical techniques in metabolomics

There are numerous analytical tools based on MS that can be used to gather information. In plant metabolomics, a combination of techniques is needed to provide the highest level of metabolite coverage because individual analytical tools are unable to identify every metabolite present in a sample (Razzaq *et al.*, 2019). Metabolomics uses a variety of analytical technologies; these include GC-MS, capillary electrophoresis–mass spectrometry (CE-MS), high-performance liquid chromatography (HPLC), liquid chromatography–mass spectrometry (LC-MS), NMR spectroscopy, liquid chromatography–nuclear magnetic resonance (LC-NMR), direct infusion mass spectrometry (DIMS), ion mobility spectrometry (IMS), Fourier-transform infrared (FTIR), and Fourier-transform–ion cyclotron resonance–mass spectrometry–matrix-assisted laser desorption/ionization (FT-ICR-MS-MALDI) (Okazaki and Saito, 2012).

GC-MS

GC-MS is a widely used analytical method for analyzing volatile and thermally stable metabolites (Fig. 18.2). In this technique, GC separates the metabolites, while MS identifies them. GC-MS is particularly useful for the analysis of primary metabolites, such as organic acids and amino acids. To extract and identify metabolites using GC-MS, two derivatization steps are necessary. First, carbonyl groups in the metabolites are converted into oximes using methoxyamine hydrochloride. The second step involves trimethylsilylation, which enhances the volatility of the metabolites by using derivatizing agents such as *N*-methyl-*N*-(trimethylsilyl)trifluoroacetamide and *N*,*O*-bis-(trimethylsilyl)-trifluoroacetamide (Harvey and Vouros, 2020).

Two primary ionization techniques are employed in GC-MS: electron ionization and chemical ionization. Of these, electron ionization is the most widely used due to its robustness, sensitivity, resolution, and repeatability, despite its limitations, such as slow scanning rates and low mass accuracy. For enhanced metabolic profiling, time-of-flight mass spectrometry (TOF-MS) analyzers combined with GC are recommended because of their faster acquisition times and improved mass accuracy (Kumar *et al.*, 2016). Metabolite profiling using GC-MS has been applied to various medicinally important plant species to assess their quality, health standards, and potential metabolic changes. For example, GC-MS analysis identified 45 metabolites in the methanolic extract and 35 metabolites in the aqueous extract of *Berginia ligulata* rhizomes, with 6 major metabolites common to both extracts, and cytotoxic analysis revealed the anticancer activity of the rhizome extract (Nadeem *et al.*, 2024). Similarly, GC-MS analysis of the methanolic extract of *Schima wallichii* revealed the presence of 14 major phytoconstituents. Among these, (Z,Z,Z)-octatriacontyl trifluoroacetate, *n*-hexadecanoic acid, and 9,12,15-octadecatrienoic acid were identified as the three major compounds. The leaf extract of *S. wallichii* was shown to exhibit hepatoprotective and renoprotective activities (Chanu *et al.*, 2024).

CE-MS

CE-MS is known for its high separation efficiency, making it particularly suitable for the analysis of charged metabolites. This technique provides fast and high-resolution analysis of charged molecules from low injection volumes and enables mass fragmentation-based metabolite characterization (Salem *et al.*, 2020), with a focus on amino acids and organic acids (Fig. 18.3). However, CE-MS has some limitations, including low sensitivity, poor repeatability, inconsistent migration times, and a lack of reference libraries, making it less ideal for analyzing metabolites from complex plant materials (Williams *et al.*, 2007). Despite these drawbacks, CE-MS is primarily used for the

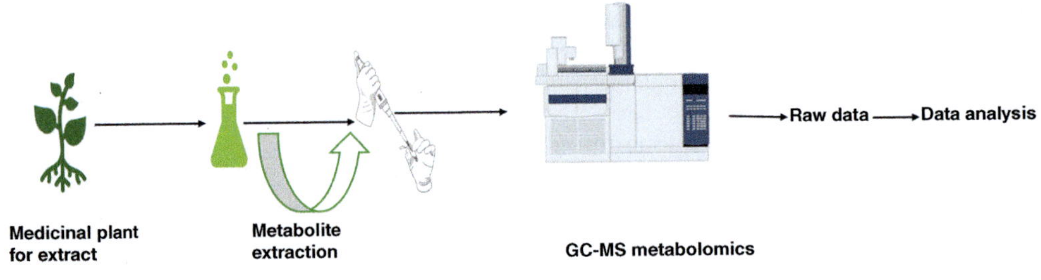

Fig. 18.2. Schematic representation of the various steps involved in metabolite analysis using gas chromatography–mass spectrometry (GC-MS).

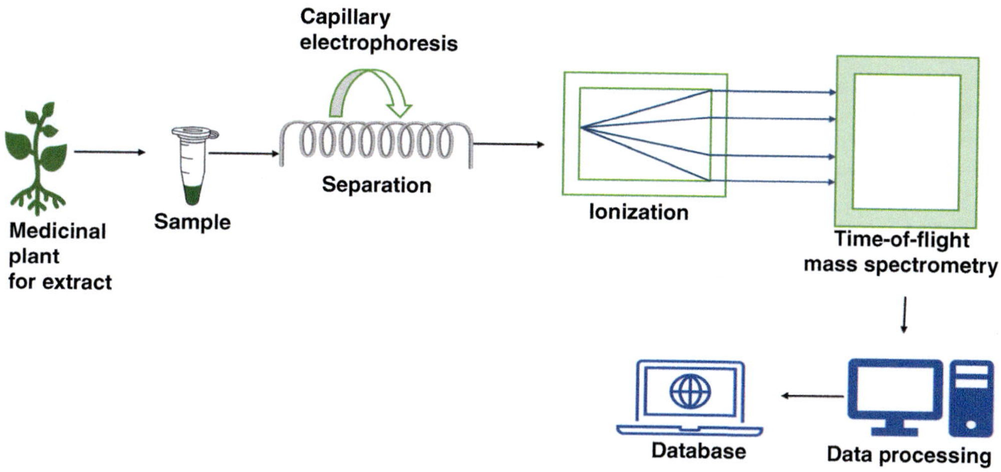

Fig. 18.3. Schematic representation of the various steps involved in metabolomic analysis using capillary electrophoresis–mass spectrometry.

separation of small amounts of compounds, making it particularly suitable for studies of plant metabolomes (Salem *et al.*, 2020).

CE-MS has been widely utilized to examine the bioactive components in various kinds of herbal medicine, including *Salvia* spp., *Hibiscus sabdariffa*, *Melissae* spp., and *Scutellaria baicalensis*. In addition, CE with ultraviolet (UV) detection has also been employed for the analysis of specific plant species. For example, it has been used for polyphenol analysis in chamomile flowers and *Cynanchum chinense* (Przybylska *et al.*, 2021), as well as for analyzing bioactive components in *Salvia* spp. (Adımcılar *et al.*, 2019), *Radix scutellariae*, *Gentiana lutea*, mate herb (Przybylska *et al.*, 2021), black tea (Memon *et al.*, 2018), rooibos, and honeybush tea (Arries *et al.*, 2017). It has also been used to study flavonoids in *Crataegus monogyna* (Przybylska *et al.*, 2021) and *Coreopsis tinctoria* (Deng *et al.*, 2017), and flavone derivatives in *Seriphidium santolinum* (Przybylska *et al.*, 2021), as well as fewflower (*Lysionotus* spp.) herb (Ren *et al.*, 2020).

Additionally, CE-MS has been used to identify organic acids in various plants, including *Pterocephalus hookeri* (Przybylska *et al.*, 2021), as well as chamomile, linden, mint, and sage. The technique has also been applied to the analysis of alkaloids in tea, *Sophora tonkinensis*, *Chelidonium majus*, *Phellodendron chinense*, and *Nelumbo nucifera*. Further applications include the detection of anthraquinones in slimming tea, coumarins in *Cacalia tangutica* (Przybylska *et al.*, 2021), and nitrate, nitrite, and bromate in tea infusions (Tembo *et al.*, 2021).

LC-MS

LC-MS is a powerful analytical technique that integrates the separation capabilities of LC with the accurate mass measurement and identification capabilities of MS (Warren, 2013). This allows the comprehensive analysis of complex mixtures, making it one of the most widely used methods in metabolomics. LC-MS is particularly advantageous because it can detect and identify a broad spectrum of metabolites, including both polar and non-polar molecules. The LC component efficiently separates these metabolites based on their chemical properties, while the MS component provides precise molecular mass information and structural details, facilitating the identification and quantification of individual compounds within a sample (Fig. 18.4). This versatility and sensitivity have made LC-MS a cornerstone in metabolomics research, enabling the detailed exploration of metabolic profiles in various biological systems.

HPLC is a widely used technique in targeted metabolomics, offering the ability to separate and quantify specific metabolites with high precision. Various detectors, such as UV, fluorescence, and diode array detectors, can be integrated with HPLC systems to enhance the detection

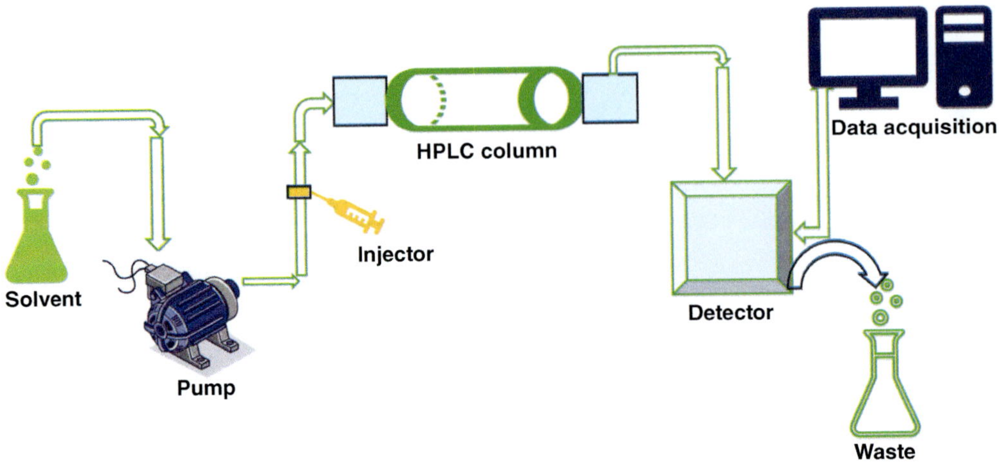

Fig. 18.4. Diagrammatic representation of the various steps involved in metabolomic analysis using liquid chromatography–mass spectrometry. HPLC, high-performance liquid chromatography.

and measurement of diverse compounds. This versatility makes HPLC particularly valuable for the comprehensive analysis of plant-derived metabolites, where it is often combined with other chromatographic methods and biological assays to screen plant extracts effectively. Historically, flavonoid analysis relied on traditional chromatographic techniques, including LC, thin-layer chromatography, column chromatography, and paper chromatography (Yonekura-Sakakibara *et al.*, 2019). However, the advent of HPLC has revolutionized this field, offering a more rapid, sensitive, and accurate approach for analyzing bioactive components. HPLC's power lies in its ability to handle complex mixtures and detect a wide range of compounds, making it indispensable for natural product research. In studies involving various plants such as *Medicago truncatula*, *Brassica rapa* L., *Lupinus albus*, *Lupinus angustifolius*, and *Sutherlandia frutescens* (L.), HPLC coupled with a UV detector has been successfully employed to isolate and characterize different flavonoid types. This technique's precision and versatility continue to drive advancements in the field, enabling researchers to explore the rich biochemical diversity of plant species and their potential applications in drug development.

NMR spectroscopy

NMR spectroscopy is a cornerstone technique in metabolomic research, offering a suite of advantages that make it indispensable for the comprehensive analysis of metabolites (Fig. 18.5). It has remarkable repeatability, is non-destructive, and facilitates rapid identification of metabolites. These features contribute to the growing preference for NMR in metabolomics, particularly in the study of plant-derived compounds (Markley *et al.*, 2017). NMR spectroscopy has emerged as a preferred method for plant metabolite analysis due to its efficiency in spectral interpretation and the simplicity of its sample preparation. The process typically requires minimal preparation, allowing preservation of sample integrity, which is crucial in metabolomic studies where the native state of metabolites is of interest. Furthermore, the technique's broad applicability extends beyond just liquid samples to tissue and biofluid extracts. One of the notable advantages of NMR is its ability to analyze solid and semi-solid materials through solid-state NMR and techniques such as magic-angle spinning. These methods enable the examination of intact tissues and other solid matrices, providing a more comprehensive metabolic profile in situations where traditional liquid-state NMR might fall short. This ability is particularly useful in clearing up spectral overlaps that can complicate the analysis of complex biological samples, leading to more accurate metabolite identification (Hong *et al.*, 2009; Diserens *et al.*, 2015; Blondel *et al.*, 2016; Jang *et al.*, 2016).

The versatility of NMR is further enhanced by the use of multi-dimensional NMR techniques, which allow the exploration of interactions

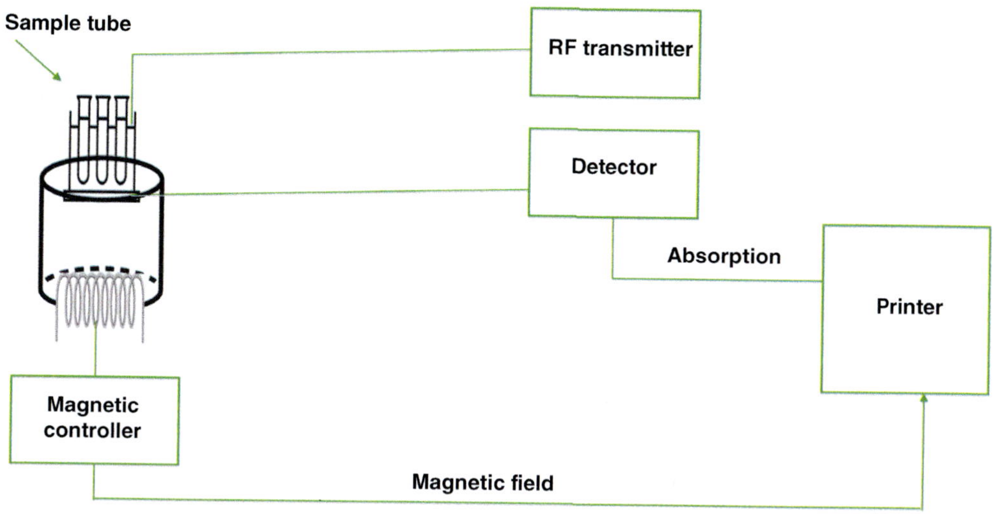

Fig. 18.5. Schematic representation of the various steps involved in sample metabolomic analysis using nuclear magnetic resonance (NMR). RF, radio frequency.

between different nuclei within a sample. This multi-dimensional approach facilitates a deeper understanding of the molecular structure and dynamics of metabolites, providing insights that are crucial for applications in drug research and development. The ability to investigate these interactions not only adds another layer of detail to metabolomic studies but also broadens the scope of research that can be conducted using NMR spectroscopy (Emwas *et al.*, 2019). For example, NMR studies on *Momordica charantia* revealed distinct variations in metabolite distribution across different parts of the fruit, including the seed, skin, and pericarp. The skin, in particular, was found to have a higher concentration of polysterols such as momordicine and charantin, which are known for their antidiabetic properties (Mishra *et al.*, 2021). Likewise, NMR spectroscopic analysis has proven effective in elucidating the structure of dalpanitin, a bioactive compound isolated from *Derris scandens* that possesses antimicrobial properties (Mohotti *et al.*, 2020).

LC-NMR

LC-NMR is a powerful analytical technique that effectively resolves complex mixtures, combining the strengths of LC for separation with the detailed molecular characterization provided by NMR spectroscopy (Fig. 18.6). This integrated approach has been widely utilized across various applications, particularly in the analysis of intricate biological samples and natural products. LC allows the separation of complex mixtures into individual components, which are then analyzed directly by NMR. This combination provides detailed structural information, making LC-NMR a valuable tool in fields such as metabolomics, natural product research, and drug discovery. When LC-NMR is combined with additional techniques, the scope of its applications broadens significantly. For example, coupling LC-NMR with SPE leads to the development of the LC-SPE-NMR system, which enhances sample preparation and enables the complete structural elucidation of sample components. This approach is particularly useful for isolating and identifying minor or unstable compounds within complex mixtures. Furthermore, integrating LC-NMR with mass spectrometry (LC-NMR-MS) allows the simultaneous acquisition of mass and fragmentation data alongside NMR spectra. This combination provides comprehensive insights into the molecular mass, structure, and fragmentation patterns of various components in a sample, making it an indispensable tool for complex mixture analysis. Additionally, the incorporation of two-dimensional NMR measurements into the LC-NMR system further enhances its analytical capabilities. This allows the detailed investigation of molecular interactions and complex structures within a sample, providing

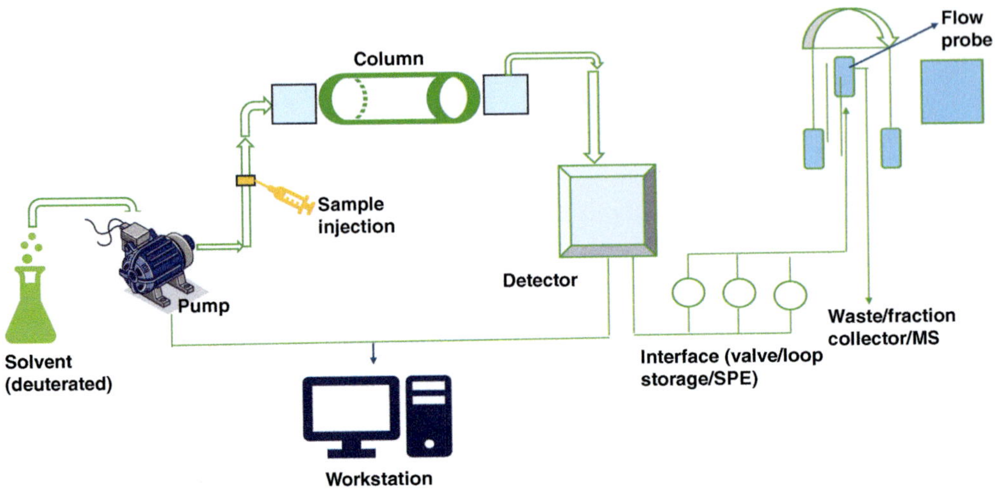

Fig. 18.6. Schematic representation of the various steps involved in sample metabolomic analysis using liquid chromatography–nuclear magnetic resonance (LC-NMR). MS, mass spectrometer; SPE, solid-phase extraction.

deeper insights into the molecular composition and dynamics of the components. Various studies have utilized these combinatorial techniques for the identification of phytoconstituents in medicinal plants. In a study on *Hypericum perforatum*, Tatsis *et al.* (2007) utilized hyphenated techniques such as LC-SPE-NMR and LC-MS to characterize naphthodianthrones (protohypericin, hypericin, protopseudohypericin, and pseudohypericin), phloroglucinols (adhyperforin and hyperforin), flavonoids, and phenolics.

DIMS

DIMS, also known as flow injection-MS, is a valuable analytical method in metabolomics research, although it is not as commonly utilized as LC-MS. DIMS enables the recording of quantitative data for all charged ions using MS, particularly high-resolution MS, without the need for chromatographic separation (Fig. 18.7). This high-throughput advantage of DIMS is notable because it eliminates the requirement for costly and time-consuming chromatographic processes. However, the direct infusion approach also presents challenges, particularly due to the significant matrix effects that can interfere with accurate measurements. For DIMS to be effectively applied in rapid metabolomics studies, two critical technical challenges must be addressed: signal alignment and structural annotation. These challenges are crucial for ensuring the accuracy and reliability of the data obtained. Traditional LC-MS or GC-MS analyses typically take 15–30 min to complete. The DIMS technique was proposed as an alternative to GC-MS or LC-MS to enhance the efficiency and capability of MS (Lichtenberg *et al.*, 2021). Despite the challenges, DIMS has been successfully applied to the investigation of herbal remedies, including *Peucedanum praeruptorum*, *Peucedanum pubescens*, *Angelica dahurica*, *Angelica sinensis*, and *Angelica decursiva*.

IMS

Untargeted metabolomics based on MS has gained popularity as a method for the in-depth characterization of metabolites within a biological system (Aksenov *et al.*, 2017). IMS is a rapid, gas-phase separation technique that employs a neutral buffer gas to separate ions based on their mobility in an electric field. An ion's mobility is influenced by its mass, charge, and collision cross-section, which is the ion's rotationally averaged surface area. IMS has been widely used around the world to evaluate a variety of toxic chemicals, such as herbicides and pesticides, which are used to control diseases, weeds, and insects in plants. This technique, along with other analytical techniques such as LC-MS and GC-MS, has been utilized for the identification of

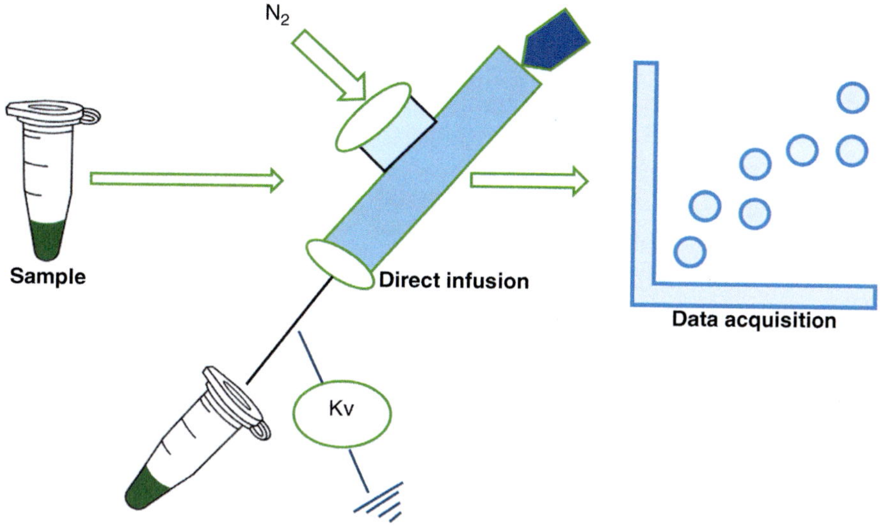

Fig. 18.7. Overview of plant metabolite analysis using direct infusion mass spectrometry. N_2, nitrogen; kV, kilovolts.

molecules with therapeutic potential. For instance, isomers of flavonoids and isoflavonoids were identified in the root extract of *Medicago truncatula* using IMS-MS (Schroeder *et al.*, 2019).

FTIR spectroscopy

FTIR spectroscopy enables the simultaneous analysis of various primary and secondary metabolites in plants, including phenolics, alkaloids, steroids, flavonoids, carbohydrates, amino acids, proteins, and polysaccharides, making it a valuable tool for metabolic fingerprinting (Manas and Marakanam, 2023). The FTIR fingerprint can identify molecular functional groups based on specific wavelengths (Fig. 18.8). Additionally, FTIR spectrometry offers several advantages, such as being non-destructive, relatively easy to use, and requiring minimal sample preparation. The technique also provides good repeatability and specificity (Tulukcu *et al.*, 2019). This technique can be utilized for the identification of bioactive compounds from medicinal plants, so they can be used for the synthesis of new drugs. For example, Sutariya *et al.* (2023) isolated bioactive compounds from the leaf and fruit extracts of *Gymnosporia montana* using FTIR. The flavonoid-rich fraction of the sample extract demonstrated strong antioxidant and anti-inflammatory properties. Similarly, FTIR was used to extract phytoconstituents from the ethanolic extract of the bark of *Prosopis cineraria*. This extract was rich in flavonoid and phenolic compounds, exhibiting significant antioxidant and free-radical scavenging activity. Furthermore, the extract showed antidiabetic potential compared with other plant extracts (Jangid *et al.*, 2023).

FT-ICR-MS-MALDI

Mass spectrometers are capable of separating metabolites at the ppb to ppm level, achieving readings with ppm accuracy, and offering the exceptional mass resolution crucial for metabolomics analysis. High-resolution mass spectrometers such as FT-ICR-MS-MALDI are among the few instruments that can deliver these qualities (Fig. 18.9). For example, dos Santos *et al.* (2019) utilized FT-ICR-MS-MALDI for the detection of cannabinoid compounds in *Cannabis* leaves.

Raman spectroscopy

Raman spectroscopy (Fig. 18.10) is an effective tool for identifying biotic and abiotic stresses in plants, as demonstrated by Sanchez *et al.* (2019). This laser-based, label-free technique analyzes plant material without the use of chemicals, resulting in a zero cost per analysis for reagents (Farber *et al.*, 2019a). In contrast, other molecular

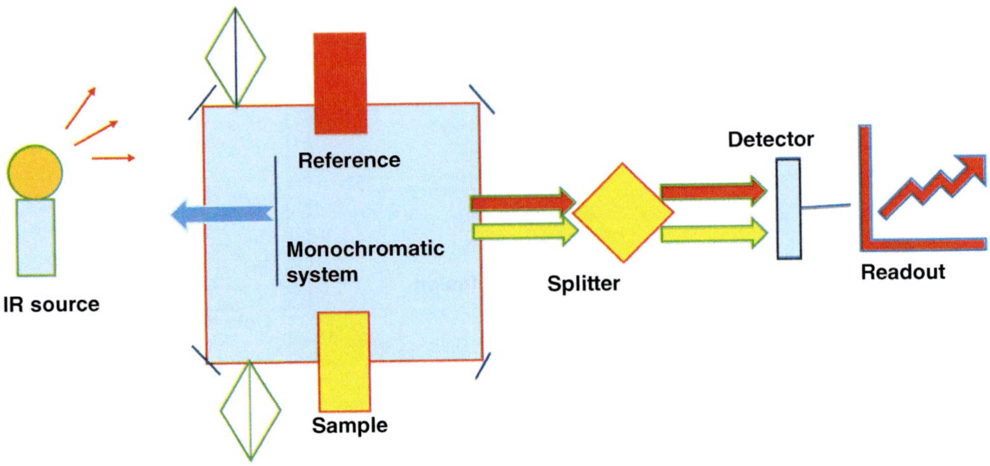

Fig. 18.8. Overview of the various steps involved in plant metabolite analysis using Fourier-transform infrared (IR).

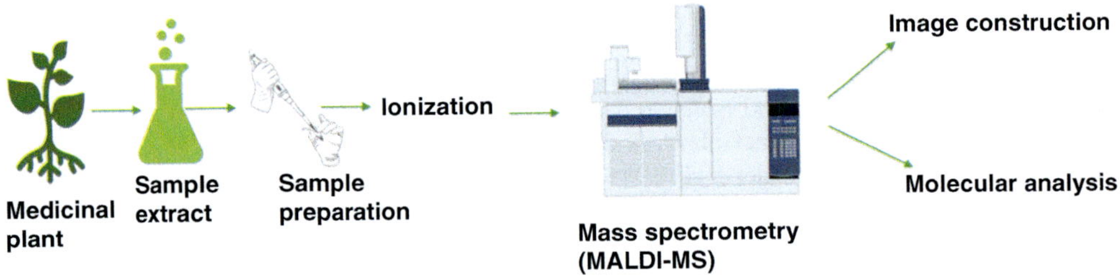

Fig. 18.9. Overview of plant sample analysis using Fourier transform–ion cyclotron resonance–mass spectrometry–matrix-assisted laser desorption/ionization. MALDI-MS, matrix-assisted laser desorption/ionization–mass spectrometry.

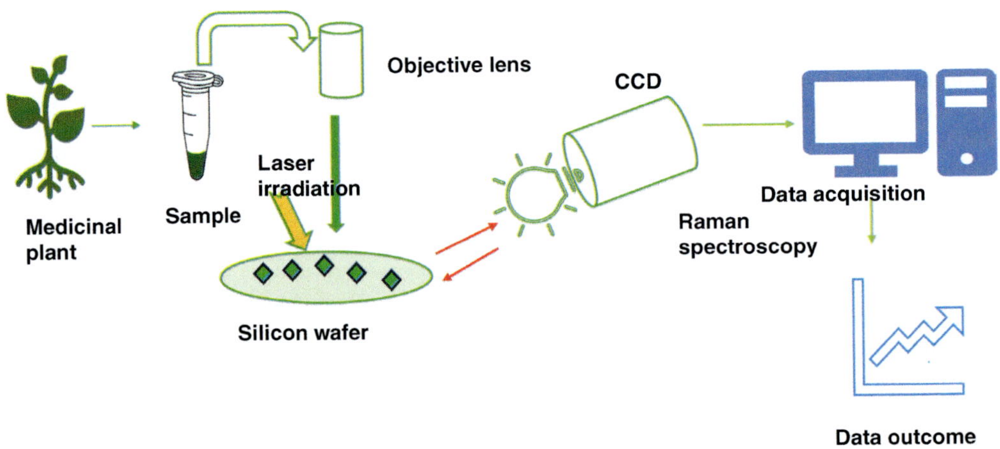

Fig. 18.10. Overview of plant sample analysis through Raman spectroscopy techniques. CCD, charge-coupled device.

analysis methods can be expensive. Raman spectroscopy is also useful for assessing the nutritional content of grains and for the rapid and accurate phenotyping of plants (Farber *et al.*, 2020).

One of the key advantages of Raman spectroscopy is its ability to analyze hydrated biological materials, such as cells and tissues, because water produces a very low Raman signal. This makes it a better option than IR spectroscopy for studying such samples (Farber *et al.*, 2019b). Rai *et al.* (2024) conducted an *in situ* analysis of secondary metabolites in *Artemisia annua* using Raman spectroscopy under exposure to red and blue light. The study revealed that combined exposure to red and blue light resulted in an increase in various bioactive compounds of medicinal importance, such as artemisinin, rutin, kaempferol, camphor, limonene, and cineole, compared with exposure to red or blue light alone.

18.4 Data Analysis in Metabolomics

18.4.1 Data preprocessing

Metabolomics research involves several key steps, beginning with data preprocessing (Sun and Xia, 2024). This step is crucial for the *in silico* examination of extensive metabolite data collected by various analytical techniques. Preprocessing involves tasks such as aligning retention times, identifying peaks, correcting baselines, and reducing noise to ensure the raw data is cleaned and prepared for further analysis. Several bioinformatics tools are available for this purpose, such as XCMS, which enhances data processing efficiency and reduces processing times, and MetaboAnalyst, a free web-based platform that improves analytical performance and user interaction (Chong *et al.*, 2019). Other platforms such as MeltDB, AMDIS, and METLIN are also widely used for data assessment, processing, and statistical analysis in plant metabolomics (Patel *et al.*, 2021). These platforms offer functionalities across four main modules: data integration and systems biology, functional analysis, data processing, and utility functions. For instance, MetAlign and MZmine 3 are open-source programs that focus on processing GC-MS and LC-MS data, respectively, providing comprehensive workflows for data analysis, including peak identification, isotope detection, and statistical analysis (Pluskal *et al.*, 2010).

18.4.2 Identification and quantification of metabolites

Once data preprocessing is complete, the next step is identification of compounds. This involves comparing experimental data, such as mass spectra and retention times, with reference spectra in metabolite databases such as METLIN and RefMetaPlant (Shi *et al.*, 2024). Isotope pattern analysis and fragmentation patterns in tandem MS (MS/MS) data can further aid in compound identification. Following identification, quantification of metabolites is performed by comparing peak areas or intensities with internal standards that were added during sample processing.

18.4.3 Data analysis and mapping

Data analysis follows, employing statistical techniques to identify significant differences between sample groups, such as treatment versus control. Common statistical methods include *t*-tests, analysis of variance, and multivariate techniques such as partial least squares discriminant analysis (PLS-DA) and principal component analysis (PCA), with the false discovery rate being used to improve multiple testing control (Jiang *et al.*, 2022). Metabolite data can also be mapped on to metabolic pathways to better understand biological processes, with tools such as Metabolomics Workbench and MetaboAnalyst aiding in pathway analysis.

Multivariate data analysis is crucial for examining complex data sets with numerous variables. Techniques such as PCA assist in dimensionality reduction and visualization of data patterns, while PLS-DA and orthogonal partial least squares discriminant analysis (OPLS-DA) are used for predictive modeling and classification. Modern machine learning approaches, including random forests, support vector machines, and deep learning methods, are increasingly applied in metabolomics for tasks such as feature selection, prediction, and classification (Chi *et al.*, 2024). Integration of metabolomic data with other omics data, such as transcriptomics, proteomics, and genomics, can provide a more comprehensive understanding of biological processes by unraveling extensive regulatory networks. Metabolite set enrichment analysis is another valuable technique offering functional insights by identifying biological processes or pathways enriched with differentially regulated metabolites (Peng *et al.*, 2024).

18.4.4 Data visualization

Effective data visualization using tools such as heatmaps, volcano plots, and metabolic circuit diagrams is essential for communicating findings clearly and efficiently. Together, these analytical methods and data-analysis techniques empower researchers to explore intricate metabolite profiles, contributing valuable insights into biological processes, potential biomarkers, and applications in health, disease, and environmental studies.

18.5 Applications of Metabolomics in Medicinal Plant Research

Metabolomics plays a pivotal role in medicinal plant research. By analyzing the complex chemical profiles of plants, metabolomics enables the identification and quantification of bioactive compounds that contribute to their therapeutic properties. This approach facilitates the discovery of novel natural products with potential pharmaceutical applications, as well as the understanding of how environmental factors, genetics, and plant physiology influence metabolite production. Additionally, metabolomics supports the quality control and standardization of herbal medicines, ensuring consistent efficacy and safety in clinical use. Through advanced analytical techniques and bioinformatics, metabolomics is revolutionizing the exploration of medicinal plants, paving the way for innovative drug development and a deeper understanding of plant-based therapies. Metabolomics has several applications in medicinal plant research, discussed below.

18.5.1 Metabolite identification and phytochemical profiling

Metabolite identification and phytochemical profiling are essential techniques in the exploration of plant-derived compounds and their potential applications in drug research. By integrating advanced analytical methods such as MS, NMR spectroscopy, and chromatographic techniques, researchers can uncover novel metabolites with potential therapeutic properties. This thorough approach not only aids in uncovering potential new drug candidates but also helps us gain insight into the medicinal properties associated with plants. For instance, integrative LC-MS and GC-MS, along with multivariate statistical analysis, were used for the identification and profiling of various compounds, including genistein, kaempferol 3-*O*-rutinoside, chlorogenic acid, and acacetin 7-*O*-rutinoside, in *Potentilla anserina* L., a plant known for its use in the prevention of premenstrual syndrome (Mari *et al.*, 2013). A similar approach involving LC-MS analysis and multivariate statistical analysis was employed to profile the phytoconstituents present in the stem bark of *Zanthoxylum* spp. with potential anti-cholinesterase activity. The results revealed a high presence of berberine and chelerythrine, which have previously been reported as cholinesterase inhibitors (Plazas *et al.*, 2019).

18.5.2 Quality control of herbal medicine

Metabolomics is a valuable tool for confirming the quality and authenticity of medicinal products. It aids in maintaining the therapeutic concentrations of bioactive components and plays a crucial role in the quality control of herbal products (Xiao *et al.*, 2022). India and other nations have traditionally used medicinal herbs to meet basic human requirements. India possesses one of the world's oldest civilizations, with a long history of using traditional medicines to manage healthcare for the entire society (Mukherjee *et al.*, 2014). The growth of the plants, environment, time of collection, drying process, and extraction techniques all affect the chemical makeup of herbal products. When evaluating the quality of Ayurvedic herbs, several elements must be taken into consideration. Modern analytical methods can be applied to medicinal plants for quality assessment and scientific validation of their formulations to meet market expectations.

Therefore, based on their primary biological reference components, the purity, quality, and various other standardized factors require immediate attention (Wang *et al.*, 2004). Research in diverse domains has focused on natural products, especially medicinal plants, to generate templates for novel chemical entities. Hyphenated technology is being used to develop several novel lead compounds. The traditional yet potent health claims of Ayurvedic medicine will be greatly supported by the biological control, standardization, and quality control of the practice. Consequently, regulatory agencies are requiring research into novel analytical techniques for greater standardization of herbal medications to increase the precision and uniformity of herbal/medicinal plant-based treatments globally. These methods must be strong and objective, and they must address the repeatability of the chemical profile content. An innovative and need-based approach called chemoprofiling is gaining traction to legitimize the use of herbal products in therapy as they are used now, specifically the extracts or marketed products.

Thus, through worldwide coordination and harmonization, the promotion and growth of herbal medicine require the utmost attention. Metabolite profiling compares the composition of compounds in addition to identifying metabolites in relation to compound distribution, and involves analyzing the results of analytical methods to estimate various secondary plant metabolites both quantitatively and qualitatively. These metabolites are evaluated in turn using a variety of statistical techniques and spectroscopic and chromatographic peak measurements. These contemporary methods can be applied to a thorough examination of the components found in the plant samples (Glassbrook and Ryals, 2001). Measurements of phenotypic and genotypic alterations in the cellular systems are part of plant metabolomics, and these methods have been used to analyze changes in genes and proteins upstream as well as changes in physiological functions downstream. Metabolomics has the potential to examine every molecule found in plant tissues and cells (Gahlaut *et al.*, 2013). Several fields of phytomics are particularly helpful for ensuring consistent quality and efficacy of herbal medicine, including chemical analysis, standardization using markers, and DNA fingerprinting (Ellis *et al.*, 2007).

18.5.3 Ethnobotanical studies

Metabolomics provides a molecular basis for the traditional use of plants in ethnobotanical research. By analyzing the metabolite profiles of plants, it helps to uncover the biochemical pathways and mechanisms underlying their therapeutic effects. This approach not only deepens our understanding of traditional remedies but also scientifically validates their effectiveness. Consequently, metabolomics bridges the gap between traditional knowledge and modern science, supporting the integration of conventional therapies into contemporary medicine. For instance, Bashir *et al.* (2021) utilized an LC-MS/MS approach for the chemical profiling of *Carissa opaca* berries. Their study revealed that the fruit extract of *C. opaca* predominantly contains flavonoids, phenolics, and triterpenoids, which exhibit α-glucosidase and urease inhibitory activities.

18.6 Challenges in Metabolomics

18.6.1 Standardization of data and quality assurance

Randomness in sample preparation, instrument performance, and data processing can result in errors and conflicts in data. For data to be consistent and repeatable across research, standardization and strong quality control methods are necessary (Di Minno *et al.*, 2022).

18.6.2 Identification of metabolites

It is still very difficult to identify and annotate metabolites in MS data, particularly for unknown or low-abundance chemicals. Ongoing activities include refining identification techniques and developing significant, carefully chosen metabolite databases (Di Minno *et al.*, 2022).

18.6.3 Quantification

Problems including matrix effects and ion suppression can make it difficult to accurately quantify metabolites, especially in complex biological matrices. It is essential to establish trustworthy quantification techniques (Di Minno *et al.*, 2022).

18.6.4 Integration of data

Merging metabolomics data with transcriptomics, proteomics, and genome data continues to be a challenging task. To gain a comprehensive understanding of biological systems, strong multi-omics integration methodologies must be developed (Di Minno *et al.*, 2022).

18.6.5 Metabolite dynamics and statistical difficulties

It is still difficult to study metabolite interactions in real time. Tracking metabolite turnover rates and achieving better time resolution are of interest. In statistical analysis, handling high-dimensional data and taking batch implications and unpredictable variables into account can be challenging. To obtain biologically significant findings, sophisticated statistical and machine learning techniques are required (Di Minno *et al.*, 2022).

18.7 Conclusion and Future Prospects

Metabolomics, combined with high-throughput technologies, has revolutionized the study of medicinal herbs, offering profound insights into their biochemical composition and therapeutic potential. By providing a comprehensive profile of metabolites, this approach enables the identification of bioactive compounds and elucidates their mechanisms of action, thereby bridging traditional knowledge with modern scientific validation. These advancements not only enhance the quality control and standardization of herbal products but also pave the way for the discovery of novel drugs derived from natural sources. The integration of metabolomics with other omics technologies, such as genomics and proteomics, holds great promise for advancing our understanding of medicinal herbs at a systems biology level. This holistic approach could lead to the discovery of new therapeutic targets and the development of more effective and personalized herbal medicines. Additionally, the application of artificial intelligence and machine learning to metabolomics data is expected to accelerate the identification of key bioactive compounds and predict their pharmacological effects. As research continues to evolve, metabolomics will play an increasingly critical role in the development of safe, effective, and scientifically validated herbal therapeutics.

18.8 Acknowledgments

The authors thank the Director, CSIR-Institute of Himalayan Bioresource Technology, Palampur, India, for infrastructure and support. S.G. thanks the University Grants Commission, Govt. of India (UGC Ref. No. 574/CSIR-UGC NET JUNE 2019) for providing a Senior Research Fellowship. The authors acknowledge the financial support provided by CSIR (MLP-201) and DBT (BT/PR45280/NER/95/1918/2022). This manuscript represents CSIR-IHBT Communication No. 5643.

18.9 References

Adımcılar, V., Kalaycıoglu, Z., Aydogdu, N., Dirmenci, T., Kahraman, A., and Erim, F.B. (2019) Rosmarinic and carnosic acid contents and correlated antioxidant and antidiabetic activities of 14 *Salvia* species from Anatolia. *Journal of Pharmaceutical and Biomedical Analysis* 175: 112763.

Aksenov, A.A., da Silva, R., Knight, R., Lopes, N.P., and Dorrestein, P.C. (2017) Global chemical analysis of biology by mass spectrometry. *Nature Review Chemistry* 1: 0054.

Arries, W.J., Tredoux, A.G.J., de Beer, D., Joubert, E., and de Villiers, A. (2017) Evaluation of capillary electrophoresis for the analysis of rooibos and honeybush tea phenolics. *Electrophoresis* 38, 897–905.

Bashir, K., Naz, S., Farooq, U., Wahid, F., Shah, A.J., *et al.* (2021) Assessing the ethnobotanical potential of *Carissa opaca* berries by merging outcomes from metabolomics profiling, enzyme assays, and *in silico* docking studies. *Food Chemistry* 363: 130259.

Blainey, P., Krzywinski, M., and Altman, N. (2014) Replication. *Nature Methods* 11, 879–880.

Bligh, E.G. and Dyer, W.J. (1959) A rapid method of total lipid extraction and purification. *Canadian Journal of Biochemistry and Physiology* 37, 911–917.

Blondel, C., Khelalfa, F., Reynaud, S., Fauvelle, F., and Raveton, M. (2016) Effect of organochlorine pesticides exposure on the maize root metabolome assessed using high-resolution magic-angle spinning ^{1}H NMR spectroscopy. *Environmental Pollution* 214, 539–548.

Bylda, C., Thiele, R., Kobold, U., and Volmer, D.A. (2014) Recent advances in sample preparation techniques to overcome difficulties encountered during quantitative analysis of small molecules from biofluids using LC-MS/MS. *Analyst* 139, 2265–2276.

Cajka, T. and Fiehn, O. (2014) Comprehensive analysis of lipids in biological systems by liquid chromatography-mass spectrometry. *Trends in Analytical Chemistry* 61, 192–206.

Cajka, T. and Fiehn, O. (2016) Toward merging untargeted and targeted methods in mass spectrometry-based metabolomics and lipidomics. *Analytical Chemistry* 88, 524–545.

Chanu, M.B., Chanu, W.K., and Chingakham, B.S. (2024) GC-MS profiling, sub-acute toxicity study and total phenolic and flavonoid content analysis of methanolic leaf extract of *Schima wallichii* (DC) Korth—a traditional antidiabetic medicinal plant. *Journal of Ethnopharmacology* 330: 118111.

Chi, J., Shu, J., Li, M., Mudappathi, R., Jin, Y., *et al.* (2024) Artificial intelligence in metabolomics: a current review. *Trends in Analytical Chemistry* 178: 117852.

Chong, J., Wishart, D.S., and Xia, J. (2019) Using MetaboAnalyst 4.0 for comprehensive and integrative metabolomics data analysis. *Current Protocols in Bioinformatics* 68: e86.

Deng, Y., Lam, S.C., Zhao, J., and Li, S.P. (2017) Quantitative analysis of flavonoids and phenolic acid in *Coreopsis tinctoria* Nutt. by capillary zone electrophoresis. *Electrophoresis* 38, 2654–2661.

Di Minno, A., Gelzo, M., Caterino, M., Costanzo, M., Ruoppolo, M., and Castaldo, G. (2022) Challenges in metabolomics-based tests, biomarkers revealed by metabolomic analysis, and the promise of the application of metabolomics in precision medicine. *International Journal of Molecular Sciences* 23: 5213.

Diserens, G., Vermathen, M., Precht, C., Broskey, N.T., Boesch, C., and Amati, F. (2015) Separation of small metabolites and lipids in spectra from biopsies by diffusion-weighted HR-MAS NMR: a feasibility study. *Analyst* 140, 272–279.

Doppler, M., Kluger, B., Bueschl, C., Schneider, C., Krska, R., *et al.* (2016) Stable isotope-assisted evaluation of different extraction solvents for untargeted metabolomics of plants. *International Journal of Molecular Science* 17: 1017.

dos Santos, N.A., de Souza, L.M., Pinto, F.E., Macrino, C.D.J., de Almeida, C.M., *et al.* (2019) LDI and MALDI-FT-ICR imaging MS in *Cannabis* leaves: optimization and study of spatial distribution of cannabinoids. *Analytical Methods* 11, 1757–1764.

Dudley, E., Yousef, M., Wang, Y., and Griffiths, W. (2010) Targeted metabolomics and mass spectrometry. *Advances in Protein Chemistry and Structural Biology* 80, 45–83.

Ellis, D.I., Dunn, W.B., Griffin, J.L., Allwood, J.W., and Goodacre, R., (2007) Metabolic fingerprinting as a diagnostic tool. *Pharmacogenomics* 8, 1243–1266.

Emwas, A.H., Roy, R., McKay, R.T., Tenori, L., Saccenti, E., and Gowda, G.A. (2019) NMR spectroscopy for metabolomics research. *Metabolites* 9: 123.

Ernst, M., Silva, D.B., Silva, R.R., Vencio, R.Z., and Lopes, N.P. (2014) Mass spectrometry in plant metabolomics strategies: from analytical platforms to data acquisition and processing. *Natural Products Reports* 31, 784–806.

Farber, C., Wang, R., Chemelewski, R., Mullet, J., and Kurouski, D. (2019a) Nanoscale structural organization of plant epicuticular wax probed by atomic force microscope infrared spectroscopy. *Analytical Chemistry* 91, 2472–2479.

Farber, C., Mahnke, M., Sanchez, L., and Kurouski, D. (2019b) Advanced spectroscopic techniques for plant disease diagnostics. A review. *Trends Analytical Chemistry* 118, 43–49.

Farber, C., Sanchez, L., Rizevsky, S., Ermolenkov, A., McCutchen, B., and Cason, J. (2020) Raman spectroscopy enables non-invasive identification of peanut genotypes and value-added traits. *Scientific Reports* 10: 7730.

Fernie, A.R., Trethewey, R.N., Krotzky, A.J., and Willmitzer, L. (2004) Innovation—metabolite profiling: from diagnostics to systems biology. *Nature Reviews Molecular Cell Biology* 5, 763–769.

Fernie, A.R., Aharoni, A., Willmitzer, L., Stitt, M., Tohge, T., *et al.* (2011) Recommendations for reporting metabolite data. *Plant Cell* 23, 2477–2482.

Folch, J., Lees, M., and Stanley, G.H.S. (1957) A simple method for the isolation and purification of total lipides from animal tissues. *Journal of Biological Chemistry* 226, 497–509.

Gahlaut, A., Vikas, D.M., Gothwal, A., Kulharia, M., Chhillar, A.K., *et al.* (2013) Proteomics and metabolomics: mapping biochemical regulations. *Drug Invention Today* 5, 321–326.

Gahtori, R., Tripathi, A.H., Kumari, A., Negi, N., Paliwal, A., *et al.* (2023) Anticancer plant-derivatives: deciphering their oncopreventive and therapeutic potential in molecular terms. *Future Journal of Pharmaceutical Sciences* 9: 14.

Giavalisco, P., Li, Y., Matthes, A., Eckhardt, A., Hubberten, H.M., *et al.* (2011) Elemental formula annotation of polar and lipophilic metabolites using ^{13}C, ^{15}N and ^{34}S isotope labelling, in combination with high-resolution mass spectrometry. *Plant Journal for Cell and Molecular Biology* 68, 364–376.

Glassbrook, N. and Ryals, J.A. (2001) Systematic approach to biochemical profiling. *Current Opinion in Plant Biology* 4, 186–190.

Harvey, D.J. and Vouros, P. (2020) Mass spectrometric fragmentation of trimethylsilyl and related alkylsilyl derivatives. *Mass Spectrometry Review* 39, 105–211.

Hong, Y.S., Coen, M., Rhode, C.M., Reily, M.D., Robertson, D.G., and Holmes, E. (2009) Chemical shift calibration of ^{1}H MAS NMR liver tissue spectra exemplified using a study of glycine protection of galactosamine toxicity. *Magnetic Resonance in Chemistry* 47, 47–53.

Jang, W.G., Park, J.Y., Lee, J., Bang, E., Kim, S.R., and Lee, E.K. (2016) Investigation of relative metabolic changes in the organs and plasma of rats exposed to X-ray radiation using HR-MAS ^{1}H NMR and solution ^{1}H NMR. *NMR in Biomedicine* 29, 507–518.

Jangid, R., Jain, S., Sharma, M.K., and Chatterjee, S. (2023) *In vitro* antioxidant and antidiabetic activity of ethanolic extract of *Prosopis* species growing in Rajasthan, India. *Vegetos* 36, 62–69.

Jiang, L., Sullivan, H., and Wang, B. (2022) Principal component analysis (PCA) loading and statistical tests for nuclear magnetic resonance (NMR) metabolomics involving multiple study groups. *Analytical Letters* 55, 1648–1662.

Kim, H.K. and Verpoorte, R. (2010) Sample preparation for plant metabolomics. *Phytochemical Analysis* 21, 4–13.

Kroymann, J. (2011) Natural diversity and adaptation in plant secondary metabolism. *Current Opinion in Plant Biology* 14, 246–251.

Kumar, M., Kuzhiumparambil, U., Pernice, M., Jiang, Z., and Ralph, P.J. (2016) Metabolomics: an emerging frontier of systems biology in marine macrophytes. *Algal Research* 16, 76–92.

Lichtenberg, S., Trifonova, O.P., Maslov, D.L., Balashova, E.E., and Lokhov, P.G. (2021) Metabolomic laboratory-developed tests: current status and perspectives. *Metabolites* 11: 423.

Lisec, J., Schauer, N., Kopka, J., Willmitzer, L., and Fernie, A.R. (2006) Gas chromatography mass spectrometry-based metabolite profiling in plants. *Nature Protocols* 1, 387–396.

Manas, R.S. and Marakanam, S.U. (2023) FTIR based metabolomics profiling and fingerprinting of some medicinal plants: an attempt to develop an approach for quality control and standardization of herbal materials. *Pharmacognosy Research* 15, 163–167.

Mari, A., Lyon, D., Fragner, L., Montoro, P., Piacente, S., *et al.* (2013) Phytochemical composition of *Potentilla anserina* L. analyzed by an integrative GC-MS and LC-MS metabolomics platform. *Metabolomics* 9, 599–607.

Markley, J.L., Bruschweiler, R., Edison, A.S., Eghbalnia, H.R., Powers, R., and Raftery, D. (2017) The future of NMR-based metabolomics. *Current Opinion Biotechnology* 43, 34–40.

Memon, A.F., Solangi, A.R., Memon, S.Q., Mallah, A., and Memon, N. (2018) Quantitative separation of hesperidin, chrysin, epicatechin, epigallocatechin gallate, and morin using ionic liquid as a buffer additive in capillary electrophoresis. *Electrophoresis* 39, 1606–1612.

Mishra, S., Ankit, Sharma, R., Gogna, N., and Dorai, K. (2021) NMR-based metabolomic profiling of the differential concentration of phytomedicinal compounds in pericarp, skin and seeds of *Momordica charantia* (bitter melon). *Natural Product Research* 36, 390–395.

Mohotti, S., Rajendran, S., Muhammad, T., Strömstedt, A.A., Adhikari, A., *et al.* (2020) Screening for bioactive secondary metabolites in Sri Lankan medicinal plants by microfractionation and targeted isolation of antimicrobial flavonoids from *Derris scandens*. *Journal of Ethnopharmacology* 246: 112158.

Mukherjee, P.K., Bahadur, S., Harwansh, R.K., and Chaudhary, S.K. (2014) Shifting paradigm for validation of medicinal plants in Indian traditional medicine. *Indian Drugs* 51, 5–15.

Mushtaq, M.Y., Choi, Y.H., Verpoorte, R., and Wilson, E.G. (2014) Extraction for metabolomics: access to the metabolome. *Phytochemical Analysis* 25, 291–306.

Nadeem, M., Zarreen, F., and Rizvi, M.M.A. (2024) GC-MS analysis of phytochemical composition and anticancer activities of methanol and aqueous extracts of *Bergenia ligulata* rhizome. *Journal of Food Science and Technology* 62, 623–634.

Okazaki, Y. and Saito, K. (2012) Recent advances of metabolomics in plant biotechnology. *Plant Biotechnology Reports* 6, 1–15.
Patel, M.K., Pandey, S., Kumar, M., Haque, M.I., Pal, S., and Yadav, N.S. (2021) Plants metabolome study: emerging tools and techniques. *Plants* 10: 2409.
Peng, Y., Xiao, X., Ji, T., Wang, X., Xu, Y., *et al.* (2024) Reveal the pharmacodynamic substances and mechanism of an edible medicinal plant *Rhodiola crenulate* in DSS-induced colitis through plasma pharmacochemistry and metabolomics. *Food Science and Human Wellness* 13, 2116–2131.
Plazas, E., Casoti, R., Murillo, M.A., Da Costa, F.B., and Cuca, L.E. (2019) Metabolomic profiling of *Zanthoxylum species*: identification of anti-cholinesterase alkaloids candidates. *Phytochemistry* 168: 12128.
Pluskal, T., Castillo, S., Villar-Briones, A., and Oresic, M. (2010) MZmine 2: modular framework for processing, visualizing, and analysing mass spectrometry-based molecular profile data. *BMC Bioinformatics* 11: 395.
Przybylska, A., Gackowski, M., and Koba, M. (2021) Application of capillary electrophoresis to the analysis of bioactive compounds in herbal raw materials. *Molecules* 26: 2135.
Rai, N., Kumari, S., Singh, S., Saha, P., Pandey, A.K., and Pandey-Rai, S. (2024) Modulation of morpho-physiological attributes and *in situ* analysis of secondary metabolites using Raman spectroscopy in response to red and blue light exposure in *Artemisia annua*. *Environmental and Experimental Botany* 217: 105563.
Raterink, R.J., Lindenburg, P.W., Vreeken, R.J., Ramautar, R., and Hankemeier, T. (2014) Recent developments in sample-pretreatment techniques for mass spectrometry-based metabolomics. *Trends in Analytical Chemistry* 61, 157–167.
Razzaq, A., Sadia, B., Raza, A., Khalid Hameed, M., and Saleem, F. (2019) Metabolomics: a way forward for crop improvement. *Metabolites* 9: 303.
Ren, X., Liu, H., and Wang, J. (2020) Determination of nevadensin in fewflower *Lysionotus* herb by capillary electrophoresis. *IOP Conference Series Earth and Environmental Science* 446: 032988.
Riekeberg, E. and Powers, R. (2017) New frontiers in metabolomics: from measurement to insight. *F1000Research* 6: 1148.
Salem, M., Bernach, M., Bajdzienko, K., and Giavalisco, P. (2017) A simple fractionated extraction method for the comprehensive analysis of metabolites, lipids, and proteins from a single sample. *Journal of Visualized Experiments* 124: e55802.
Salem, M.A., Li, Y., Bajdzienko, K., Fisahn, J., Watanabe, M., *et al.* (2018) RAPTOR controls developmental growth transitions by altering the hormonal and metabolic balance. *Plant Physiology* 177, 565–593.
Salem, M.A., Perez de Souza, L., Serag, A., Fernie, A.R., Farag, M.A., *et al.* (2020) Metabolomics in the context of plant natural products research: from sample preparation to metabolite analysis. *Metabolites* 10: 37.
Sanchez, L., Pant, S., Xing, Z., Mandadi, K., and Kurouski, D. (2019) Rapid and non-invasive diagnostics of Huanglongbing and nutrient deficits on citrus trees with a handheld Raman spectrometer. *Analytical Bioanalytical Chemistry* 411, 3125–3133.
Schroeder, M., Meyer, S.W., Heyman, H.M., Barsch, A., and Sumner, L.W. (2019) Generation of a collision cross section library for multi-dimensional plant metabolomics using UHPLC-trapped ion mobility-MS/MS. *Metabolites* 10: 13.
Shi, H., Wu, X., Zhu, Y., Jiang, T., Wang, Z., *et al.* (2024) RefMetaPlant: a reference metabolome database for plants across five major phyla. *Nucleic Acids Research* 52, D1614–D1628.
Sumner, L.W., Amberg, A., Barrett, D., Beale, M.H., Beger, R., *et al.* (2007) Proposed minimum reporting standards for chemical analysis. Chemical Analysis Working Group (CAWG) Metabolomics Standards Initiative (MSI). *Metabolomics* 3, 211–221.
Sun, J. and Xia, Y. (2024) Pretreating and normalizing metabolomics data for statistical analysis. *Genes & Diseases* 11: 100979.
Sutariya, S., Shah, A.A., Bajpai, A.B., Sharma, R.J., Pandhurnekar, C.P., and Gupta, A. (2023) Fourier transform infrared spectroscopy (FTIR) analysis, antioxidant and anti-inflammatory activities of leaf and fruit extracts of *Gymnosporia montana*. *Materials Today: Proceedings* 73, 134–141.
Tatsis, E.C., Boeren, S., Exarchou, V., Troganis, A.N., Vervoort, J., and Gerothanassis, I.P. (2007) Identification of the major constituents of *Hypericum perforatum* by LC/SPE/NMR and/or LC/MS. *Phytochemistry* 68, 383–393.
Tembo, Z.N., Aygun, F.S., and Erdoğan, B.Y., (2021) Simultaneous determination of nitrate, nitrite and bromate by capillary zone electrophoresis in tea infusions grown in the Black Sea region of Turkey. *Separation Science Plus* 4, 45–53.

Trethewey, R. (2004) Metabolite profiling as an aid to metabolic engineering in plants. *Current Opinion in Plant Biology* 7, 196–201.

Tulukcu, E., Cebi, N., and Sagdic, O. (2019) Chemical fingerprinting of seeds of some *Salvia species* in Turkey by using GC-MS and FTIR. *Foods* 8: 118.

Tweeddale, H., Notley-McRobb, L., and Ferenci, T. (1998) Effect of slow growth on metabolism of *Escherichia coli*, as revealed by global metabolite pool ('metabolome') analysis. *Journal of Bacteriology* 180, 5109–5116.

Ulrich-Merzenich, G., Zeitler, H., Jobst, D., Panek, D., Vetter, H., and Wagner, H. (2007) Application of the 'omic' technologies in phytomedicine. *Phytomedicine* 14, 70–82.

Vuckovic, D. (2012) Current trends and challenges in sample preparation for global metabolomics using liquid chromatography-mass spectrometry. *Analytical and Bioanalytical Chemistry* 403, 1523–1548.

Wang, Y., Tang, H., Nicholson, J.K., Hylands, P.J., Sampson, J., *et al.* (2004) Metabolomic strategy for the classification and quality control of phytomedicine: a case study of chamomile flower (*Matricaria recutita* L.). *Planta Medica* 70, 250–255.

Warren, C.R. (2013) Use of chemical ionization for GC-MS metabolite profiling. *Metabolomics* 9, 110–112.

Williams, B.J., Cameron, C.J., Workman, R., Broeckling, C.D., Sumner, L.W., and Smith, J.T. (2007) Amino acid profiling in plant cell cultures: an inter-laboratory comparison of CE-MS and GC-MS. *Electrophoresis* 28, 1371–1379.

Wolfender, J.L., Rudaz, S., Choi, Y.H., and Kim, H.K. (2013) Plant metabolomics: from holistic data to relevant biomarkers. *Current Medicinal Chemistry* 20, 1056–1090.

Xiao, Q., Mu, X., Liu, J., Li, B., Liu, H., *et al.* (2022) Plant metabolomics: a new strategy and tool for quality evaluation of Chinese medicinal materials. *Chinese Medicine* 17: 45.

Yonekura-Sakakibara, K., Higashi, Y., and Nakabayashi, R. (2019) The origin and evolution of plant flavonoid metabolism. *Frontiers in Plant Science* 10: 943.

19 High-throughput Metabolomics for Medicinal Plant Research

Gunna Sureshbabu Suruthi[1], Sivamani Ganesan[2], Ramakrishnan Aruna Raji[3], Shanmugam Velayuthaprabhu[4], Saradhadevi Muthukrishnan[5], Palanisamy Sampathkumar[6], Asirvatham Alwin Robert[7], and Arumugam Vijaya Anand[1*]

[1]Department of Human Genetics and Molecular Biology, Bharathiar University, Coimbatore, Tamil Nadu, India; [2]PG and Research Department of Zoology and Biotechnology, A.V.V.M. Sri Pushpam College (Autonomous), Poondi, Thanjavur, Tamil Nadu, India; [3]Department of Medicine, UV Gullas College of Medicine, Banilad, Cebu, Philippines; [4]Department of Biotechnology, Bharathiar University, Coimbatore, Tamil Nadu, India; [5]Department of Biochemistry, Bharathiar University, Coimbatore, Tamil Nadu, India; [6]Department of Chemistry and Biosciences, SASTRA Deemed University, Srinivasa Ramanujan Centre, Kumbakonam, Tamil Nadu, India; [7]Department of Endocrinology and Diabetes, Prince Sultan Military Medical City, Ministry of Defense, Riyadh, Saudi Arabia

Abstract

High-throughput metabolomics has emerged as a powerful analytical approach in medicinal plant research, offering unprecedented insights into the complex chemical compositions of plants used in traditional and modern medicine. This chapter explores the applications, advancements, and future perspectives of high-throughput metabolomics in the study of medicinal plants. We discuss how metabolomics technologies, including mass spectrometry and nuclear magnetic resonance spectroscopy, enable comprehensive profiling of plant metabolites, facilitating the identification of bioactive compounds and elucidation of their biosynthetic pathways. The integration of metabolomics with genomics, transcriptomics, and proteomics is examined, highlighting its role in providing a holistic understanding of plant biochemistry. We explore the contributions of metabolomics to drug discovery and development from medicinal plants, including lead compound identification and efficacy screening. The chapter also addresses the application of metabolomics in ethnopharmacology, validating traditional medicinal practices and uncovering novel therapeutic agents. The potential of metabolomics in personalizing plant-based therapies is discussed, along with current challenges and limitations in the field. Finally, we present emerging trends and future perspectives, emphasizing the increasing importance of high-throughput metabolomics in advancing medicinal plant research and its potential to revolutionize drug discovery and personalized medicine.

Keywords: ethnopharmacology, high-throughput, medicinal plants, metabolomics, secondary metabolites

* E-mail: avahgmb@buc.edu.in

© CAB International 2025. *High-Throughput Plant Metabolomics* (Jen-Tsung Chen ed.)
DOI: 10.1079/9781800626911.0019

19.1 Introduction to Metabolomics in Medicinal Plant Research

The study of medicinal plants has been greatly enhanced by metabolomics, a discipline that examines the full range of small molecules in living organisms. This approach has become increasingly valuable for its capacity to reveal detailed information about the chemical make-up and metabolic activities of plants (Marchev *et al.*, 2021). By examining the metabolites found in medicinal plants, researchers can gain a deeper understanding of the phytochemical components that contribute to their therapeutic effects (Gu *et al.*, 2018). This method is particularly useful for studying medicinal plants that are used in both conventional and alternative medicine because it enables the identification and quantification of the bioactive compounds that provide their therapeutic properties (Gu *et al.*, 2018).

For centuries, the use of medicinal plants has been a fundamental component of traditional medical systems, with a strong cultural and historical foundation (Gurib-Fakim, 2006; Manoharan *et al.*, 2008; Begum *et al.*, 2009; Muthumani *et al.*, 2013; Varalakshmi *et al.*, 2014; Vijaya Anand *et al.*, 2016, 2020; Bharathi *et al.*, 2018; Akilandeswari *et al.*, 2019; Divya *et al.*, 2019; Kaviya *et al.*, 2022; Manikandan *et al.*, 2022; Sangeetha *et al.*, 2022; Karthika *et al.*, 2023; Vijayakumar *et al.*, 2023). These plants are still of interest because of their therapeutic uses and have provided valuable medications (Gurib-Fakim, 2006). Researchers can find important metabolites that may be responsible for the pharmacological effects of medicinal plants by using metabolomics, a potent method for analyzing the chemical makeup of medicinal plants (Gu *et al.*, 2018). Researchers can learn more about the intricate relationships between plant metabolites and their biological functions using metabolomics, which can also help them understand the mechanisms underlying the medicinal plants' therapeutic properties (Gu *et al.*, 2018)

Enhancing the standardization and quality control of herbal materials is one of the major advantages of metabolomics in medicinal plant research (Sahoo and Umashankara, 2023). Researchers can develop strategies that ensure the consistency and effectiveness of herbal products made from medicinal plants by using methods such as Fourier-transform infrared (FTIR) spectroscopy-based metabolomics profiling (Sahoo and Umashankara, 2023). This is particularly important for the pharmaceutical sector, as it is necessary to regularly evaluate the composition and quality of herbal remedies to ensure their efficacy and safety (Sahoo and Umashankara, 2023). Moreover, metabolomics is essential for identifying the metabolic processes through which the bioactive chemicals are synthesized (Gu *et al.*, 2018). Researchers can identify important metabolic pathways that produce active chemicals such as flavonoids and alkaloids by integrating transcriptomic and metabolomic data (Gu *et al.*, 2018). This comprehensive method offers valuable knowledge about the biosynthesis of medicinal chemicals, which helps in the development of novel therapeutic compounds and the enhancement of plant-based treatments (Gu *et al.*, 2018).

In addition to its applications in traditional medicine, metabolomics has been useful in studying the potential of medicinal plants in current healthcare practices, especially in the treatment of various diseases, including gastrointestinal malignancies (Fig. 19.1). Through the application of metabolomics, scientists can examine the effectiveness of therapeutic plants from a molecular perspective, identifying significant metabolites that could have a therapeutic impact on specific diseases (Guo *et al.*, 2022). This strategy fits well with the expanding body of research on the use of natural products, such as medicinal plants, as adjuvant therapy in contemporary medicine (Guo *et al.*, 2022). For example, anti-obesity characteristics of plant metabolites have been uncovered by metabolomics, underscoring the promise of medicinal plants as sources of bioactive chemicals for treating metabolic disorders (Gooda Sahib *et al.*, 2012). Researchers can discover substances with anti-obesity effects by conducting systematic analyses of plant metabolites providing new possibilities for the development of plant-based therapies for the treatment of obesity and related disorders (Gooda Sahib *et al.*, 2012). This highlights the significance of metabolomics in investigating the various therapeutic applications of medicinal plants that go beyond their conventional medical applications.

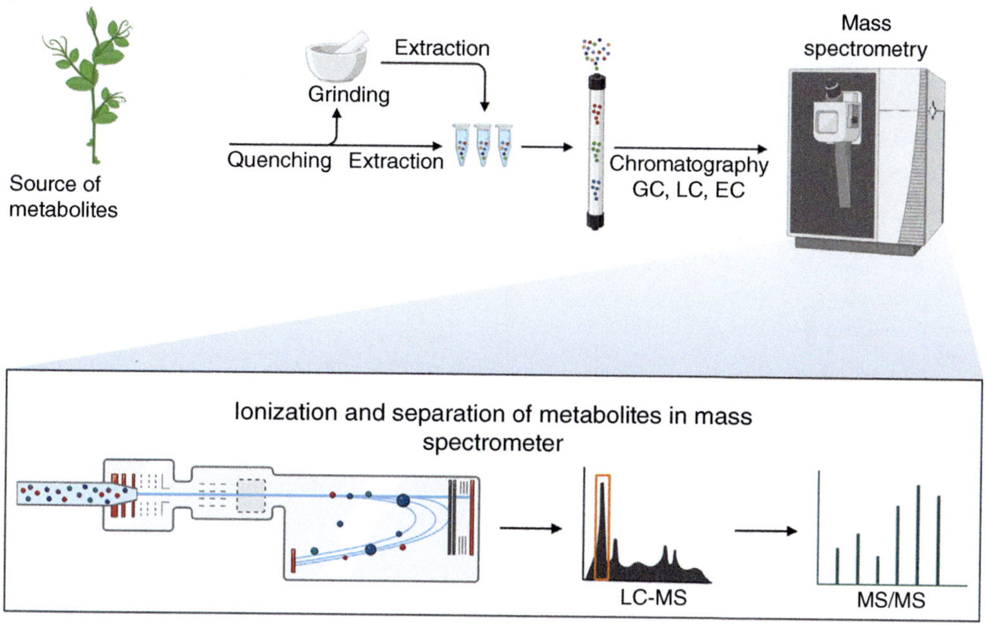

Fig. 19.1 A general overview of the use of metabolomics in plants. EC, electron capture; GC, gas chromatography; LC, liquid chromatography; LC-MS, liquid chromatography–mass spectrometry; MS/MS, tandem mass spectrometry.

19.2 High-throughput Metabolomics Technologies

High-throughput metabolomics technologies have significantly advanced the field of metabolomics, allowing researchers to analyze a large number of metabolites efficiently. Metabolite profiling can be done successfully using mass spectrometry (MS)-based techniques such as gas chromatography–mass spectrometry (GC-MS) and liquid chromatography–mass spectrometry (LC-MS) (Fig. 19.2, Table 19.1). These techniques provide vital insights into metabolic pathways and biochemical processes by enabling the sensitive and selective identification of metabolites in complex biological samples. Researchers may identify and quantify a broad range of metabolites by combining MS detection with chromatographic separation, offering a broad picture of the metabolome (Howlett *et al.*, 2017).

Another important technique for high-throughput metabolomics is nuclear magnetic resonance (NMR) spectroscopy, which offers quantitative and non-destructive examination of metabolites in biological samples (Table 19.1). To help with compound identification and the understanding of metabolic pathways, NMR spectroscopy provides structural information about metabolites. NMR spectroscopy is useful for metabolomics research because it can analyze several metabolites at once. This is particularly useful for biomarker identification and characterizing metabolic profiles under different physiological conditions (Li *et al.*, 2014). In high-throughput metabolomics research, advanced imaging methods such as direct-injection electrospray ionization–mass spectrometry (ESI-MS) have proven indispensable (Table 19.1). ESI-MS is perfect for characterizing an organism's metabolome as it enables sensitive, rapid analysis of cellular metabolites. Although there are issues with mass accuracy, ESI-MS has excellent throughput and sensitivity, making it a useful instrument for the preliminary study of many metabolites. The applicability of this technology in a variety of biological systems has been demonstrated by its success in microbial metabolomics research, such as the analysis of *Campylobacter jejuni* (Howlett *et al.*, 2017).

The development of sophisticated data-analysis tools such as MetaboReport and Protomix has been useful in handling the enormous

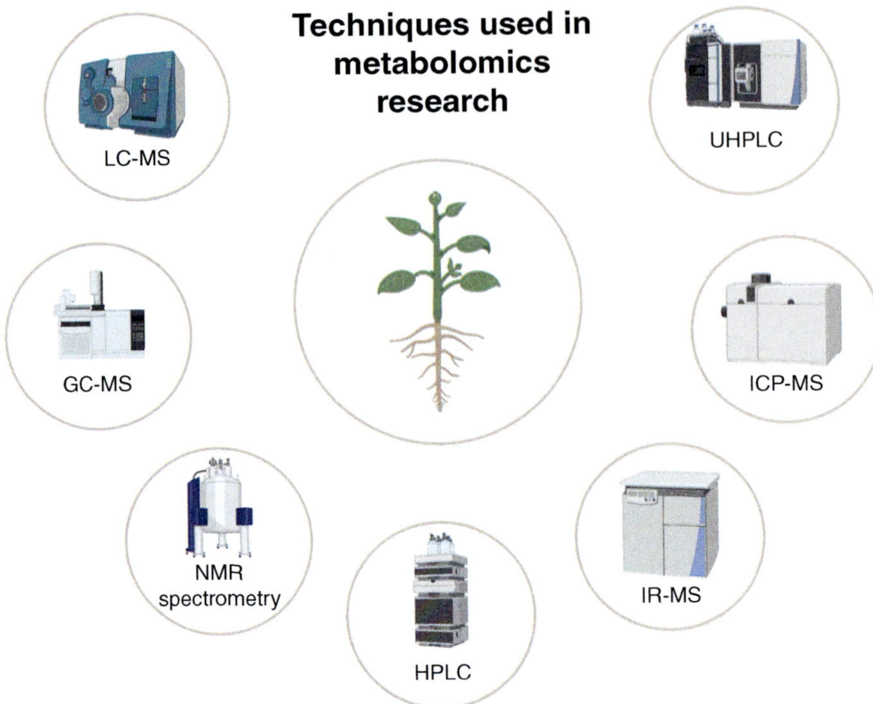

Fig. 19.2 Techniques used in metabolic research. GC-MS, gas chromatography–mass spectrometry; HPLC, high-performance liquid chromatography; IR-MS, infrared mass spectrometry; LC-MS, liquid chromatography–mass spectrometry; ICP-MS, inductively coupled plasma–mass spectrometry; NMR, nuclear magnetic resonance; UHPLC, ultra-high-performance liquid chromatography.

volume of data generated by high-throughput metabolomics technology (Zniber and Huynh, 2024). By resolving experimental variability and maintaining data quality, these technologies provide advanced data preparation abilities for NMR and MS data (Dong and Malitsky, 2024). Researchers may also import, process, and analyze metabolomics data using a variety of statistical techniques such as the metaP-Server and Omics Dashboard, which offer complete data-analysis solutions (Paley and Karp, 2024). These tools aid in the interpretation of complicated metabolomics data by facilitating multivariate analysis, metabolite identification, and metabolic pathway visualization (Kastenmüller *et al.*, 2011).

In high-throughput metabolomics research, the integration of metabolomic profiles with data from other omics technologies, such as proteomics and genomics, has become increasingly significant. Researchers can investigate the connections between various molecular levels and obtain a better understanding of intricate biological processes using the multi-omics approach, which provides a comprehensive picture of biological systems (Fiorentino *et al.*, 2021). Researchers can find biomarkers, develop personalized medical strategies, and gain new insights into disease mechanisms by combining data from multiple omics disciplines (Hasin *et al.*, 2017).

19.3 Phytochemical Profiling of Medicinal Plants

Metabolomics is a powerful tool in medicinal plant research, enabling the identification and quantification of bioactive compounds in these plants (Mohanraj *et al.*, 2018). This method is essential for studying the pharmacological effects of therapeutic plants and identifying important bioactive compounds that could be used as potential drug candidates (Mazhar *et al.*, 2022). Metabolomics in medicinal plant research significantly enhances chemotaxonomy,

Table 19.1. Techniques used in metabolomics.

Methodology	Description	Applications in medicinal plant research	Reference(s)
LC-MS	Combines chromatographic separation with mass spectrometry detection	Sensitive and specific detection of metabolites Identification and quantification of plant metabolites Biomarker discovery	Zhou and Zhong (2022); Shi (2024)
GC-MS	Separates and analyzes compounds that can be vaporized without decomposition	Analysis of volatile metabolites in medicinal plants Profiling of essential oils and other volatile compounds	Ji *et al.* (2022)
NMR spectroscopy	Provides structural information about metabolites through non-destructive analysis	Structural elucidation of plant metabolites Quantitative analysis of metabolites Metabolic profiling of plant extracts	Li *et al.* (2014)
Direct-injection ESI-MS	Allows rapid and sensitive analysis of cellular metabolites without prior separation	High-throughput screening of plant extracts Rapid profiling of plant metabolomes	Howlett *et al.* (2017)
HPLC	Separates, identifies, and quantifies components in a mixture	Phytochemical profiling of medicinal plants Quantification of bioactive compounds	Anekwe (2023)
FTIR spectroscopy	Identifies chemical bonds in a molecule by producing an infrared absorption spectrum	Metabolic fingerprinting of plant extracts Quality control and standardization of herbal materials	Sahoo and Umashankara (2023)
MS/MS	Involves multiple steps of mass spectrometry selection, with fragmentation occurring between the steps	Structural characterization of complex plant metabolites Untargeted metabolomics of medicinal plants	Kang *et al.* (2022)

ESI-MS, electrospray ionization–mass spectrometry; FTIR, Fourier-transform infrared; GC-MS, gas chromatography–mass spectrometry; HPLC, high-performance liquid chromatography; LC-MS, liquid chromatography–mass spectrometry; MS/MS, tandem mass spectrometry; NMR, nuclear magnetic resonance.

the scientific discipline that groups plants according to their chemical makeup (Xin *et al.*, 2010). The chemotaxonomic classification of medicinal plants is aided by metabolic fingerprints generated by techniques such as high-performance liquid chromatography (HPLC) and NMR spectroscopy (Xin *et al.*, 2010). These techniques make it possible to distinguish between different plant species according to their distinct chemical profiles, offering useful data for taxonomic research and the identification of chemotaxonomic markers (Hassn, 2023).

Herbal materials from medicinal plants are subjected to quality control and standardization using metabolic fingerprinting, which involves analysis of metabolite profiles in biological samples (Cardoso-Taketa *et al.*, 2008). Researchers can evaluate the authenticity and consistency of herbal products, ensuring their efficacy and quality, by comparing the metabolic profiles of plant extracts (Cardoso-Taketa *et al.*, 2008). Metabolic fingerprinting is also useful for monitoring batch-to-batch variability in herbal preparations, confirming the authenticity of therapeutic plants, and spotting adulteration (Cardoso-Taketa *et al.*, 2008). For understanding the bioactive components of medicinal plants and their therapeutic uses, phytochemical profiling is essential (Mazhar *et al.*, 2022). Techniques such as HPLC and GC-MS aid in the identification and quantification of phytochemicals in plant extracts and provide details about their biological activity (Anekwe, 2023). The characterization of bioactive compounds in medicinal plants is made possible by phytochemical profiling, which facilitates the discovery of novel drug candidates and plant-based treatments (Anekwe, 2023). Understanding the pharmacological effects and therapeutic potential of medicinal plants requires the identification and quantification of their bioactive components (Haripriya *et al.*, 2022). The chemical composition and bioactivity of medicinal plant extracts are evaluated with the aid of analytical techniques such as bioactive component identification, isolation, and extraction, maximizing the use of medicinal plants in both traditional and modern medicine by gaining an understanding of their mechanisms of action (Haripriya *et al.*, 2022).

19.4 Metabolomics in Drug Discovery and Development from Medicinal Plants

Metabolomics plays a crucial role in drug discovery and development from medicinal plants by enabling the identification of bioactive compounds and understanding their mechanisms of action (Oyenihi *et al.*, 2021).

19.4.1 Role in identifying lead compounds for drug development

Metabolomics, a powerful tool in drug discovery, plays a crucial role in identifying lead compounds for drug development from medicinal plants. Researchers are able to efficiently screen medicinal plants for anti-mycobacterial activity by using advanced techniques such as ultra-high-performance liquid chromatography (UHPLC) with high-resolution mass spectrometry (HRMS) metabolomics coupled with multivariate analysis (Moyo, 2024). This method makes it possible to identify possible bioactive compounds that could be used as leads in the development of new drugs. Similarly, metabolomics studies provide an opportunity to find new molecules using non-targeted techniques, demonstrating the potential of this strategy in identifying novel therapeutic candidates (Santacruz *et al.*, 2019). Metabolomics is a valuable tool for assessing the biological activity of medicinal plants. It aids in the quality control of natural products obtained from plants, bioactivity-guided fractionation, and the identification of bioactive leads (Marchev *et al.*, 2021). Researchers may rapidly investigate possible molecules for drug development by using computational metabolomics, which incorporates molecular networking and molecular docking, expediting the drug discovery process (Lephatsi *et al.*, 2023).

According to Kim *et al.* (2010), metabolomics is a useful analytical approach for identifying secondary metabolites from medicinal plants in addition to improving the efficiency of lead-finding from natural goods. This analytical profiling is essential for evidence-based development of novel phytotherapeutical agents and nutraceuticals (Nanjappan *et al.*, 2012). Plant natural product drug discovery has undergone a revolution due to recent advances in metabolomics, which provides insights into sample preparation methods, analytical profiling tools, and the entire bioactive molecule identification process (Salem *et al.*, 2020). Metabolomics provides drug discovery with a faster path in the search for lead compounds from medicinal plants. It is being used more frequently in bio-guided compound isolation and structural elucidation of plant extracts, demonstrating its significance in the initial phases of drug development (Umoh *et al.*, 2023). Moreover, network pharmacology's multi-targeted therapeutic assessment of medicinal plants emphasizes the potential of metabolomics in drug development and discovery. An essential technique for examining the intricate chemical composition of medicinal plants, especially those used in traditional medicine, is tandem MS (MS/MS)-based untargeted metabolomics (Kang *et al.*, 2022).

The role of metabolomics in herbal medicine and drug discovery is significant, as it offers a metabolic perspective on herbal medicines, facilitating drug development processes (Wang *et al.*, 2019). Tuyiringire *et al.* (2018) reported that metabolomics is also used to study the mechanisms of action of medicinal plants possessing anti-tuberculosis activity, indicating its versatility in examining a range of therapeutic qualities of compounds produced from plants. Plant metabolomics contributes to research on plant development and stress responses, which is important for boosting human nutrition and health (Sun *et al.*, 2020). The continued development of using plant metabolomics for drug discovery is demonstrated by the ongoing attempts to isolate and characterize pharmacologically active molecules from medicinal plants (Ojah, 2020). This area of metabolomics has developed due to technological improvements in instrumentation and data processing, which have provided a plethora of methods for the identification of bioactive compounds from plant tissues (Foito and Stewart, 2018). Anti-parasitic drug development is being driven by metabolomics in combination with other omics technologies, demonstrating its potential to identify new therapeutic molecules (Giannangelo *et al.*, 2016).

19.4.2 Screening for therapeutic potential and efficacy

Metabolomics plays a crucial role in screening for the therapeutic potential and efficacy of medicinal plants by enabling the rapid identification and characterization of phytochemicals and metabolites present in herbal medicines (Yang *et al.*, 2021). Plant metabolites can be thoroughly profiled using this advanced analytical method, which makes it easier to assess their bioactive properties and potential health benefits (Trivedi *et al.*, 2017). Researchers can identify molecules with therapeutic potential, for example those targeting type 2 diabetes, by combining metabolomic data with machine learning approaches to find particular quality markers (Q-markers) from herbal preparations (Yang *et al.*, 2021).

A comprehensive understanding of the effectiveness and molecular mechanisms of action of medicinal plants is provided by the use of metabolomics (Guo *et al.*, 2022). With its focus on the overall changes in terminal metabolites, metabolomics offers valuable information about the bioactivity and potential therapeutic uses of medicinal plants for treating a range of illnesses (Guo *et al.*, 2022). Researchers can decipher the intricate relationships between biological systems and plant metabolites using metabolomic analysis, providing insight into the mechanisms underlying the therapeutic advantages of medicinal plants (Marchev *et al.*, 2021). The study of metabolite profiles of various agricultural and medicinal plants has been made possible by metabolomics, which has enabled the identification of taxonomic and biochemical distinctions between different plant species (Marchev *et al.*, 2021). According to Bapela *et al.* (2019), this analytical method is especially useful for evaluating the quality control of plant samples used in herbal preparations, guaranteeing the uniformity and effectiveness of therapeutic plant-based products. Researchers can assess the biological activities of plant-derived compounds using metabolomics to associate the bioactivity of medicinal plants with their metabolite profiles (Gonulalan *et al.*, 2019).

The identification of compounds with particular bioactive properties is enhanced by the discovery of Q-markers for natural products using the integration of metabolomics and chemometrics techniques (Ding *et al.*, 2018). This method, which combines chemometrics with UHPLC-triple/time-of-flight-MS, enables the discovery of unique indicators that can be used to evaluate the authenticity and quality of natural products (Ding *et al.*, 2018). The therapeutic benefits of traditional herbal remedies, such as ginseng, against a range of medical diseases have been assessed effectively using such metabolomic techniques (Liu *et al.*, 2022). Researchers can determine potential therapeutic targets and understand the mechanisms of action of herbal remedies by tracking changes in metabolism and performing network pharmacology analysis (Liu *et al.*, 2022). Further evidence of use of metabolomics in quality assurance and authentication procedures comes from its application in determining the geographic origin of medicinal plants (Lee *et al.*, 2014). When determining the validity and geographic origin of plant materials, metabolomic approaches can identify changes in metabolism that result from genetic or environmental variables (Lee *et al.*, 2014). Furthermore, screening crops for yield and stress tolerance at the metabolic level is made possible by metabolomics-assisted breeding, contributing to improvements in crops (Razzaq *et al.*, 2019).

19.4.3 Case studies of successful drug discoveries

Drug discovery from medicinal plants has a long history of successful outcomes. The potential of plant-derived compounds in the fight against tuberculosis has been demonstrated by the considerable anti-mycobacterial activity of several medicinal plants against strains of *Mycobacterium bovis* and *Mycobacterium tuberculosis* (Gemechu *et al.*, 2013), highlighting how crucial it is to look for new drug leads in natural sources, especially when developing antibiotics. Balunas and Kinghorn (2005) discussed the significant contributions of drug discovery from medicinal plants, highlighting the isolation of early drugs including cocaine, codeine, digitoxin, quinine, and morphine from plant sources, some of which are still in use today. These achievements highlight the significance of natural products in drug discovery processes, as well as the influence of molecules derived from plants in contemporary pharmacology.

Another example of the use of natural product bioassays in drug development is demonstrated by how plant-based remedies have aided in the creation of vital medications such as taxol, which is derived from the Pacific yew tree and has been identified as an important anticancer agent (New *et al.*, 2003). This illustrates how thorough scientific research has led to major therapeutic improvements based on traditional knowledge of medicinal plants. Similarly, ethnomedicinal knowledge has contributed to successful drug discoveries, such as vinblastine and vincristine from *Catharanthus roseus* for treating various cancers, showcasing the valuable insights that traditional medicine can offer in identifying potent bioactive compounds (Lone *et al.*, 2015). This emphasizes how crucial it is to combine traditional knowledge with cutting-edge scientific methods to identify new therapeutic prospects derived from medicinal plants. Moreover, research on the pharmacological screening of *Tectona grandis* Linn. heartwood for antibacterial activity exemplifies the systematic approach to investigating plant extracts for potential therapeutic agents (Singh, 2022). Researchers can uncover bioactive compounds with antibacterial effects by employing techniques such as agar-well diffusion experiments, which paves the way for the development of novel drugs derived from plants.

19.5 Integration with Other Omics Approaches

Data from various omics techniques must be integrated in order to understand the synergistic interaction between metabolomics, genomics, and transcriptomics in medicinal plant research. A comprehensive understanding of the biochemical processes within medicinal plants can be obtained by combining transcriptomics, genomics, and metabolomics, which focus on the comprehensive analysis of small molecules in biological systems, effectively (Moyo, 2024). Researchers can study the genetic basis of metabolite production and discover important pathways involved in the generation of bioactive compounds by combining metabolomic data with genomic information (Santacruz *et al.*, 2019).

Conversely, transcriptomics provides information about regulatory mechanisms and patterns of gene expression in therapeutic plants. Researchers can better understand the metabolic pathways involved in the generation of bioactive compounds by linking the gene expression profiles to the production of certain metabolites through the integration of transcriptomic data with metabolomics (Marchev *et al.*, 2021). Through integration, potential genes for the development of pharmacologically significant compounds in medicinal plants can be identified (Lephatsi *et al.*, 2023). Studying the therapeutic effects of medicinal plants can be done using a systems biology approach by the integration of omics. By taking into account the networks and interactions between genes, proteins, and metabolites, systems biology seeks to understand biological systems as a whole (Oyenihi *et al.*, 2021). A thorough understanding of the therapeutic potential of medicinal plants can be obtained by researchers by combining data from different omics to decipher the intricate interactions between genes, gene products, and metabolites (Kim *et al.*, 2012).

The ability to combine transcriptomic and metabolomic studies to reveal the biosynthesis pathways of metabolites in plants has been shown in recent studies. For instance, metabolomic and transcriptome data have been merged in studies on *Polygonatum kingianum* Coll. et Hemsl and *Dendrobium huoshanense* to provide insights into the biosynthesis of flavonoids and other secondary metabolites in these medicinal plants (Salem *et al.*, 2020). Understanding the mechanisms behind plant secondary metabolism has also benefitted from the integration of omics methods. Through the integration of different omics data, researchers have expedited the identification of essential pathways implicated in the production of secondary metabolites, thereby clearing the pathway for the discovery of new bioactive substances (Gautam, 2023). Untangling the intricate regulatory networks that control secondary metabolism in plants has been made possible by this integrative approach (Kang *et al.*, 2022). Additionally, the distribution of flavonoids in various plant tissues, such as taproots and fibrous roots, has been analyzed through the integration of metabolomics and transcriptomics. Researchers can obtain a comprehensive understanding of the spatial distribution of bioactive compounds within medicinal plants by combining data from these omics techniques, providing an understanding of the physiological roles and potential therapeutic effects of these compounds (Sharma and Yadav, 2022).

19.6 Ethnopharmacology and Metabolomics

When combined with ethnopharmacology, metabolomics is essential for validating traditional medical knowledge and identifying new compounds from ethnobotanical research. Researchers can develop structure–activity relationships and advance the domains of pharmacognosy and ethnopharmacology by incorporating a strategy that involves the analyses of plant secondary metabolomes and the biological activities of plant isolates (Orlova *et al.* 2022). By bridging the gap between conventional and modern medical ideas, this comprehensive methodology offers a new perspective on how traditional medicine might be modernized (Oskouie *et al.*, 2022).

The traditional use of plants for specific therapeutic purpose, such as treating hyperglycemia, can be confirmed by ethnopharmacological examinations, such as the study of optimized *Cycas circinalis* leaf extracts for their polyphenol fingerprinting and hypoglycemic properties (Arshad *et al.*, 2021). Such validations of ethnopharmacological practices through scientific analyses contribute to the credibility and efficacy of traditional medicine. Furthermore, the translation of traditional knowledge into contemporary drug discovery methods is exemplified by the shift from ethnobotanical observations to the isolation and characterization of particular bioactive compounds for drug development (Pirintsos *et al.*, 2022).

Persian medicine has shown that metabolomics can uncover the functional features of traditional medicine (Oskouie *et al.*, 2022). This cutting-edge analytical method helps identify bioactive compounds that are consistent with traditional applications by providing insights into the chemical composition of medicinal plants. Researchers can identify novel compounds with pharmacological potential by using metabolomic investigations to understand the molecular basis of plant medicinal properties (Liu *et al.*, 2010). Metabolomics is a tool that ethnopharmacologists use to identify and authenticate medicinal plants, ensuring the efficacy, safety, and purity of medicinal products (Wangchuk *et al.*, 2011). Thus, researchers are able to evaluate the ethnobotanical applications of plants and obtain significant information for drug development by fusing traditional knowledge with scientific approaches. A thorough knowledge of the bioactive ingredients and therapeutic mechanisms of herbal medicines is made possible by the incorporation of metabolomic techniques into the research process (Banerjee *et al.*, 2021).

Moreover, the use of metabolomics in the assessment of toxicity associated with traditional Chinese medicines underscores its significance in ensuring the safety and effectiveness of herbal remedies (Duan *et al.*, 2018). Researchers can assess potential toxicities and side effects by examining the metabolic profiles of medicinal plants. This helps to standardize and ensure the quality control of medicinal products. By strengthening the pharmacological foundation of conventional drugs, this strategy increases their acceptability in contemporary therapeutic practices (Iqbal *et al.*, 2023). Studies examining the antioxidant capabilities of *Sonchus oleraceus* L. extracts have shown that metabolomics also facilitates the discovery of novel bioactive compounds from medicinal plants (Nouidha, 2023). Researchers can find compounds with specific biological functions by analyzing the metabolites found in plant extracts, which paves the way for the generation of novel medicinal treatments. According to Sun *et al.* (2012), metabolomics is a useful technique for characterizing the pharmacological properties of medicinal plants and understanding their modes of action.

19.7 Personalized Medicine and Medicinal Plants

Metabolomics plays a significant role in tailoring plant-based therapies to individual needs in the realm of personalized medicine. Metabolism allows the identification of individual differences in their metabolic profiles through advanced analytical techniques and computational tools. This allows treatment strategies to be personalized according to specific disease aspects and genetic characteristics (Jacob *et al.*, 2019). Moving away from the conventional "one-size-fits-all" approach, this personalized approach to medicine seeks to optimize medication therapy by matching it with unique patient characteristics and risk profiles (Driscoll and Ludtka, 2015). A promising approach to improving the effectiveness and precision of plant-based therapies personalized to match patients' various demands is the integration of metabolomics with personalized medicine (Zheng *et al.*, 2013).

Precision medicine, another name for personalized medicine, is the practice of personalizing medical treatment to each patient's specific requirements at every stage of care, from diagnosis and treatment to follow-up and prevention (Driscoll and Ludtka, 2015). This method emphasizes the significance of identifying variations in an individual's response to a certain medicine by taking into account their genetic profile, response to disease, and treatment (Gianazza *et al.*, 2023). As a crucial element of personalized medicine, metabolomics offers insightful information on medication management, disease monitoring, and risk assessment. This information transforms clinical practice and increases treatment efficacy (Zheng *et al.*, 2013). Plant-based therapeutics can be made more successful and personalized by using personalized medicine to analyze individual metabolomic patterns and optimize therapeutic interventions based on individual metabolic profiles (Zheng *et al.*, 2013).

The combination of plant-based medicinal research with metabolomics holds the key to developing personalized natural treatments. According to Mumtaz *et al.* (2017), metabolomics provides a thorough framework for examining the metabolite profiles of therapeutic plants, facilitating the discovery of bioactive substances and their modes of action. This analytical method advances pharmaceutical production and quality assessment of traditional treatments, in addition to helping identify new pharmaceuticals derived from natural products, all of which are ultimately beneficial to human health (Sun *et al.*, 2020). Researchers can decipher the complex metabolic pathways involved in plant-based therapies and develop personalized herbal treatments based on individual health needs by utilizing metabolomics technology and bioinformatics in the context of precision medicine (Azad and Shulaev, 2019).

The investigation of metabolomic patterns in certain diseases, including colorectal cancer, lends more support to the personalized medicine trend in the realm of natural remedies. Metabolomics presents prospects for personalized herbal treatments in oncology, including early diagnosis, drug discovery, and treatment optimization (Zheng *et al.*, 2013). Additionally, metabolomics aids in the evaluation of gastrointestinal cancers from a holistic perspective, focusing on fluctuations in terminal metabolites and offering insights into efficacy assessment and the molecular mechanisms of medicinal plants as adjuvant therapeutics for cancer therapy (Guo *et al.*, 2022). This personalized approach to plant medicine emphasizes how important it is to take into account each patient's unique metabolic profile while developing plant-based treatments to address a range of medical conditions.

LC-MS has become a potent tool for biomarker discovery and personalized treatment plans in the context of metabolomics and personalized medicine. Researchers can identify distinct biomarkers and metabolic signatures that can direct personalized herbal therapies by using LC-MS-based metabolomics to analyze metabolic profiles at the molecular level (Zhou and Zhong, 2022). In addition to complementing genomics and proteomics, this high-throughput analytical method is essential to the advancement of research into personalized medicine and customized herbal remedies (Zhou and Zhong, 2022). The combination of personalized medicine and LC-MS-based metabolomics provides a thorough foundation for understanding individual metabolic differences and tailoring plant-based medicines to the needs of individual patients.

19.8 Challenges and Limitations

High-throughput metabolomics has emerged as a powerful tool in medicinal plant research, offering the ability to analyze a vast number of metabolites in a high-throughput and comprehensive manner (Trifonova *et al.*, 2016). By utilizing advanced analytical techniques such as HRMS-LC, researchers can detect and quantify plant metabolites with unprecedented sensitivity and throughput (Shi, 2024). These methods have been pivotal in identifying differences in volatile metabolites among different species of medicinal plants, aiding in the characterization of biochemical compositions crucial for medicinal plant research (Ji *et al.*, 2022).

One of the key challenges in high-throughput metabolomics for medicinal plant research lies in the extraction of metabolites from plant tissues, which is often labor-intensive and time-consuming (Liu *et al.*, 2011). Additionally, the analysis of plant extracts can be complex, requiring sophisticated data processing due to the variability in chromatographic separations (Yilmaz *et al.*, 2016). Despite these challenges, the development of high-throughput metabolomics platforms has significantly advanced the field, allowing the identification of a large number of metabolites in plant samples (Razzaq *et al.*, 2022). This progress has been particularly beneficial in genetic studies and the improvement of plant species, offering insights into population-level variations in plant metabolites (Cao *et al.*, 2017). The integration of metabolomics with other omics technologies such as genomics, transcriptomics, and proteomics has further enhanced the scope of metabolomics studies in medicinal plant research (Razzaq *et al.*, 2022). This multi-platform approach enables researchers to achieve maximum coverage of plant metabolomes and facilitates the identification of a wide range of metabolites. Moreover, the combination of metabolomics with advanced analytical tools has opened up new avenues for diagnostic breeding and crop improvement, particularly in addressing challenges related to abiotic stress in plants (Razzaq *et al.*, 2022).

In the realm of traditional medicine, high-throughput metabolomics has been instrumental in analyzing the specialized metabolomes of medicinal plants used in East Asian traditional medicine (Kang *et al.*, 2022). MS/MS-based untargeted metabolomics has proven to be a key methodology for high-throughput analysis of the complex chemical compositions of these plants, shedding light on their therapeutic properties (Kang *et al.*, 2022). Furthermore, the creation of plant metabolite spectral libraries has been crucial for both untargeted and targeted metabolomics, facilitating the efficient handling and interpretation of metabolomics data (Li *et al.*, 2023). The continuous development of high-throughput sequencing technologies has revolutionized the study of medicinal plants, allowing in-depth analyses at the molecular level (Meng *et al.*, 2022). This shift toward microscopic molecular investigations has provided researchers with a deeper understanding of plant genomes and metabolomes, offering insights into the biosynthetic pathways of specialized compounds (Meng *et al.*, 2022). Additionally, the establishment of databases integrating traditional Chinese medicine plant genomes has further enriched the resources available for studying medicinal plants (Meng *et al.*, 2022).

19.9 Future Perspectives and Emerging Trends

High-throughput metabolomics in medicinal plant research is anticipated to advance significantly, with the integration of high-throughput metabolomics with other omics technologies becoming more prevalent. This integration is expected to enable a comprehensive understanding of plant metabolomes (Razzaq *et al.*, 2022). The multi-platform approach is projected to enhance the identification of a wide range of metabolites and facilitate the exploration of metabolic pathways and regulatory networks in medicinal plants (Razzaq *et al.*, 2022). By unraveling genetic–metabolite networks through the integration of transcriptomics and metabolomics, researchers can identify key genes and pathways involved in the biosynthesis of bioactive compounds, leading to the improved production of valuable metabolites from plants (Hirai *et al.*, 2004).

The future of high-throughput metabolomics in medicinal plant research also shows promise in personalized medicine and clinical metabolomics (Gehrke *et al.*, 2017). The application of metabolomics data alongside high-throughput measurements of low-molecular-mass substances in biosamples can pave the way for personalized metabolomics in clinical practice, offering insights into individualized drug responses and metabolic profiles (Trifonova *et al.*, 2023). This personalized approach has the potential to revolutionize clinical biochemistry and enhance the field of personalized medicine (Gehrke *et al.*, 2017). Moreover, the utilization of high-throughput metabolomics for the prioritization and targeted isolation of natural products is expected to lead to new strategies in drug discovery (Wolfender *et al.*, 2019). By integrating innovative omics-based approaches, researchers can streamline the identification and isolation of bioactive compounds from medicinal plants, leading to the discovery of novel drug candidates (Wolfender *et al.*, 2019). This shift toward targeted isolation strategies based on metabolomics data is anticipated to catalyze advancements in pharmacognosy and drug development (Wolfender *et al.*, 2019).

In the context of medicinal plant research, high-throughput metabolomics is poised to play a crucial role in elucidating the therapeutic properties of medicinal plants for the treatment of various diseases (Guo *et al.*, 2022). Metabolomics offers a holistic perspective on the biochemical compositions of medicinal plants, providing insights into the efficacy assessment and molecular mechanisms underlying their therapeutic effects (Guo *et al.*, 2022). This metabolomics-driven approach will likely enhance our understanding of the metabolite profiles of medicinal plants and their potential applications in disease management (Guo *et al.*, 2022). Furthermore, the future of high-throughput metabolomics in medicinal plant research may see an increased focus on the interaction between plant metabolomes and microbiomes (Li *et al.*, 2022). Studying the association between the alkaloidal metabolome of plants and root microbiota can offer valuable insights into the links between specific medicinal ingredients and microbial communities, providing strategies for optimizing the cultivation and application of medicinal plants (Li *et al.*, 2022). This integrated approach combining metabolomics and microbiome analysis is expected to deepen our understanding of the complex interactions shaping plant metabolomes (Li *et al.*, 2022).

19.10 Conclusion

High-throughput metabolomics has emerged as a powerful and transformative tool in medicinal plant research, offering unprecedented insights into the complex chemical landscapes of plants used in traditional and modern medicine. This advanced analytical approach has revolutionized our understanding of plant metabolomes, enabling the identification and quantification of bioactive compounds with remarkable precision and efficiency. The integration of metabolomics with other omics technologies, such as genomics and transcriptomics, has provided a holistic view of the biochemical processes within medicinal plants. This multi-omics approach has not only enhanced our understanding of biosynthetic pathways but also facilitated the discovery of novel bioactive compounds and potential drug candidates. Metabolomics has played a crucial role in validating traditional medicinal practices, bridging the gap between ethnopharmacology and modern drug discovery. It has enabled researchers to unravel the molecular basis of therapeutic properties attributed to medicinal plants, leading to the development of more targeted and effective plant-based treatments. The application of high-throughput metabolomics in personalized medicine shows great promise, offering the potential for tailored plant-based therapies based on individual metabolic profiles. This personalized approach could significantly enhance the efficacy and precision of herbal treatments, addressing the diverse needs of patients more effectively.

Despite challenges in data processing and metabolite extraction, ongoing advancements in analytical technologies and bioinformatics tools continue to expand the abilities of metabolomics in medicinal plant research. The future of this field looks promising, with emerging trends pointing toward more integrated, personalized, and targeted approaches to drug discovery and development from medicinal plants. As we move forward, high-throughput metabolomics is poised to play an increasingly vital role in unlocking the full potential of medicinal plants. By continuing to refine our understanding of plant metabolomes and their interactions with human health, we can expect significant advancements in the development of novel therapeutics, the optimization of traditional medicines, and the realization of more personalized and effective plant-based treatments.

19.11 Acknowledgment

All the authors are grateful to their respective universities and institutions for their valuable support.

19.12 References

Akilandeswari, G., Vijaya Anand, A., Sampathkumar, P., Vinayaga Moorthi, P., and Preethi, B. (2019) A prospective review on phyto-pharmacological aspects of *Andrographis paniculata*. *Systematic Reviews in Pharmacy* 10, 15–19.

Anekwe, I.I., Chikwendu, C.I., Amadi, E.S., Nwogwugwu, N.U., and Ihenetu, F.C. (2023) Gas chromatography-mass spectrometry analysis of bioactive compounds of curcuma longa leaves extract. *International Journal of Biological and Chemical Sciences* 17, 1199–1207.

Arshad, M., Chaudhary, A.R., Mumtaz, M.W., Raza, S.A., Ahmad, M., *et al.* (2021) Polyphenol fingerprinting and hypoglycemic attributes of optimized *Cycas circinalis* leaf extracts. *Journal of the Science of Food and Agriculture* 101, 1530–1537.

Azad, R.K. and Shulaev, V. (2019) Metabolomics technology and bioinformatics for precision medicine. *Briefings in Bioinformatics* 20, 1957–1971.

Balunas, M.J. and Kinghorn, A.D. (2005) Drug discovery from medicinal plants. *Life Sciences* 78, 431–441.

Banerjee, S., Kar, A., Mukherjee, P.K., Haldar, P.K., Sharma, N., and Katiyar, C.K. (2021) Immunoprotective potential of Ayurvedic herb Kalmegh (*Andrographis paniculata*) against respiratory viral infections–LC–MS/MS and network pharmacology analysis. *Phytochemical Analysis* 32, 629–639.

Bapela, M.J., Heyman, H., Senejoux, F., and Meyer, J.M. (2019) ^{1}H NMR-based metabolomics of antimalarial plant species traditionally used by Vha-Venda people in Limpopo Province, South Africa and isolation of antiplasmodial compounds. *Journal of Ethnopharmacology* 228, 148–155.

Begum, T.N., Ilyas, M.M., Kalavathy, S., Anand, A.V., Kumar, S., and Senthil, R. (2009) Effects of ethanolic leaf extract of *Azim tetracantha* Lam. on Ehrlich ascites carcinoma tumour bearing mice. *Research Journal of Medicine and Medical Sciences* 4, 351–354.

Bharathi, V., Rengarajan, R.L., Radhakrishnan, R., Hashem, A., Abd Allah, E.F., *et al.* (2018) Effects of a medicinal plant *Macrotyloma uniflorum* (Lam.) Verdc. formulation (MUF) on obesity-associated oxidative stress-induced liver injury. *Saudi Journal of Biological Sciences* 25, 1115–1121.

Cao, M., Fraser, K., Jones, C., Stewart, A., Lyons, T., *et al.* (2017) Untargeted metabotyping *Lolium perenne* reveals population-level variation in plant flavonoids and alkaloids. *Frontiers in Plant Science* 8: 133.

Cardoso-Taketa, A.T., Pereda-Miranda, R., Choi, Y.H., Verpoorte, R., and Villarreal, M.L. (2008) Metabolic profiling of the Mexican anxiolytic and sedative plant *Galphimia glauca* using nuclear magnetic resonance spectroscopy and multivariate data analysis. *Planta Medica* 74, 1295–1301.

Ding, M., Jiang, Y., Yu, X., Zhang, D., Li, J. *et al.* (2018) Screening of combinatorial quality markers for natural products by metabolomics coupled with chemometrics. A case study on pollen typhae. *Frontiers in Pharmacology* 9: 691.

Divya, N., Rengarajan, R.L., Radhakrishnan, R., Abd Allah, E.F., Alqarawi, A.A., *et al.* (2019) Phytotherapeutic efficacy of the medicinal plant *Terminalia catappa* L. *Saudi Journal of Biological Sciences* 26, 985–988.

Dong, Y. and Malitsky, S. (2024) MetaboReport: from metabolomics data analysis to comprehensive reporting. *Bioinformatics* 40: btae373.

Driscoll, J.J. and Ludtka, T. (2015) The promise and challenges of personalized medicine in hematologic malignancies. *Journal of Hematology and Thrombosis* 1: 3.

Duan, L., Guo, L., Wang, L., Yin, Q., Zhang, C.M., *et al.* (2018) Application of metabolomics in toxicity evaluation of traditional Chinese medicines. *Chinese Medicine* 13: 60.

Fiorentino, G., Visintainer, R., Domenici, E., Lauria, M., and Marchetti, L. (2021) MOUSSE: multi-omics using subject-specific signatures. *Cancers* 13: 3423.

Foito, A. and Stewart, D. (2018) Metabolomics: a high-throughput screen for biochemical and bioactivity diversity in plants and crops. *Current Pharmaceutical Design* 24, 2043–2054.

Gautam, G. (2023) Advancement of network pharmacology in multi-targeted therapeutic evaluation of medicinal plants. *Journal of CAM Research Progress* 2: 113.

Gehrke, S., Reisz, J.A., Nemkov, T., Hansen, K.C., and d'Alessandro, A. (2017) Characterization of rapid extraction protocols for high-throughput metabolomics. *Rapid Communications in Mass Spectrometry* 31, 1445–1452.

Gemechu, A., Giday, M., Worku, A., and Ameni, G. (2013) *In vitro* anti-mycobacterial activity of selected medicinal plants against *Mycobacterium tuberculosis* and *Mycobacterium bovis* strains. *BMC Complementary and Alternative Medicine* 13: 291.

Gianazza, E., Brioschi, M., Iezzi, A., Paglia, G., and Banfi, C. (2023) Pharmacometabolomics for the study of lipid-lowering therapies: opportunities and challenges. *International Journal of Molecular Sciences* 24: 3291.

Giannangelo, C.R., Ellis, K.M., Sexton, A.E., Stoessel, D., and Creek, D.J. (2016) The role of metabolomics in antiparasitic drug discovery. In: Müller, S., Cerdan, R., and Radulescu, O. (eds) *Comprehensive Analysis of Parasite Biology: From Metabolism to Drug Discovery*. Wiley, Weinheim, Germany, pp. 321–342.

Gonulalan, E.M., Nemutlu, E., and Demirezer, L.O. (2019) A new perspective on evaluation of medicinal plant biological activities: the correlation between phytomics and matrix metalloproteinases activities of some medicinal plants. *Saudi Pharmaceutical Journal* 27, 446–452.

Gooda Sahib, N., Saari, N., Ismail, A., Khatib, A., Mahomoodally, F., and Abdul Hamid, A. (2012) Plants' metabolites as potential antiobesity agents. *Scientific World Journal* 2012: 436039.

Gu, L., Zhang, Z.Y., Quan, H., Li, M.J., Zhao, F.Y. *et al.* (2018) Integrated analysis of transcriptomic and metabolomic data reveals critical metabolic pathways involved in rotenoid biosynthesis in the medicinal plant *Mirabilis himalaica*. *Molecular Genetics and Genomics* 293, 635–647.

Guo, W., Cao, P., Wang, X., Hu, M., and Feng, Y. (2022) Medicinal plants for the treatment of gastrointestinal cancers from the metabolomics perspective. *Frontiers in Pharmacology* 13: 909755.

Gurib-Fakim, A. (2006) Medicinal plants: traditions of yesterday and drugs of tomorrow. *Molecular Aspects of Medicine* 27, 1–93.

Haripriya, R., Bee, L.C., Siau, H.M., and Yin, H.C. (2022) An insight into extraction, isolation, identification and quantification of bioactive compounds from Crataegus monogyna plant extract. *Reviews in Agricultural Science* 10, 304–327.

Hasin, Y., Seldin, M., and Lusis, A. (2017) Multi-omics approaches to disease. *Genome Biology* 18: 83.

Hassn, N.T., Al-Maathide, A.M., and Al-Saffar, R.S. (2023) Phenolic compounds play a crucial role as chemotaxonomic markers in the Cucurbitaceae species cultivated in Iraq. *SABRAO Journal of Breeding and Genetics* 55, 1377–1391.

Hirai, M.Y., Yano, M., Goodenowe, D.B., Kanaya, S., Kimura, T., *et al.* (2004) Integration of transcriptomics and metabolomics for understanding of global responses to nutritional stresses in *Arabidopsis thaliana*. *Proceedings of the National Academy of Sciences USA* 101, 10205–10210.

Howlett, R.M., Davey, M.P., and Kelly, D.J. (2017) Metabolomic analysis of *Campylobacter jejuni* by direct-injection electrospray ionization mass spectrometry. *Methods in Molecular Biology* 1512, 189–197.

Iqbal, I., Saqib, F., Latif, M.F., Shahzad, H., Dima, L. *et al.* (2023) Pharmacological basis for antispasmodic, bronchodilator, and antidiarrheal potential of *Dryopteris ramosa* (Hope) C. via *in vitro*, *in vivo*, and *in silico* studies. *ACS Omega* 8, 26982–27001.

Jacob, M., Lopata, A.L., Dasouki, M., and Abdel Rahman, A.M. (2019) Metabolomics toward personalized medicine. *Mass Spectrometry Reviews* 38, 221–238.

Ji, J., Zang, L., Lu, T., Li, C., Han, X. *et al.* (2022) Widely targeted metabolomic analysis reveals differences in volatile metabolites among four *Angelica* species. *Natural Products and Bioprospecting* 15: 2.

Kang, K.B., Jeong, E., Son, S., Lee, E., Lee, S. *et al.* (2022) Mass spectrometry data on specialized metabolome of medicinal plants used in East Asian traditional medicine. *Scientific Data* 9: 528.

Karthika, P., Balamuralikrishnan, B., Yamini, K., Vijaya Anand, A., Durairaj, K. *et al.* (2023) Green synthesis, characterization of silver nanoparticles using aqueous leaf extracts of *Solanum melongena* and *in vitro* evaluation of antibacterial, pesticidal and anticancer activity in human MDA-MB-231 breast cancer cell lines. *Journal of King Saud University, Science* 35: 102663.

Kastenmüller, G., Römisch-Margl, W., Wägele, B., Altmaier, E., and Suhre, K. (2011) metaP-server: a web-based metabolomics data analysis tool. *BioMed Research International* 2011: 839862.

Kaviya, M., Balasubramanian, B., Bharathi, K., Malaisamy, A., Al-Dhabi, N.A. *et al.* (2022) Evaluation of nutritional substances and investigation of antioxidant and antimicrobial potentials of *Boerhavia diffusa* with *in silico* molecular docking. *Molecules* 27: 1280.

Kim, H.E., Choi, Y.H., Choi, K.H., Park, J.S., Kim, H.S. *et al.* (2012) Metabolic classification of herb plants by NMR-based metabolomics. *Journal of the Korean Magnetic Resonance Society* 16, 91–102.

Kim, H.K., Wilson, E.G., Choi, Y.H., and Verpoorte, R. (2010) Metabolomics: a tool for anticancer lead-finding from natural products. *Planta Medica* 76, 1094–1102.

Lee, D.K., Lim, D.K., Um, J.A., Lim, C.J., Hong, J.Y. *et al.* (2014) Evaluation of four different analytical tools to determine the regional origin of *Gastrodia elata* and *Rehmannia glutinosa* on the basis of metabolomics study. *Molecules* 19, 6294–6308.

Lephatsi, M.M., Choene, M.S., Kappo, A.P., Madala, N.E., and Tugizimana, F. (2023) An integrated molecular networking and docking approach to characterize the metabolome of *Helichrysum splendidum* and its pharmaceutical potentials. *Metabolites* 13: 1104.

Li, F., Xu, Y., Shang, D., Yang, H., Liu, W. *et al.* (2014) MPINet: metabolite pathway identification via coupling of global metabolite network structure and metabolomic profile. *BioMed Research International* 2014: 325697.

Li, H., Shi, H., Xu, P., and Yu, D. (2022) Metabolomics and microbiome reveal potential root microbiota affecting the alkaloidal metabolome in *Aconitum vilmorinianum* Kom. *BMC Microbiology* 22: 70.

Li, Y., Zhu, W., Xiang, Q., Kim, J., Dufresne, C. *et al.* (2023) Creation of a plant metabolite spectral library for untargeted and targeted metabolomics. *International Journal of Molecular Sciences* 24: 2249.

Liu, N.Q., Cao, M., Frédérich, M., Choi, Y.H., Verpoorte, R., and van der Kooy, F. (2010) Metabolomic investigation of the ethnopharmacological use of *Artemisia afra* with NMR spectroscopy and multivariate data analysis. *Journal of Ethnopharmacology* 128, 230–235.

Liu, M., Cai, Y., Li, F., Wu, Y., Zhang, M. *et al.* (2022) Metabolomics and network pharmacological analysis to explore the mechanism of ginseng treatment for spleen-qi deficiency. *Journal of Separation Science* 45, 4427–4438.

Liu, X., Zhang, L., You, L., Wu, H., Zhao, J. *et al.* (2011) Metabolomic study on the halophyte *Suaeda salsa* in the Yellow River Delta. *Clean: Soil, Air, Water* 39, 720–727.

Lone, P.A., Bhardwaj, A.K., and Bahar, F.A. (2015) Study of indigenous/traditional medicinal plant knowledge—an endeavour towards new drug discovery. *African Journal of Traditional, Complementary, and Alternative Medicines* 12, 73–95.

Manikandan, R., Balasubramanian, B., Punniyakotti, P., Vijaya Anand, A., Meyyazhagan, A. *et al.* (2022) Cardio-protective effects of *Terminalia catappa* leaves and *Terminalia chebula* fruit extract in doxorubicin-induced cardiomyopathy in rats. *Biomarkers* 27, 488–495.

Manoharan, N., Sampathkumar, P., Dheeba, B., Sheik Abdulla, S., Vinothprasanna, G. *et al.* (2008) Potential hepatoprotective effect of aqueous extract of *Gracilaria corticata* in AFB_1 induced hepatotoxicity in Wistar rats. *Journal of Biological Sciences* 8, 1352–1355.

Marchev, A.S., Vasileva, L.V., Amirova, K.M., Savova, M.S., Balcheva-Sivenova, Z.P., and Georgiev, M.I. (2021) Metabolomics and health: from nutritional crops and plant-based pharmaceuticals to profiling of human biofluids. *Cellular and Molecular Life Sciences* 78, 6487–6503.

Mazhar, M., Afzal, M., and Naveed, M. (2022) Phytochemical profiling, biological activities and *in silico* virtual screening of bioactive compounds of *Trichodesma indicum* (L) R. Br. extracts. *Chemistry Select* 7: e202203821.

Meng, F., Tang, Q., Chu, T., Li, X., Lin, Y., *et al.* (2022) TCMPG: an integrative database for traditional Chinese medicine plant genomes. *Horticulture Research* 9: uhac060.

Mohanraj, K., Karthikeyan, B.S., Vivek-Ananth, R.P., Chand, R.B., Aparna, S.R., *et al.* (2018) IMPPAT: a curated database of Indian medicinal plants, phytochemistry and therapeutics. *Scientific Reports* 8: 4329.

Moyo, P., Ofori, M., Bodede, O.S., Wooding, M., Khorommbi, N.K. *et al.* (2024) Investigation of the antimycobacterial activity of African medicinal plants combined with chemometric analysis to identify potential leads. *Scientific Reports* 14: 14660.

Mumtaz, M.W., Hamid, A.A., Akhtar, M.T., Anwar, F., Rashid, U., and AL-Zuaidy, M.H., (2017) An overview of recent developments in metabolomics and proteomics—phytotherapic research perspectives. *Frontiers in Life Science* 10, 1–37.

Muthumani, G.D., Anand, A.V., and Manikandan, R. (2013) Antioxidant, antihelmintic and antimicrobial activity of *Euphorbia thymifolia* Linn whole plant. *International Journal of Current Microbiology and Applied Sciences* 2, 66–75.

Nanjappan,.S., Narayanan, N., Nirmal, S., Nirmal, J., Jain, G.K., and Spandana, V. (2012) Analytical profiling of bioactive constituents from herbal products, using metabolomics—a review. *Natural Product Communications* 7, 1111–1115.

New, D.C., Miller-Martini, D.M., and Wong, Y.H. (2003) Reporter gene assays and their applications to bioassays of natural products. *Phytotherapy Research* 17, 439–448.

Nouidha, S., Selmi, S., Guigonis, J.M., Pourcher, T., Chekir-Ghedira, L., and Kilani-Jaziri, S. (2023) Metabolomics profiling of Tunisian *Sonchus oleraceus* L. extracts and their antioxidant activities. *Chemistry and Biodiversity* 20: e202300290.

Ojah, E.O. (2020) Medicinal plants: prospective drug candidates against the dreaded coronavirus. *Iberoamerican Journal of Medicine* 2, 314–321.

Orlova, A., Kysil, E., Tsvetkova, E., Meshalkina, D., Whaley, A. *et al.* (2022) Phytochemical characterization of water avens (*Geum rivale* L.) extracts: structure assignment and biological activity of the major phenolic constituents. *Plants* 11: 2859.

Oskouie, A.A., Aliabad, H.S., Zanjani, R.C., Kaveh, S., Sadeghi, S., *et al.* (2022) Metabolomics: a new perspective to Persian medicine. *Journal of Pharmaceutical Negative Results*, 498–504.

Oyenihi, O.R., Oyenihi, A.B., Erhabor, J.O., Matsabisa, M.G., and Oguntibeju, O.O. (2021) Unravelling the anticancer mechanisms of traditional herbal medicines with metabolomics. *Molecules* 26: 6541.

Paley, S. and Karp, P.D. (2024) The Omics Dashboard for interactive exploration of metabolomics and multi-omics data. *Metabolites* 14: 65.

Pirintsos, S., Panagiotopoulos, A., Bariotakis, M., Daskalakis, V., Lionis, C. *et al.* (2022) From traditional ethnopharmacology to modern natural drug discovery: a methodology discussion and specific examples. *Molecules* 27: 4060.

Razzaq, A., Sadia, B., Raza, A., Khalid Hameed, M., and Saleem, F. (2019) Metabolomics: a way forward for crop improvement. *Metabolites* 9: 303.

Razzaq, A., Wishart, D.S., Wani, S.H., Hameed, M.K., Mubin, M., and Saleem, F. (2022) Advances in metabolomics-driven diagnostic breeding and crop improvement. *Metabolites* 12: 511.

Sahoo, M.R. and Umashankara, M.S. (2023) FTIR based metabolomics profiling and fingerprinting of some medicinal plants: an attempt to develop an approach for quality control and standardization of herbal materials. *Pharmacognosy Research* 15, 163–167.

Salem, M.A., Perez de Souza, L., Serag, A., Fernie, A.R., Farag, M.A., *et al.* (2020) Metabolomics in the context of plant natural products research: from sample preparation to metabolite analysis. *Metabolites* 10: 37.

Sangeetha, T., Balamuralikrishnan, B., Sampathkumar, P., Velayuthaprabhu, S., Senthilkumar, N. *et al.* (2022) Characterization and phytoconstituents of *Petroselinum crispum* (Mill) and *Coriandrum sativum* (Linn) and their impacts on inflammation—an *in vitro* analysis against human adenocarcinoma cells with molecular docking. *South African Journal of Botany* 146, 776–788.

Santacruz, L., Thomas, O.P., Duque, C., Puyana, M., and Tello, E. (2019) Comparative analyses of metabolomic fingerprints and cytotoxic activities of soft corals from the Colombian Caribbean. *Marine Drugs* 17: 37.

Sharma, B. and Yadav, D.K. (2022) Metabolomics and network pharmacology in the exploration of the multi-targeted therapeutic approach of traditional medicinal plants. *Plants* 11: 3243.

Shi, H., Zhu, Y., Wu, X., Jiang, T., Li, X. *et al.* (2024) CropMetabolome: a comprehensive metabolome database for major crops cross eight categories. *Plant Journal* 119, 1613–1626.

Singh, R. (2022) Pharmacological screening of heartwood of *Tectona grandis* Linn. for antibacterial activity by agar-well diffusion method. *IOP Conference Series: Materials Science and Engineering* 1221: 012019.

Sun, H., Zhang, A., and Wang, X. (2012) Potential role of metabolomic approaches for Chinese medicine syndromes and herbal medicine. *Phytotherapy Research* 26, 1466–1471.

Sun, W., Chen, Z., Hong, J., and Shi, J. (2020) Promoting human nutrition and health through plant metabolomics: current status and challenges. *Biology* 10: 20.

Trifonova, O., Knight, R.A., Lisitsa, A., Melino, G., and Antonov, A.V. (2016) Exploration of individuality in drug metabolism by high-throughput metabolomics: the fast line for personalized medicine. *Drug Discovery Today* 21, 103–110.

Trifonova, O.P., Maslov, D.L., Balashova, E.E., and Lokhov, P.G. (2023) Current state and future perspectives on personalized metabolomics. *Metabolites* 13: 67.

Trivedi, D.K., Hollywood, K.A., and Goodacre, R. (2017) Metabolomics for the masses: the future of metabolomics in a personalized world. *New Horizons in Translational Medicine* 3, 294–305.

Tuyiringire, N., Tusubira, D., Munyampundu, J.P., Tolo, C.U., Muvunyi, C.M., and Ogwang, P.E. (2018) Application of metabolomics to drug discovery and understanding the mechanisms of action of medicinal plants with anti-tuberculosis activity. *Clinical and Translational Medicine* 7: 29.

Umoh, S.D., Bojase, G., Masesane, I.B., Majinda, R.T., and Sichilongo, K.F. (2023) Untargeted GC–MS metabolomics to identify and classify bioactive compounds in *Combretum platypetalum* subsp. *oatesii* (Rolfe) Exell (Combretaceae). *Phytochemical Analysis* 34, 127–138.

Varalakshmi, B., Anand, A.V., Karpagam, T., Bai, J.S., and Manikandan, R. (2014) *In vitro* antimicrobial and anticancer activity of *Cinnamomum zeylanicum* Linn bark extracts. *International Journal of Pharmacy and Pharmaceutical Sciences* 6, 12–18.

Vijaya Anand, A., Varalakshmi, B.P., Sampathkumar, P., Pushpa, N., and Agaath, H. (2016) *Cinnamomum zeylanicum* Linn. The spice with multi potential. *Systematic Reviews in Pharmacy* 7, 24–29.

Vijaya Anand, A., Velayuthaprabhu, S., Rengarajan, R.L., Sampathkumar, P., and Radhakrishnan, R. (2020) Bioactive compounds of guava (*Psidium guajava* L.). In: Murthy, H.N. and Bapat, V.A. (eds) *Bioactive Compounds in Underutilized Fruits and Nuts*. Springer, Cham, Switzerland, pp. 503–527.

Vijayakumar, K., Prasanna, B., Rengarajan, R.L., Rathinam, A., Velayuthaprabhu, S., and Vijaya, A.A. (2023) Anti-diabetic and hypolipidemic effects of *Cinnamon cassia* bark extracts: an *in vitro*, *in vivo*, and *in silico* approach. *Archives of Physiology and Biochemistry* 129, 338–348.

Wang, X.J., Ren, J.L., Zhang, A.H., Sun, H., Yan, G.L., *et al.* (2019) Novel applications of mass spectrometry-based metabolomics in herbal medicines and its active ingredients: current evidence. *Mass Spectrometry Reviews* 38, 380–402.

Wangchuk, P., Pyne, S.G., and Keller, P.A. (2011) Ethnobotanical authentication and identification of Khrogsman (lower elevation medicinal plants) of Bhutan. *Journal of Ethnopharmacology* 134, 813–823.

Wolfender, J.L., Litaudon, M., Touboul, D., and Queiroz, E.F. (2019) Innovative omics-based approaches for prioritisation and targeted isolation of natural products—new strategies for drug discovery. *Natural Product Reports* 36, 855–868.

Xin, Y.Y., Deng, A.J., Du, G.H., Zhang, J.L., and Qin, H.L. (2010) Fingerprinting analysis of *Saposhnikovia divaricata* using ^{1}H nuclear magnetic resonance spectroscopy and high performance liquid chromatography. *Journal of Integrative Plant Biology* 52, 782–792.

Yang, L., Xue, Y., Wei, J., Dai, Q., and Li, P. (2021) Integrating metabolomic data with machine learning approach for discovery of Q-markers from Jinqi Jiangtang preparation against type 2 diabetes. *Chinese Medicine* 16: 30.

Yilmaz, A., Rudolph, H.L., Hurst, J.J., and Wood, T.D. (2016) High-throughput metabolic profiling of soybean leaves by Fourier transform ion cyclotron resonance mass spectrometry. *Analytical Chemistry* 88, 1188–1194.

Zheng, X., Xie, G., and Jia, W. (2013) Metabolomic profiling in colorectal cancer: opportunities for personalized medicine. *Personalized Medicine* 10, 741–755.

Zhou, J. and Zhong, L. (2022) Applications of liquid chromatography-mass spectrometry based metabolomics in predictive and personalized medicine. *Frontiers in Molecular Biosciences* 9: 1049016.

Zniber, M. and Huynh, T.P. (2024) Protomix: a Python package for ^{1}H-NMR metabolomics data preprocessing. *Bioinformatics Advances* 5: vbae192.

20 High-throughput Metabolomics for the Exploration of Plant Secondary Metabolism

Gunna Sureshbabu Suruthi[1], Galeel Rahman Rihana Begum[1], Murugesan Vinothini[1], Balasubramanian Balamuralikrishnan[2], Naif Abdullah Al-Dhabi[3], Mariadhas Valan Arasu[3], Vino Palani[4], and Arumugam Vijaya Anand[1*]

[1]Department of Human Genetics and Molecular Biology, Bharathiar University, Coimbatore, Tamil Nadu, India; [2]Department of Food Science and Biotechnology, College of Life Science, Sejong University, Seoul, Republic of Korea; [3]Department of Botany and Microbiology, College of Science, King Saud University, Riyadh, Saudi Arabia; [4]Department of Botany, Sri Vidya Mandir Arts and Science College (Autonomous), Krishnagiri, Tamil Nadu, India

Abstract

Plant secondary metabolism encompasses diverse biochemical pathways that produce specialized metabolites crucial for plant defense, signaling, and interactions with the environment. High-throughput metabolomics has emerged as a transformative approach in the exploration of these secondary metabolites, providing unprecedented insights into the complex regulatory networks governing their biosynthesis and dynamics. This chapter discusses the applications of metabolomics in understanding the types of plant secondary metabolites, their biosynthetic pathways, and the integration of metabolomics with other omics technologies. The chapter highlights the utility of metabolic flux analysis and the dynamic nature of secondary metabolite accumulation in response to biotic and abiotic stressors. Furthermore, it examines the applications of high-throughput metabolomics in agriculture and industry, including crop improvement, the discovery of bioactive compounds, and food quality and safety. While the field faces challenges related to metabolite identification, data analysis, and biological variability, future innovations in analytical techniques, computational tools, and multi-omics integration are poised to advance our understanding of plant secondary metabolism and its far-reaching implications.

Keywords: bioactive compounds, biosynthetic pathways, crop improvement, high-throughput metabolomics, plant secondary metabolism, stress responses

20.1 Introduction to Plant Secondary Metabolism

Plant secondary metabolism encompasses a diverse array of biochemical pathways that produce specialized metabolites, which are not essential for basic plant growth but play crucial roles in plant defense, signaling, and interactions with the environment. These metabolites include a wide variety of compounds such as alkaloids,

* E-mail: avahgmb@buc.edu.in

© CAB International 2025. *High-Throughput Plant Metabolomics* (Jen-Tsung Chen ed.)
DOI: 10.1079/9781800626911.0020

terpenoids, and phenylpropanoids, which are synthesized through complex metabolic networks that often intersect with primary metabolic pathways (Manoharan *et al.*, 2008; Begum *et al.*, 2009; Muthumani *et al.*, 2013; Varalakshmi *et al.*, 2014; Vijaya Anand *et al.*, 2016, 2020; Bharathi *et al.*, 2018; Akilandeswari *et al.*, 2019; Divya *et al.*, 2019; Sangeetha, *et al.*, 2022; Kaviya *et al.*, 2022; Manikandan *et al.*, 2022; Karthika *et al.*, 2023; Vijayakumar *et al.*, 2023). The significance of secondary metabolites extends beyond their ecological functions; they are also of immense interest in pharmacology and agriculture due to their bioactive properties and potential applications in medicine and industry (Roy *et al.*, 2022).

The classification of secondary metabolites is typically based on their biosynthetic origins. For instance, terpenoids are derived from isoprene units, phenylpropanoids originate from the amino acid phenylalanine, and alkaloids are primarily nitrogen-containing compounds. Each class of secondary metabolites serves distinct functions: terpenoids are often involved in plant defense and attraction of pollinators, while phenylpropanoids play roles in UV protection and structural integrity of plant tissues (Dong and Lin, 2021). The intricate biosynthetic pathways leading to these metabolites are regulated by various factors, including genetic, environmental, and physiological cues, which can significantly influence the yield and composition of secondary metabolites in plants (Yang *et al.*, 2018).

Endophytes, which are microorganisms residing within plant tissues, have been shown to significantly impact the secondary metabolism of their host plants. These symbiotic relationships can enhance the production of valuable secondary metabolites by stimulating the expression of biosynthetic genes and enzymes. For example, specific endophytic fungi have been identified as key players in the biosynthesis of alkaloids in plants such as *Lycoris radiata*, where the presence of certain fungal species leads to increased alkaloid accumulation (Zhou *et al.*, 2020). This interaction highlights the potential for utilizing endophytes in biotechnological applications aimed at enhancing the production of secondary metabolites in medicinal plants (Tamang, 2023).

Abiotic stress factors, such as drought, salinity, and nutrient availability, also play a pivotal role in modulating secondary metabolite production. The carbon–nutrient balance hypothesis suggests that plants may allocate more carbon to secondary metabolites when nutrient availability is low, thereby enhancing their defensive capabilities against environmental stressors (Meng *et al.*, 2020). For instance, studies have demonstrated that salinity stress can lead to increased production of secondary metabolites in plants such as lemon verbena, indicating a stress-induced metabolic shift that enhances the plant's resilience (Ghanbari *et al.*, 2023). Furthermore, the application of exogenous substances such as selenium has been shown to alleviate stress effects and promote secondary metabolite biosynthesis, underscoring the potential for agricultural practices that incorporate such treatments to optimize plant health and metabolite production (Ghanbari *et al.*, 2023).

The regulation of secondary metabolite biosynthesis is also intricately linked to hormonal signaling pathways. Plant hormones such as abscisic acid (ABA) and jasmonic acid have been implicated in the modulation of secondary metabolite production in response to environmental stimuli. For example, ABA treatment has been shown to affect the accumulation of secondary metabolites in *Artemisia argyi*, suggesting that hormonal regulation is a critical component of the metabolic response to stress (Yang *et al.*, 2022). Additionally, microRNAs have emerged as important regulators of secondary metabolite biosynthesis, influencing gene expression and metabolic pathways involved in the production of compounds such as flavonoids and terpenoids (Khan *et al.*, 2020). The interplay between hormonal signals and microRNA regulation represents a complex network that governs the dynamic biosynthesis of secondary metabolites in plants.

Recent advances in genetic and metabolic engineering have opened up new avenues for enhancing secondary metabolite production in plants. Techniques such as virus-induced gene silencing and the manipulation of transcription factors specific to glandular trichomes have been employed to increase the yield of terpenoids and other valuable compounds. These approaches not only enhance our understanding of the genetic basis of secondary metabolism but also provide practical strategies for developing plant platforms capable of producing high-value metabolites for pharmaceutical and industrial applications (Q. Wang *et al.*, 2016). Moreover, the integration of omics technologies such as transcriptomics and metabolomics has facilitated the identification of key regulatory genes and pathways involved in secondary metabolite biosynthesis, paving the way for targeted interventions to optimize metabolite production (Shi *et al.*, 2021).

20.2 Types of Plant Secondary Metabolites

Plant secondary metabolites are a diverse group of organic compounds that play crucial roles in plant defense, ecological interactions, and human health. These metabolites are not directly involved in the primary metabolic processes necessary for plant growth and reproduction; instead, they serve specialized functions that enhance the plant's ability to survive in its environment. Plants produce secondary metabolites such as sulfur-containing chemicals, alkaloids, phenolic compounds, and terpenoids to protect themselves from exogenous limitations, and are synthesized in undifferentiated plant cells (Guerriero *et al.*, 2018). The classification of secondary metabolites typically includes terpenoids, alkaloids, phenolic compounds, and sulfur-containing compounds, each with distinct biosynthetic pathways and ecological roles (Fig. 20.1) (Erb and Kliebenstein, 2020).

Terpenoids, also known as isoprenoids, are one of the largest classes of secondary metabolites and are characterized by their diverse structures and functions. They are synthesized from five-carbon isoprene units and are involved in various ecological interactions, such as attracting pollinators and repelling herbivores. Certain terpenoids exhibit antimicrobial properties, which help protect plants from pathogens (Huang *et al.*, 2022). Additionally, terpenoids can serve as precursors for the synthesis of essential oils, which have significant applications in the food, cosmetic, and pharmaceutical industries (Guerriero *et al.*, 2018). Alkaloids represent another major class of secondary metabolites, characterized primarily by their nitrogen-containing structures. These compounds often exhibit potent biological activities, including analgesic, anti-inflammatory, and antimicrobial effects, making them valuable in traditional and modern medicine. Alkaloids such as morphine and quinine have been historically significant in pharmacology, while recent studies have identified numerous other alkaloids with therapeutic potential. The production of alkaloids is influenced by various factors, including environmental stressors, which can enhance their biosynthesis as a defensive response (Isah, 2019).

Phenolic compounds, including flavonoids and tannins, are widely recognized for their antioxidant properties and roles in plant defense mechanisms. These compounds can deter herbivory by acting as feeding inhibitors or toxins to herbivores (Maag *et al.*, 2015). Flavonoids, in particular, are abundant secondary metabolites that contribute to the coloration of flowers and fruits, thereby attracting pollinators and seed dispersers (Donadio *et al.*, 2021). Moreover, phenolic compounds have been shown to possess various health benefits for humans, including anticancer and anti-inflammatory effects, highlighting their importance beyond plant biology (Roy *et al.*, 2022). The biosynthesis of secondary metabolites is a complex process regulated by genetic, environmental, and physiological factors. For instance, the carbon–nitrogen balance hypothesis suggests that the availability of carbon and nitrogen influences the production of different classes of secondary metabolites. Under conditions of

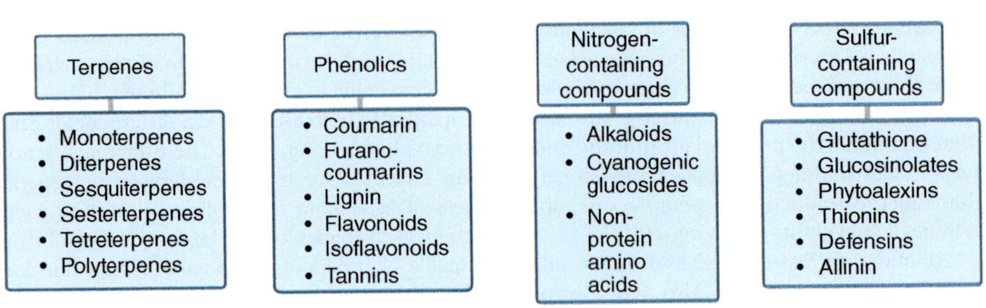

Fig. 20.1. Types of plant secondary metabolites.

nitrogen deficiency, plants may increase the synthesis of nitrogen-based metabolites such as alkaloids while reducing carbon-based metabolites such as flavonoids (Han *et al.*, 2022). This dynamic regulation underscores the adaptability of plants in response to environmental stresses and their ability to optimize resource allocation for survival.

Recent advancements in biotechnological approaches have enabled the sustainable production of plant secondary metabolites through methods such as plant tissue culture and genetic engineering. For example, the use of cyanobacteria as cell factories for the production of secondary metabolites has gained attention due to their ability to express plant biosynthetic pathways efficiently. This innovative approach not only enhances the yield of valuable metabolites but also reduces the reliance on traditional agricultural practices, which can be resource-intensive and environmentally damaging (Xue and He, 2015). Furthermore, the interactions between plants and their associated microorganisms, such as endophytic fungi, can significantly influence secondary metabolite production. Endophytes can enhance the biosynthesis of certain metabolites, providing additional defense mechanisms against herbivores and pathogens (Han *et al.*, 2024). This symbiotic relationship illustrates the complexity of plant secondary metabolism and its dependence on biotic interactions within the ecosystem.

20.3 Biosynthetic Pathways of Secondary Metabolites

Plant secondary metabolites play important roles in plant survival and in creating ecological connections between other species. Secondary metabolites act as herbivore deterrents, barriers against pathogen invasion, and mitigators of oxidative stress. The accumulation of secondary metabolites is highly dependent on environmental factors such as light, temperature, soil water, soil fertility, and salinity (Jan *et al.*,2021). Biosynthetic pathways of secondary metabolites refer to the series of enzyme-catalyzed reactions within living organisms that lead to the production of secondary metabolites. These compounds are not essential for the organism's primary metabolic processes (e.g. growth and reproduction) but often play important roles in the organism's interactions with its environment, such as defense against predators. High-throughput metabolomics is a powerful approach for exploring plant secondary metabolism, enabling the rapid analysis of a vast number of metabolites in a single experiment. One of the primary challenges in studying secondary metabolites is the intricate network of biosynthetic pathways that can lead to the production of a diverse array of compounds.

Moreover, the biosynthesis of secondary metabolites often involves a variety of enzymatic reactions, which can include polyketide synthases, non-ribosomal peptide synthetases, and other specialized enzymes. Gu *et al.* (2009) discussed the diversification of polyketide biosynthesis through the assembly of metamorphic enzymes, illustrating how variations in enzyme structure can lead to the production of diverse secondary metabolites. This enzymatic diversity is crucial for the adaptability and survival of organisms in varying environmental conditions. In plants, secondary metabolites such as flavonoids and terpenoids are synthesized through well-characterized pathways, often involving the phenylpropanoid pathway and the mevalonate pathway (Wang *et al.*, 2017). Research by Gregersen and Holm (2007) highlighted the role of specific genes in the biosynthesis of benzoxazinoids in wheat, demonstrating how plant secondary metabolites can be tightly regulated during developmental processes such as senescence. This regulation is essential for optimizing the production of these compounds in response to environmental stimuli. The potential for drug discovery from secondary metabolites is immense, particularly in the context of antibiotic resistance. The work of Ochi and Hosaka (2013) emphasized the need for new strategies to activate silent biosynthetic gene clusters in microbes, which could lead to the discovery of novel antibiotics.

20.4 Application of Metabolomics in Plant Secondary Metabolism Research

Metabolomics has emerged as a pivotal approach in the study of plant secondary metabolism, providing insights into the complex biochemical networks that govern plant responses to various biotic and abiotic stresses. This field encompasses the comprehensive analysis of metabolites, small molecules that play crucial roles in plant physiology, development, and defense mechanisms. The application of metabolomics to plant secondary metabolism research has been facilitated by advancements in analytical techniques, allowing the detailed profiling of metabolites and their interactions within the plant system. One of the significant contributions of metabolomics to plant secondary metabolism is its ability to elucidate the genetic and environmental factors influencing the production of secondary metabolites. For instance, Johnson *et al.* (2014) highlighted that the concentration of secondary metabolites exhibits substantial genetic variation within plant populations, for example in *Oenothera* spp. This genetic variability is crucial for understanding how plants adapt their secondary metabolite profiles in response to herbivory and other environmental pressures. Furthermore, Brown *et al.* (2012) emphasized that metabolomics enables researchers to build a comprehensive picture of the metabolome, moving beyond the characterization of specific classes of metabolites to a holistic understanding of plant biochemical diversity.

The integration of metabolomics with other omics technologies, such as genomics and transcriptomics, has further enhanced our understanding of plant secondary metabolism. Kumar *et al.* (2017) noted that metabolomics has established itself as a major breakthrough in plant science, allowing accurate metabolite profiling that can inform breeding programs aimed at improving crop resilience and yield. This integrative approach is particularly valuable in identifying metabolic pathways and regulatory networks that govern secondary metabolite biosynthesis, as highlighted by Piasecka *et al.* (2019) who discussed the challenges and methodologies for detecting metabolomic changes in response to various stresses. Moreover, the application of metabolomics extends to chemotaxonomy, where it aids in the classification and identification of plant species based on their secondary metabolite profiles. As noted by Cheng *et al.* (2020), chemotaxonomy leverages chemical differences among plant populations to establish phylogenetic relationships, thereby enhancing our understanding of plant biodiversity and evolution. This is particularly relevant in the context of medicinal plants, where secondary metabolites are often linked to therapeutic properties. Socaciu (2019) provided a historical perspective on the evolution of phytochemistry into metabolomics, emphasizing the interdisciplinary nature of this field and its implications for both plant biology and human health.

The ability of metabolomics to assess plant responses to abiotic stresses, such as drought and heavy-metal toxicity, has garnered significant attention in recent years. Martínez-Nicolás *et al.* (2023) demonstrated that metabolomic analyses can predict plant responses to alternative culture media and phenological stages, highlighting the potential of this approach in optimizing cultivation practices. Similarly, Chakrawarti *et al.* (2022) discussed the utility of high-throughput metabolomics in addressing a wide range of biological issues, including stress tolerance and metabolic engineering. This is particularly important as plants are subjected to increasingly variable environmental conditions due to climate change. In the context of crop improvement, Razzaq *et al.* (2019) emphasized that metabolomics serves as a powerful tool for deciphering metabolic networks associated with stress tolerance and yield. Advanced analytical techniques such as nuclear magnetic resonance (NMR) spectroscopy and mass spectrometry (MS) have revolutionized the field by enabling detailed metabolic profiling at unprecedented scales (Urakami *et al.*, 2010). These techniques allow researchers to quantify the expression of primary and secondary metabolites, facilitating the identification of genetic markers linked to desirable traits in crop species.

Furthermore, the application of metabolomics in understanding the biochemical mechanisms underlying plant responses to abiotic stress has been extensively documented. Isah (2019) showed that abiotic factors significantly influence secondary metabolite production, and the integration of molecular biology tools has enhanced our understanding of the signaling pathways involved in this process. This knowledge is crucial for developing strategies to enhance stress tolerance in crops, as demonstrated by the work of Moradi *et al.* (2017), who revealed the biochemical mechanisms underlying drought stress tolerance in thyme through metabolomic profiling.

The complexity of plant metabolomes, which can comprise thousands of metabolites, necessitates sophisticated analytical and computational approaches for data interpretation. As noted by Beniddir *et al.* (2021), metabolomics studies typically involve multiple stages, including sample collection, extraction, analytical measurement, and data processing. The integration of bioinformatics tools is essential for managing the vast amounts of data generated, enabling researchers to draw meaningful conclusions about metabolic pathways and their regulation.

20.5 Metabolic Flux Analysis and Secondary Metabolite Dynamics

Metabolic flux analysis (MFA) is an essential tool for understanding the dynamics of metabolic networks, particularly in the context of secondary metabolite production. This analysis provides insights into how metabolic pathways are regulated and how they respond to various environmental stimuli. The integration of MFA with advanced techniques such as high-performance liquid chromatography and MS has significantly enhanced our ability to study the dynamics of secondary metabolites in various biological systems. For instance, Lozano-Andrade *et al.* (2023) demonstrated a transparent soil system that allowed the exploration of secondary metabolite dynamics in *Bacillus subtilis*, revealing spatial and temporal distributions of these compounds through advanced imaging techniques. This approach overcomes traditional limitations associated with the extraction and quantification of metabolites in complex matrices, thereby facilitating a more comprehensive understanding of microbial chemical ecology.

The role of endophytes in altering the secondary metabolism of host plants is another critical aspect of metabolic flux dynamics. Tamang (2023) showed that endophytes can significantly influence the biosynthesis of secondary metabolites in plants such as *Picrorhiza kurrooa*, a Himalayan medicinal herb. The interaction between endophytes and their host plants can lead to qualitative and quantitative variations in secondary metabolites, which are crucial for the plant's medicinal properties (Tamang, 2023). This dynamic interplay emphasizes the importance of considering microbial interactions when analyzing secondary metabolite production in plants. In the context of plant stress responses, Li *et al.* (2017) explored the accumulation of secondary metabolites in licorice under ABA stress using ultra-high-performance liquid chromatography (UHPLC) coupled with LTQ-Orbitrap MS. Their findings indicated that stress conditions could lead to changes in the expression of key genes and enzymes, thereby affecting the accumulation rates of secondary metabolites. This underscores the necessity of integrating genetic and metabolic analyses to elucidate the mechanisms governing secondary metabolite dynamics in response to environmental stressors.

The intraspecific diversity in secondary metabolite production among bacterial populations, as discussed by Polano *et al.* (2019), plays a significant role in microbial community dynamics and biological control strategies. The production of secondary metabolites by *Pseudomonas protegens* and related strains can interfere with the growth of fungal and bacterial pathogens, highlighting the ecological significance of these compounds in maintaining microbial balance. This interaction not only affects the microbial community structure but also has implications for agricultural practices and biocontrol applications. The integration of MFA with genomic data is crucial for understanding the underlying mechanisms of secondary metabolite production. L. Wang *et al.* (2016) demonstrated that a systems-level analysis of cacao seed ripening revealed a sequential interplay between primary and secondary metabolism, leading to polyphenol accumulation and stress resistance preparation. This integrative approach allows a more comprehensive understanding of how metabolic pathways are interconnected and how they respond to developmental cues.

Goyal *et al.* (2022) emphasized the importance of strain-level studies in understanding the eco-evolutionary dynamics of microbial communities. They noted that transcriptional regulators and enzymes involved in central carbon metabolism play a crucial role in differentiating metabolic flux balance between strains, which can significantly influence community dynamics. This highlights the necessity of considering strain-specific metabolic capabilities when analyzing microbial interactions and their effects on secondary metabolite production.

The application of exometabolomics in soil biology, as discussed by Swenson *et al.* (2018) provides a novel perspective on linking soil chemistry and biology. Their work suggests that modeling approaches can be employed to account for metabolic state switching, which is essential for understanding the dynamics of microbial communities and their metabolic outputs. This approach can inform trait-based models that link soil metabolites with metagenomic data, thereby enhancing our understanding of microbial ecology.

In addition to the ecological and agricultural implications, understanding metabolic fluxes is crucial for biotechnological applications. Ohno *et al.* (2022) discussed the quantitative analysis of metabolic fluxes regulated by transomic networks, emphasizing the need for advanced techniques to determine metabolic fluxes accurately. This is particularly relevant for metabolic engineering, where precise control over metabolic pathways is necessary for optimizing the production of desired compounds. The dynamic nature of secondary metabolite accumulation in response to environmental factors, such as ultraviolet (UV) light exposure, was explored by Hectors *et al.* (2014). Their study on *Arabidopsis thaliana* demonstrated that secondary metabolite accumulation is a complex and dynamic process influenced by various external stimuli. This complexity necessitates a detailed understanding of the regulatory mechanisms governing secondary metabolism to harness these pathways for biotechnological applications.

20.6 Integration with Other Omics Technologies

The integration of metabolomics with other omics technologies represents a significant advancement in our understanding of plant secondary metabolism. This multi-disciplinary approach allows researchers to explore the complex biochemical pathways that govern the production of secondary metabolites, which are crucial for plant defense, adaptation, and interaction with the environment. Metabolomics, as a field, focuses on the comprehensive analysis of metabolites within biological systems, providing insights into the metabolic state of organisms under various conditions, including biotic and abiotic stresses (Manickam *et al.*, 2023). Recent studies have highlighted the utility of combining metabolomics with genomics, transcriptomics, and proteomics to create a more holistic view of plant biology. For instance, integration of these technologies enables the identification of genetic and phenotypic associations that influence metabolite production. This is particularly relevant in the context of crop improvement, where understanding the genetic basis of secondary metabolite biosynthesis can lead to the development of plants with enhanced nutritional and medicinal properties (Razzaq *et al.*, 2019). The use of high-throughput analytical tools, such as MS and NMR, has further facilitated this integration, allowing the simultaneous analysis of multiple metabolomic data sets (Yan *et al.*, 2022).

One of the key advantages of integrating metabolomics with other omics technologies is the ability to link metabolomic data to functional genomics. This connection is vital for assessing the effects of genetic modifications on secondary metabolite profiles. For example, researchers have successfully employed metabolomics to evaluate the impact of genetic engineering on the metabolic composition of plants, revealing both intended and unintended consequences of such modifications (Christ *et al.*, 2018). This approach not only enhances our understanding of plant metabolism but also informs breeding programs aimed at improving crop resilience and quality. Moreover, metabolomics plays a critical role in elucidating the molecular mechanisms underlying plant responses to environmental stressors. By analyzing metabolite changes in response to abiotic stresses, such as heavy-metal exposure or salinity, researchers can identify key metabolic pathways involved in stress tolerance. For instance, studies have shown that metabolomic profiling can reveal how plants reprogram their metabolism to cope with excess zinc or osmotic stress, highlighting the dynamic nature of plant responses to environmental challenges (Rouphael *et al.*, 2016). This information is invaluable for developing strategies to enhance crop resilience in the face of climate change and other environmental pressures.

In addition to abiotic stress responses, metabolomics is also instrumental in understanding plant interactions with pathogens and beneficial microorganisms. The integration of metabolomics with other omics technologies allows a comprehensive analysis of plant defense mechanisms, including the production of phytoalexins and other secondary metabolites that play a role in pathogen resistance. For example, metabolomic analyses have been employed to investigate the metabolic signatures associated with plant responses to pathogen attacks, providing insights into the biochemical pathways activated during these interactions (Allwood *et al.*, 2021). This knowledge can be leveraged to breed plants with improved resistance to diseases, thereby enhancing agricultural productivity.

Furthermore, the development of user-friendly computational tools for metabolomics analysis has facilitated the integration of metabolomic data with other omics data sets. Platforms such as PlantMetSuite enable researchers to visualize and interpret complex metabolomic data, making it easier to identify patterns and correlations between metabolites and other biological variables (Yu *et al.*, 2019). This ability is essential for advancing our understanding of plant secondary metabolism and its regulation, as it allows the identification of potential biomarkers associated with specific metabolic pathways or stress responses. The application of metabolomics in plant ecology is another promising area of research. By employing untargeted metabolomic approaches, researchers can gain insights into the ecological roles of secondary metabolites in plant–environment interactions (Tugizimana *et al.*, 2013). This perspective is crucial for understanding how plants adapt to their surroundings and interact with other organisms, including herbivores and pollinators. The integration of metabolomics with ecological studies can provide a more comprehensive understanding of the evolutionary significance of secondary metabolites and their contributions to plant fitness.

20.7 Applications in Agriculture and Industry

High-throughput metabolomics has emerged as a transformative tool in the exploration of plant secondary metabolism, with significant implications for agriculture and industry. This approach enables the comprehensive profiling of metabolites, which are crucial for understanding plant responses to biotic and abiotic stresses, enhancing crop quality, and discovering novel compounds for various applications (Fig. 20.2). The integration of advanced analytical techniques such as MS and NMR has facilitated the detailed characterization of complex metabolomes, allowing researchers to identify and quantify a wide array of secondary metabolites that play vital roles in plant health and productivity.

One of the primary applications of high-throughput metabolomics in agriculture is the investigation of plant responses to pathogens. For instance, Oh *et al.* (2021) demonstrated the utility of ultra-performance liquid chromatography coupled with quadrupole time-of-flight MS in identifying metabolites induced in pathogen-infected rice plants. This study highlighted the structural complexity and diversity of secondary metabolites involved in plant defense mechanisms, which have traditionally been challenging to study due to their intricate nature. By leveraging metabolomics, researchers can elucidate the biochemical pathways activated during pathogen attack, informing breeding programs aimed at enhancing disease resistance in crops.

In addition to pathogen responses, metabolomics has been instrumental in understanding the effects of environmental stressors on plant metabolism. For example, Llorens *et al.* (2019) explored how endophytes from wild cereals can protect wheat plants from drought stress by altering their physiological responses. This study found that phenolic compounds, secondary metabolites known for their role in plant defense, exhibited varying levels in endophyte-treated versus untreated plants under drought conditions. This underscores the potential of metabolomics to identify beneficial microbial interactions that can enhance plant resilience to environmental stressors, ultimately contributing to sustainable agricultural practices.

The exploration of secondary metabolites through metabolomics extends to the optimization of crop quality and nutritional value. Research by Feng *et al.* (2018) on mandarins revealed that different rootstocks significantly influenced the glycosylate secondary metabolite content in the fruit, critical for flavor and nutritional properties. Such findings emphasize the importance of metabolomics in varietal selection and breeding strategies aimed at improving

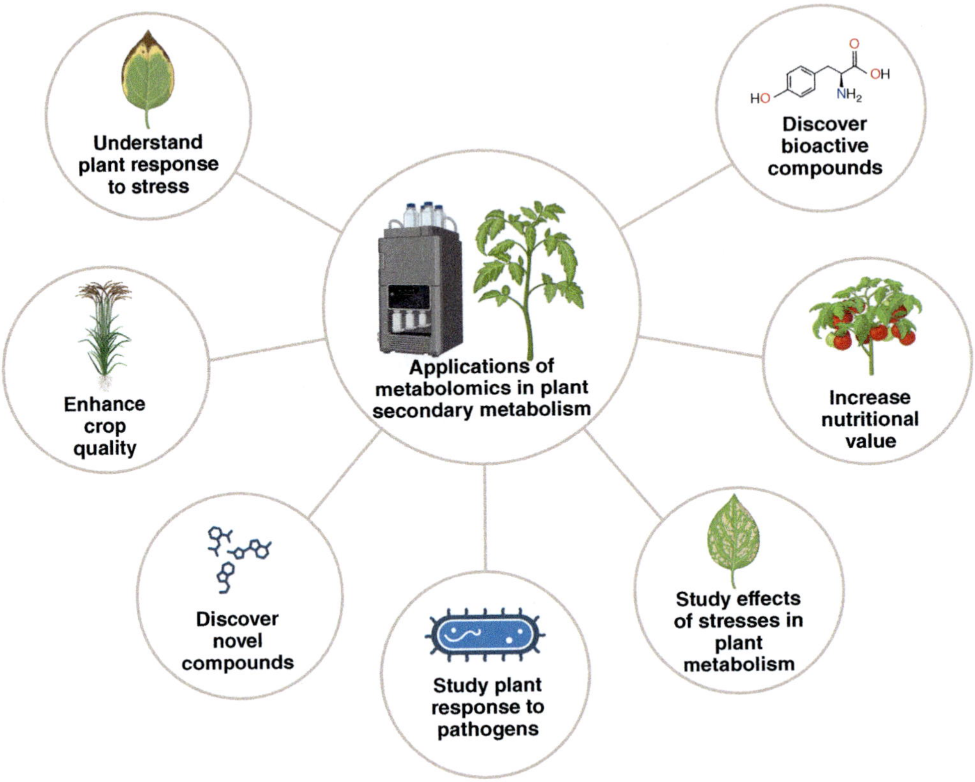

Fig. 20.2. Applications of metabolomics in the analysis of plant secondary metabolites.

crop quality. By understanding the metabolic profiles associated with desirable traits, agricultural scientists can make informed decisions that enhance the overall value of food crops. The industrial applications of high-throughput metabolomics are equally noteworthy. For instance, Padilla-González *et al.* (2023) highlighted the chemical diversity of the hemp seed metabolome, revealing new molecular families that could have implications for various industries, including pharmaceuticals and nutraceuticals. The discovery of novel metabolites through metabolomics not only expands the chemical repertoire available for industrial applications but also paves the way for innovative product development that meets consumer demands for natural and functional ingredients.

The integration of metabolomics with other omics technologies such as transcriptomics and proteomics provides a holistic view of plant responses to environmental challenges. Ramsey *et al.* (2020) conducted a longitudinal analysis of citrus plants grafted with a pathogen, revealing significant alterations in *S*-adenosylmethionine metabolism, which is crucial for the biosynthesis of secondary metabolites and plant hormones. This multi-omics approach allows a deeper understanding of the regulatory networks governing plant metabolism, facilitating the identification of key targets for genetic manipulation and breeding. The application of metabolomics is also pivotal in the context of food authenticity and safety. Erban *et al.* (2019) emphasized the potential of metabolomics to discover food identity markers, which can help in the authentication of plant-based foods and the detection of adulteration. This aspect of metabolomics is particularly relevant in an era where consumers are increasingly concerned about the provenance and quality of their food. By establishing metabolic profiles that are characteristic of specific plant varieties, stakeholders in the food industry can ensure transparency and integrity in the supply chain.

In addition to food safety, metabolomics plays a crucial role in enhancing the nutritional quality of crops. Nephali *et al.* (2020) discussed the potential of biostimulants to improve plant growth and mitigate abiotic stresses, with metabolomics serving as a valuable tool to assess the metabolic changes induced by these treatments. By identifying metabolites associated with improved stress tolerance and nutritional content, researchers can develop strategies to enhance crop resilience and nutritional profiles, ultimately contributing to food security. *Medicago sativa* L., also known as lucerne or alfalfa, is a valuable protein source for animal feed and human nutrition in developing nations and has a long history in folk medicine (Rafińska *et al.*, 2017). The exploration of secondary metabolites through high-throughput metabolomics also holds promise for the discovery of new bioactive compounds. Mishra *et al.* (2022) highlighted the role of metabolomics in understanding the interactions between plants and their endophytes, which can lead to the identification of novel bioactive metabolites with potential applications in medicine and agriculture. This aspect of metabolomics is particularly important in the search for natural products that can serve as alternatives to synthetic chemicals in pest management and disease control. Advancements in metabolomics methodologies, including the development of high-resolution analytical techniques, have significantly enhanced the capacity to analyze complex plant metabolomes. For example, González-Domínguez *et al.* (2022) used high-throughput phenolic fingerprinting to illustrate the potential of MS to provide comprehensive insights into the secondary metabolome of plant foods. Such methodologies not only facilitate the identification of a wide range of metabolites but also enable quantification of their concentrations, which is essential for evaluating their biological activities and potential health benefits.

20.8 Challenges and Limitations

High-throughput metabolomics has emerged as a powerful tool for exploring plant secondary metabolism, offering insights into the complex biochemical pathways that underpin plant responses to various environmental stimuli. However, despite its potential, the field faces numerous challenges and limitations that can hinder the comprehensive understanding of plant secondary metabolites (Table 20.1).

One of the primary challenges in high-throughput metabolomics is the inherent complexity of plant metabolomes, which contain a vast array of structurally diverse metabolites. The identification of these metabolites is often hampered by the lack of standard reference compounds and comprehensive MS spectral libraries. For instance, Shiratake and Suzuki (2016) highlighted that, while thousands of metabolite peaks may be detected in MS analyses, only a fraction can be identified reliably due to the absence of standard chemical compounds for many secondary metabolites. This limitation was echoed by Zargar *et al.* (2023), who emphasized the genetic control over metabolome diversity and the need for integrated approaches to delineate metabolic pathways effectively. The absence of a robust plant metabolite spectral library further complicates the identification process, as noted by Li *et al.* (2023), who created a library of 544 authentic standards to enhance metabolite identification efficiency.

Moreover, the analytical techniques employed in high-throughput metabolomics often present their own set of challenges. While advancements in MS, such as Fourier-transform ion cyclotron resonance and UHPLC, have improved metabolite detection capabilities, these methods can still be limited by their throughput and the complexity of sample preparation (Zhou *et al.*, 2021). For example, Yilmaz *et al.* (2016) discussed the labor-intensive extraction processes required for plant tissues, which can introduce variability and complicate data interpretation. Additionally, the integration of different analytical platforms, as suggested by Razzaq *et al.* (2022), can enhance metabolome coverage but also requires sophisticated bioinformatics tools to manage and interpret the resulting data.

Data processing and interpretation represent another significant hurdle in high-throughput metabolomics. The sheer volume of data generated from metabolomic analyses necessitates advanced computational tools for effective analysis. Allwood *et al.* (2021) highlighted the importance of understanding the timing and amplitude of signaling events during plant stress responses, which requires not only metabolomic

Table 20.1. Challenges in plant metabolomics: current hurdles in secondary metabolism research.

Challenge/limitation	Description	Reference
Metabolite identification	Lack of standard reference compounds and comprehensive mass spectrometry spectral libraries hinders reliable identification of many secondary metabolites	Shiratake and Suzuki (2016)
Metabolome complexity	The vast array of structurally diverse metabolites in plants complicates comprehensive analysis and pathway delineation	Zargar *et al.* (2023)
Sample preparation variability	Labor-intensive extraction processes for plant tissues can introduce variability and complicate data interpretation	Yilmaz *et al.* (2016)
Data integration	Integrating metabolomic data with other omics data sets (e.g. transcriptomics, proteomics) requires sophisticated bioinformatics tools and expertise	Razzaq *et al.* (2022)
Biological variability	Genetic background, environmental conditions, and developmental stages significantly influence metabolite profiles, complicating standardization	Christ *et al.* (2018)
Causal relationship identification	Determining causal relationships between metabolites and phenotypic traits remains challenging due to complex metabolic networks	Schroeder *et al.* (2019)
Technological limitations	Although advanced, current analytical techniques may still be limited in throughput and the ability to detect all metabolites present	Zhou *et al.* (2021)
Cost and accessibility	High costs associated with advanced metabolomics technologies can limit accessibility for many researchers	Implied from context
Dynamic nature of metabolomes	The rapidly changing nature of plant metabolomes in response to various stimuli makes capturing a comprehensive snapshot challenging	Allwood *et al.* (2021)
Lack of standardized protocols	Variability in experimental designs and analytical methods complicates cross-study comparisons and meta-analyses	Castro-Moretti *et al.* (2020)

data but also insights from transcriptomics and proteomics. The integration of these diverse data sets can be challenging, as each omics layer provides different perspectives on plant metabolism. Furthermore, the complexity of metabolic networks means that identifying causal relationships between metabolites and phenotypic traits remains a daunting task (Schroeder *et al.*, 2019).

The application of high-throughput metabolomics in the context of plant secondary metabolism also faces limitations related to the biological variability inherent in plant systems. Factors such as genetic background, environmental conditions, and developmental stages can significantly influence metabolite profiles. This variability complicates the establishment of standardized protocols for metabolomic analyses, as highlighted by Christ *et al.* (2018), who advocated for the use of untargeted metabolomics to assess the safety and nutritional value of crops. The dynamic nature of plant metabolomes necessitates a careful consideration of these variables when designing experiments and interpreting results.

The current state of metabolomics research often lags behind other omics fields such as genomics and transcriptomics, particularly in applications related to plant–pathogen interactions. As noted by Castro-Moretti *et al.* (2020), while metabolomics holds promise for elucidating the biochemical responses of plants to pathogens, its application in this area is still developing. The integration of metabolomic data with genomic and transcriptomic information is crucial for a holistic understanding of plant responses to biotic stressors, yet this integration is often not fully realized in practice.

In addition to these technical and biological challenges, the high costs associated with advanced metabolomics technologies can limit accessibility for many researchers. The investment required for sophisticated MS equipment and the associated operational costs can be prohibitive, particularly for smaller laboratories or institutions. This financial barrier can restrict the breadth of research MS conducted in the field and may lead to a reliance on a limited number of well-funded research groups, potentially stifling innovation and diversity in metabolomics research.

Despite these challenges, the potential of high-throughput metabolomics to advance our understanding of plant secondary metabolism remains significant. The integration of metabolomics with other omics technologies can facilitate a more comprehensive exploration of plant metabolic pathways and their regulation (Huang *et al.*, 2023). Moreover, the development of new analytical techniques, such as matrix-assisted laser desorption/ionization MS imaging, offers promising avenues for high-throughput metabolite detection directly from plant tissues, thereby overcoming some of the limitations associated with traditional extraction methods (Huang *et al.*, 2023).

20.9 Future Directions and Innovations

High-throughput metabolomics has emerged as a transformative approach in the exploration of plant secondary metabolism, providing unprecedented insights into the complex biochemical pathways that govern the production of secondary metabolites. This field has witnessed significant advancements in analytical techniques, data integration, and interpretation, which have collectively enhanced our understanding of plant responses to environmental stimuli and biotic interactions. The integration of various omics technologies—genomics, transcriptomics, and proteomics—with metabolomics has facilitated a more holistic view of plant biology, particularly in the context of secondary metabolite production.

Recent studies have highlighted the role of environmental factors, such as nutrient availability, in shaping secondary metabolism. For instance, Wu *et al.* (2023) demonstrated that nitrogen addition significantly influences the expression of secondary metabolism-related genes in *Solidago canadensis*, leading to the accumulation of specific secondary metabolites that enhance the plant's allelopathic abilities and chemical defenses against herbivores and pathogens. This finding underscores the importance of understanding how nutrient dynamics can modulate secondary metabolite profiles, which is crucial for developing strategies to enhance plant resilience and productivity.

The application of biostimulants, such as compost derivatives, has been shown to positively impact the metabolomic profiles of plants. Verrillo *et al.* (2023) used a combined metabolomic approach to assess the effects of sustainable recycled products on basil plants, revealing that these treatments not only increased the production of nutraceutical and antibacterial compounds but also provided a comprehensive characterization of the plant metabolome. This integrated methodology represents a significant innovation in metabolomics, allowing absolute quantification of metabolite changes in response to biostimulant application, thereby paving the way for sustainable agricultural practices. The exploration of epigenetic regulation in secondary metabolism has also gained traction, particularly in species such as tea plants (*Camellia sinensis*). Research by Han (2024) elucidated the role of dynamic DNA methylation in regulating seasonal variations in secondary metabolite accumulation, highlighting the intricate relationship between epigenetic modifications and metabolite biosynthesis. This underscores the potential for utilizing epigenetic markers as tools for enhancing secondary metabolite production in crops, which could have significant implications for the agricultural and pharmaceutical industries.

The integration of metabolomics with plant–microbe interaction studies has opened up new avenues for enhancing agricultural productivity. Xu and Fan (2024) emphasized that advancements in metabolomics technologies have expanded our understanding of how metabolites influence plant–microbe interactions, which are critical for plant health and resilience. This knowledge can lead to innovative strategies for sustainable agriculture, where the manipulation of microbial communities through metabolomic insights can enhance nutrient uptake and disease resistance in crops. The application of high-throughput metabolomics is not limited to understanding plant responses to abiotic factors; it also extends to the investigation of biotic interactions. For example, the study of secondary metabolites as defense compounds against herbivores and pathogens has been enriched by metabolomic approaches. Erb and Kliebenstein (2020) discussed how secondary metabolites function as multi-functional compounds that can regulate plant growth and defense mechanisms, blurring the traditional boundaries between primary and secondary metabolites. This functional diversity is crucial for developing crops with enhanced resistance to biotic stressors, thereby improving agricultural sustainability.

In addition to these advancements, the development of novel analytical techniques has significantly enhanced the capabilities of metabolomics. The use of LC-MS has become predominant in untargeted metabolomics, allowing the acquisition of thousands of metabolite signals from a single sample (An, 2024). This high-throughput capability is essential for comprehensive metabolomic profiling, enabling researchers to identify and quantify a wide range of metabolites, including those involved in secondary metabolism. The challenges associated with data analysis and interpretation in metabolomics remain a critical area of focus. As highlighted by Xu and Fan (2024), the integration of metabolomics with other omics approaches necessitates advanced data-analysis techniques to extract meaningful biological insights. The development of robust computational tools and databases, such as HMDB and METLIN, is essential for improving metabolite identification and quantification, thereby enhancing the reliability of metabolomic studies (Shi *et al.*, 2019).

The exploration of secondary metabolites in medicinal plants has gained momentum, with metabolomics serving as a valuable tool for the discovery of bioactive compounds. Puri *et al.* (2021) noted that medicinal plant metabolomics provides a wealth of bioactive phytoconstituents, facilitating the elucidation of biosynthetic pathways and the scientific validation of herbal medicines. This has significant implications for drug discovery and the development of novel therapeutic agents derived from plant secondary metabolites. As the field of high-throughput metabolomics continues to evolve, future directions will likely focus on the integration of multi-omics approaches to unravel the complexities of plant secondary metabolism. The combination of metabolomics with transcriptomics and proteomics will provide a more comprehensive understanding of the regulatory networks governing metabolite production. Additionally, the application of machine learning and artificial intelligence in metabolomic data analysis holds promise for uncovering novel patterns and relationships within complex data sets, thereby accelerating the pace of discovery in plant science.

20.10 Conclusion

High-throughput metabolomics has emerged as a transformative approach in the study of plant secondary metabolism, offering unprecedented insights into the complex biochemical networks that govern plant responses to environmental stimuli, biotic interactions, and developmental processes. The integration of metabolomics with other omics technologies has facilitated a more holistic view of plant biology, enabling researchers to elucidate the intricate relationships between genetic factors, environmental influences, and metabolic outputs. Advancements in analytical techniques and data-analysis tools have expanded our knowledge of plant chemical diversity and opened up new avenues for the discovery of novel bioactive compounds. As this field continues to evolve, high-throughput metabolomics is poised to play a crucial role in addressing global challenges related to food security, climate change adaptation, and sustainable resource management by enhancing our understanding of plant secondary metabolism and its potential applications in agriculture, industry, and human health.

20.11 Acknowledgment

All the authors are grateful to their respective universities and institutions for their valuable support.

20.12 References

Akilandeswari, G., Vijaya Anand, A., Sampathkumar, P., Vinayaga Moorthi, P., and Preethi, B. (2019) A prospective review on phyto-pharmacological aspects of *Andrographis paniculata*. *Systematic Reviews in Pharmacy* 10, 15–19.

Allwood, J.W., Williams, A., Uthe, H., van Dam, N.M., Mur, L.A., *et al.* (2021) Unravelling plant responses to stress-the importance of targeted and untargeted metabolomics. *Metabolites* 11: 558.

An, S., Lu, M., Wang, R., Wang, J., Jiang, H., *et al.* (2024) Ion entropy and accurate entropy-based FDR estimation in metabolomics. *Briefings in Bioinformatics* 25: bbae056.

Begum, T.N., Ilyas, M.M., Kalavathy, S., Anand, A.V., Kumar, S., and Senthil, R. (2009) Effects of ethanolic leaf extract of *Azim tetracantha* Lam. on Ehrlich ascites carcinoma tumour bearing mice. *Research Journal of Medicine and Medical Sciences* 4, 351–354.

Beniddir, M.A., Kang, K.B., Genta-Jouve, G., Huber, F., Rogers, S., and van der Hooft, J.J. (2021) Advances in decomposing complex metabolite mixtures using substructure- and network-based computational metabolomics approaches. *Natural Product Reports* 38, 1967–1993.

Bharathi, V., Rengarajan, R.L., Radhakrishnan, R., Hashem, A., Abd Allah, E.F., *et al.* (2018) Effects of a medicinal plant *Macrotyloma uniflorum* (Lam.) Verdc. formulation (MUF) on obesity-associated oxidative stress-induced liver injury. *Saudi Journal of Biological Sciences* 25, 1115–1121.

Brown, P.N., Murch, S.J., and Shipley, P. (2012) Phytochemical diversity of cranberry (*Vaccinium macrocarpon* Aiton) cultivars by anthocyanin determination and metabolomic profiling with chemometric analysis. *Journal of Agricultural and Food Chemistry* 60, 261–271.

Castro-Moretti, F.R., Gentzel, I.N., Mackey, D., and Alonso, A.P. (2020) Metabolomics as an emerging tool for the study of plant–pathogen interactions. *Metabolites* 10: 52.

Chakrawarti, N., Verma, R., Brahma, T., Das, A., Narzary, P.R., *et al.* (2022) Basic concepts of high throughput metabolomics in plants. *International Journal of Plant & Soil Science* 34, 365–377.

Cheng, Z., Shu, H., Zhang, S., Luo, B., Gu, R., *et al.* (2020) From folk taxonomy to species confirmation of *Acorus* (Acoraceae): evidences based on phylogenetic and metabolomic analyses. *Frontiers in Plant Science* 11: 965.

Christ, B., Pluskal, T., Aubry, S., and Weng, J.K. (2018) Contribution of untargeted metabolomics for future assessment of biotech crops. *Trends in Plant Science* 23, 1047–1056.

Divya, N., Rengarajan, R.L., Radhakrishnan, R., Abd Allah, E.F., Alqarawi, A.A., *et al.* (2019) Phytotherapeutic efficacy of the medicinal plant *Terminalia catappa* L. *Saudi Journal of Biological Sciences* 26, 985–988.

Donadio, G., Mensitieri, F., Santoro, V., Parisi, V., Bellone, M.L., *et al.* (2021) Interactions with microbial proteins driving the antibacterial activity of flavonoids. *Pharmaceutics* 13: 660.

Dong, N.Q. and Lin, H.X. (2021) Contribution of phenylpropanoid metabolism to plant development and plant–environment interactions. *Journal of Integrative Plant Biology* 63, 180–209.

Erb, M. and Kliebenstein, D.J. (2020) Plant secondary metabolites as defenses, regulators, and primary metabolites: the blurred functional trichotomy. *Plant Physiology* 184, 39–52.

Erban, A., Fehrle, I., Martinez-Seidel, F., Brigante, F., Más, A.L., *et al.* (2019) Discovery of food identity markers by metabolomics and machine learning technology. *Scientific Reports* 9: 9697.

Feng, S., Niu, L., Suh, J.H., Hung, W.L., and Wang, Y. (2018) Comprehensive metabolomics analysis of mandarins (*Citrus reticulata*) as a tool for variety, rootstock, and grove discrimination. *Journal of Agricultural and Food Chemistry* 66, 10317–10326.

Ghanbari, F., Bag-Nazari, M., and Azizi, A. (2023) Exogenous application of selenium and nano-selenium alleviates salt stress and improves secondary metabolites in lemon verbena under salinity stress. *Scientific Reports* 13: 5352.

González-Domínguez, R., Sayago, A., Santos-Martín, M., and Fernández-Recamales, Á. (2022) High-throughput method for wide-coverage and quantitative phenolic fingerprinting in plant-origin foods and urine samples. *Journal of Agricultural and Food Chemistry* 70, 7796–7804.

Goyal, A., Bittleston, L.S., Leventhal, G.E., Lu, L., and Cordero, O.X. (2022) Interactions between strains govern the eco-evolutionary dynamics of microbial communities. *Elife* 11: e74987.

Gregersen, P.L. and Holm, P.B. (2007) Transcriptome analysis of senescence in the flag leaf of wheat (*Triticum aestivum* L.). *Plant Biotechnology Journal* 5, 192–206.

Gu, L., Wang, B., Kulkarni, A., Geders, T.W., Grindberg, R.V., *et al.* (2009) Metamorphic enzyme assembly in polyketide diversification. *Nature* 459, 731–735.

Guerriero, G., Berni, R., Muñoz-Sanchez, J.A., Apone, F., Abdel-Salam, E.M., *et al.* (2018) Production of plant secondary metabolites: examples, tips and suggestions for biotechnologists. *Genes* 9: 309.

Han, D., Wang, K., Long, F., Zhang, W., Yao, X., and Chen, S. (2022) Effects of endophytic fungi on the secondary metabolites of *Hordeum bogdanii* under alkaline stress. *AMB Express* 12: 73.

Han, M., Lin, S., Zhu, B., Tong, W., Xia, E., *et al.* (2024) Dynamic DNA methylation regulates season-dependent secondary metabolism in the new shoots of tea plants. *Journal of Agricultural and Food Chemistry* 72, 3984–3997.

Hectors, K., van Oevelen, S., Geuns, J., Guisez, Y., Jansen, M.A., and Prinsen, E. (2014) Dynamic changes in plant secondary metabolites during UV acclimation in *Arabidopsis thaliana*. *Physiologia Plantarum* 152, 219–230.

Huang, W., Wang, Y., Tian, W., Cui, X., Tu, P., *et al.* (2022) Biosynthesis investigations of terpenoid, alkaloid, and flavonoid antimicrobial agents derived from medicinal plants. *Antibiotics* 11: 1380.

Huang, H., Liu, H., Ma, W., Qin, L., Chen, L., *et al.* (2023) High-throughput MALDI-MSI metabolite analysis of plant tissue microarrays. *Plant Biotechnology Journal* 21, 2574–2584.

Isah, T. (2019) Stress and defense responses in plant secondary metabolites production. *Biological Research* 52: 39.

Jan, R., Asaf, S., Numan, M., Lubna, and Kim, K.M., (2021) Plant secondary metabolite biosynthesis and transcriptional regulation in response to biotic and abiotic stress conditions. *Agronomy* 11: 968.

Johnson, M.T., Ives, A.R., Ahern, J., and Salminen, J.P. (2014) Macroevolution of plant defenses against herbivores in the evening primroses. *New Phytologist* 203, 267–279.

Karthika, P., Balamuralikrishnan, B., Yamini, K., Vijaya Anand, A., Durairaj, K., *et al.* (2023) Green synthesis, characterization of silver nanoparticles using aqueous leaf extracts of *Solanum melongena* and *in vitro* evaluation of antibacterial, pesticidal and anticancer activity in human MDA-MB-231 breast cancer cell lines. *Journal of King Saud University—Science* 35: 102663.

Kaviya, M., Balasubramanian, B., Bharathi, K., Malaisamy, A., Al-Dhabi, N.A., *et al.* (2022) Evaluation of nutritional substances and investigation of antioxidant and antimicrobial potentials of *Boerhavia diffusa* with *in silico* molecular docking. *Molecules* 27: 1280.

Khan, S., Ali, A., Saifi, M., Saxena, P., Ahlawat, S., and Abdin, M.Z. (2020) Identification and the potential involvement of miRNAs in the regulation of artemisinin biosynthesis in *A. annua*. *Scientific Reports* 10: 13614.

Kumar, R., Bohra, A., Pandey, A.K., Pandey, M.K., and Kumar, A. (2017) Metabolomics for plant improvement: status and prospects. *Frontiers in Plant Science* 8: 1302.

Li, D., Xu, G., Ren, G., Sun, Y., Huang, Y., and Liu, C. (2017) The application of ultra-high-performance liquid chromatography coupled with a LTQ-Orbitrap mass technique to reveal the dynamic accumulation of secondary metabolites in licorice under ABA stress. *Molecules* 22: 1742.

Li, Y., Zhu, W., Xiang, Q., Kim, J., Dufresne, C., *et al.* (2023) Creation of a plant metabolite spectral library for untargeted and targeted metabolomics. *International Journal of Molecular Sciences* 24: 2249.

Llorens, E., Sharon, O., Camañes, G., García-Agustín, P., and Sharon, A. (2019) Endophytes from wild cereals protect wheat plants from drought by alteration of physiological responses of the plants to water stress. *Environmental Microbiology* 21, 3299–3312.

Lozano-Andrade, C.N., Nogueira, C.G., Henriksen, N.N., Wibowo, M., Jarmusch, S.A., and Kovács, Á.T. (2023) Establishment of a transparent soil system to study *Bacillus subtilis* chemical ecology. *ISME Communications* 3: 110.

Maag, D., Erb, M., Köllner, T.G., and Gershenzon, J. (2015) Defensive weapons and defense signals in plants: some metabolites serve both roles. *BioEssays* 37, 167–174.

Manickam, S., Rajagopalan, V.R., Kambale, R., Rajasekaran, R., Kanagarajan, S., and Muthurajan, R. (2023) Plant metabolomics: current initiatives and future prospects. *Current Issues in Molecular Biology* 45, 8894–8906.

Manikandan, R., Balasubramanian, B., Punniyakotti, P., Vijaya Anand, A., Meyyazhagan, A., *et al.* (2022) Cardio-protective effects of *Terminalia catappa* leaves and *Terminalia chebula* fruit extract in doxorubicin-induced cardiomyopathy in rats. *Biomarkers* 27, 488–495.

Manoharan, N., Sampathkumar, P., Dheeba, B., Sheik Abdulla, S., Vinothprasanna, G., *et al.* (2008) Potential hepatoprotective effect of aqueous extract of *Gracilaria corticata* in AFB_1 induced hepatotoxicity in Wistar rats. *Journal of Biological Sciences* 8, 1352–1355.

Martínez-Nicolás, J.J., Hernández, F., Núñez-Gómez, D., García-Sánchez, F., Martínez-Font, R., *et al.* (2023) Metabolomic approach to study the 'Purple Queen' pomegranate cultivar response to alternative culture media and phenological stages. *Foods* 12: 352.

Mishra, S., Priyanka, and Sharma, S. (2022) Metabolomic insights into endophyte-derived bioactive compounds. *Frontiers in Microbiology* 13: 835931.

Moradi, P., Ford-Lloyd, B., and Pritchard, J. (2017) Metabolomic approach reveals the biochemical mechanisms underlying drought stress tolerance in thyme. *Analytical Biochemistry* 527, 49–62.

Muthumani, G.D., Anand, A.V., and Manikandan, R. (2013) Antioxidant, antihelmintic and antimicrobial activity of *Euphorbia thymifolia* Linn whole plant. *International Journal of Current Microbiology and Applied Sciences* 2, 66–75.

Nephali, L., Piater, L.A., Dubery, I.A., Patterson, V., Huyser, J., *et al.* (2020) Biostimulants for plant growth and mitigation of abiotic stresses: a metabolomics perspective. *Metabolites* 10: 505.

Ochi, K. and Hosaka, T. (2013) New strategies for drug discovery: activation of silent or weakly expressed microbial gene clusters. *Applied Microbiology and Biotechnology* 97, 87–98.

Oh, M., Park, S., Kim, H., Choi, G.J., and Kim, S.H. (2021) Application of UPLC-QTOF-MS based untargeted metabolomics in identification of metabolites induced in pathogen-infected rice. *Plants* 10: 213.

Ohno, S., Uematsu, S., and Kuroda, S. (2022) Quantitative metabolic fluxes regulated by trans-omic networks. *Biochemical Journal* 479, 787–804.

Padilla-González, G.F., Rosselli, A., Sadgrove, N.J., Cui, M., and Simmonds, M.S. (2023) Mining the chemical diversity of the hemp seed (*Cannabis sativa* L.) metabolome: discovery of a new molecular family widely distributed across hemp. *Frontiers in Plant Science* 14: 1114398.

Piasecka, A., Kachlicki, P., and Stobiecki, M. (2019) Analytical methods for detection of plant metabolomes changes in response to biotic and abiotic stresses. *International Journal of Molecular Sciences* 20: 379.

Polano, C., Martini, M., Savian, F., Moruzzi, S., Ermacora, P., and Firrao, G. (2019) Genome sequence and antifungal activity of two niche-sharing *Pseudomonas protegens* related strains isolated from hydroponics. *Microbial Ecology* 77, 1025–1035.

Puri, S., Sahal, D., and Sharma, U. (2021) A conversation between hyphenated spectroscopic techniques and phytometabolites from medicinal plants. *Analytical Science Advances* 2, 579–593.

Rafińska, K., Pomastowski, P., Wrona, O., Górecki, R., and Buszewski, B. (2017) *Medicago sativa* as a source of secondary metabolites for agriculture and pharmaceutical industry. *Phytochemistry Letters* 20, 520–539.

Ramsey, J.S., Chin, E.L., Chavez, J.D., Saha, S., Mischuk, D., *et al.* (2020) Longitudinal transcriptomic, proteomic, and metabolomic analysis of *Citrus limon* response to graft inoculation by *Candidatus Liberibacter asiaticus*. *Journal of Proteome Research* 19, 2247–2263.

Razzaq, A., Sadia, B., Raza, A., Khalid Hameed, M., and Saleem, F. (2019) Metabolomics: a way forward for crop improvement. *Metabolites* 9: 303.

Razzaq, A., Wishart, D.S., Wani, S.H., Hameed, M.K., Mubin, M., and Saleem, F. (2022) Advances in metabolomics-driven diagnostic breeding and crop improvement. *Metabolites* 12: 511.

Rouphael, Y., Colla, G., Bernardo, L., Kane, D., Trevisan, M., and Lucini, L. (2016) Zinc excess triggered polyamines accumulation in lettuce root metabolome, as compared to osmotic stress under high salinity. *Frontiers in Plant Science* 7: 842.

Roy, A., Khan, A., Ahmad, I., Alghamdi, S., Rajab, B.S., *et al.* (2022) Flavonoids a bioactive compound from medicinal plants and its therapeutic applications. *BioMed Research International* 2022: 5445291.

Sangeetha, T., Balamuralikrishnan, B., Sampathkumar, P., Velayuthaprabhu, S., Senthilkumar, N., *et al.* (2022) Characterization and phytoconstituents of *Petroselinum crispum* (Mill) and *Coriandrum sativum* (Linn) and their impacts on inflammation—an *in vitro* analysis against human adenocarcinoma cells with molecular docking. *South African Journal of Botany* 146, 776–788.

Schroeder, M., Meyer, S.W., Heyman, H.M., Barsch, A., and Sumner, L.W. (2019) Generation of a collision cross section library for multi-dimensional plant metabolomics using UHPLC-trapped ion mobility-MS/MS. *Metabolites* 10: 13.

Shi, X., Wang, S., Jasbi, P., Turner, C., Hrovat, J., *et al.* (2019) Database-assisted globally optimized targeted mass spectrometry (dGOT-MS): broad and reliable metabolomics analysis with enhanced identification. *Analytical Chemistry* 91, 13737–13745.

Shi, Y., Xia, H., Cheng, X., and Zhang, L. (2021) Genome-wide miRNA analysis and integrated network for flavonoid biosynthesis in *Osmanthus fragrans*. *BMC Genomics* 22: 141.

Shih, M.L. and Morgan, J.A. (2020) Metabolic flux analysis of secondary metabolism in plants. *Metabolic Engineering Communications* 10: e00123.

Shiratake, K. and Suzuki, M. (2016) Omics studies of citrus, grape and Rosaceae fruit trees. *Breeding Science* 66, 122–138.

Socaciu, C. (2019) From phytochemistry to metabolomics: eight decades of research in plant and food science. *Studia Universitatis Babes-Bolyai, Chemia* 64, 205–224.

Swenson, T.L., Karaoz, U., Swenson, J.M., Bowen, B.P., and Northen, T.R. (2018) Linking soil biology and chemistry in biological soil crust using isolate exometabolomics. *Nature Communications* 9: 19.

Tamang, A., Swarnkar, M., Kumar, P., Kumar, D., Pandey, S.S., and Hallan, V. (2023) Endomicrobiome of *in vitro* and natural plants deciphering the endophytes-associated secondary metabolite biosynthesis in *Picrorhiza kurrooa*, a Himalayan medicinal herb. *Microbiology Spectrum* 11: e0227923.

Tugizimana, F., Piater, L., and Dubery, I. (2013) Plant metabolomics: a new frontier in phytochemical analysis. *South African Journal of Science* 109, 1–11.

Urakami, K., Zangiacomi, V., Yamaguchi, K., and Kusuhara, M. (2010) Quantitative metabolome profiling of *Illicium anisatum* by capillary electrophoresis time-of-flight mass spectrometry. *Biomedical Research* 31, 161–163.

Varalakshmi, B., Anand, A.V., Karpagam, T., Bai, J.S., and Manikandan, R. (2014) *In vitro* antimicrobial and anticancer activity of *Cinnamomum zeylanicum* Linn bark extracts. *International Journal of Pharmacy and Pharmaceutical Sciences* 6, 12–18.

Verrillo, M., Koellensperger, G., Puehringer, M., Cozzolino, V., Spaccini, R., and Rampler, E. (2023) Evaluation of sustainable recycled products to increase the production of nutraceutical and antibacterial molecules in basil plants by a combined metabolomic approach. *Plants* 12: 513.

Vijaya Anand, A., Varalakshmi, B., Prasana Sampathkumar, P., Pushpa, N., and Hedina, A. (2016) *Cinnamomum zeylanicum* Linn. The spice with multi potential. *Systematic Reviews in Pharmacy* 7, 24–29.

Vijaya Anand, A., Velayuthaprabhu, S., Rengarajan, R.L., Sampathkumar, P., and Radhakrishnan, R. (2020) Bioactive compounds of guava (*Psidium guajava* L.). In: Murthy, H.N. and Bapat, V.A. (eds) *Bioactive Bioactive Compounds in Underutilized Fruits and Nuts, Reference Series in Phytochemistry*. Springer, Cham, Switzerland, pp. 503–527.

Vijayakumar, K., Prasanna, B., Rengarajan, R.L., Rathinam, A., Velayuthaprabhu, S., and Vijaya, A.A. (2023) Anti-diabetic and hypolipidemic effects of *Cinnamon cassia* bark extracts: an *in vitro*, *in vivo*, and *in silico* approach. *Archives of Physiology and Biochemistry* 129, 338–348.

Wang, L., Nägele, T., Doerfler, H., Fragner, L., Chaturvedi, P., *et al.* (2016) System level analysis of cacao seed ripening reveals a sequential interplay of primary and secondary metabolism leading to polyphenol accumulation and preparation of stress resistance. *Plant Journal* 87, 318–332.

Wang, L., Sun, X., Weiszmann, J., and Weckwerth, W. (2017) System-level and granger network analysis of integrated proteomic and metabolomic dynamics identifies key points of grape berry development at the interface of primary and secondary metabolism. *Frontiers in Plant Science* 8: 1066.

Wang, Q., Reddy, V.A., Panicker, D., Mao, H.Z., Kumar, N., *et al.* (2016) Metabolic engineering of terpene biosynthesis in plants using a trichome-specific transcription factor Ms YABBY 5 from spearmint (*Mentha spicata*). *Plant Biotechnology Journal* 14, 1619–1632.

Wu, M., Liu, H., Zhang, Y., Li, B., Zhu, T., and Sun, M. (2023) Physiology and transcriptome analysis of the response mechanism of *Solidago canadensis* to the nitrogen addition environment. *Frontiers in Plant Science* 14: 1005023.

Xu, Y. and Fan, K.T. (2024) Advancements in metabolomics: illuminating plant–microbe interactions for enhanced ecosystems and agricultural productivity. *Preprints.org*. Available at: www.preprints.org/manuscript/202406.0670/v1 (accessed 10 February 2025).

Xue, Y. and He, Q. (2015) Cyanobacteria as cell factories to produce plant secondary metabolites. *Frontiers in Bioengineering and Biotechnology* 3: 57.

Yan, S., Bhawal, R., Yin, Z., Thannhauser, T.W., and Zhang, S. (2022) Recent advances in proteomics and metabolomics in plants. *Molecular Horticulture* 2: 17.

Yang, L., Wen, K.S., Ruan, X., Zhao, Y.X., Wei, F., and Wang, Q. (2018) Response of plant secondary metabolites to environmental factors. *Molecules* 23: 762.

Yang, L., Yan, Y., Zhao, B., Xu, H., Su, X., and Dong, C. (2022) Study on the regulation of exogenous hormones on the absorption of elements and the accumulation of secondary metabolites in the medicinal plant *Artemisia argyi* leaves. *Metabolites* 12: 984.

Yilmaz, A., Rudolph, H.L., Hurst, J.J., and Wood, T.D. (2016) High-throughput metabolic profiling of soybean leaves by Fourier transform ion cyclotron resonance mass spectrometry. *Analytical Chemistry* 88, 1188–1194.

Yu, M., Xie, W., Zhang, X., Zhang, S., Wang, Y., *et al.* (2019) Arbuscular mycorrhizal fungi can compensate for the loss of indigenous microbial communities to support the growth of liquorice (*Glycyrrhiza uralensis* Fisch.). *Plants* 9: 7.

Zargar, S.M., Manzoor, M., Bhat, B., Wani, A.B., Sofi, P.A., *et al.* (2023) Metabolic-GWAS provides insights into genetic architecture of seed metabolome in buckwheat. *BMC Plant Biology* 23: 373.

Zhou, H., Zhang, Y., Liang, H., Song, H., Zhao, J., *et al.* (2021) A novel multidimensional strategy to evaluate *Belamcanda chinensis* (L) DC and *Iris tectorum* Maxim based on plant metabolomics, digital reference standard analyzer and biological activities evaluation. *Chinese Medicine* 16: 85.

Zhou, J., Liu, Z., Wang, S., Li, J., Li, Y., *et al.* (2020) Fungal endophytes promote the accumulation of Amaryllidaceae alkaloids in *Lycoris radiata*. *Environmental Microbiology* 22, 1421–1434.

21 High-throughput Metabolomics for Plant Ecology and Biodiversity Research

Sarah Jaison[1*], Chithra Devi Balasundarm Saraswathy[1], Tharani Sureshkumar[2], Ajay Samuel Siva Daniel[2], Sunitha Kumari Krishnankutty[1], Ramkumar Govindarajan[2], Muthukrishnan Arun[3], and Arumugam Vijaya Anand[4]

[1]Department of Botany, PSGR Krishnammal College for Women, Coimbatore, Tamil Nadu, India; [2]GRG BIRAC EYUVA Centre, PSGR Krishnammal College for Women, Coimbatore, Tamil Nadu, India; [3]Department of Biotechnology, Bharathiar University, Coimbatore, Tamil Nadu, India; [4]Department of Human Genetics and Molecular Biology, Bharathiar University, Coimbatore, Tamil Nadu, India

Abstract

Eco-metabolomics is an emerging field that applies metabolomics techniques to ecological research, bridging biochemistry and ecology across different spatiotemporal scales. This procedure enables the investigation of plant–organism interactions and environmental responses by analyzing changes in metabolite concentrations. Eco-metabolomics faces several challenges, including complex experimental designs, feature extraction, metabolite detection, statistical investigations, and bioinformatics analysis. However, it offers novel insights into plant resistance mechanisms against herbivores and other ecological processes. The incorporation of metabolomics with other omics techniques provides a complete understanding of biosynthetic and genetic changes during ecological responses. To facilitate successful collaborations between ecologists and metabolomics experts, strategies for project management, training, and communication of realistic outcomes are essential. As the field advances, eco-metabolomics is anticipated to play an important role in elucidating chemical mediators of organismal interactions in various ecosystems. Overcoming the challenges and improving collaboration between ecologists and metabolomics experts will enhance the potential of eco-metabolomics in understanding plant ecology and biodiversity.

Keywords: ecological responses, eco-metabolomics, integrative omics approaches, metabolite identification, plant–organism interactions

21.1 Introduction

Small compounds called metabolites are found in cells, organs, and organisms. The complete study of metabolites is known as metabolomics. The products of cellular activities are called metabolites, and a biological system's metabolic state can be precisely understood by looking at the levels of these products (Peters *et al.*, 2018). The purpose of metabolomics is to quantify and examine these metabolites to understand the connections and activities of cells in both health and illness. Eco-metabolomics is the combination of ecology and advanced metabolomics, and faces challenges in integrating the different experimental approaches, workflows, and norms of these two constituent disciplines. Technological and practical challenges in eco-metabolomics include

* E-mail: sarah@psgrkcw.ac.in

© CAB International 2025. *High-Throughput Plant Metabolomics* (Jen-Tsung Chen ed.)
DOI: 10.1079/9781800626911.0021

obtaining high-quality samples for metabolomic analysis in field conditions, and meeting requirements such as obtaining enough replicates, appropriate storage, and export regulations. Ecologists often lack knowledge about plant metabolome dynamics and have unrealistic expectations about data processing and compound identification, which can hinder productive collaborations with metabolomics experts (Uthe *et al.*, 2021)

As an example, the results of a metabolomics investigation revealed that while enriching legume abundances had conducive effects on the small-statured legume, plant diversity in the form of rising species richness had substantial impacts on small-statured non-legume herb species, suggesting light and nutrient constraints. In contrast, the tall-statured model species' metabolic profiles remained unchanged as plant diversity increased (Scherling *et al.*, 2010). Through this type of research, we hope to shed light on the mechanisms underlying the varying consequences of plant diversity on the performance of different species and identify the physiological responses of different plant species to differences in their natural abiotic and biotic environments (Hall, 2011)

The conventional method of using compatible solutes to increase the assembly of a particular compound in response to a predetermined stimulus has led to some degree of toleration in plants and shows that the ability to reroute nutrients to essential functions and the induction of appropriate metabolic adjustments are critical for plant survival in stressful environments. However, this is only one aspect of how plants control their metabolism in reaction to stressful situations. For a more thorough understanding, we need to consider that plant metabolism is a complex web of interrelated processes.

Plant metabolism is highly compartmentalized and dynamic, with metabolites varying across subcellular locations, developmental stages, and environmental conditions (Kooke and Keurentjes, 2012; Sweetlove and Fernie, 2013). This compartmentalization is crucial for metabolic network structure and regulation, involving both organelles and microcompartments formed by enzyme interactions (Sweetlove and Fernie, 2013). The plant metabolome is shaped by spatial, temporal, environmental, and genetic characteristics, which collectively determine plant development and performance (Kooke and Keurentjes, 2012). When studying plant metabolism, it is essential to consider subcellular localization, as no single method is universally optimal for metabolome analysis (Dietz, 2017). Additionally, researchers must account for the dilution effect caused by cell growth and varying vacuole sizes, which can bias metabolite concentration measurements and correlation analyses if not properly addressed (Génard *et al.*, 2014).

Understanding complex regulatory mechanisms is crucial for advancing metabolic modeling and engineering in plants. Recent improvements in plant systems biology and bioinformatics have enabled the creation of efficient plant metabolic models, enabling the study and modification of complex plant metabolisms (Baghalian *et al.*, 2014). The integration of theoretical advances in network thinking with global profiling strategies is generating system-wide molecular data sets, leading to a unique level of understanding of metabolic network principles (Sweetlove and Fernie, 2005). Knowledge-based approaches, including data mining, modeling, and genome editing, are improving the accuracy of predicting the impact of metabolic modifications (Farré *et al.*, 2015). These advancements have led to the evolution of various tools and strategies for rerouting metabolic pathways, such as genome editing and transcriptional regulation practices. Additionally, cutting-edge strategies have delivered new methods for studying plant responses to environmental stressors and associations with other species (Lankadurai *et al.*, 2013; Peters *et al.*, 2018). This approach enables the analysis of modifications in metabolite concentrations, providing evidence for biochemical procedures relevant to ecological scales (Peters *et al.*, 2018). It has been used to detect stress from abiotic factors such as exposure to xenobiotics and temperature changes, as well as biotic stressors such as herbivory and competition (Lankadurai *et al.*, 2013). Environmental metabolomics has applications across plant ecology, from understanding plant plasticity and transformation to community composition and genetic transformation in crops (Brunetti *et al.*, 2013). Different techniques are utilized, including *in situ* sampling at field sites and laboratory-based analysis of secondary metabolites (Brunetti *et al.*, 2013). This approach has been particularly valuable in assessing genetically modified plants and exploring complex plant–host interactions (Allwood *et al.*, 2008).

21.2 Plant–Environment Interactions: Abiotic Stress Responses

Metabolomics is a potent device for determining plant responses to abiotic stresses, such as salinity, drought, temperature extremes, and flooding (Uarrota, 2020). This technique enables comprehensive analysis of metabolic changes in plants under stress circumstances, revealing both shared and distinct alterations in central metabolism. Plant metabolic responses to abiotic stresses offers valuable insights into metabolic reprogramming and stress tolerance pathways (Fraire-Velázquez and Balderas-Hernández, 2013). Metabolomic studies have elucidated biochemical mechanisms underlying stress tolerance and susceptibility, providing insights into plant adaptation strategies

The combination of metabolomics with other omics platforms has further improved our understanding of complex signaling and metabolic responses to abiotic stresses (Radomiljac *et al.*, 2013). Modern advances in metabolomic methods have improved the characterization of stress-induced metabolic reprogramming, contributing to the development of techniques for enhancing plant stability in response to climate change. According to various studies (Obata and Fernie, 2012; Fraire-Velázquez and Balderas-Hernández, 2013), plants quickly modify their core metabolism, which involves amino acids, carbohydrates, and amines, to restore chemical and energy balance under stress. Secondary metabolites are essential as antioxidants, reactive oxygen species (ROS) scavengers, and regulatory molecules, and frequently show species- and genus-specific responses (Arbona *et al.*, 2013). Metabolomics techniques have revealed shared and distinct metabolic alterations under a range of abiotic stressors.

21.2.1 Drought stress

Research on metabolomics reveals intricate metabolic responses in plants under drought stress. Drought-tolerant wheat cultivars accumulate sugars, organic acids, and amino acids while exhibiting less disruption of proteins (Michaletti *et al.*, 2018; Guo *et al.*, 2020). In wheat seedlings, phenolics, flavonoids, alkaloids, thymine, and amino acids are linked to drought resistance (Guo *et al.*, 2020). Drought stress profoundly modifies mulberry lipid metabolism, resulting in a rise in prenol lipids and a decrease in fatty acyls, and polyphenols, organic acids, carbohydrates, and other metabolites also exhibit dynamic alterations (Ackah *et al.*, 2021). Through a variety of pathways, metabolic genes and primary and secondary metabolites are essential for drought tolerance (Kumar *et al.*, 2021). Gaining knowledge of these metabolic reactions can help with crop planning, shed light on pathways for drought-resistance mechanisms, and inform crop-improvement strategies for enhanced drought resistance (Michaletti *et al.*, 2018; Guo *et al.*, 2020; Kumar *et al.*, 2021).

21.2.2 Salinity stress

Salt stress induces widespread metabolic changes in plants, including alterations in amino acid and organic acid balance, accumulation of osmolytes such as myo-inositol and proline, and shifts in gluconeogenesis and the production of secondary metabolites (Sanchez *et al.*, 2008). These responses can vary between species, with some conserved and divergent metabolic adaptations observed across different plants (Sánchez *et al.*, 2008). Metabolomics approaches such as gas chromatography–mass spectrometry (GC-MS) and liquid chromatography–mass spectrometry (LC-MS) have proven valuable for unraveling these complex physiological and metabolic mechanisms (Roessner and Beckles, 2012). Additionally, seed halo priming has been shown to ameliorate salt stress effects by modulating stress-induced metabolite production, particularly in leaves (Biswas *et al.*, 2018). The ability of metabolomics to elucidate plant biochemical responses to salinity offers potential strategies for improving crop salt tolerance.

21.2.3 Temperature stress

Metabolomics is an effective method for determining plant responses to abiotic stresses, particularly temperature stress (Guy *et al.*, 2008; Obata and Fernie, 2012; Carrera *et al.*, 2021). Studies have revealed that temperature stress induces significant changes in plant metabolomes, with central carbohydrate metabolism playing a prominent role (Guy *et al.*, 2007). Heat stress in wheat, for example, leads to the accumulation of specific metabolites such as L-tryptophan and pipecolate, which negatively correlate with various traits (Thomason *et al.*, 2018). Metabolomic analyses have identified common responses across different abiotic stresses, including changes in branched-chain amino acid metabolism (Obata and Fernie, 2012). These investigations provide insights into stress tolerance mechanisms and possible biomarkers for genetic improvement (Thomason *et al.*, 2018). The comprehensive data generated by metabolomics contribute to a more holistic understanding of plant stress responses, aiding in the development of stress-resistant cultivars (Guy *et al.*, 2007; Carrera *et al.*, 2021).

21.2.4 Nutrient deficiency

Metabolomics studies have revealed distinct metabolic responses in plants under nutrient deficiencies. For example, nitrogen (N), phosphorus (P), and potassium (K) deficiencies affect energy production and amino acid metabolism in tomato plants, with specific changes in metabolite levels observed in leaves and roots. These nutrient deficiencies also impact primary metabolism and stress-related phytohormones in *Arabidopsis*, leading to nutrient-specific transcriptional responses (Forieri *et al.*, 2017). Iron deficiency, in particular, causes the strongest regulation, affecting 18% of the root transcriptome. Interestingly, cross-regulation between iron and sulfur metabolic pathways has been observed, demonstrating specific signal-transduction systems (Forieri *et al.*, 2017). Metabolomics approaches have identified numerous metabolites playing important roles in plant development and stress responses to nutrient limitations, providing insights for developing crop varieties with high nutrient content (Xue *et al.*, 2022).

Nitrogen deficiency

Metabolomics studies on N deficiency in plants have revealed significant alterations in metabolic profiles. N deficiency generally leads to increased carbohydrates and decreased amino acids in leaves (Sung *et al.*, 2018; Lin *et al.*, 2021). Nevertheless, roots may show different responses, with increased amino acids and organic acids (Sung *et al.*, 2018). These metabolic changes are often accompanied by transcriptomic and proteomic alterations, particularly in pathways related to photosynthesis, carbon (C) metabolism, and hormone signaling (Amiour *et al.*, 2012). The effect of N deficiency varies with the plant growth stage and resembles responses to other abiotic and biotic stresses (Amiour *et al.*, 2012). Incorporation of metabolomics with transcriptomics and proteomics provides a thorough knowledge of plant N metabolism, although correlations between different omics levels are not always straightforward (Amiour *et al.*, 2012). Plant N use efficiency will help in developing strategies for optimizing N management in agriculture.

Phosphorous deficiency

Metabolomic studies on plants under P deficiency have revealed significant changes in metabolite profiles across various species. Common responses include decreased levels of phosphorylated metabolites and organic acids, particularly in leaves (Muller *et al.*, 2015; Dokwal *et al.*, 2022). Conversely, amino acids and sugars often accumulate, especially in roots (Sung *et al.*, 2018; Dokwal *et al.*, 2022). These metabolic shifts reflect adaptations in carbon (C) and N metabolism, as well as energy production (Sung *et al.*, 2018; Kangi *et al.*, 2024). P-deficient plants also exhibit lipid remodeling and increased production of secondary metabolites, which may aid in stress protection and P acquisition from the soil (Muller *et al.*, 2015; Kangi *et al.*, 2024). Transcriptomic analyses show the downregulation of growth, C metabolism, and photosynthesis pathways under P deficiency (Kangi *et al.*, 2024).

Potassium deficiency

Research has shown that K deficiency significantly alters plant metabolism, affecting energy production, amino acid metabolism, and sugar accumulation (Sung *et al.*, 2018). Early responses to K deficiency involve modifications in sugar metabolism and accumulation of salicylic acid metabolites, preceding the formation of known biomarkers such as putrescine (Cui *et al.*, 2021). Metabolic profiling techniques such as GC-MS allow extensive phenotyping of genetically or environmentally altered plants. These methods can identify various compounds, including by-products of stress metabolism and signal-transduction molecules, contributing to a holistic knowledge of plant stress responses (Shulaev *et al.*, 2008). The incorporation of metabolomics with additional omics platforms provides a valuable understanding of the complex metabolic networks involved in plant adaptation to nutrient deficiencies and other stressors.

21.2.5 Nutrient excess

Metabolomics studies have revealed complex plant responses to nutrient excess. Magnesium oversupply in tomatoes alters carbohydrate and polyamine metabolism, reducing fruit quality (Kwon *et al.*, 2019), while zinc stress in tea plants affects ion homeostasis, carbohydrate metabolism, and flavonoid production. Excess molybdenum in soybeans induces organic acid production, hormone signaling, and antioxidant responses, particularly in roots (Xu *et al.*, 2018), while copper toxicity in cucumber leaves disrupts carbohydrate metabolism, antioxidant pathways, and chlorophyll degradation (Zhao *et al.*, 2018). Common themes across studies include alterations in carbohydrate metabolism, antioxidant responses, and secondary metabolite production. Plants often exhibit differential responses between shoots and roots (Xu *et al.*, 2018; Kwon *et al.*, 2019). These metabolomic approaches provide comprehensive insights into plant stress responses, revealing both shared and nutrient-specific mechanisms for coping with excess minerals.

21.2.6 Light conditions

Metabolomics studies have revealed significant changes in plant and bacterial metabolomes under different light conditions. In *Arabidopsis*, light intensity affects the accumulation of sugars, tricarboxylic acid cycle intermediates, and amino acids, with distinct patterns observed in field conditions (Jänkänpää *et al.*, 2012). In *Lactobacillus* spp., light exposure leads to significant changes in metabolic profiles, affecting multiple pathways (Xu *et al.*, 2017). Broccoli sprouts exposed to light demonstrate increased biosynthesis and metabolism of various lipids, including sterols and phenols, as well as elevated levels of polyketides and oxylipins (Maldini *et al.*, 2015).

21.2.7 Photosynthetic pigments

Recent studies have explored the metabolomics of photosynthetic pigments, particularly chlorophylls and carotenoids, in various organisms. Transcriptomic and metabolomic profiling in red maple leaves revealed key genes regulating chlorophyll and carotenoid biosynthesis (Lu *et al.*, 2020). Advanced analytical methods, including mass spectrometry (MS), have been developed for comprehensive analysis of chlorophyll and carotenoid metabolites in plants (Roca and Pérez-Gálvez, 2021). Metabolomics has also been applied to microalgae and cyanobacteria for optimizing biofuel and bio-based chemical production (Kato *et al.*, 2022). In soybean, metabolomic profiling of drought-tolerant and -susceptible cultivars showed differential phytochemical content, with the tolerant cultivar maintaining higher levels of chlorophyll-related metabolites under stress (Zemaitis *et al.*, 2021).

21.2.8 Photoprotective compounds

Recent research has explored the potential of metabolites from various organisms as photoprotective compounds. Cyanobacteria produce secondary metabolites such as ultraviolet (UV)-screening pigments and antioxidants, which can be characterized using nuclear magnetic resonance (NMR) spectroscopy (Pandey *et al.*, 2021). Lichens also synthesize UV-absorbing compounds, with some showing absorption properties such as commercial sunscreens. *Streptomyces* spp. bacteria have been recognized as a favorable source of metabolites with potential photoprotective properties, offering possibilities for developing new, eco-friendly sunscreens (Sánchez-Suárez *et al.*, 2020). The lichen *Vulpicida pinastri* produces compounds such as pinastric acid, vulpinic acid, and usnic acid, which demonstrate UV-blocking and antioxidant properties. These compounds were found to be non-cytotoxic to keratinocytes and photostable, making them attractive candidates for photoprotective applications (Legouin *et al.*, 2017).

21.2.9 Atmospheric conditions

Carbon dioxide levels

Metabolomics studies have revealed complex responses of plants and animals to elevated carbon dioxide (CO_2) levels. In *Brassica napus* guard cells, increased CO_2 triggered changes in primary and specialized metabolic pathways, with jasmonic acid signaling mediating stomatal closure (Geng *et al.*, 2017). Elevated CO_2 alleviated drought stress in cucumber seedlings by altering metabolites associated with various pathways, including amino acid metabolism and carbohydrate accumulation, while *Arabidopsis* suspension cells exposed to bicarbonate showed concentration-dependent metabolic changes, with flavonoid biosynthesis and glutathione metabolism affected under both light and dark conditions. In the Pacific oyster, an elevated partial pressure of CO_2 disturbed the tricarboxylic acid cycle and led to adjustments in energy metabolism through alanine and γ-aminobutyric acid accumulation (Jiang *et al.*, 2019). Thus, metabolomics approaches can furnish useful understandings of the molecular mechanisms underlying organisms' responses to changing CO_2 levels in both terrestrial and marine environments.

C assimilation

Recent studies have explored C assimilation pathways in photosynthetic organisms, revealing unique metabolic processes and enzymes in various phototrophic bacteria (Tang *et al.*, 2011, 2013). These investigations have used diverse techniques, including ^{13}C-based metabolomics and fluxomics, to quantify CO_2 assimilation pathways and suggest evolutionary links between photosynthetic and non-photosynthetic organisms. Research on microalgae, such as *Microchloropsis gaditana*, has shown that high CO_2 supplementation can increase biomass productivity and channel C flux toward long-chain polyunsaturated fatty acids (Kareya *et al.*, 2020). In plants, post-translational modifications play an important part in regulating C assimilation metabolism, allowing rapid and versatile responses to environmental fluctuations (Balparda *et al.*, 2023). Understanding these metabolic pathways and regulatory mechanisms is essential for improving biomass production and stress tolerance, and developing sustainable energy sources.

Ozone and pollutants

Metabolomics profiling is an effective method for investigating the effects of environmental pollutants, including ozone, on human health and plant physiology. In rice plants, ozone exposure triggers alterations in the expression of genes, protein levels, and metabolite assembly, affecting multiple cellular processes (Cho *et al.*, 2008). Similarly, in black mustard, ozone stress followed by herbivory induces changes in central metabolism and defense responses, impacting photosynthesis and stomatal closure (Papazian *et al.*, 2016). These metabolomic approaches provide an understanding of the mechanisms of pollutant toxicity and disease risks, potentially identifying early biomarkers for various environmental exposures (Deng *et al.*, 2019). In general, metabolomics offers a versatile method for characterizing the complex interactions between organisms and environmental pollutants.

21.2.10 Oxidative stress

Studies have identified potential biomarkers of oxidative stress, including malic acid, vitamin C, reduced glutathione, and tryptophan (Wang *et al.*, 2016). Oxidative stress impacts various metabolic pathways, including the urea cycle, tryptophan metabolism, and porphyrin metabolism (Doran *et al.*, 2017). The metabolomic response to oxidative stress is complex and changes depending on the basis of stress, with genetic oxidative stress causing more widespread effects than environmental stresses (Doran *et al.*, 2017). Antioxidants and redox buffers play crucial roles in mitigating oxidative stress, while ROS can act as oxidative signals (Noctor *et al.*, 2015). Exogenous administration of certain metabolites, such as pyruvate and fructose-1,6-bisphosphate, has shown potential in ameliorating oxidative stress (Liu *et al.*, 2012). Advanced analytical techniques such as NMR spectroscopy and multivariate statistical methods are being used to develop diagnostic tests for oxidative stress-related conditions (Liu *et al.*, 2012).

21.3 Plant–Plant Interactions

Metabolomics can be used to study plant–plant interactions. Recent advances in analytical methods have enhanced the sensitivity and ability of this method, allowing the investigation of hundreds to thousands of compounds in a single sample (Tenenboim and Brotman, 2016). This strategy has been used to analyze various aspects of plant biotic interactions, such as systemic acquired resistance, induced resistance, and allelopathy (Tenenboim and Brotman, 2016). Metabolomics has also been used to assess genetically altered plants and explore plant responses to pathogens and environmental stresses (Allwood *et al.*, 2008). However, despite its potential, challenges remain in metabolome coverage and compound identification (Maag *et al.*, 2015).

Allelopathy is the inhibitory effect of one plant on another through chemical compounds. Recent studies have employed metabolomics to investigate allelopathic mechanisms in various plants. For instance, research on *Eupatorium adenophorum* revealed its ability to suppress phosphate uptake and disrupt the metabolism of amino acids in *Arabidopsis thaliana*. Similarly, metabolomic analysis of canola genotypes identified specific allelochemicals, such as sinapyl alcohol and *p*-hydroxybenzoic acid, responsible for inhibiting annual ryegrass growth (Asaduzzaman *et al.*, 2015). In the case of *Wedelia trilobata*, LC-MS-based metabolomics helped identify potential allelochemical biomarkers (Azizan *et al.*, 2016). These studies demonstrate that metabolomics can elucidate the complex relationships between plants and their surroundings, providing insights into the mechanisms of allelopathy. This approach holds promise for developing organic weed control strategies and understanding plant invasiveness (Barrales-Cureño *et al.*, 2022).

21.3.2 Plant–Herbivore Interactions

Plant–herbivore relationships can be explored using metabolomics, offering an understanding of plant defense mechanisms and herbivore responses. Metabolomics has revealed induced plant responses and their effects on subsequent herbivores (Jansen *et al.*, 2009). Non-targeted approaches have revealed complex metabolic changes in plants following herbivory, including increased production of defense compounds such as 1,3-benzoxazin-4-ones and *N*-hydroxycinnamoyltyramines (Marti *et al.*, 2013). Constitutive defenses against herbivores involve elevated levels of amino acids, flavones, and saponins, while inducible defenses are associated with increases in methyl-ketone, pantothenate, and coenzyme A (Sardans *et al.*, 2021). Metabolomic analysis has also uncovered unexpected bioactive compounds, such as coumaroylquinic acids, which form a "metabolic interface" between herbivores and plants, as identified in a study on cabbage and caterpillars by Jansen *et al.* (2009). However, some induced metabolites may have contrasting effects on herbivore resistance, as demonstrated by the positive impact of *p*-coumaroyltyramine on insect growth (Marti *et al.*, 2013).

21.5 Plant–Microbe Interactions

Metabolomics has been used to study plant–microbe interactions in the rhizosphere, providing an understanding of the complex chemical communication between plants and microorganisms (Chen *et al.*, 2021; Gupta *et al.*, 2022). Flavonoids, strigolactones, coumarins, and triterpenes are just some of the many metabolites that plants release from their roots. These compounds influence the microbial community and facilitate symbiotic partnerships (Table 21.1) (Chen *et al.*, 2021). These interactions can lead to the biofilm construction of healthy microbes and induce systemic resistance in host plants (Mhlongo *et al.*, 2018). MS-based metabolomic approaches have enabled investigators to explore hundreds of compounds in a single sample, facilitating the examination of plant–microbe signaling and defense priming mechanisms (Mhlongo *et al.*, 2018; Rehman, 2020). While metabolomics has extended our knowledge of plant–microbe relationships, it remains underutilized compared with genomic and transcriptomic methods (Gupta *et al.*, 2022). Future investigation in this specialization promises to reveal further insights into plant resistance and susceptibility to various stresses.

Table 21.1. Key metabolites in plant ecology and their ecological roles.

Metabolite class	Examples	Ecological role	Significance in plant ecology	References
Primary metabolites	Glucose, amino acids, lipids	Essential for basic plant growth and development	Fundamental for energy production, structural components, and biosynthesis pathways	Stitt *et al.* (2010); Salam *et al.* (2023); Tariq *et al.* (2023)
Secondary metabolites	Alkaloids, terpenoids, flavonoids	Defense against herbivores, pathogens, and environmental stress	Crucial for plant survival, adaptation, and interactions with other organisms	Ogbemudia and Thompson (2014); War *et al.* (2020); Divekar *et al.* (2022); Jha and Mohamed (2022)
Volatile organic compounds	Isoprene, methyl jasmonate, ethylene	Attraction of pollinators, signaling to herbivores and predators	Important for plant communication, reproductive success, and defense mechanisms	Fürstenberg-Hägg *et al.* (2013); Meena *et al.* (2017)
Phenolic compounds	Tannins, lignins, anthocyanins	Ultraviolet (UV) protection, structural integrity, and pathogen defense	Protect plants from abiotic stress (UV light) and biotic stress (pathogens)	Aguirre-Becerra *et al.* (2021); Tuladhar, *et al.* (2021); Al-Khayri *et al.* (2023)
Glucosinolates	Sinigrin, glucobrassicin	Deterrent against herbivores, involved in sulfur metabolism	Specialized metabolites in *Brassicaceae*, play a role in plant defense and sulfur storage	Almuziny *et al.* (2017); Pawar (2020)
Fatty acids and lipids	Linoleic acid, phospholipids	Membrane structure, energy storage, signaling molecules	Essential for maintaining cellular integrity and responding to environmental stresses	Kumar *et al.* (2021); Ullah *et al.* (2022); Wang *et al.* (2022)

21.5.1 Symbiotic relationships

Plant–microbe relationships have been studied using metabolomics (Table 21.2), particularly in the context of arbuscular mycorrhizal symbiosis (Gupta *et al.*, 2022). This approach has revealed significant metabolomic alterations in both plants and fungi. Additionally, propionyl- and butyryl-carnitines accumulate in mycorrhizal roots and may regulate fungal gene expression *in planta* (Laparre *et al.*, 2014), as well as symbiotic associations (Ghirardo *et al.*, 2020). Arbuscular mycorrhizal fungal colonization has been shown to reprogram plant metabolic paths, affecting primary and secondary metabolites (Kaur and Suseela, 2020). For instance, structural and signaling lipid compounds increase in fungal mycelia near symbiotic protocorms, while chito-oligosaccharides are uniquely found in symbiotic protocorms (Ghirardo *et al.*, 2020). Propionyl- and butyryl-carnitines also build up in mycorrhizal roots, where they may control the expression of fungal genes in plants (Laparre *et al.*, 2014). These metabolomic changes are species specific and contribute to enhanced plant performance, particularly under stress conditions (Kaur and Suseela, 2020). Combining metabolomics with other omics approaches will provide a useful understanding of the mechanisms underlying plant–microbe relationships and their potential applications in agriculture.

21.5.2 Rhizosphere metabolomics

Plant–rhizosphere metabolomics studies the chemical exchanges that take place between soil organisms and roots. Root exudates have primary and secondary constituents that mediate interactions with microbes, insects, and other plants (van Dam and Bouwmeester, 2016). These compounds can attract beneficial microbes, deter pests, and inhibit competing plants (Mhlongo *et al.*, 2018). Metabolomics techniques allow the profiling of hundreds of compounds in root exudates, providing insights into plant defense priming and induced systemic resistance (Mhlongo *et al.*, 2018). However, challenges remain in proper sampling, analytical sensitivity, and data analysis (van Dam and Bouwmeester, 2016). Recent advances in metabolomic workflows, from sample preparation to data investigation, have improved our understanding of root exudates (Salem *et al.*, 2022). This knowledge has applications in improving crop output, developing stress-tolerant plants, and phytoremediation (Salem *et al.*, 2022). Future research may focus on generating spatially explicit metabolomes of roots and exudates at scales relevant to rhizosphere communities (van Dam and Bouwmeester, 2016).

21.5.3 Endophytes

Recent studies have explored the metabolomics of plant–endophyte interactions, revealing their complex and beneficial relationships. Endophytes, microorganisms inhabiting plants without causing harm, produce diverse bioactive compounds with potential applications in various industries (Mishra *et al.*, 2022). These microbes influence plant responses by modulating primary and secondary metabolism, enhancing stress tolerance, and promoting growth (Stranska *et al.*, 2021; Mishra *et al.*, 2022). Metabolomic approaches, associated with advanced analytical tools, provide comprehensive insights into the global metabolite profile of endophytic fungi and their relationships with host plants (Nagarajan *et al.*, 2021). Transcriptomic and metabolomic studies have deepened our understanding of plant–endophyte recognition, colonization, and protection mechanisms (X. Chen *et al.*, 2022). Specific metabolites upregulated during endophyte colonization include chlorins, phenolics, flavonoids, terpenoids, tannins, and fatty acids (Stranska *et al.*, 2021). It is necessary to find the key determinants in plant–endophyte relationships that can be utilized for future applications in agriculture and biotechnology.

21.6 Stress Response

Metabolomics is a potent tool for analyzing plant stress responses, providing insights into metabolic changes, and identifying key compounds involved in stress metabolism and signaling (Shulaev *et al.*, 2008; Jan *et al.*, 2019). This approach enables the detection of stress-induced metabolites, which can be correlated with transcriptome and proteome changes to achieve a holistic view of plant stress reactions (Shulaev *et al.*, 2008). Our

Table 21.2. Applications of high-throughput metabolomics in plant ecology and biodiversity.

Applications in plant ecology and biodiversity research	Objective	Metabolomic approach	Impacts	References
Plant–microbe interactions	Understanding the metabolic exchange between plants and their associated microbes	MS-based profiling (GC-MS, LC-MS), NMR	Discovery of metabolites that facilitate symbiosis or defense	Adeniji and Babalola (2020); Chen *et al.* (2021)
Climate change and environmental stress	Assessing plant metabolic responses to changing environmental conditions	LC-MS, GC-MS, FTIR	Identification of stress biomarkers, and adaptive metabolic shifts	Jorge *et al.* (2016b); Ahmad *et al.* (2018); Romero *et al.* (2021)
Plant biodiversity and conservation	Characterizing metabolic diversity among plant species in different ecosystems	Untargeted metabolomics (LC-MS, GC-MS), NMR	Insights into plant chemical diversity, and ecosystem functionality	Allwood *et al.* (2021); Li and Gaquerel (2021); Wolfender *et al.* (2022)
Plant–insect interactions	Investigating how plants produce metabolites in response to herbivory	LC-MS, GC-MS, targeted metabolomics	Identification of defensive compounds, signaling molecules	Maag *et al.* (2016); Macel and van Dam (2018); Parween *et al.* (2019); Allwood *et al.* (2021)
Ecosystem function and stability	Linking plant metabolic traits to ecosystem processes	Metabolomics combined with ecological modeling	Correlation between metabolic traits and ecosystem services such as carbon sequestration	Li and Gaquerel (2021); Walker *et al.* (2023); Bhattacharjya *et al.* (2024); Nizamani *et al.* (2024)
Phenotypic plasticity and adaptation	Exploring how plants adapt metabolically to different environmental conditions	LC-MS, GC-MS, CE-MS, untargeted and targeted approaches	Discovery of adaptive metabolic pathways, potential for breeding climate-resilient crops	Waseem *et al.* (2022); Roychowdhury *et al.* (2023)

CE-MS, capillary electrophoresis–mass spectrometry; FTIR, Fourier-transform infrared spectroscopy; GC-MS, gas chromatography–mass spectrometry; LC-MS, liquid chromatography–mass spectrometry; MS, mass spectrometry; NMR, nuclear magnetic resonance.

understanding of plant reactions to considerable abiotic stressors such as water, temperature, light, nutritional deficiency, and heavy-metal toxicity has been improved by the latest developments in metabolomics (You *et al.*, 2014; Jan *et al.*, 2019). Oxidative stress, characterized by increased ROS, is a common component of plant stress reactions and leads to complex metabolic alterations (Noctor *et al.*, 2015). These changes affect antioxidants, redox buffers, and primary and secondary metabolism. Lipid-derived compounds, antioxidant status, amino acid metabolism, and phytohormone pathways are particularly important in oxidative stress responses and may act as potential stress markers (Noctor *et al.*, 2015).

21.6.1 Metabolic diversity

Metabolomics has been used to explore metabolic diversity in plants and microorganisms. It enables comprehensive detection and quantification of metabolites, providing insights into biochemical processes and phenotypic traits (Lloyd *et al.*, 2015). Recent advancements in analytical methods and bioinformatics have increased the abilities of metabolomics, allowing large-scale studies of thousands of samples (van der Hooft *et al.*, 2020). In plants, metabolomics has been involved in assessing genotypic and phenotypic diversity, supporting breeding programs, and evaluating the impact of environmental factors on metabolite composition (Harrigan *et al.*, 2007; Shen *et al.*, 2023). Our knowledge of metabolic paths and gene activities has been further advanced by the incorporation of metabolomics with other omics data, such as transcriptomics and genomics (Shen *et al.*, 2023). These approaches have been successfully employed in diverse areas, including agriculture, medicine, and food science, offering new opportunities for harnessing chemical variety in organisms (Lloyd *et al.*, 2015).

21.6.2 Secondary metabolites

Liquid chromatography–tandem mass spectrometry (LC-MS/MS) is widely employed for detecting various secondary metabolites, including phytohormones, folic acid, and flavonoids. MS-based metabolomics approaches have significantly contributed to the finding and description of novel microbial natural products, bridging the gap between genome-encoded potential and known compounds (Breitling *et al.*, 2013; Krug and Müller, 2014). Integrating GC-MS and LC-MS data provides a complete view of both primary and secondary metabolisms. This integration, combined with statistical tools and metabolic modeling, allows the identification of key regulatory points between primary and secondary plant metabolism (Doerfler *et al.*, 2013). These advancements in metabolomic techniques have greatly enhanced our understanding of secondary metabolites and their relationships with primary metabolism (Table 21.3).

21.7 Biodiversity and Ecosystem Function

Metabolomics, the study of chemical phenotypes, is increasingly applied in ecology and biodiversity research. This emerging field, termed eco-metabolomics, combines traditional plant phenotypic traits with chemical analyses to understand plant–environment interactions (Uthe *et al.*, 2021). It bridges biochemistry and ecology across different spatiotemporal scales, addressing challenges such as complex experimental designs, feature extraction, and metabolite identification (Peters *et al.*, 2018). Metabolomics has revealed insights into plant responses to biodiversity, for example showing that small-statured plants exhibit metabolic changes indicating resource limitation in diverse communities, while tall-statured plants remain unaffected (Scherling *et al.*, 2010). This technique has also elucidated mechanisms of plant resistance to herbivores (Macel *et al.*, 2010). However, successful implementation of eco-metabolomics requires overcoming differences in experimental approaches between disciplines and addressing technical challenges in field sampling (Uthe *et al.*, 2021). As the field develops, it promises to provide valuable insights into plant ecology and biodiversity–ecosystem function relationships.

Metabolomics is also a potent technique for researching biogeochemical cycles in diverse environments. Soil metabolomics provides insights into nutrient cycling, C sequestration, and soil microbiome–metabolome relationships (Song *et al.*, 2024). Arsenic biotransformation studies have shown that the incorporation of metabolomics with other omics approaches helps clarify complex biogeochemical processes (Zhu *et al.*, 2017).

Table 21.3. High-throughput metabolomics technologies in plant ecology and biodiversity research.

Technology	Description	Strengths	Limitations	Applications	References
MS	An analytical technique that measures the mass:charge ratio of ions	High sensitivity, wide metabolite coverage, quantification	Requires complex sample preparation, expensive instrumentation	Identification and quantification of a broad range of plant metabolites	Allwood and Goodacre (2010); Ernst *et al.* (2014); Jorge *et al.* (2016a)
NMR spectroscopy	Technique based on magnetic properties of atomic nuclei	Non-destructive, reproducible, comprehensive metabolite profiling	Lower sensitivity requires larger sample sizes	Structural elucidation of plant metabolites, especially primary metabolites	Krishnan *et al.* (2005); Rolin *et al.* (2013); Deborde *et al.* (2017)
GC-MS	Separates and analyzes compounds that can be vaporized without decomposition	High resolution, good for volatile compounds	Limited to volatile and thermally stable metabolites	Analysis of plant volatile organic compounds, and essential oils	Allwood and Goodacre (2010); Gould *et al.* (2023)
LC-MS	Separates and analyzes compounds in the liquid phase	Versatile, high sensitivity, applicable to a wide range of metabolites	More complex data analysis, matrix effects	Profiling of polar and non-polar metabolites, secondary metabolites	La Barbera *et al.* (2017); Ali (2018); Olmo-García and Carrasco-Pancorbo (2021)
CE-MS	Separates ions based on their electrophoretic mobility	High efficiency, small sample volumes, suitable for charged metabolites	Limited metabolite coverage, less robust	Metabolite profiling of ionic species, amino acids, and organic acids	Rodrigues *et al.* (2017); Drouin and Ramautar (2021)
FTIT spectroscopy	Measures absorption of infrared radiation by molecules	Rapid, non-destructive, minimal sample preparation	Lower resolution, limited to functional group analysis	Fingerprinting of plant metabolic states, stress responses, and environmental adaptation	Cozzolino (2014); Skolik *et al.* (2019); Grabska *et al.* (2022)

CE-MS, capillary electrophoresis–mass spectrometry; FTIR, Fourier-transform infrared; GC-MS, gas chromatography–mass spectrometry; LC-MS, liquid chromatography–mass spectrometry; MS, mass spectrometry; NMR, nuclear magnetic resonance.

To facilitate such analyses, the software METABOLIC has been developed for high-throughput profiling of microbial genomes and communities (Zhou *et al.*, 2022). At the genome and community scales, this approach allows a thorough description of the processes of metabolism, exchanges of metabolites, and assistance to biogeochemical cycles. METABOLIC has been successfully applied to diverse environments, including marine and terrestrial subsurface, soil, aquatic systems, and the human gut. These advancements in metabolomics and associated computational tools are enhancing our understanding of microbial roles in biogeochemical cycling and ecosystem functioning.

21.8 Analytical Approaches

Metabolomics, the comprehensive profiling of metabolites in biological systems, employs various analytical techniques during the workflow to identify and quantify a diverse range of small molecules (Table 21.4) (Issaq *et al.*, 2009). The most used platforms are NMR spectroscopy, GC-MS, capillary electrophoresis–mass spectrometry (CE-MS), and LC-MS, each with their own strengths and limitations (Zhang *et al.*, 2012). Due to the complexity of the metabolome, no single analytical method can detect all metabolites in a sample. Therefore, combining complementary techniques has emerged as the most useful approach to maximizing metabolite coverage (Zhang *et al.*, 2012). LC-MS, in particular, has evolved into a major tool in metabolomics research, offering advantages in sensitivity and metabolite detection (Theodoridis *et al.*, 2011). However, challenges remain in data interpretation, metabolite identification, and ensuring analytical validity through proper quality assurance/quality control procedures (Issaq *et al.*, 2009; Theodoridis *et al.*, 2011). Continued expansion of these analytical outlets and their integration will accelerate the future development of metabolomics in systems biology (Fig. 21.1) (Zhang *et al.*, 2012).

21.9 Challenges

Metabolomics faces numerous challenges across its workflow, from study design to data analysis (Table 21.4). Key issues include identifying unknown metabolites, interpreting metabolite profiles, and establishing data standards (Mendes, 2006). In cancer metabolomics, distinguishing cancer-specific signals and addressing cell culture-based challenges is critical. Analytical errors can arise in the process of gathering, storing, preparing, and analyzing samples, which could jeopardize the biological interpretation and quality of the findings. Researchers must navigate the multi-disciplinary nature of metabolomics, which spans chemistry, biology, statistics, and computer science. To overcome these challenges, standardization efforts are under way to create compatible data standards across metabolomics, proteomics, and transcriptomics (Mendes, 2006). Advancements in metabolomics techniques and bioinformatics tools are essential for addressing these issues and realizing the full potential of metabolomics in understanding biological systems and improving clinical applications.

21.10 Future Directions

Metabolomics has great potential for personalized medicine and environmental monitoring. Future directions include developing standardized procedures, consistent software, and comprehensive databases to improve result interpretation and validation. The incorporation of metabolomics with other omics methods and advanced statistical techniques will facilitate the creation of detailed individual patient models, enhancing disease understanding and treatment strategies. Portable and wearable technologies, coupled with artificial intelligence, may accelerate the translation of metabolomics findings into clinical practice. As the field grows, careful consideration must be given to experimental design, sample collection and storage, metabolite extraction, analytical methods, quality control, data correlation, and statistical analysis. These advancements will contribute to the development of precision medicine and improve our ability to detect and understand the effects of environmental contaminants.

21.11 Acknowledgments

All the authors are grateful to their respective universities and institutions for their valuable support.

Table 21.4. High-throughput metabolomics workflow in plant ecology studies.

Stage	Description	Tools and techniques	Outcomes	References
Sample collection	Gathering plant samples from fields or controlled environments	Field sampling protocols, herbarium samples, freeze-drying	High-quality, representative plant material for metabolomics analysis	van der Kooy *et al.* (2009); Peters *et al.* (2023)
Sample preparation	Processing plant samples to extract metabolites	Homogenization, solvent extraction, filtration	Prepared extracts containing metabolites ready for analysis	Romanik *et al.* (2007); Salem *et al.* (2017); Roopashree and Naik (2019)
Metabolite detection	Identifying and quantifying metabolites in the samples	MS, NMR, LC-MS, GC-MS, CE-MS, FTIR	Raw data on metabolite presence and abundance	Moco *et al.* (2007); Roessner and Beckles (2009)
Data analysis	Processing and interpreting metabolomics data	Statistical analysis, bioinformatics tools, and software (e.g. MetaboAnalyst)	Identification of key metabolites, patterns, and pathways	Y. Chen *et al.* (2022); Gardinassi, *et al.* (2017)
Biological interpretation	Linking metabolomic data to ecological questions	Ecological modeling, correlation analysis	Insights into plant function, adaptation, and interactions in ecosystems	Sardans *et al.* (2011); Li and Gaquerel (2021); Walker *et al.* (2022)
Validation and integration	Confirming findings and integrating them into broader research	Targeted metabolomics, cross-validation with other omics data	Robust conclusions, potential for application in conservation and ecology	Goldansaz *et al.* (2017); Wang *et al.* (2019); Seim *et al.* (2022)

CE-MS, capillary electrophoresis–mass spectrometry; FTIR, Fourier-transform infrared spectroscopy; GC-MS, gas chromatography–mass spectrometry; LC-MS, liquid chromatography–mass spectrometry; MS, mass spectrometry; NMR, nuclear magnetic resonance.

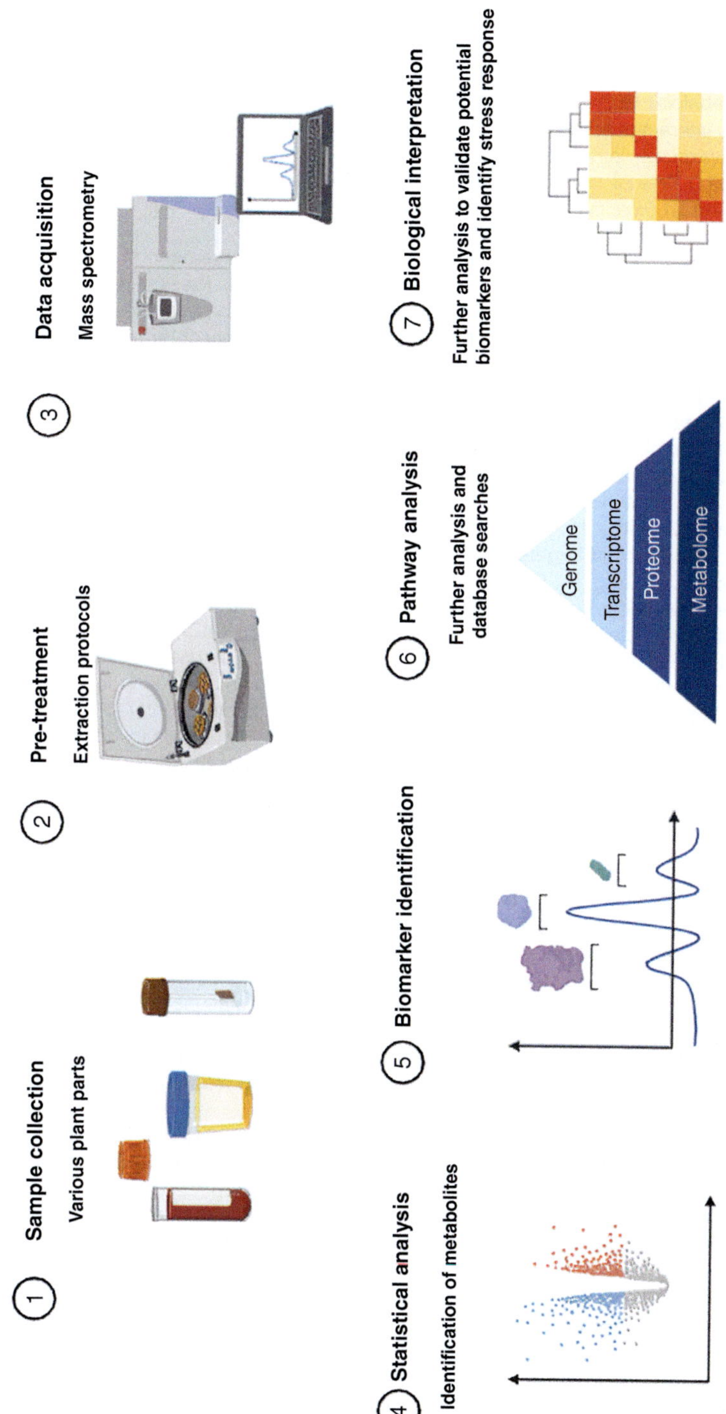

Fig. 21.1. Eco-metabolomics for understanding plant responses to environmental stresses.

21.12 References

Ackah, M., Shi, Y., Wu, M., Wang, L., Guo, P., *et al.* (2021) Metabolomics response to drought stress in *Morus alba* L. variety Yu-711. *Plants* 10: 1636.

Adeniji, A.A. and Babalola, O.O. (2020) Metabolomic applications for understanding complex tripartite plant–microbes interactions: strategies and perspectives. *Biotechnology Reports* 25: e00425.

Aguirre-Becerra, H., Vazquez-Hernandez, M.C., Saenz de la, O.D., Alvarado-Mariana, A., Guevara-Gonzalez, R.G., *et al.* (2021) Role of stress and defense in plant secondary metabolites production. In: Pal, D. and Nayak, A.K. (eds) *Bioactive Natural Products for Pharmaceutical Applications. Advanced Structured Materials, Vol.* 140. Springer, Cham, Switzerland, pp. 151–195.

Ahmad, P., Ahanger, M.A., Singh, V.P., Tripathi, D.K., Alam, P., and Alyemeni, M.N. (eds) (2018) *Plant Metabolites and Regulation Under Environmental Stress*. Academic Press, London.

Ali, A.M. (2018) Targeted and untargeted metabolomic profiling of human serum, urine and brain by liquid chromatography mass spectrometry (LC-MS). Ph.D. thesis, University of Strathclyde, Strathclyde, UK.

Al-Khayri, J.M., Rashmi, R., Toppo, V., Chole, P.B., Banadka, A., *et al.* (2023) Plant secondary metabolites: the weapons for biotic stress management. *Metabolites* 13: 716.

Allwood, J.W. and Goodacre, R. (2010) An introduction to liquid chromatography–mass spectrometry instrumentation applied in plant metabolomic analyses. *Phytochemical Analysis* 21, 33–47.

Allwood, J.W., Ellis, D.I., and Goodacre, R. (2008) Metabolomic technologies and their application to the study of plants and plant–host interactions. *Physiologia Plantarum* 132, 117–135.

Allwood, J.W., Williams, A., Uthe, H., van Dam, N.M., Mur, L.A., *et al.* (2021) Unravelling plant responses to stress—the importance of targeted and untargeted metabolomics. *Metabolites* 11: 558.

Almuziny, M., Decker, C., Wang, D., Gerard, P., and Tharayil, N. (2017) Nutrient supply and simulated herbivory differentially alter the metabolite pools and the efficacy of the glucosinolate-based defense system in *Brassica* species. *Journal of Chemical Ecology* 43, 129–142.

Amiour, N., Imbaud, S., Clément, G., Agier, N., Zivy, M., *et al.* (2012) The use of metabolomics integrated with transcriptomic and proteomic studies for identifying key steps involved in the control of nitrogen metabolism in crops such as maize. *Journal of Experimental Botany* 63, 5017–5033.

Arbona, V., Manzi, M., de Ollas, C., and Gómez-Cadenas, A. (2013) Metabolomics as a tool to investigate abiotic stress tolerance in plants. *International Journal of Molecular Sciences* 14, 4885–4911.

Asaduzzaman, M., Pratley, J.E., An, M., Luckett, D.J., and Lemerle, D. (2015) Metabolomics differentiation of canola genotypes: toward an understanding of canola allelochemicals. *Frontiers in Plant Science* 5: 765.

Azizan, K.A., Ibrahim, S., Ghani, N.H.A., and Nawawi, M.F. (2016) LC-MS based metabolomics analysis to identify potential allelochemicals in *Wedelia trilobata*. *Records of Natural Products* 10: 788.

Baghalian, K., Hajirezaei, M.R., and Schreiber, F. (2014) Plant metabolic modeling: achieving new insight into metabolism and metabolic engineering. *Plant Cell* 26, 3847–3866.

Balparda, M., Bouzid, M., Martinez, M.D.P., Zheng, K., Schwarzländer, M., and Maurino, V.G. (2023) Regulation of plant carbon assimilation metabolism by post-translational modifications. *Plant Journal* 114, 1059–1079.

Barrales-Cureño, H.J., Herrera-Cabrera, B.E., Montiel-Montoya, J., López-Valdez, L.G., Salgado-Garciglia, R., *et al.* (2022) Metabolomics studies of allelopathy: a review. *Revista Colombiana de Ciencias Químico-Farmacéuticas* 51, 243–274.

Bhattacharjya, S., Ghosh, A., Sahu, A., Agnihotri, R., Pal, N., *et al.* (2024) Utilizing soil metabolomics to investigate the untapped metabolic potential of soil microbial communities and their role in driving soil ecosystem processes: a review. *Applied Soil Ecology* 195: 105238.

Biswas, S., Biswas, A.K., and De, B. (2018) Metabolomics analysis of *Cajanus cajan* L. seedlings unravelled amelioration of stress induced responses to salinity after halopriming of seeds. *Plant Signaling and Behavior* 13: e1489670.

Breitling, R., Achcar, F., and Takano, E. (2013) Modeling challenges in the synthetic biology of secondary metabolism. *ACS Synthetic Biology* 2, 373–378.

Brunetti, C., George, R.M., Tattini, M., Field, K., and Davey, M.P. (2013) Metabolomics in plant environmental physiology. *Journal of Experimental Botany* 64, 4011–4020.

Carrera, F.P., Noceda, C., Maridueña-Zavala, M.G., and Cevallos-Cevallos, J.M. (2021) Metabolomics, a powerful tool for understanding plant abiotic stress. *Agronomy* 11: 824.

Chen, L., Schwier, M., Krumbach, J., Kopriva, S., and Jacoby, R.P. (2021) Metabolomics in plant–microbe interactions in the roots. *Advances in Botanical Research* 98, 133–161.

Chen, X., Xue, H., Zhu, L., Wang, H., Long, H., *et al.* (2022a) ERF49 mediates brassinosteroid regulation of heat stress tolerance in *Arabidopsis thaliana*. *BMC Biology* 20: 254.

Chen, Y., Li, E.-M., and Xu, L.-Y. (2022b) Guide to metabolomics analysis: a bioinformatics workflow. *Metabolites* 12: 357.

Cho, K., Shibato, J., Agrawal, G.K., Jung, Y.H., Kubo, A., *et al.* (2008) Integrated transcriptomics, proteomics, and metabolomics analyses to survey ozone responses in the leaves of rice seedling. *Journal of Proteome Research* 7, 2980–2998.

Cozzolino, D. (2014) Use of infrared spectroscopy for in-field measurement and phenotyping of plant properties: instrumentation, data analysis, and examples. *Applied Spectroscopy Reviews* 49, 564–584.

Cui, J., Nieves-Cordones, M., Rubio, F., and Tcherkez, G. (2021) Involvement of salicylic acid in the response to potassium deficiency revealed by metabolomics. *Plant Physiology and Biochemistry* 163, 201–204.

Deborde, C., Moing, A., Roch, L., Jacob, D., Rolin, D., and Giraudeau, P. (2017) Plant metabolism as studied by NMR spectroscopy. *Progress in Nuclear Magnetic Resonance Spectroscopy* 102, 61–97.

Deng, P., Li, X., Petriello, M.C., Wang, C., Morris, A.J., and Hennig, B. (2019) Application of metabolomics to characterize environmental pollutant toxicity and disease risks. *Reviews on Environmental Health* 34, 251–259.

Dietz, K.-J. (2017) Subcellular metabolomics: the choice of method depends on the aim of the study. *Journal of Experimental Botany* 68, 5695–5698.

Divekar, P.A., Narayana, S., Divekar, B.A., Kumar, R., Gadratagi, B.G., *et al.* (2022) Plant secondary metabolites as defense tools against herbivores for sustainable crop protection. *International Journal of Molecular Sciences* 23: 2690.

Doerfler, H., Lyon, D., Nägele, T., Sun, X., Fragner, L., *et al.* (2013) Granger causality in integrated GC–MS and LC–MS metabolomics data reveals the interface of primary and secondary metabolism. *Metabolomics* 9, 564–574.

Dokwal, D., Cocuron, J.C., Alonso, A.P., and Dickstein, R. (2022) Metabolite shift in *Medicago truncatula* occurs in phosphorus deprivation. *Journal of Experimental Botany* 73, 2093–2111.

Doran, O., MacLean, D.A., and Kershaw, J.A. (2017) Needle longevity of balsam fir is increased by defoliation by spruce budworm. *Trees* 31, 1933–1944.

Drouin, N. and Ramautar, R. (2021) Capillary electrophoresis-mass spectrometry for metabolomics: possibilities and perspectives. *Advances in Experimental Medicine and Biology* 1336, 159–178.

Ernst, M., Silva, D.B., Silva, R.R., Vêncio, R.Z., and Lopes, N.P. (2014) Mass spectrometry in plant metabolomics strategies: from analytical platforms to data acquisition and processing. *Natural Product Reports* 31, 784–806.

Farré, G., Twyman, R.M., Christou, P., Capell, T., and Zhu, C. (2015) Knowledge-driven approaches for engineering complex metabolic pathways in plants. *Current Opinion in Biotechnology* 32, 54–60.

Forieri, I., Sticht, C., Reichelt, M., Gretz, N., Hawkesford, M.J., *et al.* (2017) System analysis of metabolism and the transcriptome in *Arabidopsis thaliana* roots reveals differential co-regulation upon iron, sulfur and potassium deficiency. *Plant, Cell and Environment* 40, 95–107.

Fraire-Velázquez, S. and Balderas-Hernández, V.E. (2013) Abiotic stress in plants and metabolic responses. In; Vahdati, K. and Leslie, C. (eds) *Abiotic Stress: Plant Responses and Applications in Agriculture*. IntechOpen, London. Available at: www.intechopen.com/chapters/43341 (accessed 10 February 2025).

Fürstenberg-Hägg, J., Zagrobelny, M., and Bak, S. (2013) Plant defense against insect herbivores. *International Journal of Molecular Sciences* 14, 10242–10297.

Gardinassi, L.G., Xia, J., Safo, S.E., and Li, S. (2017) Bioinformatics tools for the interpretation of metabolomics data. *Current Pharmacology Reports* 3, 374–383.

Génard, M., Baldazzi, V., and Gibon, Y. (2014) Metabolic studies in plant organs: don't forget dilution by growth. *Frontiers in Plant Science* 5: 85.

Geng, S., Yu, B., Zhu, N., Dufresne, C., and Chen, S. (2017) Metabolomics and proteomics of *Brassica napus* guard cells in response to low CO_2. *Frontiers in Molecular Biosciences* 4: 51.

Ghirardo, A., Fochi, V., Lange, B., Witting, M., Schnitzler, J.P., *et al.* (2020) Metabolomic adjustments in the orchid mycorrhizal fungus *Tulasnella calospora* during symbiosis with *Serapias vomeracea*. *New Phytologist* 228, 1939–1952.

Goldansaz, S.A., Guo, A.C., Sajed, T., Steele, M.A., Plastow, G.S., and Wishart, D.S. (2017) Livestock metabolomics and the livestock metabolome: a systematic review. *PLOS One* 12: e0177675.

Gould, O., Nguyen, N., and Honeychurch, K.C. (2023) New applications of gas chromatography and gas chromatography-mass spectrometry for novel sample matrices in the forensic sciences: a literature review. *Chemosensors* 11: 527.

Grabska, J., Beć, K.B., and Huck, C.W. (2022) Current and future applications of IR and NIR spectroscopy in ecology, environmental studies, wildlife and plant investigations. *Comprehensive Analytical Chemistry* 98, 45–76.

Guo, X., Xin, Z., Yang, T., Ma, X., Zhang, Y., *et al.* (2020) Metabolomics response for drought stress tolerance in Chinese wheat genotypes (*Triticum aestivum*). *Plants* 9: 520.

Gupta, S., Schillaci, M., and Roessner, U. (2022) Metabolomics as an emerging tool to study plant–microbe interactions. *Emerging Topics in Life Sciences* 6, 175–183.

Guy, C., Kaplan, F., Kopka, J., Selbig, J., and Hincha, D.K. (2008) Metabolomics of temperature stress. *Physiologia Plantarum* 132, 220–235.

Hall, R.D. (2011) Plant metabolomics in a nutshell: potential and future challenges. *Annual Plant Reviews Online* 43, 1–24.

Harrigan, G.G., Martino-Catt, S., and Glenn, K.C. (2007) Metabolomics, metabolic diversity and genetic variation in crops. *Metabolomics* 3, 259–272.

Issaq, H.J., Van, Q.N., Waybright, T.J., Muschik, G.M., and Veenstra, T.D. (2009) Analytical and statistical approaches to metabolomics research. *Journal of Separation Science* 32, 2183–2199.

Jan, S., Abbas, N., Ashraf, M., and Ahmad, P. (2019) Roles of potential plant hormones and transcription factors in controlling leaf senescence and drought tolerance. *Protoplasma* 256, 313–329.

Jänkänpää, H.J., Mishra, Y., Schröder, W.P., and Jansson, S. (2012) Metabolic profiling reveals metabolic shifts in *Arabidopsis* plants grown under different light conditions. *Plant, Cell and Environment* 35, 1824–1836.

Jansen, J.J., Allwood, J.W., Marsden-Edwards, E., van der Putten, W.H., Goodacre, R., and van Dam, N.M. (2009) Metabolomic analysis of the interaction between plants and herbivores. *Metabolomics* 5, 150–161.

Jha, Y. and Mohamed, H.I. (2022) Plant secondary metabolites as a tool to investigate biotic stress tolerance in plants: a review. *Gesunde Pflanzen* 74, 771–790.

Jiang, Q., Wei, L., Gai, C., Yu, W., He, C., *et al.* (2019) An improved extraction method reveals varied DNA content in different parts of the shells of Pacific oysters. *Aquatic Living Resources* 32: 5.

Jorge, T.F., Mata, A.T., and António, C. (2016a) Mass spectrometry as a quantitative tool in plant metabolomics. *Philosophical Transactions of the Royal Society A: Mathematical, Physical and Engineering Sciences* 374: 20150370.

Jorge, T.F., Rodrigues, J.A., Caldana, C., Schmidt, R., van Dongen, J.T., *et al.* (2016b) Mass spectrometry-based plant metabolomics: metabolite responses to abiotic stress. *Mass Spectrometry Reviews* 35, 620–649.

Kangi, E., Brzostek, E.R., Bills, R.J., Callister, S.J., Zink, E.M., *et al.* (2024) A multi-omic survey of black cottonwood tissues highlights coordinated transcriptomic and metabolomic mechanisms for plant adaptation to phosphorus deficiency. *Frontiers in Plant Science* 15: 1324608.

Kareya, M.S., Mariam, I., Shaikh, K.M., Nesamma, A.A., and Jutur, P.P. (2020) Photosynthetic carbon partitioning and metabolic regulation in response to very-low and high CO_2 in Micro*chloropsis gaditana* NIES 2587. *Frontiers in Plant Science* 11: 981.

Kato, Y., Inabe, K., Hidese, R., Kondo, A., and Hasunuma, T. (2022) Metabolomics-based engineering for biofuel and bio-based chemical production in microalgae and cyanobacteria: a review. *Bioresource Technology* 344: 126196.

Kaur, S. and Suseela, V. (2020) Unraveling arbuscular mycorrhiza-induced changes in plant primary and secondary metabolome. *Metabolites* 10: 335.

Kooke, R. and Keurentjes, J.J.B. (2012) Multi-dimensional regulation of metabolic networks shaping plant development and performance. *Journal of Experimental Botany* 63, 3353–3365.

Krishnan, P., Kruger, N., and Ratcliffe, R. (2005) Metabolite fingerprinting and profiling in plants using NMR. *Journal of Experimental Botany* 56, 255–265.

Krug, D. and Müller, R. (2014) Secondary metabolomics: the impact of mass spectrometry-based approaches on the discovery and characterization of microbial natural products. *Natural Product Reports* 31, 768–783.

Kumar, M., Kumar Patel, M., Kumar, N., Bajpai, A.B., and Siddique, K.H. (2021) Metabolomics and molecular approaches reveal drought stress tolerance in plants. *International Journal of Molecular Sciences* 22: 9108.

Kwon, Y.S., Lee, S., Lee, S.H., Kim, H.J., and Lee, C.H. (2019) Comparative evaluation of six traditional fermented soybean products in East Asia: a metabolomics approach. *Metabolites* 9: 183.

La Barbera, G., Capriotti, A.L., Cavaliere, C., Montone, C.M., Piovesana, S., *et al.* (2017) Liquid chromatography-high resolution mass spectrometry for the analysis of phytochemicals in vegetal-derived food and beverages. *Food Research International* 100, 28–52.

Lankadurai, B.P., Nagato, E.G., and Simpson, M.J. (2013) Environmental metabolomics: an emerging approach to study organism responses to environmental stressors. *Environmental Reviews* 21, 180–205.

Laparre, J., Malbreil, M., Letisse, F., Portais, J.C., Roux, C., *et al.* (2014) Combining metabolomics and gene expression analysis reveals that propionyl- and butyryl-carnitines are involved in late stages of arbuscular mycorrhizal symbiosis. *Molecular Plant* 7, 554–566.

Legouin, B., Lohézic-Le Dévéhat, F., Ferron, S., Rouaud, I., Le Pogam, P., *et al.* (2017) Specialized metabolites of the lichen *Vulpicida pinastri* act as photoprotective agents. *Molecules* 22: 1162.

Li, D. and Gaquerel, E. (2021) Next-generation mass spectrometry metabolomics revives the functional analysis of plant metabolic diversity. *Annual Review of Plant Biology* 72, 867–891.

Lin, Z.H., Chen, C.S., Zhong, Q.S., Ruan, Q.C., Chen, Z.H., *et al.* (2021) The GC-TOF/MS-based metabolomic analysis reveals altered metabolic profiles in nitrogen-deficient leaves and roots of tea plants (*Camellia sinensis*). *BMC Plant Biology* 21: 506.

Liu, M., Tian, H., Lu, C., Xu, X., Chen, G., and Ren, W. (2012) Effects of multiple environment stresses on evapotranspiration and runoff over eastern China. *Journal of Hydrology* 426, 39–54.

Lloyd, N., Johnson, D.L., and Herderich, M.J. (2015) Metabolomics approaches for resolving and harnessing chemical diversity in grapes, yeast and wine. *Australian Journal of Grape and Wine Research* 21, 723–740.

Lu, X., Chen, Z., Gao, J., Fu, S., Hu, H., and Ren, J. (2020) Combined metabolome and transcriptome analyses of photosynthetic pigments in red maple. *Plant Physiology and Biochemistry* 154, 476–490.

Maag, D., Erb, M., and Glauser, G. (2015) Metabolomics in plant–herbivore interactions: challenges and applications. *Entomologia Experimentalis et Applicata* 157, 18–29.

Maag, D., Köhler, A., Robert, C.A., Frey, M., Wolfender, J.L., *et al.* (2016) Highly localized and persistent induction of *Bx1*-dependent herbivore resistance factors in maize. *Plant Journal* 88, 976–991.

Macel, M. and van Dam, N.M. (2018) Metabolomics of plant resistance to insects. In: Emani, C. (ed.) *The Biology of Plant–Insect Interactions*. CRC Press, Boca Raton, Florida, pp. 129–149.

Macel, M., van Dam, N.M., and Keurentjes, J.J. (2010) Metabolomics: the chemistry between ecology and genetics. *Molecular Ecology Resources* 10, 583–593.

Maldini, M., Natella, F., Baima, S., Morelli, G., Scaccini, C., *et al.* (2015) Untargeted metabolomics reveals predominant alterations in lipid metabolism following light exposure in broccoli sprouts. *International Journal of Molecular Sciences* 16, 13678–13691.

Marti, G., Erb, M., Boccard, J., Glauser, G., Doyen, G.R., *et al.* (2013) Metabolomics reveals herbivore-induced metabolites of resistance and susceptibility in maize leaves and roots. *Plant, Cell and Environment* 36, 621–639.

Meena, R.K., Jangra, S., Wadhwa, Z., Wati, M., and Wati, L. (2017) Role of plant volatiles in defense and communication. *International Journal of Current Microbiology and Applied Sciences* 6, 300–313.

Mendes, P. (2006) Metabolomics and the challenges ahead. *Briefings in Bioinformatics* 7: 127.

Mhlongo, M.I., Piater, L.A., Madala, N.E., Labuschagne, N., and Dubery, I.A. (2018) The chemistry of plant–microbe interactions in the rhizosphere and the potential for metabolomics to reveal signaling related to defense priming and induced systemic resistance. *Frontiers in Plant Science* 9: 112.

Michaletti, A., Naghavi, M.R., Toorchi, M., Zolla, L., and Rinalducci, S. (2018) Metabolomics and proteomics reveal drought-stress responses of leaf tissues from spring-wheat. *Scientific Reports* 8: 5710.

Mishra, S., Priyanka, and Sharma, S. (2022) Metabolomic insights into endophyte-derived bioactive compounds. *Frontiers in Microbiology* 13: 835931.

Moco, S., Vervoort, J., Bino, R.J., de Vos, R.C., and Bino, R. (2007) Metabolomics technologies and metabolite identification. *Trends in Analytical Chemistry* 26, 855–866.

Muller, J., Gödde, V., Niehaus, K., and Zörb, C. (2015) Metabolic adaptations of white lupin roots and shoots under phosphorus deficiency. *Frontiers in Plant Science* 6: 1014.

Nagarajan, K., Tong, W.Y., Leong, C.R., and Tan, W.N. (2021) Potential of endophytic *Diaporthe* sp. as a new source of bioactive compounds. *Journal of Microbiology and Biotechnology* 31: 493.

Nizamani, M.M., Hughes, A.C., Qureshi, S., Zhang, Q., Tarafder, E., *et al.* (2024) Microbial biodiversity and plant functional trait interactions in multifunctional ecosystems. *Applied Soil Ecology* 201: 105515.

Noctor, G., Lelarge-Trouverie, C., and Mhamdi, A. (2015) The metabolomics of oxidative stress. *Phytochemistry* 112, 33–53.

Obata, T. and Fernie, A.R. (2012) The use of metabolomics to dissect plant responses to abiotic stresses. *Cellular and Molecular Life Sciences* 69, 3225–3243.

Ogbemudia, F. and Thompson, E. (2014) Variation in plants secondary metabolites and potential ecological roles—a review. *International Journal of Modern Biology and Medicine* 5, 111–130.

Olmo-García, L. and Carrasco-Pancorbo, A. (2021) Chromatography-MS based metabolomics applied to the study of virgin olive oil bioactive compounds: characterization studies, agro-technological investigations and assessment of healthy properties. *Trends in Analytical Chemistry* 135: 116153.

Pandey, A., Kumari, N., Mishra, S., Jaiswal, J., and Sinha, R.P. (2021) NMR spectroscopy for the characterization of photoprotective compounds in cyanobacteria. *Applications of NMR Spectroscopy* 9: 1.

Papazian, S., Khaling, E., Bonnet, C., Lassueur, S., Reymond, P., *et al.* (2016) Central metabolic responses to ozone and herbivory affect photosynthesis and stomatal closure. *Plant Physiology* 172, 2057–2078.

Parween, D., Sahu, B.B., Kumari, M., and Pudake, R.N. (2019) Plant metabolites involved in plant–pathogen interactions. In: Varma, A., Tripathi, S., and Prasad, R. (eds) *Plant Biotic Interactions*. Springer. Cham, Switzerland, pp. 61–84.

Pawar, M. (2020) Understanding the chemical ecology of Brassicaceae specialist herbivore *P. xylostella* and its predators. M.Sc. thesis, IISER Pune, Pune, India.

Peters, K., Worrich, A., Weinhold, A., Alka, O., Balcke, G., *et al.* (2018) Current challenges in plant eco-metabolomics. *International Journal of Molecular Sciences* 19: 1385.

Peters, K., Poeschl, Y., Blatt-Janmaat, K.L., and Uthe, H. (2023) Ecometabolomics studies of bryophytes. In: Murthy, H.N. (ed.) *Bioactive Compounds in Bryophytes and Pteridophytes*. Springer Nature, London, pp. 637–679.

Radomiljac, J.D., Whelan, J., and van der Merwe, M. (2013) Coordinating metabolite changes with our perception of plant abiotic stress responses: emerging views revealed by integrative-omic analyses. *Metabolites* 3, 761–786.

Rehman, F., Kalsoom, M., Nasir, T.A., Adnan, M., Anwar, S., and Zahra, A. (2020) Chemistry of plant–microbe interactions in rhizosphere and rhizoplane. *Indian Journal of Pure and Applied Biosciences* 8, 11–19.

Roca, M. and Pérez-Gálvez, A. (2021) Metabolomics of chlorophylls and carotenoids: analytical methods and metabolome-based studies. *Antioxidants* 10: 1622.

Rodrigues, K.T., Cieslarová, Z., Tavares, M.F.M., and Simionato, A.V.C. (2017) Strategies involving mass spectrometry combined with capillary electrophoresis in metabolomics. *Advances in Experimental Medicine and Biology* 965, 99–141.

Roessner, U. and Beckles, D.M. (2009) Metabolite measurements. In: Schwender, J. (ed.) *Plant Metabolic Networks*. Springer, New York, pp. 39–69.

Roessner, U. and Beckles, D.M. (2012) Metabolomics for salinity research. *Methods in Molecular Biology* 913, 203–215.

Rolin, D., Deborde, C., Maucourt, M., Cabasson, C., Fauvelle, F., *et al.* (2013) High-resolution ^{1}H-NMR spectroscopy and beyond to explore plant metabolome. *Advances in Botanical Research* 67, 1–66.

Romanik, G., Gilgenast, E., Przyjazny, A., and Kamiński, M. (2007) Techniques of preparing plant material for chromatographic separation and analysis. *Journal of Biochemical and Biophysical Methods* 70, 253–261.

Romero, H., Pott, D.M., Vallarino, J.G., and Osorio, S. (2021) Metabolomics-based evaluation of crop quality changes as a consequence of climate change. *Metabolites* 11: 461.

Roopashree, K. and Naik, D. (2019) Advanced method of secondary metabolite extraction and quality analysis. *Journal of Pharmacognosy and Phytochemistry* 8, 1829–1842.

Roychowdhury, R., Das, S.P., Gupta, A., Parihar, P., Chandrasekhar, K., *et al.* (2023) Multi-omics pipeline and omics-integration approach to decipher plant's abiotic stress tolerance responses. *Genes* 14: 1281.

Salam, U., Ullah, S., Tang, Z.-H., Elateeq, A.A., Khan, Y., *et al.* (2023) Plant metabolomics: an overview of the role of primary and secondary metabolites against different environmental stress factors. *Life* 13: 706.

Salem, M., Bernach, M., Bajdzienko, K., and Giavalisco, P. (2017) A simple fractionated extraction method for the comprehensive analysis of metabolites, lipids, and proteins from a single sample. *Journal of Visualized Experiments* 124: e55802.

Salem, M.A., Wang, J.Y., and Al-Babili, S. (2022) Metabolomics of plant root exudates: from sample preparation to data analysis. *Frontiers in Plant Science* 13: 1062982.

Sanchez, D.H., Lippold, F., Redestig, H., Hannah, M.A., Erban, A., *et al.* (2008), Integrative functional genomics of salt acclimatization in the model legume *Lotus japonicus*. *Plant Journal* 53, 973–987.

Sánchez-Suárez, J., Coy-Barrera, E., Villamil, L., and Díaz, L. (2020) *Streptomyces*-derived metabolites with potential photoprotective properties—a systematic literature review and meta-analysis on the reported chemodiversity. *Molecules* 25: 3221.

Sardans, J., Penuelas, J., and Rivas-Ubach, A. (2011) Ecological metabolomics: overview of current developments and future challenges. *Chemoecology* 21, 191–225.

Sardans, J., Gargallo-Garriga, A., Urban, O., Klem, K., Holub, P., *et al.* (2021) Ecometabolomics of plant–herbivore and plant–fungi interactions: a synthesis study. *Ecosphere* 12: e03736.

Scherling, C., Roscher, C., Giavalisco, P., Schulze, E.D., and Weckwerth, W. (2010) Metabolomics unravel contrasting effects of biodiversity on the performance of individual plant species. *PLOS ONE* 5: e12569.

Seim, R.F., Glinski, D.A., Lavelle, C.M., Awkerman, J.A., Hemmer, B.L., *et al.* (2022) Using metabolomic profiling to inform use of surrogate species in ecological risk assessment practices. *Comparative Biochemistry and Physiology D: Genomics and Proteomics* 41: 100947.

Shen, S., Zhan, C., Yang, C., Fernie, A.R., and Luo, J. (2023) Metabolomics-centered mining of plant metabolic diversity and function: past decade and future perspectives. *Molecular Plant* 16, 43–63.

Shulaev, V., Cortes, D., Miller, G., and Mittler, R. (2008) Metabolomics for plant stress response. *Physiologia Plantarum* 132, 199–208.

Skolik, P., Morais, C.L., Martin, F.L., and McAinsh, M.R. (2019) Determination of developmental and ripening stages of whole tomato fruit using portable infrared spectroscopy and Chemometrics. *BMC Plant Biology* 19: 236.

Song, H., Chang, Z., Hu, X., Li, Y., Duan, C., *et al.* (2024) Combined application of chemical and organic fertilizers promoted soil carbon sequestration and bacterial community diversity in dryland wheat fields. *Land* 13: 1296.

Stitt, M., Sulpice, R., and Keurentjes, J. (2010) Metabolic networks: how to identify key components in the regulation of metabolism and growth. *Plant Physiology* 152, 428–444.

Stranska, M., Lovecka, P., Vrchotova, B., Uttl, L., Bechynska, K., *et al.* (2021) Bacterial endophytes from *Vitis vinifera* L—metabolomics characterization of plant–endophyte crosstalk. *Chemistry and Biodiversity* 18: e2100516.

Sung, J., Yun, H., Back, S., Fernie, A.R., Kim, Y.X., *et al.* (2018) Changes in mineral nutrient concentrations and C-N metabolism in cabbage shoots and roots following macronutrient deficiency. *Journal of Plant Nutrition and Soil Science* 181, 777–786.

Sweetlove, L.J. and Fernie, A.R. (2005) Regulation of metabolic networks: understanding metabolic complexity in the systems biology era. *New Phytologist* 168, 9–24.

Sweetlove, L.J. and Fernie, A.R. (2013) The spatial organization of metabolism within the plant cell. *Annual Review of Plant Biology* 64, 723–746.

Tang, K.-H., Tang, Y.J., and Blankenship, R.E. (2011) Carbon metabolic pathways in phototrophic bacteria and their broader evolutionary implications. *Frontiers in Microbiology* 2: 165.

Tang, J.K.-H., Saikin, S.K., Pingali, S.V., Enriquez, M.M., Huh, J., *et al.* (2013) Temperature and carbon assimilation regulate the chlorosome biogenesis in green sulfur bacteria. *Biophysical Journal* 105, 1346–1356.

Tariq, A., Zeng, F., Graciano, C., Ullah, A., Sadia, S., *et al.* (2023) Regulation of metabolites by nutrients in plants. In: Singh, V.P. and Siddiqui, M.H. (eds) *Plant Ionomics: Sensing, Signaling, and Regulation*. Wiley, Hoboken, New Jersey, pp. 1–18.

Tenenboim, H. and Brotman, Y. (2016) Omic relief for the biotically stressed: metabolomics of plant biotic interactions. *Trends in Plant Science* 21, 781–791.

Theodoridis, G., Gika, H.G., and Wilson, I.D. (2011) Mass spectrometry-based holistic analytical approaches for metabolite profiling in systems biology studies. *Mass Spectrometry Reviews* 30, 884–906.

Thomason, K., Babar, M.A., Erickson, J.E., Mulvaney, M., Beecher, C., and MacDonald, G. (2018) Comparative physiological and metabolomics analysis of wheat (*Triticum aestivum* L.) following post-anthesis heat stress. *PLOS ONE* 13: e0197919.

Tuladhar, P., Sasidharan, S., and Saudagar, P. (2021) Role of phenols and polyphenols in plant defense response to biotic and abiotic stresses. In: Jogaiah, S. (ed.) *Biocontrol Agents and Secondary Metabolites*. Elsevier, Duxford, UK, pp. 419–441.

Uarrota, V.G., Garcia, J., de Andrade, G.C., de Fátima Moreira de Bairros, Â., Nerling, D., *et al.* (2020) Metabolomic fingerprinting as a powerful diagnostic tool to assess the effects of abiotic stress in plants and seeds. *CABI Reviews* 2020. Available at: www.cabidigitallibrary.org/doi/10.1079/PAVSNNR202015041 (accessed 10 February 2025).

Ullah, S., Khan, M.N., Lodhi, S.S., Ahmed, I., Tayyab, M., *et al.* (2022) Targeted metabolomics reveals fatty acid abundance adjustments as playing a crucial role in drought-stress response and post-drought recovery in wheat. *Frontiers in Genetics* 13: 972696.

Uthe, H., van Dam, N.M., Hervé, M.R., Sorokina, M., Peters, K., and Weinhold, A. (2021) A practical guide to implementing metabolomics in plant ecology and biodiversity research. *Advances in Botanical Research* 98, 163–203.

van Dam, N.M. and Bouwmeester, H.J. (2016) Metabolomics in the rhizosphere: tapping into belowground chemical communication. *Trends in Plant Science* 21, 256–265.

van der Hooft, J.J., Mohimani, H., Bauermeister, A., Dorrestein, P.C., Duncan, K.R., and Medema, M.H. (2020) Linking genomics and metabolomics to chart specialized metabolic diversity. *Chemical Society Reviews* 49, 3297–3314.

van der Kooy, F., Maltese, F., Choi, Y.H., Kim, H.K., and Verpoorte, R. (2009) Quality control of herbal material and phytopharmaceuticals with MS and NMR based metabolic fingerprinting. *Planta Medica* 75, 763–775.

Walker, T.W., Alexander, J.M., Allard, P.M., Baines, O., Baldy, V., *et al.* (2022) Functional traits 2.0: the power of the metabolome for ecology. *Journal of Ecology* 110, 4–20.

Walker, T.W., Schrodt, F., Allard, P.-M., Defossez, E., Jassey, V.E., *et al.* (2023) Leaf metabolic traits reveal hidden dimensions of plant form and function. *Science Advances* 9: eadi4029.

Wang, L., Meeus, I., Rombouts, C., van Meulebroek, L., Vanhaecke, L., and Smagghe, G. (2019) Metabolomics-based biomarker discovery for bee health monitoring: a proof of concept study concerning nutritional stress in *Bombus terrestris*. *Scientific Reports* 9: 11423.

Wang, W., Pang, J., Zhang, F., Sun, L., Yang, L., and Siddique, K.H. (2022) Transcriptomic and metabolomics-based analysis of key biological pathways reveals the role of lipid metabolism in response to salt stress in the root system of *Brassica napus*. *Plant Growth Regulation* 97, 127–141.

Wang, X., Wu, Q., Wan, D., Liu, Q., Chen, D., *et al.* (2016) Fumonisins: oxidative stress-mediated toxicity and metabolism in vivo and in vitro. *Archives of Toxicology* 90, 81–101.

War, A.R., Buhroo, A.A., Hussain, B., Ahmad, T., Nair, R.M., and Sharma, H.C. (2020) Plant defense and insect adaptation with reference to secondary metabolites. In: Merillon, JM. and Ramawat, K. (eds) *Co-Evolution of Secondary Metabolites*, Springer, Cham, Switzerland, pp. 795–822.

Waseem, M., Nkurikiyimfura, O., Niyitanga, S., Nyimbo, W.J., Shaheen, I., and Aslam, M.M. (2022) Role of metabolomics and next-generation sequencing for sustainable crop production. In: Prakash, C.S., Fiaz, S., and Fahad, S. (eds) *Principles and Practices of OMICS and Genome Editing for Crop Improvement*. Springer, Cham, Switzerland, pp. 123–147.

Wolfender, J.-L., Gaudry, A., Rutz, A., Quiros-Guerrero, L.-M., Nothias, L.-F., *et al.* (2022) Metabolomics in ecology and bioactive natural products discovery: challenges and prospects for a comprehensive study of the specialised metabolome. *Chimia* 76, 954–963.

Xu, M., Zhong, F., and Zhu, J. (2017) Evaluating metabolic response to light exposure in *Lactobacillus* species via targeted metabolic profiling. *Journal of Microbiological Methods* 133, 14–19.

Xu, S., Hu, C., Hussain, S., Tan, Q., Wu, S., and Sun, X. (2018) Metabolomics analysis reveals potential mechanisms of tolerance to excess molybdenum in soybean seedlings. *Ecotoxicology and Environmental Safety* 164, 589–596.

Xue, Y., Zhu, S., Schultze-Kraft, R., Liu, G., and Chen, Z. (2022) Dissection of crop metabolome responses to nitrogen, phosphorus, potassium, and other nutrient deficiencies. *International Journal of Molecular Sciences* 23: 9079.

You, J., Zong, W., Du, H., Hu, H., and Xiong, L. (2014) A special member of the rice SRO family, OsSRO1c, mediates responses to multiple abiotic stresses through interaction with various transcription factors. *Plant Molecular Biology* 84, 693–705.

Zemaitis, K.J., Ye, H., Nguyen, H.T., and Wood, T.D. (2021) Direct infusion metabolomics of the photosystem and chlorophyll related metabolites within a drought tolerant plant introduction of *Glycine max*. *Metabolites* 11: 843.

Zhang, A., Sun, H., Wang, P., Han, Y., and Wang, X. (2012) Modern analytical techniques in metabolomics analysis. *Analyst* 137, 293–300.

Zhao, H., Chong, J., Tang, R., Li, L., Xia, J., and Li, D. (2018) Metabolomics investigation of dietary effects on flesh quality in grass carp (*Ctenopharyngodon idellus*). *GigaScience* 7: giy111.

Zhou, Z., Tran, P.Q., Breister, A.M., Liu, Y., Kieft, K., *et al.* (2022) METABOLIC: High-throughput profiling of microbial genomes for functional traits, metabolism, biogeochemistry, and community-scale functional networks. *Microbiome* 10: 33.

Zhu, Y.G., Xue, X.M., Kappler, A., Rosen, B.P., and Meharg, A.A. (2017) Linking genes to microbial biogeochemical cycling: lessons from arsenic. *Environmental Science and Technology* 51, 7326–7339.

22 High-throughput Metabolomics for Plant Natural Product Research

Gunna Sureshbabu Suruthi[1], Pandiarajan Anisha[1], Sivagaminathan Thangavalli[1], Balasubramanian Balamuralikrishnan[2], Naif Abdullah Al-Dhabi[3], Mariadhas Valan Arasu[3], Vino Palani[4], and Arumugam Vijaya Anand[1]

[1]Department of Human Genetics and Molecular Biology, Bharathiar University, Coimbatore, Tamil Nadu, India; [2]Department of Food Science and Biotechnology, College of Life Science, Sejong University, Seoul, Republic of Korea; [3]Department of Botany and Microbiology, College of Science, King Saud University, Riyadh, Saudi Arabia; [4]Department of Botany, Sri Vidya Mandir Arts and Science College (Autonomous), Krishnagiri, Tamil Nadu, India

Abstract

High-throughput metabolic profiling is an emerging transformative approach in plant natural product research, offering comprehensive insights into the complex metabolic networks of plants. This chapter explores the latest advancements in analytical technologies, data-processing tools, and integrative approaches that have revolutionized our understanding of plant metabolism, and discusses the use of high-throughput metabolomic profiling in studies of stress response, crop improvement, and natural product discovery. Incorporating metabolomic profiling with other omics is highlighted, showcasing its potential in elucidating regulatory networks and genetic bases of important plant traits. Challenges in metabolite identification and data interpretation are addressed, along with future directions including the development of more sensitive analytical methods and the application of artificial intelligence in data analysis. This chapter underscores the critical role of high-throughput metabolomics in managing global challenges associated with food safety, environmental sustainability, and human health by exploring plant natural products.

Keywords: artificial intelligence, crop improvement, data analysis, high-throughput metabolomics, multi-omics integration, plant natural products

22.1 Introduction to Metabolomics

High-throughput metabolomics is an emerging transformative approach in the plant natural product research field, enabling the comprehensive analysis of compounds that have an important role in plant biology and ecology. This methodology leverages advanced analytical techniques to detect and quantify a vast array of metabolites in living systems, providing insights into metabolic pathways, plant responses to environmental stimuli, and the biochemical basis of plant traits. Integrating high-throughput metabolic profiling with other omics technologies, such as genomics and transcriptomics, enhances our knowledge of the complex interactions within plant systems and with their environments.

* E-mail: avahgmb@buc.edu.in

© CAB International 2025. *High-Throughput Plant Metabolomics* (Jen-Tsung Chen ed.)
DOI: 10.1079/9781800626911.0022

The primary benefit of high-throughput metabolomics is its ability to analyze hundreds to thousands of samples simultaneously, which is essential for large-scale studies in plant research. This ability is facilitated by advancements in analytical technologies, particularly mass spectrometry (MS) and nuclear magnetic resonance (NMR) spectroscopy, which allow rapid identification and quantification of compounds with diverse chemical properties (Li *et al.*, 2023). However, the extraction and sample processing protocols remain critical bottlenecks in high-throughput workflows, as they must accommodate a wide variety of metabolites differing in their chemical characteristics and concentrations (Gehrke *et al.*, 2017). Zeng *et al.* (2014) highlighted the challenges associated with uniform extraction protocols in metabolomics, emphasizing that suboptimal conditions can lead to incomplete metabolite profiles.

The development of efficient extraction methods is crucial for maximizing the analytical breadth of high-throughput metabolomics. Gehrke *et al.* (2017) discussed the importance of rapid extraction protocols that minimize sample handling time while ensuring the integrity of the metabolites being analyzed. These protocols often involve sample agitation and the use of ice-cold lysis buffers to stabilize metabolites during extraction; this is particularly relevant for plant samples, which may contain a diverse array of compounds (Gehrke *et al.*, 2017). Furthermore, the integration of automated systems for sample processing can significantly enhance throughput and reproducibility, making it feasible to conduct extensive metabolomic studies in plant research (Raad *et al.*, 2016).

In addition to extraction techniques, the selection of analytical platforms plays an important role in high-throughput metabolomics. MS, especially when associated with liquid chromatography (LC-MS), has become the gold standard in metabolomic analysis due to its sensitivity and ability to analyze complex mixtures (Li *et al.*, 2023). However, NMR spectroscopy offers unique advantages, such as non-destructive analysis and minimal sample preparation, making it an attractive option for certain applications in plant metabolomics (Emwas *et al.*, 2019). The combination of these techniques allows researchers to obtain a complete picture of the metabolites, facilitating identification of novel compounds and elucidating metabolic pathways (Sun *et al.*, 2023).

High-throughput application of metabolic profiling in research of plant metabolites extends to a number of areas, such as the plant–pathogen interactions, stress responses, and the characterization of bioactive compounds. For example, Castro-Moretti *et al.* (2020) emphasized the potential of metabolomic analyses to reveal perturbations in signaling pathways during plant–microbe interactions, which can inform breeding strategies for disease resistance. Similarly, the exploration of endophyte-derived metabolites through high-throughput approaches has provided insights into the biochemical interaction of plants and microbes, highlighting the importance of these relationships in enhancing plant resilience to environmental stressors (Mishra *et al.*, 2022).

The combination of metabolomics with other omics methods has led to a more holistic understanding of plant biology. The simultaneous analysis of metabolomic, transcriptomic, and proteomic data permits the detection of key regulatory networks and metabolic pathways that govern plant traits (Li *et al.*, 2014). This integrative approach is particularly valuable in crop improvement programs, where comprehending the genetic basis of metabolic traits can enable the development of improved plant varieties with enhanced nutritional and medicinal properties (Sharma *et al.*, 2021). Statistical methods and computational tools are essential for analyzing high-throughput metabolite data, which often involves complex data sets with high dimensionality. Techniques such as Gaussian graphical modeling and machine learning algorithms have been employed to reconstruct metabolic pathways and identify key metabolites associated with specific phenotypes (Li *et al.*, 2013). These computational approaches not only enhance the interpretation of metabolomic data but also facilitate the integration of metabolomics with other biological data sets, paving the way for systems biology applications in plant research (Giacomoni *et al.*, 2015).

22.2 High-throughput Technologies in Metabolomics

High-throughput technologies in metabolomics have revolutionized the way researchers explore the vast array of small molecules found in biological systems. These developed techniques allow the comprehensive profiling of metabolites, providing crucial insights into metabolic pathways, cellular functions, and responses to environmental changes. The integration of MS, NMR spectroscopy, and imaging technologies has facilitated the rapid and precise detection of metabolites. In addition, sophisticated data acquisition and processing tools are essential for managing the large data sets generated, ensuring the accurate interpretation of complex metabolic profiles (Table 22.1).

22.2.1 MS-based techniques

Metabolomic profiling allows the analysis of an inclusive and quantitative range of metabolome in samples belonging to biological sources. These compounds have diverse physicochemical properties occurring at levels of variable abundance. Metabolome profiling requires a particular approach for preparation of the samples, the process of separation, and analysis by MS (Brown *et al.*, 2005). Multiple analytical approaches that complement the MS method, are needed to attain a detailed analysis of the metabolome. MS plays a prominent role in metabolomics investigation (Dettmer *et al.*, 2007). MS-based metabolomics in systems biology aims to amplify our understanding of the complex system of biology. This can vary from a complete study achieved via time-consuming chromatography to the rapid investigation of a few target metabolites without chromatography. Although recent MS platforms can attain throughputs of up to 0.5 s per sample, the metabolite range of these platforms is low, corresponding to low-throughputs for these separation-based MS methods (Fig. 22.1). High throughput arrives at an expense, as it is a trade-off between sample throughput and metabolite range (Wolfender *et al.*, 2015).

The use of MS in metabolite profiling has increased rapidly since the discovery and development of electrospray ionization and matrix-assisted laser desorption/ionization (MALDI) techniques. Significant advances have occurred in separation-based MS techniques as well as in separation-free MS procedures in recent decades. A reasonable sample pre-treatment process is required for metabolomics studies to confound the mismatch of the sample with the analytical technique. MS, coupled with various chromatographic processes, is suitable for wide-ranging metabolic profiling because of its huge selectivity and sensitivity, wide range of dynamics, and the information obtained (Li *et al.*, 2012).

Table 22.1. High-throughput technologies used in metabolomics.

Technology	Key points supported	Reference(s)
MS techniques	MS-based techniques such as LC-MS and GC-MS are central to metabolomics research	Dettmer *et al.* (2007)
	MS offers high sensitivity, selectivity, and a wide dynamic range for metabolite profiling	Brown *et al.* (2005); Wolfender *et al.* (2015)
	Advances in ionization methods such as electrospray and MALDI have expanded the abilities of MS	Fiehn (2008)
NMR spectroscopy	NMR provides non-destructive analysis and structural information on metabolites	Emwas *et al.* (2019)
	NMR's automation abilities facilitate high-throughput screening	Markley *et al.* (2017)
	NMR may be integrated with other techniques such as MS for comprehensive metabolite profiling	Markley *et al.* (2017)
MS imaging	MS imaging enables spatial visualization of metabolite distribution in plant tissues	Zhang *et al.* (2022)
	MS imaging complements traditional metabolomics approaches by providing spatial information	Nakabayashi *et al.* (2021)
	MS imaging can be combined with machine learning for improved metabolite annotation	Wadie *et al.* (2023)

GC-MS, gas chromatography–mass spectrometry; LC-MS, liquid chromatography–mass spectrometry; MALDI, matrix-assisted laser desorption/ionization; MS, mass spectrometry; NMR, nuclear magnetic resonance.

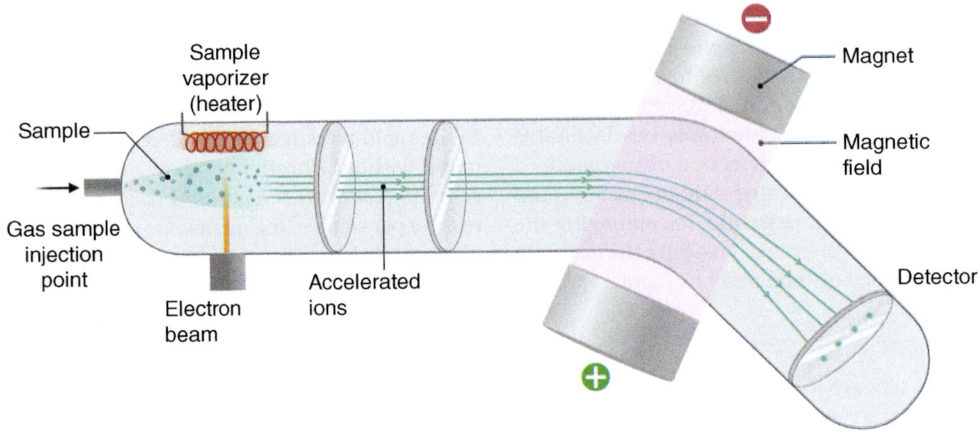

Fig. 22.1. The principles of mass spectrometry.

MS coupled with gas chromatography (GC-MS) is an important process for the investigation of metabolites and is used for the analysis of non-targeted metabolites such as metabolites of a hydrophilic nature. A high-throughput technology is required for the analysis of metabolites in non-targeted GC-MS to handle the huge number of samples, the retention time of the compounds present in the standard peak, and identification of these compounds (Fiehn, 2008). Over 400 compounds can be identified in a single-run analysis. Energetic processes such as liquid chromatography–tandem mass spectrometry (LC-MS/MS) provide an effective, economical, and rapid way to analyze changes in targeted metabolites in samples derived from biological sources (Dettmer *et al.*, 2007). Metabolomics based on LC-MS is a developing area but faces a number of different pitfalls and challenges.

22.2.2 NMR spectroscopy

NMR spectroscopy has emerged as a pivotal technology in metabolomics, particularly for the detection of plant natural products. Its unique characteristics, such as non-destructive analysis, minimal sample preparation, and apex reproducibility, make this technique an attractive alternative in high-throughput metabolomics studies. The ability of NMR to provide simultaneous structural details on a wide array of metabolites enhances its utility in understanding complex biological systems, particularly in plants where secondary metabolites play essential functions in defense and interactions with the environment (Manoharan *et al.*, 2008; Begum *et al.*, 2009; Muthumani *et al.*, 2013; Varalakshmi *et al.*, 2014; Vijaya Anand *et al.*, 2016, 2020; Bharathi *et al.*, 2018; Akilandeswari *et al.*, 2019; Divya *et al.*, 2019; Emwas *et al.*, 2019; Sangeetha *et al.*, 2022; Kaviya *et al.*, 2022; Manikandan *et al.*, 2022; Karthika *et al.*, 2023; Vijayakumar *et al.*, 2023). One important benefit of NMR spectroscopy is its non-destructive nature, allowing sample analysis without altering the sample composition. This feature is particularly beneficial in plant metabolomics, where the integrity of the sample is crucial for accurate analysis. Unlike MS, which often requires extensive sample preparation that can introduce biases, NMR can analyze crude extracts directly, significantly reducing the time and effort involved in metabolite profiling (Huang *et al.*, 2024). Additionally, the automation abilities of NMR systems facilitate high-throughput screening, making it feasible to analyze large numbers of samples efficiently (Markley *et al.*, 2017).

Recent investigations have demonstrated the efficacy of NMR in characterizing plant metabolites under various conditions. For instance, research on tomato plants treated with metabolites isolated from *Trichoderma* fungi showed that NMR could identify treatment-dependent metabolic changes with high precision. This ability is essential for understanding how plants respond to environmental stressors and for the identification of bioactive compounds that may have agricultural or pharmaceutical applications (Mazzei *et al.*, 2016). The integration of NMR with other analytical techniques, such as MS, enhances the depth of metabolomic investigations, allowing a more complete understanding of metabolic pathways and fluxes (Markley *et al.*, 2017). The benefits of NMR in plant metabolomics are not limited to research on secondary metabolites. It has also been employed to study primary compounds and their roles in plant growth and development. For example, NMR has been used to investigate the metabolome of different plant species, providing insights into the biochemical processes that underpin plant physiology (Stark *et al.*, 2020). The ability to trace metabolic pathways using isotopic labeling in NMR experiments further enriches our understanding of plant metabolism and the interactions between different metabolic networks (Dey, 2023).

The high resolution and sensitivity of NMR allow the detection of subtle changes in metabolite concentrations, which is critical in studies of plant responses to environmental changes or stressors. For instance, analysis of plants exposed to chemical pollutants using NMR has provided valuable data on the extent of damage and the metabolic shifts that appear in response to such stressors (Kim *et al.*, 2020). This information is crucial in developing a strategy for mitigating the effects of environmental pollutants on the health of plants and their productivity. In addition to its analytical abilities, NMR spectroscopy is also being explored for its potential for real-time monitoring of metabolic processes, allowing researchers to observe dynamic changes in metabolite concentrations as they occur and providing a more nuanced interpretation of plant metabolism under various conditions (Her *et al.*, 2015). The evolution of advanced NMR methods, such as high-resolution magic-angle spinning, has also enhances our ability to analyze intact plant tissues, thereby preserving the spatial and temporal context of metabolic changes (Mori *et al.*, 2015).

The integration of NMR with computational approaches, including machine learning and data mining, is another promising avenue in metabolomics research. These techniques can help in the detection of patterns and correlations within complex data sets, facilitating the discovery of novel metabolites and their potential applications (BaniMustafa and Hardy, 2020). By leveraging the strengths of NMR alongside advanced computational methods, researchers can enhance their ability to interpret metabolomic data and uncover insights that may not be apparent through traditional analytical methods.

22.2.3 MS imaging

MS imaging is a transformative method in the field of metabolic profiling, particularly for high-throughput applications in plant natural product research. This technique enables the graphic visualization of compounds in tissues of biological origin to determine their localization and distribution, which is crucial for understanding plant metabolism and the biosynthesis of natural products. The incorporation of MS imaging with traditional metabolomics approaches enhances the depth of analysis, allowing researchers to correlate metabolite profiles with specific biological functions and environmental responses. The main advantage of MS imaging is its ability to provide spatially resolved information about the metabolome without the need for extensive sample preparation or labeling. For instance, Zhang *et al.* (2022) demonstrated the incorporation of desorption electrospray ionization–MS imaging to envision the production and dispersal of metabolites in biological models, highlighting the complementary role of MS imaging in classical metabolomics studies. This ability is particularly beneficial in plant research, where the localization of metabolites can be critical for understanding their functions in plant physiology and relationships with their surroundings.

The application of MS imaging in plant metabolomics was further exemplified by Nakabayashi *et al.* (2021) who utilized MALDI–Fourier transform–ion cyclotron resonance–MS imaging to determine the localization of specific metabolites, such as asparaptine in *Asparagus officinalis*. This high-resolution technique allows the visualization of metabolite dispersal at the level of cells, providing insights into how metabolites are organized within plant tissues. Such spatial metabolomics approaches are essential for elucidating the complex biosynthetic paths that lead to the production of valuable natural products in plants. Moreover, the integration of machine learning with MS imaging has unlocked new routes for metabolite annotation and data analysis. Wadie *et al.* (2023) highlighted the development of METASPACE-ML, a machine learning framework designed to enhance the accuracy of metabolite identification in MS imaging data. This advancement is particularly relevant in high-throughput metabolomics, where the volume of data generated can be overwhelming. By improving the annotation process, researchers can more effectively interpret the spatial distribution of metabolites and their potential biological significance.

The significance of MS imaging in plant metabolomics is underscored by its ability to discover new compounds and their role in plant growth and stress responses. Dong *et al.* (2020) demonstrated that spatial metabolite mapping through MS imaging provided critical insights into gene function by revealing when and where metabolites accumulate within plant tissues. This information is invaluable for researchers aiming to understand the genetic and biochemical pathways that govern natural product biosynthesis in plants. In addition to its applications in basic research, MS imaging has potential implications for agricultural biotechnology and the evolution of new plants with enhanced traits. By elucidating the spatial distribution of metabolites associated with stress resistance or nutritional quality, researchers can determine candidate genes for breeding programs aimed at improving crop resilience and productivity. The ability to visualize metabolites *in situ* allows a more nuanced understanding of plant responses to environmental challenges, which is crucial in the context of global climate change and food safety.

The use of MS imaging in conjunction with other analytical methods such as LC-MS and GC-MS also enhances the comprehensiveness of metabolomic analyses. For example, Yang *et al.* (2019) discussed the advantages of combining different metabolomic platforms to gain a more holistic picture of metabolic changes in response to different treatments or environmental conditions. This multi-platform approach permits the identification of a broader spectrum of metabolites and their interactions, providing a more complete picture of plant metabolism. Advancements in MS imaging technology, including improvements in spatial resolution and sensitivity, have significantly expanded its applications in plant research. As noted by Alexandrov (2020), the integration of artificial intelligence and computational methods with MS-imaging data analysis is paving the way for more sophisticated interpretations of spatial metabolomics. These developments are crucial for managing the large data sets generated by MS imaging, enabling researchers to extract meaningful biological insights from complex metabolomic profiles.

22.2.4 Data acquisition and processing tools

Metabolic profiling studies produce a large amount of data, and statistical and computational methods are used for evaluation of these large data sets. The speed of data acquisition is no longer seen as the limiting factor, which has been replaced by the ability to assess the data. Skills in mathematics and statistics are essential prerequisites for the extraction of information from metabolomic profiling data. Data generated by high-throughput metabolic profiling require innovative tools for computation and statistics (Nielsen and Jewett, 2007) that can be applied consistently. These methods form omics for several molecule types, and have been developed in the fields of chemometrics, chemo-informatics, and bioinformatics. Some sources refer to tools for processing data in metabolic profiling as the process of chemometrics, while others refer to these methods as a combination of chemo-informatics and bioinformatics by making an argument that metabolic profiling is not only for metabolite identification and quantification but also for analyzing data into biological context.

After sample preparation, different analytical processes are used to separate, quantify, and identify the diverse metabolites (Allwood *et al.*, 2008). Determining the analytical approach required relies on the type of approach and also on the physical and chemical aspects of the sample. The metabolomic analysis follows the experimental procedure of sample preparation, metabolite extraction, data accession, data pre-treatment and data interpretation. For better data acquisition and quality, data development is carried out continuously in the field of metabolic profiling. Fast and sensitive MS instruments have been developed by various manufacturers with an effective range, and these improvements have allowed rapid analysis of different components and have also enabled the detection of metabolites found in low abundance in extracts prepared from plants. LC-MS is responsive to various compounds in contrast to GC-MS, which only identifies metabolites of a smaller size that are volatile or can be chemically altered to be volatile. Thus, LC-MS has the advantage of being able to detect as many compounds as possible in one analysis.

After the acquisition of LC-MS data (Kopka *et al.*, 2005), *in silico* processes are used to process the data, which depend on the software package being used, and can include mass spectral deconvolution, detection of the peak and its alignment, baseline shift adjustment, filtering of the peak by *p*-value, and generation of the final data matrices. Developments in data software packages for processing impartial peak alignment of LC-MS data, such as MetaboAnalyst, metaXCMS, MetAlign, and MZmine (Pluskal *et al.*, 2010), provide opportunities to analyze untargeted metabolic profiling across data sets from several sources, and also to collect the details of many compounds in every data set in the absence of any previous identification data (Bartel *et al.*, 2013). Additional software tools have been developed for metabolomics raw data processing, data assessment, data mining, data interpretation, statistical investigation, and mathematical modeling of metabolomic networks.

Numerous software programs are used for *in silico* data investigation of large amounts of spectrum information on metabolites using diverse analytical tools. Web-based programs are used for data processing, data mining, and integration of metabolites. Generally, acquired data are processed for the modification of baseline shifts, background noise reduction, peak detection and alignment, and, ultimately, deconvolution of the mass spectra. A number of tools for bioinformatics have been designed for preprocessing of data, including XCMS, METLIN, AMDIS (Automated Mass Spectral Deconvolution and Identification System), MZmine 2, MetaboAnalyst, MetAlign, MeltDB, and AnalyzerPro. XCMS is an online platform in bioinformatics that facilitates the direct upload of data derived from various sources and assists in the processing of data and statistical examination. For experiments using LC-MS, XCMS has been designed for the transfer of programmed data, which facilitates the data processing time and improves effectiveness (Tautenhahn *et al.*, 2012). The METLIN database is another tool that is employed in many investigations associated with plant metabolomic studies for the response to stress in plants (Smith *et al.*, 2005). It is a useful metabolome source for certain compounds and is not a time-consuming procedure for processing, mining, and annotation of data (Tsugawa *et al.*, 2011). MeltDB is a core platform web-based for the processing and assessment of data and for statistical analysis in plant metabolomic profiling (Kessler *et al.*, 2013), while MetaboAnalyst is an online platform that includes a tool for analysis of enrichment that includes possibilities for topological and visualization processes (Pang *et al.*, 2024).

22.3 High-throughput Applications of Metabolic Profiling in Plant Research

High-throughput metabolic profiling is an emerging tool that is transforming plant research, enabling scientists to delve into the complex biochemical networks that underpin plant growth and responses to the surrounding environmental stimuli. This procedure allows the complete analysis of metabolites, critical for understanding plant physiology and enhancing crop productivity. The integration of advanced analytical methods (MS and NMR spectroscopy) has significantly improved the resolution and accuracy of metabolomic studies, facilitating the detection of biomarkers and metabolic pathways relevant to plant health and resilience (Pagter *et al.*, 2017). One primary application of high-throughput metabolic profiling in research in plants is identifying metabolic signatures associated with stress responses. For instance, studies have demonstrated that metabolomic profiling can reveal distinct metabolic changes in plants subjected to abiotic and biotic stresses, such as salinity, drought, and microbe attacks (Singh *et al.*, 2024). By analyzing the metabolome, researchers can identify specific metabolites that serve as indicators of stress tolerance, thereby aiding in the selection of resilient crop varieties. This is especially significant in the context of climate adaptation, where understanding plant responses to environmental stressors is crucial for sustainable agriculture (Yandeau-Nelson *et al.*, 2016).

Moreover, metabolomics plays a necessary part in elucidating the interaction between plants and microbes. Studies of root exudates, for example, has shown that the metabolites released by plant roots can significantly impact the composition and function of the rhizosphere microbiome (Salem *et al.*, 2022). High-throughput metabolomic analyses allow researchers to characterize these exudates, providing insights into how plants communicate with beneficial microbes and how these interactions can be exploited to enhance plant health and productivity. This understanding is essential for developing strategies that enhance plant–microbe interactions, which are critical for nutrient uptake and disease antagonism (Castro-Moretti *et al.*, 2020). Additional to stress responses and plant–microbe interactions, high-throughput metabolomics is instrumental in crop-improvement efforts. By incorporating metabolomic data with genomic and transcriptomic data, investigators can identify metabolic pathways that contribute to desirable traits such as yield, nutritional quality, and disease antagonism (Razzaq *et al.*, 2019). This integrative approach facilitates the detection of candidate genes and metabolic markers that can be targeted in breeding programs, ultimately leading to the evolution of improved crops that satisfy the needs of a growing global population (Singh *et al.*, 2024).

The application of plant research metabolic profiling is expanding into the realm of plant–pathogen interactions. Metabolomic analyses can reveal metabolic alterations that occur during infection in plants, providing an understanding of the mechanisms of susceptibility and resistance (Castro-Moretti *et al.*, 2020). By analyzing the metabolomes of healthy and infected plants, researchers can determine the key compounds involved in the defense response, paving the way for the development of disease-resistant cultivars. This is particularly relevant in the context of increasing pathogen threats to global food safety, where learning about plant defenses is critical for effective management strategies (Christ *et al.*, 2018). In addition, high-throughput metabolomics is increasingly being used to assess the nutritional quality of crops. As consumer demand for healthy and nutritious food increases, understanding the biosynthesis of health-promoting metabolites in plants has become a priority (Marchev *et al.*, 2021). Metabolomic profiling enables the identification of bioactive constituents, such as flavonoids and phenolics, that contribute to the nutritional value of crops. This information can guide breeding efforts aimed at improving the nutritional range of food crops, thereby improving public health outcomes (Nalbantoglu, 2019).

In the context of precision agriculture, metabolomics offers the potential for real-time monitoring of plant health and stress status. By employing high-throughput metabolomic techniques, farmers can obtain rapid assessments of crop conditions, allowing timely interventions to mitigate stress and optimize yields (Wang *et al.*, 2023). This application of metabolomics aligns with the growing trend toward data-driven agriculture, where technology is leveraged to enhance decision making and resource management (Yan *et al.*, 2022). The advances in metabolomics approaches, including the development of high-accuracy NMR and MS, have significantly enhanced the abilities of researchers to analyze complex plant metabolomes (Augustijn *et al.*, 2021). These technologies provide high sensitivity and specificity and enable the detection of a wide range of compounds, including those present in low concentrations. As a result, metabolomics is becoming an indispensable tool for plant scientists, providing insights that were previously unattainable with traditional analytical methods (Mishra *et al.*, 2022).

22.4 Plant Natural Products: A Metabolomics Perspective

The exploration of plant natural products through a metabolomics perspective has gained significant traction in recent years, forced by advancements in genomic strategies and the increasing recognition of the therapeutic potential of these compounds (Kautsar *et al.*, 2017). Metabolomics is revolutionizing drug research by analyzing the complex chemistry of plant-derived natural products and low-level secondary metabolites, enabling screening, discovery, and quality control in phytomedicine (Salem *et al.*, 2020). The burgeoning field of plant genomics has facilitated the identification and characterization of biosynthetic gene clusters involved in the production of secondary metabolites. As highlighted by Kautsar *et al.* (2017), the rapid sequencing of plant genomes has enabled automated identification and annotation of these clusters, which represent a rich source of novel natural products. This genomic insight has been complemented by the work of Kersten and Weng (2018), who emphasized the need for gene-guided approaches to harness the potential of plant natural products, particularly in the context of the expanding genomic resources available through initiatives such as the 1000 Plant Transcriptomes project. Such advancements not only improve our knowledge of the genetic basis of metabolite biosynthesis but also increase the potential for synthetic biology applications aimed at optimizing the production of valuable phytochemicals (Liu *et al.*, 2023a).

Natural products emanating from plants have historically served as a cornerstone for drug discovery, particularly in the recognition of anticancer and anti-infective compounds. Harvey *et al.* (2015) noted that, despite the extensive use of natural products in the discovery of drugs, there remains a vast reservoir of unexplored chemical diversity within plant secondary metabolites. The *Dictionary of Natural Products* has cataloged around 300,000 unique plant secondary metabolites, underscoring the potential for discovering novel therapeutic agents (Chassagne *et al.*, 2019). Moreover, the unique structural complexity and biological activity of these compounds make them prime candidates for addressing contemporary health challenges, including antibiotic resistance and metabolic disorders (Silva *et al.*, 2016). The incorporation of metabolomics with other omics approaches has emerged as a powerful strategy for elucidating the plant metabolome and responses to environmental triggers. For instance, Wu *et al.* (2020) conducted an integrated study on *Astragalus membranaceus*, revealing a comprehensive metabolic profile that highlights the phytochemical diversity and potential health benefits of this traditional herb. Similarly, Yang *et al.* (2018) discussed how environmental factors can significantly influence the biosynthesis of secondary metabolites, which are crucial for plant defense and human health. This interplay between environmental conditions and metabolite production emphasizes the need for a holistic experience of plant metabolism, particularly in the context of climate modification and agricultural practices.

The application of advanced analytical approaches in metabolomics, such as MS and NMR, has revolutionized the field by enabling the detailed characterization of complex mixtures of phytochemicals (Marti *et al.*, 2019). These methodologies facilitate the recognition and quantification of metabolites, providing an understanding of their functional roles within the plant and their potential applications in medicine and nutrition (Turi *et al.*, 2015). The ability to profile the metabolome of various plant species not only aids in the discovery of new bioactive constituents but also improves our knowledge of the ecological and evolutionary significance of plant secondary metabolites (Shen and Hao, 2020).

In addition to their therapeutic potential, plant natural products play a vital part in sustainable agriculture and food safety. As highlighted by Butnariu and Tietel (2023), understanding the biosynthetic pathways and bioactivities of these constituents is essential for harnessing their potential in agrifood industries. The development of biotechnological approaches such as metabolic engineering and synthetic biology offers promising avenues for enhancing the production of high-value phytochemicals in sustainable ways (Wei *et al.*, 2021). For example, the engineering of terpenoid biosynthesis in *Nicotiana benthamiana* has demonstrated the feasibility of producing complex natural products in heterologous systems, thereby addressing the challenges associated with low natural abundance and difficult extraction processes (Reed and Osbourn, 2018).

The exploration of plant natural products is also critical in the context of global health challenges, particularly in the development of novel antimicrobial agents. Salam and Quave (2018) described the unique characteristics of plant-derived compounds, which often exhibit multifaceted mechanisms of action against bacterial pathogens. The revival of curiosity in natural products for infection control is driven by the urgent need for new therapeutic strategies to combat antibiotic-resistant bacteria, highlighting the crucial role of metabolites from plants as innovative drug leads (Silva *et al.*, 2016).

22.5 Integration of Metabolomics with Other Omics Technologies

The incorporation of metabolomics with other omics such as genomics, transcriptomics, and proteomics, has emerged as a transformative approach in plant natural product research (Fig. 22.2). This multi-omics strategy enables a comprehensive understanding of plant biology in context, particularly in terms of the response to environmental stress and the biosynthesis of secondary metabolites, which are crucial for plant defense and adaptation. By examining the interplay between different biological layers, researchers can elucidate the complex regulatory networks that control plant metabolism and identify key metabolites related to desirable traits in crops.

Metabolomics, as a discipline, concentrates on the comprehensive examination of metabolites within biological systems, providing an understanding of metabolic pathways and their regulation under diverse conditions. Recent advancements in MS and other analytical techniques have significantly enhanced our ability to detect and quantify a wide array of metabolites, thereby facilitating the exploration of metabolic reprogramming in plants subjected to abiotic stresses such as salinity, drought, and nutrient deficiency (Yu *et al.*, 2024). For instance, studies have shown that integrating metabolomic and

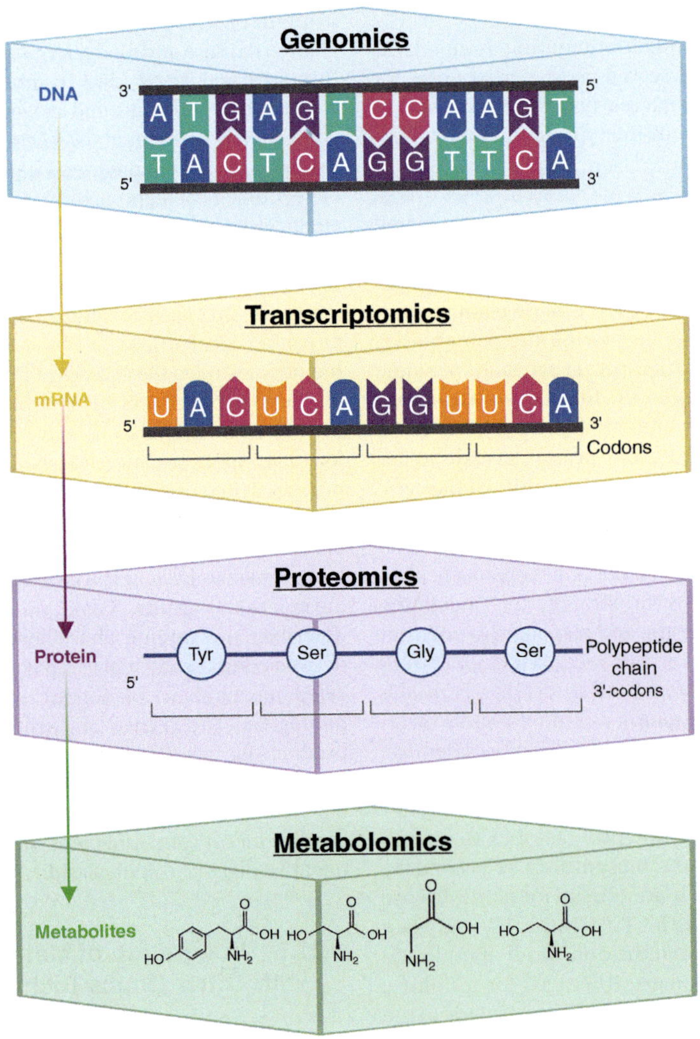

Fig. 22.2. Integrated omics.

transcriptomic data allows researchers to uncover the genetic mechanisms underlying plant responses to environmental cues, thereby revealing hidden information that can be critical for crop advancement (Zhao *et al.*, 2024).

The integration of metabolomics with transcriptomics is particularly valuable in understanding the biosynthesis of secondary metabolites that play essential roles in plant defense and have significant medicinal properties. For example, the integrated study of transcriptomic and metabolomic data has been utilized to investigate flavonoid biosynthesis pathways in various plant species, highlighting the connection between gene expression and metabolite assembly (Mo *et al.*, 2024). Such integrative approaches not only elucidate the metabolic pathways involved but also provide a perception of the regulatory mechanisms that control the expression of genes associated with metabolite biosynthesis (Yuan *et al.*, 2022).

Moreover, the benefits of multi-omics methods have been instrumental in enhancing our understanding of plant responses to biotic pressures, such as pathogen attack and herbivore feeding. By integrating metabolomic and transcriptomic data, researchers can identify specific metabolites that serve as defense compounds and elucidate the genetic basis for their production (Yang *et al.*, 2021). This knowledge is crucial for developing crops with enhanced resistance to pests and diseases, ultimately contributing to endurable agricultural practices. The role of metabolomics in plant natural product research extends beyond stress responses; it also encompasses the exploration of metabolic diversity within and among plant species. For instance, studies have demonstrated that metabolomic profiling can reveal significant variations in secondary metabolite composition among different cultivars or species, which can be leveraged for breeding programs aimed at improving the nutritional and medicinal value of crop varieties (Han *et al.*, 2023). The incorporation of metabolomics with other omics methods provides a holistic view of the genetic and environmental factors that influence metabolite production, thereby facilitating the identification of elite genotypes with desirable traits (Salam *et al.*, 2023).

In addition to enhancing our understanding of plant biology, the integration of metabolite profiling with other omics techniques has practical implications for crop improvement and biotechnology. By utilizing a systems biology approach, researchers can determine key pathways related to metabolic and regulatory networks, which can be targeted for genetic manipulation or breeding strategies (Manickam *et al.*, 2023). This integrative framework not only accelerates the selection of superior breeding stock but also aids in the development of crops with enhanced stability to environmental stresses and improved nutritional profiles (Raza, 2022). Furthermore, the challenges associated with data integration from multiple omics platforms necessitate the evolution of cutting-edge bioinformatics tools and statistical analysis methods. These tools are essential for managing the complexity of multi-omics data and for deriving meaningful biological insights from large data sets (Yang *et al.*, 2022). As the specialization of plant metabolomics continues to evolve, the integration of metabolomics with transcriptomics, proteomics, and other omics methods will undoubtedly play a pivotal role in increasing our knowledge of plant biology and addressing the challenges faced by modern agriculture.

22.6 Challenges and Limitations

High-efficiency metabolic profiling is an important tool in plant natural product research, enabling the comprehensive analysis of metabolites that play important roles in plant biology and ecology. Nevertheless, despite its potential, a number of challenges and constraints hinder the practical application of high-throughput metabolomics in this field. These challenges can be categorized into technical, analytical, and interpretative issues, each of which significantly impacts the reliability and comprehensiveness of metabolomic studies.

The primary challenge in high-throughput metabolic profiling is the accurate identification and quantification of metabolites. The complexity of plant metabolomes, which can contain thousands of metabolites with varying chemical properties, poses significant hurdles for analytical techniques. For instance, for the integration of various methodologies, such as LC-MS and NMR spectroscopy, it is often necessary to produce a comprehensive profile of plant metabolites.

However, each technique has its inherent biases and limitations, which can lead to incomplete or inaccurate metabolite identification (Salem *et al.*, 2020). Moreover, the absence of standard protocols for preparation and analysis of samples further complicates the reproducibility of results across different studies (Sousa Silva *et al.*, 2019).

Another significant limitation arises from the extensive quantity of data developed during high-throughput metabolomic analyses. Interpretation of these data is often fraught with difficulties, primarily due to the presence of non-random missing data and latent factors that can influence metabolite concentrations (McKennan *et al.*, 2020). Current analytical methods may not adequately account for these complexities, leading to potential biases in the results. Furthermore, the integration of metabolomic information with genomic and transcriptomic information remains a challenge, as the relationships between metabolites and their corresponding genes are not always straightforward (Wang *et al.*, 2023). This complexity necessitates the development of tools in advanced computational fields and statistical methods for facilitating data integration and interpretation (McKennan *et al.*, 2020).

The variability in metabolite profiles due to environmental factors also presents a significant challenge in high-throughput metabolomics. Plants respond dynamically to a myriad of biotic and abiotic stresses, which can cause fluctuations in metabolite levels. This variability complicates the establishment of baseline metabolite profiles and can obscure metabolite identification that is specifically connected with particular stress responses or developmental stages. Consequently, researchers must carefully design experiments to control for these variables, which can be resource-intensive and time-consuming (Allwood *et al.*, 2021). The issue of metabolite coverage is also a critical limitation in high-throughput metabolomics. Despite advances in analytical techniques, many metabolites, particularly those present in low concentrations or those that are highly volatile, may remain undetected (Schroeder *et al.*, 2019). This incomplete coverage can lead to a skewed understanding of the metabolomic landscape of a plant species, potentially overlooking key metabolites that could be crucial for understanding plant physiology or for applications in natural product identification (Salem *et al.*, 2020). The development of more sensitive and comprehensive analytical techniques is essential to address this burden and improve the depth of metabolomic profiling in plant investigation (Maia *et al.*, 2023).

In addition to these technical and analytical challenges, the integration of high-throughput metabolomics into plant natural product research is often limited by the availability of reference databases and libraries for metabolite identification. The identification of metabolites typically relies on comparing experimental data against existing databases, which may not include all possible metabolites found in diverse plant species (Schroeder *et al.*, 2019). This gap in reference data can hinder identification of the new compound's potential function of biological origin, thereby limiting the scope of research in plant natural products (Salem *et al.*, 2020). Moreover, the complexity of metabolite interactions within the plant system adds another layer of difficulty. Metabolites do not function in isolation; rather, they interact with one another and with various biological pathways. Understanding these interactions is essential for elucidating the roles of metabolites in plant health, stress responses, and development (Allwood *et al.*, 2021). However, current metabolomic studies often focus on individual metabolites rather than on the broader metabolic networks, which can lead to an incomplete interpretation of plant biochemistry (Razzaq *et al.*, 2019). Future research must prioritize the exploration of these interactions to fully leverage the potential of high-efficiency metabolic profiling in plant science.

22.7 Future Directions

High-efficiency metabolic profiling is emerging as a transformative approach in plant natural product research, enabling the comprehensive analysis of compounds that play important roles in plant biology and ecology, and that have potential applications in agriculture and medicine. The future directions of this field will be shaped by advancements in analytical technologies, data integration methods, and the growing understanding of plant metabolomics in response to environmental and genetic factors. One of the most influential advancements in high-throughput metabolomics has been the integration of various omics methods. This multi-omics approach allows researchers to gain a holistic picture of plant biology by correlating metabolite profiles with genetic and phenotypic data. For instance, integrating metabolomics with genomic data has enabled the identification of biomarkers associated with stress tolerance and crop improvement (Wang *et al.*, 2023). This integration is especially valuable for understanding complex traits in crops, such as yield and nutritional quality, which are influenced by a multitude of metabolites and their interactions with environmental factors (Sousa Silva *et al.*, 2019).

The evolution of user-friendly platforms for metabolomics data investigation is another critical direction for the future of this field. Tools such as PlantMetSuite (Liu *et al.*, 2023b) and MetaboAnalyst 6.0 are designed to enable the identification and visualization of metabolites in plant samples, adhering to rigorous standards for metabolite identification (Yu *et al.*, 2024). These platforms not only streamline the data-processing workflow but also enhance the accessibility of metabolomics data for researchers, enabling them to conduct more comprehensive analyses and draw meaningful conclusions from their findings. Moreover, the application of advanced analytical methods such as inclusive two-dimensional GC × GC and MS is revolutionizing the way metabolites are analyzed in plant samples. These techniques offer superior separation and detection abilities, allowing the recognition of a broader range of metabolites, including those present in low concentrations (Salem *et al.*, 2020). The ability to analyze complex mixtures of metabolites with high resolution is essential for understanding the intricate biochemical pathway responses of plant undergoing environmental stresses and for discovering new compounds from natural sources with possible therapeutic uses.

In addition to technological advancements, the integration of publicly available metabolomics databases and the use of network analysis tools are proving to be invaluable for researchers in the field. The Global Natural Product Social Molecular Networking (GNPS) platform, for example, allows the comparison of metabolite profiles across different plant species and cultivars, facilitating the identification of unique compounds and their potential role in defense mechanisms in plants (Medeiros *et al.*, 2021). This collective strategy not only improves our understanding of plant metabolomics but also fosters innovation in the discovery of new natural products. The exploration of plant–microbe relationships through metabolomics is another promising direction for future research. Understanding how plants interact with their microbial communities can provide an understanding of plant health and resilience, particularly in the context of biotic and abiotic stressors (Xu and Fan, 2024). Metabolomics can elucidate the reprogramming of metabolites that occurs when interacting with the environment, revealing how plants adapt to their environments and how beneficial microbes can enhance the growth of the plant and stress tolerance (Mashabela *et al.*, 2022).

The use of metabolic profiling in the context of climate change and environmental stressors is becoming increasingly important. As plants face challenges such as drought, salinity, and temperature fluctuations, metabolomics can help identify the metabolites implicated in stress responses and adaptation mechanisms (Wu *et al.*, 2023). By understanding these metabolic pathways, researchers can develop strategies to enhance crop resilience and productivity in the face of changing environmental circumstances. The role of metabolic profiling in drug discovery and phytomedicine is also a critical area of focus. The ability to interpret complex mixtures of compounds in plant extracts can direct the identification of bioactive constituents with therapeutic potential. As the demand for natural products in medicine continues to rise, metabolomics will play a key part in the standardization and quality control of herbal medicines, ensuring that they meet safety and efficacy standards (Salem *et al.*, 2020). Moreover, the challenges associated with data analysis and interpretation in metabolomics are being addressed by developing sophisticated computation and algorithm tools. These advancements aim to improve metabolome annotation and facilitate the integration of metabolome data with data sets of other omics studies, enhancing our overall knowledge of plant biology. The incorporation of machine learning and artificial intelligence techniques into metabolomic profiling research is also on the horizon, promising to revolutionize data analysis and interpretation (Pang *et al.*, 2024).

22.8 Conclusion

High-throughput metabolomics has revolutionized plant natural product research, offering unprecedented insights into complex biochemical networks underlying plant growth, development, and environmental responses. Advancements in analytical technologies, coupled with sophisticated data-processing tools, have significantly enhanced our ability to determine and quantify a vast array of metabolites. The incorporation of metabolomics with other omics methods provides a holistic view of plant biology, crucial for developing resilient crops, discovering novel bioactive compounds, and enhancing sustainable agriculture. Despite challenges in metabolite identification and data analysis, the field continues to evolve rapidly. Future directions include developing more sensitive analytical techniques, enhancing multi-omics data integration, and applying artificial intelligence for data interpretation. As high-throughput metabolomics advances, it will play an increasingly pivotal part in managing global challenges associated with food safety, environmental sustainability, and human health through the investigation and utilization of plant natural products.

22.9 Acknowledgment

All the authors are grateful to their respective universities and institutions for their valuable support.

22.10 References

Akilandeswari, G., Vijaya Anand, A., Sampathkumar, P., Vinayaga Moorthi, P., and Preethi, B. (2019) A prospective review on phyto-pharmacological aspects of *Andrographis paniculata*. *Systematic Reviews in Pharmacy* 10, 15–19.

Alexandrov, T. (2020) Spatial metabolomics and imaging mass spectrometry in the age of artificial intelligence. *Annual Review of Biomedical Data Science* 3, 61–87.

Allwood, J.W., Ellis, D.I., and Goodacre, R. (2008) Metabolomic technologies and their application to the study of plants and plant–host interactions. *Physiologia Plantarum* 132, 117–135.

Allwood, J.W., Williams, A., Uthe, H., van Dam, N.M., Mur, L.A., *et al.* (2021) Unravelling plant responses to stress—the importance of targeted and untargeted metabolomics. *Metabolites* 11: 558.

Augustijn, D., de Groot, H.J., and Alia, A. (2021) HR-MAS NMR applications in plant metabolomics. *Molecules* 26: 931.

BaniMustafa, A. and Hardy, N. (2020) Applications of a novel knowledge discovery and data mining process model for metabolomics. *IEEE Explore* 8, 209964–210005.

Bartel, J., Krumsiek, J., and Theis, F.J. (2013) Statistical methods for the analysis of high-throughput metabolomics data. *Computational and Structural Biotechnology Journal* 4: e201301009.

Begum, T.N., Ilyas, M.M., Kalavathy, S., Anand, A.V., Kumar, S., and Senthil, R. (2009) Effects of ethanolic leaf extract of *Azim tetracantha* Lam. on Ehrlich ascites carcinoma tumour bearing mice. *Research Journal of Medicine and Medical Sciences* 4, 351–354.

Bharathi, V., Rengarajan, R.L., Radhakrishnan, R., Hashem, A., Abd Allah, E.F., *et al.* (2018) Effects of a medicinal plant *Macrotyloma uniflorum* (Lam.) Verdc. formulation (MUF) on obesity-associated oxidative stress-induced liver injury. *Saudi Journal of Biological Sciences* 25, 1115–1121.

Brown, S.C., Kruppa, G., and Dasseux, J.L. (2005) Metabolomics applications of FT-ICR mass spectrometry. *Mass Spectrometry Reviews* 24, 223–231.

Butnariu, M. and Tietel, Z. (2023) Biosynthesis, purification, characterization and uses of natural compounds in plants. *Frontiers in Plant Science* 14: 1162676.

Castro-Moretti, F.R., Gentzel, I.N., Mackey, D., and Alonso, A.P. (2020) Metabolomics is an emerging tool for the study of plant–pathogen interactions. *Metabolites* 10: 52.

Chassagne, F., Cabanac, G., Hubert, G., David, B., and Marti, G. (2019) The landscape of natural product diversity and their pharmacological relevance from a focus on the *Dictionary of Natural Products*®. *Phytochemistry Reviews* 18, 601–622.

Christ, B., Pluskal, T., Aubry, S., and Weng, J.K. (2018) Contribution of untargeted metabolomics for future assessment of biotech crops. *Trends in Plant Science* 23, 1047–1056.

de Raad, M., Fischer, C.R., and Northen, T.R. (2016) High-throughput platforms for metabolomics. *Current Opinion in Chemical Biology* 30, 7–13.

Dettmer, K., Aronov, P.A., and Hammock, B.D. (2007) Mass spectrometry-based metabolomics. *Mass Spectrometry Reviews* 26, 51–78.

Dey, A., Charrier, B., Ribay, V., Dumez, J.N., and Giraudeau, P. (2023) Hyperpolarized ^{1}H and ^{13}C NMR spectroscopy in a single experiment for metabolomics. *Analytical Chemistry* 95, 16861–16867.

Divya, N., Rengarajan, R.L., Radhakrishnan, R., Abd Allah, E.F., Alqarawi, A.A., *et al.* (2019) Phytotherapeutic efficacy of the medicinal plant *Terminalia catappa* L. *Saudi Journal of Biological Sciences* 26, 985–988.

Dong, Y., Sonawane, P., Cohen, H., Polturak, G., Feldberg, L., *et al.* (2020) High mass resolution and spatial metabolite mapping enhance the current plant gene and pathway discovery toolbox. *New Phytologist* 228, 1986–2002.

Emwas, A.H., Roy, R., McKay, R.T., Tenori, L., Saccenti, E., *et al.* (2019). NMR spectroscopy for metabolomics research. *Metabolites* 9: 123.

Fiehn, O. (2008) Extending the breadth of metabolite profiling by gas chromatography coupled to mass spectrometry. *Trends in Analytical Chemistry* 27, 261–269.

Gehrke, S., Reisz, J.A., Nemkov, T., Hansen, K.C., and d'Alessandro, A. (2017) Characterization of rapid extraction protocols for high-throughput metabolomics. *Rapid Communications in Mass Spectrometry* 31, 1445–1452.

Giacomoni, F., Le Corguille, G., Monsoor, M., Landi, M., Pericard, P., *et al.* (2015) Workflow4Metabolomics: a collaborative research infrastructure for computational metabolomics. *Bioinformatics* 31, 1493–1495.

Han, W., Ward, J.L., Kong, Y., and Li, X. (2023) Targeted and untargeted metabolomics for the evaluation of plant metabolites in response to the environment. *Frontiers in Plant Science* 14: 1167513.

Harvey, A.L., Edrada-Ebel, R., and Quinn, R.J. (2015) The re-emergence of natural products for drug discovery in the genomics era. *Nature Reviews Drug Discovery* 14, 111–129.

Her, C., Alonzo, A.P., Vang, J.Y., Torres, E., and Krishnan, V.V. (2015) Real-time enzyme kinetics by quantitative NMR spectroscopy and determination of the Michaelis–Menten constant using the Lambert-W function. *Journal of Chemical Education* 92, 1943–1948.

Huang, Z., Bi, T., Jiang, H., and Liu, H. (2024) Review on NMR as a tool to analyse natural products extract directly: molecular structure elucidation and biological activity analysis. *Phytochemical Analysis* 35, 5–16.

Karthika, P., Balamuralikrishnan, B., Yamini, K., Vijaya Anand, A., Durairaj, K., *et al.* (2023) Green synthesis, characterization of silver nanoparticles using aqueous leaf extracts of *Solanum melongena* and *in vitro* evaluation of antibacterial, pesticidal and anticancer activity in human MDA-MB-231 breast cancer cell lines. *Journal of King Saud University—Science* 35: 102663.

Kautsar, S.A., Suarez Duran, H.G., Blin, K., Osbourn, A., and Medema, M.H. (2017) plantiSMASH: automated identification, annotation and expression analysis of plant biosynthetic gene clusters. *Nucleic Acids Research* 45, W55–W63.

Kaviya, M., Balasubramanian, B., Bharathi, K., Malaisamy, A., Al-Dhabi, N.A., *et al.* (2022) Evaluation of nutritional substances and investigation of antioxidant and antimicrobial potentials of *Boerhavia diffusa* with *in silico* molecular docking. *Molecules* 27: 1280.

Kersten, R.D. and Weng, J.K. (2018) Gene-guided discovery and engineering of branched cyclic peptides in plants. *Proceedings of the National Academy of Sciences USA* 115, E10961–E10969.

Kessler, N., Neuweger, H., Bonte, A., Langenkämper, G., Niehaus, K., *et al.* (2013) MeltDB 2.0—advances of the metabolomics software system. *Bioinformatics* 29, 2452–2459.

Kim, M., Son, J., Song, Y., and Kim, Y. (2020) Analysis of plant metabolites damaged in chemical accidents using NMR spectroscopy. *Bulletin of the Korean Chemical Society* 41, 1099–1106.

Kopka, J., Schauer, N., Krueger, S., Birkemeyer, C., Usadel, B., *et al.* (2005) GMD@CSB.DB: the Golm metabolome database. *Bioinformatics* 21, 1635–1638.

Li, Y., Ruan, Q., Li, Y., Ye, G., Lu, X., *et al.* (2012) A novel approach to transforming a non-targeted metabolic profiling method to a pseudo-targeted method using the retention time locking gas chromatography/ mass spectrometry-selected ions monitoring. *Journal of Chromatography A* 1255, 228–236.

Li, S., Park, Y., Duraisingham, S., Strobel, F.H., Khan, N., *et al.* (2013) Predicting network activity from high throughput metabolomics. *PLOS Computational Biology* 9: e1003123.

Li, F., Xu, Y., Shang, D., Yang, H., Liu, W., *et al.* (2014) MPINet: metabolite pathway identification via coupling of global metabolite network structure and metabolomic profile. *BioMed Research International* 2014: 325697.

Li, Y., Zhu, W., Xiang, Q., Kim, J., Dufresne, C., *et al.* (2023) Creation of a plant metabolite spectral library for untargeted and targeted metabolomics. *International Journal of Molecular Sciences* 24: 2249.

Liu, X., Zhang, P., Zhao, Q., and Huang, A.C. (2023a) Making small molecules in plants: a chassis for synthetic biology-based production of plant natural products. *Journal of Integrative Plant Biology* 65, 417–443.

Liu, Y., Liu, H.Z., Chen, D.K., Zeng, H.Y., Chen, Y.L., and Yao, N. (2023b) PlantMetSuite: a user-friendly web-based tool for metabolomics analysis and visualisation. *Plants* 12: 2880.

Maia, M., Figueiredo, A., Cordeiro, C., and Sousa Silva, M. (2023) FT-ICR-MS-based metabolomics: a deep dive into plant metabolism. *Mass Spectrometry Reviews* 42, 1535–1556.

Manickam, S., Rajagopalan, V.R., Kambale, R., Rajasekaran, R., Kanagarajan, S., and Muthurajan, R. (2023) Plant metabolomics: current initiatives and prospects. *Current Issues in Molecular Biology* 45, 8894–8906.

Manikandan, R., Balasubramanian, B., Punniyakotti, P., Vijaya Anand, A., Meyyazhagan, A., *et al.* (2022) Cardio-protective effects of *Terminalia catappa* leaves and *Terminalia chebula* fruit extract in doxorubicin-induced cardiomyopathy in rats. *Biomarkers* 27, 488–495.

Manoharan, N., Sampathkumar, P., Dheeba, B., Sheik Abdulla, S., Vinothprasanna, G., *et al.* (2008) Potential hepatoprotective effect of aqueous extract of *Gracilaria corticata* in AFB_1 induced hepatotoxicity in Wistar rats. *Journal of Biological Sciences* 8, 1352–1355.

Marchev, A.S., Vasileva, L.V., Amirova, K.M., Savova, M.S., Balcheva-Sivenova, Z.P., and Georgiev, M.I. (2021) Metabolomics and health: from nutritional crops and plant-based pharmaceuticals to profiling of human biofluids. *Cellular and Molecular Life Sciences* 78, 6487–6503.

Markley, J.L., Brüschweiler, R., Edison, A.S., Eghbalnia, H.R., Powers, R., *et al.* (2017) The future of NMR-based metabolomics. *Current Opinion in Biotechnology* 43, 34–40.

Marti, G., Joulia, P., Amiel, A., Fabre, B., David, B., *et al.* (2019) Comparison of the phytochemical composition of *Serenoa repens* extracts by a multiplexed metabolomic approach. *Molecules* 24: 2208.

Mashabela, M.D., Piater, L.A., Dubery, I.A., Tugizimana, F., and Mhlongo, M.I. (2022) Rhizosphere tripartite interactions and PGPR-mediated metabolic reprogramming towards ISR and plant priming: a metabolomics review. *Biology* 11: 346.

Mazzei, P., Vinale, F., Woo, S.L., Pascale, A., Lorito, M., and Piccolo, A. (2016) Metabolomics by proton high-resolution magic-angle-spinning nuclear magnetic resonance of tomato plants treated with two secondary metabolites isolated from *Trichoderma*. *Journal of Agricultural and Food Chemistry* 64, 3538–3545.

McKennan, C., Ober, C., and Nicolae, D. (2020) Estimation and inference in metabolomics with non-random missing data and latent factors. *Annals of Applied Statistics* 14: 789.

Medeiros, D.B., Brotman, Y., and Fernie, A.R. (2021) The utility of metabolomics as a tool to inform maize biology. *Plant Communications* 2: 100187.

Mishra, S., Priyanka, and Sharma, S. (2022) Metabolomic insights into endophyte-derived bioactive compounds. *Frontiers in Microbiology* 13: 835931.

Mo, X., Wang, L., Yu, C., and Kou, C. (2024) Combined metabolomics and transcriptomics analysis of the distribution of flavonoids in the fibrous root and taproot of *Polygonatum kingianum* Coll. et Hemsl. *Genes* 15: 828.

Mori, T., Tsuboi, Y., Ishida, N., Nishikubo, N., Demura, T., and Kikuchi, J. (2015) Multidimensional high-resolution magic angle spinning and solution-state NMR characterization of ^{13}C-labeled plant metabolites and lignocellulose. *Scientific Reports* 5: 11848.

Muthumani, G.D., Anand, A.V., and Manikandan, R. (2013) Antioxidant, antihelmintic and antimicrobial activity of *Euphorbia thymifolia* Linn whole plant. *International Journal of Current Microbiology and Applied Sciences* 2, 66–75.

Nakabayashi, R., Hashimoto, K., Mori, T., Toyooka, K., Sudo, H., and Saito, K. (2021) Spatial metabolomics using imaging mass spectrometry to identify the localization of asparagine A in *Asparagus officinalis*. *Plant Biotechnology* 38, 311–315.

Nalbantoglu, S. (2019) Metabolomics: basic principles and strategies. In: Nalbantoglu, S. and Amri, H. (ed.) *Molecular Medicine*. IntechOpen, London. Available at: www.intechopen.com/chapters/68486 (accessed 10 February 2025).

Nielsen, J. and Jewett, M.C. (eds.) (2007) *Metabolomics: A Powerful Tool in Systems Biology*. Springer, Berlin/Heidelberg, Gernmany.

Pagter, M., Yde, C.C., and Kjær, K.H. (2017) Metabolic fingerprinting of dormant and active flower primordia of *Ribes nigrum* using high-resolution magic angle spinning NMR. *Journal of Agricultural and Food Chemistry* 65, 10123–10130.

Pang, Z., Lu, Y., Zhou, G., Hui, F., Xu, L., *et al.* (2024) MetaboAnalyst 6.0: towards a unified platform for metabolomics data processing, analysis and interpretation. *Nucleic Acids Research* 52: W398–W406.

Pluskal, T., Castillo, S., Villar-Briones, A., and Orešič, M. (2010) MZmine 2: a modular framework for processing, visualizing, and analyzing mass spectrometry-based molecular profile data. *BMC Bioinformatics* 11: 395.

Raza, A. (2022) Metabolomics: a systems biology approach for enhancing heat stress tolerance in plants. *Plant Cell Reports* 41, 741–763.

Razzaq, A., Sadia, B., Raza, A., Khalid Hameed, M., and Saleem, F. (2019) Metabolomics: a way forward for crop improvement. *Metabolites* 9: 303.

Reed, J. and Osbourn, A. (2018) Engineering terpenoid production through transient expression in Nicotiana benthamiana. *Plant Cell Reports* 37, 1431–1441.

Salam, A.M. and Quave, C.L. (2018) Opportunities for plant natural products in infection control. *Current Opinion in Microbiology* 45, 189–194.

Salam, U., Ullah, S., Tang, Z.H., Elateeq, A.A., Khan, Y., *et al.* (2023) Plant metabolomics: an overview of the role of primary and secondary metabolites against different environmental stress factors. *Life* 13: 706.

Salem, M.A., Perez de Souza, L., Serag, A., Fernie, A.R., Farag, M.A., *et al.* (2020) Metabolomics in the context of plant natural products research: from sample preparation to metabolite analysis. *Metabolites* 10: 37.

Salem, M.A., Wang, J.Y., and Al-Babili, S. (2022) Metabolomics of plant root exudates: from sample preparation to data analysis. *Frontiers in Plant Science* 13: 1062982.

Sangeetha, T., Balamuralikrishnan, B., Sampathkumar, P., Velayuthaprabhu, S., Senthilkumar, N., *et al.* (2022) Characterization and phytoconstituents of *Petroselinum crispum* (Mill) and *Coriandrum sativum* (Linn) and their impacts on inflammation—an *in vitro* analysis against human adenocarcinoma cells with molecular docking. *South African Journal of Botany* 146, 776–788.

Schroeder, M., Meyer, S.W., Heyman, H.M., Barsch, A., and Sumner, L.W. (2019) Generation of a collision cross-section library for multi-dimensional plant metabolomics using UHPLC-trapped ion mobility-MS/MS. *Metabolites* 10: 13.

Sharma, V., Gupta, P., Kagolla, P., Hangargi, B., Veershetty, A., *et al.* (2021) Metabolomics intervention towards a better understanding of plant traits. *Cells* 10: 346.

Shen, Y. and Hao, X. (2020) Natural product sciences: an integrative approach to the innovations of plant natural products. *Science China Life Sciences* 63, 1634–1650.

Silva, L.N., Zimmer, K.R., Macedo, A.J., and Trentin, D.S. (2016) Plant natural products targeting bacterial virulence factors. *Chemical Reviews* 116, 9162–9236.

Singh, S., Karadagi, A.V., Doggalli, G., and Singh, R. (2024) Metabolomics and metabolic engineering for crop improvement: current trends and prospects. *Plant Cell Biotechnology and Molecular Biology* 25, 77–99.

Smith, C.A., O'Maille, G., Want, E.J., Qin, C., Trauger, S.A., *et al.* (2005) METLIN: a metabolite mass spectral database. *Therapeutic Drug Monitoring* 27, 747–751.

Sousa Silva, M., Cordeiro, C., Roessner, U., and Figueiredo, A. (2019) Metabolomics in crop research—current and emerging methodologies. *Frontiers in Plant Science* 10: 1013.

Stark, P., Zab, C., Porzel, A., Franke, K., Rizzo, P., and Wessjohann, L.A. (2020) PSYCHE—a valuable experiment in plant NMR-metabolomics. *Molecules* 25: 5125.

Sun, X., Jia, Z., Zhang, Y., Zhao, X., Zhao, C., *et al.* (2023) A strategy for uncovering the serum metabolome by direct-infusion high-resolution mass spectrometry. *Metabolites* 13: 460.

Tautenhahn, R., Patti, G.J., Rinehart, D., and Siuzdak, G. (2012) XCMS Online: a web-based platform to process untargeted metabolomic data. *Analytical Chemistry* 84, 5035–5039.

Tsugawa, H., Tsujimoto, Y., Arita, M., Bamba, T., and Fukusaki, E. (2011) GC/MS based metabolomics: development of a data mining system for metabolite identification by using soft independent modelling of class analogy (SIMCA). *BMC Bioinformatics* 12: 131.

Turi, C.E., Finley, J., Shipley, P.R., Murch, S.J., and Brown, P.N. (2015) Metabolomics for phytochemical discovery: development of statistical approaches using a cranberry model system. *Journal of Natural Products* 78, 953–966.

Varalakshmi, B., Anand, A.V., Karpagam, T., Bai, J.S., and Manikandan, R. (2014) *In vitro* antimicrobial and anticancer activity of *Cinnamomum zeylanicum* Linn bark extracts. *International Journal of Pharmacy and Pharmaceutical Sciences* 6, 12–18.

Vijaya Anand, A., Varalakshmi, B., Prasana Sampathkumar, P., Pushpa, N., and Hedina, A. (2016) *Cinnamomum zeylanicum* Linn. The spice with multiple potentials. *Systematic Reviews in Pharmacy* 7, 24–29.

Vijaya Anand, A., Velayuthaprabhu, S., Rengarajan, R.L., Sampathkumar, P., and Radhakrishnan, R. (2020) Bioactive compounds of guava (*Psidium guajava* L.). In: Murthy, H.N. and Bapat, V.A. (eds) *Bioactive Compounds in Underutilized Fruits and Nuts*. Springer, Cham, Switzerland, pp. 503–527.

Vijayakumar, K., Prasanna, B., Rengarajan, R.L., Rathinam, A., Velayuthaprabhu, S., and Vijaya, A.A. (2023) Anti-diabetic and hypolipidemic effects of *Cinnamon cassia* bark extracts: an *in vitro*, *in vivo*, and *in silico* approach. *Archives of Physiology and Biochemistry* 129, 338–348.

Wadie, B., Stuart, L., Rath, C.M., Drotleff, B., Mamedov, S., and Alexandrov, T. (2023) METASPACE-ML: metabolite annotation for imaging mass spectrometry using machine learning. *Nature Communications* 15: 9110.

Wang, Q., Deng, K., Ai, J., Wang, Y., Wang, Y., *et al.* (2023) Integrated transcriptomic and metabolomic analyses reveal the effects of grafting on special metabolites of *Acanthopanax senticosus* leaves. *Molecules* 28: 4877.

Wei, Y., Ji, B., Ledesma-Amaro, R., Chen, T., and Ji, X.J. (2021) Engineering yeast to produce plant natural products. *Frontiers in Bioengineering and Biotechnology* 9: 798097.

Wolfender, J.L., Marti, G., Thomas, A., and Bertrand, S. (2015) Current approaches and challenges for the metabolite profiling of complex natural extracts. *Journal of Chromatography A* 1382, 136–164.

Wu, X., Li, X., Wang, W., Shan, Y., Wang, C., *et al.* (2020) Integrated metabolomics and transcriptomics study of traditional herb *Astragalus membranaceus* Bge. var. *mongolicus* (Bge.) Hsiao reveals a global metabolic profile and novel phytochemical ingredients. *BMC Genomics* 21(Suppl. 10): 697.

Wu, C., Wang, Y., and Sun, H. (2023) Targeted and untargeted metabolomics reveals deep analysis of drought stress responses in needles and roots of *Pinus taeda* seedlings. *Frontiers in Plant Science* 13: 1031466.

Xu, Y. and Fan, K.T. (2024) Advancements in metabolomics: illuminating plant-microbe interactions for enhanced ecosystems and agricultural productivity. *Preprints.org*. Available at: www.preprints.org/manuscript/202406.0670/v1 (accessed 10 February 2025).

Yan, S., Bhawal, R., Yin, Z., Thannhauser, T.W., and Zhang, S. (2022) Recent advances in proteomics and metabolomics in plants. *Molecular Horticulture* 2: 17.

Yandeau-Nelson, M.D., Lauter, N., and Zabotina, O.A. (2016) Advances in metabolomic applications in plant genetics and breeding. *CABI Reviews* 2015. Available at: www.cabidigitallibrary.org/doi/10.1079/PAVSNNR201510040 (accessed 10 February 2025).

Yang, L., Wen, K.S., Ruan, X., Zhao, Y.X., Wei, F., and Wang, Q. (2018) Response of plant secondary metabolites to environmental factors. *Molecules* 23: 762.

Yang, Q., Zhang, A.H., Miao, J.H., Sun, H., Han, Y., *et al.* (2019) Metabolomics biotechnology, applications, and future trends: a systematic review. *RSC Advances* 9, 37245–37257.

Yang, X., Liu, C., Li, M., Li, Y., Yan, Z., *et al.* (2023) Integrated transcriptomics and metabolomics analysis reveals key regulatory network that response to cold stress in common bean (*Phaseolus vulgaris* L.). *BMC Plant Biology* 23: 85.

Yang, Y., Saand, M.A., Huang, L., Abdelaal, W.B., Zhang, J., *et al.* (2021) Applications of multi-omics technologies for crop improvement. *Frontiers in Plant Science* 12: 563953.

Yu, T., Ma, X., Zhang, J., Cao, J., Cao, S., *et al.* (2024) Progress in transcriptomics and metabolomics in plant responses to abiotic stresses. *Preprints.org*. Available at: www.preprints.org/manuscript/202406.1340/v1 (accessed 10 February 2025).

Yuan, Y., Zuo, J., Zhang, H., Li, R., Yu, M., and Liu, S. (2022) Integration of transcriptome and metabolome provides new insights into flavonoid biosynthesis in *Dendrobium huoshanense*. *Frontiers in Plant Science* 13: 850090.

Zeng, W., Hazebroek, J., Beatty, M., Hayes, K., Ponte, C., *et al.* (2014) Analytical method evaluation and discovery of variation within maize varieties in the context of food safety: transcript profiling and metabolomics. *Journal of Agricultural and Food Chemistry* 62, 2997–3009.

Zhang, D., Li, X., Zheng, W., Gui, L., Yang, Y., *et al.* (2022) Investigating the biological effect of multidimensional Ti_3C_2 (MXene)-based nanomaterials through a metabolomics approach: a multidimensional-determined alteration in energy metabolism. *Chemistry of Materials* 34, 3839–3852.

Zhao, X., Ge, W., and Miao, Z. (2024) Integrative metabolomic and transcriptomic analyses reveal the accumulation patterns of key metabolites associated with flavonoids and terpenoids of *Gynostemma pentaphyllum* (Thunb.) Makino. *Scientific Reports* 14: 8644.

Index

A
abiotic stress 80–81, 266, 392, 394
breeding 218, 224–226
challenges 262–265
conventional/genetically modified crop varieties 181–182, 185–187
crop improvement programs 221
drought 201, 224–225
environment 158, 224–226
factors 389
fundamentals and available platforms 252–256
plant biotechnology 179–180
plant–environment interactions 407–411
plant species 298–299
post-analytical steps 255
protection against 3–4
secondary metabolites 316
single-cell metabolomics 145–146
tolerance 294, 297
types of 297–299
abscisic acid (ABA) 208–209, 211, 224, 389
drought stress 294
heavy-metal stress 301
in xylem 300
accelerated solvent extraction 74
accuracy 49–50
acidification 74
agriculture 5
applications 38
bioactive compounds 6
disease prevention and health promotion 6
food processing 5
human health 5
plant–herbivore interactions 106
postharvest handling 11
quality and safety 5
alkaloids 158, 295
Lycoris radiata 389
in medicinal plants 322, 324
non-bitter steroidal 261
secondary metabolites 390
allelopathy 411
Allium sativum 304
Amaranthus cruentus 338
ambient ionization techniques 324
AmiGO 168
amino acids 25, 133, 205, 277, 294, 303
in *Arabidopsis thaliana* 411
drought stress 224, 245
durum wheat domestication 262
non-protein 158
nutritional quality 344
proline 264
salt stress 407
Solanum tuberosum 157
tyrosine 294
in YHY15 plants 104
analytical methods 428
crop protection 19–21
environmental impact 119–121
phenotyping/phenotypes 239–240
plant ecology 417
plant metabolomics 19–21
types of 319
analytical tools 9, 31, 39, 41–42, 121–123, 203, 355
annotation 42
feature extraction and 83
predictive modeling (PM) 42–46
tools and databases 103

antioxidant 263
 enzymatic and non-enzymatic 294
 glutathione 301
 in plant defense mechanisms 390
 scavenge reactive oxygen species 162
 Sonchus oleraceus L. 379
antiSMASH 53
Apolygus lucorum 106
apple scar skin viroid (ASSVd) infection 25
applications of metabolomics 179
 in medicinal plant research 364–365
 plant biotechnology 179–184
 plant secondary metabolites 395–396
Arabidopsis thaliana 3, 103, 158, 189, 218, 223, 281, 322, 343, 394, 411
arsenic biotransformation studies 415
Artemisia annua 363
Artemisia argyi 389
artificial intelligence (AI) 31–32, 39, 46, 269, 366, 400
artificial neural networks (ANNs) 48–49
Asparagus officinalis 431
Astragalus membranaceus 435
atmospheric conditions 410
atmospheric pressure chemical ionization (APCI) 20
atmospheric pressure ionization (API) 203
Atractylodes macrocephala 338
automated quantification algorithm (AQuA) 338
automation 317
 phenotyping/phenotypes 240
auxin 208, 294, 323
Avena sativa see oat *(Avena sativa)*

B
Bacillus subtilis 393
bacteria 26–29, 33–34
Beauveria bassiana 106
benzoxazinoids (BXs) 223
 maize 245
 in wheat 391
best linear unbiased prediction (BLUP) 54
bioactive compounds 350
 agriculture 6
 production 190
biochemical analyses 97
biocidal secondary metabolites 133
 see also secondary metabolites
biodiversity 57, 116
 and ecosystem function 415–417
 plant ecology and 405–419
bioinformatics 4, 38–39, 42, 83, 153–155, 256
 capillary electrophoresis-mass spectrometry (CE-MS) 159
 challenges 83
 data acquisition systems 68
 data analysis 163
 developments 53
 gas chromatography-mass spectrometry (GC-MS) 158
 genomics 160
 ion mobility spectrometry (IMS) 159
 ionomics 162–163
 liquid chromatography–mass spectrometry (LC-MS) 159
 matrix-assisted laser desorption/ionization (MALDI) 160
 metabolomics 162, 164–166
 multi-omics data analysis technologies 160–163, 167–168
 nuclear magnetic resonance (NMR) spectroscopy 159
 phenomics 162
 plant–herbivore interactions 102
 pre-processing of data 433
 proteomics 160–162
 statistical analysis 18
 under stress conditions 157–158
 and systems biology 82, 406
 transcriptomics 160
 workflow for sample preparation 155–156
biomass/tissue weight 78
biostimulants 294, 397, 399
biosynthetic pathways 101, 265, 382
 camalexin 303
 Catharanthus roseus 139
 secondary metabolites 391
 toxic metabolites 261
biotic stress 80–81, 266
 breeding 218, 223–224, 265
 challenges 262–265
 conventional/genetically modified crop varieties 181–182, 185–187
 crop improvement programs 221
 crop protection under 23
 defense response against 3
 drought 201
 environment 158, 223–224
 fundamentals and available platforms 252–256
 plant biotechnology 179–180
 post-analytical steps 255
 secondary metabolites 316
 single-cell metabolomics 145–146
 tolerance 295
bitter sugarcane (BS) 4
Botrytis cinerea 27
Bradyrhizobium japonicum 147
Brassica napus L. 301
Brassica rapa 226
brassinosteroids (BRs) 208
breeding process 177, 434
 abiotic stress 218, 224–226
 analytical technique 327
 biotic stress 218, 223–224

capillary electrophoresis-mass spectrometry (CE-MS) 219
of crop varieties 8–9
functional analysis 259–262
functional genomics 344
gas chromatography-mass spectrometry (GC-MS) 219
high-resolution magic angle spinning 220
liquid chromatography–mass spectrometry (LC-MS) 219
mass spectrometry (MS) 219
microwave-assisted extraction 220
molecular and genomic approaches 293
nuclear magnetic resonance (NMR) 218–219
phenotyping/phenotypes 244
plant–herbivore interactions 107–108
primary and secondary metabolites 216–218
supercritical fluid extraction 220
for trait selection 221–223
ultra-performance liquid chromatography tandem mass spectrometry (UPLC-MS/MS) analysis 219

brewing 9–10

C

Camelina sativa 281
CAMERA package 256
Campylobacter jejuni 373
Cannabis sativa L. 254
capillary electrophoresis–mass spectrometry (CE-MS) 73–74, 323
bioinformatics 159
biological samples 296
breeding process 219
drought stress 203
medicinal plants 356–357
phenotyping/phenotypes 240
plant ecology 416
plant metabolomics 338
single-cell 137, 141
carbon (C) assimilation 410
carbon dioxide 410
carbon–nitrogen balance hypothesis 390
carbon–nutrient balance hypothesis 389
Carissa opaca 365
Catharanthus roseus 139, 378
cellular heterogeneity 145
charge-coupled device (CCD) 143
chelating agents 300–301
chemical derivatization process 20
chemical ionization (CI) 139, 158
chemical property prediction 77
chemical shifts 100
cheminformatics tools 256
chemodiversity 56
chemometric approaches 29, 103–104, 338, 377, 432
chemoprofiling 365
chemotaxonomy 43, 374, 392
chromatography 73–74
chromatography-free approaches 42
phenotyping/phenotypes 240
see also specific chromatography
citrus fruits 8, 228
classification algorithms 47
climate change 114–115
phenotyping/phenotypes 245–247
plant stress under 114–115
current scenarios and future projections 115–118
environmental stresses 119–121
natural and anthropogenic activities 117
resilience 123–124
significance 118–119
technological advances 122–123
clustered regularly interspaced short palindromic repeats (CRISPR)/Cas gene-editing 53, 276–278, 288
acetolactate synthase *(ALS)* gene 278
case studies 340
cells and plantlet regeneration 279
characterization of 280–281
design of 279
functional studies 339
genetic segregation 288
guide RNA (gRNA) 277–278
metabolic profiling in 281–282
electrospray ionization (ESI) mass spectrometry 283
gas chromatography (GC) 283–284
high-performance liquid chromatography (HPLC) 281–283
liquid chromatography–mass spectrometry (LC-MS) 283
ultra-high-performance liquid chromatography (UHPLC) 283
non-chromatographic identification techniques 284–286
non-homologous end-joining and homology-directed repair 277
phytoene desaturase *(PDS)* 278
plant selection 278
polymerase chain reaction (PCR) 279–280, 288
principles 278–279
Sanger sequencing method 279–280, 288
sequencing and metabolite characterization 286–288
single-nucleotide polymorphisms (SNPs) 278
transcription factors 268
transcriptome profiling on 289
using sequencing technology 279–280
viral infections 277
whole-genome sequencing (WGS) 280

collaborative platforms 83
collision cross-section (CCS) database 204
computational tools 76, 139, 279, 325, 395, 397, 417
conclusive structural elucidation techniques 20
constraint-based metabolic models 341, 343
conventional "one-size-fits-all" approach 380
correlation analysis 227
correlation networks 80, 168
cotton bollworm *(Helicoverpa armigera)* 340
CottonGen 167–168
C-repeat/dehydration-responsive element-binding factor 3 (CBF3) 3
CRISPR/Cas gene-editing *see* clustered regularly interspaced short palindromic repeats (CRISPR)/Cas gene-editing
crop improvement 54–55, 177, 216, 221–229
  functional genomics 344
  metabolic engineering for 265–268
  metabolic traits 428
  multi-omics approach in 227–229
  omics technologies 293
  phenotyping/phenotypes 244
crop plants 118, 295, 345
  climate change on 125
  genetic enhancement of 288
  quality of 120
  systematic screening of 221
crop protection 18–19
  analytical techniques 19–21
  against bacteria and fungi 26–29, 33–34
  under biotic stress 23
  comprehensive analysis 21
  dynamic nature 21
  extraction methods 19
  future perspectives 31–35
  gas chromatography–mass spectrometry (GC-MS) 20, 22–24, 26–27, 29–30
  identification of differential metabolites 21
  against insects and herbivores 29–31, 34
  liquid chromatography–mass spectrometry (LC-MS) 20, 24, 27–28
  nuclear magnetic resonance (NMR) spectroscopy 18–21, 28–29
  unbiased approach 21–22
  against virus and viroid infections 22–26, 32–33
crop yield 80, 118, 221
  enhancement 188–189
  improvement 221
cross-validation strategy 49, 84
cryogenic probes 71–72
  enhanced sensitivity 71–72
  improved resolution 72
cucumber green mottle mosaic virus (CGMMV) 24
cutting-edge technologies 229, 345, 378–379, 406
cyanogenic glycosides 303
Cycas circinalis 379
cytokinins 208, 294
Cytoscape 168, 325, 342–343

D
data acquisition 75–76, 84, 318
  automated sample handling 75
  bioinformatics 68
  high-resolution 75–76
  phenotyping/phenotypes 240–241
  plant natural products 432–433
  real-time analysis 76
data analysis 76–80, 84, 325
  bioinformatics 163
  medicinal plants 363–364
  phenotyping/phenotypes 241
databased-integrated search engines 77
data integration 325, 340–342
data interpretation 76–80
  phenotyping/phenotypes 241
  plant–herbivore interactions 104
data mining 125, 220–221, 431
data preprocessing 83, 325–326
  medicinal plants 363
  plant–herbivore interactions 103
data processing 325
  plant natural products 432–433
  predictive modeling (PM) 42–46
  tools 39
data science 46
data sharing 83
data validation 84
data visualization 325
  medicinal plants 364
  plant tissues 254
  toolbox 255
decision trees 48, 79
deep learning (DL) 47–49
defense mechanisms 24–25, 221, 439
  antioxidant in 390
  biotic stress 28, 245
  chemical 264
  against drought stress 300
  indirect 106
  monocots 105
  pathogen resistance 81
  against pathogens 301–302
  plant–herbivore interactions 105
  polyamines 224
  secondary metabolites 180
derivatization process 186, 283, 296
  chemical 20, 158, 186, 188, 203
  gas chromatography–mass spectrometry (GC-MS) 179, 356
desorption electrospray ionization–mass spectrometry (DESI-MS) 137, 142, 254, 324
detoxification process 192

diet-related metabolomics 6–7
2,4-dihydroxy-7-methoxy-2H-1,4-benzoxazine-3(4H) (DIMBOA) 223
direct analysis in real time (DART) 254, 324
direct infusion mass spectrometry (DIMS) 219, 286
 medicinal plants 360–361
disease resistance 22, 28, 81, 223, 288, 343, 395, 428
distilling industry 9–10
drought stress 201–202
 abiotic factor 201, 224–225
 adaptation 209–211
 biotic factor 201
 capillary electrophoresis-mass spectrometry (CE-MS) 203
 challenge 211
 clustered regularly interspaced short palindromic repeats (CRISPR)/Cas gene-editing 288
 effects 202–205
 Fourier transform ion cyclotron resonance-mass spectrometry (FT-ICR-MS) 203–204
 future perspectives 211
 gas chromatography–mass spectrometry (GC-MS) 203, 206
 impact of 205–207
 ion mobility spectrometry (IMS) 204
 liquid chromatography–mass spectrometry (LC-MS) 203
 matrix-assisted laser desorption ionization (MALDI)-mass spectrometry (MS) 204
 nuclear magnetic resonance (NMR) 204–205
 phytohormones 207–209
 plant–environment interactions 407
 primary and secondary metabolites 205–207
 salinity and 81
 targeted and non-targeted approach 205
 tool in plant biology 209–211
drug discovery 376
 from medicinal plants 376–378
 metabolic profiling in 439
 natural products in 378
durum wheat *(Triticum durum)* 188

E
ecological responses 412
eco-metabolomics 405
 plant ecology 419
ecosystem function 56
 biodiversity and 415–417
electron ionization (EI) 139, 158
electrospray ionization–mass spectrometry (ESI-MS) 20, 101, 137, 141, 284, 324, 429
 clustered regularly interspaced short palindromic repeats (CRISPR)/Cas gene-editing 283
 drought stress 203
 medicinal plants 373
 plant biotechnology 184
element-based omics dataset analysis techniques 227–228
endophenotypes 54–55
endophytes 389, 391, 393, 395, 397, 413
environmental impact 100, 146, 390, 395
 abiotic factors 158, 224–226, 297
 beneficial effects against 267
 biotic factors 158, 223–224, 297
 climate change and 439
 conventional/genetically modified crop varieties 181–182, 185–187
 crop protection 21
 investigating responses to 180–184
 metabolite profile 120–121
 phenotyping/phenotypes 247
 plant species and analytical tool 119–121
 plant stress, under climate change 119–121
 protective compounds 267
environmental metabolomics 406
 pollution and phytoremediation 81–82
 soil microbiome 81
environmental stressors 146, 157, 183, 243, 389–390, 393–395, 433
enzymatic antioxidants 294
epigenomic data 243
ethnomedicinal knowledge 378
ethnopharmacology 379
Eupatorium adenophorum 411
exometabolomics 394
eXtensible Computational Mass Spectrometry (e-XCMS) 163
external validation 84
extracted ion chromatograms (EICs) 255
extraction method 318
 high-throughput metabolomics 428
 medicinal plants 355
 plant–herbivore interactions 99
 see also specific extraction

F
FAIR principles 43, 46
Fallopia baldschuanica 29
fatty acids 7–8, 53, 206, 262
 plant ecology and ecological roles 412
feature-based molecular networking 256
feature extraction 83
fermented beverages, sensory science in 9–10
field-based approaches 97
fingerprinting technique 254
Fisher transformation 227
flash freezing 74

flavonoids 6, 145
  biosynthesis pathways 437
  biotic stress 295
  drought stress 210
  in plant tissues 379
  secondary metabolites 390
flavor and aroma profiling 9
  advanced techniques for 9
  food design and consumer preferences 10
  food preservation and shelf life 11
  postharvest handling 11
  sensory science in fermented beverages 9–10
floral dip method 279
flow cytometry 136
flow injection-mass spectrometry (MS) *see* direct infusion mass spectrometry (DIMS)
fluorescence-activated cell sorting 136
food design and consumer preferences 10
food preservation, and shelf life 11
food processing 5, 8–9
food quality 8–9
Fourier transform infrared (FTIR) spectroscopy 361
Fourier transform ion cyclotron resonance-mass spectrometry (FT-ICR-MS) 285, 323
  drought stress 203–204
  medicinal plants 361–362
Framework Convention on Climate Change (FCCC) 116
Fritillaria unibracteata 4
FTIT spectroscopy 416
Fucus vesiculosus 145
functional foods 8
functional genomics 227, 336–339
  applications 339
  challenges and future directions 344–345
  gene discovery and functional characterization 343
  machine learning 342
  network analysis 343
  pathway modeling 343
  plant breeding and crop improvement 344
fungi 26–29, 33–34
Fusarium culmorum 28
Fusarium graminearum 3, 162, 189, 223

G
gamma-amino butyric acid (GABA) 25
Gammarus fossarum 256
gas chromatography–mass spectrometry (GC-MS) 3–4, 9, 19–20, 73, 284, 322
  bacteria and fungi 26–27
  bioinformatics 158
  biological samples 296
  clustered regularly interspaced short palindromic repeats (CRISPR)/Cas gene-editing 283–284
  drought stress 203, 206
  herbivore-induced plant volatiles 29–30
  medicinal plants 355–356, 360, 373
  melon and cotton plants 30
  non-targeted 430
  phenotyping/phenotypes 240
  plant biotechnology 179, 184, 186
  plant ecology 416
  plant–herbivore interactions 102–103, 106
  plant metabolomics 337
  plant natural products 430
  plant stress 122
  predictive modeling (PM) 41–42
  salinity stress 407
  of seeds 277
  single-cell 137, 139–141, 147
gas chromatography–time of flight–MS (GC-TOF-MS) 3
gas-phase separation technique 360
Gaussian graphical modeling 428
gene editing 227
  see also clustered regularly interspaced short palindromic repeats (CRISPR)/Cas gene-editing
gene function 168, 278, 336, 432
Gene Ontology (GO) 168
genetically modified crops 108
  under environmental conditions 186–187
  metabolite profiling of 181, 191
genetic engineering 226
genetic modifications 81, 177, 208, 268, 277, 343–344, 394
genome-editing tools 108, 191, 268, 326
genome mining 53
genome-scale modeling 228–229
genome-wide association studies (GWAS) 107, 221–222
  metabolomics-assisted breeding (MetAB) 304
  metabolomics-based 244
  phenotyping/phenotypes 243
genomic-best linear unbiased prediction (G-BLUP) 54–55
genomics 160
  biochemical pathways 316
  functional *see* functional genomics
  selection models 261
  structural 227
genomic sequencing 338–339
Gibberella pulicaris 303
Global Natural Product Social Molecular Networking (GNPS) 46, 77, 256, 439
global warming 115–116, 340
glucosinolates 303
  brassicales 105
  plant ecology and ecological roles 412
glutathione 224, 301
glycine betaine 157, 263–264, 293

glycolate oxidase 183
Golm Metabolite Database (GMD) 103
Gossypium hirsutum 267
graph-clustering approach 343
graph-guided fusion least absolute shrinkage and selection operator (GFLASSO) 227
greenhouse gasses 115
  hyperaccumulation of 116
  overproduction of 118
GreenPhylDB 167
green tea 7
guide RNA (gRNA) 277–279
Gymnosporia montana 361

H
harvesting methods 317
  medicinal plants 353–355
heavy-metal stress 226, 300–301
Helicoverpa armigera see cotton bollworm *(Helicoverpa armigera)*
herbal medicine 351, 357
  bioactive ingredients and therapeutic mechanisms 379
  and drug discovery 376
  quality control 364–365
herbivores 29–30
  crop protection against 29–31, 34
  nuclear magnetic resonance (NMR) spectroscopy against 31
  see also plant–herbivore interactions
heteronuclear single quantum coherence (HSQC) spectroscopy 21
(Z)-3-hexenyl butyrate (HB) 26
high-field nuclear magnetic resonance (NMR) spectroscopy 72–73
high-performance liquid chromatography (HPLC) 73
  chemometric techniques 338
  clustered regularly interspaced short palindromic repeats (CRISPR)/Cas gene-editing 281–283
  medicinal plants 357–358
  plant–herbivore interactions 105
  plant metabolomics 338
high-resolution magic-angle spinning 431
high-resolution mass spectrometry (HRMS) 284–285, 376
  direct infusion mass spectrometry (DIMS) 286
  predictive modeling (PM) 41
high-throughput analytical techniques (HTA) 39–40
high-throughput experimentation (HTE) 39
high-throughput metabolomics 29, 39, 55, 315–317
  abiotic stress *see* abiotic stress
  application 326–327
  benefit of 428
  biotic stress *see* biotic stress
  climate change *see* climate change
  CRISPR/Cas gene-editing *see* clustered regularly interspaced short palindromic repeats (CRISPR)/Cas gene-editing
  in crop protection 19, 21–22
  data acquisition 318–325
  data processing and analysis 325
  extraction methods 428
  future perspectives for 327–328
  with genetic analyses 56
  medicinal plants *see* medicinal plants
  omics integration 325–326
  phenotyping *see* phenotyping/phenotypes
  plant ecology *see* plant ecology
  plant–herbivore interactions *see* plant–herbivore interactions
  plant natural products 376, 427–440
  and predictive modeling 50–57
  sample preparation and automation 317–318
  secondary metabolites *see* secondary metabolites
  techniques 39–46
  technological advancements *see* technological advancements
high-throughput phenotyping *see* phenotyping/phenotypes
high-throughput technology 430
^{1}H-NMR 20–21, 25, 28
homogenization 75
hormonal signaling pathways 146
hormone profiling 80
housekeeping metabolites 78
huanglongbing (HLB) 29
human health 5
Human Metabolome Database (HMDB) 76
hydrophilic interaction liquid chromatography (HILIC) columns 41
hydrophilic solvents 355
hydrophobic solvents 355
hydroxycinnamic acid amides (HCAA) 27–28
3-hydroxy-3-methylglutaryl-CoA reductase (HMGR) 268
hyperfunctional foods 8
Hypericum perforatum 360

I
indole-3-acetic acid (IAA) 208
insects 29–31, 34
in silico fragmentation methods 43
in silico prediction tools 77
Integrated Discovery of Event-Related Metabolites (IDEOM) 163
integrative omics approaches 69, 83, 263, 345
Intergovernmental Panel on Climate Change (IPCC) 116, 118
internal standards 75, 77–78, 84

ionization 184
  chemical 139, 158
  electron 139, 158
  see also specific ionization
ion mobility spectrometry (IMS) 123, 159, 255
  drought stress 204
  medicinal plants 360–361
  untargeted metabolomics 360
ionomics 162–163
ion probe analysis *see* secondary ion mass spectrometry (SIMS)
ion trap analyzers 284
isoflavones 208, 268, 303
isoprenoids *see* terpenoids
isotopic labeling methods 101, 122, 431

J
jasmonic acid (JA) 208–209, 294, 389
  biosynthesis 281
  biotic stress 295
  pathway 264–265
  stress regulation 300

K
K-fold cross-validation 49
knowledge-based approaches 406
Kyoto Encyclopedia of Genes and Genomes (KEGG) 76–77, 228, 342

L
label-free approach 254
laser ablation ESI–mass spectrometry (LAESI-MS) 137, 141–142
laser-capture microdissection 136
leaf's metabolites 3
learning process *see* deep learning (DL); machine learning (ML)
Leptobryum pyriforme 124
leucine 28
light conditions 409
linear regression 47–48
linolenic acid 281
lipids 7
  plant ecology and ecological roles 412
lipoxygenase 3/4 (LOX3/LOX4) 30
liquid chromatography–mass spectrometry (LC-MS) 3, 8–9, 19, 73, 284, 322, 428, 433
  bacteria and fungi 27–28
  bioinformatics 159
  biological samples 296
  breeding process 219
  clustered regularly interspaced short palindromic repeats (CRISPR)/Cas gene-editing 283
  drought stress 203
  insect-resistant varieties 30
  maize 305
  medicinal plants 357–358, 360, 373
  personalized medicine 380
  plant biotechnology 179
  plant ecology 416–417
  plant–herbivore interactions 99, 102
  plant metabolomics 337–338
  plant natural products 430
  plant stress 123
  predictive modeling 39, 41–42
  salinity stress 407
  single-cell 137, 139
  untargeted metabolomics 400
  against virus and viroid infections 24
liquid chromatography-nuclear magnetic resonance (LC-NMR) 359–360
liquid chromatography–tandem mass spectrometry (LC-MS/MS) 3–4
liquid–liquid extraction 74, 355
Lycoris radiata 389
lytic metabolites 180

M
machine learning (ML) 46, 49, 79, 366, 400, 428
  algorithms 77
  artificial neural networks (ANNs) 48–49
  automated 46
  classification algorithms 47
  decision trees 48
  deep learning (DL) 48–49
  functional genomics 342
  learning process 49
  linear regression 47–48
  medicinal plants 363
  metabolomics-assisted breeding (MetAB) 305
  predictive modeling (PM) 47–49
  random forests 48, 79
  regression algorithms 47
  supervised learning 47–48
  support vector machines (SVMs) 47–48, 79
magic-angle spinning 358
maize *(Zea mays)* 3, 106, 162
Manduca sexta 259
MapMan 168, 228
marker-assisted selection programs 265, 293
mass spectrometry (MS) 8, 69–72, 284–285, 428
  advantages and disadvantages 318–324
  ambient ionization techniques 324
  biological samples 296
  characteristics 189
  crop protection 31–33
    gas chromatography–mass spectrometry (GC-MS) 19–20, 26–27, 29–30

liquid chromatography–mass spectrometry (LC-MS) 19, 24, 27–28, 30
nuclear magnetic resonance (NMR) spectroscopy 18–19, 25, 27–29, 31
MALDI-mass spectrometry (MS) *see* matrix-assisted laser desorption/ionization (MALDI)
medicinal plants 352, 355–358, 373
metabolomic profiling 429
phenotyping/phenotypes 239–240
plant ecology 416
plant–herbivore interactions 101
plant metabolomics 337–338
plant stress 119
single-cell technology 135–142
targeted metabolomics techniques 253
mathematics-based approaches 228–229
matrix-assisted laser desorption/ionization (MALDI) 101, 123, 284, 429
abiotic and biotic stress 254
bioinformatics 160
and mass spectrometry (MS) 138, 323–324
drought stress 204
single-cell technology 142–143
medicinal plants 361–362
mean absolute error (MAE) 50
mean absolute percentage error (MAPE) 50
mean squared error (MSE) 50
Medicago sativa 162, 222
Medicago truncatula 204, 268
medicinal plants 351–352
analytical methods and techniques 352–363
anti-obesity characteristics 372
anti-parasitic drug development 377
applications 372
bioactive compounds in 375–376
biochemical compositions of 382
capillary electrophoresis-mass spectrometry (CE-MS) 356–357
challenges 365–366, 381
data analysis and mapping 363–364
data preprocessing 363
data standardization and quality assurance 365
data visualization 364
direct infusion mass spectrometry (DIMS) 360–361
drug discovery and development 376–378
electrospray ionization–mass spectrometry (ESI-MS) 373
ethnobotanical studies 365
extraction techniques 355
Fourier transform infrared (FTIR) spectroscopy 361
Fourier transform ion cyclotron resonance-mass spectrometry (FT-ICR-MS) 361–362
future of 381–382
gas chromatography–mass spectrometry (GC-MS) 355–356, 360, 373
harvesting methods 353–355
high-performance liquid chromatography (HPLC) 357–358
hyphenated technology 365
identification and quantification 363, 365
ion mobility spectrometry (IMS) 360–361
limitations 381
liquid chromatography–mass spectrometry (LC-MS) 357–358, 360, 373
liquid chromatography-nuclear magnetic resonance (LC-NMR) 359–360
liquid–liquid extraction techniques 355
machine learning approaches 363
mass spectrometry (MS) 352, 355–358, 373
matrix-assisted laser desorption ionization (MALDI) 361–362
metabolite dynamics and statistical difficulties 366
Metabolomics Standards Initiative (MSI) 353
molecular mechanisms 377, 380
multi-omics approach 382
multivariate data analysis 363
natural products and 351
non-targeted methods 352, 376
nuclear magnetic resonance (NMR) spectroscopy 352, 358–359, 373
omics technologies 382
orthogonal partial least squares discriminant analysis (OPLS-DA) 363
partial least squares discriminant analysis (PLS-DA) 363
personalized medicine and 380, 382
pharmacological effects 372
phytochemical profiling of 374–376
primary metabolites 335, 361
principal component analysis (PCA) 363
quantification techniques 365–366
Raman spectroscopy 361–363
sample collection 353
samples preparation 353
screening for therapeutic potential and efficacy 377
secondary metabolites 335, 351, 361, 376, 392, 400
species 352
targeted methods 351–352
techniques 373–374
therapeutic effects of 378
Mediterranean diet 7
MetaboAnalyst 163, 256, 341, 439
metabolic analysis 21, 205, 210, 224, 338
metabolic disorders 7, 225, 372
metabolic engineering 81, 101, 107, 178, 394
for crop improvement 265–268
genetic and 389
and synthetic biology 226–227, 435

metabolic fingerprinting 102, 159, 361, 375
metabolic flux analysis (MFA) 393–394
metabolic markers 43, 57, 108, 191, 305
metabolic network reconstruction 80
metabolic pathway 6–7, 10, 53–54, 68, 80–81, 100–101, 104, 120, 133, 202, 205
  abiotic stressors 262, 298–299
  annotation 228
  biological processes 363
  drought stress on 205–207
  elucidation 245
  Fusarium graminearum 223
  genome editing 281
  lignin production 224
  nutrient composition 218
  oxidative stress 410
  under stress conditions 304
metabolic profiling 225–226
  abiotic and biotic stress 259–260
  barley 226
  beans 226
  breeding for trait selection 221–223
  in clustered regularly interspaced short palindromic repeats (CRISPR)/Cas gene-editing 281–282
    electrospray ionization (ESI) mass spectrometry 283
    gas chromatography (GC) 283–284
    high-performance liquid chromatography (HPLC) 281–283
    liquid chromatography-mass spectrometry (LC-MS) 283
    ultra-high-performance liquid chromatography (UHPLC) 283
  crops 225
  drought stress 202
  in drug discovery and phytomedicine 439
  during host–pathogen interactions 223
  in primary and secondary metabolites 223
  processing data in 432
  rice varieties 225
  tall-statured model species' 406
  see also metabolite profiling
metabolite annotation 255–259
metabolite identification 76
  databases 76–77
  in silico prediction tools 77
metabolite profiling 68, 179
  citrus 263
  environmental stresses 120–121
  genetically modified crops 181, 191
  metabolomics-assisted breeding (MetAB) 305
  phytochemicals 364
  see also metabolic profiling
metabolite set enrichment analysis (MSEA) 79
metabologenomics 53
metabolome 38, 100, 154, 315, 335
  metabolome-assisted breeding 221
  workflow 155–156
metabolomic *see* plant metabolomics
metabolomic best linear unbiased prediction (M-BLUP) 54–55
metabolomic quantitative trait locus (mQTL) 107, 190, 244, 343
metabolomics-assisted breeding (MetAB) 293
  abiotic stresses 293–294, 298–299
    drought/water-deficit stress 297–300
    heavy metals 300–301
    oxidative damage 301
    salinity 300
  biotic stresses 295
  climate-smart crops with improved stress tolerance 305–306
  crops and agroecosystems 307
  genome-wide association studies 304
  machine learning algorithms 305
  metabolite profiling 305
  multivariate statistical methods 305
  quantitative trait locus 304
  tools and techniques 295–297, 305
metabolomics-based genome-wide association studies (mGWAS) 244
Metabolomics Standards Initiative (MSI) 42, 99, 353
metal-induced stress management 226
MetExplore 163
METLIN 76, 433
mGWAS *see* metabolomics-based genome-wide association studies (mGWAS)
microbe–plant interactions 280
microcapillary extraction 136
microfluidics 135–136
micromanipulation 136
microRNAs 340, 389
microwave-assisted extraction 220
model training and validation 49–50
  accuracy 49–50
  mean absolute error (MAE) 50
  mean absolute percentage error (MAPE) 50
  mean squared error (MSE) 50
  precision 50
  predictive modeling (PM) 49–50
  root mean squared error (RMSE) 50
  sensitivity 50
  specificity 50
molecular biology techniques 29
molecular mechanisms 192
  environmental stressors 394
  medicinal plants 377, 380
  plant–herbivore interactions 97
  plant–microbe associations 223
  stress tolerance traits 293, 307
Momordica charantia 359
mQTLs *see* metabolomic quantitative trait locus (mQTL)

multi-coin inertial analysis 227
multi-omics approach 82, 178, 307, 345
  benefits of 437
  in crop improvement 227–229
  data analysis technologies 160–163, 167–168
  data integration 325, 340–342
  drought stress 211
  medicinal plants 382
  metabolomics-assisted breeding (MetAB) 304
  phenotyping/phenotypes 241–243
  plant biotechnology 177
  plant breeding and crop improvement 344
  plant natural products 436
  proline and sugar accumulation 209
  tools and software 341–342
multi-omics integration 325, 342, 366
multivariate analysis 79–80, 221, 227, 364
  medicinal plants 363
  metabolomics-assisted breeding (MetAB) 305
  phenotyping/phenotypes 241
  plant–herbivore interactions 103–104
Myricaria laxiflora 124

N
nanoLC-MS (nLC-MS) 123
network modeling 80
  functional genomics 343
  phenotyping/phenotypes 243
neural networks 48
next-generation sequencing 107, 280
Nicotiana attenuata 259
Nicotiana benthamiana 435
Nicotiana tabacum 25
Nilaparvata lugens 340
nitrogen deficiency 408–409
non-chromatographic identification techniques 284
  direct infusion mass spectrometry (DIMS) 286
  Fourier transform ion cyclotron resonance mass spectrometry (FT-ICR-MS) 285
  mass spectrometry (MS) 284–285
  nuclear magnetic resonance (NMR) 285–286
non-enzymatic antioxidants 294
non-polar metabolites 74
non-targeted approach 6, 22, 30, 120, 122, 157
  abiotic and biotic stress 253
  drought stress 205
  gas chromatography-mass spectrometry (GC-MS) 430
  ion mobility spectrometry (IMS) 360
  liquid chromatography-mass spectrometry (LC-MS) 400
  medicinal plants 352, 376
  plant–herbivore interactions 411
  plant metabolites 316
normalization techniques 78
nuclear magnetic resonance (NMR) 285–286
  advantages 285
  characteristics 189
  drought stress 204–205
  non-destructive 218
  plant biotechnology 179
  two-dimensional techniques 285–286
nuclear magnetic resonance (NMR) spectroscopy 8, 68, 324–325, 428
  against bacteria and fungi 27–29
  bioinformatics 159
  biological samples 296
  crop protection 18–19
  cryogenic probes 71–72
  against herbivores and insects 31
  high-field 72–73
  isotopic labeling in 431
  medicinal plants 352, 358–359, 373
  phenotyping/phenotypes 240
  plant ecology 416
  plant–herbivore interactions 99–101
  plant metabolomics 338
  plant natural products 429–431
  plant stress 119
  rice 305
  single-cell technology 135, 138, 142
  against virus and viroid infections 25
nutrient content 8–9
nutrient deficiency 226, 408–409
  nitrogen 408
  phosphorous 408
  plant–environment interactions 408–409
  potassium 408–409
nutrient metabolism 80
nutrigenomic diets 7
nutrition 344
  personalized 7–8, 12
  quality improvement 245

O
oat *(Avena sativa)* 223
omega (ω)-3 fatty acids 7
omics integration *see* multi-omics integration
OmicsNet 2.0 342
omics technologies 8, 11, 69, 82, 85, 105, 107, 147, 317, 345, 366, 399, 427, 435–437
  crop improvement 293
  data integration and analysis 340–342
  element-based 227–228
  medicinal plants 363–364, 381–382
  plant biotechnology 191
  plant natural products 435–437
  secondary metabolites 389–390, 392, 394–395
  see also specific omics technologies
Orbitrap MS 69–71
  phenotyping/phenotypes 239

orthogonal partial least squares-discriminant analysis (OPLS-DA) 281, 363
Oryza sativa see rice *(Oryza sativa)*
osmoprotection 81
OsTZF1 gene 340
over-representation analysis 53, 79
oxidative stress 4, 157, 188, 210, 293–294, 301, 415
  plant–environment interactions 410–411
ozone and pollutants 410

P
Pacific yew tree *(Taxus brevifolia)* 351
paclitaxel 351
Paired Omics Data Platform (PoDP) 53
partial least squares-discriminant analysis (PLS-DA) 79
  medicinal plants 363
  predictive modeling (PM) 48
Paspalum wettsteinii 263
patch-clamp technique 136
pathogen resistance 53, 81, 103, 223, 395
pathway analysis 79–80
  enrichment 79
  in plants 53–54
  tools and approaches for 79–80
pathway modeling 343
Peristenus spretus 30
personalized medicine 12, 380, 382
personalized nutrition 7–8, 12
pest management 30, 106–107
phenolic compounds 158
  plant ecology and ecological roles 412
  secondary metabolites 390, 394
phenomics 55, 124, 162, 243, 304
phenotyping/phenotypes 39, 81, 237
  analytical techniques and high-throughput capabilities 239–240
  applications 238
  automation and robotics 240
  breeding 244
  challenges and future directions 247
  climate change adaptation 245–247
  crop improvement 244
  data acquisition 240–241
  data analysis and interpretation 241
  environmental impact 247
  genome-wide association studies (GWAS) 243
  metabolic pathway elucidation 245
  multi-omics approaches 241–243
  multivariate analysis techniques 241
  nutritional quality improvement 245
  online databases/tools and software 242
  phenotype–genotype association 243
  for precision agriculture 245
  quantitative trait loci (QTLs) mapping 243
  raw data processing 240–241
  sample collection and preparation 238–239
  stress response analysis 244–245
  sustainable agriculture 247
  systems biology and network modeling 243
  trait discovery and characterization 244, 246
  univariate analysis 241
  workflow 238–243
phenylpropanoids 223
phosphorous deficiency 408–409
photoprotective compounds 409–410
photosynthetic pigments 409
phytoalexins 133, 261, 295, 303–304
phytoanticipins 133, 302
phytochemical profiling 56–57, 364, 374–376
phytohormones 80, 294, 300, 389
  abiotic and biotic stress responses 253, 262
  drought stress 202, 207–210
  single-cell technology 146
phytomedicine 434, 439
Phytophthora infestans 28
phytoremediation 81–82
phytotoxins 180
Picrorhiza kurrooa 393
Pinus radiata 263, 301
plant-based nutraceutical research 324–325
plant biotechnology 177
  applications 179–184
  bioactive compounds, production 190
  crop yield enhancement 188–189
  future directions and challenges 190–192
  gas chromatography–mass spectrometry (GC-MS) 179, 184, 186–187
  hydrophilic interaction liquid chromatography (HILIC) 184, 186
  liquid chromatography–mass spectrometry (LC-MS) 179, 184, 187
  mass analyzers 185
  mass spectrometry (MS) 178, 184–186, 189
  metabolomics and other omics areas 191
  nuclear magnetic resonance (NMR) spectroscopy 179, 186–189
  significance 177–179
  stress tolerance improvement 189–190
plant breeding *see* breeding process
plant disease 301–302
  cyanogenic glycosides 303
  glucosinolates 303
  phytoalexins 303–304
  phytoanticipins 302
  saponins 302–303
plant ecology 395, 411–412
  abiotic stress responses 407–411
  analytical approaches 417
  atmospheric conditions 410
    carbon (C) assimilation 410
    carbon dioxide 410
    ozone and pollutants 410

and biodiversity research 405–419
challenges 417
eco-metabolomics 419
endophytes 413
future directions 417
light conditions 409
liquid chromatography–mass spectrometry (LC-MS) 416–417
metabolic diversity 415
nutrient deficiency 408–409
nitrogen 408
phosphorous 408
potassium 408–409
nutrient excess 409
oxidative stress 410–411
photoprotective compounds 409–410
photosynthetic pigments 409
plant–plant interactions 411
rhizosphere metabolomics 413
secondary metabolites 415
stress response 413–415
symbiotic relationships 413
technology 416
workflow in 418
plant–environment interactions 407
abiotic stress responses 407–411
atmospheric conditions 410
carbon (C) assimilation 410
carbon dioxide 410
ozone and pollutants 410
light conditions 409
nutrient deficiency 408–409
nitrogen 408
phosphorous 408
potassium 408–409
nutrient excess 409
oxidative stress 410–411
photoprotective compounds 409–410
photosynthetic pigments 409
plant health 2–4, 100, 147, 434
plant–herbivore interactions 96–97
agriculture 106
annotation tools and databases 103
applications 105–107
biochemical analyses 97
case studies 104–105
critical concerns and best practices 99
data interpretation 104
data preprocessing 103
defense strategies 97
extraction techniques 99
field-based approaches 97
future research 107–108
gas chromatography–mass spectrometry (GC-MS) 102–103, 106
high-performance liquid chromatography (HPLC) analysis 105
liquid chromatography–mass spectrometry (LC-MS) 99, 102
mass spectrometry (MS) 101
model systems 104–105
molecular mechanisms 97
morphological strategies 96–97
non-targeted approaches 411
nuclear magnetic resonance (NMR) spectroscopy 99–101
pest management 106–107
plant breeding 107
principal component analysis (PCA) 104
sample collection and storage 99
statistical analyses 103–104
steps and types of 98
unravelling defense mechanisms 105
unveiling metabolome 97–99
plant hormones *see* phytohormones
plant metabolism 46, 53, 56, 69, 85, 101, 124, 316, 435
biotechnology 179–180
breeding programs 394
drought stress 202, 205
environmental stressors on 395
spatial and temporal changes in 139
targeted plant metabolic engineering (TPME) 265
under water-deficit stress 207–209
plant metabolome *see* metabolome
plant metabolomics 1–2, 68, 218
analytical techniques 19–21, 219
application 326–327
bioinformatics 162, 164–166
challenges 11–12, 43
cutting-edge technologies 124
data repositories and databases for 257–258
future perspectives 11–12, 327–328
gene-edited plants 280
high-performance liquid chromatography (HPLC) 338
mass spectrometry (MS) 337–338
nuclear magnetic resonance (NMR) spectroscopy 338
with other omics 82–83
to plant breeding 327
in plant health 2–4
predictive modeling *see* predictive modeling (PM)
profiling 429
sample preparation 317–318
techniques 336
workflow analysis 220–221, 259–260
PlantMetSuite 395, 439
plant–microbe interactions 4–5, 55–56, 399, 411–412, 428, 434, 439
applications of 414
endophytes 413
nuclear magnetic resonance (NMR) spectroscopy 100
rhizosphere 103, 413
symbiotic relationships 413

plant natural products 376, 427–440
application 433–434
challenges and limitations 437–438
data acquisition and processing tools 432–433
future directions 438–439
gas chromatography–mass spectrometry (GC-MS) 430
gene-guided approaches 435
liquid chromatography–mass spectrometry (LC-MS) 430
mass spectrometry (MS) 429–430
mass spectrometry (MS) imaging 431–432
nuclear magnetic resonance (NMR) spectroscopy 429–431
omics technologies 435–437
technical and analytical challenges 438
technology 429
plant–plant interactions 411
plant secondary metabolism *see* secondary metabolites
plant stress 114
bioinformatics under 157–158
under climate change 114–115
current scenarios and future projections 115–118
environmental stresses 119–121
natural and anthropogenic activities 117
resilience 123–124
significance 118–119
technological advances 122–123
single-cell metabolomics 145–146
plant tissues 19, 54, 85, 102, 119, 133, 145, 238, 254, 316–318, 323, 379, 397
plant–viroid interactions 24–25
PLAZA platform 167
polar metabolites 74
pollution 81
ozone and 410
and phytoremediation 81–82
polyamines 180, 206, 224–225, 227, 294, 300, 344
polyketide biosynthesis 391
polymerase chain reaction (PCR) 279–280, 288
polyphenols 6, 9, 105, 145, 158, 294, 357
polyunsaturated fatty acids (PUFAs) 281
Populus tomentosa 124
postgenomic approaches 107
postharvest technology 11
post-transcriptional gene silencing 339
post-translational modification (PTM) 160, 243, 268, 304, 410
potassium deficiency 408–409
potato *(Solanum tuberosum)* 106, 157, 162
potato spindle tuber viroid (PSTVd) 24
precision 50
precision agriculture 243, 434
phenotyping/phenotypes for 245
precision medicine 380, 417
predictive modeling (PM) 38–39, 47
analytical method 41
annotation 42–46
applications 50–57
artificial neural networks (ANNs) 48–49
benefits 46
best linear unbiased prediction (BLUP) 54
chemodiversity 56
classification algorithms 47
contextualization and storage 42–46
crop improvement programs 54–55
data processing 42–46
decision trees 48
deep learning (DL) 48–49
environmental interactions and ecosystem dynamics 55–57
FAIR principles 43, 46
fundamental processing steps 43
future directions 57–58
gas chromatography–mass spectrometry (GC-MS) 41–42
linear regression 47–48
liquid chromatography–mass spectrometry (LC-MS) 39, 41–42
methodologies and algorithms 47–49
model training and validation 49–50
accuracy 49–50
mean absolute error (MAE) 50
mean absolute percentage error (MAPE) 50
mean squared error (MSE) 50
precision 50
root mean squared error (RMSE) 50
sensitivity 50
specificity 50
multi-directional models 58
neural networks 48
partial least squares-discriminant analysis (PLS-DA) 48
pathway analysis 53–54
phytochemical traits 56–57
quality control 43
random forest model 48, 55
regression algorithms 47
sample preparation and processing 40–41
in silico fragmentation methods 43
software 44–45
supervised learning 47–48
support vector machines (SVMs) 47–48
workflow for 40
primary metabolites 39, 120, 133, 157, 297, 316
biotic stressors 245
breeding 216, 218
citrus rootstocks 145
drought stress 202, 205–207, 209–210
environmental stress 180–183
medicinal plants 361
metabolic profiling in 223
nutritional and medicinal value 335

plant biotechnology 179
plant ecology and ecological roles 412
for plant growth 134, 277
wheat 261
principal component analysis (PCA) 79
medicinal plants 363
plant–herbivore interactions 104
tobacco 280
proline 157, 300–301
amino acids 264
biosynthesis 263
drought stress 205
osmoprotectant activity 301
Prosopis cineraria 361
Protein Analysis Through Evolutionary Relationships (PANTHER) 167
proteomics 123, 243, 316
analyses 304
bioinformatics 160–162
matrix-assisted laser desorption/ionization (MALDI) 323
personalized medicine 380
on soybean leaves 168
protocol consistency 75
Pseudomonas protegens 393
Pseudomonas syringae 26
pseudo-targeted metabolomics 122
public repositories 83
pure shift yielded by chirp excitation (PSYCHE) method 338
pyrethrins 295
Pyricularia oryzae 304

Q
quadrupole mass spectrometry (Q-MS) 284, 323
quadrupole time-of-flight–mass spectrometry (Q-TOF-MS) 69, 71, 239
quality control (QC) 84
quantile normalization 78
quantitative trait loci (QTLs) 216
metabolomics-assisted breeding (MetAB) 304
phenotyping/phenotypes 243
see also metabolomic quantitative trait locus (mQTL)
quenching method 74, 317

R
Ralstonia solanacearum 28
Raman spectroscopy 143
medicinal plants 361–363
single-cell technology 138, 143
RAMClustR 256
random forests 48, 55, 79
random splitting 49
raw data processing 240–241
see also data processing
reactive oxygen species (ROS) 183, 263, 294
receptor for activated C-kinase 1 (RACK-1) 340
regression algorithms 47
resilience 245, 394
biotic stressors 265
to climate change 123–124
environmental stressors 124, 395, 428
resistance-related metabolites 27–28
resistant to black-streaked dwarf virus (RBSDV) 25
resveratrol 295, 304
Rhizoctonia solani 219
rice *(Oryza sativa)* 106, 124
metabolic profiling 225
receptor for activated C-kinase 1 (RACK-1) 340
salinity stress 225
rice dwarf virus (RDV) infection 24
RNA interference (RNAi) 339–340
RNA sequencing (RNA-seq) 339
robotic systems 75, 240
robots/drones 41
root exudates 55, 159, 338, 413, 434
root mean squared error (RMSE) 50

S
Saccharum officinarum 124
salicylic acid (SA) 300
biotic stress 295
pathway 264–265
willow tree *(Salix alba)* 351
salinity stress 81, 225, 263, 300
plant–environment interactions 407
secondary metabolites 389
Salix alba see willow tree *(Salix alba)*
Sanger sequencing method 279–280, 288
saponins 302–303
secondary ion mass spectrometry (SIMS) 138, 143
secondary metabolites 24–25, 39, 120, 133, 297
abiotic stress 316
abscisic acid (ABA) 389
in agriculture and industry 395–397
alkaloids 390
applications 391–393, 395–397
biological activities 390
biosynthesis 388, 391, 437
biotic stress 245, 316
breeding 216, 218
challenges 391, 397–399
chemical diversity 435
citrus rootstocks 145
defense mechanism 180
drought stress 205–207, 209–210, 294
ecological and evolutionary significance 435
ecological functions 277
environmental factors 183–184, 394
evaluation 351
flavonoids 390

secondary metabolites (*continued*)
  genetic and environmental factors 392
  genetic modifications on 394
  identification and characterization 434
  jasmonic acid (JA) 389
  liquid chromatography–mass spectrometry (LC-MS) 283
  medicinal plants 361, 376, 392, 400
  metabolic flux analysis (MFA) 393–394
  metabolic profiling in 223
  nitrogen-containing plant 158
  nuclear magnetic resonance (NMR) spectroscopy 100
  nutritional and medicinal value 335
  omics technologies 389–390, 392
  phenolic compounds 390, 394
  plant biotechnology 179
  plant ecology 412, 415
  plant–environment interactions 395
  in plant growth and development 134
  qualitative and quantitative variations in 393
  salinity stress 294, 389
  salinity tolerance 294
  significance 389, 392
  sustainable production 391
  technical and biological challenges 398
  terpenoids 390
  types of 390–391
  wheat 261
semi-targeted metabolomics 122
sensitivity 50
single-cell metabolomics 69, 82, 132–133, 345
  applications 143–147
  capillary electrophoresis–mass spectrometry (CE-MS) 137, 141
  cell types, workflow chart 140
  cellular heterogeneity 145
  desorption electrospray ionization–mass spectrometry (DESI-MS) 137, 142
  electrospray ionization–mass spectrometry (ESI-MS) 137, 141
  gas chromatography–mass spectrometry (GC-MS) 137, 139–141, 147
  hormonal signaling pathways 146
  laser ablation ESI–mass spectrometry (LAESI-MS) 137, 141–142
  liquid chromatography–mass spectrometry (LC-MS) 137, 139
  mass spectrometry (MS) 135–142
  matrix-assisted laser desorption ionization (MALDI)-mass spectrometry (MS) 142–143
  nuclear magnetic resonance (NMR) spectroscopy 135, 142
  plant species 144
  Raman spectroscopy 138, 143
  sampling types 135–138
  secondary ion mass spectrometry (SIMS) 138, 143
  stress responses 145–146
  symbiotic and pathogenic interactions 146–147
single-cell sequencing 339
soft ionization methods 20, 141
soil metabolomics 415
soil microbiome 81
Solanum lycopersicum see tomato *(Solanum lycopersicum)*
Solanum tuberosum see potato *(Solanum tuberosum)*
Solidago canadensis 399
solid-phase extraction (SPE) 74, 355, 359
Sonchus oleraceus L. 379
Soxhlet extraction 74
spatial metabolomics 69, 82, 345, 431–432
specificity 50
spectral matching 77, 259
spike-in method 78
Spodoptera littoralis 3
stable isotope labeling 54, 143
stable isotope-resolved metabolomics (SIRM) 101
standardized data formats 83
standard operating procedures 84
statistical analysis 18, 78, 154, 221, 366, 437
  data matrix for 43
  machine learning 79
  multivariate 79, 364
  online databases/tools and software for 242
  plant–herbivore interactions 103–104
  quantification and 241
  robust 155, 163
stored waveform inverse Fourier transform (SWIFT) method 285
stress response 413–415
  metabolic diversity 415
  phenotyping/phenotypes 244–245
  secondary metabolites 415
  see also specific stress
stress tolerance 259–260, 263, 265, 288, 293
  abiotic *see* abiotic stress
  biotic *see* biotic stress
  improvement 189–190
structural elucidation techniques 20
structure–activity relationships 324
supercritical fluid extraction 220
supervised learning 47
  algorithms 342
  predictive modeling (PM) 47–48
support vector classification (SVC) 48
support vector machines (SVMs) 47–48, 79
support vector regression (SVR) 48
sustainable agriculture 247
sweet sugarcane (SS) 4
symbiotic and pathogenic interactions 146–147
synthetic biology 177, 226–227, 435
systems biology 378, 437

bioinformatics and 82, 406
phenotyping/phenotypes 243

T
tailoring extraction methods 75
Tanacetum vulgare 30
tandem MS (MS/MS) 71, 203, 363
targeted approach 6, 122, 157
abiotic and biotic stress 253
drought stress 205
medicinal plants 351–352
strategies for 267
targeted plant metabolic engineering (TPME) 265, 267
Taxus brevifolia see Pacific yew tree *(Taxus brevifolia)*
technological advancements 67–85
applications 80–82
bioinformatics tools 83
chromatography 73–74
data acquisition 75–76
data analysis and interpretation 76–80
data sharing 83
extraction methods 74
integrating metabolomics with other omics 82–83
internal standards 77–78
machine learning 79
mass spectrometry 69–72
metabolite identification
databases 76–77
in silico prediction tools 77
multivariate analysis 79
normalization techniques 78
nuclear magnetic resonance (NMR) spectroscopy *see* nuclear magnetic resonance (NMR) spectroscopy
quality control 84
quantification and normalization 77–78
sample preparation 74–75
sample standardization 74–75
single-cell metabolomics 82
spatial metabolomics 82
standardization and reproducibility 83–84
standard operating procedures 84
statistical analysis 78–79
temperature stress 407–408
terpenes 73, 294, 304
terpenoids 30, 207, 210, 261, 295, 389–390, 435
Tetranychus evansi 264–265
Thellungiella halophila 3
Thermo Fisher Scientific 255
time-of-flight mass spectrometry (TOF-MS) 142, 284, 322
tobacco mosaic virus (TMV) infection 25
toggle instrumentation 285
tomato *(Solanum lycopersicum)* 162, 302
tomato mosaic virus (ToMV) infection 24–25
total ion chromatogram (TIC) normalization 78
toxic metabolites 261
traditional Chinese medicine 379, 381
trait discovery and characterization 244, 246
transcription activator-like effectors (TALEs) 268
transcription factors 207–209, 211, 267–268, 302
transcriptomic-based best linear unbiased prediction (T-BLUP) 54
transcriptomics 160, 316
transfer DNA (T-DNA) 279
tricarboxylic acid (TCA) cycle 4
triple quadrupole MS (TQMS) 122, 240, 253
Triticum durum see durum wheat *(Triticum durum)*
trophophase 277

U
ultra-high-performance liquid chromatography (UHPLC) 376
clustered regularly interspaced short palindromic repeats (CRISPR)/Cas gene-editing 283
plant stress 123
predictive modeling (PM) 41
ultra-performance liquid chromatography tandem mass spectrometry (UPLC-MS/MS) analysis 210, 219
univariate analysis 221, 241
unsupervised learning approaches 342
untargeted approach *see* non-targeted approach

V
viroid infections 22–26, 32–33
virus, crop protection against 22–26, 32–33
volatile organic compounds (VOCs) 20, 24, 101, 106, 322, 412

W
water deficit 202, 205, 207–209, 263, 297–300
waterlogging stress control 225
web-based programs 433
Wedelia trilobata 411
whole-genome sequencing (WGS) 280, 339
willow tree *(Salix alba)* 351

X
Xanthomonas oryza 304
XCMS 255, 433

Z
Zea mays see maize *(Zea mays)*
zinc-finger nucleases (ZFNs) 268

Printed and bound by CPI Group (UK) Ltd, Croydon, CR0 4YY

13/09/2026

14961116-0005